Low Cycle Fatigue and Elasto-Plastic Behaviour of Materials

Editors

K.-T. RIE
P.D. PORTELLA

Charles Seale-Hayne Library
University of Plymouth
(01752) 588 588
LibraryandITenquiries@plymouth.ac.uk

1998
ELSEVIER
Amsterdam – Lausanne – New York – Oxford – Shannon – Singapore – Tokyo

ELSEVIER SCIENCE Ltd.
The Boulevard, Langford Lane
Kidlington, Oxford OX5 1GB, UK

Library of Congress Cataloging-in-Publication Data

Low cycle fatigue and elasto-plastic behaviour of materials / editors,
K.-T. Rie and P.D. Portella.
      p.   cm.
    "Fourth International Conference on Low Cycle Fatigue and Elasto
-Plastic Behaviour of Materials was held from 7-11 September 1998 in
Gramisch-Partenkirchen, Germany"--
    ISBN 0-08-043326-X
    1. Metals--Fatigue--Congresses. 2. Elastoplasticity--Congresses.
I. Rie, K.-T. (Kyong-Tschong), 1936- II. Portella, P. D. (Pedro D.)
III. International Conference on Low Cycle Fatigue and Elasto
-Plastic Behaviour of Materials (4th : 1998 : Garmisch
-Partenkirchen, Germany)
TA460.L679 1998
620.1'66--dc21
                                                          98-34273
                                                             CIP

First Edition 1998

ISBN: 0 08 043326 X

© 1998 Elsevier Science Ltd

No responsibility is assumed by the publisher for any injury and/or damage to persons or property as a matter of products liability, negligence or otherwise, or from any use or operation of any methods, products, instructions or ideas contained in the material herein.

⊗ The paper used in this publication meets the requirements of ANSI/NISO Z39.48-1992 (Permanence of Paper).

Printed in The Netherlands.

Local Organizing Committee

Chairman: HELMUT NAUNDORF, TWP, Baldham
Members: REINHARD BARDENHEIER, Instron Wolpert Ludwigshafen
JOACHIM BERGMANN, MFPA, Weimar
KARL FEITZELMAYER, MAN, München
LEONHARD HAGN, Allianz-Zentrum für technik, Ismaning
HELMUT HUFF, MTU München
HEINZ JOAS, TÜV, Süddeutschland, München
INGRID MASLINSKI, DVM, Berlin
ANGELA OSTERMEIER, BMW, München
RUDOLF STAUBER, BMW, München

Federation of European Materials Societies

Member Societies

Benelux: Benelux Métallurgie
Czech Republic: Czech Society for New Materials and Technologies (CSNMT)
Metal Science Society of the Czech Republic (MSS)
Denmark: Danish Metallurgical Society
Estonia: Estonian Materials Science Society (EMSS)
France: Société Française de Métallurgie et de Matériaux (SF2M)
Gemany: Deutsche Gesellschaft für Materialkunde (DGM)
Deutsche Gesellschaft für Materialforschung und -prüfung (DVM)
Greece: Hellenic Society for the Science and Technology of Condensed Matter
Hungary: Országos Magyar Bányászati és Kohászati Egyesület (OMBKE)
Italy: Associaziona Italiana di Metallurgia (AIM)
The Netherlands: Bond voor Materialenkennis (BvM)
Norway: Norsk Metallurgisk Selskap (NMS)
Portugal: Sociedade Portuguesa de Materiais (SPM)
Slovak Republic: Society for New Materials and Technologies
Slovenia: Slovensko drustvo za materiale
Sweden: Svenska Föreningen för Materialteknik (SFM)
Switzerland: Schweizerischer Verband für die Materialtechnik (SVMT)
United Kingdom: Institute of Materials (IoM)

The conference is co-sponsored by:

Arbeitsgemeinschaft Wärmebehandlung und Werkstofftechnik (Germany)
ASM International (USA)
European Structural Integrity Society
Engineering Integrity Society (UK)
The Chinese Society of Metals
The Japan Institute of Metals
The Korean Institute of Metals
The society of Materials Science (Japan)
VDI-Gesellschaft Werkstofftechnik (Germany)

Preface

The Fourth International Conference on Low Cycle Fatigue and Elasto-Plastic Behaviour of Materials was held from 7 to 11 September 1998 in Garmisch-Partenkirchen, Germany, following successful conferences in Stuttgart (1979), Munich (1987) and Berlin (1992). The conference was organized by the Deutscher Verband für Materialforschung und –prüfung (DVM – German Association for Materials Research and Testing).

Following the tradition of the previous conferences, it was the intention of the organizers to provide a discussion forum for all those interested in both the fundamental aspects and the practical applications of this exciting subject.

In response to our call for papers, nearly 200 extended abstracts from 32 countries were submitted to the organizing committee. The present conference proceedings reflect the most interesting and challenging innovations in this field. These papers were presented in the conference as invited lectures or short contributions – as oral or poster presentation. All the papers were presented in poster form in extended poster sessions, a peculiarity of the LCF Conferences which allow an intense, thorough discussion of all contributions.

The papers in these proceedings were distributed in chapters according to the following main topics of the conference:

1. Isothermal low cycle fatigue; general aspects

2. Thermal and thermo-mechanical loading

3. Multiaxial loading

4. Microstructural aspects

5. Influence of environmental conditions and surface treatments

6. Advanced materials

7. Behaviour of short and large cracks

8. Crack initiation and coalescing; damage evolution

9. Constitutive equations; modelling

10. Design methods; life prediction

11. Case studies; practical experience

Each chapter provides a comprehensive overview of a materials class or a given subject. It is obvious that many contributions might be included in two or even three chapters. In order to give a better overview of the content, the reader will find a subject index, a material index and an author index in the back of the book.

On behalf of the Organizing Committee we wish to thank all the authors, invited lecturers, session chairmen, members of the International Advisory Committee and numerous others who gave an important contribution to this conference. The excellent work of the Local Organizing Committee was of fundamental importance for this event. Finally we should like to thank Mrs. Ingrid Maslinski, DVM, whose administrative and organizational capability supported once more our work from the beginning.

Kyong-Tschong Rie   Braunschweig, Germany

Pedro Dolabella Portella   Berlin, Germany

CONTENTS

Chapter 1

ISOTHERMAL LOW CYCLE FATIGUE; GENERAL ASPECTS

Chapter 2

THERMAL AND THERMO-MECHANICAL LOADING

Chapter 3

MULTIAXIAL LOADING

Chapter 4

MICROSTRUCTURAL ASPECTS

Chapter 5

INFLUENCE OF ENVIRONMENTAL CONDITIONS
AND SURFACE TREATMENTS

Chapter 6

ADVANCED MATERIALS

Chapter 7

BEHAVIOUR OF SHORT AND LARGE CRACKS

Chapter 8

CRACK INITIATION AND COALESCING; DAMAGE EVOLUTION

Chapter 9

CONSTITUTIVE EQUATIONS; MODELLING

Chapter 10

DESIGN METHODS; LIFE PREDICTION

Chapter 11

CASE STUDIES; PRACTICAL EXPERIENCE

xviii

Chapter 1

ISOTHERMAL LOW CYCLE FATIGUE; GENERAL ASPECTS

ROOM AND HIGH-TEMPERATURE LOW CYCLE FATIGUE OF LAMELLAR STRUCTURED TiAl

R. OHTANI, T. KITAMURA and M. TSUTSUMI

Department of Engineering Physics and Mechanics,
Kyoto University, Kyoto 606-8501, Japan

ABSTRACT

Experimental investigation was done to characterize the low cycle fatigue (LCF) in an as-cast Ti-34wt.%Al with near 0°-oriented lamellar structure by using the smooth bar specimens subjected to push-pull loading. Effects of test temperatures, strain waveforms and lamellar and/or grain boundaries on the crack initiation, propagation and the failure life are shown in this paper. It is pointed out that the TiAl tested is inferior to other heat resistant alloys in the LCF resistance over 0.5% total strain range at room temperature, whereas it has a satisfactory grade at high temperatures or under creep-fatigue conditions.

KEYWORDS

Intermetallic compound, TiAl, lamellar structure, low cycle fatigue, creep-fatigue, crack initiation, crack growth, high temperature

INTRODUCTION

Because of its high relative strength and heat resistance, TiAl intermetallic compound is one of the most promising materials for high-temperature use. Energies are concentrated to the development of the tougher materials and their practical usage in Japan [1]. To date, the mechanical properties of TiAl alloys with various microstructures, obtained by changing the chemical composition and heat treatment, have been intensively investigated. Among these, alloys with the $(\alpha_2 + \gamma)$ lamellar structure have attracted researchers due to their excellent mechanical performance, although the properties depend on the loading direction to the lamellae. Much research work has been carried out on the monotonic tensile and compressive properties not only at room temperature but also at high temperatures. However, studies on fatigue are still limited [2,3,4].

The objective of this study is to investigate the fundamental behavior of fatigue in a cast Ti-34wt.%Al (48at.%Al) at high as well as room temperatures. Uniaxial fatigue load is applied to the specimen in the direction nearly parallel to the lamellar boundaries. Stress is put on the effects of test temperatures, strain waveforms and lamellar boundaries for transverse cracking under relatively high stress and strain conditions in order to understand its low cycle fatigue (LCF) characteristics and to verify the applicability to the practical components subjected to thermal fatigue. Test results on the failure life are compared with those of some other heat resistant alloys.

MATERIAL, SPECIMEN AND TEST CONDITIONS

The material tested is manufactured by casting into a columnar ingot in argon gas in the plasma-skull melting process. The chemical composition (wt.%) is as follows; Al: 33.5, C: <0.01, N: 0.005, O: 0.084, H: 0.0016, Ti: bal. The grains grow from the circumference to the center in the ingot during the solidification so that the elongated grains of 3mm in the growth direction and 0.35mm in the transverse direction are formed. The lamellae composed of γ-phase (TiAl) and a small amount of α_2-phase (Ti$_3$Al) are aligned nearly perpendicularly to the grain growth direction.

Cylindrical specimens are cut from the circumference near the ingot surface so that their axes are nearly parallel to the lamellar boundaries as shown in Fig.1. Miniature specimens with 4mm-diameter are used for fatigue failure tests and standard specimens with 10mm-diameter for crack observation.

The material is neither heat treated nor HIP treated but is used as cast. It is found through the defect inspection that the specimens contain 2.5 defects per surface area of 1 mm^2 in average and 99% of the distribution of defect size is smaller than 30μm in diameter. The surface of the specimens is polished by the diamond paste before the tests to remove machine cuts and after the interrupted tests to remove oxide film and observe surface cracks.

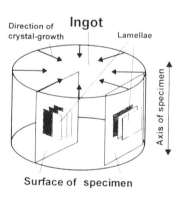

Fig.1   Direction of specimen axis. (0°-oriented lamellar structure)

In the strain-controlled failure tests, a fast–fast triangular strain-waveform is adopted at room temperature (RT) and high temperatures (600, 800 and 900℃, mainly 800℃), and fast–slow and slow–fast ones at 800℃ in air. The fast strain rate is 1 %/s and the slow strain rate is 10^{-3} or 10^{-4} %/s. The details of the test procedure have been described elsewhere [5].

TEST RESULTS AND DISCUSSIONS

Fig.2 shows the test results on the fatigue life (the number of cycles to failure, N_f) at different strain ranges for different strain waveforms. Tables 1(a) and (b) represent dominant characteristics of the fatigue cracking.

Fast–Fast Fatigue at Room and High Temperatures

The tensile rupture elongation at RT is 1% or less, being quite small as compared with that of 10% or more at 800°C, so that the fatigue life under the total strain range of 1.0% at RT is the lower bound in LCF regime as shown in Fig.2. The fatigue life at 800°C is longer than that of RT especially at high total strain ranges. Even in the LCF regime, the total strain is composed of a large amount of the elastic strain.

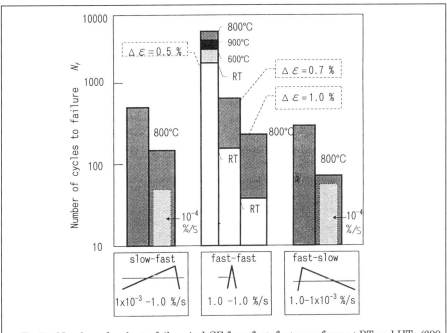

Fig.2   Number of cycles to failure in LCF for a fast–fast waveform at RT and HTs (600, 800 and 900°C) and for fast–slow and slow–fast waveforms at 800°C.

As shown in Table 1(a), the lamellar boundary cracking (Photo 1) is most frequent in the RT-fatigue, the surface defect cracking (Photo 2) is dominating below the test temperature of 800°C, and the oxide-film cracking is characteristic of the 900°C-fatigue (Photo 3). Among these fatigue cracks, the lamellar delamination grows in the direction nearly parallel to the stress axis and stops growing when it reaches the grain boundary. This is, therefore, an arrested crack for the near 0°-oriented lamellar structure [5].

On the other hand, surface defect and oxide-film cracks grow perpendicularly to the stress axis and transversely to the lamellae. Especially the surface defect crack in RT-fatigue displays a zigzag propagation accompanied with the branching due to the partial delamination along lamellar boundaries (Photo 4). Such zigzag propagation is characteristic of the fatigue, although the cracks grow rather straight at high temperatures (Photo 5).

When the strain range–fatigue life curves of the TiAl are compared with those of a Type 304 stainless steel and a Mar-M247DS nickel base superalloy, 304 SS exhibits a longer fatigue life than the TiAl at the total strain range over 0.5%. On the other hand, the fatigue life of Mar-M247DS at 900°C [6] is nearly equal to that of TiAl at 800°C.

Table 1(a). Characterization of cracking behavior in LCF at room and high temperatures

Type of Fatigue	Crack Initiation	Crack Growth
Room Temperature Fatigue (fast–fast) 1 %/s	 Photo 1 Along lamellar boundaries (delamination) blocked by GBs A few from casting defects	 Photo 4 Zigzag propagation across lamellae accompanied with micro-delamination
High Temperature fatigue (fast–fast) T = 600, 800, 900°C 1 %/s	 Photo 2 (600,800°C) Mainly from casting defects (600°C) From casting defects (800°C) Photo 3 (900°C) By oxide-film cracking (900°C)	 Photo 5 Zigzag propagation across lamellae (rather straight) $da/dN - \Delta K$ (or ΔJ_f): independent of temperature; one order faster than steels and alloys

Fast–Slow Fatigue at High Temperature

The fast–slow strain waveforms at 800°C tend to reduce the failure life as shown in Fig.2. One of the factors in decreasing the fatigue life is the increase in the maximum (peak) stress which is brought about by the unbalanced hysteresis loops.

In Table 1(b) shown are the cracks for the fast–slow strain waveform (Photo 6), indicating that they initiate at the surface oxide-film because longer test duration than the fast–fast fatigue promotes the development of oxidation. The morphology of the crack is similar to that of the crack for the fast–fast 900°C-fatigue (see Photo 3). Therefore, the fast–slow fatigue does not interact with creep but with environment (fatigue-environment interaction). Here, attention should be paid that this oxide-film crack does not always grow inward (see Photo 7) but spreads widely on the surface.

Table 1(b). Characterization of cracking behavior in fast–slow and slow–fast fatigue at 800°C

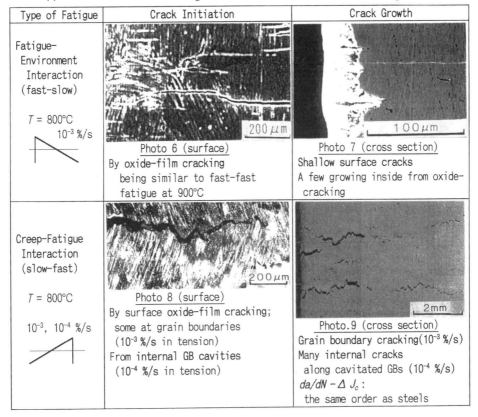

The TiAl is inferior to Mar-M247DS in the fatigue life, although the difference is not always distinguished. As thermal fatigue behavior in out-of-phase conditions (thermo-mechanical cycles combined with tension at low temperature and compression at high temperature) has a strong resemblance to the fast–slow fatigue behavior, it is necessary to consider the oxidation-induced reduction of surface resistance when the TiAl is applied to practical components such as gas turbine blades.

8

Slow–Fast Fatigue at High Temperature

The slow–fast strain waveforms also yield the reduction of the fatigue life. The lower bound of fatigue life of this type at a very low strain rate in tension of 10^{-4} %/s is nearly equal to the fast–fast fatigue life at RT as shown in Fig.2. Comparison of the failure lives with the aforementioned materials indicates that the TiAl exhibits good properties for the creep-dominating fatigue.

In this case, the crack tends to initiate on the grain boundaries at the surface and to grow inside along them. An typical example is shown in Photo 8 in Table 1(b). When the slow strain rate in tension is much smaller, 10^{-4} %/s, a number of cavities and microcracks tend to originate on the grain boundaries perpendicular to the stress axis inside the material (Photo 9). It has been clear that some austenitic and low alloy steels show a similar fracture behavior when they are subjected to slow–fast or tension-hold strain cycles [7,8]. Therefore, this slow–fast fatigue is typical of creep-fatigue, wherein the creep strain accumulated irreversibly in the tensile direction changes its fracture mechanism from fatigue transgranular cracking to grain boundary cracking or cavitation.

ACKNOWLEDGMENTS

This work was supported in part by a Grant-in-Aid for Scientific Research of the Ministry of Education. The authors acknowledge R & D Division of Daido Steel Co., Ltd., for providing the material tested.

REFERENCES

1. Izumi, S. et al (Ed.) (1996). *Intermetallic Compounds—Capabilities as Advanced High-Temperature Structural Materials* (in Japanese), Joint Committee on Intermetallic Compounds , Japan Society of Metals and other seven Societies.
2. Yamaguchi, K., Shimadaira, M. and Nishijima, S. (1992). *Iron and Steel* (in Japanese), ISJS, **78**-1, 134-140.
3. Davidson, D.L., and Cambell, J.B. (1993). *Metallurgical Transaction A*, **24A**, 1555-1562.
4. Umakoshi, Y., Yasuda, H. and Nakano, T. (1993). In: *Proc. 3rd Japan International SAMPE*, p.1329.
5. Tsutsumi, M., Ohtani, R., Kitamura, T., Takano, S. and Ohshima, T. (1995). *Journal of the Soc. of Materials Science* (in Japanese), **44**-501, 769-775.
6. Nitta, A. (1993). *Study on Thermal Fatigue in Heat Resistant Metallic Materials for Electric Power Plants* (Central Research Institute of Electric Power Industries), Ph.D thesis (in Japanese).
7. Ohtani, R. and Kitamura, T. (1994). In: *Handbook of Fatigue Crack Propagation in Metallic Structures*, A.Carpinteri(Ed.), Elsevier. Vol.2, pp.1347-1383.
8. Ohtani, R., Kitamura, T., Tada, N. and Zhou, W. (1995). In: *Proceedings of Symp. on Fatigue under Thermal and Mechanical Loading*, J.Bressers and L.Remy(Ed.), Kluwer Academic Publishers, the Netherlands, pp.199-208.

Low Cycle Fatigue and Elasto-Plastic Behaviour of Materials
K-T. Rie and P.D. Portella (Editors)

THE EFFECT OF CREEP PRE-DEFORMATION ON HIGH-TEMPERATURE FATIGUE BEHAVIOUR OF THE TITANIUM ALLOY IMI 834

S. Hardt, H. J. Maier and H.-J. Christ

Institut für Werkstofftechnik, Universität-GH Siegen,
D-57068 Siegen, F. R. Germany

ABSTRACT

The creep-induced degradation of the high-temperature fatigue properties of the titanium alloy IMI 834 was studied on samples with bimodal and equiaxed microstructure, respectively. Significant reduction of fatigue life was observed for samples pre-crept at 600°C. Moreover, in the case of a bimodal microstructure, creep pre-deformation was found to lower the cyclic yield stress leading to a level similar to that of the equiaxed condition. The reduction in cyclic life could partially be attributed to environmental effects. The higher resistance of the bimodal microstructure against fatigue crack propagation explains why the damaging effect of pre-deformation is less pronounced in this case.

KEYWORDS

creep pre-deformation, environmental degradation, low-cycle fatigue, microstructure, high-temperature titanium alloys.

INTRODUCTION

The most recent commercially available near-α titanium alloy IMI 834 was designed for high-temperature applications such as turbine blades in the compressor part of jet engines. In order to replace the heavier nickel-base superalloys an upper service temperature of 600°C was envisaged [1, 2]. As service conditions involve both monotonic and cyclic loads, excellent creep and fatigue properties are required. For optimum performance a bimodal microstructure consisting of a small volume fraction of primary α phase in a lamellar matrix of transformed β is recommended for IMI 834 [3]. Excellent mechanical properties have been obtained on IMI 834 in short-term laboratory tests, see e.g. [4]. Service conditions, however, involve long-term high-temperature exposure under combined creep and fatigue loading. Significant microstructural changes have been reported to occur in IMI 834 in creep tests run at 600°C [5]. Similar effects were expected to occur under creep-fatigue conditions, and the objective of the present paper was to study creep-induced degradation of the fatigue properties of IMI 834.

EXPERIMENTAL DETAILS

The near-α titanium alloy IMI 834 (nominal composition Ti-5.8Al-4.0Si-3.5Zr-0.7Nb-0.5Mo-0.35Si-0.06C, in wt-%) supplied by IMI Titanium Ltd. in the form of hot-rolled bars was studied with equiaxed and bimodal microstructure, respectively. The equiaxed microstructure resulted from recrystallization during solution heat treatment at 1000°C and slow air cooling. The bimodal microstructure was obtained by a solution heat treatment at 1020°C for 2h, followed by an oil quench. Finally, an additional ageing treatment (2h at 700°C/AC) was applied. Unnotched cylindrical samples with a gauge diameter of 8 mm were used for mechanical testing. To minimize surface effects, all specimens were electrolytically polished in the gauge section prior to testing. Creep behaviour was studied both in compression and tension, respectively. All creep test were performed under constant stress conditions.

10

The low cycle fatigue tests were run under closed-loop plastic strain control using a symmetrical triangular wave form. Plastic strain amplitudes used were in the range from 0.1 to 0.5%. In all fatigue tests the plastic strain rate was kept constant at $8x10^{-4}s^{-1}$.

The effect of creep pre-deformation on fatigue behaviour was studied using specimens pre-strained under constant stress conditions to different levels of monotonic plastic strain at 600°C. These tests were run in a servo-hydraulic test system, thus following creep pre-deformation of the samples the fatigue tests could be started instantaneously. In order to separate the effects of surface oxidation and pure creep damage on fatigue life, additional fatigue tests were performed on samples that had been annealed prior to fatigue testing for extended time periods in high vacuum and air, respectively.

Fracture surfaces and metallographic cross sections were studied by scanning electron microscopy (SEM) to identify the relevant damage mechanisms. The microstructural changes occurring during creep and fatigue loading, respectively, were studied by transmission electron microscopy (TEM). Thin slices were sectioned from the fractured test pieces perpendicular to the external stress axis. The final TEM samples were prepared by mechanical grinding down to 150 μm followed by conventional twin-jet polishing.

RESULTS

Creep Behaviour

As expected, creep behaviour of equiaxed and bimodal microstructures, respectively, was found to be drastically different. In essence, the bimodal microstructure displayed much higher initial strength and significantly lower minimum creep rates were observed as compared to samples with equiaxed microstructures. However, pronounced steady-state creep behaviour was observed in long-term tests performed on equiaxed microstructures, whereas a distinct minimum creep rate was monitored in samples with bimodal microstructures. The softening that set in after the minimum creep rate could be observed in both tension and compression creep tests. Thus, it was concluded that the increase in creep rate is due to microstructural changes and not caused by void formation. Indeed, creep damage as a result of void formation was observed only in samples that were deformed into the tertiary creep regime.

Fatigue behaviour without creep pre-deformation

Independent of the actual microstructure, rapid initial cyclic softening was observed in all fatigue tests. Further, continuous cyclic softening until failure occurred in tests run at high plastic strain amplitudes ($\Delta\varepsilon_{pl}/2 \geq 0.5\%$). In contrast, initial cyclic softening was followed by cyclic saturation if the alloy was tested at low plastic strain amplitudes ($\Delta\varepsilon_{pl}/2 \leq 0,2\%$).

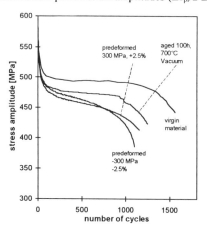

Fig. 1: Cyclic softening curves as recorded from samples with bimodal microstructure. All samples were fatigued at $T = 600°C$ and $\Delta\varepsilon_{pl}/2 = 0.2\%$.

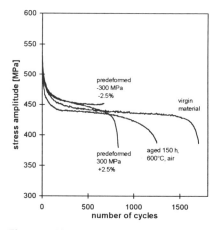

Fig. 2: Effect of creep pre-deformation and thermal exposure on cyclic stress response and fatigue life of samples with equiaxed microstructure at $T = 600°C$ and $\Delta\varepsilon_{pl}/2 = 0.2\%$.

Comparison of Figs. 1 and 2 reveals that the bimodal microstructure has a significantly higher cyclic yield stress than the equiaxed microstructure. Moreover, as soon as macro crack growth set in, the stress amplitudes dropped quite rapidly in the equiaxed microstructures, c.f. Fig. 2. In the bimodal structure, however, a rather smooth decrease in stress amplitude was observed prior to failure, c.f. Fig. 1.

Scanning electron microscopy (SEM) revealed distinctly different fracture surfaces for the two microstructures tested. As seen in Fig. 3 striations formed in samples with an equiaxed microstructure are essentially parallel to each other in all grains. In contrast, striations observed on the fracture surface of samples with a bimodal microstructure have different orientations with respect to the macroscopic crack growth direction, c.f. Fig. 4.

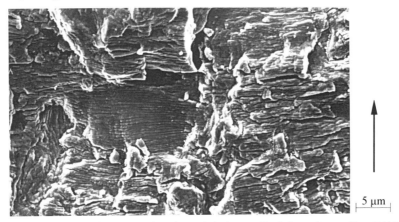

5 μm

Fig. 3: SEM micrograph of the fracture surface formed during cyclic deformation at $T = 600°C$ and $\Delta\varepsilon_{pl}/2 = 0.1\%$. Note that in this equiaxed microstructure striations are almost parallel in all grains. The arrow indicates the macroscopic crack growth direction.

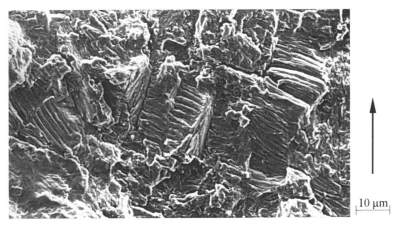

10 μm

Fig. 4: Appearance of fracture surface in a sample with bimodal microstructure fatigued at $T = 600°C$ and $\Delta\varepsilon_{pl}/2 = 0.1\%$. Note the strongly diverging striations fields.

Fatigue behaviour after a monotonic creep pre-deformation

As seen in Fig. 1, samples with a bimodal microstructure display a significant reduction in cyclic yield stress after a creep pre-deformation as compared to the virgin material. Note that both creep pre-deformation in tension and compression, respectively, cause approximately the same reduction in cyclic saturation stress amplitude and fatigue life. The effect of creep pre-deformation on fatigue life is, however, rather small.

In contrast, the effect of creep pre-deformation on cyclic stress-strain response was rather small for the equiaxed microstructure despite the high prestrains used, c.f. Fig. 2. However, fatigue life of the creep pre-deformed samples is reduced significantly as compared to the virgin material.

It has to be noted, that the creep pre-deformed samples were exposed to air for a much longer time period than the samples fatigued only. Hence, environmental degradation might have contributed to the loss in fatigue properties seen after creep pre-deformation. To separate the effects of surface oxidation and pure creep damage on fatigue life, fatigue tests were performed on samples that have been stress-free annealed in high vacuum and air. Results from such an experiment run on a specimen aged stress-free for 150h at 600°C have been included in Fig. 2. In this case, an ageing time of 150 h was chosen as this approximately equals the time elapsed during creep pre-deformation. Fatigue life of the sample that has been aged at zero stress is reduced as compared to the virgin material. However, environmental degradation alone does not fully account for the effect of creep pre-deformation on fatigue life seen in Fig. 2. A similar experiment was performed on a sample with a bimodal microstructure. In this case, however, the sample was aged at zero stress for 100h at 700°C in high vacuum. As seen in Fig. 1, this ageing treatment causes a significant reduction in cyclic stress amplitude and fatigue life is reduced only slightly.

Microstructural observations

SEM observations revealed that conventional creep damage in the form of microvoid formation was negligible in all samples that have been creep pre-deformed. Hence, the pronounced effect of creep pre-deformation on cyclic stress amplitude observed in samples with a bimodal microstructure is not a result of conventional creep damage. TEM studies indicated substantial microstructural changes after creep pre-deformation in samples with a bimodal microstructure. Figure 5 demonstrates that lamella boundaries are deteriorated after creep pre-deformation. In contrast, Fig. 6 demonstrates that the lamella boundaries in samples that have been fatigued only are significantly less affected.

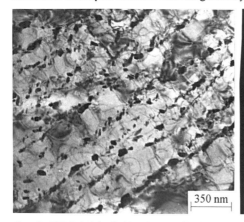

350 nm

300 nm

Fig. 5: TEM bright-field micrograph showing disintegration of lamella boundaries in creep tests run at $T = 600°C$ and 300 MPa. Creep tests was interrupted at $\varepsilon_{pl} = 2.5\%$.

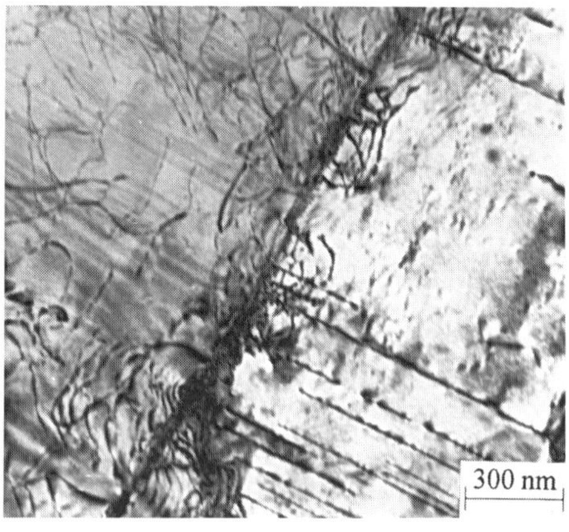

Fig. 6: Appearance of a lamella boundary in a sample that had been fatigued at $T = 600°C$ and $\Delta\varepsilon_{pl}/2 = 0.2\%$ until failure.

DISCUSSION

Fatigue behaviour without creep pre-deformation

Initial cyclic softening in age hardened titanium alloys is generally attributed to the formation of slip bands that result from shearing of Ti_3Al precipitates [6, 7]. In related work [8] slip bands in the α-phase have indeed been observed in fatigued samples tested at 600°C. The slight cyclic softening observed until macro crack growth sets in, is attributed to a loss of silicon in solid solution. It has been reported that the strength of near-α titanium alloys at test temperatures above about 500°C is governed by the amount of silicon in solid solution [9, 10]. Coarsening of silicides will reduce the amount of silicon retained in the matrix. In samples with equiaxed microstructure large silicides were present prior to testing. Hence, the lower cyclic yield stress and the higher minimum creep rate of the equiaxed microstructures results partly from the lower amount of silicon in solid solution as compared to the bimodal microstructure. TEM studies have shown that the increase in creep rate seen in long-term creep tests on samples with bimodal microstructure can be attributed both to the loss of silicon in solid solution and to the disintegration the lamella boundaries [5, 10]. The silicides seen in fatigued samples with bimodal microstructure, c.f. Fig. 6, either result from heterogeneous nucleation on dislocations or from a transformation of retained β into α plus silicides [11]. However, in samples that have been fatigued only, coarsening of silicides and disintegration of the lamella boundaries is not very pronounced, and the overall effect on cyclic stress amplitude was thus rather small.

Despite the higher stress amplitude sustained by the bimodal microstructure, fatigue life of samples with bimodal and equiaxed microstructures, respectively, was rather similar. As a result of the large plastic strain amplitude and high stresses, crack initiation occurs quite rapidly under low-cycle fatigue conditions. Hence, the superiority of the bimodal microstructure was attributed to lower crack propagation rates in this microstructure. It is well documented [12] that cracks propagate easily in equiaxed microstructure if texture is present. X-ray diffraction has indicated that a strong texture component was indeed present in the equiaxed microstructure. Hence, grain boundaries do not hinder crack propagation effectively, i.e. the crack can propagate on the same preferred plane in neighbouring grains, c.f. Fig. 3. In contrast, the bimodal microstructure provides a higher density of microstructural barriers. As seen in Fig. 4 these barriers effectively deflect the cracks, which results in lower crack growth rates. The lowest crack propagation rates have indeed been measured in tests performed on samples with bimodal microstructures, and the highest crack growth rates were obtained for equiaxed microstructures [13].

Fatigue behaviour after a monotonic creep pre-deformation

Comparison of Figs. 1 and 2 reveals that independent of initial microstructure the cyclic stress amplitude of all pre-crept samples is approximately identical. It should be emphasised that metallographic studies indicated that conventional creep damage, i.e. the formation of voids, was negligible in the pre-crept samples. Furthermore, samples pre-crept in compression, i.e. under conditions were voids should not form, showed an almost identical loss in cyclic yield strength than those pre-crept in tension. Hence, the effects of creep pre-deformation on cyclic saturation stress amplitude are attributed to one or more of the following factors: (i) a change in dislocation density, (ii) an additional precipitation of Ti_3Al, (iii) the creep-induced deterioration of the lamella boundaries, and (iv) the loss of silicon in solid solution caused by coarsening of silicides. Chu et al. [14] have observed that creep pre-deformation at room temperature is beneficial for fatigue life of titanium alloys as the increase in dislocation density promotes a more homogeneous type of slip. In contrast, the rapid initial cyclic softening observed in the present study after high temperature deformation seems to indicated that the dislocation substructure present after creep pre-deformation has no significant effect on the shearing of the Ti_3Al precipitates.

As seen in Fig. 1 the sample that has been pre-exposed at 700°C in vacuum for 100h is significantly softer than the virgin material. TEM has revealed coarsened silicides in this samples and more Ti_3Al to be present than in the virgin material. Obviously any additional hardening caused be the precipitation of Ti_3Al during ageing is compensated for by the loss in silicon in solid solution. Hence, it is concluded that the main effects of creep pre-deformation on cyclic yield stress are the loss in silicon in solid solution and the almost complete disintegration of the lamella boundaries observed after 2.5% creep strain, c.f. Fig. 5. The observation that the cyclic yield stress of creep pre-deformed samples with bimodal microstructure is almost identical to those with an equiaxed one, seems to indicate that the amount of silicon retained in the matrix is the dominant factor. Similar conclusions have been drawn from the mechanical behaviour observed in creep tests [10, 15].

The effect of creep pre-deformation on fatigue life can be rationalised as follows. Non-protective oxides form on near-α titanium alloys during exposure at 600°C. An oxygen-rich subsurface layer is built up as a result of inward diffusion of oxygen. As oxygen hardens and embrittles the α-phase, crack nucleation and earlier crack growth are promoted. It has been demonstrated that the thickness of the oxygen-enriched layer depends on the actual microstructure [16], and a somewhat thicker embrittled layer should be present in the samples with equiaxed microstructure. However, in samples pre-strained up to 2.5%, cracks that have propagated through the oxygen-rich layer were present. Thus, the different effect of creep pre-deformation on fatigue life of samples with bimodal and equiaxed microstructure, respectively, has to be attributed to differences in fatigue crack growth behaviour in the unembrittled part of the samples. Despite the deterioration of the lamella boundaries observed after creep pre-deformation,

14

diverging striation fields were still present in pre-crept and fatigued samples with bimodal microstructure. Hence, the boundaries between primary α and lamellar transformed β still deflect the fatigue cracks and thus reduce fatigue crack growth rate.

With respect to actual components, it has been reported that microstructure in IMI 834 is not stable under long-term high-temperature service conditions [9, 17]. Hence, cyclic stress-strain behaviour of components with bimodal microstructure will indeed approach that of lower-strength equiaxed ones. It should be noted, however, that the positive effect of the bimodal microstructure on fatigue crack growth is retained even under the quite severe test conditions used in the present study. In cases were crack nucleation dominates fatigue life, the formation of the brittle oxygen-rich subsurface layer seems to be the main effect of high-temperature creep pre-deformation. Earlier work [18] has demonstrated, that under such conditions, fatigue life is reduced by approximately an order of magnitude.

CONCLUSIONS

The effect of high-temperature creep pre-deformation on cyclic stress-strain response and fatigue life of the near-α titanium alloy IMI 834 was studied. The results may be summarized as follows:

- Prior creep deformation significantly reduces the cyclic yield stress of samples with a bimodal microstructure. As a result of both the creep-induced deterioration of the lamella boundaries and the loss in silicon in solid solution, the cyclic stress-strain response of pre-crept samples with a bimodal microstructure approaches that of the equiaxed ones.
- Creep pre-deformation leads to a significant reduction of cyclic life. This reduction can partially be attributed to environmental degradation giving rise to the formation of an embrittled oxygen-rich subsurface layer.
- The damaging effect of creep pre-deformation with respect to cyclic life is less pronounced in the case of a bimodal microstructure. This due to the higher resistance of this microstructure against fatigue crack propagation.

REFERENCES

[1]   Blenkinsop, P.A. (1996). In: *Titanium '95 Science and Techn.*, pp. 1-10; P.A. Blenkinsop, W.J. Evans and H.M. Flower (Eds.). The Institute of Materials, London.
[2]   Boyer, R.R. (1996). *Mater. Sci. Eng.* A213, pp. 103-114.
[3]   Neal, D.F. (1985). In: *Titanium Science and Techn.*, pp. 2419-2424; G. Lütjering, U. Zwicker and W. Bunk (Eds.). Deutsche Ges. für Metallkunde, Oberursel.
[4]   Polmear, I.J. (1989). *Light Alloys*, 2nd ed. pp. 211-273; Edward Arnold, London.
[5]   Kestler, H., Mughrabi, H. and Renner, H. (1996). In: *Titanium '95 Science and Techn.*, pp. 1171-1178; P.A. Blenkinsop, W.J. Evans and H.M. Flower (Eds.). The Institute of Materials, London.
[6]   Williams, J.C. and Lütjering, G. (1980). In: *Titanium '80 Science and Techn.*, pp. 671-681; H. Kimura and O. Izumi (Eds.). TMS-AIME, Warrendale, PA.
[7]   Beranger, A.S., Feaugas, X. and Clavel, M. (1993). *Mater. Sci. Eng.* A172, pp. 31-41.
[8]   Pototzky, P., Maier, H.J. and Christ, H.-J. (1998). submitted to: *Met. Mater. Trans. A.*
[9]   Neal, D.F. and Fox, S.P. (1992). In: *Titanium '92 Science and Techn.*, pp. 287-294; F. H. Froes and I. Chaplan (Eds.). TMS, Warrendale, PA.
[10] Kestler, H., v. Großmann, B., Höppel, H.W. and Mughrabi, H. (1998). In: *Mikrostruktur und mechanische Eigenschaften metall. Hochtemperaturwerkstoffe*, VCH Verlag, Weinheim, in print.
[11] Allison, J.E., Cho, W., Jones, J.W., Donlon, W.T. and Lasecki J.V. (1988). In: *Proc. Sixth World Conf. on Titanium*, pp. 293-298; P. Lacombe, R. Tricot and G. Béranger (Eds.). Les éditions de physiques, Les Ulis Cedex.
[12] Gregory, J.K. (1994). In: *Handbook of Fatigue Crack Propagation in Metallic Structures, Vol. 1*, pp. 281-322; A. Carpinteri (Ed.). Elsevier Science, Amsterdam
[13] Lütjering, G., Gysler, A. and Wagner, L. (1988). In: *Proc. Sixth World Conf. on Titanium*, pp. 71-80; P. Lacombe, R. Tricot and G. Béranger (Eds.).Les éditions de physiques, Les Ulis Cedex.
[14] Chu, H.P., MacDonald, B.A. and Arora, O.P. (1985). In: *Titanium Science and Techn.*, pp. 2395-2402; G. Lütjering, U. Zwicker, and W. Bunk (Eds.). Deutsche Ges. für Metallkunde, Oberursel.
[15] Paton, N.E. and Mahoney, M.W. (1976). *Met. Trans. A* 7A, pp. 1685-1694.
[16] Leyens, C., Peters, M., Weinem, D. and Kaysser, W.A. (1996). *Met. Mater. Trans. A* 27A, pp. 1709-1717.
[17] Lütjering, G., Gysler, A. and Wagner, L. (1988). In: *Proc. Sixth World Conf. on Titanium*, pp. 71-80; P. Lacombe, R. Tricot and G. Béranger (Eds.). Les éditions de physiques, Les Ulis Cedex.
[18] Specht, J.U. (1992). In: *Low Cycle Fatigue and Elasto-Plastic Behaviour of Materials 3*, pp. 19-24; K.T. Rie (Ed.). Elsevier Applied Science, London.

Low Cycle Fatigue and Elasto-Plastic Behaviour of Materials
K-T. Rie and P.D. Portella (Editors)

THE EFFECTS OF OXIDATION AND CREEP ON THE ELEVATED TEMPERATURE FATIGUE BEHAVIOUR OF A NEAR-α TITANIUM ALLOY

M.C. HARDY
DERA, Griffith Building, Farnborough, Hampshire
GU14 0LX, United Kingdom

ABSTRACT

The low cycle fatigue behaviour of near-α titanium alloy, TIMETAL 834, has been examined at 500 and 630°C. Under strain cycling, it was found that crack nucleation is controlled by inelastic strain range and mean stress, and not directly by creep effects resulting from reduced strain rates or long dwell times at maximum strain. Oxidation, however, influences fatigue life at temperatures where a relatively small (1 μm deep) oxygen rich layer forms quickly. On exposure to large inelastic strains, this layer is shown to crack and to initiate crack propagation. The implications of these observations for the use of near-α titanium alloys above 550°C and for life prediction of components with stress concentration features are discussed.

KEYWORDS

Elevated temperature, titanium alloy, oxidation, creep-fatigue

INTRODUCTION

Near-α titanium alloys are used for rotating components in the hotter compressor stages of gas turbine engines at temperatures up to 600°C. As the life of highly stressed compressor discs is limited by low cycle fatigue, an understanding of crack nucleation and growth is vital to ensure that service failures are extremely remote. At the highest operating temperatures, the component fatigue performance may be compromised by oxidation and time dependent deformation. These processes are most likely to affect stress concentration features at the rim, the hottest region of the disc, particularly during cruise or when experiencing high mean stress fluctuations that result from vibration or from changes of thrust setting.

This paper reports the results of an experimental programme aimed at investigating the effects of strain rate and dwell time on the low cycle fatigue behaviour of near-α titanium alloy, TIMETAL 834, in air at 500 and 630°C. These variations in fatigue cycle and test temperature were designed to impose different levels of time dependent deformation and oxidation. Oxidation is particularly relevant to titanium alloys at these temperatures due to the high solid solubility of oxygen and the ease of interstitial diffusion. Extensive research [1-3] has shown that exposure of near-α titanium alloys in air, at temperatures above 500°C, results in the formation of a hard surface, oxygen rich layer which reduces room temperature ductility.

MATERIAL

TIMETAL 834 (Ti-5.8Al-4Sn-3.5Zr-0.7Nb-0.5Mo-0.35Si-0.06C) is a near-α titanium alloy used in the compressor stages of aero-engines at temperatures up to 600°C.

Material for the experimental programme was cut from an isothermal forging in a tangential direction. This forging was solution treated in the α plus β phase field (α content of 12-15 %) at 1028°C for two hours and then oil quenched. Subsequently, it was aged at 700°C for two hours and finally, air cooled. This heat treatment produces a bimodal microstructure consisting of primary α grains less than 50 μm in size, surrounded by larger (50-100 μm) transformed β grains showing the basket-weave α morphology.

EXPERIMENTAL PROCEDURE

Constant amplitude, strain controlled fatigue tests were conducted on 38 mm^2, circular section test pieces in laboratory air at 500 and 630°C using a closed loop servohydraulic machine. Test control was achieved via strain measurements from an axial extensometer that was located along the parallel section of the test piece. Zero to maximum strain cycles were applied until complete test piece failure. The influence of dwell period at maximum strain (ε_{max}) was investigated by using dwell times of 1 and 120 seconds. In all these cases, fatigue cycles were based on a constant strain rate of 10^{-3} s^{-1} and included a period of 1 second at the minimum, zero strain limit. Two further cyclic strain rates, of 10^{-2} and 6×10^{-4} s^{-1}, were explored. Fatigue cycles constructed from the faster strain rate contained 1 second dwell periods at both strain limits, while triangular waveforms were applied using 6×10^{-4} s^{-1}. Maximum strain values were chosen to achieve fatigue lives of between 10^2 to 10^5 cycles.

RESULTS

Strain amplitude versus endurance data, are shown in Fig. 1 for TIMETAL 834 at 500°C. The data, presented in terms of the half life elastic and inelastic components of strain amplitude, suggest that strain rate or dwell period at maximum strain have little effect on crack nucleation.

Endurance data from 630°C, however, show two log-linear regions that coincide between 2000-3000 cycles (Fig. 2). This behaviour is apparent also at 600°C, from data published by Kordisch and Nowack [4], using a strain ratio (R_ε) of -1. To investigate these differences in behaviour with temperature, the stress histories of the test pieces exposed to 630°C were examined.

The stress response in Fig. 3 is typical of all the tests with a 120 second dwell and those others where the ε_{max} exceeded 1 %. The abscissa (cycles) axis is logarithmic enabling a linear description for the stress relaxation of post dwell stress, measured at ε_{max}. Since the material neither cyclically softens nor hardens under these conditions, the post dwell stress at minimum strain (ε_{min}) should fall away at a similar rate. This occurred until approximately 1000 cycles. After this point, there is an obvious reduction in the rate of relaxation that suggests that the test piece cracked much earlier than the detection of a macrocrack. Material ageing, *i.e.* α ordering or silicide precipitation [5,6], was not considered to be responsible for this deflection, as it is strain, and thus cycle specific, rather than time dependent.

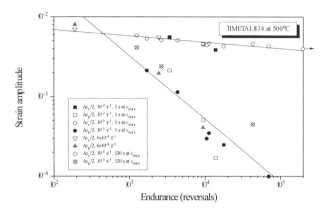

Fig. 1 Strain amplitude versus endurance data for TIMETAL 834 at 500°C. Endurance is defined by the detection of a macrocrack.

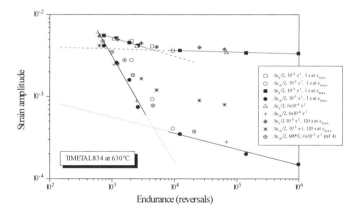

Fig. 2 Strain amplitude versus endurance data for TIMETAL 834 at 630°C. Endurance is defined by the detection of a macrocrack.

Under large inelastic strains, cracking appeared to occur after a very small number of cycles. This was confirmed from inspection of a test piece that had completed 20 cycles to a ε_{max} of 1.5 %. Acetate replicas taken over the entire circumference of the gauge section showed that many microcracks (Fig. 4) nucleated, probably to a depth of the oxygen rich layer that formed from exposure in air above 500°C. Many of these small surface microcracks subsequently propagate into thumb nail macrocracks and result in test piece failure.

Having observed early cracking from large inelastic strains, failure was re-defined as the point of deflection in the minimum stress response, as illustrated in Fig. 3, or for lower strain amplitudes, by the detection of macrocracks. Using these failure criteria, endurance data from 630°C, like those at 500°C, show the expected log-linear behaviour (Fig. 5). Although long dwell periods or slow strain rates result in mean stress relaxation, fatigue life appears to be unaffected as there is a compensating increase in inelastic strain.

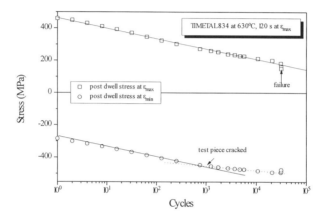

Fig. 3   Development of maximum and minimum stress from fatigue cycles with a 120 second dwell period at ε_{max} of 0.92 %.

DISCUSSION

The findings from this study have important implications for the use of near-α titanium alloys at temperatures above 550°C and for life prediction of components with stress concentration features. As cracking of the surface oxygen rich layer takes place after a small number of cycles from large inelastic strains, the total fatigue lives of components may be predicted by integrating the relevant fracture mechanics expression from an initial flaw size to the crack length associated with K_{IC}.

It is assumed that the size of the initial flaw equates to the depth of the oxygen rich layer and as such, can be determined from the rate of oxygen diffusion into the titanium alloy, given the exposure time and temperature [7]. Although, the material exhibits significant creep strain accumulation at these temperatures, linear elastic fracture mechanics may still be used for predicting the propagation lives of cracks at stress concentration features due to elastic material surrounding the localised region of plasticity. This elastic constraint imposes a strain controlled cycle on material within the plastic zone so peak stresses relax. To correlate the lives of test pieces from strain controlled fatigue tests, however, requires initial flaw sizes, significantly greater than the depth of the oxygen rich layer [8]. These flaw sizes are strongly stress dependent and show a power law increase with inelastic strain range. In extrapolating back to essentially elastic conditions, this effective initial flaw size approached the depth of the oxygen rich layer, which could be determined.

Fig. 4 Micrograph of acetate replica taken after 20, zero to 1.5 % strain cycles at 630°C. Arrow indicates loading direction.

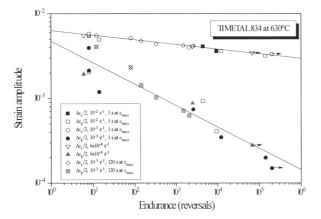

Fig. 5 Strain amplitude versus endurance data for TIMETAL 834 at 630°C.

The specimen tests described in this study suggest that the formation and subsequent fracture of a 1 μm thick layer may be sufficient to nucleate a crack. Time, temperature profiles indicate that the required oxygen rich layer forms quickly at temperatures above 550°C (*e.g.* 1.5 hours at 600°C [8]) which may limit the application of near-α titanium alloys. However, the stress-strain responses experienced in service are likely to be lower than those encountered in laboratory testing and may not be sufficient to cause fracture of the surface layer. More work is needed to establish these conditions for fixed load and displacement conditions. Published work has suggested that cracking may occur from a total static strain of 0.4 % at room temperature [7] and from inelastic strains greater than 0.5 % at 600°C [9].

CONCLUSIONS

The low cycle fatigue behaviour of near-α titanium alloy, TIMETAL 834, has been examined at 500 and 630°C. Under strain cycling, crack nucleation is controlled by inelastic strain range and mean stress, and not directly by creep effects resulting from reduced strain rates or long dwell times at maximum strain. Oxidation, however, influences fatigue life at temperatures where a relatively small (1 μm deep) oxygen rich layer forms quickly. On exposure to large inelastic strains, this layer is shown to crack and to initiate crack propagation. This has important implications for the use of near-α titanium alloys above 550°C and for life prediction of components with stress concentration features.

ACKNOWLEDGEMENTS

This work was undertaken for Brite-EuRam project 6021. As such, the author would like to acknowledge the support of the CEC and the UK Department of Trade and Industry. He also wishes to thank Turbomeca, France for conducting some of the fatigue testing.

REFERENCES

1.   Antony, K.C., (1968), *Journal of Materials*, **1**, 456.

2.   Shamblen, C.E., and Redden, T.K., (1970), In: *The Science, Technology and Application of Titanium*, R.I. Jaffee and N.E. Promisel (Eds), Pergamon Press, Oxford, pp. 199-208.

3.   Rüdinger, K., and Weigand, H.H., (1973), In: *Titanium Science and Technology*, R.I. Jaffee and H.M. Burte (Eds), **4**, Plenum Press, New York, pp. 2555-2571.

4.   Kordisch, T., and Nowack, H., (1996), In: *Fatigue '96*, G. Lütjering and H. Nowack (Eds), Pergamon Press, Elsevier Science Ltd., Oxford, UK, pp. 771-776.

5.   Cope, M.T. and Hill, M.J., (1988), In: *Sixth World Conference on Titanium*, P. Lacombe, R. Tricot and G. Béranger (Eds), Les Editions de Physique, France, pp. 153-158.

6.   Borchert, B., and Daeubler, M.A., (1988), In: *Sixth World Conference on Titanium*, P. Lacombe, R. Tricot and G. Béranger (Eds), Les Editions de Physique, France, pp. 467-472.

7.   Wallace, T.A., (1996), *Titanium '95: Science and Technology*, P.A. Blenkinsop, W.J. Evans and H.M. Flower (Eds), The Institute of Materials, The University Press, Cambridge, UK, pp. 1943-1950.

8.   Hardy, M.C., Harrison, G.F., Smith, M.E.F., and Tranter, P.H., (1997), *Influence of Creep, Fatigue and Oxidation on the Life of Fracture Critical Components*, DERA Technical Report, DERA/SMC/SM2/ TR970206/1.0, Farnborough, UK.

9.   Hardy, M.C., (1996), In: *Fatigue '96*, G. Lütjering and H. Nowack (Eds), Pergamon Press, Elsevier Science Ltd., Oxford, UK, pp. 771-776.

Low Cycle Fatigue and Elasto-Plastic Behaviour of Materials
K-T. Rie and P.D. Portella (Editors)

CYCLIC PLASTICITY OF COLD-WORKED NICKEL AT HIGH MEAN STRESSES

L. KUNZ and P. LUKÁŠ

Institute of Physics of Materials, Academy of Sciences of the Czech Republic
CZ-61662 Brno, Czech Republic

ABSTRACT

Cyclic creep behaviour of cold-worked Ni was investigated at ambient temperature. The abrupt onset of the cyclic creep accompanied by the cyclic softening acceleration was observed after the initial cyclic creep deceleration period. A unique relation between the cyclic creep rate and the plastic strain amplitude for given mean stress was found. The ductile creep fracture was not observed for mean stresses exceeding one half of the ultimate tensile strength.

KEYWORDS

Fatigue, mean stress, cyclic creep, cyclic plasticity, nickel.

INTRODUCTION

The majority of the low cycle fatigue tests has been performed as total or plastic strain controlled experiments owing to the wide-spread strain-based approach to fatigue. Stress or load controlled tests differ substantially from the strain controlled tests as they are accompanied by the cyclic creep. Cyclic creep influences both the lifetime and the dimensional stability of loaded components. Failure by cyclic creep can precede failure by fatigue at low endurance [1]. Two stage Coffin-Manson plots was reported for copper [2, 3, 5] with lower slope corresponding to the cyclic creep fracture occurrence. Fatigue loading of carbon steel with stress amplitude below yield stress σ_y and superimposed mean stress can produce cyclic creep strains exceeding the allowed limit of 0.2% for dimensioning [4]. For copper exhibiting saturating cyclic behaviour the cyclic creep curve with three stages including long secondary stage was observed [5, 6]. This makes it possible to relate the steady state creep rate and the saturated plastic strain amplitude. It was shown that a set of straight lines for different stress ratios R describes in log-log plot the dependence of those two parameters.

The cyclic creep phenomenon has not been studied in sufficient extent in materials that substantially cyclically soften. Available experimental results on carbon steels [4,7] indicate that the dependence of the plastic strain amplitude on the number of cycles is similar in shape to the dependence of the creep strain on the number of cycles. Namely, both the curves exhibit an initial phase in which the cyclic creep is negligible and the cyclic deformation is nearly elastic (i.e. the plastic strain amplitude is also almost negligible). After a certain number of loading cycles an abrupt increase of the plastic strain amplitude (i.e. abrupt cyclic softening) accompanied by an abrupt increase of the cyclic creep strain occur. The gradient of the cyclic softening curve coincides with the gradient of the cyclic creep strain.

The aim of this paper is to present experimental data on the relation between the cyclic and monotonic plasticity in the case of another type of cyclically softening material, in the case of pure cold-worked nickel.

MATERIAL AND EXPERIMENTS

Load controlled LCF tests of cold-worked Ni of 99.9% purity (σ_y = 550 MPa and σ_{UTS} = 615 MPa) were performed at ambient temperature in a servohydraulic loading system. The loading frequency was 2Hz with the exception of hysteresis loops plotting during which the frequency was lowered by factor 10. Cylindrical specimens of 4 mm in diameter and gauge length 10 mm were used. Cyclic creep strain $\varepsilon_{cc,N}$ corresponding to the number of cycles N was determined as $\varepsilon_{ccN} = (\varepsilon_1 + \varepsilon_{0,5})/2$ for N=1 and as $\varepsilon_{ccN} = (\varepsilon_N + \varepsilon_{N-0,5} - \varepsilon_1 - \varepsilon_{0,5})/2$ for N>1 and the plastic strain amplitude as a half width of (not completely closed) hysteresis loops: $\varepsilon_{ap,N} = (\varepsilon_{N-0.5} - \varepsilon_N)/2$, Fig.1. The stress-strain dependence was monitored continuously and plotted for desired number of cycles.

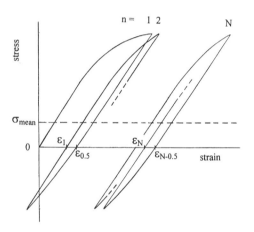

Fig.1. Schematic representation of cyclic creep determination.

RESULTS AND DISCUSSION

The cyclic creep curves do not exhibit the three-stage shape. Fig. 2 displays the set of cyclic creep curves for the mean stress 50 MPa in log-log representation. It can be seen that after a relatively short transition stage of decelerating creep a second stage characterised by accelerating creep takes place. This behaviour can be even better seen in differentiated creep curves. Fig.3 shows such curves for the mean stress 250 MPa. The initial relatively short decrease of the cyclic creep rate is followed by a relatively long continuous increase of the cyclic creep rate. It is important to note that this effect is not connected with necking of the specimen. Generally it holds that the higher the stress amplitude the shorter the stage of the decreasing cyclic creep rate. For the lowest stress amplitudes the onset of the accelerating cyclic creep was not observed at all.

An example of cyclic hardening/softening curves for the mean stress 50 MPa and different stress amplitudes is shown in Fig.4. The cyclic hardening at the beginning of all tests is replaced by softening; there is no saturation. The cyclic hardening is less pronounced for higher stress amplitudes. The plastic strain amplitude thus exhibits a similar behaviour as the cyclic creep rate. From the comparison of corresponding cyclic deformation curves and differentiated creep curves for given mean stress it follows that the onset of accelerating cyclic creep corresponds to the onset of cyclic softening. Fig.5 combines the data from the increasing branches of the curves for the mean stress 50 MPa in terms of the dependence between the cyclic creep rate and the plastic strain amplitude. It can be seen that there is a unique relation between the plastic strain amplitude and the creep strain rate for the given mean stress. Irrespective of the applied stress amplitude all the data points lie within one scatter band. Fig.6 shows (without experimental points) this dependence for three investigated mean stresses. The cyclic creep rate corresponding to the given plastic strain amplitude increases strongly with increasing mean stress.

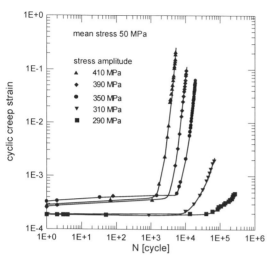

Fig.2. Cyclic creep curves for mean stress 50 MPa.

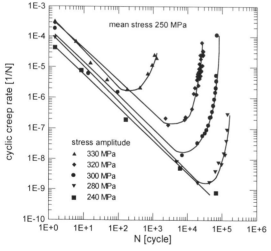

Fig.3. Differentiated creep curves for mean stress 250 MPa.

The tests presented above concern relatively low mean stresses not exceeding one half of the yield stress. As shown, in these cases there is no saturation of the plastic strain amplitude and no steady state creep. For low stress amplitudes (e.g. 240 MPa at the mean stress 250 MPa, Fig.3) the cyclic creep rate decreases below the practically measurable value. This finding can be interpreted in such a way that the prerequisite for the massive cyclic softening accompanied by cyclic creep rate increase is a high enough stress amplitude. For extremely high mean stress such an amplitude cannot be simply applied. This is documented in Fig.7 for the mean stress 500 MPa. The creep strain corresponding to the first cycle was of the order of some percent and there was naturally a severe difference for the lowest and highest (limited by σ_{UTS}) stress amplitude. In order to compare the cyclic creep strain development for different stress amplitudes the corrected creep strain ε_{ccN} - $\varepsilon_{ccN=1}$ was plotted against the number of elapsed cycles in Fig.7. Even for the highest stress amplitude 110 MPa (corresponding to 99% of the σ_{UTS}) the cyclic creep rate continuously decreases and neither creep acceleration, nor the fracture was observed (the fatigue test was continued at a resonant fatigue machine at frequency 100 Hz up to 2×10^7 cycle without failure).

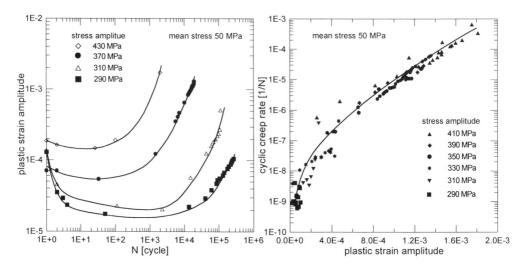

Fig.4. Cyclic deformation curves for mean stress 50 MPa.

Fig.5. The relation between the cyclic plastic rate and the plastic strain amplitude for mean stress 50 MPa.

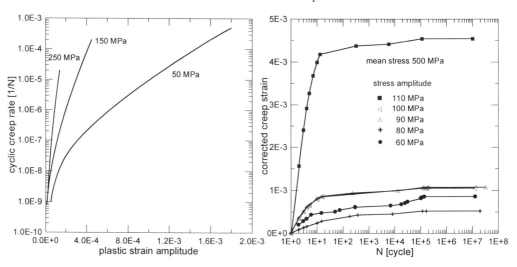

Fig.6. The dependence of the cyclic creep rate on the plastic strain amplitude.

Fig.7. Corrected creep strain rate curves for the mean stress 500 MPa.

The fractography performed by scanning electron microscopy showed that the failure is either of fatigue, ductile or mixed fatigue/ductile type. The fracture map is presented in Fig.8. Fatigue originated fractures are represented by open symbols, ductile creep fractures characterised by necking are depicted by full symbols and the transition, not entirely

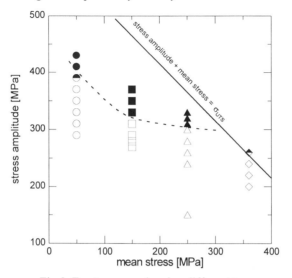

Fig.8. Fracture map showing different types of fracture in dependence on stress amplitude and mean stress.

identified fractures, by half-full symbols. The straight line represents the borderline above which ductile fracture occurs in the first half-cycle, i.e. the condition $\sigma_a + \sigma_{mean} = \sigma_{UTS}$. The ductile cyclic creep fracture is limited to the mean stresses below 300 MPa, i.e. nearly below $\sigma_{UTS}/2$. For higher mean stresses the fracture (if occurs) is of fatigue type.

CONCLUSIONS

1. The cyclic creep curves of cold worked nickel cycled under controlled stress with mean stresses lower than $\sigma_{UTS}/2$ exhibit a cyclic creep acceleration after an initial short period of cyclic creep deceleration. For mean stresses exceeding $\sigma_{UTS}/2$ the onset of cyclic creep acceleration was not observed. The ductile creep fracture was not observed for mean stresses exceeding $\sigma_{UTS}/2$.
2. The onset of the cyclic creep acceleration is related to the cyclic softening. Both the plastic strain amplitude and the cyclic creep rate are observed to increase with number of elapsed cycles.
3. For the given mean stress there is a unique relation between the cyclic creep rate and the plastic strain amplitude. Higher mean stress results in higher cyclic creep rate for a given plastic strain amplitude.

26

ACKNOWLEDGEMENT

The authors are grateful to the Grant Agency of the Czech Republic for the financial support under grant 106/96/0324.

REFERENCES

1. Oldroyd, P.W.J. (1979) *Fat. of Eng. Mater. and Structures* **1**, 297.
2. Peters, K.F., Radin, S., Radin A. and Laird, C. (1989) *Mat. Sci. Eng.* **A110**, 115.
3. Lukáš, P. and Kunz, L. (1989) *Int. J. Fatigue* **11**, 55.
4. Pilo, D., Reik, W., Mayr, P. and Macherauch E. (1979) *Fat. of Eng. Mater. and Structures* **1**, 287.
5. Eckert, R., Laird, C and Bassani J. (1987) *Mat. Sci. Eng.* **91**, 81.
6. Lukáš, P. and Kunz, L. (1989). In: *Proceeding of the 10th Int. Coll. on Mechanical Fatigue of Metals.* Technische Universität Dresden, Gaußig 1989, pp. 45 - 49.
7. Pokluda, J. and Staněk P. (1978) *Metallic materials* **16**, 583.

Low Cycle Fatigue and Elasto-Plastic Behaviour of Materials
K-T. Rie and P.D. Portella (Editors)

ISOTHERMAL HIGH TEMPERATURE FATIGUE BEHAVIOUR OF NiCr22Co12Mo9 UNDER SUPERIMPOSED LCF AND HCF LOADING

M. MOALLA, K.-H. LANG, D. LÖHE; E. MACHERAUCH

Institute of Material Science and Engineering I
University of Karlsruhe, Kaiserstr. 12, D-76128 Karlsruhe, Germany

ABSTRACT

The isothermal high temperature fatigue behaviour of NiCr22Co12Mo9 (Nicrofer 5520 Co, IN617) under superimposed LCF and HCF loading was investigated at 850°C and 1000°C. At a HCF total strain amplitude of 0.02%, the amplitude of the LCF loading was varied between 0.2 and 0.6%. In addition, at 850°C amplitudes of the HCF loading between 0.02 and 0.2% were applied at a constant LCF loading of 0.2 and 0.3%, respectively. The cyclic deformation behaviour changes from a neutral to a cyclic hardening character with increasing HCF amplitude. A low superimposed HCF amplitude of 0.02% changes the fatigue life only slightly. With increasing amplitudes of the HCF loading, however a considerable reduction of the lifetime was observed which may come to 90% of the fatigue life determined in a pure LCF tests at the same overall strain amplitude. The results from experiments with a constant LCF amplitude and varied HCF amplitudes plotted in double logarithmical scaling yield a linear dependence between the overall strain amplitude and the number of cycles to failure. The validity of this relationship for other constant LCF loadings was verified. With the combination of the relationships of Basquin and Coffin-Manson a successful lifetime prediction for the fatigue tests with superimposed LCF and HCF loading was established. To determine the effects of creep and relaxation processes, some fatigue tests with a dwell time of 50s at the maximum strain of the LCF loading were performed. Such a short hold time changes the cyclic deformation and the lifetime behaviour of the material only slightly.

KEYWORDS

Superimposed LCF and HCF loading, Nickel base superalloy, cyclic deformation, lifetime prediction.

INTRODUCTION

Repeated start up and shut down procedures of gas turbines cause a thermal induced low cycle fatigue loading in the wall of the combustion chamber. During operation, this LCF loading is superimposed by high cycle fatigue loading which results from mechanical vibrations and instationary combustion processes. Investigations of the alteration of the LCF behaviour of cast irons due to the superimposed HCF loading showed a significant change of the cyclic deformation behaviour and a considerable reduction of the fatigue life [1,2]. In the present study the cyclic deformation and the lifetime behaviour of a typical material for combustion chambers of gas turbines is investigated under superimposed LCF and HCF loading conditions. This is done in total strain controlled isothermal fatigue tests at different temperatures. The validity of the Coffin-Manson and Basquin relationship in predicting fatigue lifetime with regard to LCF and HCF loading, respectively, at very high temperatures is analyzed.

MATERIAL

The chemical composition of the investigated solid solution and carbide precipitation hardened nickel base superalloy NiCo22Co12Mo9 (Nicrofer 5520 Co, IN617) was 22.25 Cr, 11.45 Co, 8.88 Mo, 1.28 Al, 0.56 Fe, 0.4 Ti, 0.11 Si and 0.06 C (all quantities in wt.-%). The material was supplied as round bars which were annealed at 1200°C and water quenched. In the as received state, the microstructure shows grains with a high density of twin boundaries and a homogeneous distribution of primary carbides of type M_6C and Ti(C,N). The mean grain size amount to 160µm (Fig.1). From the supplied bars round specimens were manufactured with a solid

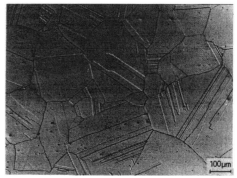

Fig. 1: Microstructure of NiCr22Co12Mo9 in the as received state

cylindrical gauge length of 10mm, a gauge diameter of 7mm and conical gripping heads.

EXPERIMENTAL PROCEDURE

The total strain controlled fatigue tests were carried out on a servohydraulic testing machine with a maximal loading capacity of 100kN. For the LCF loading, a triangular waveform with a cycle duration of 100s ($f^{LCF}=10^{-2}$ Hz) was chosen. Their amplitude $\varepsilon_{a,t}^{LCF}$ was varied between 0.2 and 0.6% for a constant superimposed HCF amplitude $\varepsilon_{a,t}^{HCF}$ of 0.02%. In addition, the HCF amplitude $\varepsilon_{a,t}^{HCF}$ was changed between 0.02 and 0.2% for a constant LCF loading of 0.2 and 0.3%, respectively. The HCF loading was sinusoidal with a frequency of 5Hz. For strain measurement a high temperature capacitive extensometer was used. The specimens were heated up to the test temperatures of 850°C or 1000°C with an induction system. The temperature was measured with a Ni-CrNi thermocouple spot welded close to the gauge length of the specimens. Some of the fatigue tests were done with a dwell time of $t_h=50s$ at the maximum strain of the LCF loading to consider the effects of creep and relaxation.

RESULTS AND DISCUSSION

Cyclic Deformation Behaviour

For both temperatures, Fig.2 shows hysteresis loops determined at N=1 (top) and just before failure at N≈N_f (bottom) from experiments with $\varepsilon_{a,t}^{LCF}$ =0.6%. On the left side, hysteresis loops from tests with pure LCF loading were plotted. The hysteresis loops on the right side in Fig.2 were determined during the LCF/HCF superposition with $\varepsilon_{a,t}^{HCF}$ =0.02%. The fluctuations of the nominal stress at 850°C in the first cycle at both loading conditions are the consequence of dynamic strain ageing effects resulting from interaction between moving dislocations and diffusing solute atoms [3]. The induced stress amplitude decreases with increasing temperature because Youngs modulus as well as the strength of materials decrease at high temperatures. Futhermore, the distinct augmentation of the plastic strain amplitude with increasing temperature results from the thermal activation of dislocation movements. For both loading conditions and both temperatures, the effects of the macro crack opening can be observed from the hysteresis loops evaluated just before failure at $N \approx N_f$. Under tension the maximum induced stress decreases significantly as a result of the decrease of the stiffness of the specimen whereas under compressive loading the hysteresis loops show an inflexion due to the crack closure effects.

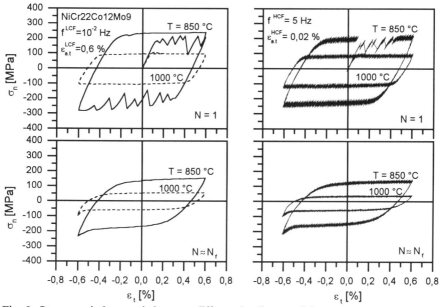

Fig. 2: Stress strain hysteresis loops at different loading conditions

Figure 3 contains the development of the plastic strain (top), the stress amplitudes and the mean stresses (bottom) during isothermal fatigue tests at 850°C with pure LCF loading (solid lines) and with superimposed LCF and HCF loading (broken lines). In experiments with pure LCF loading the cyclic deformation behaviour is neutral. Also the superposition of a low $\varepsilon_{a,t}^{HCF}$ value of 0.02% changes this behaviour hardly. Only at a superposition with $\varepsilon_{a,t}^{HCF}$ values higher than 0.05%, cyclic hardening during the first ten cycles associated with an increasing stress amplitude and up to $\varepsilon_{a,t}^{HCF} = 0.1\%$ with a decreasing plastic strain amplitude are observed. It appears that with the augmentation of $\varepsilon_{a,t}^{HCF}$ the potential for the cyclic hardening has to be characterized more and more by the higher frequent HCF loading. Therefore, in the fatigue test with the relatively high HCF amplitude of $\varepsilon_{a,t}^{HCF} = 0.2\%$ the plastic strain amplitude remains approximately constant whereas the stress amplitude increases. The fluctuations of the $\varepsilon_{a,p}$ values evaluated from experiments with the LCF/HCF superposition can be attributed to more pronounced dynamic strain ageing processes and depend clearly on the time of the occurrence of this effects during each individual cycle. For all loading conditions, the plastic strain amplitude and the stress amplitude increase with increasing overall strain amplitudes. The only exception was observed by the fatigue test with the LCF/HCF superposition of $\varepsilon_{a,t}^{HCF} = 0.1\%$. This anomaly can be attributed to the scatter because only one fatigue experiment per load combination could be performed. Concerning the induced mean stresses compressive stresses appear during all experiments. The σ_m values remain relatively small and change hardly.

Lifetime

During high temperature LCF tests, damages may occur not only due to mechanical fatigue loading but also due to the oxidation and creep processes. The superposition of LCF loading with HCF loading creates an even more complicated situation. Therefore, it is necessary to verify the already existing models for predicting fatigue life.

The total strain amplitude includes an elastic and a plastic part. In this investigation, $\varepsilon_{a,p}$ and $\varepsilon_{a,e}$ were determined at $N_f/2$ at both testing temperatures and plotted as a function of the number of cycles to failure in double

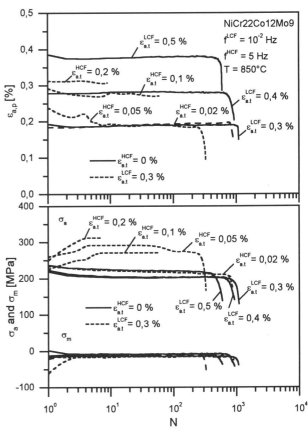

Fig. 3: Cyclic deformation curves at 850°C during fatigue tests with different loading conditions

logarithmic graphs. As an example, Fig.4 shows such an evaluation from fatigue tests with pure LCF loading and with superimposed HCF loading of 0.02% at 850°C. The $\varepsilon_{a,p}$ values at $N_f/2$ (left) appear to be linearly dependent on N_f and the determined $\varepsilon_{a,p}$ data points can be fitted in a regression straight line according to the Coffin-Manson relationship [4,5]:

$$\varepsilon_{a,p} = \varepsilon_f * N_f^{-\alpha} \tag{1}$$

where α is the fatigue ductility exponent and ε_f is the fatigue ductility coefficient.

Usually, the determined values of the elastic strain amplitude at $N_f/2$ also show a linear dependence on N_f which can be described generally with a relationship according to Basquin [6]:

$$\varepsilon_{a,e} = \sigma_f/E * N_f^{-\beta} \tag{2}$$

where β is the fatigue strength exponent, σ_f the fatigue strength coefficient and E the Young's modulus.

As can be seen in the right part of Fig.4, exceptionally in this case where the testing temperature is relatively high, it can be assumed that the elastic strain amplitude remains approximately independent of the nominal loading (β=0) for both loading conditions.

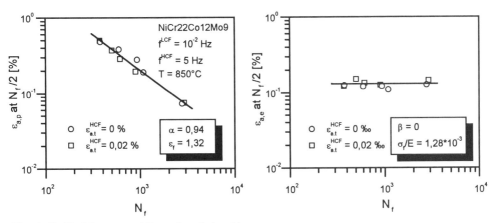

Fig. 4: Coffin-Manson and Basquin relationship

After the determination of the parameters of the relationships of Coffin-Manson and Basquin from fatigue tests with $\varepsilon_{a,t}^{HCF}$ = 0% and 0.02%, the number of cycles to failure for a given loading condition could be calculated by summing up the equation (1) and (2) to the total strain amplitude. The results are illustrated in Fig.5. They demonstrate a good agreement between the calculated lifetime (solid line) and the data points (open squares and circles) from the fatigue tests with $\varepsilon_{a,t}^{HCF}$ = 0% and 0.02%.

Also Fig.5 demonstrates that the superposition of a relatively low $\varepsilon_{a,t}^{HCF}$ =0.02% influences the fatigue life only slightly. To evaluate the effect of a variation of the superimposed HCF loading, supplementary total strain Wöhler curves at 850°C with different HCF amplitudes at a constant $\varepsilon_{a,t}^{LCF}$ =0.3% (broken line) were determined and compared to the lifetime from experiments with pure LCF loading with the same overall strain amplitude. As can be seen in Fig.5, the superposition of a suffi-

ciently high $\varepsilon_{a,t}^{HCF}$ causes a pronounced reduction of the fatigue life. For example, N_f from the fatigue test with the LCF/HCF superposition of $\varepsilon_{a,t}^{LCF}$ =0.3% and $\varepsilon_{a,t}^{HCF}$ =0.1% decrease about 92.3% compared to N_f from the fatigue experiment with a pure LCF loading of $\varepsilon_{a,t}$ =0.4%.

Furthermore, the double logarithmic Wöhler curve in Fig.5 illustrates that the overall strain amplitude $\varepsilon_{a,t}$ from fatigue experiments with the LCF/HCF superposition of a constant $\varepsilon_{a,t}^{LCF}$ =0.3% and different HCF amplitudes can be

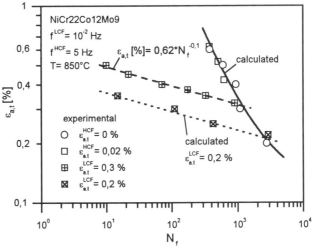

Fig. 5: Total strain Wöhler curves at 850°C from fatigue tests with different loading conditions

32

fitted with a straight line. To verify the validity of such a relationship for other superimposed loading amplitudes, supplementary fatigue tests with a constant $\varepsilon_{a,t}^{LCF}$ =0.2% and different $\varepsilon_{a,t}^{HCF}$ were carried out. To calculate the actual course of $\varepsilon_{a,t}$ as a function of N_f at $\varepsilon_{a,t}^{LCF}$ =0.2% the available data points from the fatigue tests with $\varepsilon_{a,t}^{HCF}$ =0% and 0.02% are used as a reference point to which the regression curve at $\varepsilon_{a,t}^{LCF}$ was shifted parallel. As can be seen in Fig.5, the calculated lifetime plotted as a dotted line has a good agreement with the experimental data points. So it can be assumed that the endurance line from fatigue tests with the LCF/HCF superposition of a constant LCF loading and varied HCF amplitudes are displaced only parallel to each other, i.e. the slope of the straight lines remains constant for each LCF amplitude.

Hold Time Effects

Additional fatigue tests proved that a hold time of 50s at the maximum strain in the LCF loading has a rather small influence on the cyclic deformation and the lifetime behaviour. The total strain Wöhler curves from experiments without and with a hold time in Fig.6 indicate the somewhat more pronounced lifetime reduction with decreasing $\varepsilon_{a,t}$. This can be attributed to the more pronounced diffusion controlled intergranular creep damage of grain boundaries associated with a coarsening of carbide precipitations in the vicinity of the grain boundaries which are observed by microscopic investigations after failure of the specimen.

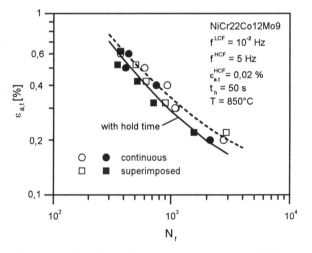

Fig. 6: Total strain Wöhler curves at 850°C from fatigue tests without and with a hold time

ACKNOWLEDGEMENTS

The support of the investigation by the "Deutsche Forschungsgemeinschaft (DFG)" within the "Sonderforschungsbereich 167" is gratefully acknowledged.

REFERENCES

1. Lang K.-H. (1991). In: Werkstoffkunde - Beiträge zu den Grundlagen and zur interdisziplinären Anwendung. DGM-Verlag, Oberursel, S. 79-88.
2. Hallstein R. (1991). Dr.-Ing. Dissertation, Universität Karlsruhe, Germany.
3. Kleinpaß B., Lang K.-H., Löhe D. and Macherauch E. (1996). In: Proceedings of the Symposium on Fatigue under Thermal and Mechanical Loading held at Petten , The Netherlands, Bressers J. and Rémy L. (Eds.), Kluwer Academic Publishers, pp. 327-337.
4. Manson S. S. (1954). Naca-Report 1170.
5. Coffin L. F. Jr. and Schenectady (1954). Trans. ASME 76, pp. 931-950.
6. Basquin O. H. (1910). Proc. ASTM 10, pp. 625-630.

Low Cycle Fatigue and Elasto-Plastic Behaviour of Materials
K-T. Rie and P.D. Portella (Editors)

LOW CYCLE FATIGUE OF SUPERALLOY SINGLE CRYSTALS CMSX-4

K. OBRTLÍK, P. LUKÁŠ and J. POLÁK
Institute of Physics of Materials, Academy of Sciences of the Czech Republic, Žižkova 22,
61662 Brno, Czech Republic

ABSTRACT

Total strain controlled tests have been performed on single crystals CMSX-4 to study the effect of orientation and temperature on LCF characteristics at asymmetrical total strain cycle. The effect of orientation was studied at 700 °C on single crystals with stress axes <001>, <011> and <111>; the effect of temperature was studied on <001> oriented single crystals in the temperature range 700 - 950 °C. The cyclic shear stress-strain curve at 700 °C was found to be independent of orientation but dependent on the strain cycle asymmetry. The Coffin-Manson curve at 700 °C is also independent of orientation. Fatigue slip bands lying along the {111} slip planes were found only at 700 °C; at higher temperatures they were not observed. The cyclic stress-strain curve of <001> oriented crystals is shifted towards lower stresses for higher temperature.

KEYWORDS

Superalloys, single crystals, cyclic stress-strain curve, fatigue life curves, strain localisation.

INTRODUCTION

Superalloy single crystals are used for the production of blades and vanes of gas turbines. Besides a good resistance against creep and thermomechanical fatigue also a high resistance against isothermal low cycle fatigue is required for this kind of material. CMSX-4 is the second generation superalloy whose creep resistance is markedly enhanced by the addition of about 3wt.% rhenium. Scarce data on LCF behaviour of CMSX-4 single crystals concern mainly crack propagation [1,2]. Orientation and temperature dependence of the LCF characteristics has been studied on several alloys [3-10] that have a similar γ/γ' two-phase structure as the alloy studied in this paper. It is generally accepted that the very strong orientation dependence of the number of cycles to fracture on the total strain range can be related chiefly to elastic modulus, low modulus orientations having longer lives. As for the cyclic plasticity the picture offered by the data in the literature is not unique. Both cyclic hardening and softening have been observed; it is important that in all the cases saturation was found to follow the hardening and/or softening [3,5,6] This makes it possible to define uniquely the cyclic stress-strain curve (CSSC). The orientation dependence of the CSSCs is weak or almost none especially when plotted in terms of resolved shear values [3,4,10]. The CSSCs are shifted towards lower stresses for higher temperatures [3,6]. At lower temperatures (at and below 650 °C) the CSSC exhibits a clear plateau [6,10]. This can be related to the persistent slip bands (PSBs) found in the structure at these lower temperatures. At higher temperatures (above 760 °C) no PSBs were found [4,11,12,13]. This is also the main difference in dislocation structures observed after cycling at different temperatures. Also the thresholds values for crack nucleation and propagation were found to be orientation and temperature dependent [7,8]. All the relevant data were obtained in symmetrical cycling. The aim of this paper is to investigate the effect of orientation and temperature on LCF characteristics of CMSX-4 single crystals at asymmetrical total strain cycle.

EXPERIMENTAL

The single crystals were kindly provided by Howmet Ltd., England in the framework of COST 501 Round III collaborative programme. They were delivered as cast rods in fully heat treated condition. The applied solution cycle resulted in solutioning of γ' better than 99% without incipient melting problems.

Table 1. Composition of CMSX-4 (wt. %)

Cr	Mo	W	Co	Ta	Re	Hf	Al	Ti	Ni
6.5	0.6	6.4	9.7	6.5	2.9	0.1	5.7	1.0	bal.

The microstructure of CMSX-4 consists of cuboidal γ' precipitates embedded in a γ matrix. The γ' particle size is about 0.5 μm and the volume fraction of the γ' phase is about 70%. The lattice parameters of γ matrix and the γ' precipitates are slightly different. The misfit parameter is negative; its value is about -1×10^{-3}.

The LCF tests were performed in air under total strain control tests on specimens having gauge length 15 mm and diameter 6 mm in a computer controlled servohydraulic testing system at total strain rate $1 \times 10^{-3} s^{-1}$ and total strain cycle asymmetry R_ε either 0.05 or 0.5.

The hysteresis loops were recorded and digitally stored at the pre-set numbers of loading cycles. This made it possible to determine the elastic moduli and the dependencies of the stress amplitude, plastic strain amplitude and mean stress on the number of loading cycles.

RESULTS AND DISCUSSION

Example of the cyclic hardening/softening curves is shown in Fig.1a. It can be seen that - with the exception of the highest amplitudes – the cyclic stress response is stable. The same holds for all the orientations and temperatures. As our cycling is asymmetrical, the mean stress response is also an important characteristic. This response is shown for the tests presented in Fig.1a in Fig.1b. It can be seen that the mean stress relaxation is more expressive for higher strain amplitudes and for higher temperatures. At 700 °C and 850 °C the mean stress levels off, at 950 °C there is no saturation. As expected, the mean stress was found to increase with increasing strain cycle asymmetry.

The stable cyclic stress response makes it possible to construct easily the CSSCs. Fig.2 shows the CSSCs for the investigated orientations at 700 °C in the shear stress vs. plastic shear strain

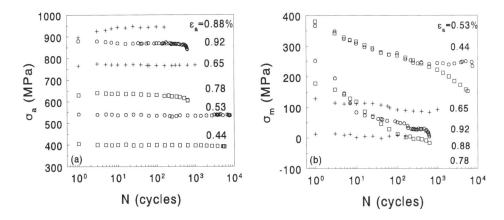

Fig.1. Cyclic stress amplitude (a) and mean stress (b) in dependence on number of cycles for
<001> crystals at $R_\varepsilon = 0.05$ and different temperatures. + 700 °C, o 850 °C, □ 950 °C.

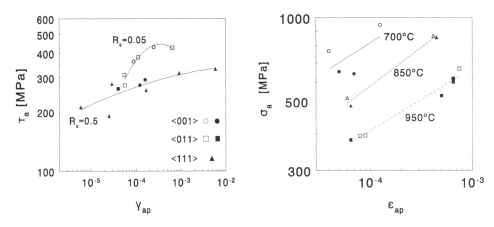

Fig.2. Cyclic stress-strain curve for three
multiple slip orientations at 700 °C.

Fig.3. Cyclic stress-strain curves for <001>
oriented crystals at three temperatures.

representation. The axial stress and plastic strain amplitudes were resolved into the slip system
{111}<011> having the highest Schmid factor. It can be seen that the CSSC does not depend
on the orientation, but it does depend on the mean strain. The CSSC is shifted towards lower
stresses for higher strain cycle asymmetry. The temperature dependence of the CSSC for
<001> oriented crystals is displayed in Fig.3. It can be seen that the CSSC depends strongly on
the temperature. For the given plastic strain amplitude the stress amplitude decreases with
increasing temperature. A slight effect of the mean strain can be again stated. The limited
amount of experimental points indicate that the effect of strain cycle asymmetry decreases with
increasing temperature. The cyclic stress-strain response is certainly closely related to the
dislocation structures. The dislocation structure in <001> oriented crystals was studied by

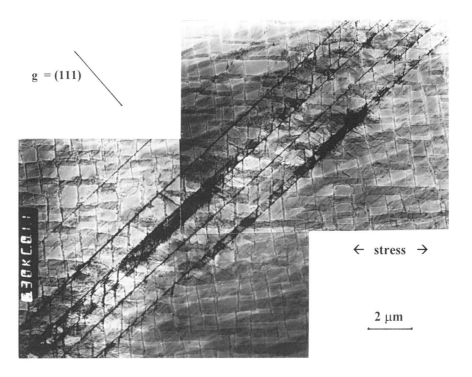

Fig.4. Dislocation structure in <001> crystal tested at 700 °C ($R_\varepsilon = 0.05$; $\varepsilon_{at} = 0.88$ %).

transmission electron microscopy. The most striking difference between the specimens tested at different temperatures is the presence of the PSBs at 700 °C and their absence at 850 °C and 950 °C. Fig.4 shows the structure in the section containing the stress axis and perpendicular to the slip plane (111) along which the PSBs run across the whole crystal. The PSBs appear here as very thin slabs (thickness below 0.1 µm) going trough both the γ channels and the γ' particles. The dislocation within the PSBs cannot be resolved. It can be only speculated that the dislocation density within them is very high. Fig.5 shows the PSBs using another technique – the scanning electron microscopy of the surface of cycled specimen. The PSBs can be seen here as white bands along the (111) plane. The much shorter and less regular lines along the trace of the (001) plane correspond to the horizontal γ channels. These lines are probably formed by the slip activity of more slip systems of the type {111}<001> operating in the γ channels. Their activity results in "squeezing out" or "sucking in" of the γ matrix in between the harder γ' particles. It should be stressed that the PSBs shown in Figs.4 and 5 can be seen only at specimens cycled at higher total strain amplitudes at 700 °C. At low amplitudes at 700 °C the PSBs have never been observed. The same holds for arbitrary amplitude at 850 °C and 950 °C. Chieragatti and Remy [10] found a clear plateau at the CSSC in the case of MAR-M200 single crystals tested in fully reversed cycle at 650 °C. The reason why the presence of the PSBs in our case does not manifest itself by the plateau must be sought either in their low density or in the fact that they harden very quickly after their formation and do not thus represent zones of substantially higher cyclic slip activity under asymmetrical cycling.

(111) plane trace   (001) plane trace

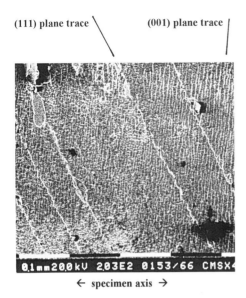

← specimen axis →

Fig.5. Surface relief of <001> crystal tested at 700 °C ($R_\varepsilon = 0.05$; $\varepsilon_{at} = 0.88$ %).

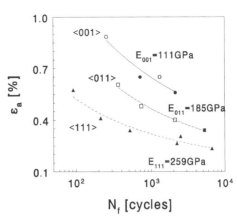

Fig.6. Total strain amplitude vs. number of cycles to fracture for three multiple slip orientations at 700 °C. Open symbols $R_\varepsilon = 0.05$; full symbols $R_\varepsilon = 0.5$.

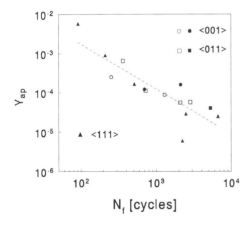

Fig.7. Coffin-Manson plot for three multiple slip orientations at 700 °C. Open symbols $R_\varepsilon = 0.05$; full symbols $R_\varepsilon = 0.5$.

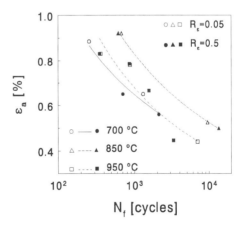

Fig.8. Total strain amplitude vs. number of cycles to fracture of <001> crystals at three different temperatures.

As expected, the fatigue life plotted in terms of the total strain amplitude vs. number of cycles to fracture at 700 °C (Fig.6) is very strongly orientation dependent. This is in agreement with the above quoted data of other authors. On the other hand, the fatigue life plotted in terms of resolved cyclic plastic strain amplitude vs. number of cycles to fracture, i.e. the Coffin-Manson diagram (Fig.7), is independent of orientation.

The temperature dependence of the fatigue life curves for <001> oriented crystals is not systematic. This is documented by Fig.8 showing the dependence of the total strain amplitude vs. number of cycles to fracture at 700 °C, 850 °C and 950 °C.

CONCLUSIONS

The results of total strain controlled tests performed on single crystals CMSX-4 at asymmetrical strain cycle can be summarised as follows.
(1) Cyclic shear stress-strain curve at 700 °C does not depend on orientation but depends on strain cycle asymmetry. The Coffin-Manson curve is also independent of orientation.
(2) Fatigue slip bands lying along the {111} slip planes were found only at 700 °C; at higher temperatures they were not observed.
(3) Cyclic stress-strain curve of <001> oriented crystals depends systematically on temperature. Temperature dependence of fatigue life curves is not systematic.

ACKNOWLEDGEMENT

This research was supported by the Grant Agency of the Academy of Sciences of the Czech Republic under the contract No. A2041704/1977. This support is gratefully acknowledged.

REFERENCES

1.   Sengupta, A., Putatunda, S.K. and Balogh, M. (1994) *J. Mater. Eng. Performance* **3**, 540.
2.   Sengupta, A. and Putatunda, S.K. (1994) *Scr. Metallurg. Mater.* **31**, 1163.
3.   Gabb, J., Gayda, J. and Miner, R.V. (1986) *Metallurg. Trans.* **17A,** 497.
4.   Gabb, J., Miner,R.V. and Gayda, J. (1986) *Scr. Metallurg.* **20**, 513.
5.   Milligan, W.W. and Jayaraman, N. (1986) *Mater. Sci. Eng.* **82**, 127.
6.   Fritzemeier, L.G. and Tien, J.K. (1988) *Acta Metallurg.* **36**, 283.
7.   Henderson, M.B. and Martin, J.W. (1992). In: *Superalloys 1992*, pp.707-716; Antolovich, S.D. et al. (Eds), TMS, Warrendale.
8.   Aswath, P.B. (1994) *Metallurg. Mater. Trans.* **25A**, 287.
9.   Li, S.X., Ellison, E.G. and Smith, D.J. (1994) *J. Strain Anal.* **29**, 147.
10.   Chieragatti, R. and Remy, L. (1991) *Mater. Sci. Eng.* **A141**, 11.
11.   Glatzel, U. and Feller-Kniepmeier, M. (1991) *Scripta Metallurg. Mater.* **25**, 1845.
12.   Decamps, B., Brien, V. and Morton, A.J. (1994) *Scripta Metallurg. Mater.* **31**, 793.
13.   Zhang, J.H., Hu, Z.Q., Xu, Y.B. and Wang, Z.G. (1992) *Metallurg. Trans.* **23A,** 1253.

Low Cycle Fatigue and Elasto-Plastic Behaviour of Materials
K-T. Rie and P.D. Portella (Editors)
39

FRACTURE MECHANISM OF FERRITIC DUCTILE CAST IRON IN EXTREMELY LOW CYCLE FATIGUE

J. KOMOTORI and M. SHIMIZU

Department of mechanical engineering, Keio University
3-14-1, Hiyoshi, Kohoku-ku, Yokohama 223-8522, JAPAN

ABSTRACT

Low cycle fatigue tests were carried out over a wide range of plastic strain levels including an Extremely Low Cycle Fatigue (ELCF) regime on two kinds of Ferritic Ductile Cast Iron (FDI) having fine- and coarse-grained graphites. To clarify the fracture mechanism in an ELCF regime, observation of microfracture behavior was also performed both on the surface and on a longitudinal section of partially fatigued specimens. In an Ordinary LCF regime, a fatal crack is formed by the propagation and frequent coalescence of small surface cracks. In an ELCF regime, however, an internal crack formed by coalescence of microvoids, which originate from the debonding of matrix-graphite interface inside the material, leads to final fracture of a specimen.

KEYWORDS

Extremely low cycle fatigue, fracture mode transition, ferritic ductile cast iron.

INTRODUCTION

Since a Ferritic Ductile Cast Iron (FDI) has excellent mechanical properties, it is now widely used in place of steel as the preferred material for various machine parts, such as crankshafts, camshafts and different types of gears. With increased demands for application of this material in various fields of industry, it has become important to clarify the fatigue fracture mechanism under severe loading conditions, such as that of a low cycle fatigue, including an Extremely Low Cycle Fatigue (ELCF) regime.

ELCF is a fatigue phenomenon at a very high level of plastic strain range $\Delta\varepsilon_p$ where fatigue life is less than about 100 cycles[1-3]. It is very important to have some basic knowledge concerning the fracture mechanism of FDI in an ELCF regime in order to prevent a catastrophic failure of machine components and structures during destructive earthquakes. However, many problems concerning the fatigue damage in an ELCF regime, remain unsolved.

The aims of the present study are (i) to clarify the fatigue fracture mechanism of FDI in an ELCF regime and (ii) to discuss the microstructural effect on fracture behavior both in an Ordinary LCF and ELCF regime.

MATERIALS AND METHODS

Two types of ferritic ductile cast iron, having fine- and coarse-graphite, were used in this study. Typical microstructures of both materials are shown in Fig. 1. Mechanical properties and metallurgical parameters are given in Table 1. Strain controlled low cycle fatigue tests were carried out over a wide range of plastic strain levels under push-pull loading condition. Cyclic strain amplitude was controlled by detecting the change in minimum diameter of the hourglass shaped specimen. To clarify the fatigue fracture mechanisms, a direct observation of a microfracture behavior was performed both on the surface and on a longitudinal section of the partially fatigued specimen using an optical microscope.

RESULTS AND DISCUSSION

Fatigue Fracture Mechanism in Extremely Low Cycle Fatigue Regime

The fracture behavior of the specimens after a given number of strain cycles were examined on its longitudinal section and on the surface using an optical microscope. As for the results, two different types of fracture modes, depending on the plastic strain range levels, were observed both in coarse- and fine- grained materials.

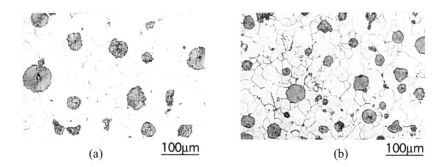

(a) 100μm   (b) 100μm

Fig. 1. Microstructures: (a) coarse-grained material; (b) fine-grained material

Table 1. Mechanical properties and metallurgical parameters

	Graphite grain size(μm)	Nodularity (%)	Vf (%)	Y.S. (MPa)	T.S. (MPa)	ε_f (%)
Fine-grained	23	78	11	282	427	29
Coarse-grained	30	76	12	272	415	30

In a small plastic strain range (OLCF regime), surface cracks originating from spheroidal graphites lead to the final fracture of the specimen. In this case, two types of crack initiation sites were observed; (i) crack initiation at the graphite in the surface microstructure and (ii) crack initiation at the subsurface adjacent to the graphite just below the surface (see Fig. 2). A fatal crack is formed by the propagation and frequent coalescence of the small cracks. This is referred to as a surface fracture mode.

In a relatively large plastic strain range (ELCF regime), however, development of internal crack leads to the final fracture of the specimen. Fig.3 shows a typical feature of an internal macrocrack observed on the longitudinal section of the specimen at the strain cycle ratio just before the final fracture. This is referred to as an internal fracture mode.

To clarify the internal crack formation process, microfracture behaviors were examined on the longitudinal section of the partially fatigued specimens using an optical microscope. As for the results, the microvoids caused by debonding of graphite-matrix interface were observed at the early stage of fatigue life, fatigue life ratio n/Nf is less than about 0.1 . The frequent coalescence of these microvoids was also observed as shown in Fig. 4. At the final stage of fatigue life, a rapid coalescence of the microvoids occurs and a macrocrack, which is shown in Fig.3, is formed inside the material . The schematic illustration of these results are given in Fig. 5, which shows the internal crack formation process.

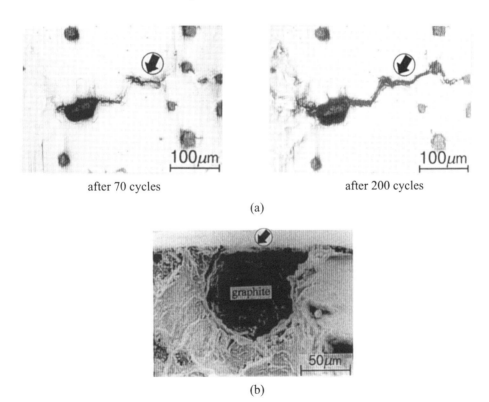

after 70 cycles   after 200 cycles

(a)

(b)

Fig. 2. Typical example of surface fracture mode (Coarse-grained, $\Delta\varepsilon_p$=0.006, Nf=573cycles): (a) successive observation of surface crack initiation behavior; (b) crack initiation site

Fig. 3. Typical feature of internal crack observed on the longitudinal section of specimen
(Coarse-grained material, $\Delta\varepsilon_p$=0.1, after 5 cycles)

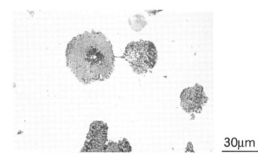

Fig. 4. Coalescence of microvoids (Coarse-grained material, $\Delta\varepsilon_p$=0.1, after 2 cycles)

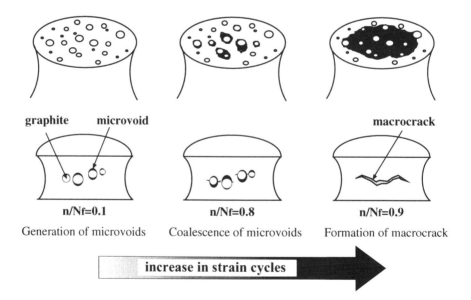

Fig. 5. Schematic illustration of internal crack formation process

Microstructural Effect on Fatigue Life of FDI

Results of low cycle fatigue tests are shown in Fig. 6, where solid and hollow marks represent the results for the specimen in which failure occurred in the internal and surface fracture modes, respectively. There is no difference of fatigue life for both materials in an ELCF regime. In an OLCF regime, however, the fatigue life of material having fine-grained graphite is greater than that of coarse-grained material. This suggests that the metallurgical parameters which determine the fatigue life change, depending on the level of $\Delta\varepsilon_p$ and on the type of fracture modes.

The solid lines in Fig. 6 are the predicted fatigue life curves which have been calculated on the basis of the law of small surface crack. The detailed procedure has been described by Harada and the others [4]. There is a good agreement with the calculated and observed fatigue life in both coarse- and fine-grained materials. This implies that the difference of fatigue life in an OLCF regime is caused by the difference of surface crack propagation rate. Since the initial surface crack length, which strongly affects the propagation of microcrack, is controlled by the size of graphite grain, the crack propagation rate of coarse-grained material is greater than that of fine-grained material.

In the case of an ELCF regime, where the specimen fails in an internal fracture mode, it has been proposed by author's [1] that a fatigue life is determined by the exhaustion of ductility under strain cycling. Accumulation of internal fatigue damage in an ELCF regime can be detected by taking a measurement of residual fracture ductility. Fig. 7 shows the relationship between $\varepsilon_{FR}/\varepsilon_f$ and n/N_f, where ε_f represents the fracture ductility of the original material and ε_{FR} is the residual fracture ductility which is defined as the static fracture ductility of a partially fatigued specimen. It should be noted that no appreciable difference of the change in the value of residual fracture ductility can be observed. This is caused by the fact that the volume fraction of spheroidal graphite which controls the fracture ductility of the material is almost the same in both materials .

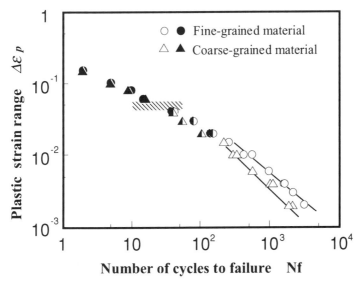

Fig. 6. Results of low cycle fatigue tests

44

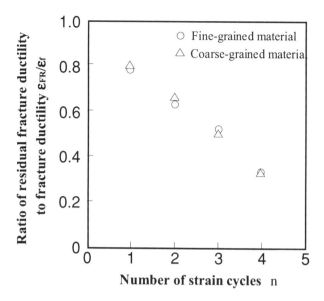

Fig. 7. Change in residual fracture ductility with strain cycling

CONCLUSION

(1) The transition of fracture mode, from surface to internal fracture, was observed with an increase in plastic strain range. In an Ordinary Low Cycle Fatigue regime, a fatal crack is formed by the propagation and frequent coalescence of small surface cracks. In an Extremely Low Cycle Fatigue regime, however, the internal crack which originates from the debonding of matrix-graphite interface inside the material leads to final fracture of a specimen.

(2) In an OLCF regime, fatigue life of fine-grained material is greater than that of coarse-grained material. This is because initial surface crack length in coarse-grained material, which controls the crack propagation rate, is much longer than that of fine-grained material. However, there is no difference in fatigue life in both fine- and coarse-grained materials in an ELCF regime. This is because the fatigue life in this regime is mainly controlled by exhaustion of fracture ductility, the volume fraction of spheroidal graphite affects the fatigue life instead of affecting size distribution.

REFERENCES

1. Kunio, T., Shimizu, M., Ohtani, N. and Abe, T. (1988) *ASTM STP* **942**, 751.
2. Shimada, K., Komotori, J. and Shimizu, M. (1987). In: *Proceedings of the second international conference on low cycle fatigue and elasto-plastic behavior of materials*, pp.680-686; Rie, K. T. (ed.). Elsevier, London.
3. Komotori, J., Adachi, T. and Shimizu, M. (1993). In: *Proceedings of the 5th international conference on fatigue and fatigue thresholds vol. 1*, pp.183-188; Bailon, J. P. and Dickson, J.I. (eds.). EMAS, West Midlands.
4. Harada, S., Murakami, Y., Fukushima, Y. and Endo, T.(1988) *ASTM STP* **942**, 1181.

Low Cycle Fatigue and Elasto-Plastic Behaviour of Materials
K-T. Rie and P.D. Portella (Editors)

LOW CYCLE FATIGUE BEHAVIOUR OF RECYCABLE
HIGH STRENGTH PM-STEELS

A. NEUBAUER, K.-H. LANG, O. VÖHRINGER, D. LÖHE
Institute of Materials Science and Engineering I
University of Karlsruhe, Kaiserstrasse 12, D-76128 Karlsruhe

ABSTRACT

The low cycle fatigue properties of copper free high strength PM-steels have been examined in the as sintered state at R = -1 under stress controled axial push-pull loading. The deformation behaviour of the materials is characterized by cyclic hardening in the first loading cycles followed by a period of nearly constant plastic strain amplitudes and finally by macroscopic cyclic softening up to the failure. Due to asymmetrical plastic deformation and microscopic crack initiation in the first cycle, generally positive mean strains $\varepsilon_{m,t}$ occur. With increasing numbers of cycles $\varepsilon_{m,t}$ rises in a degressive way during the first cycles, remains then nearly constant and grows in a progressive manner for the final cycles before fracture. The cyclic stress-strain curve proceeds above the monotonic stress-strain curve. The correlation between σ_a and $\varepsilon_{a,p}$ at $N_f/2$ can be described well with the Ramberg-Osgood-equation $\sigma_a = K' \cdot (\varepsilon_{a,p})^{n'}$. From light- and scanning-electron-microscopic analyses of the surfaces of several specimens after defined numbers of cycles a Wöhler curve for macroscopic crack initiation has been established. It was found, that the occurance of single cracks with a length of about 100 to 300 μm at the surfaces of the specimen correlate with a change in the increase of the $\varepsilon_{a,p}$-curve and in the decrease of the inverse tensile compliance plotted versus N.

KEYWORDS

PM-steels, Wöhler slopes, cyclic deformation behaviour, microscopic surface analyses, crack initiation, crack propagation

INTRODUCTION

The mass production of complex high strength parts manufactured by powder metallurgical technics increases continuously since several years, mainly because of economic reasons [1-3]. Most of these PM-parts, especially in the automobile industry, are applied under cyclic loading [4-6]. Compared to convential cast alloys, PM-steels often show worse low cycle fatigue properties because of the remaining porosity after sintering [7-8]. In order to increase the fatigue properties and to take into account modern aspects of recycling, health and environmental protection in the production process several copper free high strength PM-steels have been developed. The low cycle fatigue properties and the deformation behaviour of these copper free PM-steels are influenced by numerous factors like the density, which is the dominating factor, the pore morphology, the alloying elements, the sintering conditions, heat treatments etc. [9]. In order to improve the knowledge about these influencing parameters on the low cycle fatigue properties, three copper free PM-steels have been examined in the as sintered state.

MATERIALS AND TESTING EQUIPEMENT

The materials used, the manufacturing parameters, their chemical composition, and their density are shown in table 1. The materials consist of the base powders Fe4Ni0,5Mo (MSP4) and Fe1,5Mo (Astaloy Mo) with additions of 0,5%C in the form of graphite. After admixing lubricants and 5%Carbonyl-Iron-Powder (CEP) in the case of Fe1,5Mo+0,5%C+5%CEP specimens (see Fig. 1) were pressed at 700 MPa in a withdrawal process with filling tool to the given green densities. The following sintering happens in a walking beam furnace at 1250 °C for 40 min under a sintering atmosphere of 90% nitrogen and 10% hydrogen, followed by a slow cooling in the furnace.

Table 1: Materials, manufacturing parameters, chemical composition and density

material		Fe4Ni0,5Mo +0,5%C	Fe1,5Mo +0,5%C	Fe1,5Mo+0,5%C +5%CEP
pressing (MPa)		700	700	700
green density (g/cm³)		7,07	7,18	7,16
sintering		1250°C, 40 min, 10%H₂, 90%N₂	1250°C, 40 min, 10%H₂, 90%N₂	1250°C, 40 min, 10%H₂, 90%N₂
con-sti-tuents	C (Ma%)	0,454	0,495	0,541
	Ni (Ma%)	4,254	0,056	0,054
	Mo (Ma%)	0,524	1,517	1,416
	Cu (Ma%)	0,104	0,045	0,043
	Mn (Ma%)	0,109	0,115	0,115
	Fe (Ma%)	balance	balance	balance
density (g/cm³)		7,21	7,3	7,32

The micrographs of the three materials are given in Fig. 2. All materials show a homogeneously degenerated pearlitic microstructure. Fe4Ni0,5Mo+0,5%C additionally contains several small martensitic grains. Prior to testing the dimensions of each specimen are measured and the edges are broken in a defined way.

All tests are performed with a 100 kN Schenk-Hydropuls servohydraulic testing system at room temperature. The specimens are clamped with force-locking hydraulic axial grips and two capacitive extensometers, placed in the gauge section at opposite sides of the specimen, are used for strain measurements with high accuracy. The monotonic tests are realized with a deformation velocity of $\dot{\varepsilon} = 5\cdot10^{-2}$ 1/s. The axial push-pull fatigue tests are carried out under nominal stress control with a triangle shaped loading at 5 Hz at a stress ratio R = -1. As ultimate number of cycles $N_{ult} = 2\cdot10^6$ is choosen. The quantitative image analyses have been carried out with a Olympus CUE 2 image analyzing system. The adiabatic Youngs modulus is determined by measuring the transit velocity of ultrasonic waves. The surface deformation behaviour is examined at several polished specimen with light and scanning electron microscopes (SEM). The fracture surfaces are analysed with the SEM, too.

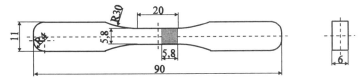

Fig. 1.   Shape and dimensions of the specimen

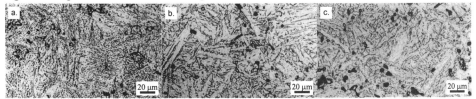

Fig. 2.   Microstructure of Fe4Ni0,5Mo+0,5%C (a), Fe1,5Mo+0,5%C (b) and Fe1,5Mo+0,5%C +5%CEP (c)

PORE MORPHOLOGY AND MONOTONIC PROPERTIES

The porosity P, the mean pore diameter d, the form factor f, the adiabatic Youngs modulus E_{adiab}, the Vickers hardness HV 30, the ultimate tensile strength R_m, the 0.2%-proof strength $R_{p0,2}$, the Youngs modulus E and the elongation at fracture A of the three materials are given in table 2.

Table 2. Porosity, pore morphology, adiabatic Youngs modulus E_{adiab}, Vickers hardness HV 30 and monotonic properties of the materials

material	P %	d μm	f	E_{adiab} GPa	HV30	R_m MPa	$R_{p0,2}$ MPa	E GPa	A %
Fe4Ni0,5Mo+0,5%C	8,4	2,99	0,665	161	188	610	440	131	3,0
Fe1,5Mo+0,5%C	7,2	2,19	0,671	171	200	663	486	142	3,8
Fe1,5Mo+0,5%C+5%CEP	7,0	1,98	0,680	172	202	653	468	148	3,6

The three materials have a total porosity between 7 and 8,4%. The mean pore diameter amounts to 3 μm for Fe4Ni0,5Mo+0,5%C and about 2 μm for the two materials with the higher density. The form factor f, defined as $f = 4\pi A/C^2$, with A as the area and C as the circumference of the pore, lies between $0{,}66 \leq f \leq 0{,}68$. The Youngs moduli E_{adiab} and E rise with the density. The 0.2%-proof strength lies between 440 and 486 MPa. R_m is 610 MPa for the nickel containing material and around 660 MPa for the FeMo-steels. The elongation at fracture was determined to 3 % and about 3,7 %, respectively. All materials show ductile deformation behaviour and ductile dimple fracture surfaces.

LOW CYCLE FATIGUE BEHAVIOUR

Under stress-controled axial push-pull loading all materials show nearly equal Wöhler slopes and an endurance limit of 158 MPa (see Fig. 3). The cyclic deformation behaviour, which is shown exemplarily for Fe4Ni0,5Mo+0,5%C in Fig. 4, is characterized by cyclic hardening in the first loading cycles followed by a period of nearly constant plastic strain amplitudes and finally by macroscopic cyclic softening up to the failure of the specimen. Because of asymmetrical plastic deformation and microscopic crack initiation in the first cycle, generally positive mean strains $\varepsilon_{m,t}$ occur (see Fig. 5). With increasing number of cycles $\varepsilon_{m,t}$ rises in a degressive way during the first cycles, remains then nearly constant and grows in a progressive manner for the final cycles before fracture. The cyclic stress-strain curve proceeds above the monotonic stress-strain curve. The correlation between σ_a and $\varepsilon_{a,p}$ at $N_f/2$ can be described well with a Ramberg-Osgood-equation $\sigma_a = K' \cdot (\varepsilon_{a,p})^{n'}$ for each materials with different K'-values (see Fig. 6).

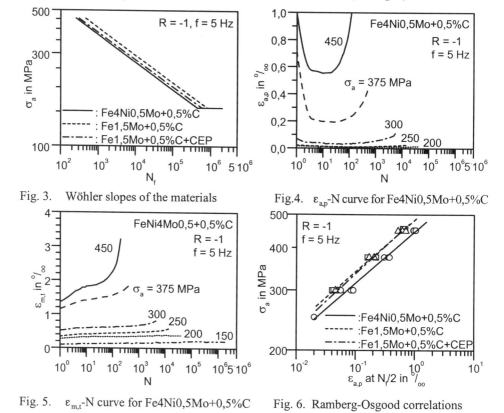

Fig. 3.   Wöhler slopes of the materials

Fig. 4. $\varepsilon_{a,p}$-N curve for Fe4Ni0,5Mo+0,5%C

Fig. 5. $\varepsilon_{m,t}$-N curve for Fe4Ni0,5Mo+0,5%C

Fig. 6.   Ramberg-Osgood correlations

Light and scanning-electron microscopic analyses of the surfaces of several specimens after defined numbers of cycles give information about slip band, micro- and macrocrack formation and the manner of crack growth. Beyond that, they allow to define a criterion for macroscopic crack initiation (see Fig. 7). Slip bands mostly are formed at sinter-necks between sharp edges of neighbouring pores or at the neighbourhood of longer cracks, especially if these cracks grow transgranularly (see Fig. 8a). Already after the first cycle microscopic cracks are detectable between or starting at sharp edges of irregular surface near pores, which lie close together (see Fig. 8b). These microcracks normally grow very slowly, but their number increases. At a certain number of cycles several of them are connected to longer cracks, which start growing faster from pore to pore. This number of cycles can be defined as a macroscopic crack initiation N_{ci}, at which single cracks with a length of more than 150 μm occur at the surfaces of the specimen. As shown in Fig. 7, N_{ci} correlates with a change in the increase of the $\varepsilon_{a,p}$-N-curve and in the decrease of the inverse tensile compliance c_t plotted versus N. Therfore $\varepsilon_{a,p}$ as well as c_t are also useful indicators for the beginning of macroscopic crack propagation. The resulting Wöhler curves for macroscopic crack initiation indicate that the ratio N_{ci}/N_f (see Fig. 9) increases with increasing numbers of cycles to fracture and by that with decreasing stress amplitudes. Scanning electron microscopic examinations of the fracture surfaces show some striations along former sintering necks (Fig. 8c) This indicates a period of stable crack growth forming an area of fatigue crack surface. The remaining fracture surface area is characterized by a dimple structure.

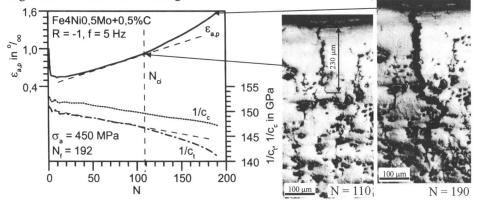

Fig. 7. Determination of the number of cycles at which macroscopic cracks initiate

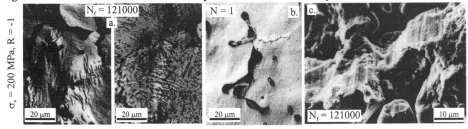

Fig. 8. Fe4Ni0,5Mo+0,5%C: slip bands (a.), microcrack (b.) and striations at sintering neck (c.)

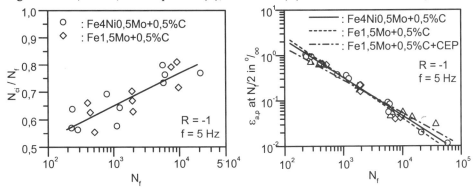

Fig. 9. N_{ci}/N_f plottet versus N_f Fig. 10. Coffin-Manson correlations

A comparison of the three PM-steels shows, that the elastic and plastic deformation behaviour of the three materials is nearly similar, although differences in density, porosity and alloying elements exist. Of course Fe4Ni0,5Mo+0,5%C shows at given stress amplitudes slightly higher plastic strain amplitudes (see Fig. 6) and a stronger increase of $\varepsilon_{m,t}$ with N than the Ni-free materials but the Coffin-Manson-curves $\varepsilon_{a,p} = \varepsilon_f'(N_f)^c$ of the three PM-steels in Fig. 10 lie close together in one scatter band. Additionally the ratio of N_{ci}/N_f plotted versus N_f in Fig. 9 also shows for all PM-steels examined that there are no obviously visible differences in macroscopic crack initiation. Regarding the two Fe1,5Mo+0,5%C-materials no significant differences in deformation behaviour, mean stresses and fatigue life could be found. The addition of 5% fine Carbonyl-iron powder improves the quality of the sintering necks but does not effect markedly the mechanical properties.

SUMMARY AND CONCLUSION

The low cycle fatigue deformation behaviour of copper free high strength PM-steels in the as sintered state at R = -1 under stress controled axial push-pull loading is characterized by cyclic hardening in the first loading cycles followed by a period of nearly constant plastic strain amplitudes and finally by macroscopic cyclic softening up to the failure. Because of asymmetrical plastic deformation and microscopic crack initiation in the first cycle, generally positive mean strains $\varepsilon_{m,t}$ occur. With increasing numbers of cycles $\varepsilon_{m,t}$ rises in a degressive way for the first cycles, remains then nearly constant and grows in a progressive manner for the final cycles before fracture. The correlation between σ_a and $\varepsilon_{a,p}$ at $N_f/2$ can be described well with a Ramberg-Osgood-equation $\sigma_a = K' \cdot (\varepsilon_{a,p})^{n'}$. The Coffin-Manson curves of the three materials lie in one scattering band. From light- and scanning-electron-microscopic analyses of the surfaces of several specimens after defined numbers of cycles a criterion for macroscopic crack initiation has been established as the occurance of single cracks with a length of more than 150 µm at the surfaces of the specimen. The number of cycles defined in that way correlates with a change in the increase of the $\varepsilon_{a,p}$-curve and with the decrease of the inverse tensile compliance plotted versus N.

The investigated high strength PM-steels in the as sintered state have improved low cycle fatigue properties compared to previous tested PM-steels with equal densities, but further improvements might be possible by additionally heat- or surface-treatments.

REFERENCES
1. Kaysser, W.A. : Leistungsfähigere Werkstoffe durch Pulvermetallurgie. Ingenieur- Werkstoffe, 1991, 3(7/8), pp. 10-16
2. Lawley, A. : PM Applications in the Automotive Industry: Case Studies; Advances in Structural PM Component Production, Proceedings of the Euro-PM 97; European Powder Metallurgy Association, München, Germany, 1997; pp. 2-10
3. Tengzelius, J. : Challenges for the Iron Powder Industry in the Next Millennium; Advances in Structural PM Component Production, Proceedings of the Euro-PM 97; European Powder Metallurgy Association, München, Germany, 1997; pp. 11-19
4. Ratzi, R. : PM Solutions for Special Requirements; Designing for High Performance PM Automotive Components, Workshop Notes of the Special Interest Session of Euro-PM 97; European Powder Metallurgy Association, München, Germany, 1997;
5. White, D.G. : Der Kfz-Markt als Wachstumsmotor für Metallpulver. Metall, 1995, 49(9), S. 564
6. Williams, B. : Wachstumstrends der Pulvermetall-Industrie. Metall, 1995, 49. Jahrgang(9), S. 563
7. Sonsino, C.M. : Fatigue Design for Powder Metallurgy. Powder Metall., 1990, 33(3), S. 235-245
8. Lindqvist, B. : Influence of microstruture and porosity on fatigue properties of sintered steels. Mod Dev Powder Metall, 1988, 21, S. 67-82
9. Koch, H.P. : PM-Werkstoffe - ihre Eigenschaften und Anwendungen. Kontakt & Studium, Pulvermetallurgie - das flexible und fortschrittliche Verfahren für wirtschaftliche und zuverlässige Bauteile, (Ed.: Esper, F.J.), 494th ed. Expert Verlag, Renningen- Malmsheim, 1997, S. 1-33

Low Cycle Fatigue and Elasto-Plastic Behaviour of Materials
K-T. Rie and P.D. Portella (Editors)

LOW CYCLE FATIGUE BEHAVIOUR OF A
POROUS PM 316L AUSTENITIC STAINLESS STEEL

U. LINDSTEDT and B. KARLSSON
Department of Engineering Metals
Chalmers University of Technology
SE-412 96 Göteborg, Sweden

ABSTRACT

The present study concerns the cyclic deformation behaviour of porous austenitic stainless steel. The material was produced from water-atomised 316L stainless steel powder, pressed and vacuum sintered at 1250°C for 1 hour. The material density after sintering was 6.9 g/cm^3. Pores enforce a strain gradient in the material resulting in a pronounced hardening during cyclic straining. Cyclic softening is hereby suppressed unlike for fully dense austenitic stainless steels. Fatigue cracks initiate at pores and grow preferentially along previous particle boundaries in the studied materials. The crack initiation is more frequent in the interior of the specimen at low strain amplitudes (<0.003) and at the surface at high strain amplitudes.

KEYWORDS

Austenitic stainless steel, fatigue cracks, low cycle fatigue, PM steel, porosity, previous particle boundaries (ppb's), stress-strain response.

INTRODUCTION

PM stainless steel is a small but growing group in the PM steel family produced by the conventional pressing and sintering technique. Low-alloyed PM steels dominate, but when there is need for special properties (e.g. corrosion and oxidation resistance, ductility, toughness, appearance, etc.), the more expensive stainless grades are selected [1]. The austenitic grades are undoubtedly the most frequently used types of PM stainless steels.

The strength of porous materials is very much dependent on the level of porosity, but also on the size and quality of sintering necks, which in turn are determined by the sintering conditions [2]. This is qualitatively understood by the fact that it is the sintering necks that transfer stresses and strains from one powder particle to the next one in a sintered porous structure [3]. Solid sintering necks with low amount of precipitates (oxides, carbides, and nitrides) are desired. The rise of precipitates can to some extent be controlled by the processing of the material [4]. Further, large sintering necks are of course better than small ones, and the growth of these necks is promoted by a high green density, and high temperature and long time at sintering.

The aim of this study is to examine the influence of pores and precipitates on the cyclic deformation behaviour of porous austenitic stainless steel. Additionally, a study on small fatigue crack initiation and propagation is provided.

EXPERIMENTALS

Materials

A water-atomised 316L stainless steel powder (particle sizes <150 µm) was pressed by uniaxial pressing (750 MPa) into rectangular blocks (120 mm x 30 mm x 11 mm) using tool lubrication (Zn-stearate suspended in acetone), resulting in a green density of ~6.8 g/cm³ for all blocks. All blocks were then sintered in vacuum at 1250°C for 1 h and allowed to cool naturally in the vacuum. The density after sintering, which was determined by Archimedes technique, was 6.9 g/cm³. The chemical compositions of the powder and sintered material are shown in Table 1.

Table 1. Chemical compositions of powder and sintered material.

	C	Si	Mn	P	S	Cr	Ni	Mo	N	O
Powder	0.018	0.85	-	<0.02	0.010	17.7	12.2	2.4	0.068	0.16
Sintered material	0.009	0.87	0.16	<0.02	0.008	18.0	12.3	2.5	0.011	0.20

Low Cycle Fatigue Testing

Low cycle fatigue (lcf) testing was performed at room temperature in laboratory atmosphere. Cylindrical specimens (∅ 6 mm) with a reduced section of 17 mm length were produced by machining and then ground and polished in several steps, the last step with 1 µm diamond paste, in order to create smooth specimens with no artificial crack initiation sites. Lcf-tests were executed in total strain control at total strain amplitudes, $\Delta\varepsilon_t/2$, between 0.0015 and 0.008. A cyclic sinusoidal strain (R_ε=-1) was applied and the strain was measured with a gauge extensometer (10 mm gauge length). During testing, peak values of strains and stresses were recorded and also hysteresis loops (stress-strain response) at some pre-selected number of cycles. During the first 25 cycles, the frequency was very low (0.01 Hz) for all tests. This procedure was selected to allow a careful recording of the initial stress-strain response and to ensure a good start up of each test. Then, the frequency was increased within five cycles. The final frequency was selected to correspond to a mean strain rate of $\bar{\dot{\varepsilon}}$ =0.005 s⁻¹ for all tests.

Longitudinal sections of fatigue-tested specimens were metallographically prepared in order to allow study of the appearance and growth of small fatigue cracks.

RESULTS AND DISCUSSION

Microstructure after Sintering

Figure 1 shows an optical micrograph of the microstructure after sintering. The density measurements, as determined by impregnation of samples, unveiled that >90% of the pores are open and thus interconnected in the structure. A high amount of interconnected pores has shown to have strongly negative effects on mechanical behaviour of PM steel generally [5]. However, in normal cases, the pores are predominately interconnected at this density level.

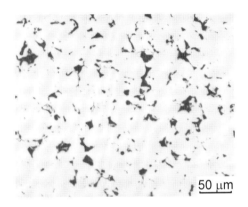

Fig. 1. The microstructure of the sintered material.

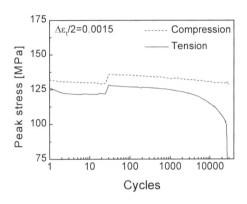

Fig. 2. The evolution of peak stresses during lcf-testing at $\Delta\varepsilon_t/2=0.0015$.

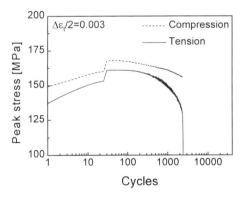

Fig. 3. The evolution of peak stresses during lcf-testing at $\Delta\varepsilon_t/2=0.003$.

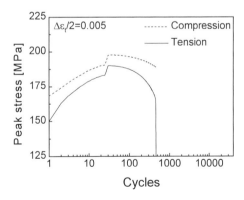

Fig. 4. The evolution of peak stresses during lcf-testing at $\Delta\varepsilon_t/2=0.005$.

Cyclic Stress-Strain Response

Figures 2 through 4 (previous page) show the development of the peak stresses during cyclic straining of the material. Three strain amplitudes are here selected for comparison: $\Delta\varepsilon_t/2$=0.0015, 0.003 and 0.005. The step-increase in peak stresses after 25 cycles is due to the increase in the applied strain rate at this point. The material shows initial cyclic hardening followed by a softening; at the lowest strain amplitude tested, $\Delta\varepsilon_t/2$=0.0015, neither cyclic hardening nor cyclic softening can be identified. Furthermore, the peak stresses in compression are higher than in tension explained by that pores are unable to carry loads in tension.

Compared to fully dense material with similar chemical composition and phase structure (single-phase austenitic structure) [6], the studied materials show a pronounced initial cyclic hardening but the following softening is less clear. The later softening is very much related to the cracking process, which causes a continuously diminishing load bearing cross-section of the specimen [7,8]. The notch effect of pores causes stress and strain gradients in the matrix. This indicates that in single-phase PM steel, a duplex matrix of plastic and elastic areas may be created during deformation. Such an inhomogeneously deformed material will affect the cyclic deformation response of the material and is responsible for the pronounced initial cyclic hardening. The stress and strain gradients are believed to cease during cyclic straining, however, resulting in a more homogeneous distribution of strains [9].

Nucleation and Growth of Small Fatigue Cracks

OM studies on longitudinal sections of lcf-specimens revealed that the strain amplitude affects the nucleation of small fatigue cracks in the specimen. At higher strain amplitudes ($\Delta\varepsilon_t/2$>0.003) surface cracking dominates and initiated fatigue cracks are more or less homogeneously distributed on the surface. At lower strain amplitudes ($\Delta\varepsilon_t/2$<0.003) internal cracking is more intense. Just a few large cracks could be seen on the surface of specimens tested at the lowest strain amplitude ($\Delta\varepsilon_t/2$=0.0015).

The cumulative plastic strain, $\varepsilon_{pl,cum}$, can be used to compare fatigue tests at different strain amplitudes [10]. This parameter is defined as

$$\varepsilon_{pl,cum} = 2N\Delta\varepsilon_{pl} \tag{1}$$

where N is the number of strain cycles and $\Delta\varepsilon_{pl}$ the plastic strain range. For example, by comparing fatigue data at $\Delta\varepsilon_t/2$=0.004 and $\Delta\varepsilon_t/2$=0.0015, the corresponding values for the cumulative plastic strain are 10 and 61 respectively [9]. Thus, at low strain amplitudes the material is exposed to a much higher cumulative plastic strain before specimen failure than at high strain amplitudes. The possibility for cracks to initiate at pores before the specimen fails is then higher at lower strain amplitudes. However, the free surfaces of smooth specimens run the biggest risk for crack nucleation because of the related stress concentration. The absolute difference between the stress in the centre and at the surface, however, is larger at high strain amplitudes. In this case, the surface runs a relatively higher risk for crack initiation than the internal of the specimen. At low strain amplitudes, this difference in stress is much less. This explains why specimens can sustain a higher cumulative plastic strain before failure at low strain amplitudes and thus allowing for internal crack initiation.

Longitudinal sections of lcf-specimens were examined qualitatively. In all cases, small cracks propagate along grain boundaries. A few small cracks, which had initiated at pores, grew transgranularly but then stopped in growth. If it is assumed that grain boundaries are unable to break through ppb's during sintering, this finding indicates that small cracks preferentially grow along ppb's.

CONCLUSIONS

The present study concerns low cycle fatigue of a porous austenitic stainless steel. The material was produced by pressing and vacuum sintering of 316L stainless steel powder. The resulted material density was 6.9 g/cm^3. The following main conclusions can be drawn:

1. Pores enforce an inhomogeneous deformation of the material when cyclically strained, resulting in a pronounced initial cyclic hardening. The distribution of strains gets, however, a more homogeneous character during further cycling.
2. Small fatigue cracks initiate at pores and grow preferentially along previous particle boundaries in the studied materials. At low strain amplitudes (<0.003), the initiation is more frequent in the interior of the specimen, and surface crack initiation dominates at higher strain amplitudes.

ACKNOWLEDGEMENTS

This work was financially supported by the Swedish Research Council for Engineering Sciences.

REFERENCES

1. Reinshagen, J.H. and Neupaver, A.J. (1989). In: *Proceedings of 1989 Powder Metallurgy Conference & Exhibition*, vol. 2, pp. 283-295, T. G. Gasbarre and W. F. Jandeska, Jr. (Eds). MPIF, Princeton, NJ.
2. Beiss, P. (1991) *Powder Metall.* **34**, 259.
3. Moon, J.R. (1996). In: *Proceedings of Deformation and Fracture in Structural PM Materials*, vol. 1, pp. 61-84, L. Parilák, H. Danninger, J. Dosza and B. Weiss (Eds). Institute of Materials Research, Kosice, Slovakia.
4. Svilar, M. and Ambs, H.D. (1990). In: *High Temperature Sintering*, pp. 75-90; H. I. Sanderow (Ed.). MPIF, Princeton, NJ.
5. Danninger, H., Spoljaric, D. and Weiss, B. (1997) *Int. J. Powder Metall.* **33**, 43.
6. Nyström, M., Lindstedt, U., Karlsson, B. and Nilsson, J.-O. (1997) *Mater. Sci. Technol.* **13**, 560.
7. Lindstedt, U. and Karlsson, B. (1996). In: *Proceedings of 1996 World Congress on Powder Metallurgy & Particulate Materials*, vol. 5, pp. 17-35 - 17-45, T. M. Cadle and K. S. S. Narasimhan (Eds). MPIF, Princeton, NJ.
8. Lindstedt, U., Karlsson, B. and Masini, R. (1997) *Int. J. Powder Metall.* **33**, 49.
9. Lindstedt, U. and Karlsson, B. (1997). In: *Proceedings of 1997 European Conference on Advances in Structural PM Component Production*, pp. 311-318. EPMA, Shrewsbury.
10. Christ, H.-J. (1996). In: *Metals Handbook*, vol. 19, pp. 73-95. ASM International, Metals Park, OH.

Low Cycle Fatigue and Elasto-Plastic Behaviour of Materials
K-T. Rie and P.D. Portella (Editors)

HIGH TEMPERATURE FATIGUE BEHAVIOUR OF X20CrMoV121 AND X10CrMoVNb91

D. RÖTTGER[1] and D. EIFLER[2]

[1] *Department of Materials Science, University of Essen, 45117 Essen, Germany*
[2] *Department of Materials Science, University of Kaiserslautern, 67635 Kaiserslautern, Germany*

ABSTRACT

Two tube segments of the steels X20CrMoV121 and X10CrMoVNb91 have been investigated with regard to their stress-strain response in isothermal stress controlled fatigue tests in the temperature range $20\,°C \leq T \leq 650\,°C$ with and without mean stress. The microstructures of the steels were characterized by means of scanning electron and transmission electron microscopy. The cyclic deformation behaviour of both materials is characterized over the whole temperature range by cyclic softening. Increasing temperatures and stress amplitudes lead to an increase in the plastic strain amplitudes and therefore to a reduction of the number of cycles to failure. Cyclic stress-strain curves will be presented and discussed. The lifetime oriented investigations are summarised in Wöhler- ($\sigma_m = 0$ N/mm²) and Haigh-diagrams ($\sigma_m \neq 0$ N/mm²).

KEYWORDS

9-12 % Cr-steels, cyclic deformation behaviour, high temperature, mean stress.

INTRODUCTION

For more than 30 years the 12 % Cr-steel X20CrMoV121 has been widely used for pipes and tubes in the chemical industry and power plants. During service life, these components are exposed to creep, fatigue, thermal fatigue and/or superposition of these loadings at temperatures of up to 550°C. In an effort to improve the efficiency of power plants by means of an increase in the service temperature up to 600 °C, the steel X10CrMoVNb91, also known as P91, was developed and is about to replace the X20CrMoV121. For the safe and economical operation of thermally and mechanically highly loaded components it is of particular interest to gain a detailed knowledge concerning the time and temperature dependent cyclic deformation behaviour of the steels used.

58

MATERIALS AND EXPERIMENTAL PROCEDURE

The chemical compositions of the steels investigated are compiled in Table 1.

Table 1: Chemical composition of the steels X20CrMoV121 and X10CrMoVNb91.

weight percent	C	Si	Mn	P	S	Cr	Mo	Ni	V	Nb	N
X20CrMoV121	0,21	0,29	0,60	0,012	0,006	10,7	0,84	0,70	0,27	not reported	not reported
X10CrMoVNb91	0,09	0,31	0,44	0,009	0,003	9,12	0,93	0,26	0,204	0,067	0,034

With 0,09 % the steel X10CrMoVNb91 has about half the carbon content of the 12 % Cr steel, 9,12 % chromium and a comparable amount of molybdenum and vanadium. 0,067 % niobium and 0,034 % nitrogen have been added in order to facilitate the precipitation of niobium/vanadium carbonitrides, conferring superior creep properties at the desired operating temperatures. The tube segments were subject to the heat treatments outlined in Table 2.

Table 2: Heat treatment of the steels X20CrMoV121 and X10CrMoVNb91.

	austenitization temp. [°C]	cooling	annealing temp. [°C]	cooling
X20CrMoV121	1040 (20 min.)	air	760 (150 min.)	air
X10CrMoVNb91	1060 (25 min.)	air	760 (120 min.)	air

Both materials show the characteristic microstructure of tempered martensite with evenly distributed carbides of the type $M_{23}C_6$ on former austenite grain boundaries and martensite laths (Fig. 1). Additionally steel X10CrMoVNb91 contains VNb-carbonitrides of the type MX [1, 2]. The transmission electron micrographs show the carbide stabilised subgrain structure for both steels.

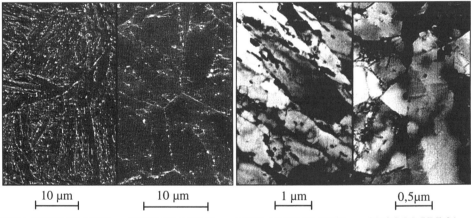

10 µm	10 µm	1 µm	0,5µm
SEM: X20CrMoV121	X10CrMoVNb91	TEM: X20CrMoV121	X10CrMoVNb91

Fig. 1: Scanning and transmission electron micrographs.

The stress controlled fatigue tests were conducted within the temperature range $20\,°C \leq T \leq 650\,°C$ with a frequency of 5 Hz and different mean stresses.

EXPERIMENTAL RESULTS

Life-Oriented Investigations without Mean Stress

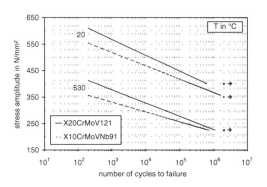

Fig. 2: Wöhler-curves for $T = 20$ °C and 530 °C.

The Wöhler-curves for 20 °C and the service temperature $T = 530$ °C are shown in Fig. 2. For constant stress amplitudes an increase in the temperature leads to significantly shorter fatigue life while a decrease in the stress amplitude leads to higher number of cycles to failure at constant temperatures. Whereas at 20 °C the steel X20CrMoV121 shows higher numbers of cycles to failure, and with $S_f = 400$ N/mm² a higher fatigue limit than the steel X10CrMoVNb91, the behaviour is different at 530 °C. At high stress amplitudes the X20CrMoV121 shows significantly longer lifetimes than the X10CrMoVNb91. With decreasing stress amplitudes, consequently longer test times, the Wöhler-curves converge and a fatigue limit of $S_f = 225$ N/mm² is observed for both steels. This is in analogy to creep tests, where the steel X10CrMoVNb91 is superior to the X20CrMoV121 for test times longer than ≈200 h only [2].

Cyclic Deformation Behaviour

In the considered temperature range both steels are characterized by cyclic softening over the whole lifetime, as shown exemplary for 530 °C in Fig. 3.

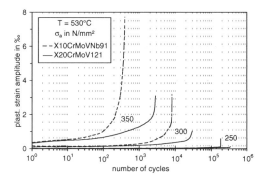

Fig. 3: Cyclic deformation curves, $T = 530$ °C.

With an increasing number of cycles the values of the plastic strain amplitude $\varepsilon_{a,p}$ rise. Stress amplitudes above 250 N/mm² lead to higher plastic strain amplitudes from the first cycle and hence to lower numbers of cycles to failure. The pronounced increase of the plastic strain amplitude towards the end of the test represents the gradual superposition of crack initiation and propagation to true, dislocation controlled plastic deformation. Due to its poorer mechanical properties, the X10CrMoVNb91 shows higher plastic strain amplitudes than the X20CrMoV121 at the same loading conditions and consequently failure occurs earlier.

In analogy to the stress-strain curves determined by tensile tests, cyclic stress-strain curves are well suited to describe the mechanical behaviour of the material under cyclic loading. As both examined materials exhibit a continuous softening behaviour and no saturation of $\varepsilon_{a,p}$, the plastic strain values were determined at half the number of cycles to failure ($N = N_f/2$).

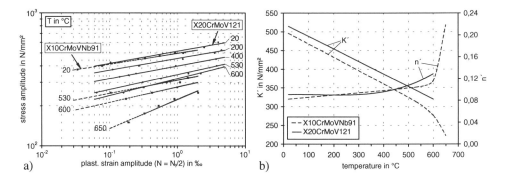

Fig. 4: a) Cyclic stress-strain curves for different temperatures,
b) Cyclic hardening coefficient and exponent as a function of the temperature.

In a log-log scale of stress amplitude vs. plastic strain amplitude determined by $N = N_f/2$, the cyclic stress-strain curves can be described by

$$\sigma_a = K' \cdot (\varepsilon_{a,p})^{n'}, \tag{1}$$

where K' is the cyclic hardening coefficient and n' the cyclic hardening exponent. For both materials higher temperatures lead to a shift of the straight lines to lower values of σ_a with an increasing slope at high temperatures (see Fig. 4 a). The values of K' and n' shown in the following Table 3 are presented as a function of the temperature in Fig. 4 b.

Table 3: Cyclic hardening coefficient K' and -exponent n'.

	T in °C	20	200	400	530	600	650
X20CrMoV121	K' in N/mm²	514	451	399	341	317	-
	n'	0,0917	0,0884	0,0956	0,113	0,1289	-
X10CrMoVNb91	K' in N/mm²	496	-	-	313	276	223
	n'	0,0821	-	-	0,1041	0,1157	0,2181

The cyclic hardening coefficient K' shows a linear relationship for both materials up to 600 °C with the exception of the steel X10CrMoVNb91, showing a deviation to smaller values at 650 °C. Due to its poorer mechanical properties, K'-value for the 9 % Cr-steel is always lower than the values for the 12 % Cr-steel. The slope of the cyclic stress-strain curves, represented by the exponent n', is for both steels nearly constant up to 400 °C, although for the 12 % Cr-steel it increases somewhat faster in the range 530 °C ≤ T ≤ 650 °C.

Life-Oriented Investigations with Mean Stress

In order to investigate the cyclic deformation behaviour of the steels when subject to mean stresses, fatigue tests were performed with constant stress ratios R = -1, -0,33 and 0. The corresponding Haigh-diagrams for both materials at T = 530 °C are shown in Fig. 5

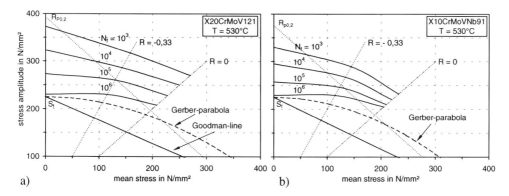

Fig. 5: Haigh-diagrams, T = 530 °C.

The number of cycles to failure decrease with increasing mean stresses. At higher mean stresses already a slight variation of the stress amplitude results in a huge difference in fatigue life, which is more pronounced for the steel X10CrMoVNb91 than for the X20CrMoV121. Therefore the mean stress sensitivity

$$M = \frac{S_f - \sigma_{a(R=0)}}{\sigma_{m(R=0)}} , \tag{2}$$

with a value of 0,25 for the 9 % Cr-steel is higher than for the 12 % Cr-steel with 0,22. For small mean stresses the Gerber-parabola describes the fatigue limit fairly well, but fails for higher values, whereas the Goodman-approximation yields always very conservative results.

Cyclic Deformation Behaviour

Cyclic deformation and plastic mean strain curves for the constant mean stress $\sigma_m = 150$ N/mm² and variable stress amplitudes σ_a at 530 °C are presented in Fig. 6. Because the maximum stress σ_O was in part beyond the yield strength $R_{p0.2}$ of the material, open hysteresis loops occurred within the first cycles. To give an impression about the deformation behaviour especially during the beginning of the tests, the values of „$\varepsilon_{a,p}$" were determined and plotted together with the plastic mean strain $\varepsilon_{m,p}$. Due to the plastic deformation during the first cycle and the fact, that the testing machine needed a few cycles to reach the desired maximum stress, the plastic strain amplitude decreases rapidly during the first 10 cycles. With increasing maximum stresses this behaviour is more pronounced.

62

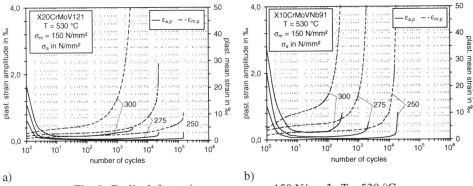

a)   b)

Fig. 6: Cyclic deformation curves, $\sigma_m = 150$ N/mm^2, $T = 530$ °C.

After this initial hardening the width of the hysteresis loops decrease more and more until a minimum value is reached, followed by a less pronounced cyclic softening behaviour towards the end of the tests. Both steels show cyclic creep behaviour; the plastic mean strain $\varepsilon_{m,p}$ increases over the whole lifetime and reaches values about 50 ‰ for high stress amplitudes towards the end of the test. Due to its poorer mechanical properties, compared to the X20CrMoV121, the steel X10CrMoVNb91 exhibits higher values of $\varepsilon_{a,p}$ and $\varepsilon_{m,p}$, consequently leading to lower numbers of cycles to failure for the same σ_a, σ_m-combinations.

CONCLUSIONS

The fatigue life decreases with increasing temperatures and stress amplitudes. At 20 °C generally and at 530 °C at high stress amplitudes only, the X20CrMoV121 shows significantly longer lifetimes than the X10CrMoVNb91. With decreasing stress amplitudes the Wöhler-curves converge and a fatigue limit of $S_f = 225$ N/mm^2 is observed. The materials show cyclic softening behaviour. For both steels the cyclic hardening coefficient K' shows a linear relationship up to 600 °C. and the cyclic hardening exponent n' is almost constant up to 400 °C. At higher temperatures it increases somewhat faster for the 12 % Cr-steel. For small mean stresses the Gerber-parabola describes the fatigue limit for $T = 530$ °C fairly well. Both steels exhibit cyclic creep behaviour.

The financial support of the *RWTÜV* (Rheinisch Westfälischer Technischer Über-wachungsverein) is gratefully acknowledged.

REFERENCES

1.   Straub, S. (1995). PhD Thesis, University of Erlangen-Nürnberg, Germany.
2.   Bendick, W., Haarmann, K., Wellnitz, G., Zschau, M. (1992) In: *Proceedings of the VGB-Conference Residual Service Life 1992*, pp. 4.1-4.21.

Low Cycle Fatigue and Elasto-Plastic Behaviour of Materials
K-T. Rie and P.D. Portella (Editors)
63

ANOMALOUS CYCLIC BEHAVIOUR OF FERRITIC STAINLESS STEELS

A. F. ARMAS*, M. AVALOS*, J. MALARRÍA*, I. ALVAREZ-ARMAS*,
and C. PETERSEN**
*Instituto de Física Rosario, CONICET-UNR,
Bv. 27 de Febrero 210 Bis, 2000 Rosario, Argentina,
**Forschungszentrum Karlsruhe für Technik und Umwelt,
Institut für Materialforschung II,
P.B. 3640, D-76021 Karlsruhe, Germany

ABSTRACT

The strain cyclic behaviour of high-chromium ferritic stainless steels (AISI 430F, AISI 420 and MANET II) has been investigated at temperatures ranging from room temperature to 823 K. Cyclic hardening curves of type AISI 430F steel show a strong temperature dependence. A pronounced cyclic hardening and an increase in saturation peak stress is observed between 623 and 773 K where the flow stress for the first tensile stroke at the maximum strain is independent of temperature. For fully annealed AISI 420 initial hardening followed by saturation stage was observed at each test temperature. A temperature independent stress saturation is the characteristical feature observed in this steel between 523 and 723 K. Normalised and tempered MANET II (a variant of the German steel DIN Nr. 1.4914) softens during cyclic loading at all temperatures. Negative strain rate sensitivity was observed plotting the peak tensile stress difference between hysteresis loops at different strain rates. It is proposed that dynamic strain aging mechanism caused by the drag of solution carbon atoms is responsible for the unusual cyclic behaviour observed in these steels.

KEYWORDS

Stainless Steels, Low Cycle Fatigue, Strengthening Mechanisms, Dynamic Strain Aging.

INTRODUCTION

Ferritic stainless steels have been increasingly used in elevated temperature applications because of their economical combination of mechanical and corrosion properties. They are currently considered to be prime structural materials for first wall and blanket of future thermonuclear fusion reactors. For these applications, structural components could be subjected to repeated thermal stresses as a result of the burning operation of fusion reactors.

Despite the importance acquired by these steels in the last years, only a few works have been found in the literature about their cyclic behaviour at elevated temperatures. Dynamic strain aging (DSA) occurs in steels owing to elastic interaction between interstitials and dislocations during straining. In this way, the effect of DSA on the cyclic behaviour of stainless steels should be considered. However, systematic investigations of DSA during fatigue deformation have been surprisingly few in stainless steels. Several authors [1,2] have reported the strong influence of DSA on the cyclic behaviour of austenitic stainless steels. Analysing the hysteresis loops, these steels show a pronounced cyclic hardening and an increase in the saturation peak tensile stress at 823K. These results together with an inverse dependence of maximum tensile cyclic stress with strain rate have been ascribed to DSA effects.

Only few works were found [3,4] in the literature about the DSA influence on the cyclic behaviour of ferritic stainless steels. Typical manifestations of DSA during monotonic tensile stress were observed in the temperature range 523 to 723 K [5,6].

As part of an extensive investigation on the low cycle fatigue at elevated temperatures of several alloys, the cyclic strain softening/hardening behaviour has been studied on steels pertaining to two ferritic steels groups: 12% chromium and high chromium (17%) steels. The purpose of the present study is to report several anomalous features observed during fatigue of these steels. It is proposed that such anomalous behaviour could be DSA manifestations.

EXPERIMENTAL PROCEDURES

Cylindrical specimens with a diameter of 5 mm and a gauge length of 18.4 mm were machined of the steels under study. AISI 420 and MANET II (a variant of the German steel DIN denomination W. Nr. 1.4914) were selected as representative of the 12% Cr group and AISI 430F for the high chromium group. Chemical compositions are given in Table 1.

Table 1. Chemical compositions of Stainless Steels

Steel	C	Si	Mn	S	P	Cr	Ni	Mo	V	Nb
AISI 420	0.27	0.36	0.35	0.02	0.03	12.3	0.13	0.07	--	0.03
MANET II	0.11	0.18	0.85	0.004	0.005	10.3	0.65	0.58	0.19	0.14
AISI 430F	0.05	0.26	0.94	0.21	0.03	17.1	0.28	0.28	--	--

According to the purpose of this paper, it is important to carry out tests with samples containing the lowest possible dislocation density to avoid eventual "hiding" effects. To obtain this condition, the steel AISI 420 was fully vacuum heat treated for 1 h at 1173 K with slow furnace cooling. The resulting structure consists of a ferrite matrix with pearlite nodules.

The vacuum heat treatment for MANET II was: austenitization 1348 K / 30 min, air cooled and tempering 1023 K / 2h, air cooled. A fully martensitic structure transforms during tempering into a ferrite matrix with carbide precipitates [7]. This structure, often called "tempered martensite", consists of elongated subgrains with walls containing a high dislocation density and mostly $M_{23}C_6$ carbides lying along prior austenite grain boundaries and tempered lath boundaries.

The steel AISI 430F was vacuum annealed at 1323 K during one hour and air-cooled. Extended grains of a ferrite matrix with some martensite islands are produced by the heat treatment and a small population of carbides has been observed in the matrix and grain boundaries.

It is important to remark that the steels under study in the present paper consists principally of a ferrite matrix with differences in dislocation density, carbide population and concentration of dissolved carbon atoms in the matrix.

Isothermal low cycle fatigue tests were carried out in air under total strain controlled conditions. Tests with total strain range 0.01 and total strain rate 2×10^{-3} s^{-1} were conducted using a fully reversed triangular wave. Optical and electron metallographic studies were carried out and both extraction-replica and thin foil electron microscopical techniques were employed.

RESULTS AND DISCUSSION

Figure 1 shows the stress response curves of the AISI 430F steel cyclically deformed at various temperatures. At the beginning of the curves in Fig. 1 one of the typical manifestations of strain aging is evident, namely the anomalous flow stress dependence on temperature. In the range 623 – 773 K, the flow stress for the first tensile stroke at the maximum strain becomes almost independent of temperature.

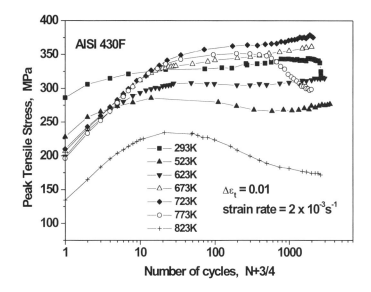

Fig. 1. Cyclic Hardening-Softening Curves of AISI 430F.

All these curves show a short, in comparison with the life of the specimen, initial cyclic hardening period. It is evident from Fig. 1 that this hardening stage is longer and much more pronounced in the temperature range between 623 K and 773 K, showing a maximum at 723 K. Below 773 K the initial hardening period is followed by a small continuous hardening that culminates in the failure of the sample.

The as-quenched AISI 430F steel consists of a highly supersaturated in interstitials carbon atoms ferritic matrix together with some martensite islands. In this work it is considered that

66

the fatigue-induced plastic deformation processes are almost completely restricted to ferrite grains. Because of the high supersaturation of carbon, the dislocation motions probably assist segregation of carbon atoms to dislocations and might even lead to precipitation of carbides. Such segregation of carbon atoms to dislocations and their resulting pinning could produce an accelerated rate of dislocation multiplication due to an additional dispersion strengthening mechanism. This effect could account for the observed pronounced cyclic hardening between 623 and 773 K in Fig. 1. Using optical microscopy, evidence of precipitation in wavy slip bands was seen in a sample fatigued up to 2500 cycles at 723 K (Fig. 2).

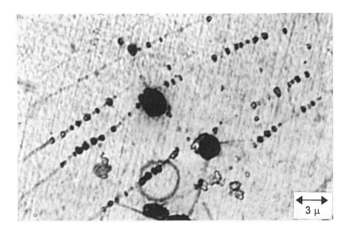

Fig. 2. Carbides in Wavy Slip Bands in a Sample Fatigued up to 2500 Cycles at 723 K.

Above 773 K the initial hardening is followed by a pronounced softening after a short stability period. Metallographical observations made at 773 K show a microstructural evolution produced by the synergism between the cycling and the elevated temperature. The cyclic process together with the higher mobility of solute atoms as a consequence of the test temperature could produce the coalescence of fine carbides in bigger ones. The softening observed above 773 K could be attributed to the coarsening of the precipitates[8].

The cyclic stress response of type AISI 420 stainless steels for strain-controlled cycling in the range 300 to 823K is shown in Fig. 3. All the curves show a short initial hardening stage followed by a saturation period independent of test temperature. This saturation stage remains up to failure, except in some cases where the stress slightly decreases. The predominant feature of the figure is the small variation among saturation stress values of each curve in the temperature range between 523 and 723 K. The fact that the saturation stress is independent of temperature suggest that DSA influences the cyclic behaviour. However, no pronounced cyclic hardening was observed in this steel in comparison with the pronounced one observed in AISI 430F. The structure of full annealed AISI 420 steel also consists of a ferritic matrix but a significant amount of carbides has precipitated during furnace cooling for this steel. In this way, the carbon concentration in the matrix will be much smaller than in the normalised AISI 430F. Because of the very low dissolved carbon content, DSA could not be strong enough to produce an enhanced dislocation multiplication during cycles.

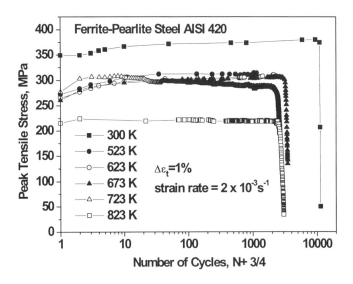

Fig. 3. Cyclic Hardening-Softening Curves of AISI 420.

Tests performed on normalised and tempered MANET II at room and high temperatures have revealed only a cyclic softening tendency. This result, according to several authors [9,10], is considered to be the result of the rearrangements of dislocations due to cyclic straining. These dislocations were previously introduced by the quenching process and could hardly disappear when tempering took place. This dislocation arrangement could "hide" the effect of dislocation-solute atom interactions in cyclic hardening curves. Negative strain rate sensitivity is also a well known manifestation of DSA. In order to examine the strain rate influence on the peak tensile stress of hysteresis loops on MANET II, strain rate changes over two orders of magnitude were performed during cyclic tests at different temperatures. Defining $\Delta\sigma_1$ as the difference between the peak tensile stress of hysteresis loops with strain rates $d\varepsilon_1/dt = 1.5 \times 10^{-3} s^{-1}$ and $d\varepsilon_2/dt = 1.5 \times 10^{-4} s^{-1}$ and $\Delta\sigma_2$ as the corresponding to $d\varepsilon_1/dt = 1.5 \times 10^{-3} s^{-1}$ and $d\varepsilon_3/dt = 1.5 \times 10^{-5} s^{-1}$, Fig. 4 shows both peak tensile stress differences as a function of temperature.

Negative values in the curves correspond to a "normal" behaviour with lower peak tensile stress for the slower strain rates. From the figure, an anomalous behaviour region is observed between 500 and 700 K with higher peak tensile stress corresponding to the slower strain rate. Both curves present a maximum at approximately 623 K. In the temperature range from 500 to 700 K solute atoms could acquire enough mobility to diffuse to the dislocation cores producing a drag effect on the mobile dislocation. This drag effect will be more pronounced for the slowest dislocation velocities.

68

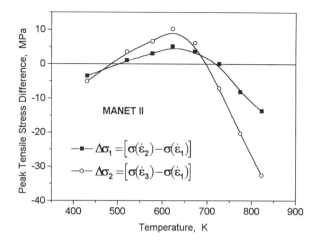

Fig. 4. Peak Tensile Stress Difference vs. Temperature. (See Text).

CONCLUSIONS

In this work it is proposed that the anomalous cyclic behaviour observed in ferritic stainless steels are produced by DSA mechanisms consequence of the interaction between dislocations and matrix dissolved carbon atoms. In AISI 430F, with a high concentration of carbon atoms dissolved in the matrix, diffusion enhanced by fatigue promotes segregation of carbon atoms to dislocations that could produce an accelerated rate of dislocation multiplication. The pronounced cyclic hardening observed in Fig. 1 could be rationalised by such dislocation multiplication. On the other hand, full annealed AISI 420 with a low concentration of solution carbon atoms, due to the enhanced carbide precipitation during furnace cooling, would not show a significant cyclic hardening. In MANET II, positive values in the peak tensile stress difference curves, as observed in Fig. 4, could be attributed to drag effects produced by carbon atoms on the mobile dislocations.

REFERENCES

1. Tsuzaki, K., Hori, T., Maki, T. and Tamura, I. (1983) Mater. Sci. Engng. **61**, 247.
2. Armas, A.F., Bettin, O.R., Alvarez-Armas, I. and Rubiolo, G.H. (1988) J. Nucl. Mater. **155-157**, 646.
3. Petersmeier, Th. and Eifler, D. (1996). In: *Fatigue '96, Proceedings of the 6th. International Fatigue Congress*, p. 753; G. Lütjering and H. Nowack (Eds.), Berlin.
4. Petersmeier, Th. and Eifler, D. (1996). VGB Kraftwerkstechnik, ISSN: 0372-5715, **76**, 321.
5. Pink, E. and Grinberg, A. (1981) Mater. Sci. Engng. **51**, 1.
6. Marmy, P., Ruan, Y. and Victoria, M. (1989) PSI-Bericht Nr. 37 Paul Scherrer Institut, Wuerenlingen, Switzerland.
7. Vitek, J.M. and Klueh, R.L. (1983) Metall. Trans. **A14**, 1047.
8. M. Avalos, M.G. Moscato, I. Alvarez-Armas and A. F. Armas. (1996). "Acta Microscopica", ISSN: 0798 - 4545, **5, Suppl. B**, 262.
9. Earthman, J.C., Eggeler, G. and Ilschner, B. (1989) Mater. Sci. Engng. **A110**, 103.
10. Chang, H.J., Tsai, C.H. and Kai, J.J. (1994) Int. J. Pres. Ves. and Piping **59**, 31.

LOW CYCLE FATIGUE OF STAINLESS STEELS;
A COMPARATIVE STUDY

A. ABEL, D. FLETCHES AND M. D. GRAHAM

Department of Civil Engineering, The University of Sydney
Sydney NSW 2006, Australia

ABSTRACT

A comparative study has been conducted on the LCF an HCF of three types of stainless steels representing ferritic, austenitic and ferritic-austenitic duplex microstructure. The initial results are presented relating to general mechanical properties, cyclic response and fatigue lives in the LCF and HCF regime. An attempt is made to resolve the cyclic state in terms of frictional and back stress resistance to dislocation movement and relate the balance of these to the fatigue performance. The results show that larger frictional component in the deformation process lowers the fatigue performance of the material and thus in LCF the best and worst results are provided by the duplex and ferritic steels respectively.

KEYWORDS

Stainless steel, LCF, microstructure, frictional stress, back stress, hysteresis

INTRODUCTION

Literature survey shows the lack of published papers relating to Low Cycle Fatigue of stainless steels generally and this applies to the LCF conferences also (1-3). Even in the recently published book dedicated to the "Stainless Steels" (4) LCF is hardly mentioned. This background provided at least one good reason to initiate a modest LCF program on stainless steels. Three types of material was selected with different metallurgical structures: 5CR12 ferritic, 316L austenitic and the SAF 2205 ferritic-austenitic duplex stainless steels.

MATERIALS AND EXPERIMENTAL PROCEDURES

The chemical composition of the tested materials are listed in Table 1.

Table 1. Chemical Compositions in Weight %

Alloy	Cr	Ni	Mo	Mn	Si	C	N	S	P
5CR12 Ferritic	10.95	0.45	-	1.12	0.4	0.011	0.011	0.002	0.018
316L Austenitic	17.1	11.1	2.11	1.3	0.36	0.024	0.022	0.005	0.028
SAF2205 Duplex	21.92	5.46	3.15	1.55	0.31	0.01	0.171	0.001	0.024

Tensile tests were carried out, according to Australian Standard 1391, in an Instron machine while high cycle fatigue tests were conducted on rotating beam specimens in Avery machines. For low cycle fatigue tests 5mm thick flat specimens were used with 12mm gauge length and 10mm gauge width using an Instron machine. As the shape of the hysteresis loops were monitored an extensometer was fitted to the specimens and the load-extension signals were processed by a computer program. The tests were controlled by a 66M Hz computer.

TEST RESULTS

The mechanical properties of the three types of stainless steel were very different as indicated by tensile test results shown in Fig. 1. The differences persist in cyclic response also where only one material shows distinct cyclic hardening while the duplex steel cyclically softens and small hardening is exhibited by the ferritic material as shown in Figs 2-3. These various cyclic tendencies are accompanied by changes in the shape of the hysteresis loop as indicated by Figs 4-5 where the Bauschinger energy parameter, β_E, is plotted against the number of applied cycles. This energy parameter for a closed hysteresis loop with full stress reversed is $\beta_E = (2\sigma_P\Delta\varepsilon_{pl} - \oint\sigma d\varepsilon)/\oint\sigma d\varepsilon$, where σ_P is the cyclic peak stress, $\Delta\varepsilon_{pl}$ is the plastic strain amplitude and $\oint\sigma d\varepsilon$ is the enclosed area of the hysteresis loop (5). Accordingly, when the hysteresis loop shape approaches that of a parallelogram the value of β_E approaches zero. Conversely, the more pointed the hysteresis loop shape the higher is the value of the energy parameter, reaching 1.0 when the loop area equals half of the area given by $(2\sigma_P\Delta\varepsilon_{pl})$. Hysteresis loop sample and the calculation method is illustrated in Fig. 6.

Table 2. Low Cycle Fatigue Results

Alloy	Plastic strain amplitude $\pm\Delta\varepsilon_{pl}$	β_E average	Number of cycles	Comment	Subsequent Tensile Test Results		
					0.2% proof MPa	UTS MPa	Total Strain %
5CR12	0.38	0.19	792	failed			
	0.56	0.175	726	failed			
316 L	0.38	0.51	1000		361(215)*	644(610)	108(74)
	0.55	0.34	1000	surface cracks	394	595	65
2205	0.25	0.38	1000		584(468)	744(730)	76(40)
	0.64	0.27	1000		556	744	70

* the bracketed numbers represent the original values before cycling

Table 3. High Cycle Fatigue Limits (at 5×10^6 cycles)

Alloy	Stresses in MPa			Fatigue Limit / $\sigma_{0.2}$
	Yield	Fatigue Limit	UTS	in %
5CR12	235	282	440	120
316 L	215	268	610	125
2205	468	389	730	83

The LCF tests were programmed for two strain amplitudes to run for 1,000 cycles but the 5CR12 material failed before this target number was reached as shown in Table 2. Included in the Table are the results of tensile data obtained subsequently on those specimens which did not fail after the 1,000 applied cycles. The corresponding stress-strain curves are presented in Fig. 7-8. Finally, the experimentally obtained fatigue limits are presented in Table 3.

DISCUSSIONS

As a starting point it can be stated: the results show that the ferritic-austenitic duplex steel is superior in respect to yield and ultimate tensile strength, LCF properties and fatigue limit while the ferritic stainless steel showed the worst performance especially in relation to LCF.

One obvious difference relates to the Nitrogen content of the duplex steel, being 0.171 in contrast to the ferritic and austenitic contents of 0.011 and 0.022 respectively. The effect of Nitrogen has been recognized and therefore it is used for both structural as well as corrosion resistance purpose (6). Degallaix *et al* (7) showed that Nitrogen favors planar slip due to the influence it has on the stacking fault energy and as a result the fatigue properties improve. The planar slip related characteristics of duplex stainless steels were further studied by Degallaix *et al* (8) by resolving the cyclic stresses into components of frictional stresses, σ_F, resulting from short range thermally activated resistance to dislocation movement and into back stresses, σ_B, resulting from long range athermal stresses. Accordingly it was shown that the austenitic steel with low Nitrogen content deforms cyclically by providing approximately equal resistance by the two components σ_F and σ_B. In the case of duplex steel with 0.11 Nitrogen content, however, the back stress resistance was found to be approximately three times the magnitude provided by the frictional resistance.

The results obtained in the present work do not support these large differences. The calculation of σ_B and σ_F leads to misleading results when only the cyclic peak and yield stresses are used. It has been pointed out (9) that the integration and averaging the stresses over the applied strain amplitude leads to a better assessment of the frictional and back stress values and this is why the Bauschinger Energy parameter becomes so useful. The value of the parameter can be experimentally determined as shown above and it can be also expressed (9) as $\beta_E = \overline{\sigma}_B / \overline{\sigma}_P$ which leads to the calculation of $\overline{\sigma}_B$ and $\overline{\sigma}_F$ as $\overline{\sigma}_B = \beta_E \overline{\sigma}_P$ or $\overline{\sigma}_F = \overline{\sigma}_P - \overline{\sigma}_B$ where the bar indicates the average values of the back-frictional and flow stresses in any half cycle (σ_P being the flow stress of the half cycle). With this approach it can be shown that the average back stress can not be larger than half the value of the peak stress in that particular cycle, but to achieve this theoretical maximum the energy parameter must have a value equal to 1.0. This approach also shows that if the hysteresis loop shape approaches that of a parallelogram, that is $\beta_E = 0$, there are no back stresses at all and the material is behaving as an ideally plastic material with 100% frictional resistance to flow.

The results shown in Table 2 therefore indicate that the proportion of frictional stresses were much higher than the back stresses. The two tests with 5CR12 material show that cycling proceeded with more than 85% frictional, therefore irreversible, deformation mechanism and both specimens failed in LCF below 800 cycles. In the case of the austenitic 316L steel the elastically stored back stresses increased in proportion and when they were at approximately 26%, with $\beta_E = 0.51$, the specimen did not fail during the 1,000 applied cycles. When however the plastic strain amplitude was set to 0.55% the average Bauschinger energy parameter dropped to 0.34 and thus the frictional stresses amounted to 78% and surface cracks were detected indicating the onset of fatigue. Results on the duplex stainless steel also indicate that the behavior in the deformation mechanism changes as the applied stain amplitude increases. At 0.25% stain amplitude, with $\beta_E = 0.38$, the average stresses are made up as: $\overline{\sigma}_P = 23.5\%\overline{\sigma}_B + 76.5\%\overline{\sigma}_F$ while at 0.64% applied strain, $\beta_E = 0.27$, $\overline{\sigma}_P = 19.7\%\overline{\sigma}_B + 80.3\%\overline{\sigma}_F$.

The tensile test results which were conducted on the "fatigued" but not failed specimens are shown in Figs 7. and 8. These results provide great difficulties when explanation is sought. The duplex steel cyclically softens while the 316L shows some degree of cyclic hardening nevertheless both show a capacity for improvement in tensile response after cycling. The 316L samples show extension and reduction in plastic response for the two tests suggesting an experimental method for the establishment of the critical point, relating to strain amplitude and number of cycles, at which a sufficient fatigue damage is altering the tensile behavior. This phenomenon offers itself for further investigation.

CONCLUSION

Ferritic, austenitic and ferritic-austenitic duplex stainless steels were low cycle fatigue tested up to a target number of 1,000 cycles and under identical test conditions the fatigue performance improved in the listed order. Some correlation has been established between fatigue performance and the balance of frictional and elastic back stresses developing during cycling. Larger is the frictional stress proportion during the cyclic deformation, shorter is the fatigue life. Evidence has been obtained that cycling with certain plastic strain amplitudes to a certain number of cycles in the LCF regime may improve the tensile properties.

ACKNOWLEDGEMENT

Thanks are due to Sandvik Steel Pty Ltd for providing the material used in the project.

REFERENCES

1. Solomon, Halford, Kais and Leis, Eds. (1988). *Low Cycle fatigue, ASTM STP 982*.
2. Rie, K. T., Eds. (1987). *The 2nd International Conference on Low Cycle Fatigue and Elastic-Plastic behavior of Materials*, Elsevier.
3. Rie, K. T., Eds. (1992). *The 3rd International Conference on Low Cycle Fatigue and Elastic-Plastic behavior of Materials*, Elsevier.
4. Lacombe, P., Beranger G., and Baroux, Eds. (1993). *Stainless Steels*, Les Editions de Physique Les Ulis, in English
5. Abel, A. and Muir, H. (1972). *Philos. Mag.*, **26**, 489.
6. Desestret, A. and Charles, J. (1990). In: *Les Aciers Inoxydables*, P. Lacombe, B. Barou and G. Beranger (Eds)., p. 633.
7. Degallaix, S., Taillard, R. and Foct, J. (1984). In: *Fatigue 84*, Beevers C. J. *et al* (Eds) p.49.
8. Degallaix, G., Seddonki, A. and Degallaix, S. (1992). In: *The 3rd International Conference on Low Cycle Fatigue and Elastic-Plastic behavior of Materials*, K. T. Rie (Ed). Elsevier, p.76.
9. Abel, A. (1993). *Mater. Sci. Eng.* **A146**, 220.

73

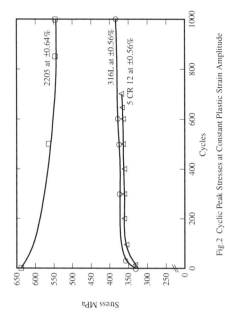

Fig.1 Tensile Response of Three Stainless Steels

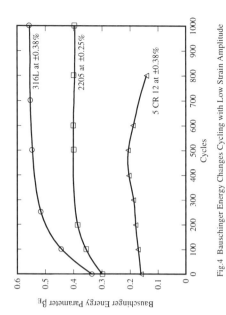

Fig.2 Cyclic Peak Stresses at Constant Plastic Strain Amplitude

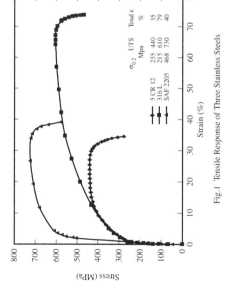

Fig.3 Cyclic Peak Stresses at Constant Plastic Strain Amplitude

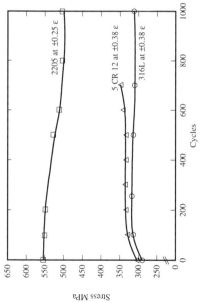

Fig.4 Bauschinger Energy Changes Cycling with Low Strain Amplitude

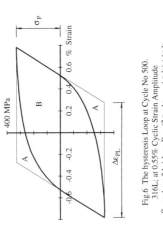

Fig.6 The hysteresis Loop at Cycle No 500.
316L; at 0.55% Cyclic Strain Amplitude
$\beta_E = $ Area 2A / Area B = $(2\sigma_p \, \Delta\epsilon_{PL} - \phi\sigma d\epsilon) / \phi\epsilon d\epsilon$

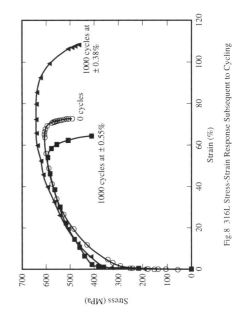

Fig 8 316L Stress-Strain Response Subsequent to Cycling

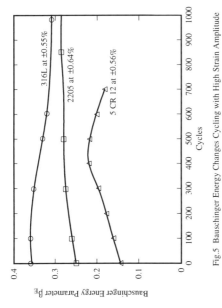

Fig.5 Bauschinger Energy Changes Cycling with High Strain Amplitude

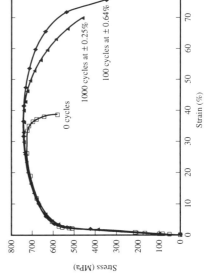

Fig.7 SAF2205 Stress-Strain Response Subsequent to Cycling

Low Cycle Fatigue and Elasto-Plastic Behaviour of Materials
K-T. Rie and P.D. Portella (Editors)

CYCLIC STRESS-STRAIN RESPONSE TO DIFFERENT STRAIN HISTORIES IN AUSTENITIC, FERRITIC AND AUSTENITIC-FERRITIC SS

E. MOUSAVI-TORSHIZI and S. DEGALLAIX
Laboratoire de Mécanique de Lille (URA CNRS 1441), Ecole Centrale de Lille
BP 48, Cité Scientifique, 59651 Villeneuve d'Ascq Cedex

ABSTRACT

The cyclic stress-strain curves of three stainless steels (austenitic, ferritic and austenitic-ferritic) assessed under different loading histories are compared. Seven methods proposed in the literature to obtain the cyclic stress-strain curve (CSSC) are applied, each of them corresponding to a specific loading history. The CSSCs are analysed and compared mutually and with monotonous tensile curves. The results show that the different loading histories generally lead to different CSSCs. The greatest difference is obtained in the austenitic stainless steel, which is the most sensitive to loading history. The sensitivity of ferritic and duplex stainless steels to loading history is much lower. The positions of the individual curves are discussed in relation to the dislocation structures developed during fatigue.

KEYWORDS

Low-cycle fatigue, cyclic stress-strain curve, loading history, austenitic SS, ferritic SS, austenitic-ferritic SS.

INTRODUCTION

A good knowledge of the behaviour of materials under cyclic loading is necessary for calculation of the stress and strain fields and for life prediction of structural elements subjected to fatigue loading. The stress-strain response to a variable loading depends on the material microstructure and on the loading conditions, in particular on the loading history. The cyclic stress-strain curve (CSSC) is defined as the plot of the stress amplitude versus the strain amplitude after stabilisation of the stress response to different imposed cyclic straining with constant strain amplitude.

Several methods have been proposed in the literature for the determination of the CSSC using short-cut precedures. The CSSCs obtained using these methods differ appreciably and depend on the type of loading applied to the specimen. Thus, each CSSC reflects the loading history to which the specimen has been subjected. The aim of the present work is to analyse the cyclic stress-strain response to different loading histories in austenitic, ferritic and austenitic-ferritic stainless steels.

EXPERIMENTAL METHOD

Three stainless steels with different cristallographic structures are studied : a type AISI 316L austenitic stainless steel with γ (fcc) structure (steel A), a type AISI 446 ferritic stainless steel with α (bcc) structure (steel F), and a type AISI 31803 austenitic-ferritic stainless steel with about 50% volume fraction of austenitic islands in a ferritic polycrystalline matrix (steel AF). The chemical composition of these steels is given in Table 1. The identification, the form of the products and the heat-treatment are defined in Table 2. Monotonic mechanical properties are shown in Table 3.

Table 1 : Chemical composition (in weight %)

Steel	C	N	Cr	Ni	Mo	Mn	Si	P	S
A	0.040	0.058	16.65	12.13	2.6	1.70	0.54	0.023	0.003
F	0.18	0.16	26.58	0.41	0.09	0.81	0.51	0.023	0.003
AF	0.02	0.158	21.99	5.25	2.8	1.33	0.44	-	-

Table 2 : Identification, shape of the product and heat treatment

Steel	AISI designation	Product shape	Heat Treatment
A	316L	forged bar Ø 20 mm	sol. annealing (1100°C)-WQ
F	446	forged bar Ø 20 mm	sol. annealing (1060°C)-WQ
AF	31803	rolled plate t 28 mm	sol. annealing (1050°C)-WQ

Table 3 : Monotonic tensile properties

Steel	E (MPa)	YS $_{0.2\%}$ (MPa)	UTS (MPa)	$A_{2.5}$ (%)
A	190000	265	587	56
F	218000	450 (R_{ei})	600	26
AF	210000	555	732	40.5

Seven different methods described in the literature [1] are used here to obtain the CSSC ; the strain histories applied to the specimens are schematically shown in Fig. 1. All the tests were performed on a servo-hydraulic testing machine INSTRON 8501 with a capacity of 100 kN. The specimens were cyclindrical and button-headed, with a 10 mm diameter and a 12.5 mm gauge length. They were mechanically polished in the longitudinal direction up to 1 μm before fatigue testing. The total axial strain was controlled in a fully reversed push-pull mode applying a triangular signal with constant strain rate ($\dot{\varepsilon}_t = 4.10^{-3}$ s^{-1}) in the range $\Delta\varepsilon_t/2 \leq 1.3$ %.

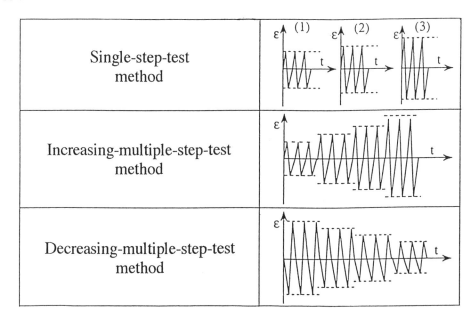

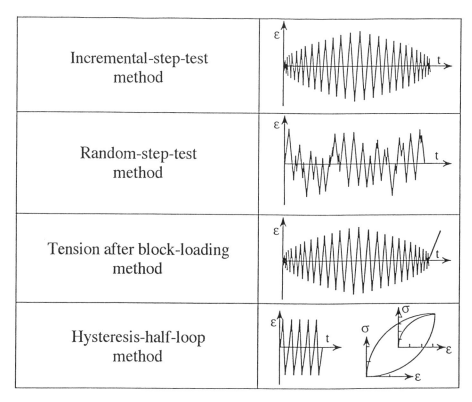

Incremental-step-test method	
Random-step-test method	
Tension after block-loading method	
Hysteresis-half-loop method	

Figure 1 : Definition of the loading history for each method used to obtain the CSSC.

In the single-step-test (SST) method, specimens were run until rupture and the stabilised stress amplitude was taken at half-life. In short-cut methods, only a single specimen was tested. In the multiple-step-test methods (IMST for increasing and DMST for decreasing steps) cyclic loading at each level continues untill stress amplitude stabilisation and the CSSC is defined as the plot of the stress amplitude at the last cycle in each step versus the imposed strain amplitude. In the incremental-step-test (IST) and random-step-test (RST) methods, successive identical loading blocks are imposed (incremental and random straining respectively). Rain Flow analysis of the stabilised block is performed. The CSSCs are defined as the plot of the stress and strain amplitudes of all closed hysteresis loops in the stabilised block. In the tension after block-loading (TABL) method, the CSSC is obtained from the monotonic stress-strain curve measured after incremental step straining up to stress stabilisation. Finally, the hysteresis half-loop (HHL) method is based on the Masing assumption, according to which tensile stress-strain curve, CSSC and hysteresis half-loops have the same shape for all strain amplitudes. The CSSC is thus derived from the increasing branch of the loop by shifting the origin to the minima of stress and strain and multiplying the coordinates by 0.5.

RESULTS AND DISCUSSION

Figures 2, 3 and 4 give, for A, F and AF steels respectively, the seven CSSCs obtained by the different methods and the monotonic tensile curve. The austenitic stainless steel is much more sensitive to the strain history than the ferritic and duplex steels. The CSSCs of steel A are well above the monotonic tensile curve, which shows that this steel hardens in cyclic loading. The CSSCs of steels F and AF do not differ appreciably from the monotonic tensile curve, which shows that these steels are cyclically stable.

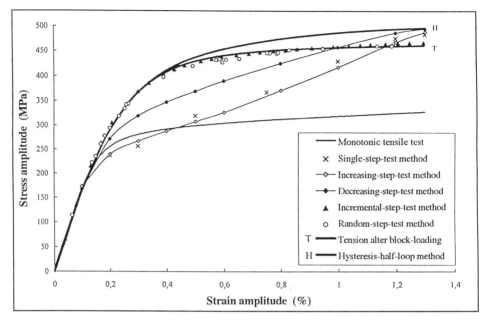

Figure 2 : CSSCs obtained by seven methods and monotonic tensile curve of steel A.

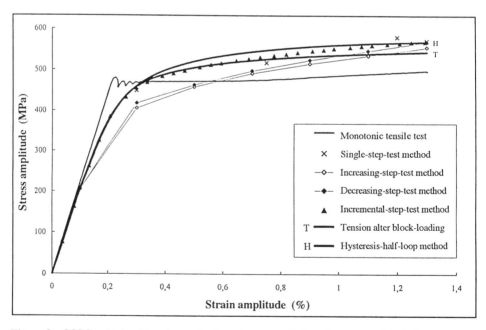

Figure 3 : CSSCs obtained by six methods and monotonic tensile curve of steel F.

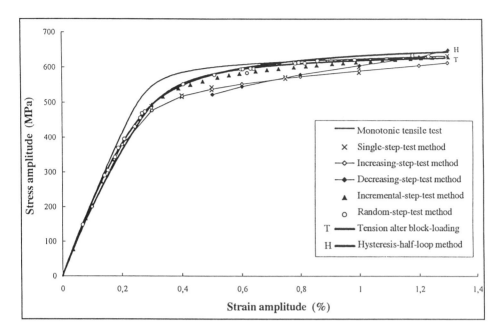

Figure 4 : CSSCs obtained by seven methods and monotonic tensile curve of steel AF.

The behaviour of the austenitic, ferritic and duplex stainless steels under various loading histories can be interpreted thanks to the observations of dislocation structures in fatigue described in the litterature. A lot of them deal with the austenitic steels, the AISI 316L more particularly, as for examples [2, 3, 4, 5], a less large number deal with the ferritic stainless steels [6, 7], and only some recent works deal with austenitic-ferritic stainless steels as [8, 9]. Detailed description of the dislocation structures in cyclic plastic deformation of bcc and fcc single- and polycristals has been published by Magnin and al. [10] some years ago, which allows to understand better the response of these materials to different cyclic strain histories.

The high sensitivity of steel A to the strain history is well known and it is explained by the strong evolution of dislocation structures during fatigue life and with strain level ; during cyclic straining, the internal structure of 316L stainless steel changes from rather planar to more complex arrangements : at low strain level, the dislocation structure remains essentially planar and homogeneous, while at high strain level, it changes to wall and cell arrangements ; the irreversibility of these arrangements increases with their complexity. On the contrary, the internal dislocation structure in ferritic steels is much more simple and does not change basically neither during cyclic straining nor with the strain level : it consists of vein and wall structures.The volume fraction of wall structure increases with the number of cycles and with the strain level ; the mixture of these two structures allows to obtain stabilised stress-strain response when the strain amplitude increases or decreases in multiple step loading. Finally, the duplex stainless steel exhibits during fatigue loading simultaneously the dislocation structures typical of each phase. Previous study [11] has shown that the behaviour of duplex stainless steels under cyclic straining is rather governed by the behaviour of its ferritic phase. The low sensibility of this steel to strain history observed here is coherent with these structure observations.

In addition to these differences, common conclusions for the three steels can be drawn from the present work :

1. The CSSC obtained by HHL method is above and that obtained by SST method is below all other curves ; this expresses the complexity, the irreversibility and the "hardness" of deformation structures increasing with the strain level.

2. The IMST and SST methods give nearly identical CSSCs ; indeed, the stress level at each step is typical of the strain level applied, also in multiple step test as the stress stabilisation is achieved.

3. IMST method leads to a CSSC below that obtained by DMST method ; that can be explained by the difficult destabilisation of a dislocation structure built at a given strain level by further cycling at lower strain level.

4. The CSSCs obtained by SST, IMST, DMST and HHL methods intersect at the same stress and strain corresponding to the highest strain level applied in each test ($\Delta\varepsilon_t/2=1.3\%$). This stress level is determined by the dislocation structure produced by this high strain level.

5. The CSSCs determined from IST, RST and TABL methods are nearly identical ; these curves correspond to the same dislocation structure. Incremental-step test can be considered as a specific case of random-step test.

6. The CSSC obtained by IST method crosses that obtained by SST method at a strain level slightly below the maximum strain ; the dislocation structure corresponds thus to a strain level slightly lower than the maximum level in the block.

REFERENCES

1. Lieurade, H.P., (1976) *Méca. - Mat. - Elec.* **29**, N° 323-324, 15.
2. Pineau, A., (1976) *Méca. - Mat. - Elec.* **29**, N° 323-324, 6.
3. Degallaix, S., (1986) Thesis, Université de Lille I, France.
4. Li, Y. and Laird, C., (1994) *Mater. Sci. Engng.*, N° A 186, 87.
5. Obrtlik, K., Kruml, T. and Polak, J., (1994) *Mater. Sci. Engng.*, N° A 187, 1.
6. Mughrabi H., Hertz H. and Stark X., (1981)*Int. J. Fatigue,* **17**, 193.
7. Lukas, P. and Kunz, L., (1994) *Mater. Sci. Engng.*, N° A 189, 1.
8. Kruml, T., Polak, J., Obrtlik, K. and Degallaix, S., *Acta Mater.,* **45**, N°12, 5145.
9. Mateo, A., Llanes, L., Iturgoyen, L. and Anglada, M., (1996) *Acta Mater.,* **44**, N° 3, 1143.
10. Magnin T., Driver J., Lepinoux J. and Kubin L.P., (1984) *Rev. Phys. Appl.,* **19**, 467 and 483.
11. Degallaix S., Seddouki A., Degallaix G., Kruml T. and Polak J., (1995) *Fat. Frac. Engng Mater. Str.,* **18**, N°1, 65.

Low Cycle Fatigue and Elasto-Plastic Behaviour of Materials
K-T. Rie and P.D. Portella (Editors)
© 1998 Elsevier Science Ltd. All rights reserved.

THE EVALUATION OF INTERNAL AND EFFECTIVE STRESSES DURING LOW CYCLE FATIGUE IN STAINLESS STEELS

F. FARDOUN*, S. DEGALLAIX* and J. POLÁK*, **

* Laboratoire de Mécanique de Lille (URA CNRS 1441), Ecole Centrale de Lille,
BP 48, Cité Scientifique, 59651 Villeneuve d'Ascq Cedex, France
** Institute of Physics of Materials, Academy of Sciences,
Žižkova 22, 616 62 Brno, Czech Republic

ABSTRACT

The hysteresis loop recorded in cyclic plastic straining on smooth specimens of an austenitic and a ferritic stainless steels cycled with constant strain amplitudes and constant strain rate were analysed using the generalised statistical theory of the hysteresis loop. The effective stress component and the probability density function of the critical internal stresses were evaluated in both materials. The effective stress changes only slightly during the fatigue life and with the strain amplitude. It is substantially higher in a ferritic steel than in an austenitic steel owing to different crystalline structures of these materials. The pronounced changes of the probability density function during the fatigue life and with the strain amplitude were discussed and related to the internal dislocation structure evolution in both materials in elastoplastic cyclic loading.

KEYWORDS

Internal stress, effective stress, low-cycle fatigue, stainless steel, austenitic steel, ferritic steel.

INTRODUCTION

The resistance of materials to fatigue loading is closely related to their cyclic stress-strain response. The analysis of the sources of the yield stress led to the identification of two additive contributions to the cyclic stress, the effective stress and the internal stress. The effective stress is related to the facility of dislocation movement in the field of short-range obstacles. It is a function of temperature and strain rate. The internal stress is caused by the obstacles producing long-range stresses and is thus determined by the internal dislocation structure.

The mesoscopic interpretation of the cyclic stress-strain response originates from the Masing hypothesis and allows formulation of the generalised statistical theory of the hysteresis loop, which considers both internal and effective stresses [1]. The usefulness of this approach has been demonstrated recently analysing the hysteresis loop shapes of 316L [2] and duplex stainless steels [3]. In this contribution, the effect of strain amplitude and the effect of number of cycles on the shape of the hysteresis loop in an austenitic and in a ferritic stainless steels is reported. The shape of the loop is analysed using second derivative, and the effective stress and the distribution of internal critical stresses are evaluated. The evolution of these quantities with the number of cycles and the strain amplitude is discussed in terms of the changes of the internal structure in cyclic straining.

EXPERIMENTAL

Two stainless steels, austenitic 316L steel alloyed with nitrogen (0.04 C, 0.056 N, 16.6 Ni, 12.1 Cr, 2.6 Mo, 1.7 Mn, 0.54 Si, and bal. Fe, all in at.%) and ferritic 446 steel (0.18 C, 0.16 N, 26.6 Ni, 0.41 Cr, 0.09 Mo, 0.81 Mn, 0.51 Si, and bal. Fe, all in at.%) were studied. Cylindrical specimens with 10 mm diameter and 12.5 mm length were tested in strain controlled cycling using 100 kN Instron computer controlled machine at room temperature. Symmetrical strain cycle with the strain rate 4×10^{-3} s^{-1} and different constant strain amplitudes was applied to the specimens. Stress and strain were recorded digitally with 300 - 1200 points per hysteresis loop. The first and the second derivatives of the loop were obtained smoothing the data by fitting a high degree polynomial to the tensile and/or to the compression reversal. The results were nearly identical for both reversals and thus only the analysis of the tensile reversal is reported.

RESULTS

Figure 1 shows the plot of the second derivative of the tensile half-loop of 316L steel vs. relative strain in a loop multiplied by the effective elastic modulus (further called fictive stress) for two strain amplitudes at different stages of the fatigue life. The initial drop of the second derivative corresponds to the drop of effective stress and to the strain relaxation and is similar for both strain amplitudes. The local minimum reaches zero only for the low amplitude in agreement with previous results [2]. The effective stress component can be found from the position of the minimum. The probability density function of the internal critical stresses, $f(\sigma_{ic})$, can be evaluated from the plot of the second derivative vs. fictive stress in the region beyond the local minimum. For the low strain amplitude (Fig. 1a) a single peak is present and does not change appreciably during the fatigue life. Its position does not change but becomes higher and narrower with increasing number of cycles. Appreciable changes of the probability density function are recorded for the large strain amplitude (Fig. 1b) and with increasing number of cycles, the function corresponds to two overlapping peaks.

Figure 2 shows analogous plot for small and large strain amplitudes applied to the 446 ferritic steel. The initial drop again corresponds to the drop of effective stress and to the strain relaxation. The effective stress is appreciably larger in this material than in austenitic steel. The probability density function is characterised by a large single peak preceded by a small peak whose position and height is nearly independent of the strain amplitude and the number of loading cycles. The main peak of the probability density function for the lowest strain amplitude undergoes characteristic changes with the number of cycles, but for the high strain amplitude does not change appreciably.

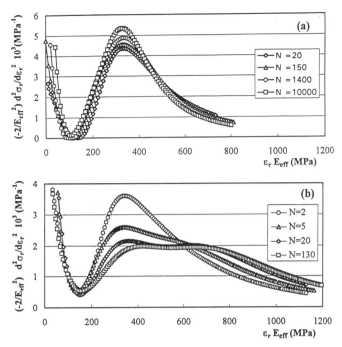

Fig. 1. Second derivative vs. fictive stress, austenitic steel, a) $\varepsilon_a = 3 \times 10^{-3}$, b) $\varepsilon_a = 1.2 \times 10^{-2}$.

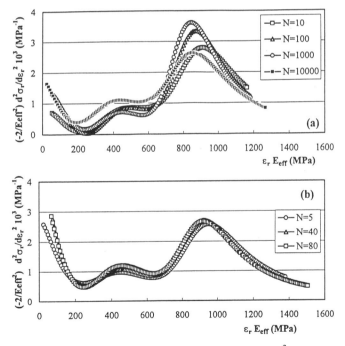

Fig. 2. Second derivative vs. fictive stress, ferritic steel, a) $\varepsilon_a = 3 \times 10^{-3}$, b) $\varepsilon_a = 1.2 \times 10^{-2}$.

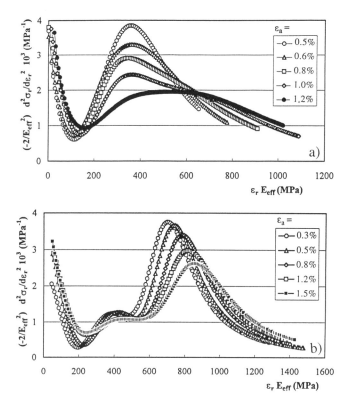

Fig. 3. Second derivative vs. fictive stress for the stabilised loops of each block in a multiple step test, a) austenitic steel, b) ferritic steel.

The effect of the strain amplitude can be studied better using single specimen subjected to multiple step test with increasing strain amplitudes. The number of cycles applied in each step was such that the total cumulative strain at each level was 600 for 316L steel and 120 for 442 steel. Figure 3 shows the second derivatives of the stabilised half-loops (the loops at the end of each block) for 316L and 446 steels. Since the scatter due to different specimens is absent, the changes of the probability density function caused by the strain amplitude changes are obvious. In austenitic steel a tendency to change the single peak to double peak distribution with increasing amplitude is confirmed. In ferritic steel a systematic shift of the position of the principal peak with increasing strain amplitude is demonstrated.

DISCUSSION

The objective of numerous experimental studies on the cyclic plastic stress-strain response in crystalline materials is to find the sources of the cyclic stress and to associate them with the pertinent material characteristics and with the strain history. This task is difficult to accomplish without a model that describes the cyclic plastic straining. One of the early models, which allows successfully to analyse and predict the hysteresis loop in elastoplastic cyclic straining, is Masing model [4]. It is based on the heterogeneous properties of the volume elements of which a material is composed. The model assumes identical deformation in all microvolumes, which are thus arranged in parallel. The microvolumes have different yield stresses and when their critical yield stress is reached, they deform without hardening. This is a reasonable approximation for fatigue loading since the strain amplitudes are generally low.

The discovery of the basic role of dislocations in plastic straining of materials led to the theory of cyclic stress that is based on the interaction of a mobile dislocation with obstacles in the crystal lattice containing other dislocations [5]. The concept of partitioning the total stress into two components, the effective (thermally activated) stress and the internal (long-range) stress was introduced. However, the movement of a typical dislocation in a stress field produced by a stabilised dislocation structure could not predict the hysteresis loop measured experimentally. The general statistical theory of the hysteresis loop [1] uses these two concepts and simulates the behaviour of a mobile dislocation in a microvolume characterised by an internal critical yield stress, under the action of an effective stress. The internal structure of a crystal is related to the distribution of the microvolumes classified according to their critical internal yield stresses σ_{ic}, which is characterised by the probability density function $f(\sigma_{ic})$.

Two stainless steels, an austenitic 316L steel (f.c.c. structure) and a ferritic 446 steel (b.c.c. structure) have been tested and their hysteresis loops were analysed. The initial drop of the second derivative in Figs 1 to 3 is related to the drop of the effective stress and simultaneous relaxation of the plastic strain in all plastically deformed microvolumes. The second derivative reaches zero at local minimum both for 316L and 446 steels cycled with low strain amplitude (Figs 1a and 2a). The position of the local minimum yields the effective stress. In 316L steel it decreases from 140 MPa to 110 MPa during cycling (Fig. 1a) for low strain amplitude. In high amplitude cycling (Fig. 1b) it is stabilised for most of the fatigue life at 140 MPa. These values correspond to the low effective stress in a material with f.c.c. structure and to the growth of the fraction of plasticised volumes with increasing strain amplitude. In 446 steel the effective stress is 220 MPa to 240 MPa, independent of the number of cycles, and increases only slightly with strain amplitude.

The probability density function (second derivative in the domain beyond the local minimum) undergoes characteristic changes, which can be lied with the changes of the internal dislocation structure. The single peak centred around $\varepsilon_r E_{eff} = 320$ MPa, whose height increases with the cycling (Fig. 1a), corresponds well to the homogeneous planar dislocation structure found after low amplitude cyclic loading [6-8]. Much profounder changes of the probability density function were observed in high amplitude loading (Fig. 1b). The single peak characterising the distribution of critical internal stresses at the onset of cycling is progressively transformed into a flat, wide distribution. This distribution is in reality a double peak distribution (one peak at $\varepsilon_r E_{eff} = 350$ MPa and the other at $\varepsilon_r E_{eff} = 700$ MPa). The observed transformation is in agreement with various types of dislocation structures found at the end of fatigue life in high amplitude cyclic straining of this material [6,8]. The position of the first peak remains unchanged, which suggests its relation to the planar structure found in about 50% of grains. The second peak can be lied to the presence of a structure of walls and cells found in the other 50% of grains. This comparison shows that at least in the case of 316L steel the generalised Masing approach yields the probability density function that reflects reasonably well the state of the internal structure of the material retrieved by transmission electron microscopy. The systematic change of the probability density function found at the end of each block in multiple step test method (Fig. 3a) corroborates the progressive irreversible changes in the internal dislocation structure of the 316L steel.

The probability density function of the 446 ferritic steel shows reproducibly one small peak at the fictive stress $\varepsilon_r E_{eff}$ = 420 MPa in both low and high amplitude loadings. It is difficult to ascribe unequivocally this peak to the presence of a specific structure or another phase. The height of the peak is low and therefore its importance is not high and we can defer to the future studies the search for the reasons of its origin. The large peak undergoes appreciable changes in cycling with low strain amplitude (Fig. 2a) but only small changes in cycling with large strain amplitude (Fig. 2b). The peak is shifted to lower fictive stress with increasing number of cycles, which corresponds to cyclic softening, and it is in agreement with the measured changes of the stress amplitude. This softening at low stress amplitude was related to the strain localisation. It results in the transformation of the small fraction of the vein structure built shortly after the onset of cyclic straining to the wall structure [9-11]. The structure of veins and cells is formed at high strain amplitude early in the fatigue life and therefore its changes during cycling are minor. Consequently neither the probability density function nor the stress amplitudes change appreciably during fatigue of 446 ferritic steel. A systematic shift of the position of the main peak of the probability density function to higher fictive stresses was found in multiple step test. This shift corresponds to cyclic hardening, which is the result of increasing volume fraction of walls and cells and decreasing fraction of the vein structure when strain amplitude increases [11]. Although the internal structures of ferritic stainless steels in cyclic straining are less documented than those of austenitic steels, the correspondence between internal structure and probability density function in this material is also reasonably established.

REFERENCES

1. Polák, J. (1991). *Cyclic Plasticity and Low Cycle Fatigue Life of Metals*. Academia-Elsevier, Praha-Amsterdam..

2. Polák, J., Fardoun, F. and Degallaix, S. (1996) *Mater. Sci. Engng* **A215**, 104.

3. Fardoun, F., Polák, J. and Degallaix, S. (1997) *Mater. Sci. Engng* **A234-236**, 456.

4. Masing, V. G. (1923) *Wisseschaftliche Verofentl. Siemens Konzern* **3**, 231.

5. Cottrell, A. H. (1953) *Dislocations and Plastic Flow in Crystals*. Clarendon Press, Oxford.

6. Gerland, M., Mendez, J., Violan, P. and AitSaadi, B. (1989) *Mater. Sci. Engng* **A118**, 83..

7. Li, Y. and Laird, C. (1994) *Mater. Sci. Engng* **A186**, 87.

8. Obrtlík, K., Kruml, T. and Polák, J. (1994) *Mater. Sci. Engng* **A187**, 1.

9. Mughrabi, H. Herz, K. and Stark X. (1981) *Int. J. Fract.* **17**, 193.

10. Magnin, T., Ramade, C., Lepinoux, J. and Kubin, L. P. (1984) *Mater. Sci. Engng* **A118**, 41.

11. Kruml, T., Polák, J., Obrtlík, K. and Degallaix, S. (1997) *Acta Mater.* **45**, 514.

Low Cycle Fatigue and Elasto-Plastic Behaviour of Materials
K-T. Rie and P.D. Portella (Editors)

VARIATION OCCURRED IN THE SECOND REGION OF DA/DN-ΔK CURVE

HAI LIN

Graduate School of Tokai University, Department of Mechanical Engineering,
1117 Kitakaname Hiratuka-shi,Kanagawa Japan
and
MORIHITO HAYASHI

Department of Mechanical Engineering, Tokai University,
1117 Kitakaname Hiratuka-shi,Kanagawa Japan

ABSTRACT

Experiments have been made to study the influence of loading frequency on fatigue crack propagation rate (FCPR), especially on variation occurred in the second region of da/dN-ΔK curve in 6/4 brass sheet. The experiments have been carried out in air and at room temperature in accordance with ASTM E647-91. Compact tension (CT) specimens with an orientation T-L and 2.75 mm in thickness were tested under three frequencies (2.5, 5, 8 Hz) of sine waveform tension-tension load. It was found that the FCPR increases with the increase of frequency in the case of R=0.11. And the FCPR decreases with the increase of frequency in the case of R=0.02. The second region of the crack propagation progress can be divided into two regions (IIa and IIb) by the change points of index m value in Paris equation. Moreover, the ΔK's at change points of m value have been found moving to the higher ΔK side as frequency increases. And the generation conditions of the change point are given.

KEYWORDS

Brass, Crack propagation rate, Fatigue, Frequency, Stress intensity factor

INTRODUCTION

In general, the crack propagation process is divided into three regions. Paris equation [1], da/dN=CΔK^m, is adapt to the second region. Index m in the Paris equation is 2-4 or a higher value [2]. This value is usually a stable and, or a constant value in the second region. However, it have been reported that m value changes by the influences of chemical environment [3] or sheet thickness [4], [5] and failure pattern [2] etc. in fatigue test. In this research, fatigue tests have been made on 6/4 brass sheet in accordance with ASTM E647-91. The change behaviors of m value by different load frequencies have been examined and the generating condition on that change of m value has been given. Furthermore, some considerations have been added.

TEST MATERIAL

The material used in this research was 6/4 brass sheet. The chemical elements of the material (mass%) are Cu: 60.51, Fe: 0.012, Pb: 0.006, Zn: Balance. And mechanical properties are as follows, Tensile strength: 410MPa, Elongation: 38.2%, 0.2% offset yield strength: 280MPa and Young's modulus: 104GPa.

To make the CT(compact tension) specimens, we got the parts of 5cm around the four sides cut out of the material with a size of $2.8 \times 365 \times 1200$mm, and let the longer sides of the specimens be parallel to the rolling direction (i.e. the rolling direction is at T-L direction in ASTM E399-90). And then made the specimens in accordance with related regulations [6]. Here, thickness t $= 2.75 \pm 0.05$mm, width W = 50mm, notch length a_n =11mm.

Notch has been processed by using a metal slitting saw, whose blade tip's angle is 30°. Moreover, because fatigue process is very sensitive to the surface qualities of specimen[7] - [10], we use specimens in the tests with the roughness (Ra) of the surfaces had been polished to 0.14μm or less.

TEST METHOD

The fatigue tests were performed in a servo-hydraulic fatigue test facility. Tension-tension loads of sine wave (P_{max}=1764N, stress ratio R=0.11 and R=0.02) were added to the CT specimens installed in the test facility and the fatigue tests were carried out in atmosphere and at the room temperature.

The test load frequencies were 2.5, 5 and 8Hz respectively and the repeating load conditions were often confirmed. The load conditions were kept constant from the precracking stage to the test end.

The propagated crack lengths were measured optically without interrupting the tests using a traveling microscope with magnification power of $50 \times$. And the measuring intervals were 1mm[6]. The crack lengths were measured only at one side for t/W<0.15. The other test methods were based on the related regulations [6].

da/dN has been calculated by using secant method [6],[11], and $\triangle$K has been calculated by using eq (1)[6].

$$\Delta K = \frac{\Delta P}{t\sqrt{W}} \cdot \frac{2+\alpha}{(2+\alpha)^{3/2}} \cdot \left(0.886 + 4.64\alpha - 13.32\alpha^2 + 14.72\alpha^3 - 5.6\alpha^4\right) \qquad (1)$$

where $\alpha = a/w$; $\Delta P = P_{max} - P_{min}$; t = thickness.

The relation between da/dN and $\triangle$K has been got by the least square method.

CHANGE POINTS OF M VALUE IN THE SECOND REGION OF DA/DN-$\triangle$K CURVE

In the Case of R=0.11. Based on the tests results, we get da/dN-$\triangle$K relations which are shown in Fig.1. From Fig.1, we can know that da/dN increases with increasing frequency in this case.

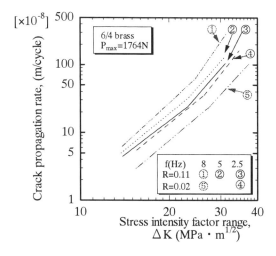

Fig.1 Results showing effects of frequency on da/dN.
(at R.T., in air, R=0.11 and R=0.02)

We can also understand that there were changing points in da/dN-ΔK curves. And when ΔK's values got larger up to about 24MPa・m$^{1/2}$, the tendency that da/dN increased sharply were seen. Therefore, the second region of crack propagation process can be divided into two regions: IIa region where da/dN rise gradually and IIb region where da/dN rise rapidly as ΔK increases. That is, the change points of m values can subdivide the second region, which has been assumed to be one region up to now, further.

By using the least square method, we can get the intersection points between each curve in IIa and curve in IIb (i.e. change points of m value) in Fig.1. Naming ΔK at change points to be ΔK_t, we show the results in Table 1.

Table 1 ΔK's at variation points in the second region of da/dN-ΔK curve.
(from results of regression analysis for Fig.1)

	R=0.11			R=0.02	
f(Hz)	2.5	5	8	2.5	8
ΔK_t	23.10	24.40	24.66	26.33	27.54

From Table1, we can understand that the change points of m value move to higher ΔK value side as load frequency f increases.

From the results of their fatigue tests, in which 6/4 brass sheets have been used as test materials, concerning the effect of sheet thickness t, Hayashi and Yano [4] have reported that the change points of m value move to higher ΔK value side as sheet thickness becomes thicker. In addition, they have also suggested that the relation between sheet thickness and the plastic zone size at the crack tip is one to cause the change of m value.

On the other hand, it was reported that plastic zone size at crack tip became smaller as frequencies increased [12], [13]. In this research, even though the sheet thickness are almost same, because the plastic zone sizes at the crack tip become smaller as the frequency increases, sheet thickness becomes thicker relatively at higher frequency. Therefore, the changing phenomenon of m value found in this research is substantially agreement to those reported by Hayashi and Yano.

Here, we try to give the generation condition of the occurrence of change points, which appeared in the second region of crack propagation process, from the relation between the sheet thickness and plastic zone size at the crack tip.

In general, the size of plastic zone at crack tip in a plane strain state is shown in equation (2) [15]

$$\gamma_p = \frac{1}{24\pi} \left(\frac{\Delta K}{\sigma_y} \right)^2 \tag{2}$$

Where, γ_p is the size of plastic zone at crack tip and σ_y is yield stress.

However, considering the influence of frequency f and sheet thickness t on the plastic zone size at crack tip, we give a constant A, relating with f and t, into equation (2), then:

$$\gamma_p = \frac{A}{24\pi} \left(\frac{\Delta K}{\sigma_y} \right)^2 \tag{3}$$

It is thought that when at the beginning of stress state shifts from the state of plane strain to the state of plane stress the change point of m value occurrences. And it is implied that, in general, when the ratio of the size of plastic zone at crack tip to sheet thickness becomes large enough to some extent, such a shift of stress state happens [4], [15].

Supposing that when the ratio of the size of plastic zone γ_p to sheet thickness t becomes ζ, the change point appears, Then, the relation between γ_p and t can be shown by eq (4).

$$\gamma_p = \zeta \cdot t \tag{4}$$

where, ζ is a constant. Substitution of eqs (4) into (3) yields:

$$\frac{A}{\zeta} = 24\pi t \left(\frac{\sigma_Y}{\Delta K} \right)^2 \tag{5}$$

When each $\frac{A}{\zeta}$ of every change points, is calculated by using eq (5), the empirical equation of the plasticity region size at the change points shown in eq (6) is obtained.

$$\gamma_p = \frac{33 f^{-0.108} \zeta}{24\pi} \left(\frac{\Delta K}{\sigma_Y} \right)^2 \tag{6}$$

Or

$$t = \frac{33f^{-0.108}}{24\pi}\left(\frac{\Delta K}{\sigma_Y}\right)^2 \qquad (7)$$

The sheet thickness condition of fracture toughness test is shown in eq (8) for one direction tension test [15][16]. Where, K_{IC} is the fracture toughness.

$$t \geq 2.5\left(\frac{K_{IC}}{\sigma_Y}\right)^2 \qquad (8)$$

Moreover, for one direction tension test, γ_p is [14]:

$$\gamma_p = \frac{1}{6\pi}\left(\frac{\Delta K_{IC}}{\sigma_Y}\right)^2 \qquad (9)$$

From eq (8) and eq (9):

$$t/\gamma_p = 47 \qquad (10)$$

That is, it can be thought that the stress state is plane strain if sheet thickness is 47 times as or much larger than the size of the plasticity region in one direction tension test.

Similarly, supposing that in fatigue test, if sheet thickness is 47 times as or much larger than the fatigue plasticity region size, the stress state is plane strain, then, the ζ value becomes 1/47. In that case, eq (6) becomes eq (11):

$$\gamma_p = \frac{0.70f^{-0.107}}{24\pi}\left(\frac{\Delta K}{\sigma_Y}\right)^2 \qquad (11)$$

This means that the change point of m value occurrences when the size of the plasticity region becomes larger to a value shown in equation (12) in the state of plane strain.

Similarly, in the case of R=0.02,

$$\gamma_p = \frac{0.53f^{-0.072}}{24\pi}\left(\frac{\Delta K}{\sigma_y}\right)^2 \qquad (12)$$

These empirical equations (11) and (12) show that the plasticity region size is direct proportion to $f^{-\lambda}$ ($\lambda > 0$) quantitatively.

The optical microscope photograph of IIa and IIb region is shown in Fig.2 and a) and b) respectively. In these photographs, both the profiles of cracks were generally straight lines and were vertical in the direction of the tension load in IIa and IIb region. These photographs suggest that the state of plane strain be always kept in the effective range of the tests microscopically. Moreover, the cracks passed through both the α phase and the β phase. And in the crack neighborhood, refined crystal grains were seen in the α phase.

Fig.2 Optical microscope photos on profile of fracture
at a) $\Delta K \fallingdotseq 17.40$ MPa·m$^{1/2}$(IIa region), b) $\Delta K \fallingdotseq 28.38$ MPa·m$^{1/2}$(IIb region).

CONCLUSION

From the results of fatigue tests on 6/4 brass, the second region of the crack propagation progress can be divided into two regions (IIa and IIb) by the change points of m value. Moreover, the ΔK's at change points of m value have been found moving to the higher ΔK side as frequency increases. The generation conditions of the change points are shown in eqs (11) and (12).

REFERENCE

1. Paris, P. C. and Erdogan: F. J. Basic Eng.,Trans ASME 85(1963), 528.
2. Kare Hellan: Introduction to Fracture Mechanics, McGraw-Hill Book Co. (1984).
3. Hiroyuki, OKAMURA: Introduction to Linear Fracture Mechanics, Baifuukann, Tokyo (1976)(Japanese).
4. Morihito HAYASHI and Ryuji, YANO: Trans. Japan Society Mech. Engineers A59 (1993), 2054-2060(Japanese).
5. J. J. McGowan and H. W. Liu: J. Eng. Mater. Technol. 102 (1980) 341-346.
6. ASTM standards: E647, (1991).
7. Motoki, YAGAWA: Fracture Mechanics, Baifuukann, Tokyo (1988), 229(Japanese).
8. Tadasi, ISIBASI: Trans. Japan Soci. Mech. Engineers I 28(1962), 1301- 1305(Japanese).
9. Tadasi, ISIBASI: Trans. Japan Soci. Mech. Engineers I 29(1963), 1693-1701(Japanese).
10. Ken-ichi, TAKAO et al: J. Soci. Mater. Sci. Japan 3(1986), 317-323(Japanese).
11. Takeshi, KUNIO, Hajime, NAKAZAWA et al: the Method for Fatigue Test, ASAKURA SHOTENN, Tokyo (1984), 164(Japanese).
12. Shigeo TAKEZONO, Masahiro SATOH et al: J. Soci. Mater. Sci. Japan 31 (1982), 76-82.
13. H. Ghonem and A. Foerch: Mater. Sci. Engng. A 138 (1991), 69-81.
14. J. F. Knott: Fundamentals of Fracture Mechanics, London Butterworths (1973).
15. Yoshihiko HAGIWARA, Toshio YOSHINO and Takeshi KUNIO: Trans. Japan Soci. Mech. Engineers A 47 (1981), 698-707(Japanese).
16. ASTM standards: E399, (1990)

Low Cycle Fatigue and Elasto-Plastic Behaviour of Materials
K-T. Rie and P.D. Portella (Editors)

DETERMINATION OF COLD WORK ENERGY IN LCF/HCF REGION

J. KALETA

*Institute of Materials Science and Applied Mechanics, Technical University of Wrocław
Wybrzeże Wyspiańskiego 27, 50-370 Wrocław, Poland*

ABSTRACT

The unit energy of cold work ΔI and its cumulated value I during the fatigue process were determined. Experiments were carried in the wide range of life times LCF/HCF. The energy ΔI was derived as the difference between the plastic strain work ΔW and the diffused from specimen heat energy ΔQ. Experiments were done for the chosen ferritic–perlitic steels. The identification of ΔI and I as the function of life times and load levels was performed. The ratio $\Delta I/\Delta W$ was determined. The value of the cumulated energy of cold work at the final failure moment was compared to other thermodynamical quantities (the work of plastic deformation for static tension test and melting energy of material).

KEYWORDS

Energy of cold work, plastic strain work, heat energy, LCF/HCF, experiment, identification.

INTRODUCTION

During fatigue process the cold work ΔI energy is accumulated in material, which, according to the I-st principle of thermodynamics is evaluated as the difference between plastic deformation energy ΔW and heat energy ΔQ. The quantities denoted by ΔW, ΔQ and ΔI, suitable are treated as the unit ones, related to the volume unit of material during single load cycle.

Till today there are no unique, commonly accepted opinion about the ratio ΔI of the total fatigue energy. Not too much papers appeared devoted to this problem [1–8].

Even quite recently, in the literature it has been stated that approximately 10% (or less) of the involved work ΔW becomes the cold work energy ΔI. That values are given for quasi–static loads as well as for cyclic load [3,11–14]. In not numerous papers from the last decade [15,16], concerning static tension, the range of values of the ratio $\Delta I/\Delta W$ has been found as from 10% to 70%.

In the case of fatigue the discrepancies of estimations of $\Delta I/\Delta W$ are especially large and cover the range from 2.5% to even 45% [4]. The authors of papers [2,17], underline that the results of experiments are function of many variables. Such parameters as cold work energy, chemical contents and grains size of a metal, the way of load control (stress or strain control), as well as temperature during experiment are currently stated. It also pays attention on the differences in materials, the way of load has been applied and the method of cold work evaluation and the class of measurement devices. For these reasons the results of $\Delta I/\Delta W$ measurements obtained by various authors are difficult to or even incomparable. Till now there are no answers for many fundamental questions. Thus it is reasonable to undertake research in this area.

Due to all said above it appears especially interesting to try to identify the cold work energy function dependence onto load level and life time, thus $\Delta I(2N_j)$ and $\Delta I(\sigma_a)$. It is also reasonable to estimate the ratio $\Delta I/\Delta W$ versus load level σ_a and life time (we have found no papers about this topic in literature). Equally important is evaluation of this ratio at the fatigue limit. It is also reasonable to compare the total accumulated energy of cold work with values of others energies characterizing thermodynamical properties of materials, as for example plastic deformation energy during static tension test or with heat energy necessary to melt material.

THE DESCRIPTION OF EXPERIMENT

The ferritic–perlitic steel or general applications will be the central object of undertaken studies. The chemical composition of the material (in wt.%) was: 0.37 C, 0.23 Si, 0.56 Mn, 0.015 P, 0.029 S. This material had the following mechanical properties: proof stress (0.2%) 365 MPa, tensile strength 560 MPa, Young's modulus $2.05 \cdot 10^5$, elongation ($\Delta l/l$) 35%, reduction of area (ψ) 50% and fatigue limit (σ_f) 220 MPa. A detailed account of the fatigue and cyclic properties of the material can be found elsewhere [23,24]. Fatigue tests were conducted on solid cylindrical 8 mm diameter section specimens.

The cold work energy was estimated with the help of so called one step method as the difference between plastic strain energy ΔW and heat energy ΔQ. To achieve this the ΔW energy was determined from the modified hysteresis loop , and the heat energy ΔQ from experiments with electrical modelling of the internal heat sources [18–20]. The details of experiment were presented in [21,22,24].

RESULTS

The Influence of the Fatigue Life Time and Load Level Upon Unit Energy of Cold Work ΔI

The results of energy measurements are completely presented below. The analysis was carried for the full range of fatigue life time. For completeness and to make the analysis possible the separate energy balance terms: mechanical ΔW as well as heat ΔQ energies are also included.

In similar way as in the literature for the case od ΔW, the power functions were applied in the identification of ΔI. The influence of life time N_f, or more accurately the number of reversals $2N_f$ is depicted in the Fig. 1. The experimental results were identified making use of the function of the form:

$$\Delta I(2N_f) = 83 \cdot (2N_f)^{-0.50} \tag{1}$$

The results of ΔW and ΔQ energies are also presented in the Fig.1.
The energy ratio $\Delta I \, \Delta W$ can be described with the help of formula:

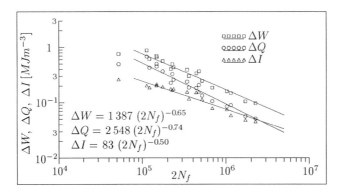

Fig. 1. The unit energies shifts dependence on the number of reversals $\Delta W(2N_f)$, $\Delta Q(2N_f)$, $\Delta I(2N_f)$.

$$\frac{\Delta I}{\Delta W}(2N_f) = \frac{83}{1387}(2N_f)^{-0.50-(-0.65)} = 5.98 \cdot 10^{-2}(2N_f)^{0.15} \qquad (2)$$

Therefore this is a monotonic increasing function of number of reversals.

For the three chosen life times the energy ratio (2) has the following values:

$N_f = 1 \cdot 10^5; \Delta I/\Delta W = 37.4\%$
$N_f = 1 \cdot 10^6; \Delta I/\Delta W = 52.8\%$
$N_f = 1 \cdot 5 \cdot 10^6; \Delta I/\Delta W = 67.3\%$

The above results correspond to the conventionally low–, high– and infinite life time.

The effect of load level onto energy of cold work unit was described by the function (3) (see Fig.2.)

$$\Delta I(\sigma_a) = 8 \cdot 10^{-20}(\sigma_a)^{7.6} \qquad (3)$$

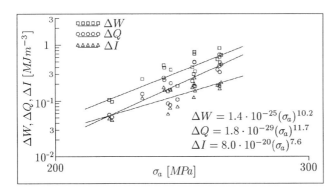

Fig. 2. The dependence of unit energies on the load level $\Delta W(\sigma_a)$, $\Delta Q(\sigma_a)$, $\Delta I(\sigma_a)$.

The ratio of cold work and mechanical energy becomes then given by

$$\frac{\Delta I}{\Delta W}(\sigma_a) = \frac{8 \cdot (10)^{-20}}{1.4 \cdot (10)^{-25}}(\sigma_a)^{7.6-10.2} = 5.71 \cdot 10^5(\sigma_a)^{-2.6} \qquad (4)$$

96

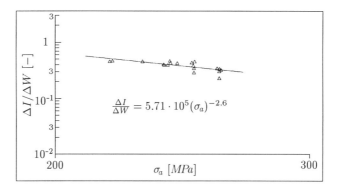

Fig. 3. The unit energies ratio dependence on the load level $\Delta I/\Delta W(\sigma_a)$.

Thus by the monotonic decreasing function (Fig.3.).
There are no data in the literature allowing for comparison of the results given by the functions (2) and (4) with results of other authors. However there is possible to compare the part of above results with the data of static tension from the last decade. Namely in the papers [15,16] it has been shown that the ΔI energy is increasing function of the load level but the ratio $\Delta I/\Delta W$ decreases for growing loads (today the second result can be treated as the classical one because first time it was presented in 1934 [11]). The range of the ratio $\Delta I/\Delta W$ changes for the two types of steels and the Armco iron covers the area from 10 to 70% depending to the load level and the type of material [15,16]. As was shown similar tendency as well as the range of changes are also obtained in this paper for cyclic loads.

The Influence of Life Time and Load Level Upon Accumulated Energy of Cold Work I

The experimentally determined values of the accumulated energy of cold work versus of the returns number to final failure were shown in the Fig.4. The identification gives

$$I(2N_f) = 70(2N_f)^{0.45} \tag{5}$$

The cumulated mechanical energy W and heat energy Q were obtained in the similar way, and the corresponding functions are also presented in the Fig.4.
Notice the the ratio of both energies can be then written in the form

$$\frac{I}{W}(2N_f) = \frac{70}{696}(2N_f)^{(0.45-0.34)} = 0.100(2Nf)^{0.11} \tag{6}$$

That is a increasing function of the number of reversals. The ratio of both energies at the fatigue fracture moment calculated from (6) varies in the range of performed identification, from 47,8% for high cycles strength to 32% in the low cycle range.
The Fig.5. presents the $I(\sigma_a)$, $W(\sigma_a)$ and $Q(\sigma_a)$ dependences. The $I(\sigma_a)$ dependence is give by the formula

$$I = 2.90 \cdot 10^{19}(\sigma_a)^{-6.33} \tag{7}$$

The ratio of both energies can be written in the following way:

$$\frac{I}{W}(\sigma_a) = \frac{2.90 \cdot 10^{19}}{3.05 \cdot 10^{14}}(\sigma_a)^{-6.33-(-4.08)} = 9.5 \cdot 10^4(\sigma_a)^{-2.25} \tag{8}$$

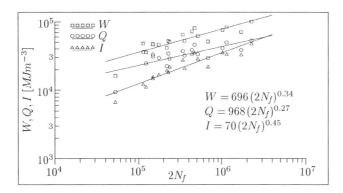

Fig. 4. The effect of reversals numbers to damage on the value of accumulated energies W, Q, I.

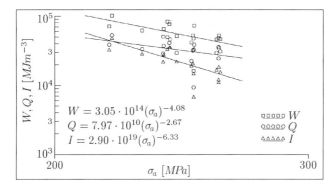

Fig. 5. The effect of load level on the value of accumulated energies at the moment of damage $W(\sigma_a)$, $Q(\sigma_a)$, $I(\sigma_a)$.

The Comparison of the Cumulated Energy I with Other Thermodynamical Characteristics

Finally lets compare the value of the accumulated energy of cold work I with the value of plastic deformation energy during the static tension test and heat energy needed to melt the examined material.

The energy I at the moment of fatigue damage, even its lowest value obtained experimentally for LCF region ($I = 9292 MJm^{-3}$), is over 20–times greater then the broke energy obtained in the static tension test ($I \cong 420 MJm^{-3}$), and 4.5–times larger then the energy necessary to melt the sample of the examined steel ($I \cong 2060 MJm^{-3}$). These ratios become $\cong 120x$ and $\cong 25x$ suitably for the greatest fatigue life time under considerations [23–24].

CONCLUSIONS

1. In the wide range of life times for the examined material the unit energy of cold work is the decreasing function of the returns number and the increasing load level function.

2. The ratio of the unit cold work energy and the plastic strain work $\Delta I/\Delta W$, in the infinite fatigue life time range is the increasing function of life times and decreasing function of the load level. The energies ratio varies from 37.4% in the low cycles area to the 67.3% in the HCF range.

3. Accumulated energy of cold work I is the increasing function of the life time but decreasing function of the load level. The ratio of accumulated energy of cold work and the total energy of plastic strain work I/W is increasing function of life time and decreasing function of the load level.

4. Accumulated energy of cold work at the final failure moment is many times greater than the plastic strain energy during static tension test but also exceeds the heat energy necessary to melt examined material.

REFERENCES

1. Welber, B. and Webeler, R. (1953). *Trans. AIME, J. Metals*, 1558–1559.
2. Bever, M.B., Holt D.L. and Titchener A.L. (1973). In: *Progress in Materials Science* **17**, Oxford, Pergamon Press.
3. Halford G.R. (1966). *J. Materials* **1**, 3–18.
4. Dillon Jr., O.W. (1962). *J. Mech. Phys. Solids*, **10**, 235–244.
5. Dillon Jr., O.W. (1963). *J. Mech. Physics Solids*, **11**, 21–33.
6. Dillon Jr., O.W. (1966). *Int. J. Solids Structures*, **2**, 181–204.
7. Clarebrough, L.M., Hargreaves, M.E., Head, A.K. and West, G.W. (1955). *Trans.AIME, J. of Metals*, 99–100.
8. Clarebrough, L.M., Hargreaves, M.E., West, G.W. and Head, A.K. (1957). *Proc. Royal Society*, Series A., **242/1228**, 160–166.
9. Troshchenko, V.T. (1971). In:*Fatigue and Inelasticity of Metals*. Kiev, Naukova Dumka (in Russian).
10. Troshchenko, V.T. (1981). In: *Deformation and Damage of Metals under Cyclic Load*. Kiev, Naukova Dumka, (in Russian).
11. Taylor, G.I. and Quinney, H. (1934). *Proc. Royal Soc. London*, **143**, 307–326.
12. Williams, R.O. (1962). *Trans. AIME*, **224**, 719–726.
13. Adams, S.L. and Krempl, E. (1984). *Res Mechanica*, **10**, 295–316.
14. Gurevič, S.E. and Gaevoy, A.P. (1966). *Zavodskaja Laboratorija* **9**, 1010–1016.
15. Chrysochoos, A. and Martin, G. (1989). *Materials Sc. and Eng.*, **A108**, 25–32.
16. Oliferuk, W., Gadaj, S.P. and Grabski, M.W. (1985). *Materials Sc. and Eng.*, **70**, 131–141.
17. Titchener, L. and Bever, M.B. (1958). In: *Progress in Metal Physics*, **7**, 247–338.
18. Błotny, R. and Kaleta, J. (1981). In:*Studia Geotechnika et Mechanica*, **3**, 45–46.
19. Błotny, R. and Kaleta, J. (1986). In:*Int. J. Fatigue*, **8**, 29–33.
20. Błotny, R., Kaleta, J., Grzebień, W. and Adamczewski, W.(1986). In:*Int. J. Fatigue*, **8**, 35–38.
21. Kaleta, J., Błotny, R. and Harig, H. (1989). In:*Proc. of ICF–7, Houston*, 1195–1202.
22. Kaleta, J., Błotny, R. and Harig, H. (1991). In:*J. Testing and Evaluation*, 325–333.
23. Kaleta, J. (1997). In:*Proc. VI Polish Conf. Fracture Mech.*, 211–219.
24. Kaleta, J. (1998). *The Scien. Papers of the Inst. Materials Sci. and Appl. Mech., Technical University of Wrocław*, **59**, Serie: Monographs No. 23.

Low Cycle Fatigue and Elasto-Plastic Behaviour of Materials
K-T. Rie and P.D. Portella (Editors)

CONTRIBUTION TO THE PHYSICAL UNDERSTANDING
OF THE MANSON-COFFIN LAW

M. MAROS

Department of Mechanical Engineering, University of Miskolc,
Miskolc-Egyetemváros, H-3515 Hungary

ABSTRACT

Mathematical statistical and thermodynamic analyses have been carried out for the verification of the supposed correlation between the empirical parameters of the Manson-Coffin relationship. Based on as much as 560 pair of LCF test data the existence of the relationship between the fatigue ductility coefficient and exponent in the form of $\ln\varepsilon_f' = A + Bc$ has been proved for different group of metals. The main results of the analysis aiming at determining the nature, strength and validity of the relationship, as well as the investigation of the parameters affecting the correlation are shown. Using the constitutive equations of the thermally activated plastic deformation a semi-empirical model is suggested for the explanation of the possible physical background of the statistically proved correlation.

KEYWORDS

Empirical power relationship, LCF, mathematical statistics, thermodynamics, thermal activation, modeling

PRELIMINARIES, OBJECTIVES

Investigations directed towards the determination of the correlation between the empirical parameters of the power type relationships describing the various damage mechanisms of metallic materials have revealed that the materials constants involved in the given relationship are not independent of each other. Existence of the correlation in the form of:

$$lgC_i = A_i + B_in$$

has been proved by mathematical statistical methods for the case of fatigue crack propagation, high cycle fatigue and creep [1], where C_i and n_i are empirical constants of the two parameter power type relationship describing the given damage process. Relating the steady state creep , creep rupture time and constant strain rate tension it has been proved [2], that the physical background of the correlation is the thermally activated nature of the plastic deformation. The present paper aims at analyzing the connection between the parameters of the Manson-Coffin relationship in the form of:

$$\varepsilon_{ap} = \varepsilon_f' \, (2N_f)^c,$$

- variation of the loading rate first of all at higher temperatures, which together resulted in the dominance of the time dependent processes, consequently the change of the damage mechanism.
- modification of the microstructure of the material by different processes (e.g. surface alloying, cold working, grain coarsening, strain aging, etc.) resulted in the change of fatigue behavior characteristic for the investigated structural group of materials.

4. The correlation has been improved by eliminating the effect of the parameters causing significant change in the LCF behavior of the given material. (Fig. 2.)

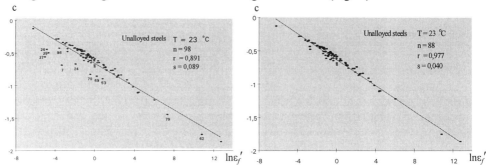

Fig. 2. Improvement of scattering of the correlation by eliminating LCF data representing significantly different behavior due to the modified test circumstances

5. The range of the fatigue ductility exponent according to the database compiled for the present analysis was much wider, namely $c = -1,8 \div -0,1$ for steels than the approximations having been formerly suggested in the literature i.e. $-0,5 \div -0,8$ for c. The $c < -1$ values were characteristic first of all for the unalloyed steels, while $c > -0,5$ values occurred mostly in case of high alloyed steels.

6. The c values showed a decreasing tendency with the increase of temperature at T>20 °C, as it is seen in Fig.3., where c_{av} represents the average value of the different c values which were processed at the given temperature.

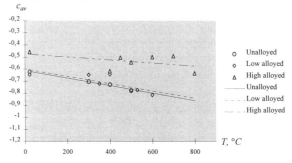

Fig. 3. Average values of the fatigue ductility exponent at different temperatures for unalloyed (o), low alloyed (◊), and high alloyed steels (Δ)

7. The temperature dependence of the correlation can not be described unambiguously on the basis of the available data. However for the unalloyed steels a clear tendency of increasing inclination with increasing temperature has been observed.

where ε_{ap} is the plastic strain amplitude, N_f is the number of cycles to failure, ε_f' and c are the materials constants. The main objectives are as follows:

- Verification of the correlation on the basis of a great number of test results;
- Determination of the nature, strength and validity of the relationship;
- Investigation of the parameters affecting the supposed correlation;
- Physical explanation of the correlation between the empirical parameters ε_f' and c.

MATHEMATICAL STATISTICAL ANALYSIS

The database was compiled on the basis of as much as 560 pair of ε_f' and c value of LCF test results, taken from the literature, see [3]. The investigated materials and temperatures were as follows:

Unalloyed steels	(23 - 500 °C)	Al alloys	(23 °C)
Low alloyed steels	(23 - 600 °C)	Ti alloys	(23 °C)
High alloyed steels	(23 - 800 °C)	Cu alloys	(23 °C)

Assuming the $ln\varepsilon_f'= A + Bc$ type correlation the equation and statistical characteristics of the regression lines for the above materials and temperatures (altogether 22 cases) have been determined. The results of the detailed analysis [3, 4] can be summarized as follows:

1. There is an unambiguous linear correlation between the logarithm of the fatigue ductility coefficient and the fatigue ductility exponent. The $ln\varepsilon_f'$ value is increasing with decreasing c values. The regression lines for the unalloyed steels at different temperatures illustrated by Fig.1.

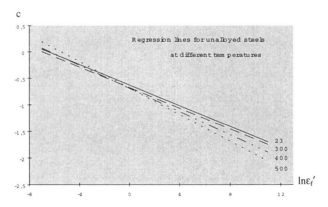

Fig.1. Regression lines for unalloyed steels at different temperatures

2. The correlation coefficient, r, characterizing the adequacy of the connection ranged from 0.504 to 0.988, however we found $r>0.9$ in the 73 % of the cases.
3. Analyzing the groups of data having $r<0.9$ correlation coefficient it was found, that the cause of the bigger scattering can be explained by the variation in the measuring circumstances, i.e. the change in the LCF behavior of the material most often due to the following reasons:

THERMODYNAMICAL ANALYSIS

Preliminaries

Investigations directed towards the physical explanation of the existing correlation between the empirical parameters of the power type relationships describing the various damage mechanisms of metallic materials revealed that in case of steady state creep, creep rupture time and constant strain rate tensile test the physical background of the correlation is a common feature of these processes. In each cases the damage is resulted by the irreversible component of the deformation [5]. Since the plastic deformation is a thermally activated process at T > 0K [6], consequently the plastic deformations during the different damage mechanisms can be described by the related constitutive equations of the thermodynamics [6, 7].

Basic Thermodynamic and Empirical Relationships

Table 1. Thermodynamic relationships

Rate of the thermally activated plastic deformation	$\dot{\varepsilon}_p = \dot{\varepsilon}_{po} \exp(-\Delta G / kT)$	(1)
Total lifetime of the thermally activated process	$t_t = t_{t_o} \exp(\Delta G / kT)$	(2)
Parameter describing the given system of obstacles	$\dot{\varepsilon}_{po} = \rho_m \cdot b \cdot s \cdot v_d$	(3)
Johnston - Gilman relationship	$m_c^* = \left(\dfrac{\partial \ln \dot{\varepsilon}_{ap}}{\partial \ln \sigma_a^*} \right)_{T,P}$	(4)
Activation enthalpy	$\Delta G = \Delta G_o - b \displaystyle\int_{o}^{\sigma^*} A^* (\sigma^*, T) d\sigma^*$	(5)
Activation area	$A^* \approx \dfrac{kTm^*}{b\sigma^*}$	(6)
Stress dependence of the activation enthalpy	$\Delta G = -kTm_c^* \ln \dfrac{\sigma_a^*}{\sigma_{a0}^*}$	(7)

Table 2. Empirical relationships describing the mechanical fatigue of smooth specimens

Manson-Coffin equation	$\varepsilon_{ap} = \varepsilon_f' (2N_f)^c$	(8)
Cyclic hardening law	$\sigma_a = K' \cdot (\varepsilon_{ap})^{n'}$	(9)
Deformation rate for simple loading functions	$\dot{\varepsilon}_p = 4 \cdot \varepsilon_{ap} \cdot f$	(10)
Effective component of the external load	$\sigma_a^* = \alpha \sigma_a$	(11)
The total time to fracture	$t_f = N_f / f$	(12)

Nomenclature:
- ΔG: activation enthalpy of the thermally activated process
- k: Boltzmann constant
- T: absolute temperature
- ρ_m: mobile dislocation density
- b: Burgers vector
- v_c: maximum velocity of the dislocations

m_c^* : strain rate sensitivity factor for cyclic deformation

σ_a^* : effective (thermal) component of the loading stress amplitude

σ_{ao}^* : resistance of the short range obstacles at T=0K

ε_{ap} : plastic strain amplitude

N_f: number of cycles to failure

ε_f' : fatigue ductility coefficient

c: fatigue ductility exponent

K': fatigue hardening coefficient

n': fatigue hardening exponent

t_f: time to fracture

f: loading frequency

α: constant

t_{to}: constant

A^*: activation area

Analysis

Assumptions:
- The dominant mechanism of the plastic deformation during LCF is the glide motion of the dislocations, which is thermally activated at $T > 0K$,
- The thermally activated dislocation motion is controlled by the short range internal (or effective) stresses (σ_a^*).

Using the constitutive equations of the thermally activated plastic deformation (Table 1.), the connection between the plastic strain rate and the effective stress component of the external load can be expressed according to the equations (1) and (7):

$$\dot{\varepsilon}_p = \dot{\varepsilon}_{po}(\sigma_{ao}^*)^{-m_c^*}(\sigma_a^*)^{m_c^*} \qquad (13)$$

On the basis of the empirical relationships describing the fatigue process (see Table 2.) from eq.(9) (10) and (11):

$$\dot{\varepsilon}_{ap} = \left(\sigma_a^*\right)^{\frac{1}{n'}}\left(K'\alpha\right)^{-\frac{1}{n'}} \cdot 4f \qquad (14)$$

The identity of eq. (12) and (14) results in

$$\frac{1}{n'} = m_c^* \qquad (15)$$

as well as:
$$\ln K' = n' \ln \frac{4f}{\dot{\varepsilon}_{po}} + \ln \frac{\sigma_{ao}^*}{\alpha} \qquad (16)$$

Eq. (16) reveals a linear correlation between the fatigue hardening coefficient and fatigue hardening exponent.

For the analogue derivation of the correlation between the $\ln\varepsilon_f'$- c parameters of the Manson-Coffin relationship the time to fracture (t_t) should be expressed in the function of the loading stress. From Eq. (8), (9), (11) and (12):

$$t_t = \frac{1}{2f}\left(\varepsilon_f' K'\right)^{-\frac{1}{c}}\left(K'\alpha\right)^{-\frac{1}{n'}} \cdot \sigma_a^{*\frac{1}{cn'}} . \qquad (17)$$

Supposing the thermally activated nature of the fracture time i. e. Arrhenius-type dependence of t_t on T, according to Eq. (2) and (7):

$$t_t = t_{to} \left(\frac{\sigma_{ao}^*}{\sigma_a^*} \right)^{m_c^*}.$$ (18)

From the identity of Eq. (17) and (18), the exponent of the effective stress:

$$m_c^* = \frac{1}{cn'}$$ (19)

and the coefficient of σ_a^*:

$$\frac{1}{2f} \cdot \left(\varepsilon_f' K' \right)^{-\frac{1}{c}} \left(K'\alpha \right)^{-\frac{1}{n'}} = t_{to} \cdot \left(\sigma_{ao}^* \right)^{-\frac{1}{cn'}}.$$ (20)

Taking the logarithm of both sides of Eq. (20) and introducing the followings:

$$A = -\left[\ln 2f + \frac{1}{n'} \ln(K'\alpha) + \ln t_{to} \right], \quad \text{and} \quad B = \frac{1}{n'} \ln \sigma_{ao}^* - \ln K' ,$$

the relationship between $\ln\varepsilon_f'$ and c can be expressed as follows:

$$\ln\varepsilon_f' = A + Bc.$$ (21)

The (21) model gives the possibility to explain the assumed and statistically proved correlation between $\ln\varepsilon_f'$ and c on a physical basis. The constants A and B are expressed in the function of characteristics related to the microstructure (ρ_m, b, σ_{ao}^*) and cyclic loading (σ_a, σ_a^*, f, m^*). The numerical verification of the model showed a good agreement between A and B values [3] based on the statistical analysis and the suggested model.

SUMMARY

Based on 560 pair of LCF test data the existence of an $\ln\varepsilon_f' = A + Bc$ type correlation between the fatigue ductility exponent and fatigue ductility coefficient was proved on the basis of statistical and thermodynamic analyses. It was concluded that the correlation is strong, and of general validity for steels, Al, Cu and Ti alloys. The correlation is strongly affected by the measuring circumstances and the microstructure of the material. The correlation can be derived also on a thermodynamic basis, supposing the thermally activated nature of the plastic deformation during LCF. The direction of the further research is the experimental verification of the physical model furthermore the expression of the parameters involved in the model in terms of basic physical quantities.

REFERENCES

1. Tóth, L. (1989) *GEP*, 16, (9), 340.
2. Krasowsky, A.J. and Tóth, L. (1994) *Strength Mater. (Probl. Proch.)*, 6, 3.
3. Maros, M.(1996). Ph.D. Thesis, University of Miskolc, Hungary
4. Maros, M. (1997) *Proc. 6th Conference on „Achievements in the Mechanical & Materials Engineering" AMME '97, Gliwice-Wisla, Miskolctapolca*
5. Krasowsky, A.J. and Tóth, L. (1997). *Metall. Trans. A.* 28A, 1831.
6. Hirth; J.P. - Lothe; J., (1982), *Theory of Dislocations, 2nd. edition*, John Wiley & Sons, Inc., New York
7. Kocks, U.F., (1973.), *Thermodynamics and kinetics of slip*, Argonne National Laboratory, Argonne

Low Cycle Fatigue and Elasto-Plastic Behaviour of Materials
K-T. Rie and P.D. Portella (Editors)
105

EFFECT OF SOME MEASURING TECHNIQUE FACTORS INFLUENCING THE LCF TEST RESULTS

GY. NAGY

Department of Mechanical Engineering, University of Miskolc
3515, Miskolc-Egyetemváros, Hungary

ABSTRACT

The uncertainty of stressing and checking may be reduced by decreasing the uncertainty of determination of the material characteristics. It is especially important during Low Cycle Fatigue (LCF) tests, when the results are influenced by several measuring-technique parameters. The present paper analyzes the effect of accuracy and reproducibility of setting the controlled parameter, as well as type of the applied extensometer and shape of the specimen on the measured characteristics.

KEYWORDS

Low Cycle Fatigue, testing, reproducibility of measurement, specimen shape, type of extensometer

INTRODUCTION

Designing and operating equipments and structures require stressing and checking. Every calculation accompanied by uncertainties (e.g. the error of modeling, uncertainty of loading and operation conditions, etc.) the effect of which are considered by using factors of safety. The reliability of the stressing and checking, the economy of production of goods can be improved by detailed analysis of the factors increasing the uncertainty as well as reducing their effects. The reliability of stressing and checking also depends on the reproducibility of the determination of material constants, which are tried to be kept constant by standardization of the test circumstances.To analyze the reliability of these material properties is especially important for such a complex tests as the Low Cycle Fatigue test is. During LCF tests the most important measuring technique factors influencing the test results are as follows:
- accuracy and reproducibility of setting of the controlled parameter;
- type of the extensometer (axial or diametral);
- shape of the specimen;
- surface-quality of specimen;
- asymmetry factor of loading;
- shape of the loading function;
- failure criteria [1]

106

The present paper aiming at presenting the results of the systematic analysis directed to determining the effect of the first three factors.

THE EFFECT OF THE ACCURACY AND REPRODUCIBILTY OF SETTING THE CONTROLLED PARAMETER AND THE TYPE OF THE EXTENSOMETER

The experiments were performed on cylindrical specimens with diameters of 8 mm according to the ASTM recommendations [2] using tangential and axial extensometers, realising R= -1 asymmetry factor up to the 25% decrement of the maximum tensile load. The tests were performed on an MTS type machine, the registered parameters were the minimum and maximum values of strain amplitude in every cycle. Fatigue tests were performed on 5-5 specimens, applying two constant (ε_{a1}=0.01 minimum, ε_{a2}=0.005 maximum) strain amplitude measured by diametral and 10 mm gauge length axial extensometers. During testing with diametral extensometers the strain amplitude was converted to axial values for the possibility of comparison.

The 200-200 maximum and minimum strain amplitudes measured on specimens fatigued with the above conditions were analysed by mathematical statistics [3]. The measured strain amplitude range has been divided into intervals and the related relative frequency bar charts have been constructed. Furthermore relative error of each measured strain amplitude has been calculated, and also plotted in bar charts. For the case of diametral extensometer Fig. 1. shows the relative frequency bar charts for ε_{a1}=0.01, while Fig. 2. illustrates the histogram for ε_{a2}=0.005 strain amplitudes. Results of measurements with axial extensometer for ε_{a1}=0.01 strain amplitude are shown in Fig. 3, and histograms for ε_{a2}=0.005 in Fig. 4.

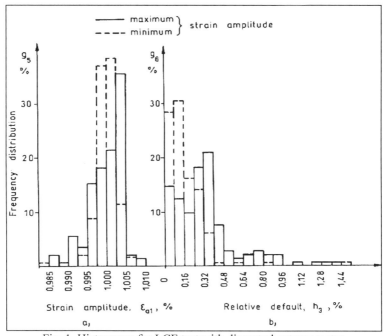

Fig. 1. Histograms for LCF test with diametral extensometer,
strain amplitude ε_{a1}=0.01

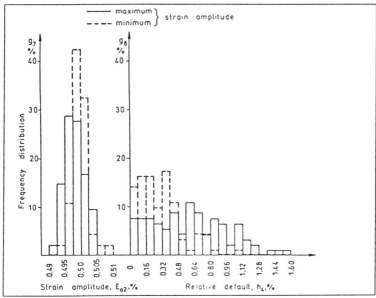

Fig. 2. Histograms for LCF test with diametral extensometer,
strain amplitude $\varepsilon_{a2}=0.005$

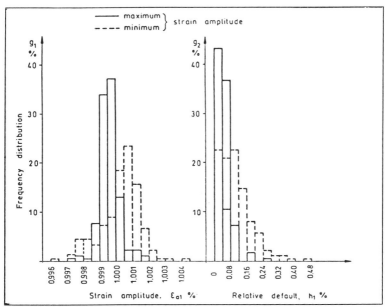

Fig. 3. Histograms for LCF test with axial extensometer,
strain amplitude $\varepsilon_{a1}=0.01$

108

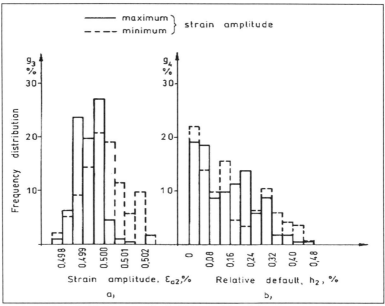

Fig. 4. Histograms for LCF test with axial extensometer,
strain amplitude $\varepsilon_{a2}=0.005$

Based on the results the following statements can be done:
- using axial extensometer the maximum and minimum strain amplitudes have been set (controlled by the testing machine) with acceptable reliability and small scattering, the relative error was less than 0.5%;

- in case of diametral extensometer the scattering of the maximum and minimum strain amplitudes is bigger than that of the axial extensometer, however the relative error is smaller than 2%, according to the standards [2].

It has to be noted that the previous statements are valid for the given machine and measuring system, because the accuracy of different machines and equipments are not necessarily the same.

THE EFFECT OF THE SHAPE OF SPECIMEN

The shape and size of the specimens are primarily determined by the shape and size of the product, from which the specimen is processed. When choosing the shape it is also important to avoid the buckling of specimens during fatigue under the usual tension and compression loading. This is why most of the LCF specimens are thick-set [2,4], although their shape can differ from each other. The circular cross-section specimens basically can be divided into two groups according to their shape. One type is when the tested, usually short, cylindrical part is connected through a radius with the head of considerably higher diameter in order to decrease the risk of buckling. The other type is when the constant diameter part is missing, i.e. the specimen is hourglass shaped. It would be proper to ask, whether the material properties determined on the two different type of specimen can be regarded as being the same or not ? To answer this question LCF tests using specimens shown in Fig. 5. have been accomplished.

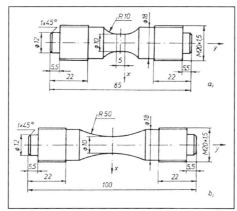

Fig. 5. Investigated types of specimens

In order to decrease the effects of microstructure inhomogenities material of specimens of X 12 CrNiTi 18 9 was chosen. The fatigue tests were performed until fracture with three different total strain amplitudes ($\varepsilon_{a1}=\pm0.0189$; $\varepsilon_{a2}=\pm0.0094$; $\varepsilon_{a3}=\pm0.0038$), on 5-5 specimens at each loading level, at room temperature in air. The number of cycles to failure (N_t) and the average value and standard deviations of the stress amplitudes belonging to the 50% of the life time (σ_{a50}) can be found in Table 1.

Table 1. Properties determined on different specimen types

Total strain amplitudes	The measured variable	Specimen type			
		average	st. dev.	average	st. dev.
	Number of cycles to failure, N_t	117	7.5	132	17.29
±0.0189	Stress amplitudes belonging to the 50% of the life time, σ_{a50}, MPa	595	4.82	534	3.35
	Number of cycles to failure, N_t	488	47.78	553	92.55
±0.0094	Stress amplitudes belonging to the 50% of the life time, σ_{a50}, MPa	425	11.97	404	5.03
	Number of cycles to failure, N_t	4609	530	5421	791
±0.0038	Stress amplitudes belonging to the 50% of the life time, σ_{a50}, MPa	317	7.33	309	8.22

The identity of the number of cycles to failure, (N_t) and the stress amplitudes belonging to the 50% of the life time (σ_{a50}) determined on different shape of specimens were checked by Wilcoxon-tests [5]. On the basis of the performed tests and the above evaluation the following statements can done:

- number of cycles to failure, (N_t) determined on the *a*, and *b*, specimens, with three different strain amplitudes are identical at 95% level of significance;
- values of stress amplitudes belonging to the 50% of the life time (σ_{a50}) in case of all the three strain amplitudes greatly differ from each other at 95% level of significance, the reason for which is most probably the difference in the shape factor of two specimens;
- the number of cycles to failure, (N_t) are less sensible to the shape of specimens than the amplitudes belonging to the 50% of the life time (σ_{a50}).

Based on the data presented in Table 1. further statements can be done:
- the standard deviation and variation coefficient (mean-square deviation) of the number of cycles to failure, exceed the values usual for the static mechanical material testing, however at the same time they are substantially smaller than those of the High Cycle Fatigue test;
- the standard deviations of the stress amplitudes belonging to the 50% of the life time (σ_{a50}) are in good agreement with those for the ultimate tensile strength and the yield strength;
- the variation, standard deviation, mean-square deviation of the average number of cycles to failure decrease with increasing strain amplitudes.

CONCLUSIONS

Analysing the effect of three measuring technique factors influencing the LCF test results the following conclusions can be drawn:
- the value of the controlled parameter has been kept constant by the testing machine with less than 0.5% relative error; in case of axial extensometer;
- the relative error of the adjustment of strain amplitudes has been smaller than 2% using diametrical extensometer;
- the shape of specimen has not influenced the number of cycles to failure, but it has had an effect on stress amplitudes belonging to the 50% of the life time.

ACKNOWLEDGEMENTS

I have the National Scientific Research Found (OTKA T 016505) to thank for its financial support.

REFERENCES

1.   Nagy, Gy. (1996) In: *Publications of the University of Miskolc, Series C., Mechanical Engineering* **46**.,University of Miskolc, Miskolc pp. 131-141.
2.   ASTM E 606-74: *Constant-Amplitude-Low-Cycle Fatigue Testing*
3.   Dowdy, S. and Wearden, S. (1983) *Statistics for research* John Wiley and Sons, New York
4.   ASTM STP 465. (1969) *Manual on Low-Cycle Fatigue Testing* ,Baltimore, USA
5.   Balogh, A., Dukati, F. and Sallay, L. (1980) *Minőségellenőrzés és megbízhatóság*, Műszaki Könyvkiadó, Budapest

Low Cycle Fatigue and Elasto-Plastic Behaviour of Materials
K-T. Rie and P.D. Portella (Editors)
© 1998 Elsevier Science Ltd. All rights reserved.   111

TESTING AND STUDY ON LOW CYCLE IMPACT FATIGUE

PINGSHENG YANG
Department of Materials Science and Engineering , Nanchang University
Nanchang 330047 , P. R . China
RONGRU GU , YONGZE WANG AND YUNFU RAO
Nuclear Industry General Co. 720 Factory
Nanchang 330101 , P. R . China

ABSTRACT

In this paper a low cycle impact fatigue testing apparatus and corresponding testing method are presented . Furthermore the work deals with some principal investigation results on low cycle impact fatigue for metallic materials .

KEYWORDS

Low cycle impact fatigue , Hopkinson's bar , Stress wave .

INTRODUCTION

Low cycle impact fatigue at high strain rate (LCIF)is a common service load and failure form which not only occurs in aerospace , weaponry and forging apparatus but is a important cause of overload damage for power plant so the investigation on LCIF is significant in theory and practice .

A push-pull impact fatigue testing apparatus has been designed by the authors . The LCIF tests have been made of a series of metallic materials and some significant results were obtained .

EXPERIMENTAL METHOD AND PRINCIPLE

Loading System and Stress Wave Propagation

In the LCIF testing apparatus the loading assembly is actually a combination of Hopkinson's pressure bar and extension bar as shown in Fig .1 [1] . A combined falling hammer consists of an outside hammer and an inside hammer so fitted together that they can slide freely relative to each other while remaining concentric . First the inside hammer collides with the wave-conducting bar , causing a compressive stress pulse in the specimen . Following that an impact of the outside hammer on the tube of specimen-weighbar assembly gave rise to a tensile stress pulse in the specimen , owing to the reflection of the incident compressive stress wave at the end of the yoke . The amplitude of the incident wave is dependent on the impact velocity and the wavelength is twice the length of the hammer . By adjusting the falling height a variable impact velocity of up to 7.7 m s^{-1} can be obtained , and the strain rate in specimens may approach 400 s^{-1} .

During testing the signals of incident wave $\varepsilon_I(t)$ and transmissive wave $\varepsilon_T(t)$ are detected by the strain gauges g_u and g_L respectively . The output of the strain gauges is amplified and fed to a storage oscilloscope , in which the signals are transformed into numeric quantities and stored . Fig . 2 shows the incident wave , from which it is apparent that the original incident wave is trapezoidal with a steep front (the time for front rise is within 10 μs) and a total duration of 120 μs .

Measure and Calculation of Dynamic Stress-Strain

The average strain and stress of specimens can be calculated from the records of the incident and transmissive wave :

$$\varepsilon_{SA} = \frac{2C_0}{L_S} \int_0^t (\varepsilon_I - \varepsilon_T)dt \dots\dots\dots\dots\dots(1)$$

$$\sigma_{SA} = E \frac{A_0}{A_S} \varepsilon_T \dots\dots\dots\dots\dots(2)$$

where ε_I and ε_T are the incident strain and transmissive strain respectively ; E is Yong's modulus ; C_0 is the velocity of elastic wave propagation ; A_0 and A_s are the cross-sectional areas of the incident bar and specimen respectively ; and L_S is the gauge length of the specimen .

For the LCIF testing apparatus the determination of strain and stress differs somewhat from ordinary SHPB . For the tensile pulse Harding's method [2] is used in principle , but an improvement has been made . Considering the effect of cross-sectional area of specimens , the authors use an elastic specimen-weighbar assembly instead of an uniform elastic bar-weighbar assembly , and regard the stress wave record given by the strain gauges g_L as the incident wave $\varepsilon_I(t)$, while the record given by the same gauges from the test specimen-weighbar assembly is taken for the transmissive wave $\varepsilon_T(t)$ (Fig . 3) . Having thus eliminated geometry effects , the stress-strain curves so obtained represent only the mechanical behaviour of the materials Based on Equations (1) and (2) , the σ-t and ε-t curves can be obtained and , in consequence , the σ-ε hysteresis loop may be drawn by a computer with time t as a parameter .

Fig . 4 represent a typical hysteresis loop of LCIF , in which there is a periodic stress fall forming an additional damped oscillation similar to the transmissive wave shown in Fig . 3 . This comes from the unloading wave reflected by the end of the specimen and has no relation to the properties of the material , so a smoothing procedure has been undertaken as shown by the dotted line in the figure .

SOME PRINCIPAL INVESTIGATION RESULTS ON LCIF FOR METALLIC MATERIALS

Overstress and Bauschinger Effect

LCIF tests have been carried out by the authors on such materials as low-carbon steel , stainless steel , medium-carbon alloy steel , aluminium , duralumin , brass and SiC/Al composites [1 , 3 , 4 , 5 , 6 , 7 , 8] . The results showed a considerable increase in flow stress for LCIF in comparison with that for LCF . This is so-called ' over-stress ' . For instance , the 0.2% proof stress in impact tests is 2.6 times as large as that in static tests for low-carbon steel , and 2.0 times for stainless steel , 0.6 times for brass , 0.25 times for aluminium respectively . However , there is no obvious overstress in 40Cr steel . The Bauschinger effect in 40Cr steel was investigated . The results showed that a obvious Bauschinger effect occurred in LCF for the steel : after an elongation of 2.3% in the positive half-cycle of the LCF test , the compression in the following half-cycle produced a round flow stress curve with yield strength decreased to 53% of the previous tension . At a stress of 880 MPa the Bauschinger strain was about 1% . However , there was no appreciable difference between the previous tension and following compression flow stress curves in LCIF tests for the steel .

Cyclic Hardening (Softening)

In both LCIF and LCF tests , a similar cyclic hardening-softening behaviours appeared for the tested materials . That is : low-carbon steel , stainless steel , duralumin , brass and SiC/Al composite appeared cyclic hardening ; 40Cr steel and aluminium appeared cyclic softening . However , there were obvious differences in hardening-softening behaviours between LCIF and LCF tests . For low-carbon steel and stainless steel the cyclic hardening in LCF was stronger than in LCIF (Fig . 5) ; On the contrary , for brass , duralumin and SiC/Al composites the cyclic hardening in LCF was weaker than in LCIF .

Fatigue Life

By plotting $\Delta\varepsilon_e/2$, $\Delta\varepsilon_p/2$ at 50% N_f against the cycles to failure (Fig . 6) , it was found that their relations in both LCF and LCIF all conformed to the Coffin-Manson law . There were similar circumstances for all the experimental materials : the $\Delta\varepsilon_e/2$ - N_f curves of LCIF are above those of LCF , which accords with the phenomenon of overstress ; by contrast , the $\Delta\varepsilon_p$ /2 - N_f curves of LCIF lay below those of LCF , showing that cyclic plastic deformation at high strain rates causes heavier fatigue damage . The transition life of LCIF inevitably shifts left , as shown in the figures . The transition life of 0.1%C steel shifts left from about 2×10^4 cycles for LCF to about 600 cycles for LCIF ; for 1Cr18Ni9Ti steel from about 4×10^3 cycles to 1×10^3 cycles ; for brass from 6×10^4 cycles to 1.6×10^4 cycles ; for aluminium from 2.6×10^4 cycles

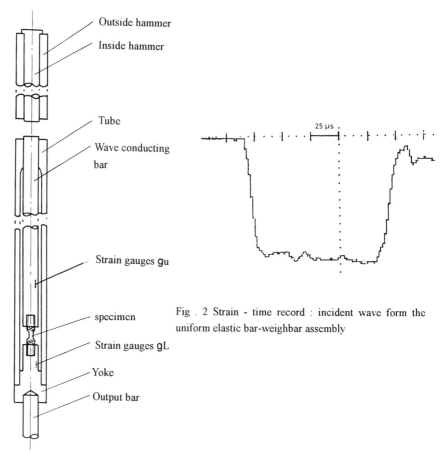

Fig . 2 Strain - time record : incident wave form the uniform elastic bar-weighbar assembly

Fig . 1 Schematic diagram of the loading system

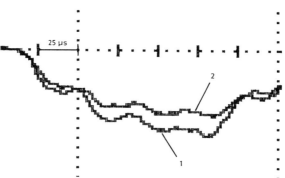

Fig . 3 Strain-time record : from the specimen - weighbar assembly ; curve 1 calulated incident wave $\varepsilon_I(t)$, curve 2 the transmissive wave $\varepsilon_T(t)$

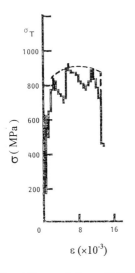

Fig . 4 Hysteresis loop of LCIF

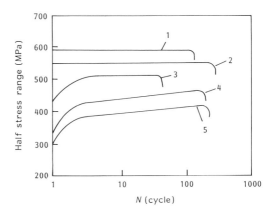

Fig . 5 Cyclic hardening curves of 0.1%C steel : 1, LCIF, $\Delta \varepsilon_T/2 = 0.009$; 2 , LCIF, $\Delta \varepsilon_T/2 = 0.007$; 3 , LCF , $\Delta \varepsilon_T/2 = 0.045$; 4 , LCF , $\Delta \varepsilon_T/2 = 0.023$; 5 , LCF , $\Delta \varepsilon_T/2 = 0.017$

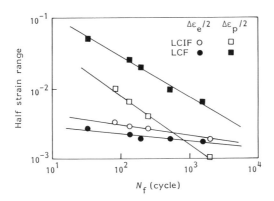

Fig . 6   Relationship between half strain range $\Delta \varepsilon_e/2$, $\Delta \varepsilon_p/2$ and cycles to failure N_f for 0.1%C steel

to 1.2×10^3 cycles ; for 40Cr steel from 10^3 cycles to 10^2 cycles ; for duralumin from 30 cycles to 13 cycles and so on . The results show that low-carbon steel and aluminium are sensitive to strain rates but stainless steel , brass and duralimin are less sensitive .

Fatigue Dislocation Structure

The TEM analysis of substructure of low-carbon steel and austenitic stainless steel shows that in LCF under the definite fatigue life condition of $N_f = 2 \times 10^3$, the dislocation density of both the materials increases with the increase in loading cycles and dislocation arrangement changes from dislocation lines through tangles to cells . However , in LCIF under the same N_f condition , the dislocation density only less increases ; the arrangements before fracture are mainly tangles but not cells . These analyses indicate that the mechanisms of crack initiation and propagation are different between LCF and LCIF .

SUMMARY

A push-pull impact fatigue testing apparatus and relevant testing procedures were developed by the authors applying a modified Hopkinson's bar technique . The LCIF tests have been made of a series of metallic materials and some significant results were obtained .

There are not only similarities but also dissimilarities in cyclic deformation and fracture behaviour between LCIF and LCF of materials ; these are strain-rate effects . Different strain-rate effects occur in each material ; therefore LCIF is an independent behaviour of materials , and it can neither be replaced by LCF test nor be estimated by impact test .

ACKNOWLEDGEMENT

The research was supported by the National Natural Science Foundation of China under Subcontract 58971075 and 59371038 .

REFERENCES

1. Yang , P. S ., Liao , X . N ., Zhu , J. H . and Zhou , H . J. (1994) Int . J. Fatigue 16 , 327 .
2. Harding , J., Wood , E . O . and Campbell , J. D . (1960) J. Mech . Eng . Sci 2(2) , 88 .
3. Yang , P. S . and Zhou , H . J. (1994) Int . J. Fatigue 16 , 567 .
4. Yang , P. S ., Tan , Y. X . and Zhou , H . J. (1995) Journal of Chinese Electron Microscopy Society 14 , 282 .
5. Yang , P. S ., Liao , X . N . and Hong , C . Y . (1994) Journal of Nanchang University (Natural Science) 18 , 397 .
6. Yang , P. S ., Liu , Y. and Zhang , M . (1996) Journal of Nanchang University (Natural Science) 20 , 156 .
7. Yang , P. S . and Zhang , M . (1996) In : Proceedings of the 6th International Fatigue Congress , G . Lütjering and H . Nowack (Eds) . Pergamon , Oxford , pp . 197-202 .
8. Yang , P. S ., Zhang , M ., Xu , F. and Gong , S . P. (1997) Materials Science and Engineering A236 , 127 .

Chapter 2

THERMAL AND THERMO-MECHANICAL LOADING

Low Cycle Fatigue and Elasto-Plastic Behaviour of Materials
K-T. Rie and P.D. Portella (Editors)

THERMAL AND THERMAL-MECHANICAL FATIGUE OF SUPERALLOYS, A CHALLENGING GOAL FOR MECHANICAL TESTS AND MODELS

L. REMY
Centre des Matériaux P.M. Fourt, UMR CNRS 7633,
Ecole des Mines de Paris, B.P. 87, 91003 EVRY Cedex, France

ABSTRACT

The complexity of thermal and thermal-loading of superalloys used in gas turbines is adressed using selected examples from recent work. Laboratory simulation is rather difficult and requires both thermal mechanical fatigue (TMF) tests on a volume element and thermal fatigue on simple structures. Modelling is most often necessary to extrapolate from laboratory to service conditions. Constitutive models with internal variables have proven to be efficient to account for stress-strain behaviour under TMF. Damage often involves oxidation, creep and fatigue interactions. Continuum damage mechanics concepts are rather powerful but one has to carefully consider synergistic effects. Some detrimental effects of coatings are discussed in the case of aluminized components.

KEYWORDS

Thermal fatigue, thermal-mechanical fatigue, nickel base superalloys, life prediction

INTRODUCTION

Many components especially in gas turbines used in aircraft or in energy production experience thermal transients during service operation. Damage under combined thermal and mechanical loading is life - limiting for parts (vanes, blades...) which are mostly made of nickel base superalloys. Many critical parts such as blades are in fact submitted to mechanical loading superimposed to thermal loading. Creep can interact with thermal fatigue and the surface of components is exposed to an aggressive environment, giving rise to oxidation mostly and to corrosion too. Long time operation and repetition of transients may in addition result in microstructure changes.

Thermal stress mainly arises in structures. Therefore the simulation of thermal loading of components is a real challenge to engineers and experimentalists. Two different kinds of tests are used in the laboratory if one excludes real component testing in an engine, which is very expensive : tests on simple structures (Glenny's discs [1], or wedges) submitted to thermal transients only and thermal mechanical fatigue (TMF) tests. The TMF test [2] is a volume element test which became mature with the development of microcomputer capabilities [3]. This test involves simultaneous temperature and strain cycling, with no stress gradient across the specimen cross section. One of the major advantages of the TMF test is that it can used to check the predictions of constitutive equations under anisothermal conditions [4]. It can provide furthermore a useful test of damage models, without relying on the predictions of a constitutive model. Thermal fatigue (TF) or thermal shock tests can simulate more closely the behaviour of static components but they are used mostly to verify the models : the life predicted depends upon constitutive models as well as upon damage models.

The purpose of this paper is to give a brief overview of thermal and thermal-mechanical loading of superalloys used in gas turbines. Testing will be discussed first emphasizing the need of tests on volume element as on structures. Modelling is difficult but essential in life assessment of components experiencing thermal transients. Constitutive equations wil be considered first. Damage modelling will then be considered for bare alloys and for coated alloys, using aluminized single crystals as an example.

THERMAL AND THERMAL-MECHANICAL FATIGUE TESTING

The major developments in testing over recent years mainly concern TMF testing which is now becoming more and more popular. Since the work of Glenny, some TF facilities were developped which use the same type of specimen geometry but different heating/cooling devices : burner rig, radio-frequency induction heater [5] or lamp furnace and removable cooling nozzle [6]. Rezai-Aria and coll. have measured the displacement along the thin edge of the wedge to check total strains [5], which can be used to validate a structure analysis.

Before the introduction of microcomputers the accuracy of the thermal cycle control in TMF was poor. Pioneer authors [2, 7] used two basic cycles, which are still very popular : "in-phase" cycles where the maximum mechanical strain occurs at maximum temperature or an "out-of-phase" cycle where the maximum mechanical strain occurs at minimum temperature. Fully reversed mechanical strain cycling is mostly used for these tests and an ISO TMF standard is in progress which uses these cycles.

A different approach is to use the TMF test to simulate more closely the strain temperature history of critical areas in parts [4]. A counter-clockwise diamond cycle has been used to simulate thermal transients occuring in blades and vanes, Fig. 1 [6, 7]. Such a cycle was shown to give rise to little mean stress in different superalloys (see Fig. 2) while the in-phase and out-of-phase cycles yield large compressive and tensile mean stress respectively [4, 8-10]. These mean stresses can explain to some extent the conflicting reports on phasing effects. Some authors are now trying to integrate the whole complexity of cycles. Meersman and coworkers use a facility which enables to simulate biaxial thermal-mechanical loading with complex path history for cast superalloys used in land - based gas turbines [11].

The concept of a volume element test imposes a constraint of no stress gradient in the gage length, which places some limitations on the temperature rate that can be achieved. It is therefore impossible in most cases to simulate the actual loading of critical parts in TMF tests, because of the temperature rate and the duration of the hold time under load during steady

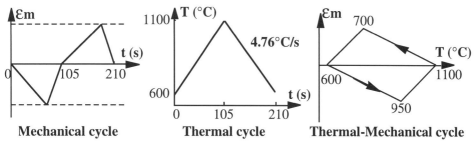

Fig. 1. A simplified thermal-mechanical fatigue cycle for blades in gas turbine :
a) temperature versus time, b) mechanical strain versus time and c) mechanical strain versus
temperature diagrams.

operation. TMF life data can seldom be used directly to design components and modelling is
therefore essential to extrapolate to service conditions or even to thermal shock test
conditions.

TMF testing is mostly useful to check the prediction of damage models and of constitutive
equations [4]. The validation of each kind of model can be made independently, which is
impossible to do with thermal shock tests : the predicted lifetime is the convolution of
predictions of both types of models and can be biased by uncertainties in the thermal analysis,
as in actual components.

CONSTITUTIVE MODELLING

Constitutive equations used for superalloys are often viscoplastic to account for creep.
Equations with internal variables were developed among others in the group of Chaboche, to
describe non linear kinematic hardening [12]. This is required by the Bauschinger effect
which is quite important to describe stress redistribution in cyclically loaded superalloy
components.The viscoplastic strain rate in uniaxial from is written as modified Norton's law :

$$\dot{\varepsilon}_v = \left\langle \frac{|\sigma - X_v| - R_v}{K} \right\rangle^n \text{sign} (\sigma - X_v) \tag{1}$$

where K and n are constant at a given temperature, < > are Mc Caulay brackets (<U> = U if
U>0 else 0), R_v is the radius of the plastic domain and X_v an internal variable that describes
the position of the centre of the elastic domain in stress space. Both R_v and X_v are given by
differential rate equations.

Such equations are identified from isothermal low cycle fatigue tests with continuous saw-
tooth cycles or including stress or strain hold and they predict fairly well the stress-strain
behaviour under varying temperature conditions in conventional superalloys [13, 4].

For anisotropic single crystal superalloys, G. Cailletaud and coll. [14, 15] have proposed to
use a crystallographic model which uses a viscoplastic phenomenological equation like Eq. 1
but at the level of the slip system. On each slip system s, Norton's law is :

122

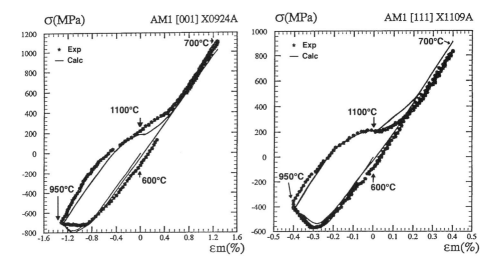

Fig. 2. Stress versus mechanical strain hysteresis loops for [001] and [111] single crystal TMF specimens of AM1 superalloys using the cycle depicted in Fig. 1 : comparison between a crystallographic viscoplastic model (solid line) and experiment (symbols).

$$\dot{\gamma}^s = \left\langle \frac{|\tau^s - x^s| - r^s}{K_I} \right\rangle^{n_I} \text{sign} (\tau^s - x^s) \qquad (2)$$

where τ^s is the shear stress on the slip system, x^s is a kinematic variable, r^s describes isotropic hardening. K_I and n_I are viscosity constant and exponent, where the index I refer to octahedral slip or to cube slip. No direct equation is postulated at the macroscopic level : the macroscopic stress is projected on to the slip systems using a generalized Schmid law, the macropic strain is the sum of all the contributions of the different slip systems. Cubic elastic anisotropy is assumed. This model which is implemented in a finite element code, is identified from isothermal tests for <001> and <111> crystals. One of the major advantages of a crystallographic model is that it can predict multiaxial stress situations and more, the active slip systems which successfully compared with experimental observations [15]. Predictions are in good agreement with experimental stress-strain hysteresis loops under TMF cycling with a diamond type cycle between 600 and 1100°C, Fig. 2 [9].

For long term operation, thermal mechanical loading can give rise to rafting of γ' precipitates, which are initially cuboids 0.5 µm in size [16, 17]. In such cases it can be necessary to incorporate a description of the evolution of microstructure. The influence of γ' rafting on the mechanical behaviour has been reported by some authors [17]. Composite models can be made using a finite element analysis of a periodic composite structure : such attempts were made recently but this will give rise to rather long structure computations for components [20]. Self consistent constitutive models can be used which simply assume a uniform stress and strain in each phase. Such an approach was recently done using experiments on various

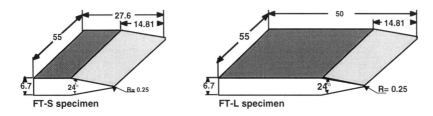

Fig. 3.   Geometry of thermal fatigue wedge specimens used for a wrought superalloy.

rafted microstructures and on single crystals of the matrix alone [18]. Good predictions of stress strain behaviour were obtained for isothermal conditions, but no attempt was made to predict TMF stress-strain loops.

A constitutive model which has been tested against TMF stress-strain loops can be used with a much better degree of confidence in a finite element analysis of a component or of a single thermal shock structure. Fig. 3 displays the geometry of specimens, standard and large, which are currently used on a thermal shock rig in our laboratory [6]. Fig. 4 shows the stress-temperature cycles as well as the stress-mechanical strain loops which can be computed after the thermal and stress analysis at the critical area of such structures with a model which has been validated using TMF experiments.

DAMAGE MODELLING

Bare Alloys

Linear accumulation of fatigue and creep damage is the simplest way of handling creep-fatigue situations under isothermal conditions as thermal-mechanical fatigue. Fatigue damage is described through Miner's rule and creep damage is estimated by the equation of Robinson and Taira. This procedure was used widely by Spera to compute the thermal fatigue life of Glenny discs and wedge specimens of superalloys submitted to thermal shocks using the fluidized bed rig as for components [19]. In a number of cases, non linear accumulation of creep and fatigue damage is definitely necessary. Continuum damage mechanics is certainly the most widely developing method over recent years [12, 20-23]. This theory uses a simplifying concept of an effective stress $\tilde{\sigma}$ which would produce the same strain in an undamaged volume element as the applied stress σ in the damaged element :

$$\tilde{\sigma} = \sigma/(1 - D) \qquad (3)$$

where damage D is such that $0 \le D \le 1$. Most authors [12, 21] assume a linear summation of creep (dD_c) and fatigue (dD_f) damage increments :

$$dD = dD_f (\Delta\sigma/2, \ \tilde{\sigma} \ , D) + dD_c (\tilde{\sigma} \ , D) \qquad (4)$$

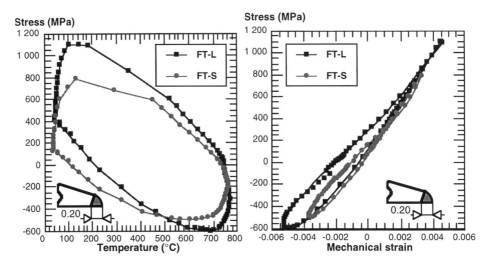

Fig. 4. Calculated stress-temperature loops (left) and stress-mechanical strain loops (right) for the first element at the edge of the two geometries FT-S and FT-L depicted in Fig. 4.

Expressions for each increment are detailed in several references and $\Delta\sigma, \bar{\sigma}$, D refer to the stress range, mean stress and damage. Multiaxial loading is taken into account in different ways by authors but damage is basically undefined. Predictions can be achieved for high temperature alloys and are used by engine manufacturers like SNECMA for various components (cooled blades, vanes, combustion chambers) [23].

Some modifications were proposed by Chaboche and coworkers to distinguish between a crack initiation and crack propagation periods : more recently the importance of oxidation in the crack initiation period was recognized in the case of coated single crystals [24]. Coffin was the first to point out at the importance of oxidation in the high temperature fatigue life of superalloys [25]. In our group, environmental attack of the surface of bare superalloys by oxidation was recognized to trigger surface crack initiation at high temperature, as shown by metallographic observations and by the reduction of life to crack initiation with respect to that in vacuum [26, 28]. The engineering life to crack initiation in cast superalloys was assumed to result from oxidation assisted microcrack growth [27, 28]. The models proposed can be written in the framework of continuum damage mechanics but here the fatigue damage is a, the length of microcracks. The oldest model [27] assumed a summation of crack growth rate increments :

$$da/dN = (da/dN)_f + (da/dN)_{ox} \qquad (5)$$

The fatigue increment was computed through Tomkins's model [29] but the oxidation contribution is a time dependent term which was deduced from metallographic measurements of localised interdendritic oxidation. This form is similar to that in Eq.4 of continuum damage mechanics. However the major difference here is that metallographic data are used in the model and not only mechanical tests. This form was used recently with some success for single crystals of AM1 [10]. If l_{ox} is the depth of oxide spikes, the kinetic equation is :

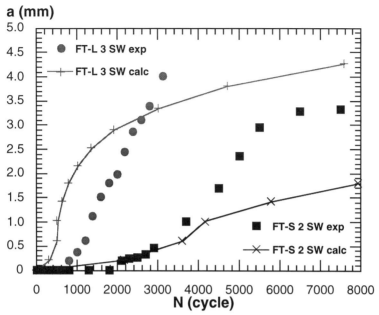

Fig. 5. Depth of the major crack as a function of cycle number for the two wedge geometry in a wrought superalloy submitted to a thermal shock between 30 and 750°C : comparison between experiment and model predictions.

$$d(l_{ox}^4) = \alpha^4 dt \qquad (6)$$

where α is a parameter which depends upon temperature according to an Arrhenius equation and which may depend upon loading. The application to thermal transients requires to integrate Eq.6 over the whole thermal-mechanical cycle.

The simple model of Eq.5 can describe TMF or TF but it cannot account for oxidation-embrittlement. Such a phenomenon was evidenced by crack growth experiments on compact tension specimens which were precracked at a temperature, then oxidized at high temperature (without applied load) and finally cracked at the test temperature. An anomalous fatigue crack growth rate behaviour was observed in various superalloys and was attributed to a local reduction in fracture toughness induced by oxidation. It is therefore necessary to use a coupled equation to describe fatigue-oxidation interactions [4, 28]. In a recent work on a wrought superalloy used for rocket engine [30], we used such a coupled model to describe oxidation-creep-fatigue interactions. Damage equation reads as :

$$da/dN = f (a, D_c, \Delta\sigma/2, \sigma_c) \qquad (7)$$

where σ_c the critical stress to break a microstructure element of size λ is given by :

$$\sigma_c = g(l_{ox}, V) \qquad (8)$$

D_c, is creep damage, l_{ox} is given by Eq. 6, V is the volume parameter which describes the size effect, to account for the differences in element size in a finite element structure analysis and laboratory specimens used to identify the model parameters.

All the parameters can be identified from isothermal tests and metallography. This model accounts for the LCF lifetime under various conditions, and can predict the lifetime under TMF or the early growth of cracks under thermal shock see e.g. Fig. 6 [30]. Such models can predict the location and size of the major crack in components [4, 28]. The prediction of the influence of microstructure on the crack growth in thermal shock on a cobalt base superalloy was a major achievement [31], using only measurements of oxidation kinetics for various microstructures without any extra mechanical test.

Coated Alloys : The Case of Aluminized Single Crystals

The intrinsic resistance of superalloys to aggressive environment (involving oxidation as well as corrosion effects) is insufficient especially with advanced high strength directionally solidified polycrystals and single crystals. Overlay coating deposited by plasma technique and aluminide coatings (chemical vapor deposition) are widely used owing to their good oxidation resistance. They prevent loss of section, which can be especially important for thin walled component operating for long service conditions. A common practice is to use a coating a few tens of µm in thickness and to design a component ignoring the presence of the coating. This simplified procedure can be conservative [32] or non conservative [33] according to the substrate / coating system and to the thermal mechanical loading path. No effect was recently obtained when a TMF cycle in the range 600 - 1100°C was applied on an aluminide coated [001] AM1 single crystal. Crack initiation in this material occurs mostly at subsurface casting pores whatever the condition, bare or aluminized [10].

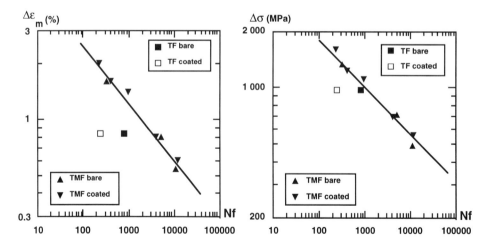

Fig. 6. Variation of lifetime under at TMF cycles between 600 and 1100°C and a severe TF cycle between 30 and 1100°C (FT-S specimen geometry) of AM1 superalloy single crystals : as function of mechanical strain range (left) and stress range (right).

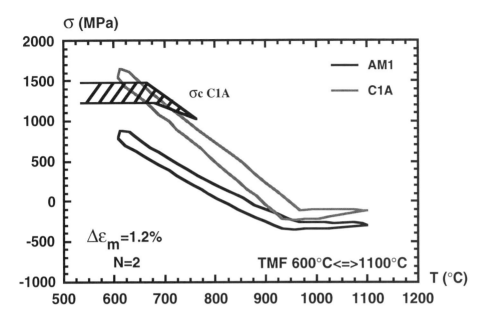

Fig. 7. Stress-versus temperature cycle of substrate (AM1) and coating (C1A) in TMF cycle. When the compressive strain range is large enough, the stress in coating exceeds the critical stress for the onset of brittle fracture.

However when a more severe TMF cycle is used as in a cycle between 30 and 1100°C, as experienced by the leading edge of wedge tested under thermal fatigue, the life of to initiate a millimetric crack is much shorter than for a bare structure Fig. [34]. Such results were reported by others authors [33] and are connected with the low temperature brittleness of the intermetallic compound coating. The ductile to brittle transition temperature can exceed 700°C for such NiAl type coating depending upon the aluminium content as well as upon alloying elements which diffuse from the substrate [35].

Brittle cracking is a primary risk for premature fatigue failure of coated components. This risk is often estimated using a measurement of tensile ductility [35], usually the strain to a macroscopic crack, as a function of temperature. For the AM1 single crystals coated with C1A, we have recently shown that this method is rather unreliable using a TMF cycle under compressive strain, see Fig. 6 [36]. Such tests can initiate large circumferential cleavage cracks around coated TMF specimens. Using constitutive equations identified from measurements on bulk NiAl with similar grain size [37], the stress cycle in the coating was estimated and found to give rise to large tensile stress in the brittle range of the coating (Fig. 7). A critical stress criterion was found to describe the onset of brittle cracking in the coating : predictions were successfully compared with different TMF loading paths.

However for conditions where the stress in the coating is high but below the threshold for brittle cracking, a crack may propagate under fatigue loading and lead to a lifetime shorter than for the bare material. A. Bickard and coll. [38] have identified isothermal fatigue life

128

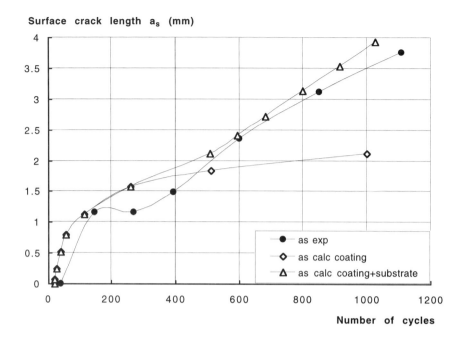

Fig. 8. Comparison between model prediction and experimental crack length (as measured on the surface) versus number of cycles between 30 and 1100°C on coated AM1. The crack growth is governed first by surface cracking in the coating and then by the substrate.

curves for the coating using specimens with an original design, in the brittle range of the coating. These S-N curves for the coating can be combined with a damage model for the substrate alloy as described earlier, which describes oxidation-fatigue damage interactions. The predictions of such a model are shown in Fig. 8, where the early surface crack growth behaviour in the structure is dominated by fatigue cracking of the coating and then cracking is governed by the growth in the substrate.

CONCLUSION

Most recent developments in testing concern thermal-mechanical fatigue (TMF) testing. Though experiments are aiming to simulate more closely service loading cycles, extrapolation is required in most cases.

Viscoplastic constitutive models with internal variables are powerful as demonstrated by the agreement with TMF stress-strain loops. Microstructure evolution during service with thermal transients remains a major challenge.

Damage models using the concept of continuum damage mechanics are promising. However the synergy between oxidation, creep and fatigue has to be considered carefully. Coated superalloys are still more complex and will require further improvement of the models.

REFERENCES

1.   Glenny, E., Northwood, J.E., Shaw, S.W.K. and Taylor, T.A. (1958-1959), *Journal of the Institute of Metals* **87**, 284-302.
2.   Hopkins, S.W. (1976). In: *Thermal Fatigue of Materials and components*, ASTM STP 612, D.A. Spera and D.F. Mowbray (Eds), ASTM, Philadelphia, pp. 157-169.
3.   Koster, A., Fleury, E., Vasseur, E. and Rémy, L. (1994). In: *Automation in Fatigue and Fracture: Testing and Analysis*, ASTM STP 1231, C.Amzallag (Ed), ASTM, Philadelphia, pp. 563-580.
4.   Rémy, L. (1990). In: *High temperature fracture mechanisms and mechanics*, EGF6, P. Bensoussan and J.P. Mascarell (Eds), Mechanical Engineering Publications, London, pp. 353-377.
5.   Meyer-Obersleben, F., Rezai-Aria, F. and Ilschner, B. (1990). In: *High temperature alloys for gas turbines and other applications*, W. Betz, R. Brunetaud, D. Coutsouradis et al. (Eds), II, Reidel, Dordrecht, pp. 1121-1129.
6.   Koster, A., Laurent, G., Cailletaud, G. and Rémy, L. (1995). In; *Fatigue under Thermal and Mechanical Loading*, J. Bressers, L. Rémy, M. Steen and L. Valles (Eds), Kluwer, Amsterdam, pp. 25-35.
7.   Taira, S. (1973). In: *Fatigue at elevated temperature*, ASTM STP 520, ASTM, Philadelphia, 239-254.
8.   Malpertu, J.L. and Rémy, L. (1990) *Metall Trans.* **21A**, 389-399.
9.   Fleury, E. and Rémy, L. (1994) *Metall. Trans.* **25A**, 99-109.
10.   Chataigner, E. and Rémy, L. (1996). In: *Thermo-mechanical fatigue behavior of materials 2nd volume,* ASTM STP 1263, M.J. Verrilli and M.G. Castelli (Eds), ASTM, Philadelphia, pp.3-23.
11.   Meersman, J., Ziebs, J. and Kühn, H.J. (1994). In: *Materials for Adanced Power Engineering,* D.Coutsouradis, J.H.Davidson, J.E. Ewald (Eds), Kluwer, Dordrecht, pp. 841-852.
12.   Lemaitre, J. and Chaboche, J.L. (1985). *Mécanique des Matériaux Solides*, Dunod, Paris.
13.   Cailletaud, G., Culie, J.P. and Kaczmarek, H. (1983). In: *Mechanical Behaviour of Solids IV*, J. Carlsson and N.G. Ohlson (Eds), Pergamon Press, Oxford, **1,** pp. 255-261.
14.   Méric, L., Poubanne, P. and Cailletaud, G. (1991) *ASME J. of Eng. Materials and Technology* **113**, 165-170.
15.   Hanriot, F., Cailletaud, G. and Rémy, L. (1991). In: *High Temperature Constitutive Modeling* , A.D. Freed and K.P. Walker (Eds), ASME, New-York, MD 26/ AMD 121, pp. 139-150.
16.   Arrel, D.J., Ostolaza, K.M., Vallès, J.L. and Bressers, J. (1996). In: *Fatigue under Thermal and Mechanical Loading*, J. Bressers, L. Rémy, M. Steen and J.L. Vallès (Eds), Kluwer, Ansterdam, pp. 295-304.
17.   Fredholm A. and Strudel, J.L. (1984); In: *Superalloys 1984*, M. Gell and C.S. Kortovich (Eds), AIME, p. 211.
18.   Espié, L. (1996) Thesis, Ecole des Mines de Paris.
19.   Bizon, P.J. and Spera, D.A. (1976). In: *Thermal Fatigue of Materials and Components,* ASTM STP 612, D.A. Spera and D.F. Mowbray (Eds), ASTM, Philadelphia, pp. 106-122.
20.   Rabotnov, Y.N. (1969). *Creep problems in structural members,* North Holland, Amsterdam.
21.   Leckie, F.A. and Hayhurst, D.R. (1977) *Acta Metall.* **25**, 1059.

22. Chaboche, J.L. (1982). In: *Low cycle fatigue and life prediction*, ASTM STP 770, C. Amzallag, B. N. Leis and P. Rabbe (Eds), ASTM, Philadelphia, pp. 81-104.

23. Dambrine, B. and Mascarell, J.P. (1990). In: *High temperature fracture mechanisms and mechanics*, EGF 6, P. Benssoussan and J.P. Mascarell (Eds), Mechanical Engineering Publications, London, pp. 195-210.

24. Gallerneau, F. (1996) Thesis, Ecole des Mines de Paris.

25. Coffin, L.F. (1973). In: *Fatigue at elevated temperatures*, ASTM STP 520, ASTM, Philadelphia, pp. 5-34.

26. Reger, M. and Rémy, L. (1988) *Mater. Sci. Engng.* **A101**, 47-54; 55-63.

27. Reuchet, J. and Rémy, L. (1983) *Metall. Trans.* **14A**, 141-149.

28. Rémy, L. (1993). In: *Behaviour of defects at high temperatures*, ESIS 15, R.A. Ainsworth and R.P. Skelton (Eds), Mechanical Engineering Publications, London, pp. 167-187.

29. Tomkins, B. (1968) *Phil. Mag.* **18**, 1041-1066.

30. Koster, A. (1997) Thesis, Ecole des Mines de Paris.

31. François, M. and Rémy, L. (1990) *Metall. Trans.* **21A**, pp. 949-958

32. Bernard, H. and Rémy, L. (1990). In: *Advanced Materials and Process, EUROMAT 89, Edited by H.E. Exner and V. Schumacher, DGM, oberursel*, **1**, pp. 529-534.

33. Bressers, J., Timm, J., Williams, S., Bennet, A. and Affeldt, E.E. (1995); In: *Thermo-Mechanical Fatigue Behavior of Materials*, **2**, ASTM STP 1263, American Society for Testing and Materials, Philadelphia.

34. Koster, A., Chataigner, E., and Rémy, L. (1996). In: *Thermal Mechanical Fatigue of Aircraft Engine Materials*, AGARD-CP-569, pp. 8-1/8-8.

35. Duret-Thual, C., Morbioli, R. and Steinmetz, P. (1986). *A guide to the control of high temperature corrosion and protection of gas turbine materials*, Report EUR 10682 EN CEC, Luxembourg.

36. Chataigner, E and Rémy, L. (1995). In: *Fatigue et Traitements de surface*, SF2M, Revue de Métallurgie, Paris, pp. 11-22.

37. Whittenberger, J.D. (1988) *J. of Materials Science* **23**, 235-240.

38. Bickard, A., Chataigner, E. and Rémy, L. (1996). Unpublished results, Centre des Matériaux.

Low Cycle Fatigue and Elasto-Plastic Behaviour of Materials
K-T. Rie and P.D. Portella (Editors)

THERMO-MECHANICAL FATIGUE (TMF) IN TURBINE AIRFOILS:
TYPICAL LOADING, 3D EFFECTS AND ADJUSTED TESTING

J. ESSLINGER and E.E. AFFELDT

MTU, Motoren- und Turbinen-Union, München GmbH
Postfach 500640, D-80976 München, Germany

ABSTRACT

Thermo-mechanical loading are caused by three effects: centrifugal forces due to turbine rotation, bending due to the gas stream, and thermal stresses which are caused by stationary and transient inhomogeneities in the temperature distribution. Typical loading conditions were evaluated by FEM and define the parameter for TMF testing. Consequences for the requirements to TMF testing are drawn and possible solutions are shown and discussed in detail.

KEYWORDS

Thermo-mechanical loading analysis, TMF testing, test performance, 3D thermal analysis.

INTRODUCTION

Due to the requirement of increased efficiency, and due to the increased temperature capability of new materials, turbine airfoils are exposed to higher and higher temperatures. Therefore, to ensure the structural integrity of the design, the computation and evaluation of thermal loading and viscoplasticity has become more and more important.

In this presentation, it is intended to focus on the aspects leading to fatigue by thermo-mechanical loading. Typical loading conditions of turbine airfoils are presented. New calculation results, aspects of three-dimensional effects and resulting typical loading condition, as well as testing procedures which are adjusted to the requirements of the application are presented.

LOADING ANALYSIS

The origin of the mechanical loading relevant for TMF may be divided into three parts: centrifugal forces due to turbine rotation, bending due to the gas stream, and thermal stresses which are caused by stationary and transient inhomogeneities in the temperature distribution. In particular the thermal stresses tend to have peaks located in areas where, due to the geometry, the development of meshing for the finite-element (FE) analysis used requires much of experience and sophisticated procedures to determine the stress levels with sufficient accuracy.

132

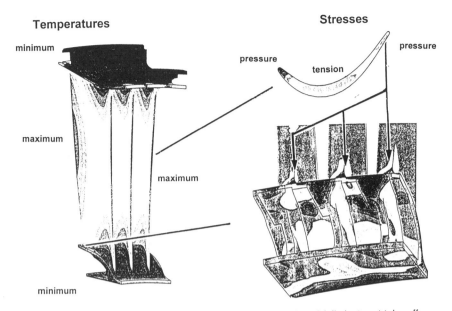

Temperatures

minimum

maximum

maximum

minimum

Stresses

pressure

pressure

tension

Fig. 1: Transient thermal gradients and stresses in a turbine airfoil cluster at take-off

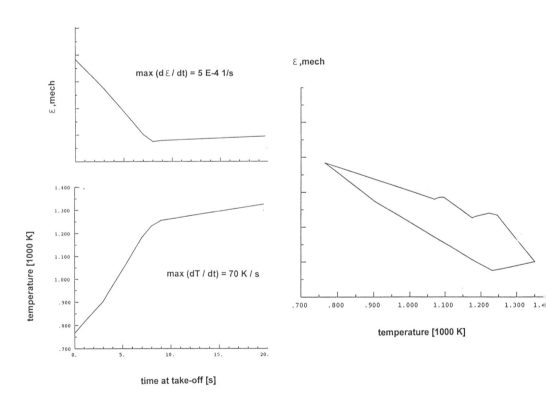

ε,mech

max (d ε / dt) = 5 E-4 1/s

ε,mech

max (dT / dt) = 70 K / s

temperature [1000 K]

time at take-off [s]

Fig.2 : Mechanical strain and temperature with respect to time,
airfoil midspan leading edge (compare fig.1)

As an example fig.1 shows for a low pressure turbine vane cluster the take-off transient temperature distribution together with its resulting stress field. Critical positions for TMF are typically located at leading and trailing edges of the airfoils' midspan regions, where peak stresses are mainly driven by the non-uniform warming up of the airfoil section by the hot gas stream. The higher temperatures at the leading and trailing edges give rise to high pressure at this positions because of the geometrical restrictions of the airfoil and its shroud. As a reaction to this the fillet radii of the airfoil to the shroud and the shroud itself show stress concentrations, which, together with shroud unbending effects due to thermal gradients in the shrouds (not treated here in more detail), may cause critical positions too. (See also fig. 1)

The typical take-off time-dependence of thermal strain and temperature at the leading edge of the airfoil's midspan region is plotted in fig.2, where the maximum time gradients with respect to time are approximately 5 E-4 1/s and 70 K/s respectively. The resulting strain / temperature cycle (now for the whole mission from take-off to landing) show that the assumption of a phase shift of -135° for TMF-testing is a good approximation.

TESTING REQUIREMENTS

Whereas the above mentioned analytical procedure describes one essential part of the design, it defines the input data of the second part, which is to calculate the life of the component. To do that reference data are necessary. One important aspect which was emphasised in the last years is that damage mechanisms change with temperature, thus anisothermal testing is necessary. The best compromise between a component test in a real engine and specimen testing, that focuses on the understanding of isolated damage mechanisms, is a thermomechanical fatigue test, with conditions held as closely to real engine conditions as possible without overly exacerbating the performance of the test results. The exceptional advantage of the TMF test is that temperature and mechanical strain can be controlled separately.

The following parameters have been found to be critical :

- upper and lower test temperature:
    The upper temperature controls the amount of stress relaxation and oxidation. The lower temperature should be low enough to enable the possibility of a brittle cracking of the coating, which is often observed with typical coating compositions.
- heating and cooling rate:
    The heating and cooling rates influence the loading rate, as in a TMF test both signals have the same period. Thus again stress relaxation and oxidation are influenced.
- phase shift:
    The phase shift defines the correlation between the strain and the temperature and thus controls whether high stresses are experienced at high, low or intermediate temperature. This is obviously of great influence.
- influence of a coating:
    Typical coatings applied to protect against oxidation are often observed to be very brittle at low temperatures (lower than 600°C) . If these experience sufficiently high stresses the coatings exhibits brittle cracking which can drastically reduce the life of the component.

■ loading amplitude and ratio:

> The loading amplitude can influence the damage mechanism. It is necessary to define loading conditions which result in a TMF test life which is closely related to that component life which would be realistic under typical service conditions.

TMF TEST PROCEDURE

The aim of a thermomechanical fatigue test is to simulate the conditions of a highly stressed volume elements of a turbine blade or vane. That means the whole gauge length of a TMF specimen experiences the conditions of an often much smaller volume element on the blade. Precautions must be done to avoid high temperature gradients (longitudinal and transverse) in the gauge length. That enables to measure the stress response of the material under the applied conditions and thus confirm the calculated stresses. As TMF conditions are thought to be mainly caused by thermal gradients the test is strain controlled.

Specimen design

Three of the above mentioned requirements are important in the specimen design. Typical heating and cooling rates of aero engine blades and vanes are in the range of 5-100°C/s. To achieve high rates in a specimen but avoid unacceptable high temperature gradients in the gauge length a high surface to volume ratio is necessary. Two specimen designs are known to fulfil that condition:
• round hollow cylindrical specimen, typically with a wall thickness of about 1mm
• flat rectangular specimens with rounded corners
both with threaded ends and smooth radii from the gauge section to the threads.
As coating (can) influence(s) the TMF life and almost all turbine blades/vanes are coated the whole surface of specimen in the gauge length should be coated. Some coating procedures are only applicable on outer surfaces. Therefore we prefer the rectangular specimen as given in fig. 3. It also works under compression conditions ($R=-\infty$).

Heating and cooling rates

Whereas components approach the upper temperature asymptotically, which means the heating rate lowers with increasing temperature, TMF tests are performed with constant rates to have well-defined conditions. Some effort enables TMF test heating rates as high as 25°C/s with local differences below 10°C. As the rate is important in a temperature range which shows temperature dependant behaviour (i.e. at the high temperature branch of the cycle) a test condition of 10°C/s or 25°C/s is a good simulation of realistic conditions.

Phase shift

As pointed out in the description of typical loading conditions for turbine airfoils a phase shift of -135° (fig. 4) is often found at critical locations of blades and vanes.
That is based on in-house design experience, as exemplified above, but can be confirmed by the phase shift applied in the literature. Very similar cycles [1-5] are used from different authors mainly if they are or cooperate with design engineers.

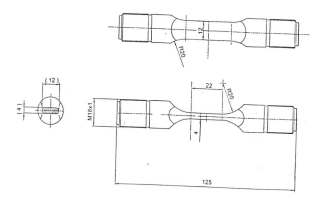

fig. 3: TMF specimen

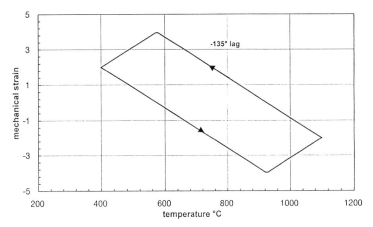

fig.4: TMF test cycle with -135° lag phase shift

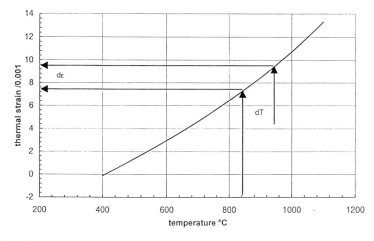

fig.5: Correlation between deviations in strain and temperature.

Test performance

The most critical point of a TMF test is the compromise between realistic parameters and high precision test control. One possible comprise was discussed above for the heating rate applied. Another one is not to focus on one parameters, but to weight with respect to the other ones. For TMF tests that means not to focus for high precision temperature control and underestimate the errors caused by deviations in phase shift and strain control. Almost all testing laboratories apply a test procedure which measures the thermal strain response of each specimen by application of the thermal test cycle under load control. Then the total strain cycle is calculated by adding the thermal strain to the intended mechanical strain cycle and testing is performed under total strain control. A typical value for the maximum thermal strain is 1.5% (temperature cycle 400°C - 1100°C). A strain controlled test with an error of 2% of the maximum strain (i.e. 0.03% strain) will cause an erroneous stress of 30 MPa (Young's modulus about 100 000 MPa), which in turn corresponds to an temperature deviation of 2% or 14°C or, (s. fig. 5) based on a heating rate of 10°C/s, a synchronisation error of 1.4s.

Good practice is it to conduct a performance test. For TMF a strain controlled test with a mechanical strain of 0% will evaluate the erroneous stresses produced by the whole control cycle. Examples of typical result are given in [1, 2]

LITERATURE

1. Bressers, J.; Timm, J.; Williams, S.J.; Affeldt, E.E.; Bennett, A. (1996) In:*"Thermal Mechanical Fatigue of Aircraft Engine Materials"*, AGARD-CP-569, Advisory group for aerospace research & development, Neuilly-sur-Seine, France, pp. 9-1 - 9-10
2. Bressers, J.; Timm, J.; Williams, S.J.; Bennett, A.; Affeldt, E.E. (1996) In: *Thermomechanical fatigue behavior of materials: Second Volume"*, Ed. M.J. Verrilli, M.G. Castelli, ASTM STP 1263, American Society for Testing and Materials, Philadelphia, pp. 56-67
3. Chataigner, E.; Remy, L. (1996) In: *"Thermomechanical fatigue behavior of materials: Second Volume"*, Ed. M.J. Verrilli, M.G. Castelli, ASTM STP 1263, American Society for Testing and Materials, Philadelphia, pp. 3-26
4. Fleury, E.; Remy, L. (1990) In: *"High Temperature Materials for Power Engineering 1990"*, Ed. E. Bachelet et al., Kluwer Academic Publishers, London, pp. 1007-1016
5. Heine, J.E.; Ruano, E.; DeLaneuville, D.E. (1996) In: *"Thermal Mechanical Fatigue of Aircraft Engine Materials"*, AGARD-CP-569, Advisory group for aerospace research & development, Neuilly-sur-Seine, France, pp. 20-1 - 20-6

LOW CYCLE FATIGUE AND THERMOMECHANICAL FATIGUE BEHAVIOUR OF COATED AND UNCOATED IN738 SUPERALLOY

V. BICEGO and P. BONTEMPI
ENEL-SRI-PDM, Via Reggio Emilia, 39 - 20090 Segrate (Milan) - I

N. TAYLOR
CISE Milan - I, presently at JRC-IAM Petten - NL

ABSTRACT

Results of isothermal Low Cycle Fatigue (LCF) tests made at 850°C on uncoated IN738LC samples are compared with similar tests carried out on specimens with two types of coating, namely a platinum aluminide coating (RT-22) and a plasma sprayed MCrAlY coating (A995). When compared on the basis of identical values of applied strain range, it is found that the presence of the MCrAlY coating did not significantly lower the cyclic life with respect to that obtained for the un-coated material.

The Thermo-Mechanical Fatigue - Linear Out of Phase (TMF-LOP) cycling mode was also applied, in tests on the coated and uncoated specimens of IN738. Interestingly, the TMF-LOP lives for the coated samples were considerably inferior than the TMF-LOP lives of the uncoated specimens. A modest diminution of life for the RT22 coated specimens was also confirmed. The crucial feature, leading to the observed dramatic life shortening of the coated samples of RT22 in TMF-LOP tests with respect to the isothermal ones was attributed to be the reduction of the elastic-plastic ductility of the coating, occurring in the cold part of the thermomechanical cycle when a large tensile stress is occurring with a low value of the temperature.

KEYWORDS

Low cycle fatigue, thermo-mechanical fatigue, nickel base superalloy, coatings.

INTRODUCTION

The high temperature mechanical strength properties of IN738LC, a Ni-base superalloy extensively used for industrial gas turbine blading, have been the subject of numerous investigations over the last few years. In order to improve its endurance and temperature capability, traditional and advanced coating technologies are now being widely applied. In this situation there is concern about the reliability of the coating processes with regard to retain the alloy's good resistance to elastic-plastic straining cycles resulting from the main transient operation of the gas turbine.

The prime objective of this study was to compare isothermal LCF strength at 850°C of uncoated IN738LC samples with similar tests carried out on specimens with two types of coating, namely a platinum aluminide coating (RT-22) and a plasma sprayed MCrAlY coating (A995).

On the other hand it is widely recognised that the traditional LCF characterisation of materials, made at the maximum temperature reached by the component during transients, may be poorly representative of the actual damaging effects occurring on superalloy blades for the most severe transients, namely cold starts and full load trips. The simplest idealised thermo-mechanical history to be considered in laboratory tests is provided by the Linear Out of Phase cycle (TMF-LOP: maximum tensile stress at the lowest temperature). Therefore the second objective of this study was to apply such a TMF-LOP cycling mode, on the coated and uncoated specimens of IN738.

MATERIAL AND EXPERIMENTAL

IN738LC superalloy has been produced by Thyssen and supplied as cast bars with 22 mm diameter x 152 mm long, with the chemical composition given in Table 1.

Table 1. Chemical composition of IN738LC superalloy (weight %).

C	Cr	Co	Mo	W	Ti	Al	B	Zr
0.9-0.13	15.7-16.3	8.0-9.0	1.5-2.0	2.4-2.8	3.2-3.7	3.2-3.7	0.007-0.012	0.03-0.08

LCF and TMF specimens have been coated with:
- NiCoCrAlY Vacuum Plasma Spray coating named AMDRY 995, by ATLA (I), with tickness of 200÷300 μm,
- Chemical Vapour Deposition Pt-Al named RT22 pack cementation diffusion coating, by Chromalloy (UK), with tickness of 100÷130 μm.

The isothermal LCF tests of this study were carried out with the following modalities:
- on servohidraulic units, with 100 KN capacity,
- under total strain range control condition,
- using resistive furnaces,
- at 850°C,
- triangular strain cycling, with strain rate = $3 \ 10^{-4} \ s^{-1}$.

The TMF-LOP tests were carried out with the following characteristics:
- servohidraulic unit with 100 KN capacity,
- under total mechanical strain control,
- temperature and mechanical strain imposed by a computer,
- induction furnace,
- max and min temperature equal to 850 and 500°C respectively.

RESULTS AND DISCUSSION

Results of isothermal LCF tests made at 850°C on uncoated IN738LC samples are compared in Fig. 1 with similar tests carried out on specimens with the two types of coating.

When compared on the basis of identical values of applied strain range, it is found that the presence of the MCrAlY coating did not significantly lower the cyclic life with respect to that obtained for the un-coated material; the lives of the specimens coated with RT22 were lower than the base material data by an average factor of 2, i.e. an appreciable effect but not really dramatic.

On the other hand the TMF-LOP lives for the coated samples were considerably inferior than the TMF-LOP lives of the uncoated specimens, Fig. 2. A modest diminution of life for the RT22 coated specimens with respect to isothermal tests was also confirmed.

The crucial feature, leading to the observed dramatic life shortening of the coated samples of RT22 in TMF-LOP tests with respect to the isothermal ones was attributed to be the reduction of the elastic-plastic ductility of the coating, occurring in the cold part of the thermomechanical cycle when a large tensile stress is occurring with a low value of the temperature, below the ductile-to-brittle transition temperature (DBTT) of this RT22 coating, as shown in Fig. 3 [1÷5].

CONCLUSIONS

Results of isothermal LCF tests made at 850°C on uncoated IN738LC samples were compared with similar tests carried out on specimens with two types of coating, namely a platinum aluminide coating (RT-22) and a plasma sprayed MCrAlY coating (A995). On the basis of identical values of applied strain range, it was found that the presence of the MCrAlY coating did not significantly lower the cyclic life with respect to that obtained for the un-coated material.

On the other hands, results from TMF-LOP cycles for the coated samples were considerably inferior than the TMF-LOP lives of the uncoated specimens. A modest diminution of life for the RT22 coated specimens was also confirmed, with respect to isothermal LCF data on these same specimens.

The crucial feature, leading to the observed dramatic life shortening of the coated samples of RT22 in TMF-LOP tests with respect to the isothermal ones was attributed to be the reduction of the elastic-plastic ductility of the coating, occurring in the cold part of the thermomechanical cycle when a large tensile stress is occurring with a low value of the temperature.

REFERENCES

1. D.H. Boone, Technical Data Sheets, Airco Temescal, Inc., Berkeley, CA, Jan. 1976, in A. Strang and E. Lang, Effect of Coatings on the Mech. Prop. of Superalloys, in High Temperature Alloys for Gas Turbines, R. Brunetaud Ed., Proc. of Conf. in Liege, Belgium, 1982, D. Riedel Pub., Dordrecht, 1982, 469-506.

2. D.H. Boone, Overlay Coatings for High Temperature Applications, Airco Temescal, Inc., Berkeley, CA, Jan. 1976, in Ref. 1.

3. G.W. Goward, Journal of Metals, Oct. 1970, 31-39.

4. G.W. Goward, in Symposium on Properties of High Temperature Alloys, Las Vegas, Oct. 1976, 806-823, in Ref. 1.

5. R. Viswanathan, Damage Mechanisms and Life Assessment of High Temperature Components, ASM International, 1989, ISBN 0-87170-358-0, 447.

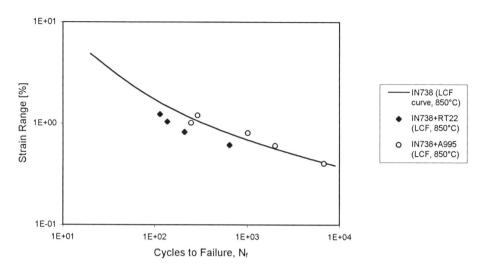

Fig. 1. LCF strength data for coated and uncoated IN738.

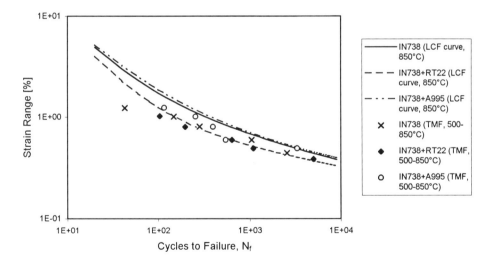

Fig. 2. Comparison of LCF e TMF-LOP strength data for coated and uncoated IN738.

142

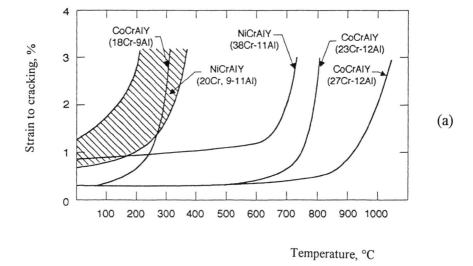

(a)

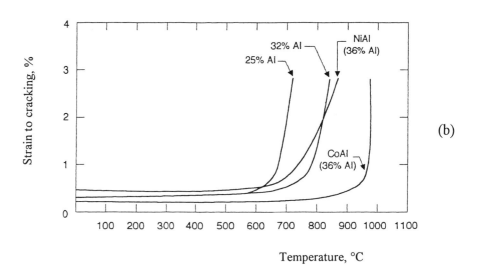

(b)

Fig. 3. Ductility characteristics of (a) MCrAlY coatings and (b) aluminide coatings [5].

HIGH TEMPERATURE FATIGUE BEHAVIOUR OF COATED IN738 LC

L. LINDÉ* and P. J. HENDERSON+
*Swedish Institute for Metals Research, Drottning Kristinas väg 48,
S-114 28 Stockholm, Sweden
+Vattenfall Energisystem AB, P.O. BOX 528,
S-162 16 Stockholm, Sweden

ABSTRACT

In this study, the effects of a plasma sprayed and an aluminised coating on the fatigue behaviour of a Ni-base superalloy have been investigated. The coated material was subjected to two different fatigue conditions; low cycle fatigue (LCF) at 850°C or thermo-mechanical fatigue (TMF) between 400 and 850°C. It was found that for the aluminised material, the difference in fatigue life between the LCF and TMF tests was quite small. However, for specimens with the plasma sprayed coating, the LCF tests resulted in a fatigue life a factor of two or more greater than for the TMF tests. It was also found that coating type and fatigue conditions had a decisive effect on the mechanisms for coating related crack initiation. For the plasma sprayed material, LCF testing resulted in a higher number of cracks in the coating compared to the TMF tests, where cracks also initiated between the coating and substrate. In the aluminised material, there was no difference in cracking mechanism between the two testing types, although the number of cracks was larger for the TMF tests. Further, all observed cracks in the aluminised material initiated in the coating but a majority was arrested in the coating or the transition zone.

KEYWORDS

Thermo-mechanical fatigue, low cycle fatigue, aluminised, plasma sprayed, coating, Ni-base superalloy, crack initiation.

INTRODUCTION

Coatings are often used on components for gas turbines and aero engines as a protection against aggressive environments. The coating must be compatible with the substrate alloy and at the same time be able to withstand the strains it is exposed to during service. One important aspect of coatings, which affect their mechanical properties, is the ductile to brittle transition temperature (DBTT) which has been defined as the temperature corresponding to a fracture strain of 0.6% [1]. Below this temperature the brittle behaviour reduces the ability of the coating to endure plastic deformation. Mechanical loads, *e.g.* during fatigue and especially below the DBTT, may cause the coating to form cracks and thereby expose the base metal to environmental attacks. Cracks in the coating can also propagate into the substrate alloy, causing fracture. The effect of coatings on the fatigue life is therefore an important field of study for applications where fatigue is a major life-limiting factor.

EXPERIMENTAL

Investigated material in this study was the conventionally cast Ni-base superalloy IN 738LC with a plasma sprayed NiCoCrAlY coating, PWA 1365-2, or a diffusion aluminide coating, ELCOAT 101. The thickness of the coatings / interdiffusion zones were 100-130 µm / 15-35 µm for the plasma sprayed coating and 75-85 µm / 25-30 µm for the aluminised coating respectively. All specimens were solid with a parallel gauge length of 27 mm and a diameter before coating of 8 mm. The chemical composition of the substrate alloy and the NiCoCrAlY coating is shown in Table 1.

Table 1. Chemical composition of the investigated IN738 LC alloy and NiCoCrAlY coating, in wt%.

Material	Al	B	C	Co	Cr	Fe	Mo	Nb	Ni	Ta	Ti	W	Y	Zr
IN738 LC	3.6	0.01	0.10	8.10	16.0	0.06	1.7	0.80	bal.	1.7	3.4	2.6	-	0.03
NiCoCrAlY	12.5	-	0.01	22.5	17.1	-	-	-	bal.	-	-	-	0.52	-

The coated material was tested in thermo-mechanical fatigue (TMF) between 400 and 850°C with temperature and strain cycled in phase, *i.e.* with maximum tensile strain at maximum temperature. Isothermal low cycle fatigue (LCF) testing of the material at 850°C was also carried out. Testing were performed in total strain control. For the TMF tests, the total strain was equal to the sum of total mechanical and thermal strain, while for the LCF tests it was equal to the total mechanical strain. LCF of the plasma sprayed material was performed on as sprayed as well as on specimens where the plasma sprayed coating had been polished. Before polishing the R_a value was 5.8 µm compared to 0.36 µm after polishing. TMF and LCF testing of aluminised material was only performed on as coated material. Testing temperatures were chosen to be above the DBTT during LCF and cycled through during TMF. The transition temperature was assumed to be less than 800°C but above 500°C for both coatings. After testing, some specimens were subjected to metallographic investigation and sectioned longitudinally, polished to 1 µm and etched in glyceregia, which consists of 10 ml HNO_3, 20 ml HCl and 20 ml glycerol. During etching, the solution was heated to 50-60°C. The specimens were examined by light optical or scanning electron microscopy (SEM).

RESULTS

Fatigue Testing

The results from the LCF and TMF testing are shown in Fig. 1, where N_{90} was determined as a 10% drop from the linear behaviour of the maximum tensile stress versus number of cycles. As can be seen in Fig. 1 there was a significant difference in fatigue life between LCF tests performed on plasma sprayed and aluminised specimens at similar total mechanical strain ranges. It can also be seen that polishing the plasma sprayed LCF specimens had a small but beneficial effect on fatigue life compared to as sprayed tests. The results for the TMF tests show a much smaller difference between the two coating types. One aluminised TMF test (within brackets) failed prematurely while another was interrupted before failure. For the aluminised specimens a cross-over in fatigue life was observed for the two test types where the TMF tests, although the difference was quite small, had longer fatigue life at low strain ranges. The lines correspond to a summation of the Coffin-Manson and Basquin equations, Eqn. 1 and 2 respectively. These equations are:

Coffin-Manson
$$\Delta\varepsilon_p = A \cdot N_{90}^{\alpha} \tag{1}$$

Basquin
$$\Delta\varepsilon_e = B \cdot N_{90}^{\beta} \tag{2}$$

where $\Delta\varepsilon_p$ and $\Delta\varepsilon_e$ are the plastic and elastic strain ranges respectively and A, α, B and β are material constants. The values of the constants (in the same order) for the aluminised material were 12.42, -0.66, 1.53, -0.16 and 9.38, -0.72, 1.54, -0.14 for the LCF and TMF series respectively. For the plasma sprayed material the corresponding values were 14.2, -0.58, 1.77, -0.15 for the LCF and 11.6, -0.76, 1.59, -0.14 for the TMF series respectively, for the polished LCF material the values were 23.6, -0.64, 1.85, -0.15.

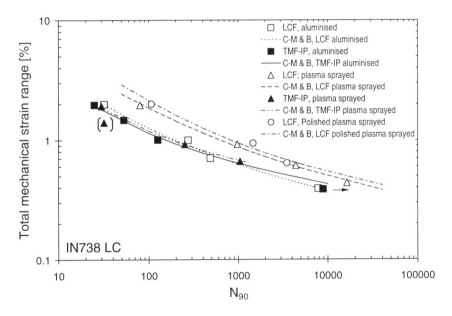

Fig. 1. Total mechanical strain range versus number of cycles to 10% stress reduction for coated IN 738LC. The lines (C-M & B) correspond to a summation of the Coffin-Manson and Basquin equations respectively.

Metallography after Testing

In the plasma sprayed as well as in the aluminised coating, cracking was intergranular in all investigated specimens. The substrate IN 738LC material showed intergranular cracks after TMF testing but transgranular in the LCF specimens. Some of the intergranular cracks in the TMF specimens were internal and showed no signs of oxidation, similar to cracks observed after creep testing.

In the plasma sprayed material, defects in the interdiffusion layer, *e.g.* inclusions and pores, were found [2]. The number of cracks in the coating was generally lower after TMF testing than after LCF testing at corresponding strain ranges, except at the largest strain range where it was similar. Cracks in the plasma sprayed LCF specimens all initiated at the coating surface but only a minor fraction propagated into the substrate alloy. The behaviour differed from that observed in the TMF tests, where no cracks in the coating were found that had not reached the substrate. Further, in one of the TMF specimens, cracks had evidently initiated in the interdiffusion zone, sometimes with the coating more or less intact, see Fig. 2. There were also signs of delamination with cracks growing between the coating and substrate alloy [2].

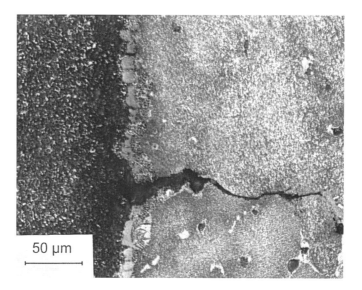

Fig. 2.   Crack initiated in the interdiffusion zone of the plasma sprayed NiCoCrAlY coating after TMF testing between 400 and 850°C with a total mechanical strain range of 0.92%. Stress direction vertical.

The behaviour of the aluminide coating was quite different from that of the plasma sprayed. Here the TMF specimens showed more extensive crack initiation, more than for corresponding LCF tests. All cracks observed in the substrate alloy had initiated in and propagated through the coating, although a majority of the observed cracks did not propagate through the interdiffusion zone, see Fig. 3. No crack initiation in the interdiffusion zone or delamination was observed. However, in TMF specimens tested at a total mechanical strain range of 0.71% or above, cracks were observed in the coating perpendicular as well as parallel to the stress axis, which was more unexpected [3, 4]. The parallel cracks were probably formed during the compressive, low temperature, part of the TMF cycle, during which the diameter of the gauge length increased inducing tensile strains in the brittle coating.

DISCUSSION

LCF testing of the plasma sprayed material resulted in a fatigue life a factor of two or more greater than for TMF, probably due to the fact that the temperature was cycled through the DBTT of the coating during the TMF-tests. At lower temperatures in the TMF cycle, where the coating was brittle, it was exposed to strains that may have caused it to form early cracks. This resulted in few cracks that propagated through the coating. On the other hand, the isothermal LCF tests were performed above the DBTT where the coating was ductile and resulted in a notable increase in the number of cracks and fatigue life. However, specimens with an aluminised coating exhibited a much smaller difference between the TMF and LCF tests. The fatigue lives for LCF and TMF tests of the aluminised material were similar to the TMF results for the plasma sprayed material. One reason for this could be the fact that during LCF, the plasma sprayed coating was more effective in arresting the growth of cracks, *e.g.* because it was more ductile than the aluminised coating. Apparently the difference between the two test types had a much smaller effect and was equally damaging for the aluminised material.

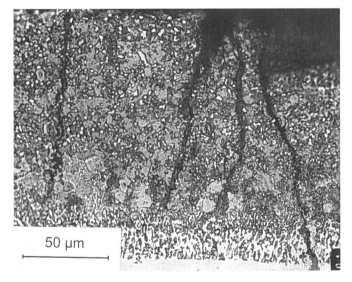

Fig. 3.   Cracks in the aluminised coating after LCF testing at 850°C with a total mechanical strain range of 0.39%. Stress direction horizontal.

It has been considered before that the most important mechanical property of a coating is the resistance to thermal-induced cracking [5, 6]. These results emphasise the fact that isothermal fatigue testing cannot always include all factors that affect the component life and may result in lifetime predictions that are non-conservative. The error may be as large as for the presented results for the plasma sprayed material or smaller but still significant, as observed for the aluminised material. Also, the wrong conclusions could be drawn when comparing the performance of different materials. From the LCF results, it would have been easy to assume that the plasma sprayed coating would perform better than the aluminised in an application. The TMF results show that this is not necessarily true.

Polishing of the rough plasma sprayed coating surface increased the fatigue life during LCF, which could be expected. However, since the polishing reduced the surface roughness considerably and all cracks were found to initiate at the coating surface for the LCF tests, one might have expected an even larger increase in fatigue life. This implies that the roughness of the original surface did not have a very large effect on fatigue crack initiation during LCF, probably because the roughness was caused by pits rather than grooves [2]. The same was probably true for the TMF tests as well, since fewer cracks initiated at the coating surface during TMF testing compared to LCF, despite the fact that the coating was subjected to tensile strains at lower temperatures where it might be more brittle.

CONCLUSIONS

The nickel based superalloy IN 738LC with a plasma sprayed NiCoCrAlY or aluminised coating, was tested in TMF between 400 and 850°C or LCF at 850°C. From the results it was found that for the plasma sprayed material, LCF tests had a fatigue life a factor of two or more longer than for corresponding TMF tests. Polishing of the coating before LCF testing increased the fatigue life slightly. The aluminised material showed a crossover in fatigue life between TMF and LCF tests, but the difference was much smaller than for the plasma sprayed material. LCF testing of the plasma sprayed material resulted in more extensive cracking compared to the TMF tests. Cracks also initiated in the interdiffusion zone in one of the TMF specimens. In the aluminised coating, the crack mechanism was the same for both testing types, but the TMF specimens showed the larger amount of cracks. All observed cracks initiated at the coating surface and a majority was arrested in the transition zone.

ACKNOWLEDGEMENTS

The authors wish to thank Dr Jacek Komenda for metallographic assistance. Financing was provided by ABB STAL AB, Avesta Sheffield AB, CSM Materialteknik AB, AB Sandvik Steel, Volvo Aero Corporation and ELFORSK. The support is gratefully acknowledged.

REFERENCES

1. Lowrie, R. and Boone, D. H. (1977). In: *International Conference on Metallurgical Coatings*, San Francisco, USA.
2. Lindé, L. and Henderson, P. J. (1997). In: *High Temperature Surface Engineering* (to be published).
3. Lindé, L. and Henderson, P. J. (1995). In: *8th CIMTEC*, pp. 313-320, P. Vincenzini (Ed.). Techna Srl, Faenza.
4. Lindé, L and Henderson, P. J. (1994). In: *Materials for Advanced Power Engineering, Part II*, pp. 1367-1376, D. Coutsouradis et al (Eds). Kluwer Academic Publishers, Dordrecht.
5. Patnaik, P. C. and Immarigeon, J. P. (1989) *Mat. & Man. Proc.* **4**, 347.
6. Strang, A. and Lang, E. (1982). In: *High Temperature Alloys for Gas Turbines*, pp. 469-506, R. Brunetaud et al (Eds). D. Reidel Publishing Co, Dordrecht.

Low Cycle Fatigue and Elasto-Plastic Behaviour of Materials
K-T. Rie and P.D. Portella (Editors)
© 1998 Elsevier Science Ltd. All rights reserved.

THERMOMECHANICAL FATIGUE OF A COATED DIRECTIONALLY SOLIDIFIED NICKEL-BASE SUPERALLOY

Dedicated to Professor Albrecht Gysler on the occasion of this 60th birthday

R. KOWALEWSKI[*)] and H. MUGHRABI

Institut für Werkstoffwissenschaften, Lehrstuhl I, Allgemeine Werkstoffeigenschaften,
Universität Erlangen-Nürnberg, Martensstr.5, D-91058 Erlangen, F.R. Germany
[*)] now at: Siemens AG, ASI 7, QM 2, D-90441 Nürnberg, F.R. Germany

ABSTRACT

The termomechanical fatigue (TMF) behaviour of the directionally solidified (DS) nickel-base superalloy CM 247 LC, coated with a plasma-sprayed NiCrAlY-coating (PCA-1), was investigated experimentally. The tests were carried out, using a particular out-of-phase cycle, at different total strain ranges in the temperature range between the lower temperature of 400°C and the upper temperature of 1000°C. For a better fundamental understanding, corresponding tests were performed not only on the uncoated substrate material but also on bulk specimens of the coating material. Detailed microstructural studies were carried out in order to characterize the microstructural changes occurring in the substrate and in the coating material during a TMF cycle and in order to elucidate the basic damage mechanisms. It was concluded that fatigue cracks initiate in the coating as a consequence of the high tensile stresses prevailing at the lower temperature and that the fatigue life of the coated alloy is governed by the mechanical strain amplitude experienced by the coating.

KEYWORDS

Directionally solidified nickel-base superalloy, CM 247 LC, NiCrAlY-coating, thermomechanical fatigue, microstructure, fatigue damage, fatigue life

INTRODUCTION

The objective of the present work was to study in some detail the thermomechanical fatigue (TMF) behaviour of the directionally solified (DS) nickel-base superalloy CM 247 LC, coated with a plasma-sprayed NiCrAlY-coating (PCA-1). And to explore the underlying microstructural processes. A special effort was made to investigate the mechanical properties and the thermomechanical fatigue of the pure coating material, as opposed to the mere study of the thermomechanical fatigue of the uncoated and the coated substrate materials. In an earlier publication, work on the isothermal high-temperature low-cycle fatigue behaviours of the coated DS alloy and the bulk coating material was reported [1]. This paper deals with selected results on the TMF behaviour, obtained i a more comprehensive study [2]

150

EXPERIMENTAL DETAILS

The directionally solidified DS superalloy CM 247 LC with the nominal composition (in wt.%): Ni bal., Cr 8.1, Co 9.2, Mo 0.5, W 9.5, Ta 3.2, Ti 0.7, Al 5.6, Zr 0.01, B 0.01, C 0.07, Hf 1.4 was provided in the form of casting slabs. Cylindrical specimens with a diameter of 8 mm and a gauge length of 14 mm were taken from these casting slabs in such a manner that the crystallographic <001> directions of the columnar grains were aligned within 15° parallel to the specimen axes. These specimens were coated over their gauge length by low-pressure plasma spraying with a NiCrAlY-alloy, called PCA-1, consisting of (in wt. %): Ni bal., Cr 25, Al 5, Y 0.5 and unspecified amounts of Ta and Si in a thickness of 200 μm. Also, cylindrical specimens of the bulk coating material with a diameter of 5 mm and a gauge length of 14 mm were prepared from continuously deposited coating material. Both types of specimens were standard heat-treated, and their surfaces were finally mechanically polished.

The fatigue tests were performed on a closed-loop servohydraulic testing machine (MTS 810 with a digital controller) which is equipped with a 200 kHz induction furnace. Strain was measured with a high-temperature axial extensometer with 12 mm gauge length. Temperature control was conducted with a strip thermocouple, which was wound around the circumference of the specimen. The thermomechanical fatigue test cycle was selected so as to simulate the critical conditions prevailing in a near-surface volume element of a turbine blade in a land-based gas turbine during start-up and shut-down operation. The tests were carried out in a particular unsymmetrical out-of-phase cycle at different total strain ranges $\Delta\varepsilon_t$ in the temperature range ΔT from the lower temperature 400^0C to the upper temperature 1000^0C with a frequency of $v = 3.17\times10^{-3}$ s^{-1}. The tests were started at the lower temperature of 400^0C and at zero total strain ε_t.

A schematic representation of the out-of-phase cycle used in this study is shown in Fig. 1 a which displays the time dependence and the phase relationships of temperature T and of the strain amplitudes ε_t, ε_{th} and ε_{mech}, where the total strain ε_t is the sum of the mechanical strain $\varepsilon_{mech} = \varepsilon_{el} + \varepsilon_{pl}$ (ε_{el}: elastic strain, ε_{pl}: plastic strain) and the thermal strain ε_{th}. It should be noted that this TMF cycle is unsymmetrical with respect to ε_{mech}. For comparison, the more

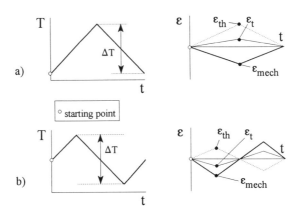

Fig. 1: Comparison between (a) the unsymmetrical (with respect to the mechanical strain $\Delta\varepsilon_{mech}$) out-of-phase TMF test cycle used in this work and (b) the more common symmetrical out-of-phase TMF cycle. Time is denoted by the symbol t.

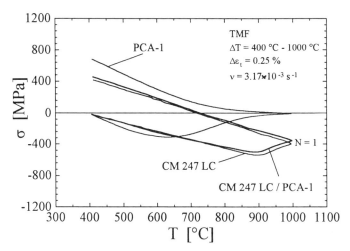

Fig. 2: Comparison of first stress versus temperature TMF cycle for the bulk alloy CM 247 LC, the bulk NiCrAlY coating PCA-1 and the coated alloy composite CM 247 LC/PCA-1. The data of the TMF cycle are listed in the figure.

commonly used symmetrical out-of-phase TMF cycle, compare [2-6], is shown in Fig. 1 b. For the unsymmetrical out-of-phase TMF cycle applied in this study, the mechanical strain ε_{mech}, given by the difference between ε_t and ε_{th}, is shifted with respect to the temperature by a phase angle of 180^0 and varies between zero and some negative value throughout the test. As a consequence, the specimen (blade) experiences a high (damaging) tensile stress at the lower temperature during this out-of-phase TMF cycle.

In addition, the elastic and the thermal properties of the bulk coating and substrate materials were investigated. Furthermore, detailed microstructural studies were conducted on the specimens in the initial state and after fatigue by standard metallographic techniques.

RESULTS AND DISCUSSION

Valuable information is obtained from a comparison of the first stress vs. temperature cycles of the uncoated substrate material CM 247 LC, the bulk coating material PCA-1 and the coated directionally solidified alloy CM 247 LC, as shown in Fig. 2 for experiments conducted at a total strain range of $\Delta\varepsilon_t = 0.25\%$. The curve obtained for the substrate-coating composite closely resembles that of the uncoated substrate material, since the cross section of the coating material is only a small fraction of the total cross section. Furthermore, one recognizes that the mechanical responses of the substrate and the coating material differ significantly under identical thermomechanical straining conditions. Thus, whereas the substrate material experiences significant compressive stresses at the upper temperature throughout the TMF cycle, the compressive stresses acting on the coating material at the upper temperature are very small as a consequence of severe stress relaxation, associated with reduced strength, of the coating material. On the other hand, the coating material experiences increasing larger tensile stresses during cooling than the substrate material. It could be shown [2] that the stress-temperature response of the coated material corresponds closely to that predicted by the superposition of the responses of the substrate and the coating materials in a simple composite model [7].

152

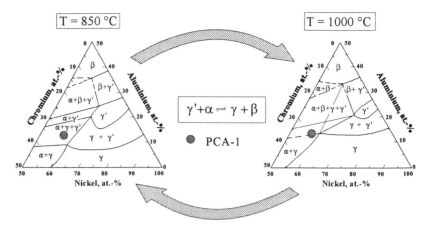

Fig. 3: Isothermal sections of the ternary nickel-chromium-aluminium system at 850°C and 1000°C. The drawn-in point represents the composition of the NiCrAlY-alloy PCA-1, lying in the (α+γ+γ') phase field at 850°C and just in the (α+β+γ+γ') phase field at 1000°C.

It is emphasized that severe microstructural changes occur in the coating material during the temperature cycles and that the TMF behaviour of the coating material cannot be described on the basis of the isothermal behaviour. Thus, measurements of the thermal expansion behaviour of the bulk coating material in a temperature cycle between 400°C and 1000°C yielded a much stronger thermal expansion than in the case of the substrate material CM 247 LC and a pronounced hysteresis [2]. The latter is related to a time-dependent phase reaction, associated with a volume change which can be understood by inspection of isothermal sections of the ternary NiCrAl phase diagram, shown in Fig. 3 for temperatures of 850°C and 1000°C, compare [8]. The composition according to the major constituents of the coating material PCA-1 is indicated in the figure. At the lower temperature of 850°C and at still lower temperatures, the coating material PCA-1 consists of the following three phases (disregarding the yttrium-rich M_5Y-phase, where M stands for Ni+Cr+Al, cf. [9]): the body-centred cubic α-phase, the face-centred cubic γ-phase and the ordered γ'-phase with $L1_2$-structure (cf. Fig. 3). At about 1000°C during the heating branch of the TMF cycle a fourth phase, the ordered face-centred cubic β-phase with B2-structure appears.

Due to the fact that the alloy composition lies in a three-phase region at lower temperatures and in a four-phase region at the highest temperatures of the TMF cycle, respectively, the coating material undergoes a change of its microstructure for every single TMF cycle. Since the variation in the content of the coating material goes along with volume changes, an opened hysteresis loop (thermal expansion vs. temperature) occurs during thermal cycling, which is coupled with formation and dissolution of the γ- and the β-phase (according to the equation: γ' +α↔γ+β) depending on the rate of heating and cooling (cf. Fig. 3).

Fig. 4 shows some TMF life data for the substrate-coating composite and, for comparison, for the bulk substrate material in a double-logarithmic plot of the mechanical strain amplitude, $\Delta\varepsilon_{mech}/2$, as a function of twice the number of cycles to failure, $2 N_f$. Under the TMF conditions described, the fatigue life of the coated alloy is smaller than that of the uncoated material. The reason lies in the initiation of fatigue cracks in the coating, very probably as a conse-

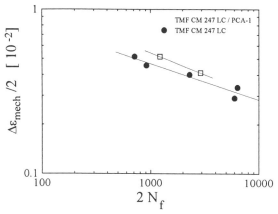

Fig. 4: Double-logarithmic plot of the mechanical strain amplitude $\Delta\varepsilon_{mech}/2$ versus twice the number of cycles to failure, $2\,N_f$, for the uncoated alloy CM 247 LC and for the coated alloy composite CM 247 LC/PCA-1 for TMF tests at different total strain ranges $\Delta\varepsilon_t$ under otherwise identical conditions as listed in Fig. 2.

quence of the high tensile stresses experienced at the lower temperature, cf. Fig. 2. These microcracks propagate subsequently into the substrate material.

Generally, crack initiation and propagation are governed by the interaction between oxidation processes at the higher cycle temperatures and the high tensile stresses occurring at the lower cycle temperatures (cf. Fig. 2) under the given thermomechanical (out-of-phase) testing conditions. This statement applies to all types of specimens (the coated and uncoated substrate and the bulk coating material). All cracks originate from the surface of the thermomechanically fatigued specimens. In the cases of the CM 247 LC/PCA-1-composite and the bulk coating material, the main initiation sites for surface cracks are grains of the oxidized yttrium-rich M_5Y-phase (cf. Fig. 5). The detailed metallographic observations show that, under the conditions of TMF cycling, a number of cracks grow perpendicularly to the stress axis towards the transition zone of coating and substrate and enter preferentially at oxidized carbides into the base material. The oxidized carbides play also a crucial role as crack initiation sites during thermomechanical fatigue of the uncoated substrate.

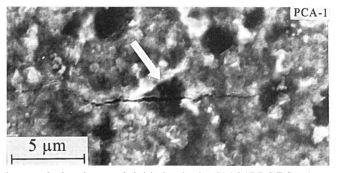

Fig. 5: SEM micrograph showing crack initiation in the CM 247 LC/PCA-1 composite at the yttrium-rich phase M_5Y (arrow) of the oxidized surface. Stress axis perpendicular. Test conditions: $\Delta\varepsilon_t = 0.25$ % ($\Delta\varepsilon_{mech} = 0.80\%$), $\Delta T = 400°C - 1000°C$, $\nu = 0\ 3.17 \times 10^{-3} s^{-1}$.

154

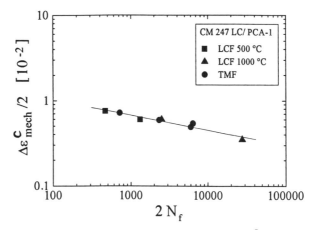

Fig. 6: Double-logarithmic plot of mechanical strain amplitude $\Delta\varepsilon^C_{mech}/2$ in the coating against twice the number of cycles to failure, $2 N_f$. All data of the composite CM 247 LC/PCA-1 fall on one common line for both TMF tests and for isothermal LCF tests at 500°C and 1000°C.

Finally, it is interesting to note that, in a double-logarithmic plot of the mechanical strain amplitude experienced by the coating, $\Delta\varepsilon^C_{mech}/2$, against $2N_f$ for the coated material. the fatigue life data for the coated composite material for isothermal fatigue tests performed at the lower and upper temperatures of the TMF cycle, respectively, and for thermomechanical fatigue tests fall on the same common line [2]. This is shown in Fig. 6. It is concluded that the TMF life of the coated alloy is dictated by that of the coating. This finding corresponds precisely to the earlier results obtained for the fatigue lives in isothermal LCF tests on the composite CM 247 LC/PCA-1, when compared with the data for the bulk coating material PCA-1 [1,2].

ACKNOWLEDGMENTS

Sincere thanks are extended to ABB Corporate Research Centre, Heidelberg, for partial financial support and to Dr. M. Bayerlein and Dr. C. Sommer for their continued interest.

REFERENCES

[1] Kowalewski, R. and Mughrabi, H. (1998). *Mater. Sci. Eng.*, in print.

[2] Kowalewski, R. (1997). Doctorate Thesis, University of Erlangen-Nürnberg.

[3] Kraft, S.A. and Mughrabi, H. (1996). In: *Thermomechanical Fatigue Behaviour of Materials: Second Volume*, M.J. Verrilli and M.G. Castelli (Eds.). ASTM STP 1263, pp. 27-40.

[4] Chen, H., Chen, W., Mukherji, D., Wahi, R.P. and Wever, H. (1995). *Z. Metallkde.* **86** 423.

[5] Boismier, D.A. and Sehitoglu, H. (1990). *Trans. ASME* **112**, 68.

[6] Skelton, R.P. (1994). *Fatigue Fract. Mater. Struct.* **17**, 479.

[7] Heine, J.E., Warren, J.R. and Cowles, B.A. (1989). Report, Wright-Patterson Air Force Base, Ohio, WRDC-TR-894027.

[8] Taylor, A. and Floyd, R.W. (1952). *J. of the Institute of Metals* **81**, 451.

[9] Frances, M., Vilasi, M., Mansour-Gabr, M., Steinmetz, J. and Steinmetz P. (1987). *Mater. Sci. Eng.* **88**, 89.

Low Cycle Fatigue and Elasto-Plastic Behaviour of Materials
K-T. Rie and P.D. Portella (Editors)
155

VISCOPLASTIC MODEL FOR THERMOMECHANICAL FATIGUE

MARK L. RENAULD
RAVINDRA ANNIGERI
SAM Y. ZAMRIK
Department of Engineering Science and Mechanics
The Pennsylvania State University, University Park, PA, USA

ABSTRACT

A viscoplastic model was used to characterize the time-dependent stress-strain response of IN-738LC gas turbine material with NiCoCrAlY overlay coating. The model is based on a flow rule which describes the inelastic strain rate as a function of the applied stress and a set of three internal stress variables known as back stress, drag stress and limit stress. It is hypothesized that only part of the applied stress is responsible for the inelastic deformation and hence, an effective stress is responsible for the inelastic deformation. To validate the use of the viscoplastic analysis, thermomechanical fatigue (TMF) tests with and without hold-time period were conducted on IN-738LC at temperatures of 482-871°C. Test results show that the viscoplastic model can reasonably predict the cyclic stress-strain response of the material from the first cycle to the saturation stage that occurs at mid-life. The model can also simulate the stress-strain hysteresis loop variation as a function of number of cycles.

KEYWORDS

Viscoplastic model, flow rule, mid-life cycle, internal stress variables, TMF cycle

INTRODUCTION

Life prediction techniques for cyclic loading are usually based upon the cyclic stress-strain response of the material at mid-life. The mid-life cycle is considered to be the material's average deformation. The mid-life stress-strain cycle, called the hysteresis loop, is used in relations such as the Manson-Coffin life prediction relation. However, in case of thermomechanical (TMF) type cycle, one needs to describe the stress-strain-cycle evolution, which is different from response obtained under isothermal condition. Most of the evolution techniques are based on viscoplastic modeling. The advantage of using a viscoplastic model is by treating a polycrystalline material similar to fluids and by writing governing equations for flow rate or inelastic strain rate. The viscoplastic analysis assumes that the inelastic strain rate is a function of stress, temperature, and a set of internal variables that defines the inelastic deformation and describes the kinematic and isotropic hardening behavior and can be express as :

$$\dot{\varepsilon} = f(S, \kappa, T) \tag{1}$$

The concept of internal variables characterizes the inelastic deformation and has been applied in the viscoplastic models proposed by a number of researchers [1-12]. The model of Freed [1] was applied to a nickel-based superalloy IN-738LC and overlay coating NiCoCrAlY materials.

VISCOPLASTIC MODEL

The basic components of the viscoplastic model are strain rate, stress and temperature. The inelastic strain rate is assumed to be a function of deviatoric stress S, internal stress variables represented by κ, and temperature T, as expressed in eq. (1). The internal stress variables are a deviatoric tensor back stress $\mathbf{B}_{ij}$ and two positive scalars representing drag stress D, and limit stress L. The back stress accounts for kinematic hardening and the drag and limit stresses together account for isotropic hardening.

The uniaxial deviatoric stress S_x, can be defined in terms of the applied stress, σ_x as:

$$S_x = \frac{2}{3}\sigma_x \tag{2}$$

The model also uses the concept of an effective stress, Σ, which is responsible for inelastic deformation. The effective stress for the uniaxial case, is defined as:

$$\Sigma_x = S_x - B_x \tag{3}$$

For non-linear state condition (when hysteresis loops are not stabilized or for cycles less than mid-life) and under varying temperature such as TMF strain cycle, the three internal variables are defined by evolutionary differential equations which, for the uniaxial condition, are written as:

$$\dot{B}_x = H_b \left[\dot{\varepsilon}_x^{in} - \frac{B_x}{L}\left|\dot{\varepsilon}_x^{in}\right| \right] + \frac{B_x}{H_b}\frac{\partial H_b}{\partial T}\dot{T} \tag{4}$$

$$\dot{D} = \frac{H_d}{G}\left[1 - \theta\,R(G)\right]\left|\dot{\varepsilon}_x^{in}\right| + \frac{D}{H_d}\frac{\partial H_d}{\partial T}\dot{T} \tag{5}$$

$$\dot{L} = \frac{H_l\,D}{L}\left[F - G\right]\left|\dot{\varepsilon}_x^{in}\right| + \frac{L}{H_l}\frac{\partial H_l}{\partial T}\dot{T} \tag{6}$$

Where H_b is the kinematic hardening modulus and H_d and H_l are the isotropic hardening moduli. Under isothermal condition, the temperature rate is zero and T is constant. Therefore, the scalar variables L and D become constant at mid-life cycle. The inelastic strain rate under uniaxial loading is expressed as:

$$\dot{\varepsilon}_x^{in} = \theta\,Z(F)\frac{\Sigma_x}{\Sigma_2} \tag{7}$$

Where θ is the thermal diffusivity that characterized the temperature-dependent flow response of the material expressed as:

$$\theta = \begin{cases} \exp\left\{\dfrac{-Q}{RT}\right\} & for \quad T \geq \dfrac{T_m}{2} \\[2em] \exp\left[\dfrac{-2Q}{RT_m}\left\{1n\left(\dfrac{T_m}{2T}\right)+1\right\}\right] & for \quad T < \dfrac{T_m}{2} \end{cases} \tag{8}$$

Z(F) is the Zener-Hollomon parameter which accounts for the stress-dependence of the inelastic strain rate and maybe defined by two equations, a power law of low stresses and an exponential relation for high stresses as expressed by:

$$Z(F) = A\, F^n \qquad\qquad for \quad F \leq 1 \tag{10a}$$

$$= A\, \exp\left[n\,(F-1)\right] \qquad\qquad for \quad F > 1 \tag{10b}$$

Where A and n are temperature-independent material constants and the flow function, F, for the uniaxial situation are defined by:

$$F = \frac{\Sigma_2}{D} = \left|\frac{\Sigma_x}{D}\right| = \frac{|S_x - B_x|}{D} \tag{11}$$

The thermal recovery function R(G) accounts for transient material response that describes the strain-cycle evolution from first fatigue cycle to saturation and is defined as:

$$R(G) = A\, G^n \qquad\qquad for \quad G \leq 1 \tag{12a}$$

$$= A\, \exp\left[n(G-1)\right] \qquad\qquad for \quad G > 1 \tag{12b}$$

and $\quad G = \dfrac{L}{C-D}$ \qquad\qquad\qquad\qquad\qquad\qquad (13)

C is a temperature-independent material constant that represents the transition stress from power law to exponential function, and it is always greater than the drag stress D. C is determined from steady state condition or during stabilized cycle where the inelastic strain rate can be assumed to approach the steady state creep rate $\dot{\varepsilon}_{ss}$. Using steady state creep data at several stress levels and temperature, the ratio of $\dot{\varepsilon}^{in}/\theta$ is plotted versus the deviatoric stress as shown in Figure 1 where the drag stress, D, has evolved to the steady stress levels and an exponential relationship is typically observed. This change of flow behavior has been attributed to a change in recovery mechanisms (dislocation movement) where diffusional recovery dominates at low stress/ high temperature conditions and dynamic recovery dominates at high stress/ low temperature combinations.

MODEL SIMPLIFICATION

Fatigue life prediction is often based on mid-life hysteresis loops since they represent the average material deformation over the life of the specimen. Because the evolution of mid-life behavior is not required, several simplifications can be made in the model. Under saturated conditions at mid-life, the drag and limit stresses have evolved to a steady value (no longer cyclic hardening or softening). As such, F (eq. 11) and G (eq. 13) become equal and by eq. (10) and eq. (12), Z(F) equals R(G). With hardening and recovery equal, the evolution equations for the drag and limit stress rate approach zero, and the drag and limit stresses reach saturated values. By neglecting transient material behavior, the evolutionary equations for L and D, the two hardening moduli, H_l and H_d, the two initial conditions, L_0 and D_0, and the two equations for G and R(G) are all eliminated. The steady state or saturated values of L and D are assumed to be a function of temperature only for the slow temperature variations and small inelastic strain ranges observed in the materials of interest. The value of the third internal variable B is not constant within the mid-life cycle since the back stress is a function of inelastic strain rate and H_b, both of which generally will change within the mid-life hysteresis loop. This simplified viscoplastic model requires two elastic constants, E and α (thermal expansion coefficient) and six inelastic constants, A, n, C, H_b, D and L. The temperature dependence of $\dot{B}$ is neglected in eq. (4) since $B_x << H_b$ and $\dfrac{\partial H_b}{\partial T}$ is small and $\dot{T}$ is also small, therefore the product of these three parameters is negligible. Therefore for the uniaxial case, eqs. (4-6), are simplified as:

$$\dot{B}_x = H_b \left[\dot{\varepsilon}_x^{in} - \frac{B_x}{L} \left| \dot{\varepsilon}_x^{in} \right| \right] \tag{14}$$

$$Hb = f(T), \qquad\qquad L = f(T), \qquad\qquad D = f(T)$$

APPLICATION OF THE MODEL TO TMF

To apply the model to TMF strain controlled cycle, a number of constants have to be identified. The elastic moduli for IN-738LC, NiCoCrAlY overlay and coefficients of thermal expansion are needed as a function of temperature. The melting temperature of IN-738LC was found to be 1232°C and the melting temperature of the overlay coatings was assumed to be 1288°C although coatings have been subjected to this temperature without incipient melting [13]. This activation energy for creep of a NiCoCrAlY coating has been reported in [14] to be 290 kJ/mol. The creep activation energy of IN-738LC of 544 kJ/mol was calculated from the Dorn equation:

$$\dot{\varepsilon}_{ss} = A\sigma^n \exp\left[\frac{-Q_c}{RT} \right] \tag{15}$$

Where $\dot{\varepsilon}_{ss}$ is steady state creep rate. Q_c may also be dependent upon the applied stress. Representative TMF experimental and predicted hysteresis loops are shown in Figure 2 for the OP TMF at a temperature range from 482-871°C. The model predictions are quite accurate although at 0.8% without the hold-time, the inelastic strain range is slightly over-predicted and with a hold-time the compressive stresses are over-predicted. Figure 3 shows the coating test data and predicted hysteresis loops using Walker's and Freed's models. The Walker model slightly over-predicts the stress response at the high temperature and severely over-estimates the tensile stress at the low temperature of the fatigue cycle. The model was found numerically to be more efficient and its advantages lie in its ability to analyze both the creep and the pure inelastic response without any distinction. It also accounted for the hardening process. The mean stress effect in the OP TMF was considerable at all strain ranges. This effect is attributed due to the difference in the inelastic flow properties and Young's modulus as the test temperature changed within the TMF cycle. The tensile mean stress increased rapidly in first cycles to practically constant value at mid-life cycle. Also, the material was found to exhibit kinematic hardening in LCF cycle and TMF cycle with hold-time period of 90 seconds. Consequently, the hysteresis loop shifted progressively with cycles.

The internal stress variables showed different responses in TMF. The back stress was saturated after a few cycle at high mechanical strain range test; however, at low mechanical strain range test, many cycles were needed before back stress saturation. This effect at low strain ranges caused considerable upward (tensile side) shift in the hysteresis evolution of hysteresis loops because they remained constant indicating negligible isotropic hardening for IN-738LC material.

SUMMARY AND CONCLUSIONS

The viscoplastic model has been used for uniaxial TMF cycle to characterize the evolution of strain from first cycle to mid-life cycle. The stabilized mid-life cycle was in a life prediction model. The TMF cycle is more complicated than the isothermal cycle particularly if hold-time period was injected. As a result of the OP TMF compressive cycle, a strong tensile mean stress was developed due to Young's modulus and coefficient of thermal expansion variation with temperature and the mismatch of properties between the substrate and coating. Test results showed that the OP TMF cycle is the most damaging cycle for the coated IN-738LC material. Also the model in its simplified form was successfully applied to TMF for both the substrate and the overlay coating.

REFERENCES

1. Freed, A.D. (1988). "Structure of Viscoplastic Theory", NASA-TM-100794, pp. 1-26.
2. Geary, J.A. and Onat, E.T. (1974). ORNL-TM-4525.
3. Bodnar, S.R. and Partom, Y. (1975). *Journal of Applied Mechanics* **42**, 385.
4. Hart, E.W. (1976). *Journal of Engineering Materials Technology* **98**, 193.
5. Miller, A. (1976). *Journal of Engineering Materials Technology* **98**, 97.
6. Ponter, A.R.S. and Leckie, F.A. (1976). *Journal of Engineering Materials Technology* **98**, 47.
7. Chaboche, J.L. (1977). "Viscoplastic Constitutive Equations for the Description of Cyclic and Anisotropic Behavior of Metals", *Bull. Acad. Pol. Sci., Ser. Sci. Tech.* **25**, 33.
8. Krieg, R.D., Swearengen, J.C. and Rhode, R.W. (1978). "Inelastic Behavior of Pressure Vessel and Piping Components", ASME-PVP-PB-028. ASME, New York, pp. 15-28.
9. Robinson, D.N. (1978). "A Unified Creep-Plasticity Model for Structural Materials at High Temperature", ORNL-TM-5969.
10. Walker, K.P. (1981). NASA Contractor Report CR-165533.
11. Chaboche, J.L. and Rousselier, G. (1983). *Journal of Pressure Vessel Technology* **105**, 153.
12. Schmidt, C.G. and Miller, A.K. (1981). *Research Mechanics* **3**, 109.
13. Wood, M.I. and Goldman, E.H. (1987). "Chapter 13, Protective Coating", *Superalloys II*, pp. 359-383; C.T. Sims et al. (Eds.). Wiley & Sons, New York.
14. Hebsur, M.G. and Miner, R.V. (1986). "High Temperature Tensile and Creep Behavior of Low Pressure Plasma-Sprayed NiCoCrAlY Coating Alloy", *Materials Science and Engineering* **83(2)**, 239-245.

160

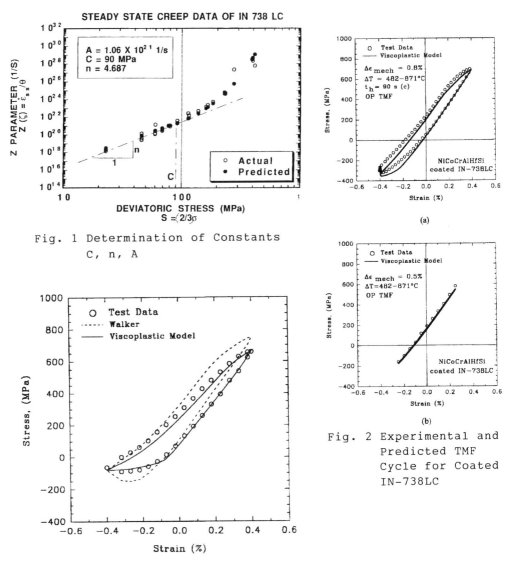

Fig. 1 Determination of Constants
C, n, A

(a)

(b)

Fig. 2 Experimental and
Predicted TMF
Cycle for Coated
IN-738LC

Fig. 3 Experimental and Predicted Mid-Life
Cycle for Stand Alone Coating

Low Cycle Fatigue and Elasto-Plastic Behaviour of Materials
K-T. Rie and P.D. Portella (Editors)

FEATURES ON THERMAL FATIGUE
OF FERRITE MATRIX DUCTILE CAST IRON

MORIHITO HAYASHI

Department of Mechanical Engineering, Tokai University,
1117 Kitakaname Hiratsuka,Kanagawa Japan

ABSTRACT

In the study, the thermal fatigue tests on ferrite matrix ductile cast iron were carried out as the specimens were axially constrained completely at room temperature. Then, the topics were studied as follows, the dependence of thermal fatigue life on the peak temperature of thermal cycle, the ordinary serration and the peculiar serration of thermal stress cycle, change of visual fatigue cracks and fracture pattern, the influence of dynamic strain aging, and microstructure change caused by polygonization or by $\alpha \Leftrightarrow \gamma$ transformation on thermal fatigue. Furthermore, the activation energy for thermal fatigue in Arrhenius equation in each temperature range was investigated and discussed.

KEYWORDS

Ductile cast iron, Thermal fatigue test, Fatigue life, Microstructure, Thermal stress, Thermal activation energy, Ordinary serration, Peculiar serration

INTRODUCTION

As known, ductile cast iron is stronger than conventional gray cast iron and it is superior in wear-resistance, shock, heat, cast-ability and price etc. what make it lead industrial material used widely. For example, it is utilized as heat-resisting material for some machinery components such as exhaust pipe and other engine parts of automobile. So, several studies on its strength at elevated temperature were reported [1~4]. Whereat, the phenomenon of stress serration versus strain appeared at temperatures from 423K to 673K and the brittleness at 673K were observed at ferrite cast iron. Regards to thermal fatigue on ductile cast iron, there were a few researches reported, so far [5,6,7]. In the report [5], the thermal fatigue test under the condition of completely constrained strain was carried out on ferrite matrix ductile cast iron and the Manson-Coffin's equation was introduced to conjecture the thermal fatigue life. It was reported [6] that fatigue crack growth rate in out-of-phase type was faster than that in in-phase type. The another report [7] showed that the thermal fatigue life was affected by the peak temperature of thermal cycle, by heating duration and peak-temperature holding duration, and the effect of phase type of temperature versus stress, and then proposed the life prediction equation.

In the study, thermal fatigue tests were carried out on ferrite matrix spheroidal graphite cast iron under the condition of complete constrained strain, which was set up at zero at room temperature, and the lowest temperature was kept at 323K in each thermal test cycle. And the several features on thermal fatigue were studied.

EXPERIMENTAL PROCEDURE

Specimen

Chemical composition and static mechanical properties of experimental material, which was classified into FCD400 in JIS, are shown in Table 1 and Table 2 respectively. The diameter of graphite and that of ferrite grain are about 22μm and about 27μm, respectively.

Table 1 Chemical composition of specimen of ductile cast iron (mass%).

C	Mg	Mn	Si	P	S	Fe
3.68	0.041	0.14	3.09	0.065	0.007	Residue

Table 2 Static mechanical properties on specimen of ductile cast iron.

Tensile strength (MPa)	0.2%Proof strength (MPa)	Elongation (%)	Reduction rate (%)
408	320	20	17

Each specimen for thermal fatigue test was machined from a cast rod, with 30 mm in diameter and 170 mm in length. And the specimen configuration was with gage length of 15 mm and central diameter of 10 mm. The central part 35 mm long was axially machined with the curvature radius of 60 mm. And its surface roughness was finished below 0.2 μm. To avoid the breaking out of gage length by bulging phenomenon and/or by high temperature brittleness during thermal fatigue test, such curvature was given. In this case, the stress concentration factor was about 1.035 [8].

Experimental Equipment

The test machine utilized in the thermal fatigue tests was a hydraulic servo-pulsar, Shimazu EHF-ED100kN-TF-20L, with maximum load capacity of $\pm$10t. Temperature of specimen was measured by an R-type thermocouple with 0.3 mm in diameter, which was spot-welded at a point on the central circumferential surface of specimen. The specimen was repeatedly heated by high frequency induced alternate current and cooled by compressed air blowing. The scattering of temperature along gauge length was within $\pm$5K. During thermal fatigue test, the output of displacement of gage length from strain transducer was controlled to be zero, so the test was constrained completely. In this way, the thermal stress was generated repeatedly by thermal cycles. The load was detected by load-cell. And all data was treated by microcomputer.

Experimental Condition

The repetition of thermal cycle, heating from lowest temperature up to peak temperature then cooling down to lowest temperature in triangle wave, was given to the completely constrained specimen. Heating and cooling speed was kept at 3.1 K/s and holding duration was zero at the peak or at the lowest temperature. In all thermal fatigue tests, the lowest temperature was fixed at 323K and peak temperature was selected as variable parameter, from 673K to 1223K. The thermal fatigue life was decided by the number of thermal cycles, when the peak stress was equal to or less than the value of 3/4 the maximum of peak tensile thermal stress.

EXPERIMENTAL RESULTS AND DISSCUSSION

Thermal Stress Cycle

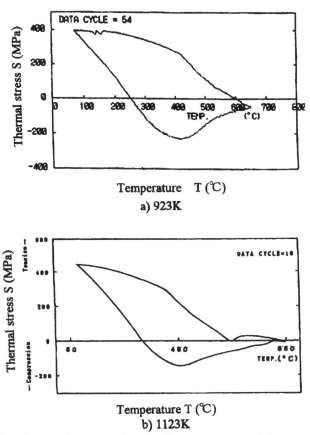

Fig.1 Examples of thermal stress cycle of peak temperature at a) 923K and at b) 1123K and at the lowest temperature of 323K in thermal fatigue test.

One example of thermal stress cycle on the thermal fatigue test of peak temperature at 923K was shown in Fig.1a), in which the ordinate showed thermal stress generated in gauge length of specimen and abscissa showed the temperature. The cycle was described in counter clockwise in the thermal fatigue tests. While temperature ascending from the lowest temperature, the thermal stress descended due to compression of thermal expansion of specimen; as the temperature ascending further until the peak temperature, then reversely the stress ascended due to reducing of yield stress of specimen. While the temperature started falling, the thermal stress ascended due to contraction of specimen; with further falling of temperature, the thermal stress increased. And when the thermal stress became larger than the yield point of specimen, then the curve rose gently until the lowest temperature. Namely, the thermal stress was compression as heating and in the upper half of cycle toward to tension as cooling which meant that the thermal fatigue test was belong to out-of -phase type. In this case of the peak temperature at 923K, the serration appeared at peak left side of the curve. Because the flow resistance of specimen became smaller at elevated temperature, the absolute value of the maximum thermal stress in tension was larger than that of the minimum in compression in each thermal fatigue cycle.

The shape of the thermal stress cycle at the peak temperature above 1098K was different from others at the other peak temperatures. Because the microstructure of specimen transformed from ferrite to austenite, the sudden shrink of specimen happened to increase the thermal stress, and vice versa.

Ordinary Serration in Thermal Stress

There was no any special phenomenon found in the thermal stress cycle at the peak temperature from 673K to 753K. At the peak temperature from 773K to 823K, the small serration like ripple was found near the peak stress in the initial life stage, but the serration disappeared in intermediate life stage. And the ordinary serration appeared at the peak temperature from 843K to 1078K. The serration appeared within the temperature range from 333K to 458K in each cycle. As known, the ordinary serration was caused by the dynamic interaction between the migration of dislocations and the interstitial atoms in crystal.

Visual Fatigue Crack

After the visual crack was found, it took tens to hundreds cycles for the specimen to fail, in the peak temperature range from 673K to 773K. However, in the peak temperature range from 798K to 1223K, the specimen broke just soon after the crack was detected. In the case from 673K to 773K, the material was strengthened by the solid solution that easily initiated a leading crack and the propagation of the crack took a large part of fatigue life. As for the case from 798K to 1223K, the specimen became easy to flow by the softening of crystals and grain boundaries themselves, and easy to cause brittle oxide layer at specimen surface, so that many micro-voids and micro-cracks were generated and grew. The crack coalescences occurred. And the cracks became visual at the final stage of fatigue life just before its failure. A great deal of hair like cracks appeared on the surface of specimen above 1023K.

Peculiar Serration

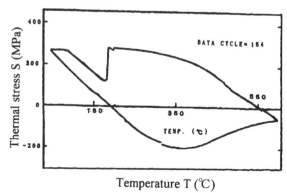

Fig.2 An example of the peculiar large serration appears in thermal stress cycles of thermal

fatigue of the peak temperature at 843K.

Here, particularly being worthy of note, a peculiar large serration was discovered at the peak temperatures above 798K, as shown in Fig.2, at several final cycles, namely at the final stage of fatigue life. The peculiar serration led to the rupture of specimen. This kind of serration appeared at temperature range from 333K to 538K and stress range from 468MPa to 670MPa, in the thermal cycle. But this kind of peculiar serration disappeared at the peak temperature below 798K or above 1098K.

As mentioned above, the fact of the phenomenon of the peculiar serration coincided with the appearance of ordinary serration. So the peculiar serration might be related to dynamic strain aging. The peculiar serration may be caused by the formation of macro-crack through the process of initiation, growth, coalescence of micro-voids or micro-cracks and the dynamic strain aging mechanism. Particularly, it could be construed that the "abrupt big fall", peculiar serration, was caused by the abrupt coalescence of micro-cracks and micro-voids. And/or the peculiar serration was caused by the great deal of instantaneous migration of a large number of dislocations, which were multiplied and stored in material and were adhered by activated interstitial atoms at the final stage of fatigue life. In the final fatigue stage these adhered dislocations were released abruptly.

Fatigue Life versus Peak Temperature

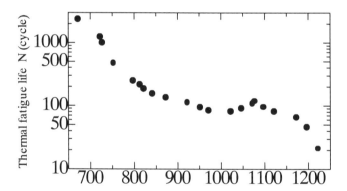

Fig 3 The curve of the thermal fatigue life versus the peak temperature
on ferrite ductile cast iron.

As shown in Fig.3, the relationship of thermal fatigue life to the peak temperature could be divided into 3 ranges, the first range was α range from 673K to 1023K where the life decreased with the increase of peak temperature, the second was $\alpha \Leftrightarrow \gamma$ range from 1023K to 1073K where the life increases with increase of temperature and the third was γ range above 1078K that was the same as the first range. The relationship in each range could be represented by the equation (1),

$$N = A \cdot \exp(Q/T) \tag{1}$$

where, N: thermal fatigue life, R: gas constant, 8.31451 J/mol·K, Q : thermal activation energy, KJ/mol and A: coefficient. The relationship of fatigue life versus the peak temperature was linear in each range. But in the first range the activation energy was large at low temperature and became smaller at high temperature. The values in each range are as shown in Table 3.

The equation (1) was the same form as Ahhrenius equation that clearly showed the thermal fatigue was affected by the micro-mechanism of thermal activation process.

Table 3 The values of activation energy and coefficient of each ranges in equation (1).

	Lower α range	Higher α range	All α range	$\alpha \Leftrightarrow \gamma$ range	γ range
Q:	83	26.7	58	-63	109
A:	9.7×10^{-4}	0.3	5.5×10^{-2}	1.3×10^{5}	6.5×10^{-4}

SUMMARY

By performing thermal fatigue tests on ferrite matrix ductile cast iron, the main results are attained as follows.
1. The shape of thermal stress cycle is affected by the transformation of microstructure.
2. The ordinary serration appears at peak temperatures from 773K to 1078K.
3. The peculiar serration appears at peak temperatures from 843K to 1078K.
4. Visual crack is found early and lasts long at low peak temperatures in the test. And at high temperatures, visual crack is found at final stage just before the failure of specimen.
5. The relationships of peak temperature to fatigue life are attained.

REFERENCE

1. Chijiiwa, K., Hayashi, M. (1979) Journal of The Faculty of Engineering, The University of Tokyo (B), Vol. 35, No.2, pp.205-230.
2. Chijiiwa, K., Hayashi, M. (1979) Imono, Vol. 51, pp.395-400.
3. Chijiiwa, K., Hayashi, M. (1979) Imono, Vol. 51, pp.513-518.
4. Yanagwasawa, O. (1986) Science of Machine, Vol. 38, pp.915-920.
5. Yasue, K., Wasotani, M., Kondo, Y., Kawamoto, N. (1982) Imono, Vol. 54, pp.739-743.
6. Nakashiro, M., Kitagawa, M., Hukuhara, Y., Oohama, S. (1984) Transaction of Iron and Steel, Vol. 70, pp.1230.
7. Takeshige, N., Uosaki, Y., Asai, H. (1994) Prepr. of Jpn. Soc.Mech.Eng. No.940-10, pp.56-58.
8. Peterson, R. E. (1965) Stress Concentration Design Factors, John Wiley & Sons, Inc.New York.

Low Cycle Fatigue and Elasto-Plastic Behaviour of Materials
K-T. Rie and P.D. Portella (Editors)
167

THE INFLUENCE OF SDAS ON TMF RESPONSE OF AL 319 ALLOYS

TRACY J. SMITH, HUSEYIN SEHITOGLU, XINLIN QING, AND HANS J. MAIER*

Department of Mechanical and Industrial Engineering
University of Illinois at Urbana-Champaign
1206 West Green Street, Urbana, IL 61801, USA
*Institut f. Werkstofftechnick, FB 11, Universitat-GH-Siegen, Paul-Bonatz-Str. 9-11
57068 Siegen, Germany

ABSTRACT

The stress-strain behaviors of cast aluminum 319-T6, with small and large Secondary Dendrite Arm Spacing (SDAS), are studied under conditions of thermal exposure at high temperature and thermo-mechanical fatigue loading. A two state variable unified inelastic constitutive model was proposed with explicit dependence on dendrite arm spacing. The results showed that the model provided successful predictions and described the marked softening under thermo-mechanical fatigue. The material behavior was described by using a unified theory in which the creep and plastic strains are combined as inelastic strains.

KEYWORDS

Thermo-mechanical fatigue, secondary dendrite arm spacing, constitutive modeling, aluminum-copper-silicon alloy.

INTRODUCTION

The Al 319 class of alloys are being used in cylinder head applications in automotive engines. For a constitutive model to be both useful and applicable for a number of different deformation histories the model should incorporate the most dominant microstructural features which affect the material's mechanical response. The two most important microstructural features considered in this study and incorporated into the model for Al 319 were (a) dendritic microstructure and (b) coarsening of the precipitates.

The dendritic microstructure is an inherent result of solidification and has been shown to have an effect on the mechanical properties of various aluminum alloys. Flemings et al. [1] showed that the yield strength, ultimate strength and ductility were all inversely affected by increasing SDAS, which results from increasing solidification times. The effect of SDAS has also been studied in isothermal fatigue and low cycle fatigue experiments by Gundlach et al.[2] and Stephens et al.[3]. They found that increasing SDAS decreased the fatigue life of the aluminum alloys. Samuel and Samuel [4] endorsed the observation that ductility decreased with increasing SDAS by analyzing the fracture surfaces of test specimens. They found that the smaller SDAS exhibited fracture surfaces indicative of ductile failure, whereas the larger SDAS fracture surface was characteristic of brittle fracture. Consequently, SDAS has a significant impact on the mechanical properties of a material and should be incorporated into a constitutive model.

The second significant microstructural feature of this precipitation hardened alloy is the coarsening of the precipitates during thermal exposure. Calabrese and Laird [5] studied the effect of the coarsening of θ´ precipitates on the cyclic response of an aluminum-copper alloy. They found that the finer θ´ structure increased cyclic saturation and monotonic flow stresses. Since the finer θ´ structure possesses smaller interparticle spacings, which are less than the self-trapping distance of dislocations, the particles control the cyclic response. The coarser θ´ structure, however, has interparticle spacings which are greater than the self-trapping distance of the dislocations and the subsequent cyclic response approaches that of the single phase material.

The unified constitutive model proposed in this work incorporates both the dendritic microstructure and the coarsening of the microstructure due to thermal exposure. Combining both of these features results in a constitutive model which is both accurate and applicable to numerous deformation histories over a range of SDAS.

EXPERIMENTAL PROCEDURE

A cast aluminum 319 alloy with a T6 (8h/495°C;BWQ;5h/190°C) heat treatment, of composition shown in Table 1, was used in the present study. The solutionizing temperature is 495°C and the aging temperature is 190°C. This heat treatment produces a peak aged microstructure which results in a material of maximum strength. At room temperature, a fine θ´ structure with disk shaped precipitates forms. The samples for thermo-mechanical fatigue testing were prepared from a sand-cast wedge with a copper chill positioned at the apex of the wedge. The wedge geometry results in different solidification rates based upon a similar principle as the varied cooling rate castings used in early work[6, 7]. This solidification control permits the machining of samples with controlled secondary dendrite arm spacing (SDAS). For the purpose of this study, small SDAS represents a sample with a SDAS of approximately 30μm and large SDAS denotes a sample with a SDAS of approximately 90μm.

Table 1. Chemical composition of Al 319

wt%	Si	Cu	Mg	Mn	Fe	Ti	Zn	Cr	Al
min.	7.2	3.3	0.25	0.20	-	-	-	-	balance
max.	7.7	3.7	0.35	0.30	0.40	0.25	0.25	0.05	balance

The thermo-mechanical fatigue tests were conducted using a servo-hydraulic Instron test machine in conjunction with a 15kW induction heater. The thermo-mechanical fatigue experiments were conducted in total strain control with independent closed-loop temperature and strain control. The experiments were controlled under out-of-phase (OP) conditions in which the maximum mechanical strain coincided with the minimum temperature. The strain and temperature waveforms followed a triangular wave-shape. The experiments were controlled at a mechanical strain rate of $5 \times 10^{5} s^{-1}$ and were conducted over a temperature range of 100-300°C and a strain range of 0.6%.

CONSTITUTIVE MODEL

The model proposed in this study, to predict the influence of SDAS size on the TMF response of the 319 Al alloy, is an experimentally based unified constitutive model. The model was first proposed by Slavik and Sehitoglu [8] to model the mechanical behavior of 1070 steel. The model has since been modified to predict both the mechanical behavior of Al 319 and to also include the explicit influence of SDAS on the mechanical behavior and even more importantly on the TMF response of the material.

Two state variables, back stress and drag stress, are incorporated into the model to predict various material responses of the material such as strain hardening, cyclic softening or hardening, strain rate sensitivity, stress relaxation etc. The state variables evolve using a hardening and recovery format which permit the state variables to simulate the previously described material reponses under varying temperature and loading histories. A kinetic form of the flow rule is used to determine the inelastic strain rate components as a function of the two state variables. A detailed description of the physical interpretation of the state variables in deviatoric stress space has been described in previous work [8]. Previous work by Smith et al [9] details the determination of model parameters from experiments which capture the essential features of the material's mechanical response.

Flow Rule

The purpose of the flow rule is to determine the inelastic strain rate components based on the evolution of the two state variables. The functional form of the flow rule used in the model is given below as

$$\dot{\varepsilon}_{ij}^{in} = \frac{3}{2} A f\left(\frac{\overline{\sigma}}{K}\right) \frac{S_{ij} - S_{ij}^c}{\overline{\sigma}} \tag{1}$$

where $\dot{\varepsilon}_{ij}^{in}$ is the inelastic strain rate, K is the drag stress, $\overline{\sigma}$ is the effective stress, S_{ij} is the deviatoric stress, and S_{ij}^c is the deviatoric back stress. The flow function, $f(\overline{\sigma}/K)$, depends upon the active deformation mechanism such as power law creep, plasticity, or diffusional flow. This flow function is well motivated by Ashby's deformation mechanism maps[10]. The second term of $S_{ij} - S_{ij}^c$ determines the direction of the strain rate. The constant, A, is a function of temperature and SDAS and can be expressed as follows:

$$A = A_c \left(\frac{L}{L_o}\right)^{m_o} exp\left(-\frac{\Delta H_c}{RT}\right) \tag{2}$$

where L is the size of SDAS and L_o is a normalization constant.

Back Stress Evolution

The back stress, otherwise known as internal stress, describes the increase in a material's resistance to deformation with straining. For example, the material's resistance may be due to dislocation pile-up resulting from dislocation interaction with barriers such as precipitates and other dislocations , in addition to other internal stresses at the micro-level.

The evolution of the back stress state variable follows a hardening and recovery formulation and can be modeled as

$$\dot{S}_{ij}^C = \frac{2}{3} h_\alpha(\alpha, L) \dot{\varepsilon}_{ij}^{in} - [r_\alpha^D(\alpha, T, \dot{\overline{\varepsilon}}^{in}, L) \dot{\overline{\varepsilon}}^{in} + r_\alpha^s(\alpha, T, L)] S_{ij}^C \tag{3}$$

where $\dot{S}_{ij}^c$ is the back stress rate, $\dot{\overline{\varepsilon}}_{ij}^{in}$ is the effective inelastic strain rate, h_α is the back stress hardening function, and r_α^D and r_α^s are the dynamic and static back stress recovery functions, respectively. The hardening function, h_α, describes the evolution of the back stress during high strain rate experiments and acts to increase the back stress. In a converse manner, r_α tends to decrease the back stress rate to zero with increasing temperature or time.

The forms for the back stress hardening and recovery terns were determined from experiments. The hardening term was ascertained from room temperature experiments and the recovery term from high temperature experiments with temperatures ranging from 150-250°C. For the present study, the static recovery term, $r_\alpha^s(\alpha, T, L)$, was assumed to be equal to zero and the dynamic recovery term adequately described the recovery behavior of the material.

Drag Stress Evolution

The cyclic softening and hardening behaviors of the material are governed by the evolution of the drag stress. In terms of the microstructure, changes in the drag stress represent the effect of a developing microstructure on initial yielding behavior. The drag stress characterizes the initial resistance to slip from precipitates, grains, and point defects in the initial microstructure and also cyclic hardening that arises from a dislocation structure, such as cells and subgrains in fatigue.

The evolution of the drag stress also follows a hardening-recovery format and can be modeled as:

$$\dot{K} = h_k(K, T, L) - r_k(K, T, L) + \theta(K, T, L)\dot{T} \qquad (4)$$

where $\dot{K}$ is the drag stress rate and $\dot{T}$ is the temperature rate. The drag stress hardening term h_k represents hardening or softening due to plastic deformation. The drag stress recovery term r_k governs the change in K caused by microstructural changes. The theta-term, θ, represents the variation of the initial drag stress with temperature.

The evolution of the drag stress, K, is determined by examining the cyclic material response. Measurements of the current yield stress range, $(2\bar{\sigma})$, saturated yield stress range, $(2\bar{\sigma}_{sat})$, and the yield stress after thermal exposure, established the corresponding levels of drag stress, K, K_{sat}, and K_{rec}. Subsequently the hardening function, h_k, and the recovery term, r_k, were established.

RESULTS

The constitutive model was checked with thermo-mechanical loading experiments which were independent of determining model parameters. The comparisons of simulation and experiment for small SDAS and large SDAS are given respectively in Figs. 1a and b. The mechanical strain ranges in the two TMF experiments was 0.6%. The experimental results are denoted by the solid line and the dashed line gives the model predictions. Note that the stress levels obtained under small SDAS conditions exceed the large SDAS case by as much as 30%. The model is capable of accurately predicting the decrease in stress with cycling and temperature, in addition to the increase in plasticity. The failure criterion for theses experiments was taken to be the cycle in which the maximum tensile load was 50% of the maximum value of the tensile load averaged over cycles 2 and 3. The specimen with the large SDAS failed before attaining the 50% load drop which would be expected due to the decrease in ductility with increasing SDAS, hence leading to brittle failure. On the other hand, the sample with small SDAS (N_f=741) reached the 50% load drop and had a life which was 25 times greater than that of the large SDAS (N_f=29) sample.

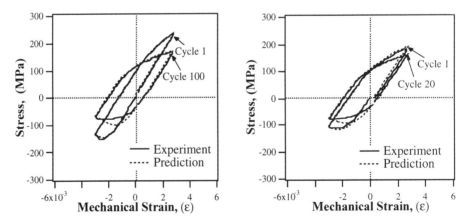

Fig. 1: Predicted and experimental results for an OP TMF experiment with $\Delta T=300°C$, $\Delta\varepsilon=0.6\%$ at a strain rate of $5\times10^{-5}s^{-1}$ for (a) small SDAS and (b) large SDAS.

TEM images obtained from the undeformed material and after thermal exposure, 1000hrs at 250°C, show that extreme coarsening and dissolution of θ' precipitates have occurred, as seen in Figs. 2a and b. The coarse precipitates are no longer effective barriers to dislocation motion and the material could, in the limit, reach an unprecipitated structure. Consequently, the coarsening microstructure acts to degrade the mechanical properties of the material. The effect of thermal exposure has been considered in the constitutive model as influencing both the drag stress and the back stress.

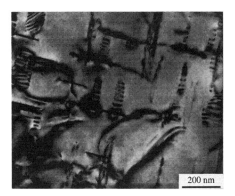 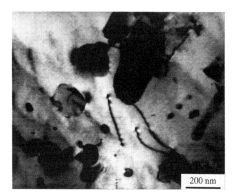

Fig. 2: TEM images showing coarsening and dissolution of precipitates due to thermal exposure at 250°C for 1000hrs: (a) undeformed material and (b) after 250°C for 1000hrs.

Figs. 1 and 2 demonstrate the importance for incorporating both dendritic structure and the coarsening of the precipitates into the constitutive model since both microstructural features have a large impact on the mechanical response of the material during an experiment. The cyclic softening due to the coarsening of the microstructure is shown below in Figs. 3a and b for both experiment and prediction. The stresses in compression are lower than those in tension because of the TMF loading. The dashed line again represents the prediction and the markers display the experimental points. The small SDAS showed significantly more softening over the life of the experiment but with the majority of the softening occurring within the first 200 cycles. The model slightly over predicts the softening at low temperature in the small SDAS case while the predictions are very accurate for the large SDAS case. For the most part, the model accurately predicted the cyclic softening of the material.

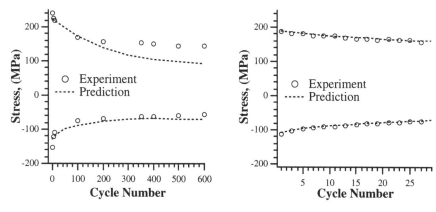

Fig. 3: Predicted and experimental results demonstrating the softening behavior during an OP TMF experiment (300°C, $\Delta\varepsilon$=0.6%, $5\times10^{-5}s^{-1}$): (a) small SDAS and (b) large SDAS

CONCLUSIONS

The preceding work supports the following conclusions:

(1) Cyclic stress-strain response of Al 319 strongly depends on solidification conditions. Rapidly solidified material with small SDAS displayed higher cyclic stresses than materials with large SDAS.

(2) Due to the significant metallurgical changes of the material's microstructure due to thermal exposure at high temperatures and extended periods of time, the variation of drag stress due to temperature and time should be considered in the constitutive model as well as the variation of back stress. The functions were readily established from experiments and predicted the correct stress-strain behavior.

(3) The cyclic softening of drag stress occurs rapidly followed by saturation. The recovery processes were accelerated during deformation and this observation was reflected in the model. For example, the saturated value of drag stress caused by cyclic softening is much smaller than the fully recovered value of drag stress caused by thermal exposure at high temperature.

ACKNOWLEDGMENTS

The work is supported by Ford Motor Company, Dearborn, Michigan. The support of Dr. John Allison and Mr. John Lasecki, Materials Science Dept., Ford Research Laboratories, is appreciated.

REFERENCES

1. Flemings, M. C., Kattamis, T. Z. and Bardes, B. P. (1991). *AFS Trans.* **99**, 501.
2. Gundlach, R. B., Ross, B., Hetke, A., Valtierra, S. and Mojica, J. F. (1994). *AFS Trans.* **102**, 205.
3. Stephens, R. I., Berns, H. D., Chernenkoff, R. A., Indig, R. L., Koh, S. K., Lingenfelser, D. J., Mitchell, M. R., Testi, R. A. and Wigant, C. C. (1988). *SAE Technical Paper* **No. 881701**.
4. Samuel, A. M. and Samuel, F. H. (1995). *Metall. Mater. Trans. A* **26A**, 2359.
5. Calabrese, C. and Laird, C. (1974). *Mater. Sci. Eng.* **13-14**, 159.
6. Wickberg, A., Gustafson, G. and Larson, L. E. (1984). *SAE Trans.* **93**, 728.
7. Vorren, O., Evensen, J. E. and Pedersen, T. B. (1984). *AFS Trans.* **92**, 459.
8. Slavik, D. and Sehitoglu, H. (1987). *ASME PVP* **123**, 66.
9. Smith, T. J., Maier, H. J., Sehitoglu, H., Fleury, E. and Allison, J. (1998). *Submitted to Met. Trans.* .
10. Frost, H. J. and Ashby, M. F., (1982). *Deformation-Mechanism Maps*. Pergamon, Elmsford.

Low Cycle Fatigue and Elasto-Plastic Behaviour of Materials
K-T. Rie and P.D. Portella (Editors)
173

THERMOMECHANICAL FATIGUE OF SURFACED AND SPRAYED ELEMENTS

J. DZIUBIŃSKI and M. CIEŚLA

*Faculty of Material Science, Metallurgy and Transport, Silesian Technical University,
PL-40 019 Katowice, Poland*

ABSTRACT

Methodology and results of examining thermomechanical fatigue of surfaced and sprayed elements are discussed. Mechanisms of cracking of surfaced elements with 18-8 and Cr-Mo electrodes and plasma sprayed elements with Al_2O_3 powder are determined. A relation between cracking mechanisms and life of tested elements is shown. The elements surfaced with Cr-Mo electrodes showed longer life in relation to the elements surfaced with 18-8 electrodes, which was connected with a different mechanism of their fatigue cracking. In the case of austenitic padding welds an intensive destruction process was observed on the border between the padding weld and the base material. The process of cracking of elements surfaced with Cr-Mo electrodes is initiated on the surface of the padding weld. In the conditions of thermomechanical fatigue, the processes of cracking of sprayed elements Al_2O_3+Ni with a NiAl precoat and without this layer are initiated both in the ceramic layer and in the precoat layer, however, the sprayed elements where the NiAl precoat layer was applied show greater fatigue life than the elements without this layer.

KEYWORDS

Thermomechanical fatigue, surfacing, thermal spraying.

INTRODUCTION

A most often way of material damage in metallurgy, chemical and power industry is fatigue caused by cyclic changes of temperature [1÷3]. Therefore the elements working in mentioned branches of industry are often regenerated by means of welding technologies or protected against high temperatures influence (in a form of so called thermal barriers coating -TBC) by the way of plasma spraying technology. Thus, there is a need of estimation of an influence of these technologies on changes of regenerated objects properties. In the present research work a trial

174

of assessment of influence of surface layers obtained by means of surfacing or spraying on elements life in thermomechanical fatigue process is being undertaken.

TESTING MATERIALS

Two series of round surfaced samples made of 45 steel grade of 12mm diameter were prepared for tests. The first series was being surfaced by means of 18-8 electrodes (chemical composition of weld metal 0.07%C 1.2%Mn, 19.5%Cr, 9%Ni) whereas the other one by means of Cr-Mo electrodes (chemical composition of weld metal 0.07%C, 0.7%Mn, 0.9%Cr, 0.6%Mo). Longitudinal padding welds of thickness equal 2mm were made in a form of outer ring situated along the sample.

There were also prepared two series of round samples made of 45 steel grade of 12 mm diameter on which two types of heat-resistant adhesive coatings were constituted (Table 1). The heat resistant coatings were obtained by means of plasma spraying with using METCO 7MB device under following parameters J=560A, U=60V, where the distance between the plasma-arc gun and the surface of the sample was equal to 110 mm and hydrogen with argon were applied as carrier gas.

Table 1. Types of protective coatings created on samples made of steel 45

Marking	Type of coating	Method of obtaining
„A"	adhesive - thickness approx. 200μm	plasma spraying of Al_2O_3 powder+Ni on the sample made of steel 45
„B"	adhesive - thickness approx. 230 μm	plasma spraying of NiAl powder (30% Al, 70%Ni) (approx. 80μm) on the ground made of steel 45 and next plasma spraying of Al_2O_3 + Ni powder (approx. 150 μm)

THERMOMECHANICAL FATIGUE AND METALLOGRAPHIC TESTS

Tests of fatigue durability determined by the number of cycles until a failure of test piece were carried out by means of MTS-810 servo-hydraulic machine. During tests the samples were heated and cooled in cycles in the range of temperatures between $200 \div 700^0$C, $200 \div 650^0$C and $200 \div 600^0$C. Heating times depended on the maximal cycle temperature and were equal respectively: 40s, 30s and 20s. Similarly the cooling process down to the minimal temperature of the cycle (T_{min} = 200^0C) was carried out within a period of 40 to 60 seconds. Thermomechanical fatigue tests were carried out with the strain controlling on the machine where the factor $K = \varepsilon_M/\varepsilon_t = 1$, where ε_M- mechanical strain, ε_t - sample's thermal strain. During tests stabilised hysteresis loops were recorded (σ,T) which were a ground for $\Delta\varepsilon_{pl}$ determination.

For the surfaced samples average values of the number of cycles till failure N_f of samples and also plastic strain values $\Delta\varepsilon_{pl}$ of samples with defined type of padding welds are shown in Table 2. Ranges of plastic strains $\Delta\varepsilon_{pl}$ corresponding with the hysteresis loop width at zero value of strain were calculated on the basis of thermal linear expansion coefficients values β of the padding weld materials in appropriate temperature ranges [4]. On the ground of received results and on the Coffin's criterion [5] a mathematical model was constructed describing life of layer-heterogeneous materials (Table 3) in the form: $\Delta\varepsilon_{pl} \cdot N_f^k = M$, where k, M - material constants referring to both the basis material and the padding weld material. A graphical form of this model is shown in Fig. 1a. The life characteristics of tested materials in function of maximal cycle temperature are given in Fig. 1b.

Table 2. Results of surfaced specimens thermomechanical fatigue tests

Padding weld type	Temperature range in thermal cycle (°C)	$\Delta\varepsilon_{pl}$	Number of cycles till appearance of first cracks on the padding weld *N_f
18-8	200 ÷ 700	0.00450	190
	200 ÷ 650	0.00200	330
	200 ÷ 600	0.00140	555
Cr-Mo	200 ÷ 700	0.00280	260
	200 ÷ 650	0.00133	465
	200 ÷ 600	0.00088	715

* N_f - average values from two samples testing

The thermomechanical fatigue tests of sprayed samples were carried out according to methodology similar to the one accepted for surfaced samples. The only difference was that the tests were being carried out without the first half-cycle of load in which so called plastic adjustment of material appears. This adjustment takes place in the range of very high compressive stress. In the result of that a cracking of friable (by nature) ceramic heat-resistant coatings that was noticed during initial tests was eliminated already in the first cycle of heating. The results of thermomechanical fatigue tests of samples covered with heat-resistant layers obtained by means of plasma spraying are set together in Table 4.

Table 3. Mathematical model of tested samples life under conditions of thermomechanical fatigue. Values of correlation and regression functions coefficients

Padding weld type	Life criterion: $\Delta\varepsilon_{pl} \cdot N_f^k = M$
18 - 8	M = 1.3 ; k = 1.093 ; r = 0.97
Cr-Mo	M = 1.9 ; k = 1.170 ; r = 0.98

176

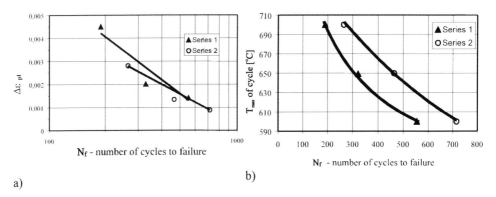

a)

b)

Fig. 1. Life characteristics of tested materials according to following formula: $\Delta\varepsilon_{pl} = f(N_f)$ - (a) and T_{max} of cycle = $f(N_f)$ - (b). Series 1 - samples surfaced with electrodes of 18-8 type. Series 2 - samples surfaced with Cr- Mo electrodes.

Table 4. Results of sprayed specimens thermomechanical fatigue tests.

Layer type	Range of temperature changes in thermal cycle (^{0}C)	Number of cycles till appearance of first cracks in the sprayed layer *N_f
(Al_2O_3 + Ni)	200 ÷ 600	70
„A"	200 ÷ 700	44
(NiAl)+(Al_2O_3+Ni)	200 ÷ 600	108
„B"	200 ÷ 700	63

*N_f - average values from two samples testing

Metallographic observations of the surfaced and sprayed elements after thermomechanical fatigue tests have been carried out. Cracks on the border of penetration of padding weld made by means of 18-8 electrode and cracks from the surface of padding weld made by means of Cr-Mo electrode have been found. In the ceramic layer (Al_2O_3 + Ni), cracks in the ceramic layer and on the border between the layer and the ground have been observed. In the ceramic layer (Al_2O_3) with precoat layer (NiAl), cracks in the ceramic layer itself and on the border between the precoat layer and the base material have been observed too.

ANALYSIS OF TEST RESULTS

As one can see on the basis of presented results (Fig. 1, Table 2), samples of steel 45 with padding weld of Cr-Mo type are characterised by greater durability in assumed conditions of thermomechanical fatigue examinations. High durability of this type of layer-heterogeneous materials can be explained by their high resistance to variable cyclic plastic strains the range of which $\Delta\varepsilon_{pl}$ for the same thermal cycles is significantly lower than that for samples surfaced with 18-8 electrodes (Table 2).

Received results show simultaneously that due to a great surface fraction of padding weld in the sample's cross-section in the tested material the decisive factor of cracking and surface layer degradation processes in the tested layer-heterogeneous materials are first of all mechanical and physical properties (β, σ_y, σ_{UTS}) of the surfaced layer. For example for austenitic padding welds the linear expansion coefficient β in the range of temperatures $20 \div 600^0$C equals $18.7 \cdot 10^{-6}$ 1/K and for the ground material (steel of 45 grade) β equals $14.1 \cdot 10^{-6}$ 1/K [4]. That means that the difference $\beta_{18-8} - \beta_{45} = 4.6 \cdot 10^{-6}$ 1/K. However for padding welds of bainitic structure made with Cr-Mo electrodes for which $\beta_{20 \div 600^0C} = 15 \cdot 10^{-6}$ 1/K [4], the difference of appropriate coefficients equals $\beta_{Cr-Mo} - \beta_{45} = 0.9 \cdot 10^{-6}$ 1/K.

Microscope observations have shown that in case of austenitic padding welds application the cracking process is initiated in a fused zone (the border between the core material and the padding weld material). The cracking in the material of the core probably happens along the former austenite grains boundaries. This austenite came into being in the heat-affected zone during the surfacing process. In case of padding welds made by means of Cr-Mo electrodes the cracks are being generated on the padding weld surface. These diverse mechanisms of materials degradation in the fatigue process determine the life of surfaced elements.

The results of thermomechanical fatigue examinations (Table 4) and structural metallurgy investigations of sprayed layers show that fatigue processes of layers cracking are stimulated mainly by simultaneous influence of high thermal stresses and plastic strains of metallic base. In the analysed case the thermal stresses are determined by basically different values of linear thermal expansion coefficient of the base made of steel 45 ($\beta_{20 \div 700^0C} = 14.1 \cdot 10^{-6}$ 1/K) and ceramic ($Al_2O_3 + Ni$) heat resistant coating ($\beta_{20 \div 700^0C} = 7.4 \cdot 10^{-6}$ 1/K) [6]. At the same time an increase of durability of heat resistant coatings in which precoat layers (NiAl) were applied should be connected with their compensating influence on the thermal stresses pattern, especially in the outer, friable ceramic layer [7]. Indeed, the value of β coefficient of the precoat layer which is equal $\approx 13.6 \cdot 10^{-6}$[1/K] in the range of temperatures between $20 \div 700^0$C (the value of coefficient determined by the authors for synthetically prepared NiAl phase) is contained in the range the borders of which determine β coefficients values for steel 45 and oxide layer $Al_2O_3 + Ni$.

CONCLUSIONS

1. In case of austenitic padding welds an intensive destruction process was observed on the surfaced elements being submitted to thermomechanical fatigue. This process manifested itself in cracks appearances on the border between the padding weld and the base material. The process of cracking of elements surfaced with Cr-Mo electrodes is initiated on the surface of the padding weld.

2. The elements surfaced with Cr-Mo electrodes showed longer life in relation to the elements surfaced with 18-8 electrodes, which was connected with a different mechanism of their fatigue cracking.

3. In the conditions of thermomechanical fatigue, the processes of cracking of sprayed elements $Al_2O_3 + Ni$ with a NiAl precoat and without this layer are initiated both in the ceramic layer and in the precoat layer, however, the sprayed elements where the NiAl precoat layer was applied show greater fatigue life than the elements without this layer.

4. The layers on the base of Al_2O_3 due to their low durability should not be used for elements working under heavy thermomechanical loads.

REFERENCES

1. Żuchowski, R. (1982). *Thermal Fatigue of Metals* (in Polish), Wyd. Politechniki Wrocławskiej, Wrocław.
2. Weroński, A. and Hejwowski, T. (1992). *Thermal Fatigue of Metals*, Marcel Dekker Inc., New York.
3. Okrajni, J. (1986). *ZN Politechniki Śl., Seria Mechanika*, **83**, 145.
4. Beitz, W. and Kuttner, K. H. (1987). *Taschenbuch für den Maschinebau*, Springer - Verlag, Berlin.
5. Kocańda, S. (1985). *Fatigue Cracks of Metals* (in Polish), Wyd. Nauk.-Techn., Warsaw.
6. Nadasi, E. (1975). *Modern Methods of Thermal Spraying* (in Polish), Wyd. Nauk.-Techn., Warsaw.
7. Rhys-Jones, T. N. (1989). *Cor. Science*, **29**, 234.

Low Cycle Fatigue and Elasto-Plastic Behaviour of Materials
K-T. Rie and P.D. Portella (Editors)
179

ASPECTS OF THERMAL FATIGUE DEFORMATION PROCESS OF WROUGHT NI-BASE SUPERALLOY

P. HORŇAK, J. ZRNÍK, V. VRCHOVINSKÝ

Department of Material Science, Technical University of Košice,
042 00 Košice, Slovakia

ABSTRACT

The wrought nickel base superalloy has been exposed in conditions of thermal fatigue. The thermal fatigue resistance was investigated under repeated thermal shock cycles and thermal cycles with introduction of a dwell time at maximum temperature. The different cycling loading regimes defined by the heating and cooling rate and dwell times at maximum heating temperature were used to simulate the different experimental cycles of thermal fatigue conditions, giving a faithful simulation of thermal transients imposed on components or structures. The effect of thermal cycling was related with structural changes and crack nucleation process. The advances in structural damage were studied by surface metallography and related to the number of thermal cycles. The progress in damage process due to build-up of an internal stresses, considering progress in thermal cycling, was for individual testing procedure investigated by TEM. It was found that whereas the deformation proceeded throughout the specimen cross section the crack nucleation was just surface effect. The development in deformation process strongly depended on thermal fatigue procedure, so did the first crack appearance. The hold time introduction had significant effect on damage process (appeared to heal the induced as well the cooling rate). The structural stability of alloy varied in dependence of thermal cycling conditions applied. The number of cycles was taken as the thermal fatigue life criterion.

KEYWORDS

Nickel base superalloy, thermal fatigue, hold time, microstructural characteristics, crack nucleation, failure process

INTRODUCTION

The potential failure modes for gas turbine components are creep, high cycle fatigue, low cycle fatigue, thermal fatigue, and surface attack. To understand the damage evolution process under both thermal and mechanical loading an improved durability of components exposed might be achieved. Knowledge on the thermal fatigue (TF) behaviour can provide a valuable information for the understanding of material behaviour under real thermomechanical fatigue.

The thermal fatigue failures are caused by the repeated application of stress that is thermal origin. Thermal fatigue stresses are originating due to steep temperature gradients in a component or across a section, but TF can develop even under conditions of uniform specimen temperature in result of internal constrains or can be product of various thermal expansion coefficients of phases the alloy consists.

The studies on TF have been conducted in different wrought, cast, or directionally solidified nickel base superalloys and composites, examining the role of structural parameters of alloys and evaluating their TF resistance under various test conditions, including heating and cooling rate, temperature range, duration of thermal cycles, cycling frequency, and environmental effect [1-7]. The resistance of some nickel base superalloys to the initiation and crack propagation under thermal cycling, or recently, in more complex thermomechanical fatigue (TMF) conditions have been also reported [7-10]. Research on TF has also described structure degradation caused in TF that significantly influence deformation behaviour and TF lifetime of nickel base superalloys. However, for stress-strain behaviour there is still not enough information on TF damage model in sophisticated conditions of TF simulation.

In the present paper special attention is paid to study the microstructural changes, and degradation process evolution in wrought nickel base superalloy, using light metallography, scanning electron microscopy and transmission electron microscopy (using various metallography techniques) in dependence of various thermal fatigue loading.

MATERIAL AND PROCEDURE

The investigated material was wrought nickel base superalloy EI 698 VD the chemical composition of which in weight % is: 0.08 C, 13.0 Cr, 2.8-3.8 Mo, 1.3-1.7 Al, 2.3-2.7 Ti, 1.8-2.2 Nb, 2.0 Fe, 0.007 S, 0.015 P, and balance Ni. To obtained the required mechanical properties at room and higher temperature three step heat treatment was performed: 1. Solution annealing at 1100 °C for 8 hours/air cooling; 2. Ageing at 1000 C for 4 hours/air cooling; 3. Ageing at 775 °C for 16 hours/air cooling. The initial structure of the alloy after this defined heat treatment is presented in

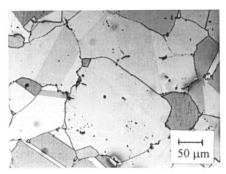

Fig. 1. It can be shown that structure is equiaxed and of uniform grain size. Two types of carbides is possible to observe in matrix. The coarse, bulky MC carbides were randomly distributed in matrix and along grain boundaries. The finer $M_{23}C_6$ carbide particles, the result of ageing at higher temperature, are precipitated along grain boundaries and these are supposed to provide sufficient grain boundary creep strengthening of the alloy.

Fig. 1. The initial structure of EI 698 VD alloy.

To investigate the structural stability of alloy and the mechanism of the failure process under different thermal fatigue conditions the four following experimental thermal cycles were used:

1. heating at 650 °C / rapid cooling by water; 2. heating at 650 °C / slow cooling by airstream; 3. heating at 650 °C / 1,5 hour hold time / rapid cooling by water; 4. heating at 650 °C /1,5 hour hold time / slow cooling by airstream. The diagrams of used TF cycles are presented on Fig. 2.

The temperature was levelled in 10 seconds, and was controlled with thermocouple welded to the specimen. The specimens were heated by passage of electric current. The free dilatation of specimen, to eliminate additional effect of longitudinal specimen thermal expansion/constrain/, during heating and cooling, was maintained. The thermal fatigue resistance and the changes of microstructural characteristics follow to rapid degradation of structure characteristics. After 12500 cycles is can to see expressive after considering number of thermal cycles were continuously controlled by light and scanning electron microscopy. A deformation progress, in dependence of number of applied thermal cycles, until first appearance, was analysed in thin foils by TEM.

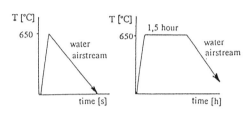

Fig. 2. Thermal fatigue cycle diagrams crack.

RESULTS AND DISCUSSION

1. heating at 650 °C / rapid cooling by water

The initial attributes of the damage were observed already after 200 cycles. In the individual coarse carbides were locations where the first thermal fatigue cracks appeared. The growth of already nucleated cracks was realised by the process of continuous coalescence of the individual cracks nucleated preferentially by carbide cracking or by matrix-carbide interface decohesion, Fig. 3. In Fig. 4 shows the microstructure of alloy after 1000 cycles. The developed network of the cracks on the specimen surface is corresponding already

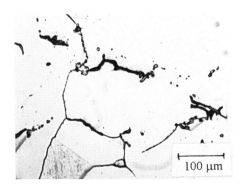

Fig. 3. The growth coalescence process of the individual cracks.

Fig. 4. The network of the cracks on the specimen surface.

182

pronounced damage process. By increasing the number of cycles these cracks have weak grown up to critical size and their further propagation was governed by structure's weak sites. he grains The individual cracks followed up not only the grain boundaries but also row arrangement of bulky MC carbides inside the grains. However, consequent microstructure analysis of specimen cross section confirmed the fact, that process of crack nucleation was surface only process, as can be seen in Fig. 5. TEM observations showed that local stresses nucleated in the exposed material leads to formation of relatively developed slip bands throughout the matrix, Fig. 6. Such development of dislocation structure manifests the stresses induced in matrix due to thermal cycling can respond in inelastic deformation and to start the slips on the different slip systems.

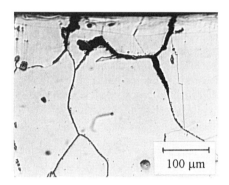
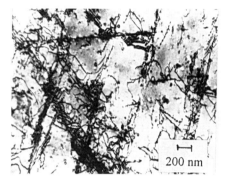

Fig. 5. The surface nucleated intergranular cracks.

Fig. 6. TEM micrograph of exposed specimen.

2. heating at 650 °C / slow cooling by air stream

The first visible changes in the microstructure of the exposed specimen were observed after 10000 cycles. The thermal cycling caused crack initiation either by cracking of coarse carbides, or by decohesion on carbide-matrix interface, Fig. 7. This number of cycles appeared to be critical for progress of damage process because additional cycling followed to rapid degradation of alloy structure. The specimen exposed to 12500 cycles shows excessive cracks formation, which preferentially copy grain boundaries, Fig. 8.

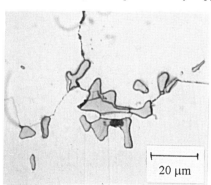

Fig. 7. The cracking of coarse MC carbide.

Fig. 8. The micrograph of the exposed specimen after 12 500 cycles, 2. regime.

The extent of deformation process was also confirmed by the study of deformation structure by means of TEM.

3. heating at 650 °C / 1,5 hour hold time / rapid cooling by water

Evaluation of the influence of the hold time introduction at maximum temperature the degradation process started again by cracking of coarse MC carbides (34 cycles) or by cavitation process on interface carbide-matrix (233 cycles). The extension of thermal cycling time (up to 500 cycles) caused the further propagation of cracks preferentially by intergranular mechanism, Fig. 9. In case of cycling with hold time introduction at upper temperature level the number of cycles to initiate the crack nucleation was increased and resulted in less impressive deformation in interior of grains. The possible stress increase, rising in specimen due to rapid cooling and successive heating cycle, elimination by thermal stress relaxation during the hold time is the most probable reason to delay the deformation process. This relaxation effect significantly influenced the deformation behaviour of alloy and development of dislocation structure as it is manifested in Fig. 10. The extent of deformation process in individual grains, and comparing it with previous cycling regimes, was not so expressive in this time. The stresses, which originated on specimen rapid cooling were probably relaxed by consecutive long-duration heating, and it resulted in less width and density of slip bands. However, more important is fact that long-duration heating did not eliminate the evolution of thermal stresses either, which probably originate following different coefficients of thermal expansion of individual alloy structural constituents and these might have a degradation influence at lifetime of alloy.

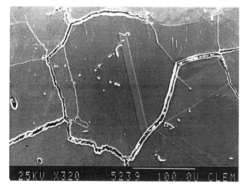
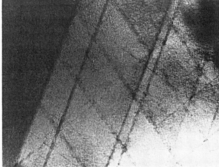

Fig. 9. SEM micrograph of intercrystalline cracks

Fig. 10. The slip bands structure of exposed specimen, 500 cycles, 3. regime

4. heating at 650 °C / 1,5 hour hold time / slow cooling by air stream

The obtained results showed that this soft regime caused only slightly changes in the microstructure of the alloy. The realised number of thermal cycles (333) was not to cause a visible degradation of structural characteristics. This regime of thermal cycling is from view of degradation of structure the most acceptable.

The obtained results showed that the structure degradation effect in dependence of thermal cycles number was evident for all regimes, except the last one, where hold time and slow cooling were involved. The degradation process is time dependent and the considered extent of deformation was specific corresponding to the progress of thermal fatigue. As implies from obtained results the crucial for deformation behaviour of nickel base superalloy under conditions of thermal fatigue is the presence of structure constituents of different physical properties (λ, α), which due to originating of critical stresses in local sites of the structure. From this point of view in nickel base superalloy the coarse MC carbide particles present in matrix showed the main harmful effect on the behaviour of the alloy in conditions of thermal fatigue. This conclusion is in good agreement with the results of calculation of local thermal stress for given temperature range due to thermal expansion nucleated at carbide-matrix interface. Results proved that theoretically calculated stresses could reach the level higher than tensile yield stress at corresponding cycling temperature, and caused mechanical inelastic strain. The local stresses in specimen are built up as a results of thermal gradient across the sample and can have additive effect on deformation process. The peak stress elimination by thermal stress relaxation during the hold time is the most probable reason to delay the deformation process. The introduction of hold time and air stream cooling provided the softer conditions of thermal fatigue and reflected in delay of degradation process, however, the mechanism of damage was identical as for severe regime when water cooling was applied.

CONCLUSIONS

1. Damage process in superalloy was evident due to thermal fatigue regardless the procedure.
2. Degradation process evolution strongly depends on applied regime of thermal fatigue.
3. Deleterious damage resulting in critical size crack growth was surface effect. Crack nucleation was due to carbide cracking or matrix-carbide interface decohesion. Critical intergranular crack propagation was localised on specimen surface.
4. The introduction of hold time at cycling regime and low rate cooling reflected in delay of degradation process.

REFERENCES

[1]   Glenny, E., Taylor, T.A. (1959) A Study of the Thermal Fatigue Behaviour of Metals, *J.Inst. Met.* **88** 449
[2]   Tilly, G.P. (1971) Thermal Stresses and Thermal Fatigue, Ed., D.J. Littler, Butterworth, London 47
[3]   Woodford, D.A., Mowbray, D.F. (1974) Mater. Sci. Eng. **16** 5
[4]   Bizon, P.T., Spera,D.A. (1976) In: *Thermal Fatigue of Materials Components*, STP 612, Ed. Spera, D.A. and Mowbray, D.F. 106
[5]   Ford, D.A., Arthey, R.P. (1984) In: *Superalloys 1984, Proceedings of the Fifth Int. Symposium on Superalloys*, Ed. Gell, M. et al, Met. Soc. AIME, Warrendale
[6]   Hopkins, S.W. (1976) In: *Thermal Fatigue of Materials Components*, STP 612, ed. Spera, D.A. and Mowbray, D.F. 157
[7]   Varin, J.D. (1972) In: *The Superalloys*, Ed. by Sims, C.T., Hagel, W.C., Willey, J., Interscience Publications, Wiley, J. & Sons, New York 231
[8]   Malpertu, J.L., Remy,L. (1990) *Met. TransA*. **21A** 2 389
[9]   Chataigner, E., Fleury, Remy, L. (1995) In: *Proc. Fatigue under Thermal and Mechanical Loading*, Ed. Bressers, J., and Remy, L. European Commission, Petten
[10]   Howes, M.A.H. (1973) In: *Fatigue at Elevated Temperatures*, STP 520, Ed. Carden, A.E., McEvily, A.J. and Wels, C.H., ASTM, 242

Low Cycle Fatigue and Elasto-Plastic Behaviour of Materials
K-T. Rie and P.D. Portella (Editors)
© 1998 Elsevier Science Ltd. All rights reserved.

DEFORMATION BEHAVIOUR OF WROUGHT NICKEL BASE SUPERALLOY SUBJECTED TO ISOTHERMAL CYCLIC CREEP AND THERMOMECHANICAL FATIGUE

J. ZRNÍK*, P. WANGYAO[a], V. VRCHOVINSKÝ*, P. HORŇAK* and
E. NISARATANAPORN[a]
*Technical University of Košice, Department of Materials Science , Park Komenského 11,
040 01 Košice, Slovak Republic
[a]Chulalongkorn University Bangkok, Department of Metallurgical Engineering ,
Bangkok 10300 Thailand

ABSTRACT

The deformation and damage mechanisms in wrought nickel base superalloy EI 698 VD subjected to isothermal and anisothermal cycling loading were investigated by examining the deformation characteristics, microstructure, crack nucleation and crack propagation mechanisms of representative specimens. All tests, high temperature creep, isothermal cyclic creep, including pure fatigue and cyclic creep with additional thermomechanical fatigue stress component were load controlled and have been conducted at temperature of 650 °C. The effect of individual testing hold periods as well as combined effect of hold periods and temperature change in loading cycle were considered to evaluate the role of superimposed fatigue stress component onto creep in deformation process. Depending on testing conditions the controversial cycling frequency effect on deformation behaviour with respect to isothermal cyclic creep and thermomechanical fatigue process was found. The results of microstructural analysis may suggest that deformation behaviour would depend on stress level build-up in single loading cycle which in turn promote the dislocation mechanism to become operative. Fractography examination have shown that introduction of fatigue stress onto creep stress had been clearly proved to affect the crack nucleation process only for thermomechanical fatigue testing with shortest hold period and for pure fatigue.

KEYWORDS

Nickel base superalloy, isothermal cyclic creep, thermomechanical fatigue, creep-fatigue interaction, microstructure and fracture analysis

INTRODUCTION

The wrought nickel base superalloy EI 698 VD, the one of nickel base superalloys grades developed in Russian aircraft industry, is being used due to its attractive high temperature properties for turbine shafts and discs.

High temperature components assembled in aircraft turbine engines are subjected to complex stresses, including creep stress, fatigue stress and thermal stress, originating from centrifugal force, high frequency vibrations and temperature transients. A distinction must be drawn between isothermal high temperature fatigue (IF), and/or isothermal cyclic creep (ICC), as cycling straining under constant nominal temperature conditions, and thermomechanical fatigue (TMF), where strain-temperature waveform is classified according to the phase relation between mechanical strain and temperature.

Increased attention in last decade has been paid to study the creep-fatigue interaction in different either isothermal or anisothermal fatigue conditions. The introduction of hold times in a low cycle fatigue test at high temperature can be considered a frequency effect, and is the widely used method of studying creep-fatigue interaction in high temperature alloys. The net effect from introducing hold time is to systematically impose a creep component on the fatigue load cycling. There is abundance of information on introducing hold time of different periods under IF conditions. The introduction of tensile hold times during low cycle fatigue test has been shown to result in a decrease in the numbers of cycles to failure relative to continuous cycling [1,2,3], [4,5]. Only in rare cases does the hold times have no effect [3], [6,7], or introducing hold periods increases fatigue life [8,9]. But there is less information on thermomechanical testing carried out under load control where hold periods have been introduced either at maximum tensile or compressive stress. More of a valuable information are accessible on the strain controlled TMF.

In present study the isothermal cyclic creep, and thermomechanical fatigue tests with introduced hold time in tension were carried out in order to investigate the high temperature stability of nickel base superalloy and its deformation behaviour. The variables of load controlled tests were the cycling frequency versus hold times introduced at tensile amplitude peaks, whereas the stress range interval was kept constant. The TMF fatigue stress component was developed as the result of a forced temperature reduction between the individual hold periods when load was on and off. The high temperature deformation characteristics and damage mechanism were explained with respect to the additional effect of creep stress and temperature changing.

EXPERIMENTAL DETAILS

The alloy studied was a commercial wrought nickel base superalloy. Its chemical composition (in wt.%) is: 0.08 C, 13.0-16.0 Cr, 2.8-3.8 Mo, 1.3-1.7 Al, 2.3-2.7 Ti, 1.8-2.2 Nb, maximum 2.0 Fe, and balance Ni. The following heat treatment was performed: Solutioning at 1100 °C for 8 h and air cooling + Ageing at 1000 °C for 4 h and air cooling + Ageing at 775 °C for 16 h and air cooling. The heat treatment produced a uniform equiaxed grain structure. The gamma prime precipitates were uniformly distributed in the matrix as shown in Fig. 1. The sphere shaped precipitates of 40 to 80 nm in diameter had a full coherency with matrix. The volume fraction of gamma prime precipitates was determined to be over 40 %.

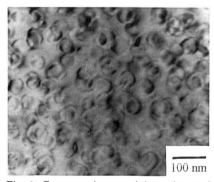
100 nm

Fig. 1. Gamma prime precipitates in matrix.

Solid, uniform gauge section, specimens with diameter of 5.0 mm were machined for all testing types. The pure creep, isothermal cyclic creep including fatigue, and cyclic creep with thermomechanical fatigue stress component were performed in air in creep testing machines, allowing the applied stress to remain constant during the testing. The testing temperature was 650 °C and was measured by a thermocouple located at the middle of specimen's gauge length. The strain was measured by means of an extensometer fixed on the specimen. The testing conditions for creep, cyclic creep and thermomechanical fatigue were following:

a) *Creep*. Creep tests were performed under applied stress of R = 740 MPa and test duration's were 41.6 h until rupture.

b) *Isothermal Cyclic Creep*. The load wave diagram as a function of time was of trapezoidal shape and unchanged until rupture, Fig. 2. Four different hold times $\Delta t = 0$ (pure fatigue), 1, 3, and 5 hours respectively at peak stress $R_{max} = 740$ MPa were introduced. The load rise and load fall time were 1 minute respectively. Cycling frequency range was between 9 x 10^{-3} and 3 x 10^{-5} Hz and stress ratio R = 0.027. The stress after load reduction was maintained at $R_{min} = 20$ MPa to keep the loading system in tension.

c) *Thermomechanical Fatigue*. The cycling schedule was the same for thermomechanical fatigue loading as for ICC testing, only the thermal fatigue stress component was introduced into the load waveform due to the specimen's forced air cooling prior to load reduction and reheating prior to reloading, Fig. 3. The testing procedure was as follows: heating (zero load) $\rightarrow$ load rising (01) $\rightarrow$ hold time (12) $\rightarrow$ cooling (prior to load reduction) $\rightarrow$ load reduction (23). The time for cooling and unloading in each cycle was 3 minutes, and for heating and loading was 2 minutes. The thermal cycles were generated by passage of an electric current, and the temperature was monitored throughout the test with spot-welded thermocouples. The longitudinal deformation was measured with a strain gauge extensometer fixed on the specimens. All tests were conducted until fracture. The specimen elongation, the fracture mode and fracture life, expressed in the numbers of cycles to fracture of individual specimens, were the criteria for the evaluation of the cyclic creep resistance in different conditions and compared with either creep or conventional fatigue characteristics.

The microstructural characteristics of fractured specimens were studied by transmission electron microscopy (TEM) of thin foils prepared from the gauge sections. The TEM analyses were used to reveal a dislocation structure and dislocation deformation mechanisms with respect to deformation characteristics obtained for applied testing procedure. The fractured surfaces were examined using scanning electron microscopy (SEM) with aim to evaluate the participation of fatigue stress component superimposed onto creep stress in fracture process.

RESULTS AND DISCUSSION

Creep Test

The creep tests were performed at five different stress levels. To carry out the ICC tests and thermomechanical fatigue tests the maximum stress level R = 740 MPa was chosen on the basis of creep results. The strain - time curve for pure creep is presented in strain - time diagrams for ICC and TMF either. The results on creep behaviour of the alloy are presented in Ref [10].

Isothermal Cyclic Creep

The strain - time curves for the isothermal cyclic creep tests for various hold times are presented in Fig. 2. The results might be treated and discussed either from creep behaviour or isothermal fatigue point of view. Considering isothermal fatigue behaviour, the obtained results are in conformity with results in [1-5], confirming the number of cycles to failure decreases as the hold time is increased. A possible mechanistic approach on the hold time introduction considers the conversion of elastic to plastic strain, i.e., stress relaxation occurs. In addition, there is a transition in fracture mode from transgranular to intergranular with introduction of tensile hold times [2]. On the other hand, evaluating the results by considering creep behaviour of the alloy the obtained strain - time dependencies are controversial as to the effect of hold time. The introduction of fatigue stress component (the hold times introduction effect) superimposed onto the stationary creep resulted in a positive change of fracture life and in strain rate decrease. The time to fracture is proportionally extended with duration of hold time. It would be rather assumed, if reduction in creep stress component , the more net fatigue behaviour would be expected, and that ought to result in life time extension and relative hardening too. But contrary to this the softening behaviour was found for shorter hold times (1 and 3 h), whereas the longer creep exposure resulted in life extension and strain rate reduction as well. The possible explanation of softening behaviour can be attributed to dislocation - precipitate interaction, as it can be seen in Fig. 3. The gamma prime precipitate shearing and Orowan bowing participated in dislocation surpassing the dispersed gamma prime for shorter hold times, and dense dislocation clustering was observed in specimen cycled with longer hold time. Creep - fatigue interaction and fracture behaviour depend on deformation process operating during cycling, which is influenced by the testing and alloy structure conditions. To explain the deformation behaviour of alloy during IC creep the introduction of the hold time at maximum strain can be considered as either stress relaxation [2], [11], i.e., when conversion of elastic to plastic strain takes place, or as cyclic creep acceleration based on on-load time, when hardened microstructure is formed due to repeated periods of primary creep. More frequent introduction of on-load periods might cause the creep acceleration [12,13].

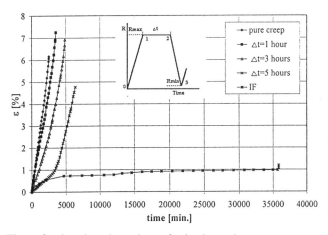

Fig. 2. Strain - time dependence for isothermal creep.

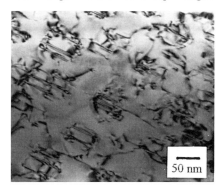

Fig. 3. Gamma prime precipitates shearing.

But relying on the to date information which literature review could provide there was not reporting on the cyclic strengthening effect occurred at very long hold times.

Thermomechanical Fatigue

Different trends in deformation behaviour were observed in the type of creep-fatigue interaction where both, the change of a thermal and mechanical stress were introduced into the deformation cycle. The results on deformation behaviour of alloy subjected to TMF are presented in Fig. 4. The temperature change which preceded to the mechanical strain change in loading cycle together with hold times introduction resulted in significant cyclic creep strain rate lowering for all introduced hold periods related to creep. The strengthening effect is becoming more pronounced as the frequency of cycling is risen. This fact could be employed as a proof of the reverse behaviour of alloy comparing IC creep with behaviour of alloy. The

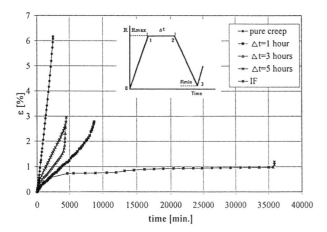

Fig. 4. Strain - time dependencies for thermomechanical
fatigue testing

more frequent temperature and strain change the higher reduction in the cyclic creep strain rate was observed. The additional internal stress component arising in specimen from the temperature change, either as result of temperature gradient across a section of specimen or caused by internal constraining, such as anisotrophy of the thermal expansion coefficient, appeared to have not detrimental effect relating the results with alloy's creep lifetime as well comparing them with ICC results. However, the introduction of thermal fatigue stress component together with mechanical strain, on the other hand, reduced the deformation resistance and lifetime considerably comparing it with isothermal fatigue.

Damage mechanism

The fracture analysis of the failed specimens in the case of isothermal cyclic creep with the different hold periods indicates that the dominant mechanism of the failure was governed by creep. All fractures, either crept or failed under ICC have characteristic intergranular crack initiation starting at the specimen surface. The propagation of the crack continued by intergranular mechanism till abrupt failure of specimen. The morphology of the fracture surfaces did not provide any evidence on the cyclic stress participation in crack nucleation or propagation process, except of continuous cycling without hold periods, where the fatigue stress component considerably affected the fracture process by presence of transgranular cleavage mode.

190

The similar fracture behaviour was observed at cycling with the thermomechanical simulation. The only fatigue stress participation at the crack nucleation process was observed in case

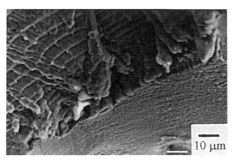

where 1 hour hold time was introduced, after that the fatigue striations on the fracture facet were clearly displayed, as shown in Fig. 5. Relating the fracture characteristics with mechanical testing results the appeared strengthening in individual testing, did not resulted in any change of fracture mode. The fracture mechanism for crack propagation and the final rupture was similar as for creep deformation.

Fig. 5. Fatigue crack nucleation at fracture surface subjecting to TMF 1 h hold time.

CONCLUSIONS

1. Introducing the cyclic stress component in creep process the controversial effect of hold time on deformation behaviour was found in ICC testing.
2. Introduction of tensile hold times in low cycle fatigue test has been shown to result in decreease in the number of cycles to failure relative to isothermal fatigue.
3. The reduction in the strain rate at isothermal cyclic creep is proportional to hold times. Small reduction in envelope strain was observed.
4. The reduction in the strain rate and the envelope strain was more pronounced in case of TMF cycling and it was proportional to the duration of hold times.
5. The fatigue stress component, resulted from introduced thermomechanical stress, was evidenced in fracture nucleation process, only for the shortes hold time.

REFERENCES

[1] Leveillant, C., Rezgui, C., and Pineau, A. (1977) *Mechanical Behaviour of Materials*, Pergamon Press, Oxford, 163
[2] Wareing, (1977) *J.Met. Trans. A*, **8A** 711.
[3] Rezgui, B., Petrequin, P. and Mottot, M. (1982) *Advances in Fracture Research*, Pergamon Press, New York, 2393.
[4] Day, M.F. and Thomas, G.B. (1979) *Met. Sci.* **13** 25.
[5] Zrník, J.,.Wang, Z.G., Žitňanský, M. and Hazlinger,M. (1995) *J. of Mater. Sci.* **11** 1 5.
[6] Remy, L., Rezai-Aria, F.,.Danzer, R. and Hoffelner, W. (1988) In: *Low Cycle Fatigue*, STP 942, H. Solomon et al.(Ed.), ASTM 1115.
[7] Bill, R.C., Verrilli, M.J., McGaw, M.A. and Halford, G.R. (1984), In: *Preliminary Study of MAR-M 200 Thermomechanical Fatigue of Polycrystalline* , TT-2280, NASA.
[8] Antolovich, C.D., Baur, R. and Liu, S. (1980) In: *Superalloys 1980*, ASM 605.
[9] Antolovich, C.D., Liu, S. and Baur, R. (1981) *Met. Trans A*, **12A** 473.
[10] Zrník, J., Wangyao, P., Vrchovinský, V. and Horňak, P. (1997) *Metalurgija* **36,** 4, 225.
[11] Ostergren, W.J. (1976) In: *Proceedings of the Symposium on Creep Fatigue Interaction*, R.M. Curran (Ed.), ASME-MPC, New York, 87.
[12] Evans, J.T. and Parkins, R.N. (1976) *Acta Met.*, **24** 511.
[13] Morris, D.G. and Harries D.R. (1978) *J. Mat. Sci.*, **13** 985.

Chapter 3

MULTIAXIAL LOADING

Low Cycle Fatigue and Elasto-Plastic Behaviour of Materials
K-T. Rie and P.D. Portella (Editors)

MICROMECHANICAL SIMULATIONS OF STAGE I FATIGUE CRACK GROWTH UNDER MULTIAXIAL LOADING.

V.DOQUET
Laboratoire de Mécanique des Solides.UMR-CNRS 7649
Ecole Polytechnique.91128 Palaiseau cedex. France.

ABSTRACT

Simulations of dislocations nucleation, glide and annihilation ahead of a stage I crack are performed for push-pull and reversed torsion, taking into account the presence of grain boundaries. An influence of the normal stress on crack flanks friction as well as on the condition for dislocation emission is introduced.The crack growth rates are deduced from the dislocation fluxes at the crack tip. A comparison between the loading modes is made.

KEYWORDS

Stage I, multiaxial loading, micromechanics, dislocations dynamics, grain boundaries.

INTRODUCTION

It is widely acknowledged that the reversed shear strain is the driving force for stage I fatigue crack growth but that the presence of an opening stress on highly sheared facets can assist the development of microcracks. The absence of such an opening stress in torsion is often considered responsible for slower stage I growth and longer fatigue life compared to push-pull, for equivalent strain ranges [1]. But the way the normal stress plays during Stage I is not clear, from the point of view of physical mechanisms.The aim of the present work is thus to suggest mechanisms by which the normal stress could influence Stage I crack growth and illustrate them through micromechanical simulations of dislocation glide in the plastic zone.

PHILOSOPHY OF THE SIMULATIONS

Stage I fatigue cracks grow along localized slip-bands. They are submitted to either pure shear loading (but mixed mode II+III anyway) in torsion or combined opening and shear in push-pull. But plasticity - that is dislocation nucleation and glide - is related only to the shear components and is restricted to the slip band colinear to the crack, otherwise (that is : if noncoplanar slip is activated) there is a transition towards Stage II propagation. As a consequence, and unlike the shear stresses, the singular opening stress is not relaxed (or shielded) by the dislocations stress field [2].

In the present simulations, only mixed mode I+II loading has been considered for sake of simplicity. It means that the dislocations that are emitted and glide along the coplanar slip plane have a pure edge character . They are supposed to be perfect dislocations. However, cross slip is not envisaged. Dislocations are supposed to be emitted from the crack tip as it has been observed during in situ tests performed in a transmission electron microscope [3]. In a first approach, decribed in details in [4] and briefly recalled here, no microstructural obstacle to dislocation glide was considered. In the present paper, grain boundaries will be introduced.

The author is aware of the limitations of Linear Elastic Fracture Mechanics as concerns short cracks. But, considering "long stage I cracks " though still in the size range where it undergoes the influence of the microstructure, like those encountered in torsional fatigue or in cyclically softening materials, stress intensity factors will be used. However, a "correction for plasticity" will be introduced, as shown below, through the stress field of dislocations .

Concerning the mode II stress intensity factor, allowance has to be made for the friction forces distributed along the crack flanks that make the **effective** stress intensity factor, K_{II}^{eff}, different from its **nominal** value, K_{II}^{nom} [5]. These friction forces exist, even in pure mode II, because of the asperities of the crack flanks that can form either when the crack adopts a zig-zag path between several parallel slip bands of a same grain or when it crosses a grain boundary and is thus slightly tilted (fretting debris have been observed to come out Stage I crack flanks [4]). Friction forces will of course be intensified in presence of a normal compressive stress and reduced if there is an opening stress.

Let us now consider that n edge dislocations with Burgers vectors $b_i = \pm b$ are present in the plastic zone. It can be shown [2] that they have a shielding effect on the stress distribution at the crack tip, characterised by

$$K_{II}^{tip} = K_{II}^{eff} - \sum_{i=1}^{n} \frac{\mu b_i}{(1-v)\sqrt{2\pi x_i}} \qquad (1)$$

where x_i stands for the distance of the ith dislocation to the crack tip. μ and v are the shear elastic modulus and Poisson's ratio respectively.

The criterion for dislocation nucleation at a crack tip was recently analysed by Rice on the basis of the Peierls concept. A tension-shear coupling equation, based on atomic models, was later established by Sun, Beltz and Rice [6]. According to them:

$$K_{II}^{nucl} = \sqrt{\frac{2\mu}{(1-v)}[\gamma_{us} - \alpha.(\gamma_{us}^u - \gamma_{us})(\frac{\pi}{2} - \psi)]} \qquad (2)$$

where $\psi = \arctan(\frac{K_{II}}{K_I})$ is the phase angle between mode I and II loadings.(In torsion, a pure shear case, $\psi = \frac{\Pi}{2}$, whereas for push-pull, at any time, the normal and shear stresses are equal in magnitude, but since K_I is zero when the normal stress is compressive, $\psi = \frac{\Pi}{4}$ during the tensile phase and $\frac{\Pi}{2}$ in the compressive phase). α is a positive constant that depends on the crystallographic and electronic structure of the material. γ_{us} is termed "the unstable stacking energy". It is defined as the energy increase per unit area of slip plane when the lattice on one side of the plane is shifted relative to the lattice on the other side, to the unstable equilibrium position at or near to a sliding displacement of b/2. Now, this can be achieved in two ways, either letting a certain amount of opening displacement occur (this is the **relaxed** case) or without any opening displacement (this is the **unrelaxed** case). the latter case corresponds to a value γ_{us}^u of γ_{us} which is higher than γ_{us} in the previous case. Equation 2 thus predicts a lower threshold stress intensity factor for dislocation nucleation when ψ decreases, that is when an opening stress is present.

In a first approach, only fully reversed loadings are envisaged , so that K_{II} changes sign. As a consequence, negative as well as positive edge dislocations can be emitted when, respectively

$$K_{II}^{tip} \leq -K_{II}^{nucl} \text{ or } K_{II}^{tip} \geq K_{II}^{nucl} \qquad (3)$$

Once emitted, these dislocations glide or rest according to the law:

$$v_i = v_0.\text{sign}(b_i).\text{sign}(\tau_i).\langle|\tau_i| -\tau_f\rangle^m \text{ where } \langle x \rangle \text{ is zero if } x \leq 0 \text{ and } x \text{ otherwise} \qquad (4)$$

where v_i is the velocity of the ith dislocation, τ_f is the lattice friction on the dislocations, v_0 and m two constants. τ_i is the shear stress on the ith dislocation, that is, according to Ohr [3]:

$$\tau_i = \frac{K_{II}^{eff}}{\sqrt{2\Pi x_i}} - \frac{\mu b_i}{4\Pi(1-\nu)x_i} - \sum_{j\neq i} \frac{\mu b_j}{2\Pi(1-\nu)} \sqrt{\frac{x_j}{x_i}} \cdot \frac{1}{x_j - x_i} \qquad (5)$$

The second term on the right handside of equation 5 represents the image stress on the dislocation due to the free surface of the crack, the third term is the shear stress due to the other dislocations (and their own image dislocations, whence the $\sqrt{\frac{x_j}{x_i}}$ term).

Annihilation between a positive and negative dislocation is allowed when their mutual distance is less than a critical value y_e. A dislocation will return to the crack tip and then annihilate if it approaches closer than a critical distance y_r. The cyclic loading path is followed by incremental time steps, Δt, small enough for the velocity of each dislocation to be considered constant over Δt. Two successive cycles only are simulated because the second cycle is representative of the steady-state .

The crack is considered to grow by one Burgers vector each time a pair of positive-negative dislocation has been emitted at the crack tip or when a positive (or negative) dislocation returns to the crack tip. In the latter case, it is assumed that even though the crack tip geometry before the dislocation nucleation is, in principle, recovered when this dislocation comes back, the free surface increment created at nucleation, that has been exposed to environment and gas adsorption in the meantime, cannot be rewelded afterwards. Anyway, both events (positive-negative pair emission or emission and return of a dislocation) correspond to some **cyclic** plastic flow at the crack tip and should thus contribute to its growth.

Grain boundaries (G.Bs.) are often considered as good sinks for lattice dislocations as well as dislocation sources. If the plastic zone at the crack tip reaches a grain boundary, dislocations are assumed to pile-up there, until the shear stress on the leading dislocation reaches a value allowing slip transfer into the next grain, on a slightly misoriented plane, this involving energetically costly dissociation reactions and leaving geometrically necessary dislocations at the G.B. that will not be taken into account. This naïve scheme avoids introducing an arbitrary position for a dislocation source in the next grain. The critical shear stress for slip transfer, $\tau_{G.B.}$, should depend on the crystallographic misorientation of the adjacent grains, and on the angle of incidence of the slip plane on the G.B. plane.

IMPLEMENTATION OF THE SIMULATIONS

The nominal stress intensity factors are calculated at each time-step as:

$$K_I = \sigma_{ncp}.\sqrt{\pi a} \text{ when } \sigma_{ncp} \geq 0 \qquad K_{II} = \tau_{cp}.\sqrt{\pi a} \qquad (6)$$

where σ_{ncp} and τ_{cp} are the current values of the normal stress and shear stress on the critical plane (i.e. the plane where the shear stress range is maximum).

Experimental data concerning crack flanks frictional interactions for a long crack loaded in mode II has been obtained in the framework of an other study [5]. It was shown that a the presence of a uniform friction stress, c, along the flanks could account for the measured displacement profiles. In the present study, an attempt is made to reproduce frictional effects **qualitatively** through empirical equations.The influence of the normal stress, σ_n, on the friction shear stress, c, distributed along the whole crack length: 2a, is expressed by:

$$c = c_0.\exp(-k_+.\sigma_n) \text{ if } \sigma_n \geq 0$$
$$c = c_0.\exp(-k_-.\sigma_n) \text{ if } \sigma_n \leq 0 \qquad (7)$$

c_0 is a constant which characterises the friction stress in the absence of any normal stress and is thus related to the tortuosity of the crack path . k_+ and k_- are two constants, the latter connected with the friction coefficient of the material. Most of the simulations will be performed with k_+ =0.057, k_- =0.014. (it means that c is divided by one hundred for $\sigma_n = 80$MPa, but it is multiplied by three only for $\sigma_n = -80$MPa)

It comes: $\qquad K_{II}^{eff} = K_{II}^{nom} \pm c\sqrt{\Pi a} \qquad (8)$

the sign affecting the second term depending on the loading (-) or unloading (+) situation. Of course, when loading changes direction, there will be periods where K_{II}^{nom} changes but K_{II}^{eff} remains constant, until the frictional resistance to reverse crack flanks slip is overcome.

Figure 1 shows how K_{II}^{nom} and K_{II}^{eff} vary in time during a reversed torsion and a push-pull cycle described respectively by: $\tau = \tau_0.\sin\omega t$ and $\sigma = \sigma_0.\sin\omega t$ with $\sigma_0 = 2\tau_0$ (equivalent stress ranges in the sense of Tresca) assuming that the crack lies along the maximum shear stress plane in each case. Note that ΔK_{II}^{nom} is the same in both cases, but that the effective loading cycle is asymmetrical in the push-pull case, due to enhanced friction in the compressive part of the cycle.

Sun, et al. [6] have calculated numerical values of γ_{us}^u, γ_{us}^u and α for various metals. For Al and Ni, $\frac{\gamma_{us}^u}{\gamma_{us}^u}$ is approximately 0.866 and $\alpha = 1.2$ so that equation 2 gives a nucleation stress intensity factor in the tensile phase of the push-pull cycle that is 92% of its value in torsion or in the compressive part of the cycle. (Note that the effect would be more pronounced for 90° out-of-phase tension and torsion loading - a case for which ψ varies continuously between 0 and $\frac{\Pi}{2}$, and consequently, K_{II}^{nucl} can be as low as 84% of its value in pure shear-. This case thus deserves a special study.)

K_{II}^{nucl} for pure mode II was set to 0.3 MPa√m in accordance with the order of magnitude given in [3].

The critical distance for positive-negative dislocations annihilation, y_e, was taken as 16 nanometer, which is the annihilation distance for edge dipoles in copper .The critical distance for dislocation annihilation at the crack tip, y_r, was chosen as 10^{-10} m, which is the order of magnitude of the core radius. A lattice friction τ_f of 20MPa, close to the local shear stress measured in the channels of persistent slip bands in copper was chosen.The constants in the expression of dislocations velocity (eq. 4) are chosen as $v_0 = 13 ms^{-1}$ and m=0.88 which are typical experimental values for FCC metals.

RESULTS

Whithout any obstacle to dislocation glide.

The dislocation dynamics during push-pull and reversed torsion loadings is analysed in details in [4]. Only global results will be given here. Simulations have been carried out for reversed torsion for various loading ranges and various friction stresses c_0.The results are drawn on a bilogarithmic da/dN versus ΔK_{II}^{nom} plot on Fig.2 (curves labelled B,E,F).

It can be seen that the existence of a threshold stress intensity range below which the crack does not propagate for lack of cyclic plasticity at the tip, as well as the well known shape of da/dN versus ΔK curves in the vicinity of this threshold are qualitatively reproduced by the simulations. In addition, the influence of the friction stress along the crack flanks is illustrated: quite similarly to the closure effects for mode I propagation, the friction effects shift the da/dN versus ΔK_{II}^{nom} curve compared to the "intrinsic" curve corresponding to zero friction. In the absence of quantitative data on the friction stresses along the flanks of a stage I crack, a discussion on the calculated values of the threshold stress intensity factors would be premature, and what is more, experimental values of mode II thresholds in single crystals, that could be compared to the calculated values, seem to be rare.

Several simulations were performed for push-pull with various loading ranges and friction stresses, c_0. The results are compared to torsional results on Fig.2 (curves labelled A,C,D). Comparing the curves labelled A and B, it appears that even without friction, the crack growth rates are higher in push-pull than in reversed torsion, but the difference, appreciable for small loading ranges, vanishes as the loading range increases. This is probably because close to the threshold, the critical stage for crack propagation is dislocation **nucleation**, which is made easier by the opening stress, in push-pull, whereas for higher loads, nucleation is no more critical, but **reverse dislocation glide**, more natural under reversed torsion because of the symetry of the effective loading, is. According to this trend, the ratio between torsional and tensile fatigue lives should thus increase as the loading range decreases. This corresponds to experimental observations [1] and it is thus very encouraging. When the influence of the normal stress on crack flanks friction is introduced, the difference in Stage I kinetics between the two loading modes is increased, and this difference is amplified when the friction stress c_0 increases.

Interactions with grain boundaries

Simulations of stage I growth at a constant stress range (and thus increasing ΔK_{II}^{hom}) have been performed, in reversed torsion, for a crack of initial length $2a = 200\mu m$ in a polycrystal with a $50\mu m$ mean grain size. The results are plotted on Fig.3 for various values of the critical stress for slip transfer at grain boundaries (these values where chosen only to illustrate qualitatively different behaviours but are not considered meaningful themselves). If the value of $\tau_{G.B.}$ is small enough, given the initial ΔK_{II}^{hom}, the crack growth rate is not influenced by the microstructure. If $\tau_{G.B.}$ is high enough, the crack growth rate first increases, as long as the plastic zone does not reach the grain boundary. Then it decreases, since the dislocations piled-up against the G.B., closer and closer from the crack tip, inhibit the emission of new dislocations due to their shielding effect (see equation 1). Finally, the crack becomes non propagating. If $\tau_{G.B.}$, has a medium value, the deceleration is only temporary and followed by an acceleration,when the conditions for slip transfer are met. This occurs earlier and earlier in successive grains because of the increase in ΔK_{II}^{hom} associated with an increase in the number of dislocations emitted and piled-up against the G.B., so that the fluctuations in the crack growth rate progressively disappear, and that the kinetics becomes independent of the microstructure, as in the first case.
It has been observed, on a Co45Ni alloy, that for equivalent stress ranges, it takes approximately twice as long for microcracks to cross the first G.B. encountered in reversed torsion than in push-pull [1]. This can be explained by the simulations. Figure.4 compares the calculated shear stress concentrations due to the dislocations emitted by a microcrack and piled-up against the G.B.(of infinite resistance) at $5\mu m$ from the G.B.in the next grain, for reversed torsion and push-pull of equivalent stress ranges. Much more dislocations are emitted in push-pull, because of the influence of the opening stress on both K_{II}^{eff} and K_{II}^{nucl}, so that the stress concentration is much higher, this making slip transfer, and thus G.B. crossing easier.

CONCLUSIONS

The development of Stage I cracks by dislocations emission/annihilation at the tip has been simulated, taking into account their interactions with grain boundaries. da/dN versus ΔK curves were obtained for push-pull and reversed torsion . The existence of a threshold stress intensity factor for crack growth, as well as the shape of the da/dN versus ΔK curves in the vicinity of the threshold are qualitatively reproduced. The influence of friction stresses along the crack flanks is also illustrated: it shifts the da/dN versus ΔK_{II}^{hom} curve compared to the "intrinsic" curve corresponding to zero friction. The periodic decelerations, sometimes leading to crack arrest typical of microstructure sensitive Stage I propagation were simulated. Stage I is predicted to be slower under reversed torsion than under push-pull for equivalent stress (or strain) ranges in the sense of Tresca, because of slower transgranular growth rate and more difficult G.B. crossing. The ratio between torsional and tensile fatigue lives is predicted to increase as the loading range decreases, consistently with experimental data on many materials. It should also depend on the roughness of Stage I cracks and the frictional properties of the material.

REFERENCES

1.   Doquet ,V.(1997) Fat.Fract.Engng.Mater.Struct.**20**, 227
2.   Lin, I.H. and Thomson, R. (1986) Acta Metall.**34**, 187
3.   Ohr, S.M.(1985). Materials Science and Engineering,**72**, 1
4.   Doquet, V. (1998) Fat.Fract.Engng.Mater.Struct.**21**,
5   Pinna,C.,Doquet,V.(1997) Proc.5th Int.Conf.Biaxial/Multiaxial Fat/Fract.Cracow,Sept 8-12, Vol.2,pp 97-113, E.Macha, Z.Mroz eds
6.   Sun,Y.,Beltz,G.E. and J.R.Rice,J.R.(1993) Mat. Sci. Eng.,**A170**, 67

198

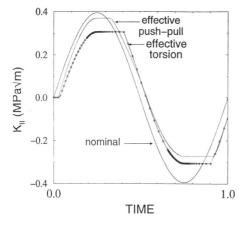

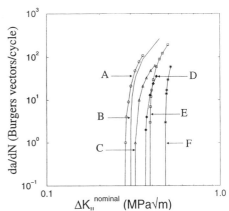

Fig.1 Variation of nominal and effective mode II stress intensity factors for a crack lying along the maximum shear stress plane under reversed torsion and push-pull loadings of equivalent amplitude in the sense of Tresca.

Fig.2. Comparison of the calculated crack growth rates versus nominal K_{II} curves (A) push-pull, no friction (B) reversed torsion, no friction (C) push-pull, c_0=5MPa (D) push-pull, c_0=10MPa (E) reversed torsion, c_0=5MPa (F) reversed torsion, c_0=10MPa.

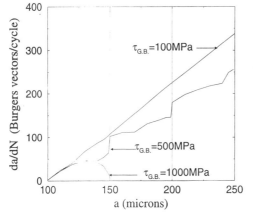

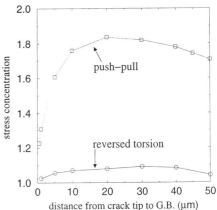

Fig.3. Evolution of the crack growth rate of a torsional stage I crack of initial length 200μm in a polycrystal of 50μm mean grain size for various critical stress for slip transfer at G.Bs.

Fig.4. Stress concentration in the next grain, at 5μm from the G.B., due to the dislocations piled-up at the G.B., versus the distance between crack tip and G.B.

Low Cycle Fatigue and Elasto-Plastic Behaviour of Materials
K-T. Rie and P.D. Portella (Editors)

INFLUENCE OF STRESS CONCENTRATION AND MULTIAXIAL STRESS STATE ON LOW CYCLE FATIGUE

S. BARAGETTI and L. VERGANI

Dipartimento di Meccanica, Politecnico di Milano
Piazza Leonardo da Vinci 32, 20133 Milano, Italy

ABSTRACT

Torsion tests were performed on thin tubular specimens made of 34CrMo4 steel. The low-cycle curves obtained from torsional tests were compared with the ones obtained from axial tests. Experimental tests were executed on notched specimens made of the same material and the fatigue behaviour was studied. Microscopic observations and profile measurements of the fracture surfaces permitted to put in evidence the different kind of fractures and the fracture surface changes induced by the shape of the specimen and the load amplitude. The experimental tests together with a numerical analysis (conducted by using a finite element model) enabled the determination of all the useful parameters to perform a low-cycle fatigue life prediction. The calculation of the multiaxial stress and strain state near the notch was performed by the Neuber's rule and the ESED method.

KEYWORDS

Low cycle fatigue, stress concentration, multiaxial stress state, life prediction

INTRODUCTION

Low cycle fatigue loaded mechanical components present plasticisations that especially occur near notches, where the stress state is multiaxial. This provokes crack initiation and propagation until rupture after few load cycles. Due to the plasticisation the strain amplitude has to be taken into account to estimate the mechanical components life. The method usually adopted is the one based on the local strain value and on the Manson and Coffin curves, experimental curves obtained from smooth axially loaded specimens [1], [2]. The employment of this method on notched components is not trivial. In fact the determination of the stress and strain state on notched components can be evaluated only by means of finite element analyses or by using approximated literature methods as Neuber's rule and the *ESED* method that are the extreme limits of the effective behaviour. Besides, once known the strain amplitude to introduce in the Manson-Coffin curves, the fatigue life that is determined is often not similar to the experimental values due to the presence of the multiaxial stress state and of the strain and stress gradient. The fatigue lives instead are obtained from analyses carried out on smooth specimens. In a previous work [3] was shown that if the experimental strain-amplitude values were entered in the Manson-Coffin curves, the corresponding lives did not agree with the experimental values. In this work the same analysis is conducted on smooth and notched torsional loaded specimens in order to consider a multiaxial stress state too.

Another method is the energy based one; in particular in [4], [5] the total deformation work W_t is used as a parameter characterising the low-cycle fatigue loaded components [6], [7]. This criterion assumes the employment of a correction coefficient in order to take the multiaxial stress and strain state into account; besides it can be applied both to low cycle and to high cycle fatigue. Other researches have tried to define a critical plane or, better, a plane in which the cracks were observed to form and grow [8], [9]. In [8] the strain energy density W^* is related to the critical plane which is assumed to be the plane of maximum shear strain.

The aim of this work is to study the problem of the prediction of the low cycle fatigue life of notched components; the work develops through several experimental tests executed on specimens made of 34CrMo4 steel which has a martensitic structure. During the torsional experimental tests the principal strains were measured while a numerical analysis permitted to know the complete stress state. Besides it was possible to apply the literature criteria to these results in order to obtain the fatigue life of the notched specimens.

MATERIAL AND EXPERIMENTAL PROCEDURES

The material tested in the experimental analyses is the 34CrMo4 steel (EN 10083), furnished in 100 mm diameter bars, and it was quenched at 850 °C for 1h and tempered at 600 °C for about 1.5h. Tensile tests were carried out according to the EN 10002 standard in order to determine the static properties of this material, while hardness measurements enabled to obtain the surface hardness values. The values obtained are the following: R_m=1074 MPa (ultimate tensile strength), $R_{p0.2}$=1022 MPa (0.2% offset yield strength), E=206000 MPa (Young modulus), ν=0.3 (Poisson coefficient), 32 HRC (hardness).

Smooth specimens, according to the ASTM E606-92 Standard, were used to determine the cyclic behaviour of the material which was established by carrying out experimental tests in deformation control on a Instron 8501 machine. The values obtained are the following: K'=1200 MPa (hardening coefficient), n'=0.098 (hardening exponent), σ_f'=1730 MPa, b=-0.104, ε_f'=0.893, c=-0.749 (Manson-Coffin parameters).

In Figure 1 the comparison between the cyclic (a) and the Manson-Coffin (b) curves obtained from the tensile and the torsional tests is reported.

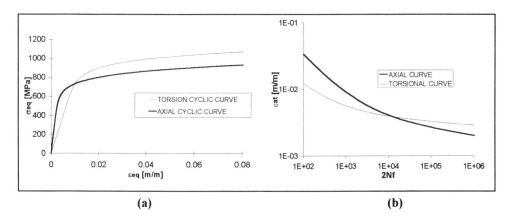

(a)   **(b)**

Fig. 1. (a) Cyclic and (b) Manson-Coffin curves obtained from the tensile and the torsional tests

Figure 2 shows the smooth tubular and the notched specimens employed in the experimental analyses.

All the specimens were polished in order to obtain a mean surface value of roughness equal to R_a = 0.2 μm.

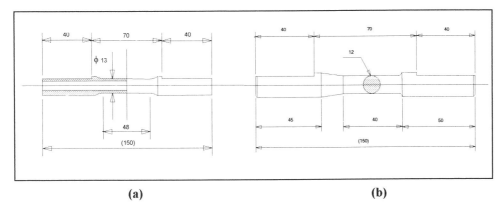

(a)   **(b)**

Fig. 2.   Smooth tubular (a) and notched (b) specimens used in the experimental analyses

The torsional alternating load is applied by using an axial test machine (opportunely modified in order to enable the execution of the torsional fatigue tests).The specimens were equipped with both axial and torsional strain gages positioned in the middle of the smooth specimens and near the notch in the notched specimens. Six smooth tubular specimens were tested and the results obtained are listed in Table 1 (τ is the applied stress, N_f is the number of cycles to failure, γ_{at} is the total deformation amplitude, γ_{ap} is the plastic deformation amplitude, ε_{ateq} is the equivalent total strain amplitude, W_t is the total deformation work, W^* is the strain energy density).

Table 1.   Results of the torsional tests carried out on the smooth specimens

Spec. N°	τ [MPa]	N_f	γ_{at} [μm/m]	γ_{ap} [μm/m]	ε_{ateq}[μm/m]	W_t[MJ/m³]	W^* [MJ/m³]
1	477	74	14124	6692	8769	10.790	6.800
2	500	154	14037	7229	8664	7.830	7.209
3	492	205	15300	8900	9464	9.948	7.856
4	373	1380	5946	619	3898	1.625	2.290
5	437	1970	6875	1220	4458	2.308	2.935
6	357	30043	4668	166	3660	0.924	1.648

The parameters K' and n' of the cyclic curve $\gamma_a = \dfrac{\tau_a}{G} + \left(\dfrac{\tau_a}{K'}\right)^{1/n'}$, obtained from the torsional tests

carried out on the smooth specimens, are respectively equal to 747 MPa and 0.087. In Figure 1 the cyclic and the Manson-Coffin curves obtained from the torsional tests are compared with the ones previously obtained from the axial tests.

The stress intensity factor for the notched specimens, calculated by using the maximum value of deformation in the notched section, is equal to K_t=1.26. Literature data [10] furnish the value K_t=1.24.

The experimental tests on the notched specimens resulted to be very expensive. At the moment only one specimen was tested and the results are reported in Table 2 (γ_{ae} is the elastic shear strain amplitude).

Table 2.   Results of the torsional tests carried out on the notched specimen

Spec. N°	τ [MPa]	N_f	γ_{at} [μm/m]	γ_{ap} [μm/m]	γ_{ae} [μm/m]
1	373	4490	6319	623	5696

The cracks observed on the smooth specimens were all along the maximum shear strain direction (both longitudinal and circunferential) while on the notched specimen only one circunferential crack initiated and propagated from the surface in correspondence of the notched section. Figure 3 shows the fracture surfaces of the smooth specimen N° 3 (a) and of the notched one (b). The observation of the fracture surface of the smooth specimen, subjected to a high load, puts in evidence regions of different crack propagation speed (longitudinal crack). Figure 3b, in which the fracture surface of the notched specimen is reported, enables the observation of three distinct regions: the first one, near the surface of the specimen, is limited to a thin layer of material while the second one interests almost half the fracture surface area; the third region, positioned in centre of the specimen, corresponds to the zone of static rupture. In the first thin region the crack initiated, even if a clear initiation point was not observed, and propagated in the second internal region.

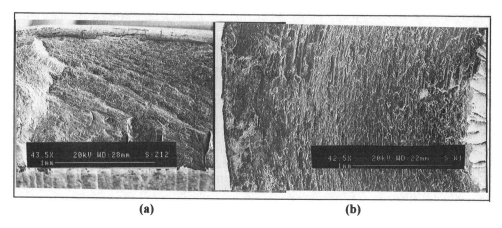

(a)   (b)

Fig. 3.   Fracture surfaces of a smooth specimen (a) and of the notched one (b)

In order to determine the different mechanisms of initiation of the fatigue crack in the surface and of propagation in the internal region of the specimen, roughness measurements of the fracture surface were carried out. In Figure 4 two samples of the profile of the fracture surface concerning the two external regions of the specimen, recorded by means of a profilometer, is reported.

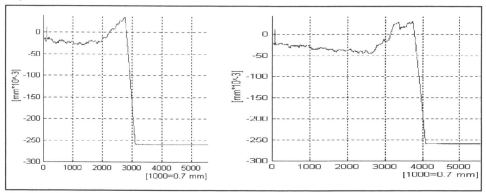

Fig. 4. Profiles of the fracture surface, recorded by means of a profilometer, concerning the two external regions of the notched specimen

From these observations of the profiles of the fracture surface it is evident that the external thin region is positioned on a different layer with respect to the internal region of propagation of the crack. Besides the external region presents an higher roughness, even if it was not possible to carry out a surface roughness measurement because of the limited dimensions of the area of interest. Probably the crack initiates on planes of maximum shear stress (external region of the specimen) and propagates on a different plane normal to the specimen axis (internal region of the specimen).

NUMERICAL ANALYSES

A finite element model (see Figure 5) of the notched specimen was executed in order to evaluate the results of the experimental torsional tests. The mesh was refined in correspondence of the notch and the analyses carried out with a linear elastic behaviour of the material furnished a stress intensity factor equal to $K_t=1.25$ which is in a good agreement with experimental literature values.

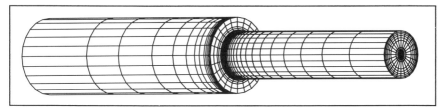

Fig. 5.   Finite element model of the notched specimen employed in the numerical analyses

In the elasto-plastic analyses the monotone stress-strain curve was used only in the first load cycle, while the cyclic curve (kinematic hardening) was used for the following cycles. Two kind of analyses were carried out: in the first one the cyclic behaviour was assigned to the whole model after the first cycle while in the second one the cyclic behaviour was assigned only to the region of the specimen which had plasticised during the first cycle. The results obtained from the numerical analyses of the two different approaches are almost the same and only the values related to the second one are considered. The *ESED* (Equivalent Strain Energy Density) method [11,13] and Neuber's rule were employed to evaluate the stress intensity factor in correspondence of the notch for different values of the torsional applied load. Figure 6 shows the results obtained from the numerical analysis, the experimental tests and from the application of the *ESED* method and of the Neuber's rule respectively for maximum applied torsional moment 112 Nm.

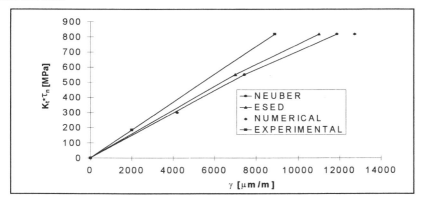

Fig. 6. Comparison of the numerical and the experimental results

The results of the numerical analyses are near to the ones obtained by applying Neuber's rule; for higher torsional loads the difference becomes greater. This is probably due to the fact that the literature methods are utilisable only in the cases of rather limited plasticised regions and, for high torsional applied moments, the plasticised zone is extended. In the same figure the experimental trend measured on the notched specimen is shown. It seems that the shear strain values are lower than the ESED ones, but the results are related to only one specimen and they are meaningless.

CONCLUSIONS

Experimental axial and torsional low cycle fatigue tests were carried out on 34CrMo4 steel (EN 10083). The stress-strain and the cyclic curves of the material were obtained and the material showed a cyclic softening behaviour. The Manson-Coffin curve was drawn out too.

The torsional tests carried out both on smooth tubular specimens and on notched ones enabled the determination of the torsional cyclic curve of the material and permitted to put in evidence the influence of the stress concentration on the fatigue behaviour of the specimens. The observation of the fracture surfaces and the fracture surface profile measurement (by means of a profilometer) enabled to put in evidence the different phases of crack initiation and propagation both on smooth and on notched specimens. Numerical analyses were carried out in order to evaluate the stress concentration in correspondence of the notch of the notched specimens. The *ESED* method and Neuber's rule were employed to evaluate the stress intensity factor in correspondence of the notch for different values of the torsional applied load and permitted to put in evidence that the results of the numerical analyses are near to the ones obtained by applying Neuber's rule.

BIBLIOGRAPHY

[1] You, B.R., Lee, S.B. (1996). *Int. J. Fatigue* **18**, N° 4, pp 235-244.
[2] Sehitoglu, H., Gall, K., Gargia, A.M. (1996). *Int. J. Fracture* **80**, pp. 165-192.
[3] Giglio, M., Vergani, L. (1995). *Trans. ASME „J. Engng. Mater. Technol.* **117**, pp. 50-55.
[4] Ellyin, F., Golos, K., (1991). *Trans. ASME, J. Engng. Mater. Technol.* **113**, p. 112.
[5] Ellyin, F., Xia, Z., (1993). *Trans. ASME, J. Engng. Mater. Technol.* **115**, p. 411.
[6] Liu, K.C., (1993). *Advances in Multiaxial Fatigue*, ASTM STP 1191, D.L. McDowell and R. Ellis. Eds, American Society for Testing and Materials, Philadelphia, pp. 67-84.
[7] Doquet, V., Pineau, A., (1991). *Fatigue Under Biaxial and Multiaxial Loading*, ESIS10 (Edited by K. Kussmaul, D. McDiarmid, D. Socie), Mechanical Engineering Publications, London, pp. 81-101.
[8] Glinka, G., Shen, G., Plumtree, A., (1995). *Fatigue Fract. Engng. Mater. Struct.* **18**, N° 1, pp. 37-46.
[9] Socie, D. (1993). *Advances in Multiaxial Fatigue*, ASTM STP 1191, D.L. McDowell and R. Ellis. Eds, American Society for Testing and Materials, Philadelphia, pp. 7-36.
[10] Peterson, R. E., (1974). *Stress Concentration Factors*. John Wiley & Sons, New York.
[11] Moftakhan, A., Buczinski, A., Glinka, G. (1995). *Int. J. Fracture* **70**, pp. 357-373.
[12] Glinka, G. (1985). *Engng. Fract. Mech.* **22**, N° 3, pp. 485-508.
[13] Glinka, G., Sheen, G., Plumtree, A., (1995). *Fatigue Fract. Engng. Mater Struct..* **18**, N° 1, pp. 37-46.

AN EVALUATION OF METHODS FOR ESTIMATING FATIGUE LIVES UNDER MULTIAXIAL NONPROPORTIONAL VARIABLE AMPLITUDE LOADING

DARRELL SOCIE
Mechanical Engineering Department
University of Illinois at Urbana-Champaign
1206 West Green
Urbana, Illinois 61801, USA

ABSTRACT

Estimating fatigue lives under variable amplitude multiaxial nonproportional loading requires consideration of a number of issues. There are four components needed to estimate fatigue lives for variable amplitude nonproportional multiaxial loading just as there are in uniaxial loading:

cyclic stress strain model,
cycle counting method,
damage model, and a
damage accumulation model.

This paper focuses on the entire process and makes an assessment of the life estimation procedure and identifies areas that are reasonably well understood and other areas where our understanding and modeling need improvement..

KEYWORDS

multiaxial, fatigue damage, nonproportional loading, cyclic plasticity

CYCLIC PLASTICITY MODEL

A complete cyclic plasticity model will be composed of three major components: a *yield function* to describe the combinations of stress that will lead to plastic flow, a *flow rule* to describe the relationship between the stresses and plastic strains during plastic deformation, and a *hardening rule* to describe how the yield criterion changes with plastic straining. Most of the multiaxial models use similar yield functions such as Mises and flow rules such as the normality postulate by Drucker. They all work reasonably well for proportional loading without mean stresses even though they have different hardening rules. Histories with combined cyclic torsion and static axial mean stresses cause problems for many models and the solution is unstable. An example of such a loading history is shown in Fig.1. A static axial strain is combined with a cycle of shear strain and unloaded back to zero strain. Then an axial strain cycle is combined with a static shear strain. The loading history is repeated in compression. This loading history involves nonproportional hardening, ratcheting and stress relaxation leading to the complex stress response shown in the figure. Stresses were computed from the Jiang and Sehitoglu[1] plasticity model. Six loading cycles are shown before the stress response stabilizes.

206

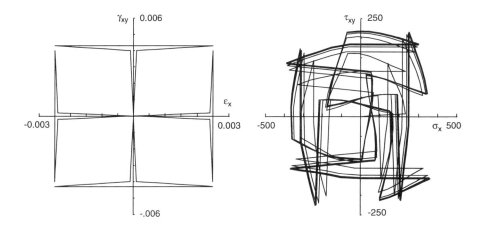

Figure 1 Strain history and computed stresses

One good way to judge a plasticity model is to use the same model in both stress and strain control. That is, use the model to estimate the stress history from the input strains. Then use these computed stresses as input to a stress controlled version of the same model and make an estimate of the strains. The output strains of the stress controlled model should be identical to the input strains to strain controlled model. Results of this procedure for the Jiang and Sehitoglu model are shown in Fig. 2 and the ability of the model to predict the original loading history is good. This simple technique provides a way to quantify the accuracy of the stress strain model for a complex loading history. Real materials can be loaded in either stress or strain control and plasticity models that can only be formulated in either stress or strain control should be used with caution.

CYCLE COUNTING

The most widely used cycle counting scheme for uniaxial loading is the rainflow method. Two multiaxial variations of the method have been proposed. Bannantine and Socie[2] proposed that any nonproportional loading could be resolved into stresses and strains acting on a plane in the material. Once the stresses and strains have been resolved on a plane, conventional rainflow counting procedures that can account for the complex stress response can be used to compute fatigue damage on each plane. Wang and Brown[3] take a different approach where they use an equivalent strain. To overcome the problem with the sign of the equivalent strain they compute a relative equivalent strain by using a reference strain. Both methods have the ability to identify cycles and reduce to rainflow counting for proportional loading. Both methods have some difficulty in defining the mean stress for a cycle. How to determine the mean stress for a asymmetric hysteresis loop is an open question. For proportional loading both cycle counting methods will give the same cycles. Both methods were used to count cycles for the loading history shown in Fig. 1. The shear strain for the 20° plane is shown in Fig. 3 where the major cycle and four subcycles are easily identified.

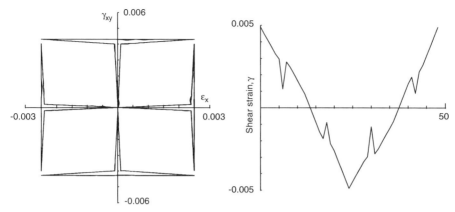

Figure 2 Computed strain history   **Figure 3** Shear strain on the 20° plane.

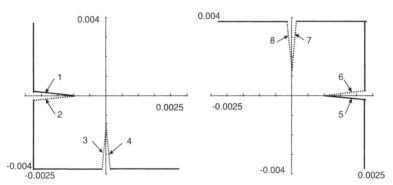

Figure 4 Cycles identified by Wang and Brown method

Reversals identified by the Wang and Brown method are shown in Fig. 4. The solid line is counted as one reversal. Smaller amplitude reversals are shown with the dotted lines. Both methods identify the same major cycle and differ in the number and magnitude of the smaller cycles identified. Most of the fatigue damage is done by the larger cycles and both methods give similar cycle counts for typical service loading histories.

DAMAGE MODEL

Most of the damage parameters have been originally formulated to provide correlation with combined tension torsion fatigue data for simple loading paths. Damage is defined on the plane experiencing the largest range of shear or normal strain. They are then modified for either normal stresses or strains on this plane. Chu[4] notes that all critical plane parameters should be based on the maximum value of the damage parameter. If the effect of a normal stress or strain is to enhance shear crack growth, then this effect should be present on all shear planes. A plane with a slightly lower cyclic shear strain could have a much larger normal stress and strain and thus have more damage and be more likely to form and grow a crack.

208

During nonproportional loading many planes are damaged, not just the plane experiencing the maximum shear strain range. Figure 5 shows the distribution of fatigue cracks for an in-phase and 90° out-of-phase nonproportional loading history obtained by Ohkawa et. al [5] for 1045 steel with a tensile strength of 590 MPa. The solid lines in Fig. 5 show the relative magnitude and distribution of normal strain with orientation and the dashed line shows the magnitude and distribution of shear strain. The bar chart shows the measured distribution of small cracks at about 25% of the fatigue life.

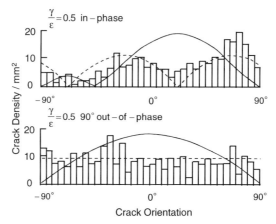

Figure 5 Distribution of fatigue damage

For in-phase proportional loading, there is a tendency for more cracks to form on planes where the cyclic shear strain is highest. There is no preferred orientation of cracks for 90° out-of-phase nonproportional loading. These observations show that a single loading cycle damages many planes during both proportional and nonproportional loading. The magnitude of the damage as evidenced by the number of cracks is dependent on orientation of the plane with respect to the loading. This leads to a concept of computing fatigue damage on all planes for each loading cycle and then defining the critical plane as the plane experiencing the maximum damage. Models that define fatigue damage on a single plane do not match the physics of the process. All of the critical plane models can be formulated in terms of the maximum value of the damage parameter and these formulations should be used in methods that search a variable amplitude loading history to find the critical plane.

DAMAGE ACCUMULATION MODEL

Fatigue damage must be accumulated for each cycle in the loading history. Miner's linear damage rule is used with all of the methods that are currently in use. Its limitations are well known. High amplitude cycles followed by small amplitude cycles are more damaging than small amplitude cycles followed by high amplitude ones. In nonproportional loading, additional effects have been observed for tension torsion loading sequences as shown in Fig. 6. Harada and Endo [6] tested 1045 steel under combinations of rotating bending and cyclic torsion. Stress levels for the tests are given in the figure. For example, the solid circle denotes a test conducted with an initial tension cycling at 250 MPa followed by torsion cycling at 125 MPa. Fatigue damage for each part of the test is determined from the linear damage rule and plotted in the figure. The linear damage rule is shown by the straight line. Data to the right side of the line have cycle ratio's greater than one and the use of Miner's rule results in conservative life estimates. These results show that torsion followed by tension ($\tau \rightarrow \sigma$) is much more damaging than tension followed by torsion ($\sigma \rightarrow \tau$).

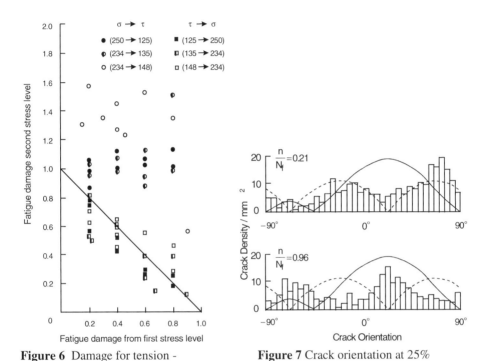

Figure 6 Damage for tension - torsion sequences

Figure 7 Crack orientation at 25% and 95% of the fatigue life

Figure 7 shows the distribution of cracks at both 21% and 96% of the fatigue life. The majority of the cracks nucleate on planes of with the largest shear strain ranges. Later in the life, more cracks are found on planes of the larger principal strains. The number of cracks on the maximum shear strain planes actually decreases as the life becomes longer. Final failure results from the growth and coalescence of many small cracks. Initially these cracks grow and coalescence on the maximum shear strain planes. Early in the life, more cracks nucleate than coalesc and the overall crack density increases. Later in life, the shear cracks turn to become mode I tensile cracks. The cracks in the lower half of the figure on planes of maximum principal strain actually started out as shear cracks.

The test data in Fig. 6 can be explained by noting that torsion loading nucleates many small cracks on planes where subsequent tensile loads cause them to grow. Tensile loads do not nucleate cracks on planes where torsion loads would cause them to propagate.

There have been a few attempts to model this type of behavior. Robillard and Cailletaud[7] are one of the few to propose a multiaxial damage model based on these observations.

Damage $D(\theta)$ on each plane θ is defined by an incremental growth law. Two types of damage are considered, shear damage D_τ and tensile damage, D_σ.

$$dD(\theta) = dD_\tau (\theta) + dD_\sigma(\theta) \qquad (1)$$

Damage is directionally dependent; each plane has a fatigue damage associated with it. On each plane, the critical variable in the beginning of the life is shear stress or strain and the propagation of microcracks at the end of the life is governed by the normal stress or strain.

Failure occurs when damage reaches a critical value, D_c. They propose specific functions for D_τ and D_σ but they are complex and require many material constants.

SUMMARY

This paper summarizes the primary issues involved in assessing fatigue damage for variable amplitude nonproportional loading. Space limitations prevent a detailed discussion of each issue but I believe that the basic physics of the failure process is reasonably well understood from a qualitative viewpoint. For time independent behavior, quantitative models of deformation are adequate for estimation fatigue damage. However, there is little quantitative modeling of the fatigue damage accumulation process that considers both shear and tensile damage and the transition between the two.

REFERENCES
1. Jiang and Sehitoglu "Modeling of Cyclic Ratcheting Plasticity, Part I: Development of Constitutive Equations" Journal of Applied Mechanics, Vol. 63, 720-725, 1996
2. Bannantine, J.A. and Socie, D.F, "A Variable Amplitude Multiaxial Fatigue Life Prediction Method" Fatigue Under Biaxial and Multiaxial Loading, ESIS10, 1991, 35-51
3. Wang, C.H. and Brown, M.W. "Life Prediction Techniques for Variable Amplitude Multiaxial Fatigue - Part 1: Theories" *Journal of Engineering Materials and Technology*, Vol. 118, 1996, 367-370
4. Chu, C.-C., "Fatigue Damage Calculation Using the Critical Plane Approach" Journal of Engineering Materials and Technology, Vol. 117, 1995, 41-49
5. Ohkawa, I, Takahashi, H., Moriwaki, M., Misumi, M., "A Study Of Fatigue Crack Growth under Out-of-phase Combined Loadings" *Fatigue and Fracture of Engineering Materials and Structures*, Vol. 20, No. 6, 1997, pp. 929-940.
6. Harada, S. and Endo T, "On the Validity of Miner's Rule under Sequential Loading of Rotating Bending and Cyclic Torsion, ESIS 10, 1991, 161-178
7. Robillard, M., and Cailletaud, G., "Directionally Defined Damage in Multiaxial Low Cycle Fatigue: Experimental Evidence and Tentative Modeling", ESIS 10, 1991, 103-130

Low Cycle Fatigue and Elasto-Plastic Behaviour of Materials
K-T. Rie and P.D. Portella (Editors)

INFLUENCE OF LOAD- AND DEFORMATION-CONTROLLED MULTIAXIAL TESTS ON FATIGUE LIFE TO CRACK INITIATION

C.M. SONSINO
Fraunhofer-Institut für Betriebsfestigkeit (LBF)
Dept. Component Related Material Behaviour
Bartningstr. 47, D-64289 Darmstadt / Germany

ABSTRACT

Generally, areas of components with notches or geometrical transitions are critical because of the resulting stress/strain concentrations. In these areas due to the stress-gradients and constraint local deformations are displacement controlled even if the material's yield stress is exceeded, as long as the deformations are below the structural yield point. Therefore, load controlled tests in the elasto-plastic region with unnotched specimens from ductile materials under combined axial loading and torsion are not suitable for the interpretation of component's behaviour because of uncontrolled local deformations. Thus, the influence of multiaxial stress/strain states on the fatigue behaviour of a component under elasto-plastic deformations can be determined reliably with unnotched specimens only by deformation controlled tests, if cyclic creep is not expected in critical areas. The modified effective equivalent strain hypothesis (MEES) for ductile materials renders a satisfactory evaluation of the in- and out-of-phase test results.

KEYWORDS

Multiaxial fatigue, load and deformation controlled tests, in- and out-of-phase loading, fine grained steel FeE 460, fatigue life to crack initiation.

INTRODUCTION

The influence of multiaxial, in-phase or out-of-phase stresses and strains on the fatigue life and the corresponding strength hypotheses are usually investigated with the aid of specimens. However, contrasting results can occur in the areas of low-cycle and finite fatigue life region depending on the form and loading mode of the specimens: While load controlled tests with unnotched hollow specimens from different ductile steels Ck45 (SAE 1045), 30CrNiMo8 (AISI 4330), X 10 CrNiTi 18 9 (AISI 321) show an increase of fatigue life to crack initiation for a phase shift of 90° between axial loading and torsion [1, 3], deformation controlled tests reveal a reduction of fatigue life [3, 5]. The contradictions with regard to the influence of phase shifting on the life duration can only be resolved by the results of load and deformation-controlled tests using the same material, specimen shape and loading mode (axial loading and torsion) in the same test equipment, under the same test conditions (temperature, alternating stress and strain (R = -1), frequency) whilst registering the local deformations and loads. The following report deals with tests of this kind on the cyclic softening fine grained steel FeE 460 [2, 5] which is commonly used in steel, piping and pressure vessel constructions.

MATERIAL DATA, SPECIMEN FORM AND TEST PROCEDURE

The chemical composition, the mechanical and cyclic properties of the ferritic-pearlitical fine grained steel FeE 460 are shown in Table 1.

Table 1. Data of the steel FeE 460

a.   Chemical composition in %

C	Si	Mn	P	S	Cr	Mo	Ni	V	Cu	Sn	Nb	As	N
0.15	0.15	1.52	0.004	0.002	0.020	0.005	0.54	0.15	0.020	0.002	0.002	0.004	0.020

Heat treatment: normalized, 940° C/20 min in air
Microstructure: ferrite + 30 % pearlite + traces of bainite

b.   Mechanical properties

			axial	torsion
Tensile strength	R_m	[MPa]	670	-
Yield strength	$R_{p,0,2,monotonic}/\tau_{p,0,2,monotonic}$	[MPa]	467	227
Stress coefficient	$K_{\sigma,monotonic}/K_{\tau,monotonic}$	[MPa]	477	227
Hardening exponent	$n_{\sigma,monotonic}/n_{\tau,monotonic}$		0.0186	0
Elongation	A_5	[%]	25	-
Area reduction	Z	[%]	65	-
Young's / shear modulus	E/G	[GPa]	206	80
Impact energy	A_v (bending)	[J]	(214)	-
Poisson's constant	μ_{el}		0.28	-

c.   Cyclic properties (f = 1s^{-1})

Yield strength	$R_{p,0,2,cyclic}/\tau_{p,0,2,cyclic}$	[MPa]	455	162
Stress coefficient	$K_{\sigma,cyclic}/K_{\tau,cyclic}$	[MPa]	508	319
Hardening exponent	$n_{\sigma,cyclic}/n_{\tau,cyclic}$		0.2161	0.1733
Constants of	$\varepsilon'_f / \gamma'_f$		0.5227	1.8697
Basquin-Coffin	c_ε / c_γ		0.5945	0.6882
equations	σ'_f / E / τ'_f / G		0.0032	0.0073
	b_ε / c_γ		0.0588	0.0900

The hollow cylindrical specimens, Fig. 1, of forged plate material, were taken mainly parallel to the rolling direction but also transversal to it. The hollow cylindrical specimens were tested under deformation and load control with a maximum load dependant frequency of 5 s^{-1} in a biaxial torsion/tension/compression test rig consisting of a 25 kN hydraulic actuator for torsion and a 100 kN cylinder for axial loading. The deformation was controlled by a self-developed transducer, Fig. 1, with which both, the axial strain and the shear strain could be simultaneously measured or applied.

For the tests with non-synchronous combined loading and deformation, a phase shift of $\delta = 90°$ was chosen between axial loading and torsion, as experience shows that the greatest reduction of life duration in ductile materials is achieved at this value [1, 4]. The failure criterion was defined as the breakthrough of the specimens with a wall thickness of 1.5 mm as a technical crack, which was determined by a vacuum system.

RESULTS OF EXPERIMENTAL INVESTIGATIONS

The results of the load-controlled tests are shown in Fig. 2 in form of S-N curves. The tests under sinusoidal combined loading were performed in such a way that the normal stress σ_x and the shear stress τ_{xy} were equally damaging according to the von Mises distortion energy hypothesis (DEH), i.e. the particular amplitudes had the ratio $\sigma_{xa}/\tau_{xya} = \sqrt{3}$. An increase of the fatigue life occurs under out-of-phase loading, Fig. 2. This could be explained by measurements of the local axial strains and shear strains [2, 5]. The local deformation amplitudes increase more strongly under in-phase load-controlled loading and lead to earlier failures than under out-of-phase load-control.

The tests under combined deformation control were carried out with a ratio of $\gamma_{xya}/\varepsilon_{xa} = 2(1+\mu)/\sqrt{3}$. According to the DEH, equal proportions of damaging axial strain and shear strain are assumed by this ratio [1, 2, 5]. In contrast to the load controlled multiaxial tests, the deformation controlled tests, Fig. 3, reveal a reduction of fatigue life for out-of-phase loading [2, 5]. Comparing the hysteresis loops ($\sigma_x = f(\varepsilon_x)$ and $\tau_{xy} = f(\gamma_{xy})$) for in- and out-of-phase deformation combinations [2, 5], it becomes apparent that under a phase shift of 90° higher stresses must be exerted and therefore more energy must be applied to the material in order to reach the stipulated amplitudes of axial and shear strain than under synchronous strains. Due to the changing directions of principal stresses and principal strains, dislocations are initiated in differing directions which block each other. In order to overcome the resulting obstacles, higher loading and so higher energy levels must be exerted on the material. For this reason, the phase shift under deformation control produces the observed reduction of fatigue life in ductile materials [2, 5].

EVALUATION OF THE TEST RESULTS

Only the multiaxial strain states under deformation control are evaluated because they are component and so application relevant. The multiaxial strain state is transformed by the

214

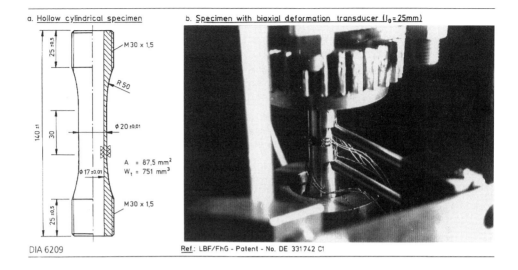

Fig. 1 Hollow cylindrical specimen with biaxial deformation transducer

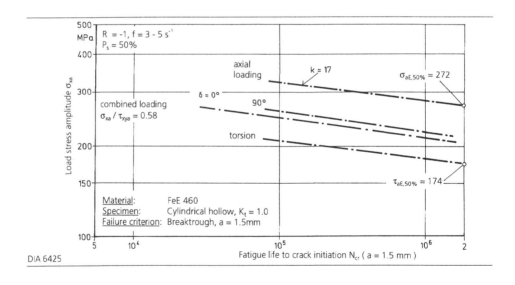

Fig. 2 S-N curves for uniaxial and combined load controlled loading

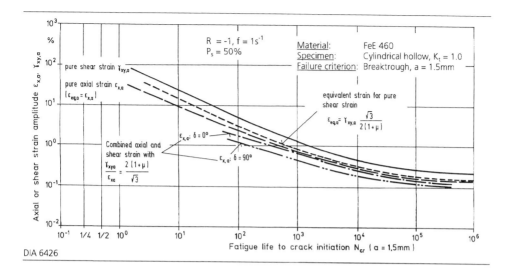

Fig. 3 Strain controlled determined S-N curves

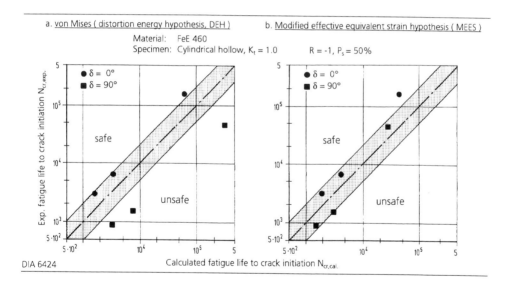

Fig. 4 Evaluation of test results obtained under deformation controlled combined axial and shear strain according to different hypotheses

interference plane based modified effective equivalent strain (MEES) hypothesis [2, 5]. For each interference plane φ the shear strain $\gamma(\varphi)$ is calculated. Its arithmetical mean value γ_{arith} over all planes takes into account the before explained influence of changing principal stress/strain directions. The equivalent strain is calculated for any phase angle δ by

$$\varepsilon_{eq}(\delta) = \varepsilon_{eq}(\delta = 0°) \frac{\gamma_{arith}(\delta)}{\gamma_{arith}(\delta = 0°)} \sqrt{G \exp\left[1 - \left(\frac{\delta - 90°}{90°}\right)^2\right]} \quad \text{with}$$

$$\varepsilon_{eq}(\delta = 0°) = \frac{1}{\sqrt{2}(1+\mu)} \sqrt{(\varepsilon_x - \varepsilon_y)^2 + (\varepsilon_y - \varepsilon_z)^2 + (\varepsilon_x - \varepsilon_z)^2 + f_G^2 \frac{3}{2} \gamma_{xy}^2} \quad .$$

$G = (1 + K_{ta}) / (1 + K_{tt})$ and $f_G = \varepsilon_{eq,von\ Mises}$ (pure axial strain) $/ \varepsilon_{eq,von\ Mises}$ (pure shear strain) consider gradient and size effects [2, 5, 6, 7]. The MEES hypothesis is the strain based version of the stress based effective equivalent stress (EES) hypothesis [2, 6, 7].

For the assessment of the fatigue life to crack initiation the calculated equivalent strain amplitude is related to the S-N curve for pure axial strain. Fig. 4 shows the comparison of calculated and experimental fatigue axial strain. Fig. 4 shows the comparison of calculated and experimental fatigue lives according to von Mises (DEH) and MEES. While the von Mises criterion renders for out-of-phase loading an unsafe assessment, the MEES criterion delivers results within an acceptable scatter range.

CONCLUSIONS

When elasto-plastic deformations occur in load-controlled tests with unnotched specimens made of ductile materials under combined axial loading and torsion, hypotheses based on such results cannot be applied to components, because of the occuring cyclic creep while for components the local stress/strains remain deformation-controlled due to strain gradients until the structural yield point has been reached or exceeded. Therefore, the testing as well as the development and application of multiaxial criterion must consider the real local deformation conditions of components.

REFERENCES

1. Simbürger, A. (1975). Fraunhofer-Inst. f. Betriebsfestigkeit, Darmstadt. Rep. FB-121.
2. Sonsino, C.M. (1997). EUR-Report No. 16024, Luxembourg.
3. Sanetra, C. (1991). Dissertation, TU Clausthal.
4. Brown, M. W. and Miller, K. J. (1982). In: ASTM STP 770, pp. 482 - 499.
5. Sonsino, C.M. (1995). Mat. u. Werkstofftechnik 26, 8, pp. 425 -441.
6. Sonsino, C.M. (1995). Int. Journal of Fatigue 17, 1, pp. 55 - 70.
7. Sonsino, C.M. (1997). In: Proceedings of the 5th Int. Conf. on Biaxial/Multiaxial Fatigue and Fracture, Cracow, Vol. I, pp. 395-419.

Low Cycle Fatigue and Elasto-Plastic Behaviour of Materials
K-T. Rie and P.D. Portella (Editors)

Influence of Multi-Axial Loading on the Lifetime in LCF Strength

J. HUG and H. ZENNER

Institut for Plant Engineering and Fatigue Analysis, Technical University of Clausthal.

D-38678 Clausthal-Zellerfeld, Germany

ABSTRACT

Three different steels were used to investigate the influence of multi-axial loading on the damage process and the subsequent effect on the LCF-lifetime. Plain cylindrical hollow specimens of the three technically relevant materials X6 CrNiTi 18 10, Ck 15, 30 CrNiMo 8 were subjected to normal- and shearstresses which were induced either in-phase, out-of-phase and consecutively, figure 1.

The experiments were carried out on a multi-axial servohydraulic testdevice capable to perform torsion- and tension/compression-loads. The characterisation of the cyclic stress-strain behavior of the specimens subjected to various load histories was realized with a multiaxialextensometer. Microscopic surface investigations are used to determine the formation and propagation of microcracks.

The experimentally achieved data form the basis for the evaluation and verification of the stresshypotheses. The parameters which decisively influence the process of damage were analyzed. Laws of microcrack-initiation and microcrackgrowth were also investigated.

KEYWORDS

LCF, multiaxial, lifetimeprediction, damage process, in-phase, out-of -phase, microcracks

INTRODUCION

The parameters which influence the fatigue process are manifold and often interfere with each other so that a reliable lifetimeprediction of components and machines becomes difficult. A main portion of dynamically loaded components is subject to multiaxial loads and is mostly combined with complex geometries. The current state in the field of lifetimeprediction of multiaxial loaded components is rather incomplete and unsatisfying. A reason for this is that the process of damage under multiaxial loading is not yet thoroughly investigated, (1).

218

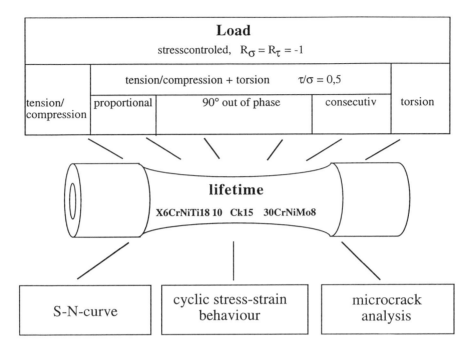

Figure 1: General view of the project

The mechanism of damage in LCF is initialized by plastic deformations. Alternating sliding in grains leads to the formation of microcracks.

Under cyclic loading X6 CrNiTi 18 10 hardens, Ck 15 shows to have a neutral behavior and 30 CrNiMo 8 softens. The degree of cyclic stress-strain behavior (magnitude of the variation of the cyclic stress-strain behavior) depends on the material as well as on the load amplitude and history. The respective cyclic stress-strain behavior under pure torsion loading and combined torsion-tension/compression loading is different. Simultaniously, the lifetime is affected. There is no damage parameter which sufficiently and acurately describes the cyclic stress-strain behavior for multiaxial loading. A lifetime prediction on the basis of the cyclic stress-strain behavior under pure ten sion/compression loading is therefore insufficient, figure 2, 3, 4.

The cyclic stress-strain behavior has a major influence on the lifetime. In a stresscontrolled experiment with a cyclic hardening material the multiaxial loading results in a favourable lifetime extension as opposed to uniaxial loading; in the case of cyclic softening materials multiaxial loading leads to a shorter lifetime. The declination of the S-N-curve is not significantly changed by multiaxial loading. The evaluation of the microcrack investigations showed that the microcrackorientation is parallel to the direction of maximum shear stresses. The microcrack formation and microcrack propagation are to a large degree dependent on the material. The underlying laws are not generally describable.

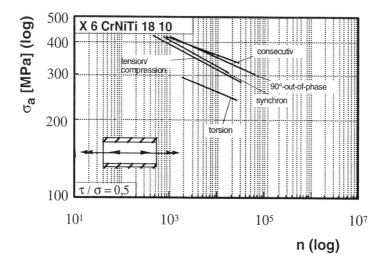

Figure 2: S-N-curve, material X6 CrNiTi 18 10

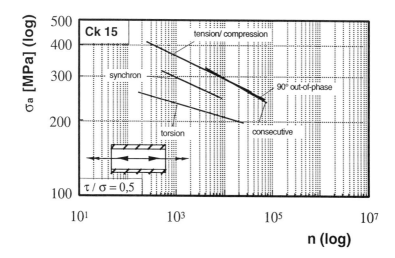

Figure 3: S-N-curve, material Ck 15

220

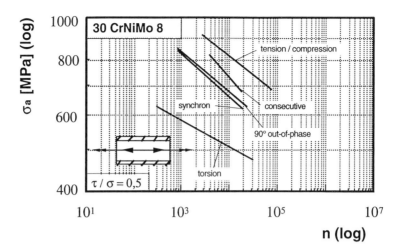

Figure 4: S-N-curve, material 30 CrNiMo 8

Microcracks often already start at the specimen surface after a low number of cycles. The density of microcracks increases in an S-shaped curve with increasing numbers of cycles. A saturation is possible when microcrack coalescence and microcrack growth prevail.

For nonproportional loading (90°out-of-phase) the crackdensity is higher as compared to proportional loading provided that the same local stressdistribution prevails; this is because all plains are evenly stressed and the probability of the formation of a microcrack is higher. The crack orientation under 90° out-of-phase loading scatters across a wide area since the shearstresses are of equal magnitude in any direction, figure 5.

Five different strain damage parameters (max. strain, max. shearstrain, von Mises, Kandil-Brown-Miller, Lohr-Ellison) and one stress damage parameter (plastic energy) were investigated which could be distinguished by a various evaluation of the stress-strain correlation as well as the shear-stress / shear-strain correlation, (2). A comparison of the multiaxial forms of loading shows that the lifetime of specimens subjected to proportional syncronic loading can be reliably predicted through most of the damage parameters; the nonproportional 90°out-of-phase loading, however, leads to an unreliable and unfavourable lifetime prediction in all cases. The lifetime of the cyclic softening material 30 CrNiMo 8 is rather unfavourably predicted. A lifetime prediction using the parameter of plastic energy generally shows to be reliable for LCF loadings. In the case of a high 90°out-of-phase loadamplitude the calculated value largely varies from the experimentally achieved result. As entrydata the stress-strain hystersis has to be in the saturated range. A lifetime prediction in LCF is only possible applying the local stress-strain concept.

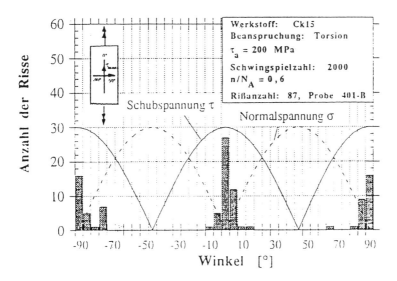

Figure 5: Crackorientation, torsion loading

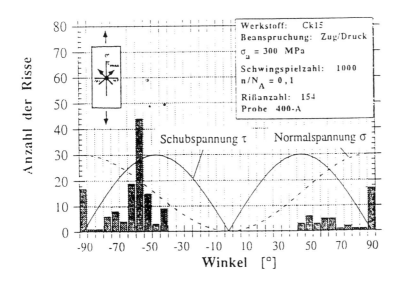

Figure 6: Crackorientation, tension/compression loading

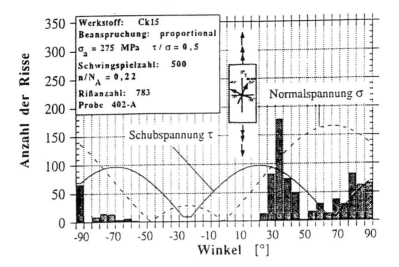

Figure 7: Crackorientation, proportional loading

Presently, no damage-parameter exists that is capable to sufficiently analyse the influence of a complex multi-axial load for various materials. The explanation for that is that the damageparameters are materialindependent, the lifetime, however, as well as the damageprocess is to a great extend materialdependent under various states of stress. The lifetimeshortening influence of nonproportional loading in combination with cyclic softening material behavior has to be accounted for with safetyfactors. A logical consequence of these investigations is that the cyclic stress-strain behavior has to be quantified taking the materialspecific properties into account and has to be considered in the damage parameters. Finally, a proposal for a safe lifetimeprediction for steels under multiaxial loading is introduced.

ACKNOWLEDGEMENTS

The work was supported by Deutsche Forschungsgemeinschaft (Ze 248/4, "Mehrachsige Schädigung")

REFERENCES

1. Hug, J. (1994) Einfluß mehrachsiger Beanspruchung auf die Lebensdauer im Zeitfestigkeitsgebiet. Dissertation Technische Universität, Clausthal.

2. Liu, J. (1991) Beitrag zur Verbesserung der Dauerfestigkeitsberechnung bei mehrachsiger Beanspruchung. Dissertation Technische Universität, Clausthal.

Low Cycle Fatigue and Elasto-Plastic Behaviour of Materials
K-T. Rie and P.D. Portella (Editors)

ON THE UTILITY OF COMPLEX MULTIAXIAL CYCLIC LOADINGS IN TENSION-TORSION-INTERNAL AND EXTERNAL PRESSURES, TO IMPROVE THE FORMULATION OF THE CONSTITUTIVE EQUATIONS

L. BOCHER, D. JEUNE, P. ROBINET AND P. DELOBELLE

Laboratoire de Mécanique Appliquée R. Chaléat, UMR CNRS 6604
UFR Sciences et Techniques, 24 rue de l'Epitaphe, 25030 Besançon Cedex, France

ABSTRACT

This paper is concerned with the experimental behaviour of a 316 austenitic stainless steel at room temperature and under non proportional cyclic strainings in tension-torsion-internal and external pressures. The two or three sinusoïdal strains were applied both in and out-of-phase or with different pulsations. The main investigations deal with the additional hardening due to multiaxiality of the loadings. With respect to the increasing maximum additional hardening, the different tests can be classified as follows : in phase tests $\varphi = \delta = 0$, out-of-phase internal-external pressure tests with $\varphi = 90°$ or $60°$, out-of-phase tension-torsion tests with $\delta = 90°$ and finally tension-torsion pressure tests with significant phase angles and tension-torsion tests with different pulsations n = 2. Concerning the ratchetting effect, from tests performed under in or out-of-phase cyclic tension-torsion plus a static stress due to internal pressure, it was shown that the rate of diametral ratchetting is an increasing function of the phase angle between the cyclic components.
Concerning the modeling, certain observations described in this paper are not predictable neither by the phenomenological approaches nor by the micromechanical approaches. Hence, if the results are confirmed certain improvements in the model formulation will be necessary.

KEYWORDS

Multiaxial plasticity, tension-torsion-internal-external pressure tests, ratchetting effect; 316 austenitic stainless steel.

INTRODUCTION

Since the advent of hydraulic tension-torsion machines twenty years ago, experimental studies of the behaviour of metallic materials under cyclic biaxial loadings have developed significantly. However, only few studies have been realized under tension-torsion-internal and external pressure loadings. So, this paper is concerned with the experimental behaviour of an austenitic stainless steel under non-proportional cyclic strainings combining these three components (cyclic tests and ratchetting tests). In the case of a 316 stainless steel at ambient temperature it has been shown that a very significant supplementary hardening appears under cyclic biaxial loadings, that is directly due to the phase difference between the components of the strain and to the ratio between the maximum amplitudes of these components [1,4]. In the same way, for this steel, the description of the progressive strain phenomena under mechanical uniaxial and biaxial loadings has been extensively described [5,6]. In the present study, we propose to analyse qualitatively and quantitatively the effect on the tension-torsion behaviours in the case of tension-pressure and tension-torsion-pressure where these components are out-of-phase with one another.

EXPERIMENTAL METHODS

The various biaxial or triaxial tests have been conducted on a modified Schenck hydraulic tension-torsion machine [7]. A device mounted on this machine is realized and which allows cyclic tests to be performed on tubes for loadings in tension-torsion-internal and external pressures. It is composed of a medium pressure chamber enclosing the gage thickness of the test specimen, directly mounted on the specimen and connected to two pressure regulators. The specimen is also connected to the jaws of the hydraulic machine through two extension rods. The entire device is controlled with the aid of strain gauges fixed directly on the gage thickness of the test specimen. The two or three control signals are delivered by a H.P. function generator which itself can be controlled by a micro-computer. In this way, any kind of signal can be generated.

THE TWO OR THREE DIMENSIONAL CYCLIC TESTS

The behaviour of this type of steel under uni and bidirectional loadings in tension-torsion has been very thoroughly studied at ambient temperature which allows, given a few calibrations and verification tests, to establish a reliable and complete experimental basis for these two types of loadings. These classical tests will not be examined here. The tests presented below consist in the applications and control of either two strains, $\left\{\varepsilon_{zz}^{T}, \varepsilon_{\theta\theta}^{T}\right\}$, $\left\{\varepsilon_{zz}^{T}, \varepsilon_{z\theta}^{T}\right\}$ or three strains $\left\{\varepsilon_{zz}^{T}, \varepsilon_{z\theta}^{T}, \varepsilon_{\theta\theta}^{T}\right\}$. Superscript T denotes total values of the axial (zz), hoop ($\theta\theta$) and shear ($z\theta$) strains.

Before continuing, it is important to specify the exact nature of these tests. The three controlled components are planar and the resultant stresses are tri-dimensional (Eq. 1) :

$$\sigma_{ij}=\begin{pmatrix}\sigma_{zz} & \sigma_{z\theta} & 0 \\ \sigma_{z\theta} & \sigma_{\theta\theta} & 0 \\ 0 & 0 & \sigma_{rr}\end{pmatrix} \quad \text{with} \quad \begin{array}{l} \sigma_{zz}=\dfrac{F}{\pi\left(r_{e}^{2}-r_{i}^{2}\right)}+\dfrac{1}{2}\dfrac{\left(P_{1}r_{i}-P_{2}r_{e}\right)}{r_{e}-r_{i}}, \quad \sigma_{\theta\theta}=\dfrac{P_{1}r_{i}-P_{2}r_{e}}{r_{e}-r_{i}} \\[3mm] \sigma_{rr}=-\dfrac{P_{1}+P_{2}}{2} \quad \text{and} \quad \sigma_{z\theta}=\dfrac{3C}{2\pi\left(r_{e}^{3}-r_{i}^{3}\right)} \end{array} \tag{1}$$

In these equations, F is the value of the force read by the cell, C the torque exerted at one extremity of the tube and P_1, P_2 respectively the applied pressures on internal (radius r_i) and external (radius r_e) surfaces of the specimen.

Thus, we can speak of tri-axial tests in the strict sense of the word. However, it is true that σ_{rr} is small in comparison with the other components and that it intervenes more as a parasitic stress than as a test parameter. The two or three imposed signals have the form :

$$\varepsilon_{zz}^{T}=\varepsilon_{zz_{M}}^{T}\sin\omega t, \quad \varepsilon_{z\theta}^{T}=\varepsilon_{z\theta_{M}}^{T}\sin\left(n\omega t-\delta\right), \quad \varepsilon_{\theta\theta}^{T}=\varepsilon_{\theta\theta_{M}}^{T}\sin\left(\omega t-\varphi\right) \tag{2}$$

The values of $\varepsilon_{zz_{M}}^{T}$, $\varepsilon_{z\theta_{M}}^{T}$ and $\varepsilon_{\theta\theta_{M}}^{T}$ condition the importance of each one of these components, φ, δ set the phase shift between them, and n impose the difference between the pulsations of ε_{zz}^{T} and $\varepsilon_{z\theta}^{T}$ when $\varepsilon_{\theta\theta}^{T}=0$. In the general case of tension-torsion-pressure loadings with n = 1, each one of these components express themselves as a function of the maximal equivalent von Mises strain, $\bar{\varepsilon}_{M}^{T}$ (Eq. 3), namely :

$$\varepsilon_{zz_{M}}^{T}=\frac{\sqrt{3/2}\ \bar{\varepsilon}_{M}^{T}}{\sqrt{A-\sqrt{B}+C}}, \quad \varepsilon_{z\theta_{M}}^{T}=r_{2}\frac{\sqrt{3}}{2}\varepsilon_{zz_{M}}^{T}, \quad \varepsilon_{\theta\theta_{M}}^{T}=r_{1}\varepsilon_{zz_{M}}^{T}, \quad \text{with :}$$

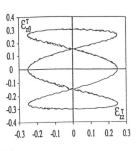

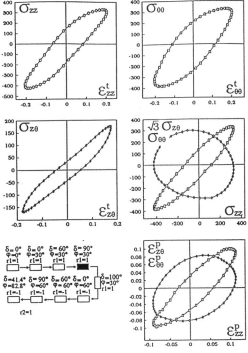

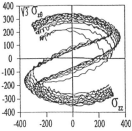

•Fig. 1 : Responses at the stabilized cycle of the
test where the three components are out-of-phase :

$r_1 = r_2 = 1$, $\delta = 90°$, $\varphi = 30°$. $\left(\sigma_{zz}, \sigma_{\theta\theta}, \sigma_{z\theta}\right) = f\left(\varepsilon_{zz}^T, \varepsilon_{\theta\theta}^T, \varepsilon_{z\theta}^T\right)$

$\left(\sqrt{3}\sigma_{z\theta}, \sigma_{\theta\theta}\right) = f(\sigma_{zz})$ and $\left(\varepsilon_{z\theta}^P, \varepsilon_{\theta\theta}^P\right) = f\left(\varepsilon_{zz}^P\right)$

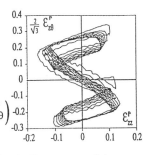

•Fig. 2 : Responses of the
tension-torsion test where $n = 3$,
$\varphi = \delta = 0$.

Imposed path : $\varepsilon_{z\theta}^T = f\left(\varepsilon_{zz}^T\right)$

responses : $\sqrt{3}\,\sigma_{z\theta} = f(\sigma_{zz})$,

$2/\sqrt{3}\ \varepsilon_{z\theta}^P = f\left(\varepsilon_{zz}^P\right)$.

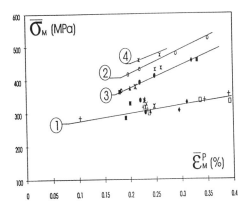

•Fig. 3 : $\overline{\sigma}_M$ versus $\overline{\varepsilon}_M^P$.
Situation of the totality of the
experimental points ($n = 1$).

$$A = 1 + r_1^2 + \frac{3}{4} r_2^2 + r_1 \cos\varphi, \quad B = \left(\frac{3}{4} \ r_2^2 \sin 2\delta + r_1 \sin\varphi + r_1^2 \sin 2\varphi \right)^2$$

and $\quad C = \left(1 + \frac{3}{4} \ r_2^2 \cos 2\delta + r_1 \cos\varphi + r_1^2 \cos 2\varphi \right)^2$ \hfill (3)

In this way, working with $\overline{\varepsilon}_M^T$ fixed, it is possible to calculate, using Eq(3), the three amplitudes $\varepsilon_{zz_M}^T$, $\varepsilon_{z\theta_M}^T$ and $\varepsilon_{\theta\theta_M}^T$ for r_2, r_1, φ and δ imposed.

For the performed tests we imposed $\overline{\varepsilon}_M^T = 0.4\,\%$, $\omega = 1.88 \ 10^{-2} \ S^{-1}$, with $n = 1.$, $r_1 = 0., \pm 1.$, $r_2 = 0,1.$ and $0. \langle \varphi \langle 180°.$, $0. \langle \delta \langle 90°.$ or when $r_1 = 0.$, $\delta = 0.$, we imposed $n = 1., 1.5, 2.$ and $3.$. On each specimen, several incremental or decremental phase shifts are imposed and for each step 50 cycles are performed.

Figures 1 and 2 give examples of the loops obtained respectively with three out-of-phase components and two components with $n = 3$ and $\delta = 0$.

In order to be able to perform a quantitative analysis of the totality of these tests, the maximal equivalent stress $\overline{\sigma}_M$ as well as the maximal plastic strain $\overline{\varepsilon}_M^P$ are calculated numerically and for each one of the imposed sequences. The set of the calculated pairs $\left(\overline{\sigma}_M, \overline{\varepsilon}_M^P \right)$ are reported in figures 3 and 4 respectively for $n = 1$ and $n \neq 1$. Reference points from the literature obtained in tension-compression, alternating torsion, tension-torsion in and out-of-phase $(\delta = 90°)$ are also reported in Fig. 3. These figures allow the different types of loading to be classified with respect to the supplementary hardening.

This classification is established as follows :

- The in-phase tests with two or three components $(\delta = \varphi = 0°)$: non supplementary hardening, curve 1, Fig. 3.

- The tension-pressure tests such as $r_1 = 1$, $\varphi = 90°$ and $r_1 = -1$, $\varphi = 60°$, the hardening is slightly lower than that of tension-torsion tests, curve 3, Fig. 3.

- The tension-torsion tests such as $r_2 = 1$, $\delta = 90°$, curve 2, Fig. 3 and curve 1, Fig. 4.

- The tension-torsion-pressure tests where the three components are strongly shifted, namely : $r_1 = r_2 = 1$, $\delta = 90°$, $\varphi = 60°$

and $r_2 = 1$, $r_1 = -1$, corresponding to the two previous cases, $\delta = 41.4°$ and $\varphi = 82.8°$, corresponding to a strain path following the von Mises ellipsoid, curve 4, Fig. 3. In this case, the hardening is slightly greater than the one recorded in tension-torsion $(\delta = 90°)$.

- The tension-torsion tests with $n \cong 2$, curve 2, Fig. 4. The hardening is greater than the one obtained in tension-torsion $(\delta = 90°)$.

Note that this last observation corroborates those of Calloch [8] obtained with the same loading path.

THE MULTIAXIAL RATCHETTING TESTS

In order to extend the validity of the observations under two dimensional loading to the case of multiaxial loading where two components are cyclic, several ratchetting tests involving three stress components are performed. The primary loading is applied by internal pressure and the cyclic secondary loading is in tension and torsion, either in or out-of-phase. The characteristics of the tests are :

$$\theta_{\theta\theta} = 50\,\text{MPa}, \quad \varepsilon_{zz}^T = \varepsilon_{zz_M}^T \sin\omega t \quad \text{and} \quad \varepsilon_{z\theta}^T = \varepsilon_{z\theta_M}^T \sin(\omega t - \delta) \quad \text{with} \quad 0 \langle \delta \langle 90°. \quad (4)$$

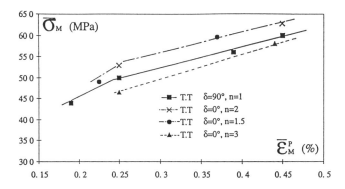

•Fig. 4 : $\overline{\sigma}_M$ versus $\overline{\varepsilon}_M^P$. Situation of the experimental points corresponding to different values of the n parameter $(\delta = 0)$.

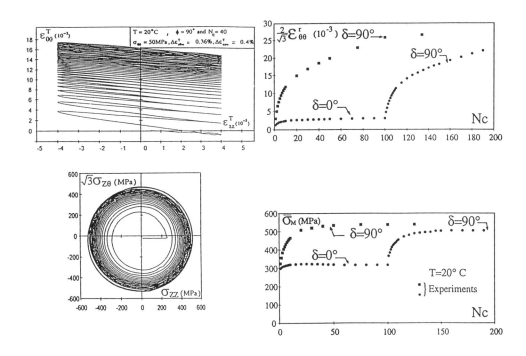

•Fig. 5 : Ratchetting test where the tension-torsion secondary loading is out-of-phase. The static internal pressure primary loading is such that $\sigma_{\theta\theta} = 50\,\text{MPa}$.

•Fig. 6 : Variations of the ratchetting and of the cyclic hardening for different values of δ $(\delta = 0$ and $\delta = 90^\circ\text{C})$ and $\sigma_{\theta\theta} = 50\,\text{MPa}$.

228

When the secondary loading path is radial, the diametral ratchetting amplitude and the maximum cyclic stress level are nearly identical (in the von Mises sense) to those obtained in the case of tension-torsion ratchetting.

However, as soon as the secondary loading path is non proportional, a very large increase of the diametral ratchetting rate and of the maximum equivalent cyclic stress is observed. An example of results obtained with $\delta = 90°$ is given in Fig. 5. Fig. 6 shows the increase of the progressive strain and of the equivalent stress when the phase shift increase from $\delta = 0°$ to $\delta = 90°$. Other tests with intermediate values of δ confirm that the ratchetting rate is an increasing function of the phase lag.

Concerning the modelling, certain observations described previously are not predictable by the phenomenological approaches [7,8] and only qualitatively by the micromechanical approaches [9,10]. Hence, if all the observations are confirmed certain improvements in the model formulation will be necessary.

CONCLUSIONS

A device has been designed which allows cyclic tests to be performed on tubes for loading in tension-torsion-internal and external pressures. Different tests have been performed at ambient temperature on an austenitic stainless steel which has the particularity of presenting a strong supplementary hardening related to the non-radiality of the loadings. We study the influence of the phase shift parameters, namely δ and φ, and of the difference between the pulsations, namely n, for two or three sinusoïdal components.

The results are very rich in informations and allow the different types of loading to be classified with respect to the observed supplementary hardening.

Concerning the ratchetting test with three independent components it was shown that the rate of diametral ratchetting is an increasing function of the phase angle between the cyclic components.

The presented complex results are very difficult to be modelized by phenomenological or micromechanical models.

REFERENCES

1. Cailletaud, G., Kaczmarek, H. and Policella, H. (1984) *Mech. Mat.* **3**, 333.
2. Benallal, A. and Marquis, D. (1987), *J. Eng. Mat. Techn.* **109**, 326.
3. Murakami, S., Kawai, M., Aoki, K. and Ohmi, Y. (1989), *J. Eng. Mat. Techn.* **111**, 32.
4. Bouchou, A. and Delobelle, P. (1996), *Nucl. Eng. Design.* **162**, 21.
5. Chaboche, J.L. and Nouailhas, D. (1989), *J. Eng. Mat. Techn.* **111**, 384.
6. Delobelle, P., Robinet, P., Bocher, L. (1995), *Int. J. Plast.* **114**, 295.
7. Delobelle, P., Bocher, L. (1997), *J. Phys. III France* **7**, 1755.
8. Calloch, S. (1997), Thesis, ENS Cachan, France.
9. Pilvin, P. (1994). In: *Proc. Int. Conf. on Biaxial/Multiaxial Fatigue*, pp. 31-46, ESIS/SF2M, Saint-Germain en Laye, France.
10. Pilvin, P. and Geyer, P. (1997). In: *Trans. of the 14th Int. Conf. on Struct. Mech. in Reactor Techn.*, pp. 277-284, SMIRT 14, Lyon, France.

DESIGN CRITERIA FOR MULTIAXIAL CREEP-FATIGUE SUPPORTED BY ADVANCED MULTIAXIAL EXPERIMENTS

MASATERU OHNAMI[*] ,MASAO SAKANE[*] ,SHOICHI MUKAI[**]
and TAKAFUMI TSURUI[***]

[*] Department of Mechanical Engineering, Faculty of Science and Engineering, Ritsumeikan University,Noji-cho,Kusatsu,525,Japan.
[**] Power Reactor & Nuclear Fuel Develop.Corp.,Ibaraki-gun,Ibaraki, 311-13,Japan.
[***]Kobe Material Testing Laboratory Ltd.,Harima-cho,Kako-gun,Hyogo, 675-01,Japan.

ABSTRACT

This paper first shows the result of biaxial tension creep rupture test using cruciform specimen of type 304 stainless steel at 923K and that of triaxial creep rupture test using triaxial cubic specimen of the material at 923K. The biaxial creep rupture time increases with increasing principal stress ratio, λ ,in the range of $0 \leq \lambda \leq 1$ under the fixed Mises' equivalent stress. The equivalent stress based on crack opening displacement, COD stress, gives the best stress parameter to correlate the biaxial and triaxial creep rupture lives among five stress parameters. Second, this paper describes that the biaxial creep-fatigue lives in a full range of principal strain ratio of $-1 \leq \phi \leq 1$ at 873K is successfully predicted from both uniaxial tension creep rupture and uniaxial tension-compression low cycle fatigue test data, using both criteria of COD stress/strain and linear damage rule. This will be the most practical conclusion supported by the advanced multiaxial experiments.

KEYWORDS

COD stress/strain, multiaxial creep-fatigue, type 304 stainless steel.

INTRODUCTION

Practical applications operated at high temperatures receive multiaxial creep-fatigue loadings and multiaxial structural design is required. The typical case of equi-biaxial creep-fatigue loadings is known in the nuclear reactor vessel of type 304 stainless steel near moving liquid sodium surface in fast breeder reactor (FBR) [1]. Type IV crack in high-temperature boiler piping portion is related to equi-triaxial loadings which affects creep cavitation. It is clear that Mises stress/strain used in the design code takes a zero under equi-triaxial tension/ compression loadings and Mises' basis is insufficient. The studies of the authors [2-9] have revealed that Mises stress/strain is not effective in correlating multiaxial stress/strain controlled LCF life and multiaxial creep-fatigue life of heat resistant steels.

The aim of this paper is to describe the design criteria for multiaxial creep-fatigue supported by the following advanced multiaxial experiments of type 304 stainless steel:
(1)Biaxial tension creep rupture tests using cruciform specimen under multiaxial stress states of $0 \leqq \lambda \leqq 1$ at 923K.
(1)Equi-triaxial tension creep rupture tests using triaxial specimen at 923K.
(2)Biaxial strain-controlled creep-fatigue tests using tubular and cruciform specimens under a full range of multiaxial stress states of $-1 \leqq \phi \leqq 1$ at 873K.

BIAXIAL TENSION CREEP RUPTURE CRITERIA
(a)Comparison of Stress Parameters to Correlate Multiaxial Tension Creep Rupture Life Being Relevant to Multiaxial Creep-Fatigue

Table 1 shows the stress parameters correlating multiaxial tension creep rupture life of materials. They are:① Mises stress,② maximum principal stress,③ COD stress [2-9],④ Joshi stress [10],⑤ Huddelston stress [11]. Fig.1 shows the comparison of these parameters and the coordinate is normalized by the maximum principal stress in y-direction, σ_1. The criteria ③ and ⑤ take the intermediate location between the classical ones ① and ② under the multiaxial stress states of $-1 \leqq \lambda \leqq 0$, where λ is principal stress ratio and is defined as σ_3/σ_1.Under the stress states of $0 \leqq \lambda \leqq 1$, only COD stress ③ decreases monotonously with increasing λ, where $\lambda = \sigma_2/\sigma_1$, in comparison with the variation of the stresses ④ and ⑤. COD stress is written by two equations as shown in the table under $-1 \leqq \lambda \leqq 0$ and $0 \leqq \lambda \leqq 1$, respectively, irrespective of the inelastic constitutive relation of the material. In the analysis, the positive principal stress parallel to mode I crack significantly decreases the crack propagation rate, and this was historically confirmed in the experiments by Hopper and Miller [12] and the authors [2,3] .

(b) Experiment of Biaxial Tension Creep Rupture of Type 304 Cruciform Specimen.

The shape and dimensions of the cruciform specimen for biaxial tension creep rupture test requires uniform stress distribution within the gage area and no crack initiation besides the gage area. 3D FE creep analyses by using MARC-K5 program were made to determine the shape and dimensions of the specimen as shown in Fig.2 [13]. From the analysis of stress at the center of the gage area of $16\text{mm} \times 16\text{mm}$ with time under the equi-biaxial tension, it was found that the normal stress components, σ_x, σ_y reached a steady state in a short elapsed time and there is no change in the value of the stress components with time. The effectiveness of COD stress ③ in Table 1 correlating biaxial tension creep rupture life was shown in Fig.3 [13] and most of the data are within a factor of two scatter band. Huddelston stress ⑤ in Table 1,on the other hand, gives a scatter depending on λ in the data correlation, and it is not suitable for correlating high-temperature creep rupture lives under multiaxial stress states.

TRIAXIAL TENSION CREEP RUPTURE CRITERIA

Newly developed testing machine using triaxial specimen can make equi-triaxial tension creep rupture and equi-triaxial tension-compression LCF tests. The load capacity is 49KN and the maximum temperature is 923K for performing the test of heat resistant steels. The apparatus is an electric-hydraulic servo machine and has six actuators and six servo controllers to generate widely ranged trixial stress states. Analogue control for the stroke of the six actuators was adopted and additionally a digital feedback control circuit was used for performing the stress/strain-controlled test. Fig.4 [14] shows the shape and dimensions of the triaxial specimen, in which six grips are prepared to perform perpendicularly triaxial loadings on the cubic portion of $10mm \times 10mm \times 10mm$. Shape of the specimen was determined from 3D FE elastic and creep analyses so that the stress and strain are distributed uniformly in the region of $5mm \times 5mm \times 5mm$ in the gage part. The measurement of displacement of the cubic gage part of the triaxial specimen was made by using tri-couples of extensometers of contact type having a resolution of $1\mu m$. Quartz glass rod of the extensometer was inserted into the bottom of the groove which has slope of 45 degrees to x,y,z-axes as shown in Fig.4.

Fig.5 [14] shows the equi-triaxial tension creep rupture test result in comparison with uniaxial and equi-biaxial tension test data on the common basis of the maximum principal stress of 196 MPa. In the figure, the abscissa shows the first invariant of stress, $J_1 = \sigma_1 + \sigma_2 + \sigma_3$. The triaxial tension creep rupture test was interrupted after the elapsed time of 135 hrs, because one groove portion of the specimen was ruptured. However,the grain boundary creep cavities were observed in the central portion of the gage part of the specimen crept during 135 hrs under the equi-triaxial tension, irrespective of taking zero of Mises stress. Therefore, it is considered that the triaxial specimen tested will be failured after longer time than 135 hrs if the specimen groove is not ruptured. We found that the increase of positive value of J_1 has an effect to increase high-temperature creep rupture life of type 304 stainless steel from the comparison of three kinds of test data. From the comparison of three kinds of creep rupture curves under uniaxial, equi-biaxial and equi-triaxial tensions, it is clear that creep deformation under equi-triaxial tension becomes remarkably smaller than those under uniaxial and equi-biaxial tensions and that creep decreases with increasing positive value of J_1.

MULTIAXIAL CREEP-FATIGUE CRITERIA

The shape and dimensions of the cruciform specimen used was determined from FE analyses so that the stress and strain are distributed uniformly within the $8mm \times 8mm$ gage area [4]. The variation of stress and strain within the gage area was less than 10%. The experimental apparatus used was an electric-hydraulic servo machine for the cruciform specimen [4]. The apparatus has four actuators and four servo controllers to generate widely ranged biaxial stress states. Strain-controlled proportional multiaxial LCF tests having dwell time, tH, of 10min. and 30min.,respectively,at the maximum tensile strain of y-direction in each cycle, were carried out for type 304 stainless steel at 873K. The tests were performed under the fixed Mises total strain range, $\Delta \varepsilon_{eq}$,of 1% and the principal strain ratio, $\phi = \varepsilon_x / \varepsilon_y$,of -0.5,0 and 1, respectively. Fig.6 [15] shows the effect of principal strain ratio, ϕ, on the reduction of failure life ratio, Nf/Nf*,with dwell time, where Nf* denotes the failure life for tH =0.It is clear from the figure that the life ratio decreases with increasing tH and the reduction of multiaxial creep-fatigue life becomes larger with increasing ϕ in a full range from -1 to 1.

In order to evaluate the creep damage of type 304 stainless steel in biaxial creep-fatigue tests, the biaxial stress relaxation of the cruciform specimen during dwell time was calculated from 3D FE creep analyses using MARC-K5 program. Fig.7(a) ~(c)[15] show the evaluation results of both creep damage,Dc,and fatigue damage, Df, of type 304 cruciform specimen based on three kinds of stress parameters, respectively. They are:(a) Mises stress,(b) COD stress and (c) Huddleston stress as previously shown in Table 1. Both Df and Dc were calculated from the following equations;

$$Dc = Nf \int_0^{tH} \frac{dt}{tr} \ , \ Df = \frac{Nf}{Nf^*}$$ (1)

where Nf^* and Nf are the failure lives of multiaxial LCF at 823K and those of multiaxial creep-fatigue with tH =10 and 30min. at 873K,respectively, on the basis of COD strain, $\varepsilon^* = 1.83 \ (2-\phi)^{-0.66} \ \varepsilon_y$. COD strain is the strain intensity of COD as well as COD stress [2]. In the figure, the solid line denotes the linear damage rule of Df+Dc=1 and the broken line shows the straight line which goes through the coordinate (0.3,0.3). It was found from the figure that the data points based on Mises stress and Huddleston stress locate outside of the linear damage rule line but those based on COD stress were arranged well by the linear damage rule. It was clarified by Hamada [16] that the creep-fatigue lives of type 304 tubular specimen under $-1.0 \leqq \phi \leqq -0.5$ at 923K were also arranged by the linear damage rule on the base of COD stress and COD strain. These findings reveal : The multiaxial creep-fatigue failure life of type 304 stainless steel at high temperatures in a full range of principal strain ratio, ϕ ,from-1 to 1 is analytically predicted from both data of uniaxial LCF test and uniaxial tension creep rupture test, by using the COD stress and COD strain parameters based on mode I crack and the linear damage rule.

REFERENCES

1. Yamauchi,M.,Ohtani,T.and Takahashi,Y.(1996).In:*Thermomechanical Fatigue Behavior of Materials*, 2,*ASTM STP* 1263.ASTM.
2. Hamada,N.,Sakane,M.and Ohnami,M.(1984) *Fatigue Engng Mater.Struct.*,7,85.
3. Hamada,N.,Sakane,M.and Ohnami, M.(1985) *Bull.JSME*,28,1341.
4. Sakane,M.,Ohnami,M.and Sawada,M.(1987) *J.Engng.Mater.Technol.,Trans. ASME*,109,236.
5. Sakane,M.and Ohnami,M (1989).In:*Proc.7th Inter. Conf. on Fracture (ICF7)*, K. Salma,K.Ravi-Chandar, D.M.R.Taplin and P.Rama Rao (Eds). Pergamon, Oxford,pp.1667-1674.
6. Sakane,M.,Ohnami,M.and Sawada,M.(1991) *J.Engng.Mater.Technol.,Trans. ASME*,113,244.
7. Itoh,T.,Sakane,M.Ohnami, M.Takahashi,Y.and Ogata,T.(1992).In:*Proc.5th Inter. Conf. on Creep of Mater.*, ASM International, pp.331-339.
8. Ohnami,M. and Sakane,M.(1992). In: *Proc. 5th Inter. Conf. on Creep of Mater.*, ASM International,pp.13-22.
9. Itoh,T.,Sakane,M.and Ohnami,M.(1994) *J.Engng.Mater.Technol.,Trans.ASME*, 116,90.
10. Joshi,S.R.and Shewchuk,J.(1970) *Exper.Mech.*,10,529.
11. Huddleston,R.L.(1985) *J. Pressure Vessel Technol.Trans.ASME*,107,421.
12. Hopper,C.D.and Miller,K.J.(1977) *J.Strain Analysis*,12,23.
13. Sakane,M.and Ohnami,M.(1996). In: *Proc.6th Inter.Conf.on Creep and Fatigue*, I Mech E.pp.143-152.
14. Mukai,S.,Komatsu,Y.,Ueda,D.,Sakane,M.,Ohnami,M.and Tsurui,T.(1997) *J.Soc.Mater.Sci.,Japan*,46,1374.
15. Mukai.S.,Sakane,M.,Ohnami, M.and Ogata,T.(1997) *J.Soc.Mater.Sci.,Japan*, 46, 1083.
16. Hamada,N.(1992) *J.Soc.Mater.Sci.,Japan*,41,1799.

Table 1 Stress parameters correlating multiaxial creep rupture life

Mises stress	①	$\sigma_{eq} = \sigma_1 (1 - \lambda + \lambda^2)^{0.5}$	
Max. principal stress	②	σ_1	
COD stress	③	$\sigma' = 0.91 \, \sigma_1 (2 - \lambda)^{0.14}$	for $-1 \leqq \lambda \leqq 0$
		$\sigma' = 0.71 \, \sigma_1 (2 - \lambda)^{0.5}$	for $0 \leqq \lambda \leqq 1$
Joshi stress	④	$\sigma_e = \sigma_1 (1 - 0.5 \, \lambda + \lambda^2)^{0.5}$	for $-1 \leqq \lambda \leqq 0$
		$\sigma_e = \sigma_1$	for $0 \leqq \lambda \leqq 1$
Huddelston stress	⑤	$\sigma_H = \sigma_{eq} \exp \left\{ 0.24 \left(\dfrac{1 + \lambda}{\sqrt{1 + \lambda^2}} - 1 \right) \right\}$	for $-1 \leqq \lambda \leqq 1$

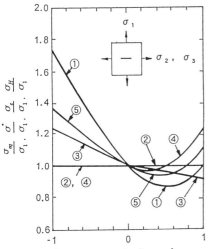

Fig. 1 Comparison of stress parameters.

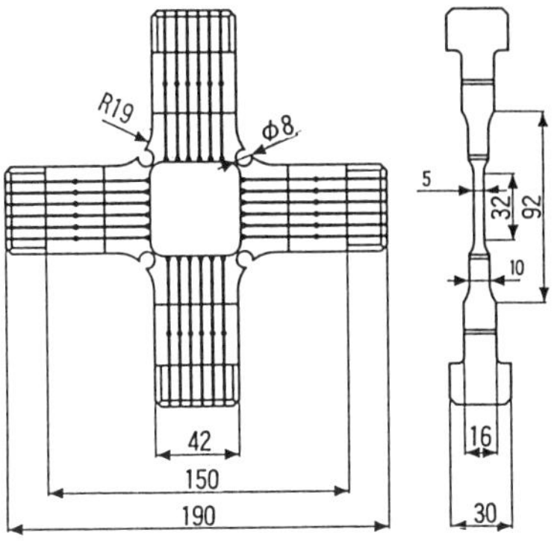

Fig. 2 Shape and dimensions (mm) of cruciform specimen for biaxial tension creep rupture test.

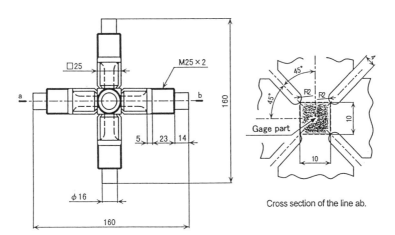

Fig. 4 Shape and dimensions (mm) of triaxial specimen for triaxial tension creep rupture test.

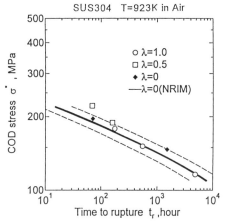

Fig. 3 Correlation of time to rupture with COD stress.

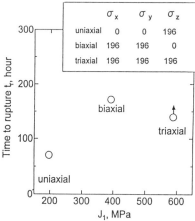

Fig. 5 Relationship between creep rupture time and J_1.

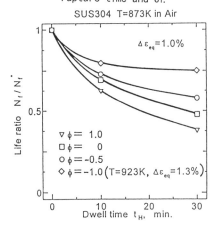

Fig. 6 Reduction of biaxial LCF failure lives with dwell time.

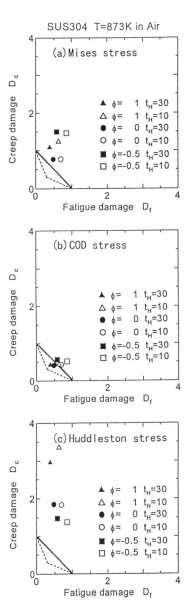

Fig. 7 Creep-fatigue damage diagram by linear damage rule on the basis of (a) Mises stress, (b) COD stress and (c) Huddleston stress.

Low Cycle Fatigue and Elasto-Plastic Behaviour of Materials
K-T. Rie and P.D. Portella (Editors)

COMPARATIVE TECHNIQUES FOR MULTIAXIAL TESTING

R. D. LOHR
Department of Engineering, Instron Ltd,
High Wycombe, England

ABSTRACT

The progression in design of testing machines leading to testing of single geometry specimens over the full range of strain ratios is reviewed. The optimisation of machines and specimens for studies of low cycle fatigue life including initiation, or crack propagation from a starter feature, is examined. Design approaches which have led to improvements in testing facilities, so that engineers and researchers are better able to simulate real world environments and loading conditions, are highlighted. The overall choice of system specification, taking into account both performance and inherent cost is analysed using a semi-graphical tabular presentation. References are made throughout the paper to actual multiaxial fatigue testing systems, some arising from the author's employment with a major materials testing machine manufacturer.

KEYWORDS

Multiaxial, Torsion, Internal pressure, External pressure, Biaxial strain ratio, Cruciform, Thin-wall tube, Modal control, LCF, Crack propagation

SYMBOLS

ϕ = Biaxial strain ratio = $\varepsilon_2 / \varepsilon_1$ where $\varepsilon_1 > \varepsilon_2 > \varepsilon_3$
ν = Poisson's Ratio

INTRODUCTION

Interest in multiaxial testing has been driven by recognition of the inadequacy of uniaxial data plus classical yield criteria to predict fatigue failures in real engineering components such as pressure vessels and rotating elements. Further stimulus has come from the high biaxial strain fields introduced by thermal transients during startup and rapid shutdown of steam and gas turbine power plant which can lead to LCF failures. The design of testing systems has progressed from simple mechanical rigs with open loop analogue control to the complex multi-actuator systems of today featuring digital closed loop control. Ever closer simulation of real world mechanical and thermal environments is being achieved through the development of advanced application software for multi-axis waveform generation, test control and data analysis.

BEAM AND PLATE BENDING

In 1950 a research program sponsored by the Pressure Vessel Research Committee in the USA led to investigations based on variable width cantilever bend [1], anticlastic bending of rhombic plates [2] and reversed pressurisation of circular and oval plates [3]. Such systems have the advantages of relatively simple mechanical design, no tendency to buckle and ease of specimen observation, however they suffer from high through-thickness strain gradients, limited achievable ϕ and the need to vary specimen geometry in order to change ϕ. Nevertheless the large uniform stress field associated with anticlastic bending has enabled it to continue to be useful in crack initiation and propagation studies [4].

CRUCIFORM SYSTEMS

Development [5] in 1963 at Chance Vought Co. of a rig enabling the application of tension to the ends of a cross shaped plate specimen was a great step forward since in principle all ϕ can be achieved with this single geometry. At Cambridge, orthogonal mounting of 4 hydraulic actuators in an annular frame (Fig 3) enabled fully reversed fatigue testing of a dished cruciform specimen [6], however it was the addition of closed loop control [7] that realised testing for $-1 \le \phi \le +1$. Problems include: load paths around the dished centre so that gauge area stresses are not known explicitly; total strains can be measured but plastic strains can only be estimated; buckling in compression limits maximum strains especially if the dished centre is flat bottomed.

A significant improvement in specimen design at Sheffield [8] introduced slotted arms and a large square flat bottomed gauge area with radiused edges which works well in tension-tension applications, minimising cross-coupling of forces between adjacent arms of the specimen and creating a large essentially constant biaxial strain field. These features, coupled with ease of visual observation, have enabled the cruciform to become a standard for crack propagation studies and, because of the flat specimen geometry, an ideal choice for testing sheet and plate materials. Recent advances include:

- a facility at JUTEM in Japan (Fig 6) to test CMCs at temperatures up to 1800°C by RF heating plus susceptor in a vacuum chamber with in-situ biaxial extensometry. Crack monitoring by laser scanning microscope is enabled by modal control (derived from sum and difference signals of opposing actuator transducers) holding the centre of the specimen stationary within microns [9].

- a large system at NASA (Fig 9) for testing aerospace materials including intermetallics [10] at temperatures up to 1500°C using the latest radiant furnace technology.

- at LMT, Cachan, a six actuator system provides the ability to test not only cruciform specimens, but also cubic triaxial specimens (Figs 10 & 11) enabling studies of hydrostatic tension and compression together with shear on all possible 45° planes [11].

THIN WALL TUBES

Tubular specimens have the advantage that all loads, axial, torsional or pressure, are fully carried by the gauge area. As a result stresses and plastic strains can be determined explicitly with benefit to fundamental studies of multiaxial fatigue, while buckling can be better controlled than with flat plate geometries. On the other hand, specimen observation is often obscured by a pressure vessel and elevated temperatures limited by the pressurisation medium. At Kyoto, in the

1960s, axial + torsion tests ($-1 \leq \phi \leq -\nu$) were performed at ambient and elevated temperature [12] while from the 1970s onwards closed loop servo-hydraulic systems have been widely utilised for such tests including the effects of out-of phase loading.

The effects of combining axial + internal pressure loading at elevated temperature (815°C) were investigated in the 1960s at Oakridge [13], however development at Waterloo [14] of a system combining axial with internal and external pressure, theoretically permitting all ϕ to be achieved, was an important advance. During the 1970s, work at Bristol [15,16] resulted in a system with closed loop servo-hydraulic control of axial force and differential pressure, a specimen design to minimise buckling, and capacitive axial and diametral extensometry (Figs1 & 2). This realised the goal of fully reversed strain controlled LCF, including stress-strain hysteresis loop generation on both axes, for all $-1 \leq \phi \leq +1$. In the 1980s, Sheffield developed a system which combined axial force, torsion, internal and external pressure, enabling all ϕ to be achieved but with the ability (through torsion) to rotate the principal axes of strain [17]. Systems, described above, employing external pressure used oil as the pressurising medium and were limited to testing at ambient or up to 140°C [15].

Figures 4 & 5 show an axial force + cyclic internal pressure (argon) system at CEAT, Toulouse, for testing gas turbine materials. A combination of direct current and RF heating is used to attain 1100°C and to control temperature gradients. Axial and transverse extensometry is used for strain control. Figures 7 & 8 show an axial force + torsion + internal pressure (oil) system at NTH, Trondheim. An internal mandrel eliminates axial force due to pressure while an axial/torsional extensometer provides strain control.

Thin wall tubes have also been favoured for TMF studies where their ease of rapid heating by RF induction and cooling by gas blow-down through the bore, or conduction into water cooled grips, is a great advantage. Recently, TMF studies at BAM, Berlin, have been published for axial + torsion loading of nickel based superalloys including single crystal material [18].

LCF FAILURE CRITERIA

In uniaxial fatigue studies, failure is frequently characterised by a percentage drop in peak tensile stress, which is closely related to the crack area in relation to the specimen cross sectional area. However in multiaxial studies other criteria are often more relevant and may be unavoidable.

Cruciform specimens may show little loss of peak force, even with quite large cracks, because load is shunted around the gauge area by thicker material. As a result, specified surface crack length is often used as the end-of-test criterion. Tests on thin wall tubes under internal or external pressure need to be terminated rapidly once the crack has penetrated the wall because of loss of pressurising medium or control if internal and external pressure volumes become interconnected. However the surface crack lengths associated with penetration of the wall vary considerably with ϕ as recent studies have shown [19]. Specifically, equibiaxial ($\phi = +1$) cracks had short surface lengths (0.6mm) at penetration, whereas pure shear cracks ($\phi = -1$) were much longer (5mm).

The above variation in crack morphology with ϕ is not restricted to pressurised tubes, indeed similar comments were made in respect of cruciform tests [6]. However it may help to explain differences in ascribed fatigue lives between research groups and the relevance of different endurance models in the LCF regime.

CONCLUSIONS

Techniques for multi-axial testing have been reviewed.

Table 1 overleaf provides an assessment of ten design approaches by ten major features, including the range of biaxiality (ϕ) achievable, which should facilitate system selection for the application need.

The last column in the table gives the number of actuators, and therefore control channels, required by the system. This number is related to the system cost, although issues like force capacity, special grips, extensometry and environmental requirements will act as modifiers.

It is perhaps persuasive to conclude that for studies of biaxial fatigue crack propagation, especially of materials in sheet or plate form, cruciform systems are the preferred approach.

Studies of LCF endurance and modelling , which need explicit knowledge of stresses and plastic strains, point to the specification of systems based on thin walled tubes subject to axial force, internal and external pressure.

REFERENCES

1.   Sachs, G., Gerberich, W.W., Weiss, V. and Latorre, J.V. (1961) *Proc. ASME 60, 512.*
2.   Zamrik, S.Y., (1967) *Third Annual Progress Report, Penn. State University, October.*
3.   Zamrik, S.Y., (1968) *Fifth Annual Progress Report, Penn. State University, October.*
4.   Zamrik, S.Y., (1997) *Proceedings of the 5th International Conference on Biaxial / Multiaxial Fatigue and Fracture,* pp. 167-187, Technical University of Opole, Poland.
5.   McLaren, S.W. and Terry, E.L., (1963) *ASME Paper, 63-WA-315.*
6.   Pascoe, K.J. and de Villiers, J.W.R., (1967) *Journal of Strain Analysis,* 2, 117.
7.   Parsons, M.W. and Pascoe, K.J., (1975) *Journal of Strain Analysis, 10, 1.*
8.   Brown, M.W. and Miller, K.J., (1985) *Multiaxial Fatigue, ASTM STP 853,* pp. 135-152.
9.   Masumoto, H. and Tanaka M., (1995) In: *Ultra High Temperature Mechanical Testing,* pp. 193-207, R.D.Lohr and M.Steen (Eds.) Woodhead, Cambridge, UK.
10.   Bartolotta, P.A., Kantzos, P.and Krause, D.L., (1997) *Proceedings of the 5th Int. Conf. on Biaxial / Multiaxial Fatigue and Fracture,* pp. 389-402, Tech. University of Opole.
11.   Calloch, S., (1997) PhD Thesis, E.N.S. de Cachan, Paris.
12.   Taira, S., Inoue, T. and Takahashi, M., (1967) *Proceedings of the 10th Japan Congress on Testing Materials,* pp.18-23.
13.   Kennedy, C.R., (1963) *Fatigue of Aircraft Structures, ASTM STP 338,* pp. 92-104.
14.   Havard, D.G. and Topper, T.H., (1969) *Proc. Soc. Expl. Stress Analysis,* 16, 305.
15.   Andrews, J.M.H. and Ellison, E.G. (1973) *Journal of Strain Analysis,* 8, 168.
16.   Lohr, R.D. and Ellison, E.G., (1980) *Fatigue of Engineering Materials and Structures,* 3, 1.
17.   Found, M.S., Fernando, U.S. and Miller, K.J., (1985) *Multiaxial Fatigue, ASTM STP 853,* pp.11-23.
18.   Meersman, J., Frenz, H., Ziebs, J., Klingelhoeffer, H. and Kuehn, H.-J., (1997) *Proceedings of the 5th Int. Conf. on Biaxial / Multiaxial Fatigue and Fracture,* pp. 303-322, Tech. University of Opole.30
19.   Shatil, G., Smith, D. and Ellison, E.G., (1994) *Fatigue and Fracture of Engineering Materials and Struct*ures,17, 159.

Table 1 Biaxial Specimen Schematics and Modes of loading	Range of surface Principal Strains	Single geometry	Immune to buckling	Invariant σ,ε on gauge area	Minimal thro. 't' ε gradient	Monitoring of biax σ and ε_p	Specimen observation	High temp capability	Rotation of principal axes	Actuator no. prop. to cost
Cantilever Bend		N	Y	Y	N	Y	Y	Y	N	1
Anticlastic Bend		N	Y	Y	N	Y	Y	Y	N	1
Plate Pressurisation		N	Y	Y	N	Y	N	N	N	1
Cruciform LCF		Y	N	N	Y	N	Y	Y	N	4
Cruciform crack propagation		Y	N	Y	Y	Y	Y	Y	N	4
Axial + Torsion		Y	N	Y	Y	Y	Y	Y	Y	2
Axial + P_{int}		Y	N	Y	Y	Y	Y	Y	N	2
Axial + P_{int} + Torsion		Y	N	Y	Y	Y	Y	Y	Y	3
Axial + P_{int} + Torsion const P_{ext}		Y	N	Y	Y	Y	N	Y	N	2
Axial + P_{int} + Torsion + P_{ext}		Y	N	Y	Y	Y	N	N	Y	4

240

Figs 1 & 2: Bristol, 1970s, Axial Force + Internal + External Pressure, Capacitive Extensometry

Fig. 3: Cambridge Cruciform, 1960s and 1970s   Figs. 4 & 5: CEAT, Toulouse, 1980s, Axial + Internal Pressure (Argon), 1100°C in air

Fig. 6: JUTEM, Japan, 1990s, Cruciform, 1800°C in vacuum

Figs 7 & 8: NTH, Trondheim, 1990s, Axial + Torsion + Internal Pressure

Fig 9: NASA, 1990s, Cruciform, 1500°C, radiant furnace

Figs 10 & 11: LMT, Cachan, 1990s, 6 Actuator Triaxial system and specimen

Low Cycle Fatigue and Elasto-Plastic Behaviour of Materials
K-T. Rie and P.D. Portella (Editors)

AN ALGORITHM FOR SOLVING NON-LINEAR PROBLEMS IN MECHANICS OF STRUCTURES UNDER COMPLEX LOADING HISTORIES

Markus Arzt

GKSS Forschungszentrum Geesthacht GmbH
Institut für Werkstofforschung
Max-Planck-Straße, 21502 Geesthacht, Germany

ABSTRACT

An algorithm in the framework of the Large Time Increment method is presented. This computational algorithm allows to deal with non-linear material behaviour, such as plasticity and viscoplasticity, for quasi-static problems under cyclic loading histories. The method differs from the common integration methods because it is an iterative method which takes into account the whole loading history in one single increment at once. A significant advantage of this algorithm is that no hypothesis of the interpolation time functions is needed. At each iteration, these functions depend on material properties algorithm parameters and previous iterations.

KEYWORDS

Structural analysis, numerical simulation, cyclic loads, constitutive equations, viscoplasticity

INTRODUCTION

Structures under cyclic loading conditions are studied. The behaviour under cyclic loadings is an interesting task since the failure of structures is often caused by repeated cycles of loadings, e.g. an aircraft turbine disk, see ref. [11] and [14]. To simulate the phenomena of material behaviour in case of cyclic loadings, such as multiaxial and out-of-phase loading conditions, non-linear constitutive relations have to be applied, see ref. [2], [3] and [8]. The material behaviour is described by so-called internal variables, see ref. [17], [19] and [20]. These constitutive relations are strongly non-linear, see ref. [4], [6] and [7]. Therefore, the numerical analysis of structures is complex and the numerical algorithms are difficult to develop, see ref. [9]. The required processing time and memory for storage, to solve these problems, is still important. Even today, there are few complex computations carried out in industry, although the performances of computers is increasing. In order to reduce processing time and memory for storage, the costs of a simulation, efficient algorithms have to be developed, see ref. [5], [10] and [18]. In this paper the principles of an algorithm to solve the problems under complex loading histories is given.

PROBLEM TO BE SOLVED

The problem to be solved is quasi-static, isothermal conditions and small strains are assumed. Hence, the equilibrium equation and the kinematic condition are linear.

242

MATERIAL MODEL

The material model chosen is the model based on the works of Lemaitre and Chaboche. A standard and normal formulation of this model is applied. The constitutive equations are strongly non-linear. The state equations are linear except the relation of the isotropic hardening. A transformation is carried out to obtain a linear relation of the isotropic hardening. This transformation leads to a normal formulation, see ref. [15]. Therefore, the state equations are linear.

THE ALGORITHM

An algorithm in the framework of the LArge Time INcrement method, LATIN method, see ref. [15] and [16], is adapted to elasto-viscoplasticity, see ref. [2]. The method has been developed by Ladevèze, see ref. [13]. This algorithm differs from the common integration methods, see ref. [1] and [12], since it is an iterative method which takes into account the whole loading history at once in one single time increment. No a priori hypothesis is necessary. The first principle of the method is to separate the set of equations to be solved into two groups. Therefore, the iterative method consists of two stages, the local and the global stage. The first group of equations are linear which could be global and are solved at the global stage, e.g. equilibrium equation. The second group of equations are solved at the local stage, at an integration point, and these could be non-linear, e.g. constitutive equations. Each of the two groups of equations defines a subspace of solutions. The subspace of the solutions which verify the linear equations is denoted Ad. The subspace of the solutions which verify the local equations is denoted Γ. And again, the first principle of the method is to determine successive a solution of the two subspaces defined by the two groups of equations, until convergence is reached. The solution of the problem verifies the two groups of equations. The solution is the intersection of the two curves of the two subspaces Ad and Γ, see Fig. 1. The separation depends on the problem and the material model.

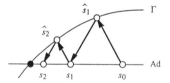

Fig. 1. Large Time Increment method

The second principle of the method is the appropriate choice of the directions E^+ and E^-, see Fig. 2. The choice has an important influence on the number of iterations required to reach convergence. The aim is to diminish the number of iterations. Unlike the first principle, the directions depend on the material model only.

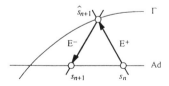

Fig. 2. Directions

The third and last principle is the key to decrease numerical computations and memory needed for storage of the unknowns. The storage depends on the problem to be solved.

Finally, the equations are separated, the first principle. The equilibrium equation, kinematic condition and state equations are solved at the global stage, since these equations are linear. The equations to be solved at the local stage are the constitutive equations, these equations are nonlinear.

The second principle consists in the choice of the directions, see Fig. 3. The direction is obtained by differentiation of the constitutive equations. Since a standard formulation is applied, the tangent operator is symmetric.

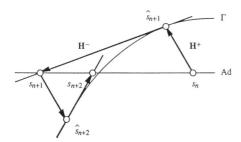

Fig. 3. Directions appropriate to the material model

The last principle is the important one, because the memory to store the unknowns is limited. In practice, not solutions are determined at the global stage. Instead corrections are calculated, and determined as a sum of products of space fields and scalar time functions. At each iteration of the method a space field and a time function is determined for each unknown. This storage allows the problem to be separated in two simple ones. An iterative procedure at the global stage is carried out. At convergence the optimum of space fields and scalar time functions is determined to store the unknowns. This procedure is well adapted to quasi-static problems, see ref. [10]. The advantage is obvious, especially when the number of degrees of freedom is large. Instead of storing a space field at each time step, which could exceed the memory capacity of a computer, one space field is stored only. The correction at a given time step is obtained by multiplying with the scalar time function.

The time function can be calculated efficiently by introducing two time scales. Since the loading is cyclic, the variation over the cycle length is nearly periodic. Therefore, one time scale is defined over the period of a cycle. The second time scale is related to the number of the cycle. This interpolation allows the calculation of the unknowns over the period of some cycles. A parabolic interpolation is applied over an interval which could contain hundreds of cycles.

244

EXAMPLE

As a complex numerical simulation is carried out the rotation of an aircraft turbine disk. The total number of degree of freedom is about 3000 and the structure is analysed when submitted to a 1000 cycle loading history. The mechanical solicitation is the centrifugal force. The temperature field is homogeneous, and the temperature is of 600° C. The material is a stainless steel called AISI 316 or Z6CNDT17-11.

The shape of one cycle of loading is presented in fig. 4.

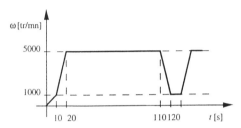

Fig. 4. Shape of cycle of loading

The mesh of the turbine disk is presented in Fig.5, it is an axisymmetric problem. Elements of three nodes are used. The number of elements is 2464, and the number of nodes is 1357.

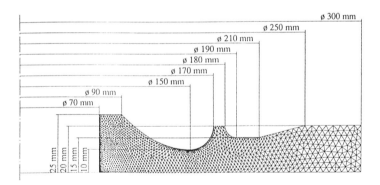

Fig. 5. Dimensions and mesh of the aircraft turbine disk

A sufficient accuracy is reached at iteration 7, the error with respect to the constitutive equations is less than 10 %.

The simulation is carried out by using the two time scales, since the variation over the period of a cycle length is nearly periodic. Using an interpolation, the initial values of a given cycle are calculated. The unknowns are then determined over the period of 20 cycles. It is not necessary to determine the unknowns over the period of all cycles. In fact of the range of the graph, the history of the stress appears as vertical lines, fig. 6. Although, at the scale of a cycle, the history is complete.

In fig. 6 is shown the stress versus time for a node at the bore. An important variation of the stress is noticed.

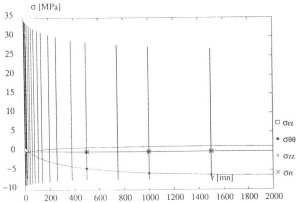

Fig. 6. Stress versus time

Figures 7 and 8 give the stress distribution at the 1000th cycle. Figure 7 shows the stress in circumferential direction, and fig. 8 shows the stress in radial direction. A high stress concentration is observed at the bore and at the neck.

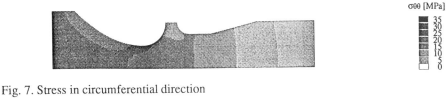

Fig. 7. Stress in circumferential direction

Fig. 8. Stress in radial direction

REFERENCES

1. Aravas, N., (1987). On the Numerical Integration of a Class of Pressure-Dependent Plasticity Models, *International Journal for Numerical Methods in Engineering*, vol. 24, pp. 1395–1416, John Wiley & Sons, Ltd.
2. Arzt, M., (1994). Approche des phénomènes cycliques par la méthode à grand incrément de temps, Ph.D. Thesis, Ecole Normale Supérieure de Cachan, France
3. Benallal, A., (1987), On the Stability of Time-Integration Schemes in Quasi-Static Hardening Elasto-Viscoplasticity, *Engineering Analysis*, volume 4, no. 2, pp. 95–99, Computational Mechanics Publications
4. Benallal, A. and Marquis, D., (1987). Constitutive Equations for Non-Proportional Cyclic Elasto-Viscoplasticity, *Journal of Engineering Materials and Technology*, vol. 109, pp. 218–228, Transactions of the ASME

246

5. Boisse, Ph.; Ladevèze, P. and Rougée, P., (1989). A Large Time Increment Method for Elastoplastic Problems, *European Journal of Mechanics*, A/Solids, volume 8, no. 4, pp. 257–275, Gauthier-Villars, Paris

6. Chaboche, J.-L., (1993). Cyclic Viscoplastic Constitutive Equations, Part I: A Thermodynamically Consistent Formulation, *Journal of Applied Mechanics*, vol. 60, pp. 813–821, USA

7. Chaboche, J.-L., (1993). Cyclic Viscoplastic Constitutive Equations, Part II: Stored Energy—Comparison Between Models and Experiments, *Journal of Applied Mechanics*, vol. 60, pp. 822–828, USA

8. Chaboche, J.-L. and Cailletaud, G., (1985). Sur le calcul de structures en viscoplasticité cyclique, *La Recherche Aérospatiale (French and English Edition)*, no. 1, pp. 41–54, Gauthier-Villars, Paris

9. Cognard, J.-Y. and Ladevèze, P., (1991). A Parallel Computer Implementation for Elastoplastic Calculations with the Large Time Increment Method, *Non-linear Engineering Computations*, pp. 1–10, Swansea

10. Cognard, J.-Y. and Ladevèze, P., (1993). A Large Time Increment Approach for Cyclic Viscoplasticity, *International Journal of Plasticity*, vol. 9, pp. 141–157, Pergamon Press Ltd.

11. Dambrine, B. and Mascarell, J. P., (1987). About the Interest of Using Unified Viscoplastic Models in Engine Hot Components Life Prediction, *Proceedings of MECA-MAT '87, International Seminar on High Temperature Fracture Mechanics*, vol. II, pp. 1–15, Dourdan

12. Hibbitt, Karlsson & Sorensen, Inc.: ABAQUS, (1995). *Theory Manual*, Version 5.5

13. Ladevèze, P., (1985). Sur une famille d'algorithme en mécanique des structures, *Comptes Rendus de l'Académie des Sciences*, tome 300, série II, no. 2, pp. 41–44, Gauthier-Villars, Paris

14. Ladevèze, P. and Rougée, P., (1985). Plasticité et viscoplasticité sous chargement cyclique : propriétés et calcul du cycle limite, *Comptes Rendus de l'Académie des Sciences*, tome 301, série II, no. 13, pp. 891–894, Gauthier-Villars, Paris

15. Ladevèze, P., (1989). La méthode à grand incrément de temps pour l'analyse de structures à comportement non linéaire décrit par variables intermes, *Comptes Rendus de l'Académie des Sciences*, tome 309, série II, pp. 1095–1099, Gauthier-Villars, Paris

16. Ladevèze, P., (1992). New Advances in the Large Time Increment Method, *New Advances in Computational Structural Mechanics*, pp. 3–21, Ladevèze, P and Zienkiewicz, O. C., Elsevier, Amsterdam

17. Lemaitre, J. and Chaboche J.-L., (1994). *Mechanics of Solid Materials*, Cambridge University Press, UK

18. Lesne, P. M. and Savalle, S., (1989). An efficient Cycles Jump Technique for Viscoplastic Structure Calculations Involving Large Number of Cycles, 2nd *International Conference on Computational Plasticity*, pp. 591–602, Barcelona

19. Nouailhas, D., (1989). Unified Modelling of Cyclic Viscoplasticity: Application to Austenitic Stainless Steels, *International Journal of Plasticity*, vol. 5, pp. 501–520, USA

20. Watanabe, O. and Atluri, S. N., (1986). Internal Time, General Internal Variable, and Multi-Yield-Surface Theories of Plasticity and Creep: A Unification of Concepts, *International Journal of Plasticity*, vol. 2, no. 1, pp. 37–57, Pergamon Press Ltd.

Low Cycle Fatigue and Elasto-Plastic Behaviour of Materials
K-T. Rie and P.D. Portella (Editors)

FATIGUE IN BIAXIAL STRESS STATE OF PRESTRAINED ALUMINIUM

W. JERMOŁAJ

*Department of Applied Mechanics, Bialystok Technical University,
15-463 Bialystok, Czysta 11, Poland*

ABSTRACT

In the paper is presented a review of fatigue problems of metal alloys prestrained plastically. The results of own investigation on fatigue in a biaxial stress state for an aluminium alloy noprestrained and prestrained statically are also placed. Tests were carried out in a range limited fatigue strength. It was made the results of testing for stage of a failure by means of modified strength criterion. The limited fatigue strength of aluminium alloy prestrained statically shows a decrease in comparison with a determined value of fatigue strength for noprestrained material.

KEYWORDS

Prestrained statically, aluminium, fatigue in biaxial stress state.

INTRODUCTION

A plastic deformation can influence really on some properties as for example a creep strength or fatigue strength of metals. This influence was tested under simple loads in numerous papers. For example this effect in fatigue process is shown for low carbon steels in the papers [1, 2] and for certain aluminium alloys [3, 4, 5]. In this papers were observed a decrease of fatigue strength by small (ca 2-3%) initial plastic deformations for both mentioned metal alloys.
The authors of paper [3] give that at higher amounts of an initial deformation of metals the fatigue strength is higher than that for the virgin material. In this papers by fatigue testing was kept a direction of uniaxial stress initiating plastic deformation. In the paper [6] are presented the results of fatigue testing in uniaxial stress state by pulsating stresses of different magnitudes σ_{max} and σ_{min} on flat specimens prestrained plastically. This specimens were prestrained statically in different complex stress states. The authors [6] gave that they have used the special way of statically loading. It give an equivalent results of static investigations on tubular specimens. Here was also stated that a variability, i.e. decrease or increase of a fatigue strength depend from magnitude and direction initial deformation.
A behaviour during of a fatigue in complex stress state of metals prestrained plastically can determine by fixing [7] that each of the identically prestrained plastically specimens of a set is loaded by a different cyclic complex stress states. The measured characteristic, exemplary fatigue strength, is shown in a stress space by means of a radius vector oriented in the same

manner as the cyclic stress vector. This procedure [7] is called the loading scheme 1.

An effect of behaviour of material prestrained plastically can also determine by means of procedure [7] called the loading scheme 2. Here each of specimen of the set is plastically prestrained under a different stress states in the direction of the proportional loading path to time obtainment of an equivalent stress $\sigma = \left(1.5\,S_{kl}S_{kl}\right)^{1/2}$. S_{kl} denotes the deviator of the stress tensor σ. All prestrained specimens are loaded a cyclic stress in the same manner under same level and stress state. The determined fatigue strength is shown in stress space by a radius vector oriented in the same manner as the prestressing stress vector.

The aim of this paper is to investigate the effect of behaviour under fatigue metal alloy prestrained statically. For describing of fracture the fatigue strength criterion is introduced.

EXPERIMENT

The fatigue tests were carried out at temperature 293K on two series of tubular specimens of an aluminium alloy. For the specimens of both series were made a stress relieving after a machining. The tempered specimens of second set, in contradistinction to first set, were prestrained statically by tension in direction $\varepsilon_{11} = \varepsilon_{11}^{e} + \varepsilon_{11}^{p} = 1.5$ percent. Tests were made on the pulsator type EDZ40Dyn adapted to testing in biaxial stress states, i.e. under cyclic tension at frequency of 15Hz and also static torsion. The specimens of both series were loaded in an asymmetrical cycle under stress intensity amplitude coefficient $A_{\sigma_i} = \sigma_i^{a}/\sigma_i^{m} = 0.25$ (σ_i^{a} - stress amplitude intensity, σ_i^{m} - mean stress intensity). The stress state was characterised by $\lambda = \sigma_{12}/\sigma_{11} = 0,\ 0.43,\ 0.73$ (σ_{12} - shear stress by torsion, σ_{11} - normal stress on direction of tension). During tests were recorded the number cycles to failure.

TEST RESULTS

The test results are made for stage of a failure of specimens noprestrained [8] and prestrained statically investigated in biaxial stress states on fatigue in an asymmetrical cycle.

The limited fatigue strength was describe by means of a modified creep strength criterion [9] in the form

$$\sigma_{red}^{c} = \beta\sigma_{max} + \left(1-\beta\right)\sigma_i \qquad (1)$$

where: σ_{red}^{c} - reduced stress in conditions of creep, σ_{max} - maximal principal stress, σ_i - stress intensity, β_c - material constant for data creep.

The periodically variable stress in a direction of tension in an asymmetrical cycle is presented in the form

$$\sigma_{11}\left(t\right) = \sigma_{11}^{m} + \sigma_{11}^{a}\,sin\,2\pi ft = \sigma_{11}^{m}\left(1 + A_{\sigma}\,sin\,2\pi ft\right) \qquad (2)$$

where: σ_{11}^{m} - mean stress, σ_{11}^{a} - stress amplitude, f - frequency, $A_{\sigma} = \sigma_{11}^{a}/\sigma_{11}^{m}$ - stress amplitude coefficient.

The amplitude σ_{11}^{a} of cyclic stress overlapping the static stress σ_{11}^{m} changes a kinetics of the creep process. In this paper is taken into account an effect stress amplitude σ_{11}^{a} and stress σ_{11}^{m}

on this changed process, i.e. dynamic creep. It is applied a substitute static stress tensor σ_{ij}^s in the following form

$$\sigma_{ij}^s = \sigma_{ij}^m + p\sigma_{ij}^a \tag{3}$$

where: $\sigma_{ij}^m, \sigma_{ij}^a$ - mean and amplitude stress tensor co-ordinates, p - parameter determined experimentally, determined value of parameter $p \approx 0.5$.

The modification of criterion (1) consists in dependence of the constant β on the stress intensity amplitude coefficient $A_{\sigma_i} = \sigma_i^a / \sigma_i^m$ (σ_i^a - stress intensity amplitude, σ_i^m - mean stress intensity).

For calculating of the magnitude of σ_{max} and σ_i eq. (1) is used the substitute stress tensor eq. (3). Finally is obtained the fatigue criterion in the form [8]

$$\sigma_{red}^t = \beta_f\left(A_{\sigma_i}\right)\sigma_{max} + \left(1 - \beta_f\left(A_{\sigma_i}\right)\right)\sigma_i \tag{4}$$

where: σ_{red}^t - reduced stress for data of fatigue, β_f - material constant in conditions of fatigue by steady of A_{σ_i}.

The discrete value of constant β_f independent from stress states, is obtained from eq. (4) by the method of the sum of least squares, respectively for both series of specimens by $A_{\sigma_i} = 0.25$.

The durability of aluminium alloy as a function of reduced stresses σ_{red}^t by $A_{\sigma_i} = 0.25$ is shown respectively, for noprestrained specimens, i.e. by $\varepsilon_{11} = 0$ in Fig.1a and for prestrained specimens, i.e. by $\varepsilon_{11} = 1.5$ percent in Fig. 1b.

The straight line of regression of the variability of durability t_r from σ_{red}^t is described by the dependence

$$ln\left(t_r/t_0\right) = p_2 + p_1\left(\sigma_{red}^t/\sigma_0\right) \tag{5}$$

where: $t_0 = 1h$, $\sigma_0 = 1MPa$, p_1, p_2 - coefficients determined statistically.

This coefficients are calculated by value of confidence level equal 95% and coefficient of correlation $R = 0.94$.

The values of coefficients p_1, p_2 and material constant β_f determined for fatigue of two series specimens, i.e. noprestrained and prestrained statically are presented in table 1.

Table 1. Calculated values of coefficients p_1, p_2 and constant β_f

Aluminium	$A_{\sigma_i} = \sigma_i^a / \sigma_i^m$	$\lambda = \sigma_{12}/\sigma_{11}$	p_1	p_2	β_f
noprestrained $\varepsilon_{11} = 0$	0.25	0 0.43 0.73	-0.0395	12, 223	0
prestrained $\varepsilon_{11} = 1.5\%$	0.25	0 0.43	-0.0333	12, 414	0.4

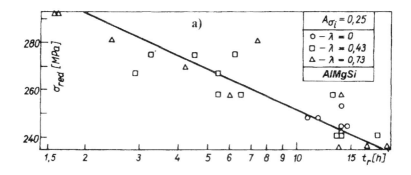

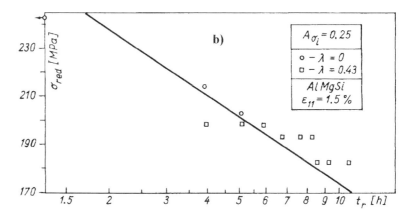

Fig. 1. Durability of aluminium alloy as a function of reduced stress by $A_{\sigma_i} = 0.25$: a - noprestrained alloy $\varepsilon_{11} = 0$, b - prestrained alloy $\varepsilon_{11} = 1.5\%$

For series of specimens noprestrained statically the failure goes on plastic state, $\beta_f = 0$ and σ_{red} depend from σ_i. The specimens prestrained statically fracture in elasto-plastic state, here $\beta_f = 0.4$ and σ_{red} depend from σ_i and σ_{max}.

CONCLUSIONS

The static deformation of aluminium alloy under fatigue in biaxial stress state induce:
1. decrease of value of the reduced stresses in comparison with value of noprestrained alloy,
2. change of value of the material constant β_f in the range $0 \le \beta_f \le 1$,
3. transformation of failure criterion by fatigue eq. (4), on $\sigma_{red}^f(\sigma_i)$ for noprestrained and $\sigma_{red}^f(\sigma_i, \sigma_{max})$ for prestrained alloy.

REFERENCES

1. Dyląg, Z. and Orłoś, Z. (1960) Testing of Behaviour of Plastic Strain on Fatigue Strength Some Low Carbon Steel [in Polish]. Biuletyn WAT **11**, 50.
2. Ingerma, A. and Rannat, E. (1974) On the Influence of Small Plastic Deformations on the Fatigue Strength of Metals [in Polish]. Czasopismo techniczne **70**, 1.
3. Arora, P.R. and Raghavan, M.R. (1973) Effect of Tensile Prestrain on Fatigue Strength of Aluminium Alloy in High Cycle Fatigue. Trans. ASME, J. Engng. Materials and Technology **95**, 76.
4. Frost, N.E. (1958) Fatigue Strength of Specimens Cut from Preloaded Blanks. Metalurgia **57**, 279.
5. Vitovec, F.H. (1958) Effect of Static Prestrain on the Fatigue Properties under Uniformly Increasing Stress Amplitude. Proc. ASTM **58**, 552.
6. Miastkowski, J. and Szczepiński, W. (1983) An Experimental Study of the Effect of Biaxial Plastic Prestrain on the Fatigue Strength of an Aluminium Alloy. Archives of mechanics **35**, 241.
7. Szczepiński, W. Ditrich, L. and Miastkowski, J. (1981) Plastic Properties of Metals. Experimental Methods of Mechanics of Solid Bodies [in Polish]. Mechan. Techn. (Ed.). PWN Warsaw.
8. Jermołaj, W. (1992) In: Proceedings of 3rd International Conference on Low Cycle Fatigue and Elasto-Plastic Behaviour of Materials (Ed.). Elsevier Applied Science, London & New York. **3**, pp 375-380.
9. Sdobyrev, W.P. (1959) Long Strength Criterion for Some Heat-Resisting Alloy by Complex Stress State [in Russian]. ANSSR. Mechan. Mashinostrojenye **6**, 12.

Low Cycle Fatigue and Elasto-Plastic Behaviour of Materials
K-T. Rie and P.D. Portella (Editors)
253

A NEW NONLINEAR DAMAGE CUMULATIVE MODEL
FOR MULTIAXIAL FATIGUE

D. G. SHANG, W.X.YAO and D.J.WANG*

*Department of Aircraft Engineering, Nanjing University of Aeronautics and Astronautics,
Nanjing,210016, P.R. China*

*Institute of mechanical Engineering, Northeastern University, Shenyang,110006,P.R.China

ABSTRACT

A nonlinear uniaxial fatigue damage cumulative model is proposed based on the continuum
fatigue damage theory, and then a nonlinear multiaxial fatigue damage cumulative model was
developed on the basis of combining multiaxial fatigue damage characteristics and the critical
plane principle for multiaxial fatigue. The recurrence formula of fatigue damage model was
derived under multilevel loading. The experimental results show that this model can be used
to predicted multiaxial fatigue life, and good agreement is demonstrated with experimental
data.

KEYWORDS

Nonlinear fatigue cumulative, damage, multiaxial fatigue, nonproportional loading, life
prediction.

INTRODUCTION

The basis and frame of continuum damage mechanics are gradually formed since
L.M.Kachanov[1] firstly presented the concepts of "continuum factor" and "effective stress".
J.Lemaitre and J.L.Chaboche [2] take the concept of damage to use in the research of low
cycle fatigue that make the analysis method of fatigue damage be of the theoretical basis, and
solve many problems of which only dependent experimental data can not deal with, for
example, nonlinear damage cumulation. But how to give multiaxial fatigue damage
cumulative model, the research works for this aspect very few and lack the experimental
verification [3]. It is further need to study and verify.
The aim of the present paper is that a nonlinear uniaxial fatigue damage cumulative model is
proposed, and then according to the results of the critical plane approach by author [4,5],
combining the behavior of multiaxial fatigue damage, a multiaxial fatigue damage cumulative
model is developed. The proposed multiaxial fatigue damage cumulative model may be used
in both proportional and nonproportional loading.

UNIAXIAL NONLINEAR FATIGUE DAMAGE CUMULATIVE MODEL

On the basis of J. Lemaitre and J.L.Chaboche model [3], Refer[4] presented a fatigue damage cumulative model to describe the process of material degeneration gradually.

$$\frac{dD}{dN} = (1-D)^{\alpha(\Delta\sigma/2,\sigma_m)} \left[\frac{\sigma_{max} - \sigma_m}{M(\sigma_m)}\right]^{\beta} \tag{1}$$

Where the function $(\alpha\,\Delta\sigma/2,\sigma_m)$ is taken as follows

$$\alpha\,(\Delta\sigma/2,\ \sigma_m) = 1 - \frac{H(\Delta\sigma/2 - \sigma_l(\sigma_m))}{a\ln\left|\Delta\sigma/2 - \sigma_l(\sigma_m)\right|} \tag{2}$$

Where $H(x)$ is Heaviside function, i.e. for $x > 0$, $H(x)=1$; for $x \le 0$, $H(x) = 0$, a, M, β are the material constants. For $(\Delta\sigma/2 - \sigma_l(\sigma_m)) > 1$, it is thought to be loading above fatigue limit.

For the constant loading parameter $\Delta\sigma/2$, Equation (1) is integrated that may obtain the number of cycle to failure and the evolution equation, respectively

$$N_f = \frac{1}{1-\alpha} \left[\frac{\sigma_{max} - \sigma_m}{M(\sigma_m)}\right]^{-\beta} \tag{3}$$

$$D = 1 - (1 - \frac{n}{N_f})^{\frac{1}{1-\alpha}} \tag{4}$$

It may be shown from Equation (2), for $\Delta\sigma/2 \le \sigma_l(\sigma_m)$ i.e. the cyclic loading below the fatigue limit, $\alpha = 1$, it is given from Equation (3), $N_f = \infty$.

For $\Delta\sigma/2 > \sigma_l(\sigma_m)$, $H(\Delta\sigma/2 - \sigma_l(\sigma_m)) = 1$, Equation (3) is become as

$$N_f = aM_0^{\beta} \ln\left|\Delta\sigma/2 - \sigma_{-1}(1-b\sigma_m)\right| \left[\frac{\Delta\sigma/2}{1-b'\sigma_m}\right]^{-\beta} \tag{5}$$

It may be seen from Equation (5) that the constants β and aM_0^{β} may be determined by S — N curve with the mean stress $\sigma_m = 0$, and b' may be determined by S — N curve with the mean stress $\sigma_m \ne 0$.

If the loading parameters are the strain control, the stress may be transformed to the strain by means of the cyclic stress-strain relationship. It is gotten by the strain hardening rule (for steady state).

$$\Delta\sigma/2 = K(\Delta\varepsilon/2)^n \tag{6}$$

Thus Equation (1),(2)and (3) are as follows, respectively

$$dD = (1-D)^{\alpha(\Delta\varepsilon/2,\sigma_m)} \left[\frac{K(\Delta\varepsilon/2)^n}{M_0(1-b'\sigma_m)}\right]^{\beta} dN \tag{7}$$

$$\alpha = 1 - \frac{H(K(\Delta\varepsilon/2)^n - \sigma_{-1}(1-b\sigma_m))}{a\ln\left|K(\Delta\varepsilon/2)^n - \sigma_{-1}(1-b\sigma_m)\right|} \tag{8}$$

$$N_f = aM_0^\beta \frac{\ln\left|K(\Delta\varepsilon/2)^n - \sigma_{-1}(1-b\sigma_m)\right|}{H(K(\Delta\varepsilon/2)^n - \sigma_{-1}(1-b\sigma_m))}\left[\frac{K(\Delta\varepsilon/2)^n}{1-b'\sigma_m}\right]^{-\beta} \quad (9)$$

MULTIAXIAL FATIGUE DAMAGE CUMULATIVE MODEL

Multiaxial fatigue damage cumulative model under proportional loading

It may show from Equation (7), which the damage control parameter is $\Delta\varepsilon/2$ for uniaxial fatigue damage rate equation. Owing to the fatigue behaviour of the multiaxial proportional cyclic loading and the uniaxial cyclic loading being identical, thus $\Delta\varepsilon/2$ of uniaxial may be replaced by von Mises equivalent strain amplitude $\Delta\varepsilon_{eq}/2$, then multiaxial fatigue cumulative model may be gotten. Because the Sines criteria corresponds to von Mises ellipse on the plane $\sigma_3 = 0$, the multiaxial fatigue limit may take the Sines criteria[5], thus Equation (7) and (9) may be rewritten as follows

$$dD = (1-D)^{\alpha(\Delta\varepsilon_{eq}/2,\overline{\sigma}_H)}\left[\frac{K(\Delta\varepsilon_{eq}/2)^n}{M_0(1-b'\overline{\sigma}_H)}\right]^\beta dN \quad (10)$$

$$N_f = aM_0^\beta \frac{\ln\left|K(\Delta\varepsilon_{eq}/2)^n - \sigma_{-1}(1-3b\overline{\sigma}_H)\right|}{H(K(\Delta\varepsilon_{eq}/2)^n - \sigma_{-1}(1-3b\overline{\sigma}_H))}\left[\frac{K(\Delta\varepsilon_{eq}/2)^n}{1-b'\overline{\sigma}_H}\right]^{-\beta} \quad (11)$$

Where the material constants, $aM_0^\beta, \beta, M_0, K, n, \sigma_{-1}$ may be determined by uniaxial fatigue test.

Multiaxial fatigue damage cumulative model under nonproportional loading

Refer [6] indicated, for tension-torsion thin-walled specimens, the normal strain excursion ε_n^* between adjacent turning points of maximum shear strain γ_{max} on the critical plane is very little under proportional loading, and ε_n^* equals the biggest range of ε_n. For the nonproportionl loading of the phase difference $\varphi = 90°$ between tension and torsion, the normal strain excursion ε_n^* between adjacent turning points of maximum strain on the critical plane has reached its maximum value. Because the additional hardening is the most serious, results in fatigue life to be the shortest. It is indicated that γ_{max} and ε_n^* on the critical plane are the two parameters of controlling multiaxial fatigue damage.

If ε_n^* and γ_{max} are combined as an equivalent strain by von Mises criterion and it is used as fatigue damage controlled parameter on the critical plane, a new multiaxial fatigue damage parameter will be given as:

$$\Delta\varepsilon_{eq}^{cr}/2 = \left[\varepsilon_n^{*2} + 1/3(\Delta\gamma_{max}/2)^2\right]^{1/2} \quad (12)$$

$$\varepsilon_n^* = 1/2\Delta\varepsilon_n\left[1 + \cos(\xi + \eta)\right] \quad (13)$$

Where the range of $\xi + \eta$ is between $-\dfrac{\pi}{2}$ and $\dfrac{\pi}{2}$.

By relating Equation (12) with Manson-Coffin equation, the multiaxial fatigue damage formula may be given:

$$\Delta\varepsilon_{eq}^{cr}/2 = \frac{\sigma_f'}{E}(2N_f)^b + \varepsilon_f'(2N_f)^c \qquad (14)$$

Where σ_f', ε_f', b and c are the material constants of uniaxial fatigue, N_f is fatigue life.

In the case of the proportional loading, it may be derived by Equation (12)

$$\Delta\varepsilon_{eq}^{cr}/2 = \Delta\varepsilon_{eq}/2 \qquad (15)$$

I.e., equation (14) reduces the equivalent strain approach form. In the case of the uniaxial loading, equation (14) become as:

$$\Delta\varepsilon_{eq}^{cr}/2 = \left[\Delta\varepsilon_n^{\ 2} + \tfrac{1}{3}(\Delta\gamma_{max}/2)^2\right] = \Delta\varepsilon/2$$

I.e., equation (33) reduces Manson-Coffin equation in uniaxial form.

A large number of experimental verification showed that the more accurate prediction for multiaxial fatigue life may be gotten by equation (14)[4,6].

In the case of nonproportional loading, if von Mises equivalent strain is directly used in multiaxial fatigue damage model, the large errors will be created due to no considering the additional hardening from nonproportional loading. It may be known through above analysis that the maximum shear strain amplitude $\Delta\gamma_{max}/2$ and normal strain excursion $\varepsilon_n^{\bullet}$ on the critical plane are two important parameters controlled the fatigue damage. The maximum shear strain amplitude $\Delta\gamma_{max}/2$ and the size of the normal strain excursion between adjacent turning points the maximum strain on the critical plane are combined as an equivalent strain $\Delta\varepsilon_{eq}^{cr}/2$ to replace $\Delta\varepsilon_{eq}/2$ of multiaxial proportional fatigue damage model, which may consider the additional hardening effect of nonproportional loading. Thus multiaxial fatigue damage cumulative model under nonproportional loading may be gotten as follows

$$dD = (1-D)^{\alpha(\Delta\varepsilon_{eq}^{cr}/2,\overline{\sigma}_H)}\left[\frac{K(\Delta\varepsilon_{eq}^{cr}/2)^n}{M_0(1-b'\overline{\sigma}_H)}\right]^\beta dN \qquad (16)$$

$$N_f = aM_0^\beta \frac{\ln\left|K(\Delta\varepsilon_{eq}^{cr}/2)^n - \sigma_{-1}(1-3b\overline{\sigma}_H)\right|}{H(K(\Delta\varepsilon_{eq}^{cr}/2)^n - \sigma_{-1}(1-3b\overline{\sigma}_H))}\left[\frac{K(\Delta\varepsilon_{eq}^{cr}/2)^n}{1-3b'\overline{\sigma}_H}\right]^{-\beta} \qquad (17)$$

$$D = 1 - (1-\frac{n}{N_f})^{\frac{a\ln|K(\Delta\varepsilon_{eq}^{cr}/2)^n - \sigma_{-1}(1-3b\overline{\sigma}_H)|}{H(K(\Delta\varepsilon_{eq}^{cr}/2)^n - \sigma_{-1}(1-3b\overline{\sigma}_H))}} \qquad \mathbf{(18)}$$

In the case of proportional loading, $\Delta\varepsilon_{eq}^{cr}/2$ is von Mises equivalent strain amplitude $\Delta\varepsilon_{eq}/2$, and $\Delta\varepsilon_{eq}^{cr}/2$ is the nominal strain amplitude $\Delta\varepsilon/2$ under uniaxial loading. Thus the uniaxial loading and proportional loading may be thought to be the special cases of the proposed multiaxial nonproportional fatigue damage model in this paper. From feature of this model , owing to the principle of the critical plane being combined to multiaxial fatigue damage model, thus the multiaxial fatigue damage behaviour is considered.

Owing to this model suiting either uniaxial loading or multiaxial loading, all material constants in model may be determined by the uniaxial fatigue experiment, and may avoid making the expensive multiaxial experiments.

Application of multiaxial fatigue damage model under multilevel loading

It is easy through sequential calculation to get multiaxial fatigue damage cumulative formula, i.e. by using some variable change, $Y_i = 1 - (1 - D_i)^{1-\alpha_i}$, where D_i is the value of damage at end of the i th loading level, then damage cumulative cyclic ratio

$$Y_i = 1 - 1 - D_{i-1})^{1-\alpha_i} + \frac{n_i}{N_{fi}} = 1 - (1 - Y_{i-1})^{\frac{1-\alpha_i}{1-\alpha_{i-1}}} + \frac{n_i}{N_{fi}} \qquad (i = 2,3,4,.....n) \qquad (19)$$

The integration is pursued until $Y_i = 1$, that corresponding to the fatigue life may be gotten.

EXPERIMENTAL VERIFICATIONS

In order to verify the proposed multiaxial fatigue damage model, it is needed to determine the material constants aM_0^β, β. After Equation (9) is rewritten, the two sides of equation are taken log (as taken $\sigma_m = 0$)

$$\lg(N_f / \ln|K(\Delta\varepsilon/2)^n - \sigma_{-1}|) = -\beta\lg[K(\Delta\varepsilon/2)^n] + \lg(aM_0^\beta)$$

The material constants may be determined by means of uniaxial S — N curve. For the normalized 45 steel, $K = 448.3 MPa, n = 0.209, \sigma_{-1} = 241.8 MPa$, the material constants may be gotten by fitting uniaxial experimental data. $aM_0^\beta = 1.23566 \times 10^{38}$, $\beta = 13.62$, the interrelated coefficient $\gamma = 0.9837$. For the material constants b, b', they basically are equal from Refer[7], which may take $1/\sigma_b$ on the basis of the Goodman Equation. where σ_b is ultimate tensile strength.

The comparisons between the predicted results by Equation (17) and experimental life under both proportional and nonproportional loading are shown Refer[4]. The errors are within a factor of 1.5. In order to verify the description ability of Equation (18) under multiaxial multilevel loading, the experimental data from Refer [8] are used to verify in the case of the two-level loading, the results are shown in Table 1. which the predicted values of n_2 / N_{f2} are gotten by Equation (19). It is shown in Table 1 that the predicted results in this paper are satisfactory, and are better than Miner' rule and the predicted results of Refer[8].

Table1. Comparisons between prediction and the experimental values under two-level loading*.

Strain ratio λ	Sequence	n_1	$\frac{n_1}{N_{f1}}$	n_2	$\frac{n_2}{N_{f2}}$	Prediction $\frac{n_{2p}}{N_{f2p}}$	Miner $\frac{n_2}{N_{f2}}$	Refer [8] $\frac{n_2}{N_{f2}}$	$\frac{n_1}{N_{f1}} + \frac{n_2}{N_{f2}}$	$\frac{n_1}{N_{f1}} + \frac{n_{2p}}{N_{f2p}}$
0	L-H	43000	0.41	1122	1.0	0.71	0.59	0.85	1.41	1.12
1	L-H	56000	0.49	885	0.62	0.645	0.51	0.75	1.11	1.14
∞	L-H	42000	0.5	572	0.64	0.637	0.5	0.7	1.14	1.14
0	H-L	550	0.49	33055	0.3	0.35	0.51	0.2	0.79	0.84
1	H-L	700	0.48	39690	0.35	0.36	0.52	0.21	0.83	0.84
∞	H-L	450	0.5	18641	0.22	0.34	0.5	0.28	0.72	0.84

*Two-level loading Mises equivalent strain amplitude are 0.22 %,1.0 %, respectively.

CONCLUSIONS

1. On the basis of the research results of uniaxial fatigue damage cumulative model and considering the mechanism of multiaxial fatigue failure, a multiaxial fatigue damage cumulative model is proposed by combining the critical plane approach to the uniaxial fatigue damage cumulative model, and the multiaxial fatigue damage evolution formula and the multiaxial fatigue life formula are given.

2. The application of the proposed multiaxial fatigue damage cumulative model is discussed , and the recurrence formula are derived under multilevel loading.

3. The uniaxial fatigue and multiaxial fatigue experimental verification show that the proposed multiaxial fatigue model is of the good precision of prediction. When this model is used to multiaxial cyclic loading, there is not to need any additional material constant. thus it is convenient for engineering application.

REFERENCES

1 .   Kachanov, L. M.(1958).Time of the rupture process under creep condition , TVZ Akad Nauk, S.S.R. Otd   Tech , Nauk,8.

2 .   Lemaitre,J. and Chaboche, J. L.(1974). In: Proceedings of IUTAM, Symp on Mechanics of Viscoelastic Media and Bodies , Springer-Verlag, Getheubury,.

3.   Lemaitre, J. and Chaboche, J. L.(1990). Mecanique des materiaux solides, Dunod , and Cambridge University Press,(English Edition).

4.   Shang, D.G. (1996).Ph.D, Dissertation, Northeastern University,.(in Chinese)

5.   Sines, G. (1956).Metal Fatigue , G. Sines J. L. Wailsman(Ed.), Mc-Graw, Hill, London , 138.

6.   Shang, D. G. and Wang D.J. (1998) Int. J. Fatigue 20 (in press).

7.   Wu, H.Y., (1990). The damage mechanics, National defense industry press, (in Chinese).

8.   Hua, C. T.Socie .D. F.(1984).Ph.D. Dissertation,Illinois University,.

9.   Gail.E. L. and Jodean, M. (1985). ASTM STP 853, pp482-496.

Low Cycle Fatigue and Elasto-Plastic Behaviour of Materials
K-T. Rie and P.D. Portella (Editors)
259

EXPERIMENTAL VERIFICATION OF SIMILARITY OF CREEP VELOCITY CURVES

ROBERT UŚCINOWICZ
Chair of Applied Mechanics, Bialystok Technical University,
15-483 Bialystok, Czysta 11, Poland

ABSTRACT

This paper presents experimental results concerned with the verification of similarity of creep curves. The test device, specimens and experimental procedure are expounded. The experimental tests were made at an elevated temperature (823K) for stationary creep process and simple monotonic loading process (non-stationary creep). Verification of similarity of creep curves was carried out for biaxial and multiaxial stress states for few discrete values of time. In the paper the essence of similarity assumption is presented.

KEYWORDS

Creep process, anisotropy, complex stress state, metal properties, proportional load

INTRODUCTION

Metals and alloys obtain anisotropy of some properties in the manufacturing (technological) process. Directional changes of the metal properties are often the effect of stretching, cold and warm multistage rolling and drawing. Understanding the behaviour of anisotropic processes in metals and the development of constitutive relationships are very important for numerical simulations, design and forming technological process. It is able to predict the occurrence of technological defect like necking and tearing. The creep process, which is particularly concerned with utilisation properties of metals and alloys, must be put to theoretical study. This study should respect anisotropy process. From the historical point of view, the equations of anisotropy creep were formulated and based on isotropic creep equation by their generalisation. These methods were established from theory of plasticity.
A lot of works on this subject present the basic equation as similarity of creep velocity tensor $\mathbf{d}$ and tensor of auxiliary stresses μ, i.e.

$$\mathbf{d} = \mathbf{S}\,\mu ,\qquad\qquad (1)$$

where: S-function of invariant of stress, strain and velocity of strain tensors.
Similar equations have been found in the research of Goldenblat-Kopnov [1], Betten [2], Sobotka [3], Baltov and Sawczuk [4]. The first anisotropy creep equations were formulated on the basis of existence of a creep (strain) velocity potential. These equations needed very restrictly assumptions. For this reason and other, the simpler solutions of this problem without using potential form were analysed. In the Mieleszko's work [5] simple, quasi-linear equation of creep anisotropy is presented in the form.

$$\mathbf{d} = \mathbf{F}^{(IV)}(\sigma)\,\sigma\,, \tag{2}$$

where: $F^{(IV)}$ – tensor, coefficient of proportionality.

Non-linearity is occurring in the coefficient $\mathbf{F}^{(IV)}(\sigma)$, because relationship among strain velocity tensor and stress tensor is quasi-linear. Transformation of above equation to constitutive law requires fulfilment of the assumption of creep curves similarity. This assumption is not easy to verified in experiment. It should be noted that similitude of the creep curves is usually observed in creep processes in metals and alloys.

For various stress states, if the proportionality of stress tensor components σ_{ij} to components of stress state vector λ_{ij} in the creep process will be kept, the components of velocity strain tensor d_{ij} will be proportional to components of strain vector χ_{ij}. This relation is given by equation (3).

$$\sigma_{ij} = \lambda_{ij}\,a \;\Rightarrow\; d_{ij} = \chi_{ij}\,a^1, \tag{3}$$

where: σ_{ij} - components of stress tensor, d_{ij} - components of velocity tensor, χ_{ij} - components of strain state vector ($\chi_{ij} = [\chi_1, \chi_2, \chi_6] = [d_{11}/d_{11}, d_{22}/d_{11}, d_{12}/d_{11}]$), λ_{ij} - components of stress state vector ($\lambda_{ij} = [\lambda_1, \lambda_2, \lambda_6] = [\sigma_{11}/\sigma_{11}, \sigma_{22}/\sigma_{11}, \sigma_{12}/\sigma_{11}]$), a, a^1 -parameters of proportionality, σ_{11}, σ_{22} - normal stress component, σ_{12} - shear stress component.

In this way suitable experimental verification of this assumption by using relationship (3), gives possibility to write equation (2) in below form (4):

$$\mathbf{d} = F_A^{(IV)}(\sigma)\,\mathbf{A}\,\sigma\,, \tag{4}$$

$F_A^{(IV)}(\sigma)$ is superposition of linear transformation and conversion space of stresses to space of strain velocities by using homothety operation. Tensor $\mathbf{A}$ is linear operator of this transformation, interpreted as tensor of strain anisotropy. The final, transforming equation (4) takes the form [5,6]:

- for secondary creep: $\mathbf{d}(\sigma) = G(\sigma_{red})\,\mathbf{A}\,\sigma,$ \hfill (5)
- for first and third period of creep: $\mathbf{d}(t', \sigma) = G(\sigma_{red}, t')\,\mathbf{A}(t')\,\sigma,$ \hfill (6)

For non-stationary creep (monotonic loading process) this equation is replaced by:

$$\mathbf{d}^f(t', \sigma^f) = G^*(\sigma^f_{red}, t')\,\mathbf{A}(t')\,t^f\,\underline{\sigma}^f, \tag{7}$$

where: $\mathbf{d}$-creep velocity tensor , $\mathbf{A}$-anisotropy tensor, $G(\sigma_{red}, t')$, $G^*(\sigma^f_{red}, t')$ - non-linearity functions of equations, σ_{red}, σ^f_{red}-reduced stresses, σ-stress tensor, $\underline{\sigma}^f$- rupture velocity tensor, t' - normalised time parameter; $t' \in \langle 0,1 \rangle$, t^f -time to rupture.

The time t', which is shown in equation (6-7), describes creep process in following way: value of zero (t'=0) describe the moment in which the specimen obtains the last load (σ_i) and value of one (t'=1) when the specimen is ruptured.

The aim of this paper is the verification of similarity of creep velocity curves, as necessary condition for describing the evolution of strain anisotropy i.e. components of anisotropy tensor.

PROGRAM OF INVESTIGATIONS

Material, Specimens & Testing Device

The investigated material was low-allowed, molybdenum-chromium 15HM steel (Cr-0.9%, Mo -0.55%, Mn-0.5%, C-0.15%). Creep investigations were carried out on thin-walled tubular specimens with external diameter, thickness and gauge length 12.0, 0.5 and 50.0 mm, respectively. After machining process the samples were subjected to heat treatment, it means, they were annealed for one hour at 950K and air-cooled to achieve a uniform ferritic-pearlite structure.

The main experiments were performed in special designed creep stands [7,8]. The testing system was designed to impose tension, torsion and internal pressure simultaneously on thin-walled tubular specimens. The software of personal computer controlled processes of heating (to 823K) and loading of samples and automatically transferred and registered all data. The loading process was realised with constant velocity of stress intensity by using the proportional loading. The displacements (strains) were measured directly from the sample by three various, independent channels by using seven inductive detectors. The first group of detectors was used for measuring the axial strains (ε_{11}), second for shear strains ($2\varepsilon_{12}$) and the last diametrical inductive transducer for the hoop strains (ε_{22}). All these three strains were measured in the range of near 0% to sample rupture, during the experiment. The data (electrical signal) from three different measuring channels (tension, torsion and internal pressure) were gathered in hard disc of computer in the creep process in real time.

Experimental Procedure

The investigations described two kinds of processes: stationary creep (process with constant load of specimen), non-stationary creep (proportional, monotonic loading process).
The specimens were loading a combination of axial load, torsional moment and internal pressure. In this way the simple and complex stress states were obtained which could be described by using stress state vector λ. In this way it was obtained three kinds of loading states, conforming to three kinds of stress states, such as:
- uniaxial and plane stress state (tension and tension with torsion),
- "specific" multiaxial stress state (tension and internal pressure),
- multiaxial stress state (tension, internal pressure and torsion).
The temperature was estimated at 832K. Scheme of loading of specimens in stationary creep (plane stress state) is shown in Fig. 1.

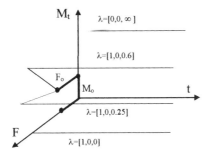

Fig. 1. Scheme of loading the specimens in the creep tests (plane stress state)

EXPERIMENTAL RESULTS

Stationary Creep Tests

Twenty-seven creep tests were made for description the stationary creep by using constant load. Seventeen of them were made in plane stress state and ten in multiaxial stress state. Tests were carried out in the 117- 258 MPa range of stress intensity and in the time to specimen rupture up to 1890 hours. All the curves displayed primary for $t' \in \langle 0, 0.04 \rangle$, secondary for $t' \in \langle 0.04, 0.3 \rangle$ and tertiary for $t' \in \langle 0.3, 1 \rangle$ creep periods. The primary creep period on the creep curves was very short, but tertiary - long (70% of specimen lifetime). Experimental data were described by using general equations in the form:

$$\varepsilon_{ij}(t/t_0) = p_1(t/t_0)^{p_2} + p_3 + p_4(t/t_0) \quad (8)$$

where: $\varepsilon_{ij}(t/t_0)$- components of strain tensor, p_i - coefficients of equation (i=1,..,4), t_0=1 hour, t-real time. The constant values of stress tensor components σ_{ij} were kept during tests and because of them the proportionality of components of creep velocity tensor dij took place. Above statement were proceeded for secondary creep i.e. $t' \in \langle 0.04, 0.3 \rangle$ and for other creep periods, too. The experimental values of d_{ij} on the planes d_{22}-d_{11} and d_{12}-d_{11} should fulfilled conditions: $d_{22} = \chi_2 d_{11}$ and $d_{12} = \chi_6 d_{11}$ and should lain in one straight line in one direction (χ_2 i χ_6 are slops of straight line). Verification of assumption of creep velocity curves similarity was made for a few discrete parameter time t' for three periods of creep. Exemplary graphical verifications of similarity creep velocity curves are presented on Fig.2. and Fig.3.

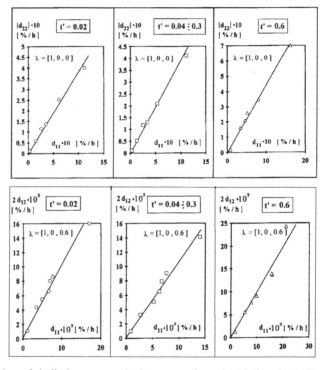

Fig.2. Verification of similarity creep velocity curves for uniaxial ($\lambda = [1,0,0]$) and biaxial stress state ($\lambda = [1,0,0.6]$) for various values of normalised time t'

Non-stationary Creep Tests

Description of non-stationary creep process was based on the results of investigation of 37 samples, twenty-one in plane stress state and twenty three in multiaxial stress state. Elementary increment of loading was at the average $\Delta\sigma_i = 15$ MPa. The tests were realised with three basic loading velocities: 1.8, 11.3, 30.0 MPa/min. In this case the limit of time t' was described in this way: values of zero describing the moment in which the sample obtained the limit of proportionality and value one-when the sample was ruptured. The results of examinations gave

263

possibility to determine strain velocity curves for the parameter t'. The samples were loading proportional (value of λ was constant), so increments of values of strain velocities were proportional, too. It is described in Fig.4. and Fig.5.

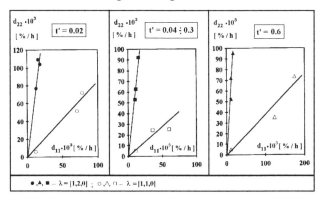

Fig.3. Verification of similarity creep velocity curves for multiaxial stress state (λ=[1,2,0], [1,1,0]) for various values of normalised time t'

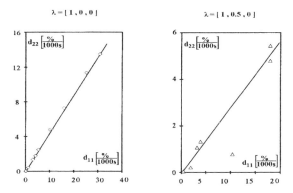

Fig.4. Verification of similarity strain velocity curves for biaxial stress state for time t'=0,6

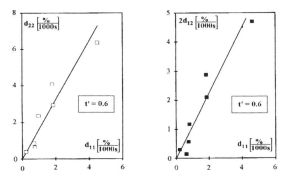

Fig.5. Verification of similarity strain velocity curves for biaxial stress state (λ = [1, 1, 0.5]) for time t'=0,6

264

DISCUSSION OF RESULTS

It could be state that the creep velocity curves for realised stress states (simple and complex stress states) are similar, because the relation (3) is fulfilled. Hence, the equation could be applied in quasilinear form for description examined processes. It is related to stationary and non-stationary creep process in an equal degree. For the group of samples which exposed to the same load, tangents of inclination angle of experimental straight-line to horizontal line achieved similar value for different time t' in plane: d_{11}- d_{12} and d_{11}-d_{12}. The best results were obtained in plane and uniaxial states. The great scatter of experimental values is observed in the multiaxial stress state. Alike tendency is observed in non-stationary creep (proportional, monotonic loading process) where relation between strain velocity tensor components at constant increment of loading is constant. In examined case the scatter of values is significant greater. The causes of this fact are linked to difficulty of description the creep in the initial and final phases of process, when the strain velocity is high. The measuring and approximation errors should be taken into account for description the curves. The same results were presented by Jakowluk & Mieleszko[5,9].

CONCLUSIONS

- The lower scatter of experimental points was observed in second period of creep process.
- Verification of assumption of similarity creep velocity curves was carried out for biaxial and multiaxial stress states for a few discrete values of normalised time t' gave positive result.
- Hypothesis of geometrical similarity of creep velocity curves gave possibility to use quasi-linear anisotropic theory in the research of creep process in metals and alloys.

REFERENCES

1. Goldenblat I.I., Kopnov V.A., (1965), Kriterij procnosti anizotropnych materialov. Izv. AN SSSR, Mech. 6, , 77-83.
2. Betten J.,(1981), Creep theory of anisotropic solids, J. Rheol., 6, **25**.
3. Sobotka Z.,(1969), Theorie der plastischen Fliessens von anisotropic Korpern, Z. Angen Math. Mech., 1-2, **49**, 25-32.
4. Baltov A. and Sawczuk A. ,(1965), A rule of anisotropic hardening, Acta Mech.,2, 1, 81-92.
5. Mieleszko E., (1987), Study about equation of creep constitutive laws (in Polish), Bialystok
6. Jakowluk A. & Mieleszko E., (1985), No-potential theory of the construction of anisotropic creep constitutive laws, Res. Mechanica, **2**, 16.
7. Uścinowicz R., (1995), The influence of stress state on the anisotropy creep in the ferritic steel (in Polish), Doctoral Thesis, Technical University of Białystok, Białystok.
8. Uścinowicz R., (1996), Influence of kind of loading state on the anisotropy development in the 15HM steel at an elevated temperature, Proceedings of 9th Int. Symp. -Creep Res. Met. Mater., Publishers Vitkovice a.s. Research & Development Division, Ostrava, s. 400-405.
9. Jakowluk A. , (1993), Creep and fatigue processes in materials, (in Polish), PWN, Warsaw.

Chapter 4

MICROSTRUCTURAL ASPECTS

Low Cycle Fatigue and Elasto-Plastic Behaviour of Materials
K-T. Rie and P.D. Portella (Editors)

MICROSTRUCTURAL ASPECTS OF LOW CYCLE FATIGUE

P. LUKÁŠ

*Institute of Physics of Materials, Academy of Sciences of the Czech Republic,
Žižkova 22, 61662 Brno, Czech Republic*

ABSTRACT

Several microstructural aspects of low cycle fatigue of metals are reviewed. The possibility to divide the metallic materials into different categories according to the fraction of loaded volume bearing the cyclic plastic deformation is discussed. Further it is shown that as the LCF requires activity of more slip systems, the cyclic stress-strain curves of multiple-slip oriented single crystals and polycrystals can be mutually transformed using the Taylor factor. Finally, the role of the cyclic slip localisation in cyclic plasticity and in initiation of microcracks is examined.

KEYWORDS

Cyclic plasticity, stress-strain curve, transformation, Taylor factor, strain localisation, dislocation structures, microstructure, microcrack initiation.

LOW-CYCLE-FATIGUE CYCLIC PLASTICITY

Low-cycle-fatigue properties of metals – as well the mechanical properties in general - are strongly dependent on the starting structural and microstructural parameters. Moreover, the microstructure undergoes changes during cyclic loading. Low cycle fatigue is defined on the basis of the number of cycles to failure. Usually the range of 10^2 to 10^5 cycles to failure is considered to be the LCF region, while the range beyond 10^5 cycles to failure is understood as the high-cycle-fatigue region. The obvious difference between these two regions is the extent of cyclic plasticity, being larger for the LCF region than for the HCF region. The extent of the cyclic plasticity within the LCF region depends strongly on material and less strongly on testing parameters like temperature. Fig.1 shows Coffin-Manson plot for Cu-22%Zn single crystals oriented for single slip cycled at room temperature [1] and for superalloy single crystals CMSX-4 of <001> orientation cycled at 950 °C [2]; in both the cases the axial values of the plastic strain amplitude are used in the plot. It can be clearly seen that the plastic strain amplitudes for a constant number of cycles to failure differ considerably, namely by two orders of magnitude. The range of the plastic strain amplitudes for the CMSX-4 single crystals at

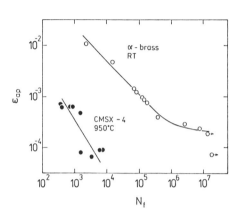

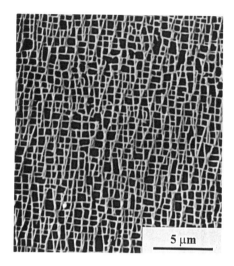

Fig.1. Coffin-Manson plot for single crystals of α-brass (RT) and superalloy CMSX-4 (950 °C).

Fig.2. SEM micrograph showing two-phase structure of superalloy CMSX-4.

the LCF region roughly corresponds to the range of the plastic strain amplitudes for Cu-22%Zn single crystals at the HCF region. Thus the magnitude of the cyclic plastic strain itself cannot be used for the delimitation of the LCF region. From the point of view of the LCF cyclic plasticity the metallic materials can be divided into two broad categories: (i) materials in which the whole loaded volume deforms plastically and (ii) materials in which only a part of the total volume bears the cyclic plastic deformation. In the latter case the "macroscopic" plastic strain amplitude measured over the whole gauge length of the specimen is lower than the local plastic strain amplitude in the plastically deforming regions. With a certain inevitable imprecision we shall call the first category homogeneous materials and the second category composite materials.

Homogeneous materials

The inspection of the huge amount of existing data on Coffin-Manson curves shows that for all materials of this category the number of cycles to failure corresponding to the plastic strain amplitude $\varepsilon_{ap} = 10^{-3}$ lies within a relatively narrow interval of about 5×10^4 to 2×10^5, most frequently within even a narrower interval around the value 1×10^5. This holds both for polycrystalline materials and for single crystals (see the curve for α-brass in Fig.1). Generally, the cyclic plastic strain is larger in the LCF region than in the HCF region. Consequently a more expressive effect of cycling on the microstructure can be expected in the LCF region. The small plastic strain in the HCF region can be well accommodated by the slip on one slip system with only a minor contribution of the secondary slip systems. The larger plastic strains corresponding to the LCF region ($\varepsilon_{ap} > 10^{-3}$) certainly require extensive activity of more slip systems both in single crystals and polycrystals. Moreover, in the case of polycrystals an indispensable activity of the secondary slip systems is always needed due to the compatibility requirements at the grain boundaries. Thus the LCF cyclic plasticity of homogeneous materials is always conditioned by multiple slip. This offers an excellent possibility to mutually transform the LCF stress-strain characteristic of single crystals and polycrystals using a mean orientation factor. This will be discussed later.

Composite materials

To this category belong not only the composites, as they are nowadays understood, but generally all the materials in which only part of the volume contributes to the cyclic plastic deformation like some high strength metallic materials. Representative of this category is superalloy CMSX-4, the Manson-Coffin curve of which is shown in Fig.1. The microstructure of the single crystals CMSX-4 is a two-phase one, namely it consists of a γ matrix in which cuboidal γ' precipitates are coherently embedded. Fig.2 shows this microstructure. The volume fraction of γ' precipitates is about 70%, the size of γ' cubes (measured along the cube edge) is about 0.4 to 0.5 μm and γ/γ' interfaces are aligned to {001} planes. The main bearers of the cyclic plastic deformation are the γ channels, as the γ' particles are the harder phase. The misfit between the two phases results in the presence of the misfit stresses. The stress acting on the dislocations is given by the sum of the externally applied stress and the misfit stress. Due to the fact that the CMSX-4 single crystal is a negative misfit alloy, the stress acting in the horizontal γ channels (channels perpendicular to the stress axis) is higher then the stress acting in the vertical γ channels (channels parallel with the stress axis). The cyclic plastic deformation takes place on all the eight available {111}<011> slip systems just in these horizontal γ channels. As the amplitude of the cyclic plastic deformation is small (below 0.1% - see Fig.1), the activity of the vertical γ channels is not necessary at all. The total volume fraction of the horizontal γ channels is about 15%. Moreover, not every horizontal γ channels undergoes cyclic plastic deformation. This point was proved experimentally by TEM and SEM observation. Zhang et al. [15] found in a similar superalloy after cycling bands of dislocations in some of the γ channels. Obrtlík et al. [9] observed by SEM of cycled CMSX-4 crystals that approximately every 20th horizontal γ channel produces a visible surface relief. Summing up we can state that only horizontal γ channels can contribute to the plastic deformation (they represent altogether about 15% of the total volume) and roughly every 20th horizontal γ channel is really active. Thus we come to the conclusion that only 1% of the total volume bears the cyclic plastic deformation. This is the reason why the Coffin-Manson plot of this material is shifted by two orders of magnitude towards lower plastic strains with respect to the homogeneously deformed material, as shown in Fig.1.

STRESS-STRAIN BEHAVIOUR OF SINGLE CRYSTALS AND POLYCRYSTALS

The relation between polycrystal deformation and single-crystal deformation has been subject of investigation for many years. In 1928 Sachs [3] formulated the polycrystal plasticity model based on single slip in individual grains, in 1938 Taylor [4] published another model assuming multiple slip in individual grains. Since that time a considerable amount of experimental data on the monotonic deformation has been gathered and several theoretical extensions of the original models have been presented. Both the experiments and the theoretical work have also been reviewed [e.g. 5,6]. The quoted models make it possible to relate the axial flow stress for the polycrystal σ and the resolved shear flow stress for the single crystal τ and the axial plastic strain for the polycrystal ε and the resolved plastic shear strain for the single crystal γ by a

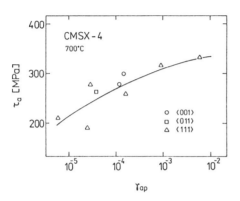

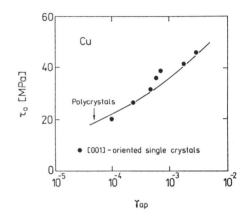

Fig.3. Cyclic stress-strain curve for three multiple slip orientations of superalloy single crystals cycled at 700 °C [9].

Fig.4. Cyclic stress-strain curve for <001> oriented single crystals [11] and polycrystals [12] of pure copper. The polycrystalline data were converted by the Taylor factor.

mean orientation factor M. Thus it holds $\sigma = M\,\tau$ and $\varepsilon = \gamma / M$, where M = 2.24 (Sachs) or M = 3.06 (Taylor). The Sachs model assumes that the grains in the polycrystalline aggregate deform independently by single slip on their most highly stressed slip systems, i.e. on the slip systems having the highest Schmid factor. This kind of single slip deformation would lead to material separation at the grain boundaries unless the compatibility were accommodated elastically. This is possible only for extremely small plastic strains and the validity of the Sachs model is thus a priori limited. The Taylor model assumes that the strains are the same in all grains and equal to the macroscopic strain. In order to satisfy these conditions every grain embedded in a polycrystalline aggregate must in general deform by slip on five independent slip systems to maintain macroscopic compatibility with its neighbours.

Let us first mention the possibilities of the transformation in the case of monotonic deformation. The monotonic stress-strain curves (tensile diagrams) of single crystals depend very strongly on the orientation. For the same flow stress the strain may differ by two orders of magnitude. This holds for the diagram both in terms of axial stress/strain values and in terms of shear stress/strain values resolved on the most stressed slip system using the Schmid factor. This fact itself limits the transformation to special cases only. Moreover, the monotonic stress-strain curves of polycrystals depend on the grain size. This dependence is not so strong as the orientation dependence in the case of single crystals. For example, for small plastic strains the flow stress in copper polycrystals with grain sizes differing by a factor of 16 differ roughly by a factor of 3 [7]. Both Sachs and Taylor model give the possibility to transform mutually the monocrystalline and the polycrystalline monotonic stress-strain curves. The question is which orientation should be chosen on the side of single crystals and which grain size should be chosen on the side of the polycrystals. As the above mentioned compatibility requirements lead to the necessity of multiple slip in polycrystals, it is plausible to adopt the point of view that a proper relation can be established only if the single crystals are also deformed under similar conditions of multiple slip. It means that multiple slip oriented single crystals (i.e. orientations <001> and <111>) must be considered for comparison with polycrystals. Experiments confirm that the tensile monotonic stress-strain curves measured on polycrystals of medium grain size and transformed by Taylor factor roughly agree (up to about ±30% as for the flow stress) with the tensile monotonic stress-strain curves measured on <111> or <001> oriented single crystals [6,8]. It means that Taylor factor can serve for the approximate evaluation of the monocrystalline tensile behaviour on the basis of the polycrystalline data (and vice versa) provided the "proper" orientations and grain sizes are compared.

The transformation of the CSSCs between single crystals and polycrystals is much more general and substantially better provided the multiple-slip single monocrystalline data are used. All the above quoted models of transformation implicitly assume homogeneous deformation. As already mentioned, the most prominent feature of the cyclic plasticity is formation of persistent slip bands in some metals under suitable conditions. This directly contradicts the requirement of homogeneous deformation. Fortunately, the experimental data show that this is a disturbing factor only when the degree of cyclic strain localisation is so expressive that there is a plateau on the CSSC. In all other cases the strain localisation – if there is any - does not adversely affect the possibility of transformation. The plateau on the CSSCs occurs only in some single-slip oriented single crystals (see the following section). If we use in these cases the CSSCs of multiple-slip oriented single crystals we can successfully perform the transformation using the Taylor factor. This is true not only for the conversion of the CSSCs, but also for the conversion of fatigue life curves. Contrary to monotonic plasticity, there is no or only very weak orientation dependence of the CSSCs of multiple-slip oriented single crystals and no or only very weak dependence of the CSSCs of polycrystals on the grain size. The first point is documented by Fig.3 [9] showing the CSSCs of superalloy single crystals CMSX-4 with the orientations <001>, <011> and <111> cycled at 700 °C in the shear stress vs. plastic shear strain representation. The direcly measured axial stress and plastic strain amplitudes were resolved into the slip system {111}<011> having the highest Schmid factor. It can be seen that the CSSC does not depend on the orientation of the stress axis. The effect of grain size on cyclic plastic deformation is considerably less expressive than the grain size effect in the case of monotonic plastic deformation. For example, in polycrystalline copper specimens with grain sizes differing by a factor of 16 the saturation stress amplitudes differ at most by 20% [10]. The grain size effect is generally due to the compatibility requirements leading to secondary slip in the vicinity of the grain boundaries. Equilibrium requirements lead to an internal stress field with a characteristic range for the grain size. The fact that the plastic deformation within one stress cycle is small suggests that the extent of the grain boundary induced secondary slip is also small. Moreover the fact that the plastic deformation changes its sign from tension to compression and vice versa allows for the relaxation of the internal stresses. These are probably reasons why the effect of grain size on CSSC is weak.

Figs.4 and 5 show examples of CSSCs transformation. Fig.4 displays the results for copper. The data for the <001> oriented copper single crystals (full points) are taken from the paper by Gong et al. [11]. The directly measured axial stress and plastic strain amplitudes were resolved into the mostly stressed slip systems {111}<011>. Sixty seven specimens of medium grain size copper (mean grain size 0.07 mm) were used to determine the polycrystalline CSSC [12]. The directly measured axial stress and plastic strain amplitudes were converted using the

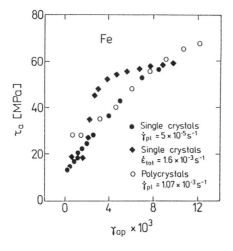

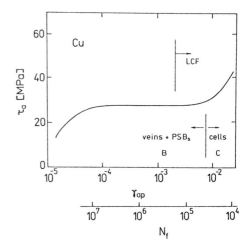

Fig.5. Cyclic stress-strain data for single crystals [13,14] and polycrystals of pure iron. The polycrystalline data were converted by the Taylor factor.

Fig.6. Cyclic stress-strain curve of single slip oriented copper single crystals [16]. Regions of different dislocation structures are also marked.

Taylor factor. Due to the high number of experimental data the CSSC is not presented by the individual points. An excellent agreement between the monocrystalline and the polycrystalline data can be stated. The second example shown in Fig.5 concerns pure iron. Both the monotonic and the cyclic plastic deformation of bcc metals is strongly temperature and strain rate sensitive. Mughrabi et al. [13] cycled crystals at a constant plastic shear strain rate $d\gamma_{pl}/dt = 5 \times 10^{-5}$ s^{-1} . Šesták et al. [14] used constant axial total strain rate $d\varepsilon_{tot}/dt = 1.6 \times 10^{-3}$ s^{-1}, which yields (for the range of strain amplitudes used) the plastic shear strain rate in the range from 2×10^{-5} to 2×10^{-3} s^{-1}. The CSSC for polycrystals shown in Fig.4 was determined also by Mughrabi et al. [13] at a constant axial plastic strain rate. This CSSC was converted by the Taylor factor; the converted value of the plastic shear strain rate is $d\gamma_{pl}/dt = 1.07 \times 10^{-3}$ s^{-1}. It can be seen that the agreement between the data for single crystals and for polycrystals is good. It is important that no PSBs can be observed at room temperature and at the strain rates used.

CYCLIC STRAIN LOCALISATION

The most prominent feature of the cyclic plasticity is formation of persistent slip bands (PSBs) in some metals under suitable conditions. It seems that in the past too a great importance has been attributed to these bands. The conditions for the formation of the PSBs (type of metal, stress and strain requirements, temperature etc.) will be shown and the role of the PSBs in the overall cyclic plasticity (including their effect on the cyclic stress-strain curve) and in the initiation of microcracks will be examined in this part of the paper.

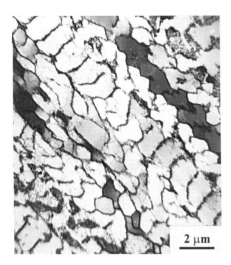

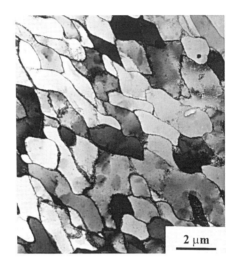

Fig.7. Set of PSBs in copper single crystal. Section perpendicular to primary slip plane.

Fig.8. Cell structure in copper single crystals.

The structure of the PSBs and their relation to the CSSCs have been most extensively studied on copper single crystals. Fig.6 shows the CSSC of the copper single crystals [16] in semi-logarithmic plot together with further relevant data. The N_f –scale is based on the extensive fatigue life data of copper single crystals obtained in strain controlled tests by Cheng and Laird [17]. They found a power law relation between γ_{ap} and N_f ; therefore the N_f –scale is also logarithmic. The dislocation structure produced by cycling with the shear plastic strain amplitude $\gamma_{ap} < 8 \times 10^{-3}$ (region B) consists of veins and PSBs. Example of this structure is shown in Fig.7. Cycling with $\gamma_{ap} > 8 \times 10^{-3}$ results in cell structure of the type shown in Fig.8. Adopting the above introduced definition of the LCF as cycling with $\varepsilon_{ap} > 1 \times 10^{-3}$ (i.e. $\gamma_{ap} > 2 \times 10^{-3}$ corresponding here to $N_f > 1.7 \times 10^5$), it can be seen that the LCF covers both the end of the structural region B (veins + PSBs) and the region C (cells). In region B the plastic strain is localised in the PSBs, in region C the whole volume deforms homogeneously. In both the cases there are coarse slip bands on the specimen surface. In region B they can be correlated with the PSBs, in region C there is no correlation to the cell structure beneath the surface. Nevertheless in both the cases the microcracks initiate in the surface intrusions. It means that the plastic strain localisation into the PSBs is not prerequisite for the microcrack initiation.

The ladder-like or cell-like type of the PSBs shown in Fig.7 are typical only for metals with easy cross slip. In fcc metals the easiness of the cross slip decreases with decreasing stacking fault energy γ_{SFE}. The observations performed on single crystals of the Cu-Zn [19] and Cu-Al systems [20] with a range of the alloying elements Zn or Al (the γ_{SFE} decreases with increasing Zn or Al content) make it possible to delimitate the region of the γ_{SFE} in which the PSBs can be

274

formed. It can be stated that the PSBs were observed for $\gamma_{SFE} > 0.020$ J/m^2. For systems having lower stacking-fault energy the PSBs were not observed.

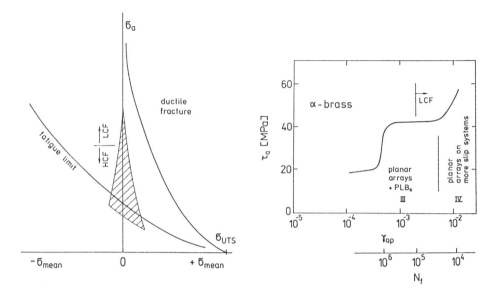

Fig.9. Schematic diagram showing the stress requirements for the formation of PSBs. The PSBs are confined to the hatched area.

Fig.10. Cyclic stress-strain curve of α-brass single crystals [18]. Regions of different dislocation structures are also marked.

Temperature has also a clear effect on the occurrence of the PSBs. Lisiecki et al. [21] found in copper cycled at 250 °C (T/T$_m$ = 0.39) and below well defined PSBs; at 405 °C (T/T$_m$ = 0.5) no PSBs were found in the structure. In superalloy single crystals having the two-phase γ/γ′ structure of the type shown in Fig.2 the PSBs running along the {111} slip planes across both the phases were detected for lower temperatures only. More exactly, no PSBs were detected at temperatures above 760 °C (T/T$_m$ ≈ 0.6) [22,23,24,25].

In precipitation hardened alloys the cycling can lead to the formation of precipitate-free zones; the slip activity in such zones is higher than that in the surrounding matrix and they are permanent. Therefore they can be also called PSBs. There are several excellent experimental works performed on the precipitation hardened alloys. The general conclusion is that the PSBs are typical for alloys containing shearable precipitates. There is no slip localisation in alloys containing non-shearable precipitates [26-29].

Stress-controlled cycling with non-zero mean stress affects adversely not only the lifetime, but also the formation of the PSBs. For example, Holste et al. [30] found in polycrystalline nickel the ladder-like PSBs only at symmetrical or nearly symmetrical cycling. Fig.9 shows schematically the range of stress amplitude and mean stress (hatched area) at which the PSBs can be found in easy-cross-slip materials like copper and nickel. It can be seen that this range trenches both into the LCF and HCF region; the higher the stress amplitude the narrower the range of the PSB occurrence.

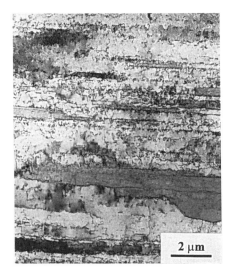

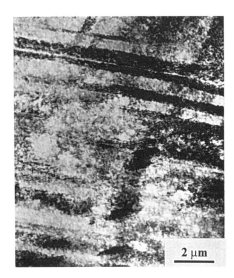

Fig.11. Dislocation arrangement into one-system planar arrays in α-brass single crystals (region III of Fig.10).

Fig.12. Dislocation arrangement into two-system planar arrays in α-brass single crystals (region IV of Fig.10).

Another type of cyclic slip localisation and transition to homogeneous slip is shown in Fig.10 for the low-stacking-fault energy material α-brass (Cu-22%Zn) [18]. The relation between ε_{ap} and N_f is not of a power law type (see the curve for α-brass in Fig.1); therefore the N_f –scale is not logarithmic. The dislocation structure produced by cycling with the shear plastic strain amplitude $\varepsilon_{ap} < 5x10^{-3}$ (region III) consists of planar arrays of edge dislocations on the primary slip plane. Example of this structure is shown in Fig.11. This TEM micrograph was taken using a foil perpendicular to the primary slip plane and containing the primary Burgers vector. The edge dislocations forming the arrays are perpendicular to the micrograph and thus appear as black points or short dislocation segments. It can be seen that there are denser and less dense slabs; this witnesses a certain inhomogeneity of the cyclic slip. Cycling with $\varepsilon_{ap} > 5x10^{-3}$ results in a very similar structure, the only difference being a clearly higher activity of secondary slip systems (Fig.12). The slabs of activated primary slip planes (such as are in Fig.11) are often called "persistent Lüder's bands" (PLBs) [31]. These bands are assumed to represent zones of localised cyclic slip. Contrary to the PSBs, PLBs are not permanent and do not represent zones of dislocation structure distinctly different from surrounding matrix. The PLBs are not stable – the localised strain moves around the gauge length and the active life of the slip bands is short. The idea of PLBs as bearers of the cyclic plastic strain is very attractive as it explains the existence of the plateau in the CSSC in a very natural way. On the other hand, the fact that the plateau has not been observed in the case of all planar slip alloys and that the PLBs structure is not distinguishable from the matrix structure throws some doubts on this idea. The plateau simply means that the plastic strain is being spread without increase in stress. How far specific zones are needed for this process remains to be answered.

276

A very important question is the relation between the slip activity within the crystal and the surface morphology. The surface slip relief in easy-cross-slip metals (see Fig.6) is related to the slip activity of the PSBs for lower strain amplitudes (region B in Fig.6) and most probably to the slip of the whole layers of the cells for higher strain amplitudes (region C in Fig.6). In some cases a one-to-one correspondence between PSBs and the surface extrusions was proved. In planar slip metals a regular hill and valley surface morphology was found many times after sufficiently high number of loading cycles both at lower strain amplitudes (region III of Fig.10) and at higher strain amplitudes (region IV of Fig.10), the only difference between the two regions being the occurrence of the slip markings corresponding to secondary slip systems. The mechanism of the surface relief formation is here probably related to the active slip bands that repeatedly move around the gauge length, i.e. the slip bands can cease their activity and be reactivated again [32,33].

Summing up this section it can be state that the occurrence of the permanent bands of the higher cyclic slip activity (PSBs) is a very frequent but certainly not the general phenomenon. General phenomenon is initiation of microcracks in surface intrusions. Thus the necessary prerequisites for the microcrack initiation are (i) expressive notch-peak topography; the way how it is formed is less important, (ii) locally higher cyclic plastic strain at the intrusion root; this might be due to the localised slip or simply to the concentrating effect of surface profile and (iii) hardening of material around the surface intrusions due to dislocation structure built up by cycling.

CONCLUSIONS

1.   From the point of view of the microstructural changes taking place during LCF the metallic materials can be divided into two broad categories: (i) materials in which the whole loaded volume deforms plastically and (ii) materials in which only a part of the total volume bears the whole cyclic plastic deformation. The macroscopic characteristics are adjusted accordingly.

2.   The CSSCs of multiple-slip oriented single crystals and polycrystals can be mutually transformed using the Taylor factor. Transformation of the CSSC of the single slip oriented single crystals exhibiting PSBs is not possible due to the massive slip localisation.

3.   The PSBs play often, but not always, the crucial role in the initiation of fatigue microcracks. The necessary prerequisites for the microcrack initiation are (i) expressive notch-peak topography, (ii) locally higher cyclic plastic strain at the intrusion root, and (iii) hardening of material around the surface intrusions. These prerequisites can be reached by different ways, one of them being formation of the PSBs.

REFERENCES

1.   Lukáš, P., Kunz, L. and Krejčí, J. (1992) *Mater.Sci.Engng.* **A158**, 177.
2.   Obrtlík, K. and Lukáš, P. , unpublished results.
3.   Sachs, G. (1928) *Z.Ver.Dtsch.Ing.* **72**, 734.

4. Taylor, G.I. (1938) *J.Inst.Met.* **62**, 307.
5. Macherauch, E. (1964) *Z.Metallkde.* **55**, 60.
6. Kocks, U.F. (1970) *Metal.Trans.* **1**, 1121.
7. Thompson, A.W. and Baskes, M.I. (1973) *Phil.Mag.* **28**, 301.
8. Chin, G.Y. (1977). In: *Work Hardening in Tension and Fatigue*, A.W.Thompson (Ed.). AIME, New York, pp. 45-66.
9. Obrtlík, K., Lukáš, P. and Polák, J. (1998), this volume.
10. Lukáš, P. and Kunz, L. (1987) *Mater.Sci.Engng.* **85**, 67.
11. Gong, B., Wang, Z. and Wang, Z.G. (1997) *Acta Mater.* **45**, 1365.
12. Lukáš, P. and Kunz, L. (1985) *Mater.Sci.Engng.* **74**, L1.
13. Mughrabi, H., Herz, K. and Stark, X. (1981) *Int.J.Fract.* **17**, 193.
14. Šesták, B., Novák, V. and Libovický, S. (1988) *Phil.Mag.* **A57**, 353.
15. Zhang, J.H., Hu, Z.Q., Xu, Y.B. and Wang, Z.G. (1992) *Met.Trans.* **23A**, 1253.
16. Mughrabi, H. (1978) *Mater.Sci.Engng.* **33**, 207.
17. Cheng, A.S. and Laird, C. (1981) *Mater.Sci.Engng.* **51**, 55.
18. Lukáš, P., Kunz, L. and Krejčí, J. (1992) *Mater.Sci.Engng.* **A158**, 177.
19. Lukáš, P., and Klesnil, M. (1972). In: *Corrosion Fatigue*, pp. 118-132; A.J.McEvily and R.W.Staehle (Eds.), NACE, Houston.
20. Abel, A., Wilhelm, M. and Gerold, V. (1979) *Mater.Sci.Engng.* **37**, 187.
21. Lisiecki, L.L., Boehme, F. and Weertman, J.R. (1987). In: *Fatigue 87*, pp. 1047-1056; R.O.Ritchie et al. (Eds.), Emas, Warley.
22. Gabb, J., Miner, R.V. and Gayda, J. (1986) *Scripta Metall.* **20**, 513.
23. Glatzel, U. and Feller-Kniepmeier, M.(1991) *Scripta Metall. Mater.* **25**, 1845.
24. Decamps, B., Brien, V. and Morton, A.J. (1994) *Scripta Metall. Mater.* **31**, 793.
25. Zhang, J.H., Hu, Z.Q., Xu, Y.B. and Wang, Z.G.(1992) *Metallurg. Trans.* **23A**, 1253.
26. Horibe, S, Lee, J.K. and Laird, C. (1984) *Mater.Sci.Engng.* **63**, 257.
27. Wilhelm, M. (1981) *Mater.Sci.Engng.* **48**, 91.
28. Steiner, D. and Gerold, V. (1986) *Mater.Sci.Engng.* **84**, 77.
29. Calderon, H.A., Vogel, P. and Kostorz, G. (1991). In: *Strength of Metals and Alloys*, ICSMA 9, pp. 529-536,; D.G. Brandon et al. (Eds.). Freund Publishing Company, London.
30. Holste, C., Kleinert, W., Gurth, R. and Mecke, K. (1994) *Mater.Sci.Engng.* **A187**, 113.
31. Hong, S.I. and Laird, C. (1991) Fatigue *Fract.Engng.Mater.Struct.* **14**, 143.
32. Hong, S.I. and Laird, C. (1990) *Mater.Sci.Engng.* **A128**, 55.
33. Lukáš, P., Kunz, L. and Krejčí, J. (1992) *Scripta Metall. Mater.* **26**, 1511.

Low Cycle Fatigue and Elasto-Plastic Behaviour of Materials
K-T. Rie and P.D. Portella (Editors)

RELATIONSHIP BETWEEN CYCLIC STRESS-STRAIN RESPONSE AND SUBSTRUCTURE

A. PLUMTREE and H.A. ABDEL-RAOUF*

Department of Mechanical Engineering
University of Waterloo, Waterloo, ON Canada
*Now, Department of Materials Engineering
Zagzig University, Egypt

Abstract

A series of fully reversed cyclic strain tests have been conducted on a range of ferrous and non-ferrous metals. For metals with finely dispersed particles and single phase low stacking fault metals the stabilized hysteresis loops coincided with the cyclic stress-strain curve magnified by two (Masing Behaviour). In this case the Bauschinger strain increased linearly with cyclic plastic strain. For high stacking fault energy metals where the cyclic deformation was matrix controlled, the cyclic stress-strain response was Non-Masing. However, Masing behaviour was observed below a threshold strain level. Above this threshold where Non-Masing behaviour occurred a dislocation cellular microstructure formed. The Bauschinger strain increased non-linearly with increasing strain.

Keywords

Masing behaviour, Bauschinger strain, Cyclic stress-strain loops, dislocation cells.

INTRODUCTION

The need for analyzing the cyclic inelastic response and fatigue damage is encountered in various applications. Masing's hypothesis [1] that the shape of the cyclic stress-strain hysteresis loop is geometrically similar to the monotonic stress-strain curve magnified by a scale factor of two has been extensively used in modelling the inelastic stress-strain response of metals. In this case, the monotonic stress-strain curve is replaced by either a transient or steady state cyclic stress-strain curve. Jhansale an Topper [2] noted that for some metals the hysteresis loop plotted with matched compressive tips fell on each other and coincided with the cyclic stress-strain curve magnified by two. This was termed "Masing" behaviour. For other metals, however, coincidence occurred only when the hysteresis loop branches were translated downwards along the elastic slope from the position of matched compressive tips. The common curve was not the cyclic stress-strain curve magnified by two. This "Non Masing" behaviour was expressed by introducing a strain increment to denote the amount by which each branch must be translated in order to coincide with the upper part of the loop branches. Whereas the shape of the nonlinear part of the hysteresis loop remained constant for metals exhibiting Non Masing behaviour, the proportional limit and the linear part of the stress-strain loop increased with cyclic strain range.

Abdel-Raouf et al. [3] showed that the Bauschinger strain (defined as the plastic strain in the reverse direction at 75 percent of the prestress in the forward direction could also be used to distinguish between Masing and Non Masing behaviour. For the former, the Bauschinger strain increased linearly with plastic strain range while for the latter it increased more slowly in a parabolic manner. A variety of metals displaying Non Masing behaviour was examined and it was noted that below a certain threshold strain level these metals exhibited Masing behaviour. Abdel-Raouf et al [4] showed that an aluminum-4% copper alloy with small interparticle spacing exhibited Masing behaviour. However, Non Masing behaviour was observed in the same alloy with a larger interparticle spacing allowing the plastic deformation to be controlled by the matrix rather than the particles.

The purpose of the present paper continues to investigate more fully the relationship between the microstructure and the cyclic hysteresis loop shape. In doing so a more fundamental explanation of Masing or Non-Masing behaviour may be given.

MATERIALS AND EXPERIMENTATION

The metals investigated are given in Table I. All cyclic tests were fully reversed and performed on round cylindrical specimens using an electrohydraulic servocontrolled test system operating under strain control. The specimens were first cycled to saturation using an incremental step test consisting of a sequence of strain cycles of increasing and then decreasing magnitude. Afterwards, the specimens were cycled until the loop became completely stable at each different strain range. The corresponding steady state hysteresis stress-strain loops were recorded.

RESULTS AND DISCUSSION

The steady state stress-strain hysteresis loops were translated to a common point of maximum compressive stress, shown in Figures 1 and 2. The group of metals represented by AISI 1018 steel containing about 30 percent martensite is shown in Figure 1. This group contained the dispersion hardened aluminum alloys 2024-T6 and Al-4% Cu with the smaller interparticle spacing ($\lambda = 0.53\,\mu m$) as well as AISI type 304 stainless steel. All these metals exhibited Masing behaviour. For each metal the hysteresis loop branches fell on the doubled cyclic stress-strain curve for all strain levels.

Table I Materials and Heat Treatment

	Material	Heat Treatment
1.	Al-4% Cu	Solution heat treated, aged 6 hr. at 400° C. Average interparticle spacing, λ, of 1.4 µm.
2.	Al-4% Cu	Solution heat treated, aged 25 min. at 400°C. Average interparticle spacing, λ, of 0.5 µm.
3.	2024-T6	Solution heat treated, aged 12 hr. at 190°C.
4.	CDA 102 copper	Annealed ½ hr. at 650°C.
5.	AISI 1018 steel	Hot rolled.
6.	AISI 1018 steel	Heat treated 30 min. at 760°C. Water quenched (30% martensite).
7.	CSA G40.11 steel	Hot rolled C = 0.22%.
8.	A36-70A	Hot rolled C = 0.11%.
9.	AISI 304 stainless steel	Water quenched from 1000°C.

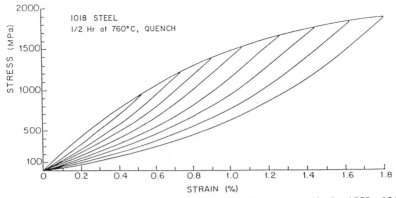

Figure 1. Stable hysteresis stress-strain loops. Water quenched AISI 1018 steel

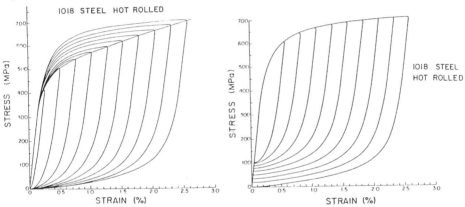

Figure 2. Stable hysteresis stress-strain loops shifted to common zero. Hot rolled AISI 1018 steel.

Figure 3. Loops shown in Figure 2 moved on to common cyclic stress-strain curve.

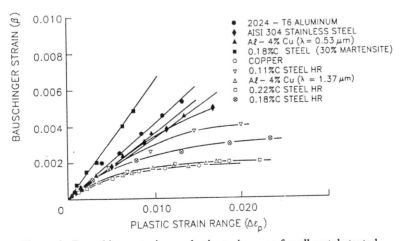

Figure 4. Bauschinger strain vs. plastic strain range for all metals tested

Materials represented in Figure 2 by the ferritic-pearlitic hot rolled AISI 1018 steel exhibited Non-Masing behaviour at the higher strain levels. This group also contained the 0.11%C A36 and 0.22%C G40.11 hot rolled steels, CDA102 copper and Al-4% copper alloy with the larger interparticle spacing ($\lambda = 1.37 \, \mu m$). Above a critical strain amplitude the hysteresis branches for each of these metals deviated from a common cyclic stress-strain curve. The hysteresis loop branches for those strain levels above the critical for Masing behaviour when translated downward along the elastic slope fell on a common curve for each of these metals as represented in Figure 3. Hence, the shape of the nonlinear portion of the hysteresis branch curve was the same at all strain levels but the proportional limit stress or strain increased with increasing strain range. The critical plastic strain range representing the onset of Non-Masing behaviour was found to be 0.0043 for the ferritic-pearlitic steels, 0.0024 for the Al-4% Cu alloy ($\lambda = 1.37 \, \mu m$) and 0.0052 for the commercial purity copper. It is noteworthy that below these small strain ranges the substructure contained patches of dislocation dipoles whose density governed the friction stress. Figure 2 shows that at strain amplitudes higher than the critical, the loop branches rose above the doubled cyclic stress-strain curve only to join it at the loop tips. At these higher cyclic strain amplitudes the substructure consisted of dislocation cells.

In high stacking fault energy materials the formation of a cellular structure is an efficient means of storing the appropriate dislocation density and the cell size determines the average slip distance of dislocations. This explains the rapid hardening and softening during each stable half cycle of a material exhibiting non-Masing behaviour, while the microstructure is still able to maintain an insignificant change in the dislocation structure from cycle to cycle. On loading, a higher stress must be imposed for dislocation sources to operate in the cell walls. Once these sources are operational however, there will be a lower resistance to dislocation motion in the relatively dislocation-free cell until the adjacent walls are encountered where interactions and annihilations occur. This process will repeat itself on each strain reversal.

When dislocation cells are formed, the flow stress of the hysteresis branches lies above the cyclic stress-strain curve except at the loop tip which, by definition, lies on the curve. In the absence of cells on the other hand, the hysteresis branches follow the cyclic stress-strain curve.

An alternative method of describing an hysteresis loop shape is by using the Bauschinger strain measured at a given fraction (7/8) of the peak stress in a hysteresis loop branch [5]. For Masing materials all loops followed the doubled cyclic stress-strain curve which was expressed by

$$\Delta \sigma = K'(\Delta \varepsilon_p)^{n'} \qquad (1)$$

where $\Delta\sigma$ is the steady state stress range, K' is the cyclic strength coefficient, $\Delta\varepsilon_p$ is the plastic strain range and n' is the cyclic work hardening exponent. Equation (1) represents loop tip and branch behaviour.

The Bauschinger strain β may be expressed by

$$\beta = R\Delta\varepsilon_p \qquad (2)$$

where $R[= (0.875)^{-n'}]$ is a function of the cyclic strain hardening exponent. The Bauschinger strain was plotted against plastic strain for the four materials exhibiting Masing behaviour to yield a linear relationship, as seen in Figure 4.

A similar plot for the Non-Masing group of metals is also shown in Figure 4, but after an initial linear region at low strain levels the curve became convex and the rate of increase in Bauschinger strain with plastic strain range decreased continuously. Hence, for the same plastic strain level, Non-Masing behaviour was associated with smaller Bauschinger strains.

The larger Bauschinger strains and Masing behaviour was observed in alloys containing strong non-shearable second phases. This effect is attributed to the long range internal stresses due to dislocation loops associated with the Orowan bowing mechanism. On the other hand single phase alloys containing shearable precipitates had smaller Bauschinger strains than those with strong second phase particles [6]. Masing behaviour gives rise to large Bauschinger strains because plastic deformation is controlled by the structure which prevents the formation of a dislocation cellular structure.

The hysteresis branch curves for the 2024-T4 alloy and the aluminum 4% Cu alloy with close interparticle spacing ($\lambda = 0.53$ µm) all fall on the cyclic stress-plastic strain curve over a wide range of strain amplitudes indicating good agreement with the Masing hypothesis. When the interparticle spacing of a two-phase alloy is smaller than the self- trapping distance of dislocations, deformation is controlled by the microstructure rather than by the matrix, leading to a stress response dependent upon only strain and interparticle spacing. This is consistent with the observed Masing behaviour in which the hysteresis branch stress is dependent upon the strain (measured from the branch origin) but independent of the strain range of the branch. A similar argument can be advanced to explain the Masing behaviour of the water quenched AISI 1018 steel where the fine martensite provides barriers to dislocation slip.

The Masing behaviour in the AISI 304 stainless steel can be accounted for the initially high cyclic strains in this low stacking fault energy metal. The absence of cross slip inhibits softening and the dislocation substructure formed during the first application of the high cyclic strains in the incremental step test remains throughout cycling at progressively lower strain levels during the descending sequence of the test. Since this structure changes extremely slowly during strain cycling the stress-strain hysteresis branches for all subsequent smaller cycles continue to follow the curve of this largest cycle.

CONCLUSIONS

1.   Metals in which cyclic plastic straining gives rise to dislocation cells exhibit Non-Masing behaviour with a progressive increase in the proportional limit and flow stress of stress-strain hysteresis loop branches plotted from their compressive tips with increasing strain range for those above a threshold strain level. At strain ranges below this level the hysteresis loop branches lie along the doubled cyclic stress-strain curve (Masing behaviour). The coincidence of the strain threshold with that for dislocation cell formation and the characteristic increase of flow stress with increasing strain range and decreasing cell size provides a basis for describing this behaviour in terms of dislocation cell formation.

2.   Masing behaviour was observed in metals where deformation is controlled by the presence of fine particles and in low stacking fault energy metals which normally, do not form dislocation cells. In AISI 304 stainless steel the difficulty of cross slip retarded softening at lower cyclic strain levels. Consequently the dislocation substructure formed at the highest strain level broke down very slowly.

3.   The strain increment and Bauschinger strain provide measures of the degree to which a metal deviates from the Masing hypothesis. The former progressively increased with increasing cyclic strain range while the latter increased linearly with cyclic plastic strain for Masing behaviour yet increased at a progressively slower rate with plastic strain for Non-Masing behaviour.

ACKNOWLEDGEMENTS

The authors would like to thank the Natural Sciences and Engineering Research Council of Canada for financial support and Marlene Dolson for typing the manuscript.

REFERENCES

1.   Masing, G. (1926), In Proceedings 2nd International Congress of Applied Mechanics, Zurich.
2.   Jhansale, H.R. and Topper, T.H., Topper, (1973). In STP 519, ASTM, Philadelphia, pp. 246-270.
3.   Abdel-Raouf, H., Topper, T.H. and Plumtree, A., (1977), In Proceedings 4th International Conference on Fracture, Waterloo, pp. 1207-1216.
4.   Abdel-Raouf, H. Topper, T.H. and Plumtree, A., (1979), Met. Trans. A 10A, pp. 449-456.
5.   Wooley, R.L., (1953), Phil. Mag. 44, p. 597.
6.   Wilson, D.V., (1965), Acta. Met. 13, p. 807.

Low Cycle Fatigue and Elasto-Plastic Behaviour of Materials
K-T. Rie and P.D. Portella (Editors)
285

CHARACTERIZATION OF PLASTICITY-INDUCED MARTENSITE FORMATION
DURING FATIGUE OF AUSTENITIC STEEL

H.-J. BASSLER and D. EIFLER

University of Kaiserslautern, Department of Materials Science

Gottlieb-Daimler-Straße,D-67663 Kaiserslautern, Germany

ABSTRACT

The assessment of the actual fatigue damage and thus the remaining lifetime of materials is a subject of enormous scientific and economical relevance. This paper describes some aspects of the cyclic deformation behaviour of the metastable austenitic steel X6CrNiTi1810 which is often used in power stations and chemical plant constructions. The main aim of the present investigation is to determine the fatigue damage of cyclically loaded austenitic steel specimens by stress-strain and thermometric measurements as well as the detection of the plasticity-induced martensite content with nondestructive magnetic measurements [1-3]. Several specimen batches were tested with constant stress amplitudes at room temperature and T = 300°C. The experiments yield characteristic cyclic deformation curves and martensite fractions corresponding to the actual fatigue state. The fatigue behaviour of the investigated material is characterized by cyclic hardening and softening effects, which are strongly influenced by plasticity-induced martensite formation.

KEYWORDS

Fatigue, metastable austenitic steel, plasticity-induced martensite, magnetic measurements

INTRODUCTION

Metastable austenitic steel tends to transform from austenite to α`-martensite. The transformation depends on kinetic factors and requires a driving force from cooling or quasi-static and cyclic plastic deformations. In order to initiate the phase transformation a certain amount of cumulated plastic strain has to be exceeded [4,5]. The stability of austenitic steels, i.e. the resistance against martensitic transformation, mainly depends on their chemical composition. In the Schaeffler-Diagram the investigated steel X6CrNiTi1810 is placed in the region of austenite and δ-ferrite quite close to the region of stable martensite. Thus, due to local chemical inhomogeneities, martensitic transformation is to be expected. But even in the case of complete homogeneous distribution of the alloying elements and a position outside the region where austenite, martensite and δ-ferrite exist together martensitic transformation may occur as a consequence of sufficiently high plastic deformation.

MATERIAL AND EXPERIMENTAL PROCEDURE

The material investigated is a metastable austenitic steel of the type X6CrNiTi1810 corresponding to the US-grade AISI 321. The microstructure is characterized by a mean grain size of about 45 µm, a small amount of Cr- and Ti-carbides and a δ-ferrite content of about 2 %. A typical microstructure, the chemical composition and the mechanical properties of the tested material are shown in Fig. 1 and Tab. I and II. The fatigue tests were performed on a 100 kN servohydraulic testing machine under stress control at room temperature and T = 300°C with different R-values, a frequency of 5 Hz and triangular load-time functions.

Table I. Chemical composition (wt-%).

C	Si	Mn	P	S	Cr	Ni	Ti
0,025	0,44	1,76	0,026	0,020	17,15	9,83	0,16

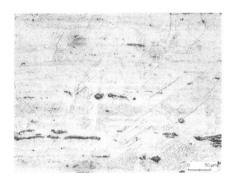

Figure 1. Micrograph of X6CrNiTi1810 (AISI 321), virgin state.

Table II. Mechanical properties.

$R_{p0,2}$ [MPa]	R_m [MPa]	A [%]	$R_{p0,2}/R_m$ [%]	Hardness HV40	Magnetic Phase [%]
449	691	43,6	65	272	≤ 2

EXPERIMENTAL RESULTS

Cyclic deformation behaviour

Characteristic results of stress controlled fatigue tests at room temperature are shown in Fig. 2. The development of the plastic strain amplitude as a function of the number of cycles at constant stress amplitudes represents the microstructural changes in the tested material. The cyclic deformation curves reveal three different states: first a quasi-saturation state with nearly constant $\varepsilon_{a,p}$, followed by cyclic softening with increasing $\varepsilon_{a,p}$ and finally cyclic hardening with decreasing $\varepsilon_{a,p}$. Generally the material reveals characteristic plastic strain amplitudes at the beginning of the test. With increasing stress amplitudes these plastic strain amplitudes increase. Consequently the hardening and softening effects are more pronounced at higher stress amplitudes. In Figure 3 the development of the magnetic fraction, detected with an Eddy Current sensor, is plotted. The martensitic transformation and thus the increase of the magnetic fraction starts at about $2 \cdot 10^2$ -$3 \cdot 10^4$ cycles if a certain plastic strain is accumulated. With increasing stress amplitudes the martensitic transformation starts sooner. Obviously the cyclic hardening is caused by the martensite formation.

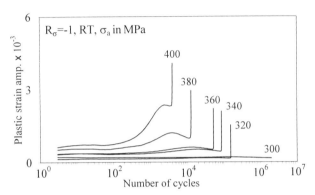

Figure 2. Cyclic deformation curves.

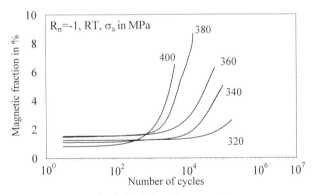

Figure 3. Magnetic fraction due to plasticity induced martensite.

Temperature changes caused by heat dissipation due to plastic deformation can be used to describe the cyclic deformation behaviour. Figure 4 shows cyclic temperature curves corresponding to the cyclic deformation curves in Fig. 2. The temperature changes and the plastic strain amplitudes develop qualitatively in a similar manner. During the cyclic hardening phase the temperature decreases as a consequence of smaller plastic deformations. Figure 5 contains cyclic deformation curves for T=300°C. As expected the hardening is less pronounced due to the higher stability of the austenitic phase. The plastic strain amplitude increases progressively and hardening effects are hard to see. The amount of plasticity-induced martensite in this case is extremely small and not detectable with the Eddy Current technique.

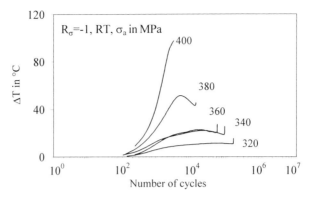

Figure 4. Cyclic stress-temperature curves.

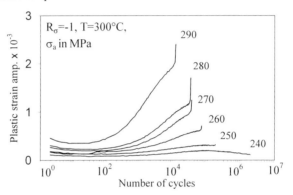

Figure 5. Cyclic deformation curves at T = 300°C.

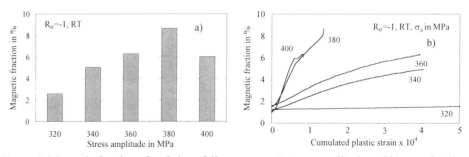

Figure 6. Magnetic fraction after fatigue failure versus a) stress amplitude and b) cumulated plastic strain.

The magnetic fraction after fatigue failure (Fig. 6a,b) is directly related to the martensite content and depends on the stress amplitudes (Fig. 6a) and the cumulated plastic strain ε_{cum} (Fig. 6b).

$$\varepsilon_{cum}(N) = N \cdot \frac{1}{N_B} \int_1^{N_B} \varepsilon_{a,p}(N)dN \qquad (1)$$

Increasing stress amplitudes lead to increasing plastic strain amplitudes, but at the same time to smaller numbers of cycles to failure. There exist counterrotating processes of enhanced martensite formation rate and reduced lifetime with increasing stress amplitudes. For the investigated material state the optimum conditions to form martensite are reached at a stress amplitude of 380 MPa. Consequently the cyclic hardening effect in the cyclic deformation (Fig. 2) and in the cyclic stress-temperature curve (Fig. 4) is most pronounced for this stress amplitude.

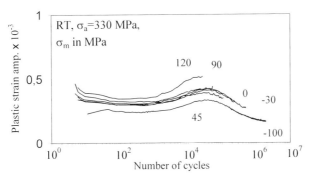

Figure 7. Influence of different mean stresses on cyclic deformation curves.

The influence of mean stresses on the cyclic deformation behaviour and the development of the magnetic α`-martensite, detected with an Eddy Current sensor, is shown in Fig. 7-9. Several tests were carried out with a constant stress amplitude of 330 MPa and different mean stresses in the range -100 to 120 MPa. The cyclic deformation curves are arranged in a relatively narrow scattering band of $\Delta\varepsilon_{a,p} < 0,25\cdot10^3$ (Fig. 7). Positive mean stresses lead to increasing plastic strain amplitudes and positive mean strains at decreasing numbers of cycles to failure, negative mean stresses lead to decreasing plastic strain amplitudes and negative mean strains and consequently increasing numbers of cycles to failure (Fig. 8). The martensitic transformation starts at about 10^4 cycles and is enhanced by positive mean stresses as can be seen in Fig. 9.

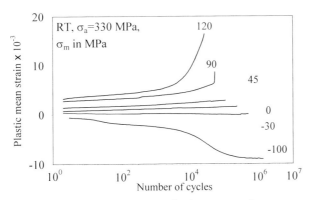

Figure 8. Influence of different mean stresses on plastic mean strain curves.

290

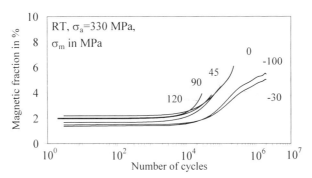

Figure 9. Influence of different mean stresses on plasticity induced martensite formation.

SUMMARY

Stress controlled fatigue tests with the metastable austenitic steel X6CrNiTi1810 were performed at room temperature and T = 300°C. At both temperatures the cyclic deformation behaviour is characterized by cyclic softening and hardening phases. A certain amount of cyclically accumulated plastic deformation has to be exceeded to initiate the martensite formation. The martensitic transformation varies with the stress amplitude, mean stress, testing temperature and number of cycles. Microscopic investigations confirm the influence of the martensitic transformation on the cyclic deformation behaviour. The experiments yield characteristic cyclic deformation curves and corresponding Eddy Current and temperature values related to the actual fatigue state. At T = 300°C, only low fractions of plasticity-induced martensite were formed and therefore highly sensitive magnetic sensors are necessary to detect phase transformations in fatigued specimens [1-3] under the used loading conditions.

REFERENCES

1. M. Lang, J. Johnson and H.-J. Bassler, Fatigue characterization of AISI 321 austenitic steel by means of magnetic materials characterization using a HTSL-SQUID, Proceedings of COFREND Congress, Nantes (1997).
2. M. Lang, J. Johnson and H.-J. Bassler, Fatigue-Characterization of AISI 321 austenitic steel by means of HTSL-SQUID, Proceedings of the 8. International Symposium on Nondestructive Characterization of Materials, Boulder, USA (1997).
3. H.-J. Bassler, D. Eifler, M. Lang, G. Dobmann, Characterization of the fatigue behavior of austenitic steel using HTSL-SQUID, 24th Annual Review of Progress in Quantitative Nonedestructive Evaluation, San Diego, California, July 27-August 1, 1997.
4. G. B. Olsen, M. Cohen, Kinetics of Strain Induced Martensitic Nucleation, Metall. Trans. 6A (1975) p. 791-795.
5. M. Bayerlein, H.-J. Christ, H. Mughrabi, Plasticity induced martensitic transformation during cyclic deformation of AISI 304L stainless steel , Mater. Sci. Eng. A114 (1989), L11-L16.

INFLUENCE OF 160 000 SERVICE HOURS ON THE HIGH TEMPERATURE LCF BEHAVIOUR OF 2.25Cr-1Mo STEEL

S. ARGILLIER[1,2], J. LEON[2]. V. PRUNIER[1], J. P. MASSOUD[1] and J.B. VOGT[2]

1- EDF Centre des Renardières. Département Etude des Matériaux.
77818 Moret sur loing Cedex, France.
2- Université de Lille 1. Laboratoire de Métallurgie Physique de l'URA CNRS 234
Bât C6 59655 Villeneuve d'Ascq, France.

ABSTRACT

The influence of ageing on the high temperature low cycle fatigue (LCF) strength is studied on a ferrito-bainitic 2.25Cr-1Mo steel. Microstructure of aged steel (after 160 000 hours service exposure) and of the same steel after regeneration (by a normalizing-tempering heat treatment) have been studied before and after LCF tests. The long service exposure has induced microstructural evolutions (regarding dislocation arrangement and precipation) especially in bainite. As a consequence this ageing leads to a decrease of the cyclic stress, to a softening instead of a hardening at the beginning of fatigue life and thus to an increase in fatigue resistance. The cyclic plasticity is controlled by the ability to soften the bainite which depends on its density of mobile dislocations before LCF tests. The partition of deformation between bainite and ferrite can be evaluated by the dislocation arrangements in the ferrite after LCF tests.

KEYWORDS

Ferrito bainitic steel, ex service materials, precipitation, high temperature fatigue, dislocation structure

INTRODUCTION

Due to its excellent mechanical properties especially creep resistance at elevated temperature, 2.25Cr-1Mo steel (10CrMo9.10) has been extensively used for many years for high temperature components in power generation. The recent trend in some countries to adopt a semi-basis load for the conventional power plants may accelerate the thermal fatigue damage. After ageing in service, critical components of conventional power plants headers are further damaged by thermal fatigue due to cyclic operation, load transients or fluctuations. This paper presents the comparison between high temperature low cycle fatigue behaviour of a 160000 service hours 2.25Cr-1Mo and of the same steel after regeneration. In order to understand the influence of thermal ageing on fatigue strength, transmission electron microscopy was used to characterize the microstructure of both steels and to reveal their mechanisms of fatigue damage.

EXPERIMENTAL

Material

The 2.25Cr-1Mo steel was taken from a header of a conventional power plant which was retired after 160000 h of service at the nominal temperature of 565°C. A part of this material was tested in this condition and is called aged steel. Another part was normalized (900°C/1.5h) and tempered (750°C/3h) according to the original heat treatment applied to the header at its end of fabrication. It is-called regenerated steel. So both steels have the same chemical composition and a similar ratio of ferrite (Table 1).

Table 1.   Chemical composition (wt %) of the aged and regenerated 2.25Cr-1Mo.

C	S	P	Si	Mn	Ni	Cr	Mo	Cu	V	Fe
0.11	0.026	0.014	0.35	0.48	0.11	2.24	1.05	0.19	<0.01	bal

Fatigue testing

High temperature low cycle fatigue tests were performed on aged and regenerated 2.25Cr-1Mo. Fully reversed uniaxial total strain controlled push-pull fatigue tests were carried in an MTS servohydraulic fatigue machine. Heating of the specimens was obtained using a radiation furnace. Tests were conducted in air at 565°C under total strain control. The total strain variation were $\Delta\varepsilon_t = 0.4 ; 0.8 ; 1.2 ; 2$ % with a strain ratio $R_\varepsilon = -1$ and the total strain rate was kept constant 4.10^{-3} s^{-1}.

The life time (N_f) was defined as the number of cycles which corresponds to a stress reduction of 25 % compared to the saturated region. Some tests on the aged steel were interrupted for $N/N_f = 5$ or 50 % to examine the microstructure by TEM.

Transmission Electron Microscopy

The microstructure of aged and regenerated steels has been studied before fatigue on carbon extraction replicas for the identification of carbides in a PHILIPS EM 430 (300kV) transmission electron microscope. Carbon extraction replicas were prepared by vacuum depositing thin carbon films into polished and lightly etched specimen. Electron microdiffraction and energy dispersive X-ray were used to determine the cristallographic structure and the chemical composition of the carbides.

Thin foils were observed in order to characterize dislocation structures before and after fatigue tests. They were prepared by electropolishing at 15°C in an acetic acid electrolyte with 10% perchloric acid. The variation of microstructure due to sampling makes the use of any single micrograph as unique characterisation difficult. Actually the micrographs shown in this work represent the most frequently observed substructure for each condition.

RESULTS

Microstructural observations before LCF tests

Regenerated steel. Optical microscopy reveals a microstructure of approximatively 45% bainite and 55% proeutectoide ferrite with an average grain size about 45µm (Fig 1a). Numerous carbides are observed in bainite grains and along all types of grain boundaries. There are mainly little spheroidally shaped carbides and coarse carbides identified respectively as M_7C_3 and $M_{23}C_6$. Few needlelike M_2C and rare coarse M_6C carbides were also observed.

In ferrite grains there is a fine and dispersed precipitation of short M_2C platelets aligned along <100> directions (Fig 1b).

The organisation of dislocations in ferrite and bainite is very different (Fig 2). Shear stresses developped during the bainitic transformation have induced a large density of tangled dislocations whereas ferrite grains contain only rare isolated dislocations.

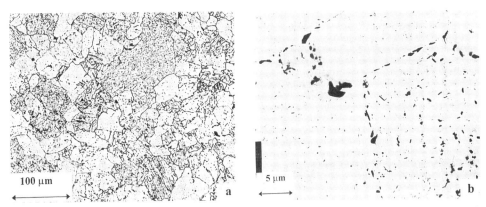

Fig 1. (a) Optical micrograph and (b) TEM micrograph on carbon replica for the regenerated steel.

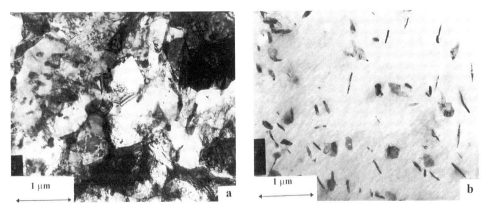

Fig 2. TEM micrographs on thin foils (a) in bainite and (b) in ferrite for the regenerated steel.

294

<u>Aged steel</u>. The optical micrograph of this steel is similar to the one of the regenerated steel, with a smaller grain size (30 μm) (Fig.3a). The service exposure promoted carbide growth, coalescence and enrichment of some elements from the solid solution (Cr, Mo) (Fig.3b). In bainite grains and at grain boundaries, many M_7C_3 carbides are still present and a lot of coarse M_6C carbides have grown at the expense of $M_{23}C_6$ and M_2C, as reported in the literature [1,2]. In ferrite, it should be noted that contrary to A. Latif [1] et al but in agreement with T. Wada [2] proeutectoide ferrite still contains a numerous precipitation of M_2C platelets but no M_6C. M_2C carbides have grown and have been enriched in Mo.

The main difference of the dislocation structure between the regenerated and the aged steels concerns bainite grains (Fig 4). Here they contain high density dislocation networks organised in cells, the walls of which are pinned on carbides. This evolution into cell walls indicates that extensive polygonisation has occured during high temperature service exposure.

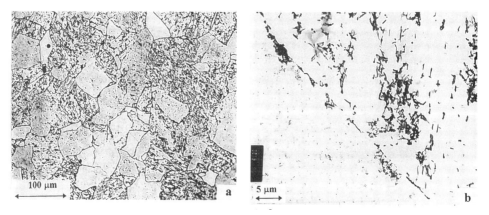

Fig 3. (a) Optical micrograph and (b) TEM micrograph on carbon replica for the aged steel.

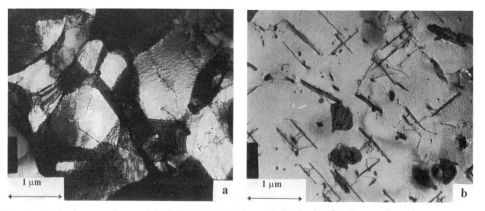

Fig 4. TEM micrographs on thin foils in (a) bainite and (b) ferrite for the aged steel.

High temperature LCF behaviour

The evolution of the stress amplitude with the life fraction is reported figure 5a for the regenerated steel and in figure 5b for the aged steel. During the first cycles, a hardening corresponding to the increase of the cycling stress is observed for the regenerated steel while a strong softening is observed for the aged steel before a stabilisation of the stress. This softening has already been described for virgin steels [3, 4] but not for the aged condition. In addition, the stress amplitude is smaller for the aged steel than that for the regenerated one. Thus the aged steel exhibits higher fatigue lives.

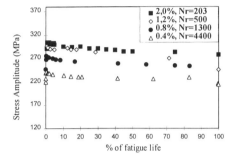

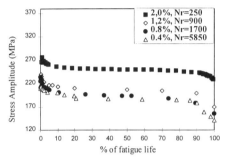

Fig 5a. Stress amplitude versus fraction of lifetime for the regenerated steel.

Fig 5b. Stress amplitude versus fraction of lifetime for the aged steel

Microstructural observations after LCF tests

Though both types of grains have been examined with the same attention, the micrographs presented in this paper concern only ferrite grains where there is the most noticeable microstructural evolution. In ferrite of both steels fatigue damage leads to an increase of the dislocation density. The dislocations tend to arrange into walls but their formation is more significant and rapid in the regenerated steel than in the aged steel. Ferrite grains present well formed little cells for $\Delta\varepsilon_t$=1.2% after fracture in the regenerated steel whereas subgrains boundaries or cells formation is in progress in the aged steel (Fig 6). Contrary to Yoshida et al [3], no dipolar structure was observed at the walls.

Fig 6. TEM micrographs in ferrite after $\Delta\varepsilon_t$=1.2% LCF tests for (a) regenerated, (b) aged steel.

DISCUSSION

The long term service exposure at high temperature has induced some important microstructural modifications in the bainite. The evolution of the nature and chemical composition of carbides has caused a decrease of the interstitial and substitutional elements in the matrix. The decreases of these solid solution elements and of the carbide density (due to coarsening) are beneficial to the dislocation mobility [5]. This increase of dislocation mobility is emphasized by the arrangement of pre-existing dislocations : dislocations in poligonized structure are more mobile than entangled dislocations. Thus bainitic grains in the aged steel are more able to soften. The ferrite is much less affected by the 160 000h service exposure. It still contains numerous fine acicular M_2C carbides, which provide its good mechanical strength, and rare free dislocations.

In such steels, plastic deformation will be accomodated at first by grains with a high density of mobile dislocations, here bainite. The contribution of the bainite in the cyclic plasticity essentially during the first cyles leads to a softening of the bainite. This has been confirmed by microhardness measurements [6]. The softening rate is higher in the aged steel than in the regenerated steel. Then, bainite having less and less capacity to accommodate the strain, ferrite hardens as a consequence of a progressive load transfer from bainite to ferrite. The bainite of the regenerated steel having a poor capacity to accomodate, the contribution of the ferrite occurs earlier in the fatigue life compared to the aged steel. This explains why there is a hardening at the beginning of the LCF cycling while the aged steel softens. Moreover this earlier and greater contribution of ferrite during LCF test for the regenerated steel explains that well formed fatigue cells are observed at fracture while cells in ferrite are under formation in the aged steel.

CONCLUSION

The LCF behaviour at 565°C of an aged 2.25Cr-1Mo steel (after a 160 000 h service exposure) has been compared to that of the same steel after regeneration treatment. The long term service exposure leads to a decrease of the cyclic stress, to a softening instead of a hardening at the beginning of fatigue life and then to an increase in fatigue resistance.

The differences have been explained on the basis of microstructural analysis. We have shown that the macroscopic fatigue behaviour of a ferrito-bainitic steel is strongly dependent on the initial microstructure of the bainite, which governs its possibility to accomodate the plastic deformation. The observation of dislocation arrangements in ferrite allows to follow the load transfer between bainite and ferrite.

REFERENCES

1.   A.M. Abdel-Latif, J.M. Corbett, D.M.R. Taplin.(1982) *Metal Science* **16**, 90.
2.   T Wada.(1985) 8505-052 *Metals / Materials Technology Series*, ASM, Metal Park.
3.   T. Yoshida, H. Umaki, M. Nakashiro.(1996) *Conf: Creep and fatigue Design and Life.*
4.   J. Polak, J. Helesic, M. Klesnil. (1985) *Conf: Low Cycle Fatigue, New-York, USA.*
5.   R.L. Klueh. (1978) *Materials Science Engineering*, 35, 239.
6.   S. Argillier, J. Leon, V. Prunier, J. P. Massoud, J.B. Vogt, (1998) *to be presented at Conf : Lifetime Management and Evaluation of Plant, Structures and Components, U.K.*

Low Cycle Fatigue and Elasto-Plastic Behaviour of Materials
K-T. Rie and P.D. Portella (Editors)
297

CYCLIC DEFORMATION BEHAVIOUR OF A SUPERAUSTENITIC STAINLESS STEEL AND THE ROLE OF EMBRITTLING PRECIPITATES FORMED DURING WELDING

S. HEINO and B. KARLSSON
Department of Engineering Metals, Chalmers University of Technology
SE-412 96 Göteborg, Sweden

ABSTRACT

The monotonic and cyclic deformation behaviour, including the role of grain boundary precipitates, have been studied for a superaustenitic stainless steel with nominal composition Fe-0.01C-24Cr-22Ni-7.3Mo-3Mn-0.5N (wt%). Pronounced planar slip modes, caused by the high N content, were identified as well as indications of persistent slip band formation during cyclic deformation. Cyclic softening at the initial part of the life-time followed by fairly constant saturation stresses was observed. Grain boundary precipitate structures, resembling those found in heat-affected zones of welds, greatly affect the fracture mode in tensile straining but have a minor influence on the flow stress levels and elongation to fracture. Cyclic straining causes an earlier onset of micro cracking in material containing grain boundary precipitates.

KEYWORDS

Superaustenitic, austenitic, stainless steel, cyclic deformation, low cycle fatigue, precipitates.

INTRODUCTION

The deformation behaviour of austenitic stainless steels has been subject to many studies, both regarding the stress-strain response in cyclic and monotonic deformation and its relation to the microstructure.

Recently, several highly alloyed austenitic stainless steels have been developed which show very high resistance to localised and crevice corrosion in a wide variety of environments. The high corrosion resistance is achieved mainly by high levels of Cr, Mo and N, which is balanced by a high Ni level to keep the austenitic microstructure. In addition to the excellent corrosion properties, the mechanical properties are improved, mainly due to the high nitrogen levels, but the high Cr and Mo levels may also have some influence. High nitrogen content has been found to increase strength and promote planar slip without significant losses in ductility [1-4].

As a consequence of high alloying levels, the materials become more or less prone to formation of intermetallic precipitates when exposed to elevated temperatures, typical examples being the heat-affected zone of welds or high temperature use in heat exchangers. In the higher alloyed, superaustenitic grades, observed intermetallic phases include the σ, χ and R phases [5,6]. These phases are hard and brittle and may act as crack nucleation sites.

Thus, two counteractive phenomena may be observed: Improved mechanical properties due to higher alloying levels and possible embrittlement caused by precipitates. The aim of this paper is to study the tensile and cyclic deformation behaviour of a high Mo, high N superaustenitic stainless steel and to quantify the influence of grain boundary precipitates. The emphasis is on short time precipitation.

EXPERIMENTAL DETAILS

The tested material was a commercial superaustenitic stainless steel, Avesta 654 SMO (Table 1). The material was manufactured in electro slag remelted (ESR) condition as 16 mm thick plate, subsequently cold rolled to 12 mm thickness and finally recrystallised at 1200°C for 5 min in argon atmosphere. To resemble the precipitate structure found in the HAZ of welded material [6], isothermal heat treatments were performed at 1050°C for 5 min.

Table 1. Chemical composition of test material (wt%)

	Fe	C	Cr	Ni	Mo	N	Mn	Other
Avesta 654SMO	Bal.	0.012	24.7	21.6	7.6	0.46	3.29	0.38Cu

Tensile tests were performed on test bars 5 mm in diameter and in accordance with ASTM standards. The testing machine was an Instron 4505 operated in computer controlled mode. Low cycle fatigue specimens were manufactured with a diameter of 5 mm and a gauge length of 35 mm. The specimens were ground and polished with diamond paste down to 1 μm grain size. Testing was performed in an Instron 8500 series servohydraulic testing machine operated in computer controlled mode using Instron Wavemaker/Waverunner computer software. Constant amplitude of total strain ($\Delta\varepsilon_{tot}/2$) was the controlling parameter and the total strain amplitude levels employed were 0.4, 0.6, 0.8 and 1.0%. The waveform was sinusoidal with a mean strain rate of $5 \cdot 10^{-3}$ s^{-1}.

Specimens for transmission electron microscopy (TEM) were prepared from slices cut perpendicular to the fatigue specimen axis. Thinning was performed by electropolishing in 15% perchloric acid in ethanol at 25V and −30°C followed by ion thinning in a Gatan PIPS ion mill. TEM studies were performed with a Zeiss 912 OMEGA energy filtering microscope operating at 120 kV, which provides possibilities to image and analyse electrons of specific energy losses. Scanning electron microscopy (SEM) was performed with a JEOL 733 scanning electron microprobe.

RESULTS AND DISCUSSION

The investigated material was initially free of grain boundary precipitates. After heat treatment, plate-like grain boundary precipitates were found and the appearance is shown in Fig. 1a. Compared to the structure found in the heat-affected zone of welded material (Fig. 1b) these precipitates were slightly more elongated in the direction of the grain boundaries and were wider, but still in the same size range [6]. Typical sizes were 1000 nm long by 300 nm wide. Electron diffraction studies showed that the dominating precipitate type was σ-phase.

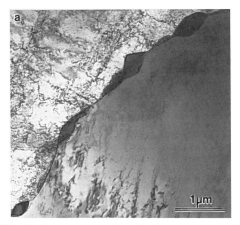 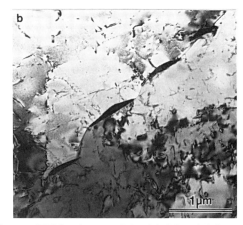

Fig. 1. Appearance of grain boundary precipitates (TEM micrographs): (a) Material heat treated at 1050°C for 5 min. (b) From the heat-affected zone of welded material [6].

The grain size was initially 180 μm in the hot-rolled condition. Later cold rolling and recrystallisation lead to a decrease in grain size to 80 μm, which was unaltered by the later heat treatment to create precipitates (Table 2).

Tensile tests revealed no significant differences in yield stress, $R_{p0.2}$, ultimate tensile stress, R_m, and uniform elongation, A_g, when comparing recrystallised and recrystallised+heat treated material (Table 2). It was however found that the reduction of area at fracture, Z, decreased significantly with heat treatment and the fracture mode changed from dimple type transgranular to a partly intergranular type of fracture, Fig. 2. Note that the type of fracture shown in Fig. 2b is promoted by the grain boundary precipitates and the degree of intergranular fracture has been found to be dependent on the amount of precipitates [3]. As the precipitates nucleate cracks in the grain boundaries, there is also simultaneous growth of secondary cracks (Fig. 2b). These cracks are judged to easily create macrocracks, which limit the further straining in the triaxially loaded volume in the necking zone, and therefore the area reduction at fracture is decreased.

In the present study intergranular fracture was observed already after 5 min of ageing. Degallaix et al. [3] observed that the area of intergranular fracture increased with ageing time (2000 and 10000 hours at 600°C) for a 316L type of stainless steel. However, differences in ageing temperature do not allow any direct comparison, and it should be noted that intergranular fractures in the present study did not give any drop in ductility as observed by Degallaix et al. [3].

Table 2. Tensile properties of recrystallised and recrystallised+heat treated material.

	Grain size [μm]	E [GPa]	$R_{p0.2}$ [MPa]	R_m [MPa]	Uniform elongation, A_g [%]	Elongation to fracture [%]	Reduction of area, Z [%]
HR, //	180	182	409	832	63	74	76
CR + Recr., //	80	181	439	875	56	72	73
CR + Recr., ⊥	80	179	433	879	55	69	72
CR + Recr. + HT, //	80	178	447	882	56	69	60

Grain size= Mean intercept length
HR= Hot rolled
Recr.= Recrystallised at 1200°C, 5 min   CR= Cold rolled, 25% reduction of thickness
//= Longitudinal direction   HT= Heat-treated at 1050°C, 5 min
                               ⊥= Transverse direction

300

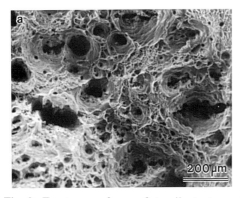

Fig. 2. Fracture surfaces of tensile test specimens (SEM micrographs): (a) Recrystallised material with dimple-type, transgranular fracture, and (b) Recrystallised + heat treated material, giving partly transgranular, partly intergranular fracture.

Cyclic hardening curves are shown in Fig. 3 for two total strain amplitudes. The cyclic peak stresses are fairly constant during the first 10-15 cycles at the high strain amplitude, $\Delta\varepsilon_{tot}/2 = 1.0\%$, followed by a marked cyclic softening until final fracture. At the lower strain amplitude, $\Delta\varepsilon_{tot}/2 = 0.4\%$, there is a continuous softening from the beginning which ends as a plateau region is reached.

The pronounced cyclic hardening during the first 20-30 cycles observed in lower alloyed, nitrogen containing 316LN type stainless steels [4] was not observed in the present superaustenitic material. This deviation from the 316LN steel is likely to be caused by the high N level in the present case, and a similar behaviour has been observed for a Cr-Mn austenitic stainless steels containing 0.9 wt% N [2]. The gradual softening prior to fracture in heat treated material (Fig. 3) may arise from crack initiation at grain boundary precipitates which lowers the load bearing area in the same manner as has been observed for the growth of short fatigue cracks [7].

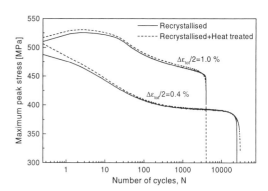

Fig. 3. Cyclic stress strain behaviour of recrystallised and recrystallised+heat treated material. $\Delta\varepsilon_{tot}/2=0.4\%$ and $\Delta\varepsilon_{tot}/2=1.0\%$.

The life length in strain amplitude control for the studied material is shown in the Coffin-Manson diagram in Fig. 4, where a comparison is made with the conventional austenitic material 316LN, containing less nitrogen and having the same ductility but lower yield strength. It is seen that the superaustenitic material shows larger lifetime at larger strain amplitudes, while the situation is reverse at low strain amplitudes. It is believed that this difference is caused by the more pronounced planar slip in the superaustenitic material with its higher nitrogen content (cf. [3,4]).

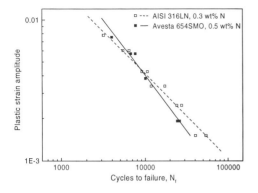

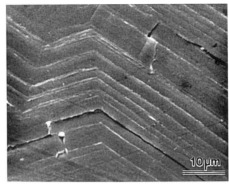

Fig. 4. Coffin-Manson diagram of the superaustenitic Avesta 654SMO (this investigation) and of a N containing AISI 316LN austenitic stainless steel [4].

Fig. 5. SEM micrograph of low cycle fatigue specimen surface cycled at 0.8% total strain amplitude, $N=N_f/2$.

On the surfaces of fatigued specimens very pronounced planar glide could be observed (Fig. 5) in accordance with earlier studies on nitrogen containing austenitic stainless steels [4]. The planar glide may also be represented in terms of dislocation structures as shown in Fig. 6. Depending on the grain orientation, differences in the developed structures may be observed in austenitic stainless steels [8-10]. Figure 6a shows very pronounced planar glide and Fig. 6b shows an onset of formation of persistent slip bands (PSBs) which represent an equilibrium condition as the plateau stress region is reached.

Regarding the surface crack morphology and the dislocation structures no general differences between recrystallised and recrystallised + heat treated material could be observed. However, TEM observations indicate pile up of dislocations at precipitates, which implies that dislocation glide across adjacent grains is hindered in material containing precipitates.

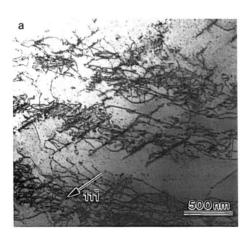

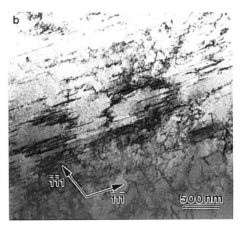

Fig. 6. Dislocation structure in cyclically deformed material ($\Delta\varepsilon_{tot}/2=0.8\%$) at $N=N_f/2$. TEM micrographs, beam direction $\gamma<011>$: (a) Planar deformation. Zero-loss image. (b) Onset of PSB-formation. Image formed by 80eV-loss electrons.

CONCLUSIONS

The superaustenitic stainless steel Fe-0.01C-24Cr-22Ni-7.3Mo-3Mn-0.5N (wt%) was studied regarding monotonic and cyclic deformation behaviour. The material was studied in virgin state as well as in heat-treated condition where intermetallic phases were precipitated in the grain boundaries. Following conclusions could be made:

1. The material is prone to precipitation of intermetallic phases during heat treatment cycles corresponding to those found in welding.
2. A slight increase in yield and tensile strength results after heat treatment, while the elongation to fracture is virtually unaffected. On the contrary, the reduction of area at fracture is markedly reduced. This reduction is associated with intergranular cracking caused by the grain boundary precipitates.
3. The precipitates cause a minor hardening at cyclic straining, while the life time is virtually unaffected.
4. Compared to conventional austenitic stainless steel, the superaustenic materials exhibit longer life lengths in lcf at higher strain amplitudes, while the opposite is the case at low strain amplitudes (long life times). This is related to the marked planar slip occurring in the superaustenic material with its high nitrogen level.

ACKNOWLEDGEMENTS

Avesta Sheffield AB (Mr. A. Brorson) is acknowledged for providing the material and performing the heat treatments. Financial support was given by the Swedish Research Council for Engineering Sciences.

REFERENCES

1. Vogt, J.B., Magnin, T. and Foct, J. (1993) *Fatigue Fract. Eng. Mater. Struct.* **16**, 555-564.
2. Vogt, J.-B., Messai, A. and Foct, J. (1996) *ISIJ Int.* **36**, 862-866.
3. Degallaix, S., Degallaix, G. and Foct, J. (1988). In: *Low Cycle Fatigue*, pp. 798-811; H.D. Solomon, G.R. Halford, L.R. Kaisand and B.N. Leis (Eds.), ASTM, STP 942, Philadelphia, PA.
4. Nyström, M., Lindstedt, U., Karlsson, B. and Nilsson, J.-O. (1997) *Mater. Sci. Technol.* **13**, 560-567.
5. Charles, J., Pugeault, P., Soulignac, P. and Catelin, D. (1987). In: *Proceedings of Eurocorr '87*, pp. 601-606, Karlsruhe.
6. Heino, S., Knutson-Wedel, E.M. and Karlsson, B. (1998) Accepted for publication in *Mater. Sci. Technol.*
7. Lindstedt, U., Karlsson, B. and Nyström, M. (1998) *Fatigue Fract. Eng. Mater. Struct.* **21**, 85-98.
8. Li, Y. and Laird, C. (1994) *Mater. Sci. Eng.* **A186**, 65-86.
9. Li, Y. and Laird, C. (1994) *Mater. Sci. Eng.* **A186**, 87-103.
10. Kruml, T., Polák, J., Obrtlík, K. and Degallaix, S. (1997) *Acta Materialia* **45**, 5145-5151.

Low Cycle Fatigue and Elasto-Plastic Behaviour of Materials
K-T. Rie and P.D. Portella (Editors)

THE EFFECT OF δ-FERRITE ON LOW CYCLE FATIGUE BEHAVIOUR AT HIGH TEMPERATURE IN TYPE 304L STAINLESS STEEL

SOO WOO NAM*·HYUN UK HONG·BYUNG SUP RHO, YONG DEUK LEE·SOO CHAN LEE**

*Department of Materials Science and Engineering, Korea Advanced Institute of Sci. & Tech.,
373-1 Kusong-dong Yusong-gu, Taejon, 305-701, Korea
*Jointly Appointed at the Center for the Advanced Aerospace Materials
Pohang Univ. of Sci. and Tech., San 31 Hyoja-dong Nam-gu, Pohang, 790-784, Korea
**POSCO Technical Research Laboratories, Pohang 790-785*

ABSTRACT

Type 304L stainless steel has been investigated at 600℃ for the effect of δ-ferrite on the continuous fatigue and creep-fatigue behaviours. Under continuous fatigue test, specimens with a larger amount of δ-ferrite had shorter lives than that of the specimens with a smaller amount of δ-ferrite, whereas under the condition of creep-fatigue interaction they showed same creep-fatigue behaviour regardless of the amount of δ-ferrite. From microstructural observation it can be said that under continuous fatigue test the interface between δ-ferrite and matrix provides the surface crack initiation site as a result of stress concentration which results in fatigue failure predominantly. But under creep-fatigue interaction test there is little influence of δ-ferrite on the creep-fatigue behaviour, because the main damage mechanism in this condition is the grain boundary cavitation. It can be concluded that in case of fatigue predominant condition, the interface between δ-ferrite and matrix is the site of the fatigue crack, however, in case of creep dominant condition, grain boundary cavitation is more significant than the crack initiation.

KEYWORDS

Type 304L stainless steel, δ-ferrite, continuous fatigue, crack initiation site, creep-fatigue interaction, damage mechanism, grain boundary cavitation

INTRODUCTION

Austenitic stainless steels are widely used for architectural, consumer and industrial applications because of their excellent corrosion resistance, toughness, formability and weldability. Type 304L stainless steel has been considered as one of the strongest candidate materials for the high temperature structural component. But due to the limitation of manufacturing process, the detrimental δ-ferrite remains in γ phase unavoidably [1]. According to many reports for the δ-ferrite, large amounts of δ-ferrite (usually more than 10 vol.%) in the austenitic stainless steels give rise to decrease the hot workability [2]. And δ-ferrite can be transformed to the hard brittle σ phase by long exposure at the elevated temperature range of $500 \sim 900\,°C$ [3]. Whereas the absence of δ-ferrite can be the cause of longitudinal facial crack and hot shortness of continuously cast slab of 304L stainless steel [4-5]. But there are little reported papers related with the effect of δ-ferrite on fatigue properties at high temperature.

The purpose of this study is to investigate how δ-ferrite affects the continuous fatigue and creep-fatigue properties at high temperature in Type 304L stainless steel.

EXPERIMENTAL DETAILS

The chemical composition and heat treatment condition of Type 304L stainless steel investigated are given in Table 1. The control of the amount of δ-ferrite has been accomplished by solution annealing treatment at $1050\,°C$ for 1hr which can lead to decrease the amount of δ-ferrite without distinct changes in other microstructures. Table 2 shows the amount of δ-ferrite qualitatively measured by ferrite scope.

Table 1. The chemical composition of Type 304L stainless steel. (in wt.%)

Alloy	C	Si	Mn	P	S	Ni	Cr	Mo	Cu	N	Co
304L	0.029	0.57	1.05	0.020	0.003	10.13	18.01	0.07	0.19	0.034	-

Solution treated at $1050\,°C$ for 30 min.

Table 2. The amount of δ-ferrite of Type 304L stainless steel at t/2 site. (in vol.%)

Alloy	As-received	After solution annealing treatment
304L	0.97	0.23

Cylindrical specimens having 8mm gage length and 7mm diameter were cyclically strained at $600\,°C$ in an air under axial total strain control in a servohydraulic testing machine. Symmetric wave shapes were used with a total strain rate of 4×10^{-3}/s to conduct continuous fatigue test. Creep-fatigue tests were also conducted by applying 20min hold times at the maximum strain in tension.

RESULTS

Continuous fatigue testing

The results of continuous fatigue tests are plotted in Fig. 1 in terms of hysteresis loop energy with the number of cycles to failure. It is found that under continuous fatigue test, specimens with a smaller amount of δ-ferrite have somewhat longer lives than specimens with a larger amount of δ-ferrite, which can be explained that the δ-ferrite has some influence on the continuous fatigue properties.

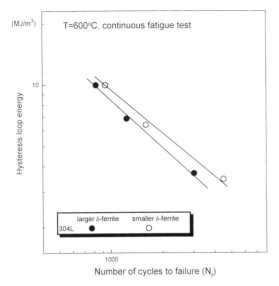

Fig. 1. Comparison of continuous fatigue test results in two alloys which have a different amount of δ-ferrite respectively.

Creep-Fatigue testing

Creep-fatigue test results are presented in Fig. 2. It is seen from this plot that the effect of δ-ferrite is not so significant as in the case of fig. 1. Therefore, they show same creep-fatigue behaviour regardless of the amount of δ-ferrite.

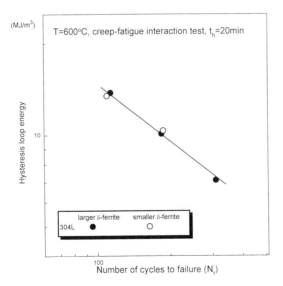

Fig. 2. Comparison of creep-fatigue interaction test results in two alloys which have a different amount of δ-ferrite respectively.

306

Microstructure and Discussion

It is generally accepted that under continuous fatigue the fatigue life is determined mainly by crack initiation stage. As shown in fig. 3 the δ-ferrite acts as the crack initiation site. Those surface cracks are interlinked each other to be the main crack which results in fatigue failure. It is believed that the local stress concentration occurs at the interface between δ-ferrite and matrix due to the strain incompatibility[6], which provides the surface crack initiation site. The reason why specimens with a smaller amount of δ-ferrite has somewhat longer lives than ones with a larger amount of δ-ferrite can be explained partially in the view point of the number of crack initiation site. Therefore, it can be said that specimens with a smaller amount of δ-ferrite have less number of possible crack initiation sites than specimens with a larger amount of δ-ferrite.

When the tensile hold is imposed at the maximum tensile strain on each fatigue cycle, the damage mechanism changes. Therefore, unlike in the case of continuous fatigue, the major factor which dominates creep-fatigue life is cavitation at grain boundary associated with intergranular fracture mode rather than surface crack initiation and growth. For this reason, density of grain boundary carbide which can be beneficial site for cavity nucleation is most significant under creep-fatigue interaction test[7-9], as shown in fig. 4.

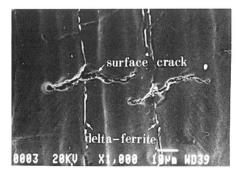

Fig. 3. Crack initiation on the interface between δ-ferrite and matrix during continuous fatigue test of as-received 304L.

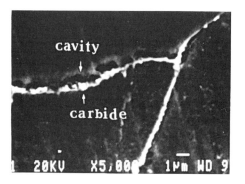

Fig. 4. SEM micrograph showing grain boundary carbide and cavity of as-received 304L after creep-fatigue interaction test.

Fig. 5 shows the influence of δ-ferrite on the grain boundary crack propagation during creep-fatigue interaction test in 304L. As mentioned above, the crack formed by cavity is essentially intercrystalline. According to Hull and Rimmer[10] cavity growth is most active in grain boundary facet which is perpendicular to loading axis. So it can be inferred that there are little possibility to form cavity in δ-ferrite which is parallel to loading axis. Furthermore, δ-ferrite is isolated in the interior of grain. For those reasons cracks formed by cavity interlinkage propagate along grain boundary irrelevant to presence of δ-ferrite, as shown in fig. 5.

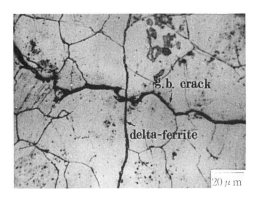

Fig. 5. Crack propagation mode during creep-fatigue interaction test in as-received 304L.

SUMMARY

1. Under continuous fatigue test, there are strong effect of δ-ferrite on fatigue properties to reduce the life. Interface between δ-ferrite and matrix provides the surface crack initiation site as a result of stress concentration.
2. Under creep-fatigue interaction test, the effect of δ-ferrite on the creep-fatigue behaviour is negligible. This can be explained by the change of the damage mechanism from fatigue dominant crack to creep dominant grain boundary cavitation.

ACKNOWLEDGEMENTS

The authors are grateful to POSCO for their financial support and supply of specimens.

REFERENCES

1. F. Matsuda and Y. Arata (1976) Trans of JWRI, **5**, p. 35.
2. B. Ahlblom and R. Sandstrom (1982) Int. Metals Rev., 1 p. 1.
3. M. O. Malone (1967) Weld. J., 46, p. 241.
4. N. Suutala and T. Takalo (19 9) Metall. Trans., 10A, 4, p. 512.
5. S. David (1981) Weld. Res. Supp., p. 63.
6. J. P. Hirth (1972) Metall. Trans., 3, p. 3047.

7. P. S. Maiya and S. Majumdar (1977) Mat. Trans., 8A, p. 1651.
8. S. W. Nam, Y. C. Yoon, B. G. Choi, J. M. Lee and J. W. Hong (1996) Metall. Mater. Trans. A, 27A, p. 1273.
9. B. G. Choi, Y. C. Yoon, J. J. Kim and S. W. Nam (1996) J. Mat. Sci., 31, p. 4957.
10. D. Hull and D. E. Rimmer (1959) Phil. Mag., 4, p. 673.

Low Cycle Fatigue and Elasto-Plastic Behaviour of Materials
K-T. Rie and P.D. Portella (Editors)

LOW-CYCLE FATIGUE BEHAVIOUR OF A 316LN STAINLESS STEEL AT 77 K AND ASSOCIATED STRUCTURAL TRANSFORMATION

M. BOTSHEKAN[1], J. POLÁK[1,2], Y. DESPLANQUES[1] and S. DEGALLAIX[1]

[1] *Laboratoire de Mécanique de Lille (URA CNRS 1441), Ecole Centrale de Lille,*
BP 48, 59651 Villeneuve d'Ascq Cedex, France
[2] *Institute of Physics of Materials, Academy of Sciences,*
Žižkova 22, 616 62 Brno, Czech Republic

ABSTRACT

Low-cycle fatigue tests were performed on a 316LN austenitic stainless steel at liquid nitrogen temperature. The presence of α'-martensite was detected using a ferromagnetic-fluid method and evaluated quantitatively using magnetic saturation method. The changes of the internal structure were studied using transmission electron microscopy. The stress amplitudes are doubled compared to room temperature and cycling with the highest strain amplitudes results in secondary hardening stage. A high density of thin lamellae of twins and α'-martensitic islands embedded in the austenitic matrix were identified. The results are discussed in terms of the effect of high amplitude cyclic straining on the transformation of austenite to martensite and the effect of strain induced martensite on the fatigue life of 316LN stainless steel.

KEYWORDS

Low-cycle fatigue, 316LN stainless steel, low-temperature strain-induced martensitic transformation.

INTRODUCTION

Good mechanical properties as well as structural stability are required for materials used at cryogenic temperatures. AISI 300 series stainless steels have been extensively used at low temperatures [1]. However, the austenitic structure of most of these alloys is metastable and martensitic transformation can occur during cooling, plastic straining or fatigue. Carbon and nitrogen play an essential role in plastic behaviour and in stability of the austenite. The nitrogen favours the formation of stacking faults and promotes the nucleation of hexagonal ϵ-martensite facilitating thus the formation of α'-martensite, which negatively affects the magnetic properties of these alloys, used for the construction of toroidal field coils.

The aim of the present work is to study the low-cycle fatigue behaviour and the associated microstructural transformations in an AISI 316LN austenitic stainless steel at 77 K.

EXPERIMENTAL PROCEDURE

The AISI 316LN austenitic stainless had the following chemical composition: 17.2 Cr, 13.5 Ni, 2.5 Mo, <0.03 C, 0.16 N, <1 Si, <2 Mn, 2.5 Mo, <0.045 P, <0.03 S, bal Fe, all in wt.%. It was cross-rolled and the plates were solution treated at 1050 °C and water-quenched. During heat-treatment the interstitial (C and N) and the substitutional (Cr, Ni, Mo, Mn, ...) alloying elements are brought in solid solution. Low carbon content is compensated by a relatively high nitrogen content allowing to stabilise the austenitic structure and to improve the mechanical properties. The cylindrical, button-headed specimens with 7 mm diameter were taken at an angle 45° of the rolling direction. The surface of the specimens was mechanically polished up to 1 μm. In different sections of the specimen a fully austenitic microstructure with an average grain size of (80 ± 4) μm in each plane was found.

The low-cycle fatigue tests were performed on a computer controlled servohydraulic testing machine (INSTRON 8501) with a capacity of 50 kN in a cryostat designed in the laboratory. Symmetrical loading with constant imposed total strain amplitude (ε_a) and with constant total strain rate $\dot{\varepsilon}_t = 4 \times 10^{-3}\ s^{-1}$ was applied. The total strain amplitudes imposed were in the range $1\% \leq \varepsilon_a \leq 2.5\%$.

Ferrofluid micrographic method and magnetic saturation method were used to detect and evaluate the content of martensite. By magnetic saturation method, the volume fraction of α'-martensite in a specimen, f, can be evaluated from the following relation:

$$f = \frac{(\sigma_s)_f}{(\sigma_s)_{f=1}} \tag{1}$$

where $(\sigma_s)_{f=1}$ is the saturated specific magnetic moment for a purely martensitic state ($f = 1$) evaluated by Didieux theory [2] as $(\sigma_s)_{f=1} = 143$ emu/g, and $(\sigma_s)_f$ is the saturated specific magnetic moment of the specimen analysed. This magnetic moment was measured by a FONER magnetometer at the University of Nancy I. The absolute error in the evaluation of the α'-martensite content using the magnetic saturation method was estimated to be less than 0.5%. The shape of the martensitic islands in the fatigued specimens was observed in a Hitachi S2500 scanning electron microscope (SEM) using secondary electrons on longitudinal sections of the fatigued specimens. The sections were mechanically polished and chemically etched in regal water [3].

Thin foils for electron microscopy were prepared from lamellae cut parallel to the specimen axis. They were thinned by mechanical grinding and by double jet electropolishing. The foils were observed using a Philips CM-12 transmission electron microscope.

EXPERIMENTAL RESULTS

The evolution of the stress amplitude with the number of cycles for different strain amplitudes (cyclic hardening/softening curves) are shown in Fig. 1. The stress amplitudes are about double of those at 300 K [3,4]. The stress evolution at the lowest strain amplitude consists of a rapid hardening followed by a softening and by a quasi-stabilisation phase. Similar behaviour has been also observed by Vogt et al. [5] on a 316LN austenitic stainless steel alloyed with 0.235% N. Secondary hardening, whose intensity increases with the strain amplitude, is observed. at high strain amplitude. This secondary hardening stage was previously found by other authors [6,7] on metastable austenitic stainless steels.

Low temperature cyclic straining results in partial transformation of austenite in α'-martensite. In order to find the dependence of the α'-martensite content on the number of cycles, companion specimens were cycled with the same strain amplitude, and when a desired number of cycles was reached, the α'-martensite volume fraction was measured. The α'-martensite volume fraction is shown simultaneously with the cyclic hardening/softening curves in linear coordinates in Fig. 1b. The α'-martensite volume fraction f is plotted vs. the cumulative plastic strain in Fig. 2. α'-martensite content increases with the strain amplitude and with the cumulative plastic strain.

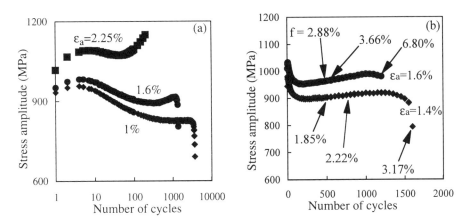

Fig. 1. Cyclic hardening/softening curves (a) and induced α'-martensite contents (b) at 77 K.

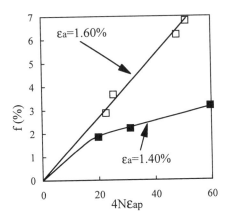

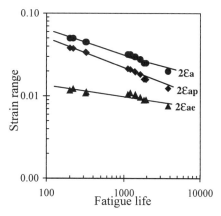

Fig. 2. Evolution of α'-martensite volume fraction during cycling at 77 K versus accumulated plastic strain .

Fig. 3. Fatigue life curves in strain controlled cycling at 77 K.

The low-cycle fatigue life curves showing the fatigue life as a function of total ($2\varepsilon_a$), elastic ($2\varepsilon_{ae}$) and plastic ($2\varepsilon_{ap}$) strain ranges at half-life are shown in a double logarithmic diagram in Fig. 3. Manson-Coffin and Basquin laws expressed as:

$$\varepsilon_a = \varepsilon_{ap} + \varepsilon_{ea} = \varepsilon'_f \left(2N_f\right)^c + \frac{\sigma'_f}{E}\left(2N_f\right)^b \qquad (2)$$

where ε'_f is the fatigue ductility coefficient, c the fatigue ductility exponent, σ'_f the fatigue strength coefficient and b the fatigue strength exponent, can be fitted to the experimental data. Straight lines which are fitted to the plastic and elastic strain ranges correspond to $\varepsilon'_f = 0.200$, $c = -0.384$, $\sigma'_f = 2504$ MPa, $b = -0.119$.

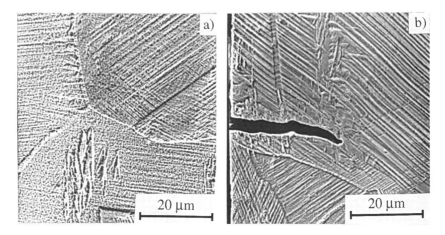

Fig. 4. SEM observations of etched longitudinal sections of specimens fatigued with $\varepsilon_a = 1.6\%$ at 77 K.

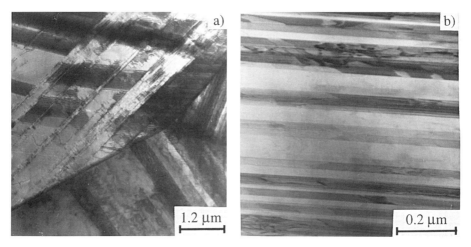

Fig. 5. Internal dislocation structures in specimens cycled with $\varepsilon_a = 1.6\%$ at 77 K up to rupture, $N_f = 1214$ cycles; a) intersecting bands close to triple junction; b) high magnification of thin lamellae.

Figure 4 shows the etched longitudinal sections of a specimen cycled at 77 K. The presence of transformed phases is distinguished by an irregular shape of martensitic islands in a regular parallel pattern of etched planar dislocation arrangements in austenitic grains. The islands are usually in the interior of a grain (Fig. 4a). A short crack starting from the specimen surface is temporarily stopped when it encounters a martensitic island (Fig. 4b).

Internal dislocation structures as observed by transmission electron microscopy after cycling are shown in Fig. 5. The structure is characterised by planar slip, twinning and/or martensite formation. Planar dislocation arrangements, stacking faults, microtwins and/or ε-martensite platelets are present in the majority of grains. The slip bands are very pronounced and the intersection of systems systems of slip results in appreciable localised deformations (Fig. 5a). The high magnification micrograph in Fig. 5b reveals the presence of spatially distributed bands with different contrast. In analogy with similar bands observed after unidirectional straining [8], these bands can be ascribed to the presence of microtwins and/or ε-martensite platelets. The nuclei of α'-martensite can be initiated in these bands or at the intersection of the intense slip bands [9].

DISCUSSION

The measurement of α'-martensite content during cyclic loading revealed no α'-martensite formation in tests run with strain amplitude $\varepsilon_a = 1\%$. It agrees with the findings of Baudry and Pineau [6] concerning the critical total or plastic strain amplitude, which must be achieved in order to induce α'-martensite formation in cyclic straining. In our steel, using the α'-martensite measurements on fatigued specimens, the critical amplitude for α'-martensite formation can be evaluated as $1\% < \varepsilon_{a,c} < 1.6\%$.

Since only magnetic measurements were used for the identification of martensite during cycling, the presence of ε-martensite could not be unequivocally verified. In a similar study on AISI 304 stainless steel fatigued at 77 K, Sadough [7] observed the presence of both α'- and ε-martensites. At the start of cycling, both types of martensite were produced, however, while the fraction of α'-martensite increased until saturation, ε-martensite progressively disappeared, presumably transforming to α'-martensite. In analogy with these results, we judge that irregularly shaped formations in longitudinal sections of broken specimens after cycling with $\varepsilon_a = 1.6\%$ (Fig. 4) are α'-martensite.

The evolution of α'-martensite volume fraction in our steel (Fig. 2) can be compared with that formed in 304L stainless steel as observed by Sadough [7]. An initial increase and later a tendency to saturation are observed. The saturated value depends appreciably on the type of steel and on the applied strain amplitude. 316LN stainless steel is appreciably more stable than 304L stainless steel and the fraction of α'-martensite is much lower.

The experimental data concerning the low-cycle fatigue life fit well Manson-Coffin and Basquin laws and the relevant parameters were evaluated. The comparison with the fatigue life at room temperature [3,4] shows equal fatigue life in terms of plastic strain and higher fatigue life in terms of total strain and stress. A higher fatigue life suggests a higher resistance of the material to the initiation and early growth of fatigue cracks at 77 K. The martensite islands could serve as effective obstacles to the early growth of short fatigue cracks. It can be deduced from Fig. 4b, which shows an early crack blocked by martensite islands.

The low temperature cyclic loading produces planar dislocation structures, numerous twins and/or bands of ε-martensite (Fig. 5). The presence of these structures shows that besides the movement of perfect and partial dislocations, other straining mechanisms as mechanical twinning, ε-martensite and α'-martensite formation appear in low temperature cyclic straining. The presence of other deformation mechanisms is due to a high effective stress for dislocation glide. The internal structure becomes very complex and results in a high cyclic yield stress and secondary cyclic hardening.

CONCLUSIONS

1) 316LN austenitic stainless steel shows high resistance to low-temperature low-cycle fatigue straining. Cyclic stress at 77 K is doubled compared to room temperature and fatigue life is longer.
2) Formation of strain induced α'-martensite results in secondary cyclic hardening.
3) Low temperature cyclic straining is characterised by planar dislocation arrangements, microtwins and martensite lamellae. Observed internal structures correspond to the high cyclic stress at 77 K.
4) The higher low-cycle fatigue life at 77 K relative to room temperature can be explained by a high resistance to early crack growth due to partial blocking of short crack path by martensitic islands.

REFERENCES

1. McHenry, H. I. (1983). In: *Austenitic Steels at Low Temperature*, R. P. Reed and T. Horiuchi (Eds.), Plenum Press, New York, pp. 1-27.
2. Didieux, J. (1952) Thése de Docteur-Ingénieur, Université Paris VI, France.
3. Botshekan, M. (1997) Thèse de doctorat, Université Lille I, France.
4. Botshekan, M. Degallaix, S., Desplanques, Y. and Polák J. (1998) *Fatigue Fract. Engng Mater. Struct.*, in print.
5. Vogt, J. B., Magnin, T. and Foct, J. (1993) *Fatigue Fract. Engng Mater. Struct.* **16**,555.
6. Baudry, G. and Pineau, A. (1977) *Mater. Sci. Engng* **28**, 229.
7. Sadough, A. (1985) Thèse de doctorat, Université Paris VI, France.
8. Thomas, B. and Henry, G. (1990) In: *Les aciers inoxydables*. P. Lacombe, B. Baroux and G. Béranger (Eds.) Les Èditions de Physique, Les Ulis, pp. 61-107.
9. Maier, H. J., Schneeweis O. and Donth B. (1993) *Scripta Metall.* **29**, 521.

Low Cycle Fatigue and Elasto-Plastic Behaviour of Materials
K-T. Rie and P.D. Portella (Editors)
315

INFLUENCE OF THE BAINITIC MICROSTRUCTURE ON SHORT FATIGUE CRACK GROWTH IN HSS STEEL BUTT-WELDED JOINTS

C. BUIRETTE *, G. DEGALLAIX * and J. MENIGAULT **

* *Laboratoire de Mécanique de Lille (URA CNRS 1441), Ecole Centrale de Lille,
BP 48, F-59655 Villeneuve d'Ascq Cedex*
** *SOLLAC, CRDM, BP 2508, F-59381 Dunkerque Cedex 1*

ABSTRACT

Fatigue failure of welds generally starts from weld toes mostly due to geometrical stress concentration effect. Nevertheless, the bainitic HAZ microstructure at the weld toe plays an important role in short fatigue crack initiation and growth. Short crack kinetics is assessed on arc-welded joints of HSS steel tested in repeated tensile loading. Crack growth is monitored by the means of replica technique. Bainitic lath packets and prior austenitic grain boundaries act as major microstructural barriers which slow down short crack growth rates especially in the very early stage. Crack growth is satisfactorily modelled according to Brown's laws for short crack propagtion stage and to a Paris-type law for long crack propagation stage. The predicted S-N_R curve and conventional fatigue limit are in good agreement with experimental data. Moreover, the model is used to predict a S-N_i curve in respect of the Microstructurally Short Crack threshold condition.

KEYWORDS

High Strength Structural Steel, Arc-Welding, Bainite, Short Crack, Crack Propagation Model, Life Prediction, Fatigue Limit, Replica technique.

INTRODUCTION

Numerous studies have shown the interest in applying post-weld treatments to improve the fatigue life of welded structures made of High Strength Structural (HSS) steel sheets. This improvement is obtained by rounding off the local weld toe geometry, and/or introducing compressive residual stresses. Due to the weld toe geometry, fatigue cracks generally initiate in the Heat Affected Zone (HAZ). Nevertheless, only little work has been done concerning the role of the HAZ microstructure on short crack growth in welded joints [1,2]. The aim of the present work is to evaluate the influence of the bainitic HAZ microstructure on short crack growth and to identify a crack propagation model leading to fatigue life prediction.

EXPERIMENTAL CONDITIONS

The material under investigation is a C-Mn micro-alloyed HSS hot-rolled steel. The steel grade is S550MC and is provided by SOLLAC. The sheet thickness is 6 mm. The mechanical properties (YS≈ 600 MPa, UTS ≈ 700 MPa, A = 25%) are obtained for a grain size of 5 to 10 µm with a ferrite-bainite microstructure. The sheets are butt-welded using a MAG welding process in the flat position, following a X-type edge preparation. More details are given elsewhere [3,4]. The welding procedure generates an upper bainitic HAZ microstructure with a prior austenitic grain size of 80-100 µm, and a bainitic lath packet size of about 10 µm. It has been fully demonstrated earlier [3-5] that fatigue crack initiation generally takes place in the bainitic coarse grain HAZ close to the fusion line as it coincides with the weld toe.

Fatigue tests were carried out on a 100 kN servo-hydraulic testing machine. Tests are performed under nominal stress control, with a stress ratio $R_\sigma = 0.1$ and a sinusoidal waveform (frequency < 15 Hz). The tests are conducted until specimen failure or stopped at 2.10^6 cycles. Specimens were machined and grooved in order to locate crack initiation in the bainitic coarse grain HAZ. The damage history at the groove surface was recorded using replica technique. The specimen preparation consists in removing the weld seams on both sides by grinding and machining a groove along the second pass fusion line (Fig. 1). The groove surface is then polished and etched to reveal the microstructure. This procedure proves to be quite suitable to show the influence of the microstructure on the cracking process, the geometrical effect being controlled by the shape of the groove and not imposed by the weld toe itself [4].

The fatigue tests are regularly interrupted under maximum load so that the cracks are opened for replica application purposes. For each specimen, the main crack and secondary cracks are identified through observations of the replicas using light microscopy starting with the last. The smallest crack detected with this method is a crack of surface length of about 10 µm. The surface length $2c$ of a crack is taken as the distance between its two ends. When two cracks coalesce, the new crack length is the sum of their lengths. Moreover, fatigued specimens were cut perpendicularly to the direction of the crack at the crack initiation site. The polished cross-sections were observed in light or scanning electron microscopy in order to analyse the in-depth propagation of the cracks [4].

INFLUENCE OF THE MICROSTRUCTURE ON CRACKING PROCESS

Figure 2 shows the S-N curves obtained for the following conditions: base metal (BM), as-welded (AW) and the machined and grooved condition (MG). The fatigue limits $\Delta\sigma_{2.10^6}$, identified using a simplified Bastenaire's model, are respectively 450, 250 and 396 MPa. The differences between these values can be explained in terms of microstructure (ferritic-pearlitic for BM, bainitic for AW and MG) and of detrimental geometrical effect ($K_t = 2$ at the weld toe for AW, $K_t = 1.5$ in the groove for MG), despite the beneficial effect of residual stresses ($\sigma_R = -300$ MPa for AW, $\sigma_R = -350$ MPa for MG, measured by X-diffraction in the longitudinal direction).

Whatever the loading level, multiple short crack initiation is clearly evidenced from replica observations. For each specimen, the main crack (defined as the crack leading to the final fracture) and, in some cases also the largest secondary crack, are followed during the fatigue life. Figure 3 shows the development of the surface length $2c$ of such cracks for three different stress levels. A common feature can be observed: the crack length varies very slightly at the beginning, then increases more rapidly, with some jumps due to coalescence, and finally increases very drastically. Of course, the initiation point and subsequent development depend on the stress level, the whole process being favoured by a high stress level.

From these results, the surface crack growth rate $d2c/dN$ can be deduced using a difference method. Thus, Fig. 4 shows the crack growth rate $d2c/dN$ versus the length $2c$ of the main crack for three stress levels. Microstructural analysis based on the observations of the replicas on the one hand, and of the cross-sections on the other hand, allows a good understanding of the cracking process [4]. The three crack growth stades, as described in [6-10], are identified:
i) Microstructurally Short Crack (MSC) stade for $2c \leq d$,
ii) Physically Short Crack (PSC) stade for $50 < 2c \leq 2c_{th}$,
iii) Long Crack (LC) stade for $2c > 2c_{th}$.

Since there is only a slight dependance of these thresholds d and $2c_{th}$ on the stress level, the following values are adopted: $d = 50$ µm and $2c_{th} = 750$ µm.

Strong drops of crack growth rate occur in the MSC stade and are caused by the presence of microstructural barriers. These barriers are sufficiently strong to stop temporarily the microstructurally short cracks. The average distance at the surface between two consecutive drops is 12 to 15 µm, which corresponds to a crack depth of 6 to 8 µm, according to the relationship between surface and in-depth crack lengths [4]. Thanks to the microstructural observations, the microstructural barriers are identified as the boundaries between bainitic lath packets. The microstructural threshold d corresponds to half of the primary austenitic grain size in the HAZ.

In the PSC stade, the crack growth rate exhibits large fluctuations but they decrease progressively as the crack length increases. The slow-downs are due to the presence of bainitic lath packets and prior γ-grain boundaries, which act as strong microstructural barriers. The physically short crack goes around lath packets, following the packet boundaries, or passes through the lath packets, depending on lath orientation relatively to that of the advancing crack. Crack branching from the main crack can occur in the short crack growth domain (MSC and PSC) and generates secondary cracking. The secondary cracks can either follow the packet boundaries or pass through the lath packets. It must be noticed that the crack propagation mode depends strongly on the relative orientation of the lath packet ahead of the crack front. As an illustration, Fig. 5 shows a replica taken at a life fraction of 64 % and relative to a slow-down, in the PSC stade, of a surface crack of a length 173 µm. It was documented later that this crack passed through misoriented lath packets (on the left of the picture) and followed a prior γ-grain boundary (on the right), before it links with the visible secondary cracks.

When $2c > 750$ µm ($2c_{th}$), which corresponds to an in-depth length a greater than 190 µm (a_{th}), the crack has passed over about ten primary γ-grains. Then, the crack path becomes rectilinear, without no more clear influence of the microstructure, and goes in the long crack propagation regime (tensile mode I regime).

The crack front shape changes during the fatigue process: after initiation, the numerous short cracks are nearly semi-circular. Then the surface crack growth is predominant, due to microcrack coalescences. Finally, a single long crack propagates both in depth and at the surface, with a front turning from a semi-elliptical to a more semi-circular shape. At failure, the fatigue crack front covers, in general, at least 12 mm at the surface and about 5 mm in depth.

MODELLING OF THE CRACK PROPAGATION

The crack growth rate, in terms of surface crack length $2c$ is modelled according to Brown's laws [7] in the MSC and PSC stades and to a Paris-type law in the LC stade. The respective equations are:

$$\frac{d2c}{dN} = A.\left(\Delta\sigma^{\alpha}\right).(d - 2c) \qquad \text{for } 2c \le 50 \text{ µm } (= d) \qquad \text{in MSC stage} \qquad (1)$$

$$\frac{d2c}{dN} = B.\left(\Delta\sigma^{\beta}\right).(2c - d) \qquad \text{for } 50 < 2c \le 750 \text{ µm } (= 2c_{th}) \qquad \text{in PSC stage} \qquad (2)$$

$$\frac{d2c}{dN} = C.\left(\Delta\sigma.\sqrt{\pi.2c}\right)^{\gamma} \qquad \text{for } 2c > 750 \text{ µm} \qquad \text{in LC stage} \qquad (3)$$

The parameters A, B, C, α, β and γ are evaluated from the experimental data reported in Fig. 4. The following values are obtained: $A = 5.8 \ 10^{-36}$, $B = 6.8 \ 10^{-41}$, $C = 5.8 \ 10^{-25}$, $\alpha = 11.8$, $\beta = 13.5$ and $\gamma = 5.2$, with stress range in MPa and crack length in mm. Figure 6 shows the as-modelled propagation kinetics, in the three crack growth stages, for stress ranges of 500 and 430 MPa. These curves are compared to the experimental data (the main crack for $\Delta\sigma = 500$ MPa, and the main and the largest secondary crack for $\Delta\sigma = 430$ MPa). There is a good

318

agreement between predicted values and experimental ones, so that the modelling of the crack propagation process is quite satisfactory.

The integration for a given stress level of equations (1), (2) and (3) on their respective domains of validity gives the durations of each propagation stade and consequently an estimation of the fatigue life: $N_R = N_{MSC} + N_{PSC} + N_{LC}$. The upper limit for the LC crack propagation stade is here taken equal to 12 mm, the typical crack length measured in surface at failure. It is then possible to predict a whole fatigue life curve. Figure 7 compares the as-predicted S-N_R curve with experimental points: the predicted curve is in very good agreement with experimental results. Moreover, fatigue limits, as defined in [6], can be derived at the conventional fatigue life of 2.10^6 cycles, as follows:

i) $\Delta\sigma_{2.10^6}$ (MSC) = 325 MPa which is related to the threshold under which cracks remain microstructurally short and do not propagate (MSC threshold condition),

ii) $\Delta\sigma_{2.10^6}$ = 376 MPa, which is the conventional *engineering* fatigue limit. Once again, this value is very close to the one determined experimentally, i.e. 396 MPa (Fig. 2).

Finally, the crack initiation phase can be defined as the period during which the cracks can grow but are arrested before they can penetrate the dominant microstructural barrier (see d) while the stress level is insufficient to propagate them across that barrier. Based on this definition, a S-N_i curve can be consequently deduced from integration of the equation (1) of the MSC stade. This curve, presented in Fig. 7, shows that the ratio N_i/N_R represents about 22% at $\Delta\sigma = 500$ MPa and 19% at $\Delta\sigma_{2.10^6} = 376$ MPa.

REFERENCES

1. Baudry G. and Amzallag C. (1987). *Ann. Chim. Fr.* **12**, 529.
2. Dziubinski J. (1992). In:*Welding International* (Abington Publishing) **6**, 56.
3. Buirette C., Degallaix G., Dauphin J.Y. and Ménigault J. (1996). *La Revue de Métallurgie - CIT* **JA96**, 127.
4. Buirette C., Degallaix G., Dauphin J.Y. and Ménigault J. (1998). *Welding in the World* **IIW-Doc. XIII-1693-97**, under press.
5. Tricoteaux A., Fardoun F., Degallaix S. and Sauvage F. (1995). *Fat. Fract. Engng. Mat. Str.* **18**, 189.
6. Miller K. J. (1993). *Mat. Sc. Tech.* **9**, 453.
7. Brown M. W. (1986). In: *The behaviour of short fatigue cracks*, pp. 423-439; K. J. Miller & E. R. de los Rios (Eds.) M.E.P., London.
8. Turnbull A. and de los Rios E. R. (1995). *Fat. Fract. Engng. Mat. Str.* **18**, 1455.
9. Beretta S. and Clerici P. (1996). *Fat. Fract. Engng. Mat. Str.* **19**, 1107.
10. Demulsant X. and Mendez J. (1995). *Fat. Fract. Engng. Mat. Str.* **18**, 1483.

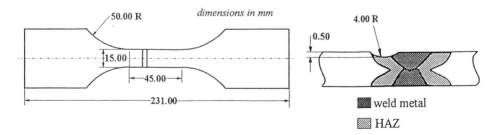

Fig 1. Specimen geometry (MG-type specimen).

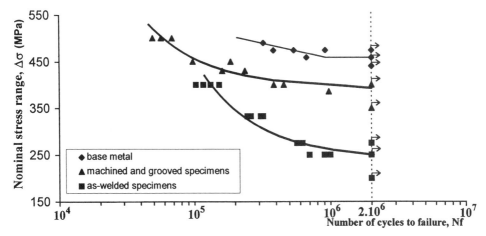

Fig. 2. S-N$_R$ curves for the different types of specimens

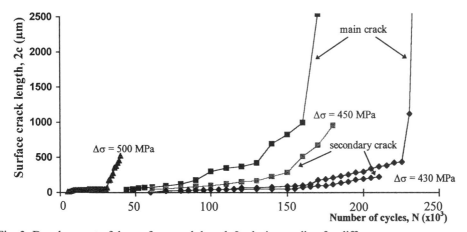

Fig. 3. Development of the surface crack length 2c during cycling for different stress ranges

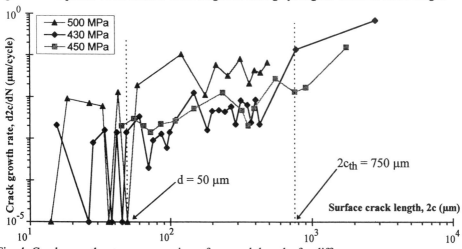

Fig. 4. Crack growth rates versus main surface crack lengths for different stress ranges

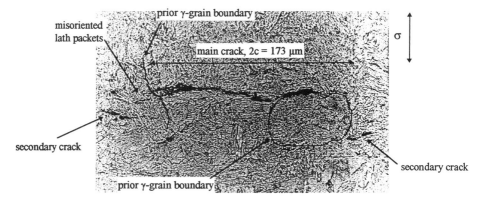

Fig. 5. Observation of the main crack from the replica taken at 64 % of life (specimen tested at $\Delta\sigma = 500$ MPa)

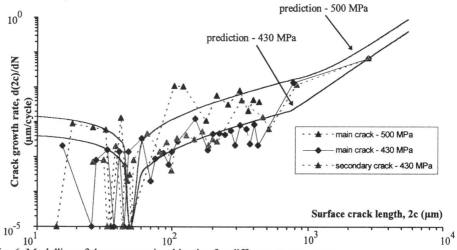

Fig. 6. Modelling of the propagation kinetics for different stress ranges

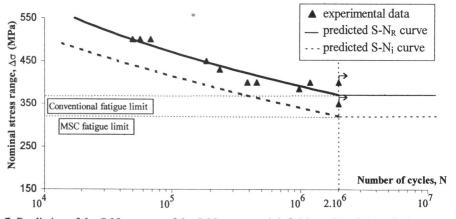

Fig. 7. Prediction of the S-N_R curve, of the S-N_i curve and definition of the fatigue limits

Low Cycle Fatigue and Elasto-Plastic Behaviour of Materials
K-T. Rie and P.D. Portella (Editors)

FATIGUE-INDUCED MARTENSITIC TRANSFORMATION IN METASTABLE STAINLESS STEELS

R.G. TETERUK, H. J. MAIER, and H.-J. CHRIST
Institut für Werkstofftechnik,
Universität-GH-Siegen, 57068 Siegen, Fed. Rep. Germany

ABSTRACT

In metastable stainless steels deformation-induced formation of martensite gives rise to drastic changes in the material's monotonic and cyclic deformation properties. The objective of the present study was to characterize the effect of sample geometry, austenite stability and testing mode on the cyclic deformation-induced austenite/martensite transformation in AISI 304 (German designation X5 CrNi 18 9, 1.4301) and AISI 304L (X2 CrNi 18 11, 1.4306) at room temperature. For this purpose the behaviour of bulk samples was compared with that of thin foils of thickness down to 25 μm. These tests indicated that in both geometries a similar tendency to form martensite exists. Furthermore, the carbon content was changed by a carburization and decarburization annealing treatment, respectively. The corresponding results confirmed the austenite-stabilizing effect of carbon. Tests which were carried out under block loading conditions showed that a linear martensite accumulation rule based on data from corresponding constant amplitude fatigue tests gives reasonable results in certain cases only.

KEYWORDS

metastable stainless steels, low-cycle fatigue, cyclic block loading, austenite/martensite transformation, influence of carbon content

INTRODUCTION

The transformation from austenite to martensite in ferrous systems can be accomplished either by heat treatment or by deformation of the austenite. The deformation-induced martensite may be classified in two categories: stress-assisted or strain-induced [1]. Maxwell et al. [2] have indicated that stress-assisted martensite is formed during deformation if the stress level is sufficiently high to provide the energy required for the transformation from austenite to martensite. Stress-assisted transformation involves the spontaneous nucleation and growth of martensite in the form of plates [2, 3], usually in a manner similar to that obtained by thermal quenching of austenite. However, strain-induced martensite is a direct consequence of plastic deformation and can be morphologically different from stress-assisted or thermally produced martensite. The formation of lath-type α'-martensite by strain-induced processes appears to be common in ferrous systems [1]. Plastic deformation of the austenitic matrix leads to a proper defect structure initiating the transformation process. The potential nucleation sites are deformation twins, stacking faults, and ε martensite [1].

The mechanical behaviour has been found to be changed drastically by martensitic transformation. Martensite formation reduces crack growth [4] and leads to rapid hardening in fatigue tests [5, 6], i.e. it increases the resistance to plastic flow. In tensile tests the transformation is trigged by a certain amount of plastic strain which depends on the material and on the temperature [7]. In fatigue tests at room temperature the plastic strain amplitude can be considered to be the decisive parameter that triggers the martensitic transformation [5, 8].

In the present study the effect of sample geometry, austenite stability and testing mode on the cyclic deformation-induced austenite/martensite transformation in the steels AISI 304 and AISI 304L at room temperature is addressed.

MATERIALS AND EXPERIMENTAL PROCEDURES

The chemical compositions of the two materials used in the present study are shown in Table 1. This table indicates that a carbon content of 0.05 wt.% is present in AISI 304, while AISI 304L contains only 0.015 wt.%. The mean grain size of the as-received AISI 304L stainless steel was determined to be 150 μm and that of AISI 304 which was obtained in the form of thin foils was about 30 μm (twins not taken into account).

Table 1. Chemical composition of AISI 304 and AISI 304L (wt.%)

Steel	C	Si	Mn	Cr	Ni	Mo	N	Cu
304L	0.015	0.19	1.25	18.35	10	0.1	0.045	0.05
304	0.05	0.5	1.5	18	9	–	–	–

All fatigue tests were performed in symmetrical push-pull using plastic or total strain control at room temperature by means of a closed-loop servohydraulic testing machine (SCHENK S56). Fatigue tests were conducted on bulk samples and on thin foils. For fatigue testing of bulk samples, cylindrical specimens with a diameter of 12 mm within the gauge length were machined from AISI 304L-type stainless steel bars and electropolished prior to cyclic loading. Fatigue testing of AISI 304 thin foils of thickness t down to 25 μm was accomplished by glueing each foil onto the gauge length of an aluminium specimen that had a rectangular cross section. Strain gauges attached to both the aluminium sample and the thin foil indicated that as long as the strain amplitudes were not too high, the total strain is almost completely transferred from the aluminium substrate to the foil. The plastic strain amplitude ($\Delta\varepsilon_{pl}/2$) was varied between 0.185% and 1.1% in the tests on the bulk samples. The experiments on the thin foils were carried out in total strain control with amplitudes varying between 0.2% and 0.52%. Two different testing modes were applied. In constant amplitude fatigue tests plastic or total strain amplitude was kept constant throughout each test. Fatigue tests with block loading were used to study the way in which the martensite formation is affected by a changing amplitude. The test frequency was kept constant at 0.2 Hz to avoid an increase in temperature of the sample because the rate of deformation-induced martensite formation depends very sensitively on the temperature.

The α'-martensite formed in an austenitic matrix (γ) is ferromagnetic and can be detected by magnetic permeability measurements. A commercially available measuring instrument (Fischer Feritscope) was used. This instrument was carefully calibrated with standards of known martensite content. It should be noted that the output signal is dependent on the thickness of the sample studied. To compare the measurements obtained on the bulk samples with these on the thin foils a correction function $K(t)$ was determined. The difference between displayed and true martensite content was found to be negligible only above a thickness of $t \approx 200\mu m$.

RESULTS AND DISCUSSION

An example of the temperature effect on the deformation-induced martensitic transformation caused by variations in test frequency is represented in Fig. 1. The cyclic hardening curve was measured at a plastic strain amplitude of 1.2%, applying step-wise constant frequencies.

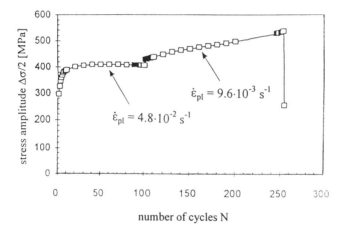

Fig. 1. Cyclic hardening curve of the austenitic steel AISI 304L. The increase of the stress amplitude after the reduction of frequency can be attributed to the formation of martensite. See text for details.

The higher deformation energy per time introduced into the material at higher frequency causes a slight increase of the temperature of the plastically deforming sample (by about 10°C) at 1 Hz as compared to 0.2 Hz. As a consequence of this temperature rise at the higher frequency, the martensite-induced cyclic hardening is restricted to the low-frequency part in the cyclic hardening curve depicted in Fig. 1, whereas at the faster strain rate a stable cyclic deformation behaviour is observed.

The change in the volume fraction of α'-martensite with the number of cycles during constant amplitude loading at different plastic strain amplitudes is shown in Fig. 2a and Fig. 2b for bulk samples and for thin foils, respectively. In either case the direct correlation between the amount of the martensite formed, the number of cycles, and the strain amplitude is clearly visible. The volume fraction of α'-martensite increases continuously with the number of cycles and the slopes of the curves are strongly affected by plastic strain amplitude. Furthermore, the difference in the content of martensite at fracture decreases with increasing

324

plastic strain amplitude. For example, the bulk specimens (AISI 304L) fatigued at $\Delta\varepsilon_{pl}/2 =$ 0.8% and at 1.1% failed at martensite volume fractions of 19% and 19.5%, respectively. The main reason for this effect is that a saturation content of martensite is not reached because of premature failure.

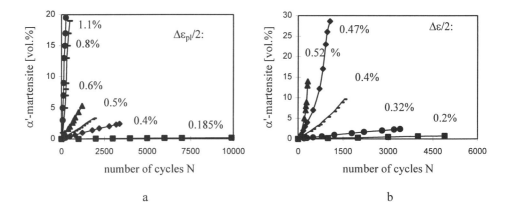

a   b

Fig. 2. Volume fraction of α'-martensite as a function of the number of cycles in bulk samples of AISI 304L (a) and in thin foils of AISI 304 (b) for different plastic and total strain amplitudes, respectively.

As already mentioned above, the tests on the thin foils were performed under total strain control to ensure that the strain amplitude acting in the thin foil is identical to that of the supporting aluminium specimen. The maximum total strain amplitude that could be applied was $\Delta\varepsilon/2 = 0.52\%$. At higher values the adhesive bond did not transfer the strain in a sufficient way. However, the maximum volume fraction of α'-martensite in thin foils was not observed at the maximum tolerable total strain amplitude of 0.52% but at $\Delta\varepsilon/2 = 0.47\%$. This is a consequence of the drastically reduced number of cycles to failure at $\Delta\varepsilon/2= 0.52\%$ which is about one third of that obtained at $\Delta\varepsilon/2 = 0.47\%$. Consequently the final α'-martensite content at $\Delta\varepsilon/2 = 0.52\%$ is relatively small.

At amplitudes smaller than those represented in Fig. 2, martensite formation in both bulk samples and thin foils was not observed. In agreement with observations in a previous study [5], a threshold of the plastic strain amplitude $\Delta\varepsilon_{pl,th}/2$ of about 0.3% for martensite formation could be confirmed.

In order to study the effect of changing amplitudes on the martensitic transformation, block loading tests were run. In these tests each block consisted of nine cycles with $\Delta\varepsilon_{pl,1}/2$ and one overload cycle with $\Delta\varepsilon_{pl,2}/2$, where $\Delta\varepsilon_{pl,1}/2 < \Delta\varepsilon_{pl,2}/2$. The blocks were continuously repeated until failure. The plastic strain amplitudes $\Delta\varepsilon_{pl,1}/2$ for the tests on bulk samples were chosen in such a way that values below and above $\Delta\varepsilon_{pl,th}/2$ resulted. Similar tests were performed on the thin foils. However, instead of the plastic strain amplitudes the total strain amplitudes were varied accordingly.

The increase in volume fraction of α'-martensite in a thin foil as a function of the number of cycles at $\Delta\varepsilon_2/2 = 0.52\%$ during block loading is presented in Fig. 3a as solid symbols. In this test the value of $\Delta\varepsilon_1/2$ was lower than the threshold value for martensite formation. Obviously this threshold value has not changed during cyclic loading as can be seen by the comparison with the martensitic transformation curve from the constant amplitude test at $\Delta\varepsilon/2 = 0.52\%$ (Fig. 3a, open symbols). The agreement of both curves documents that in tests with changing amplitudes martensite transformation occurred only during those cycles where the strain amplitude was larger than $\Delta\varepsilon_{th}/2$.

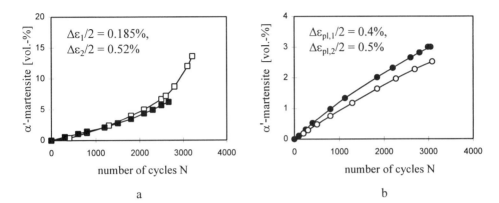

Fig. 3. Comparative analysis of the development of α'-martensite as a function of the total number of cycles in block loading tests (full symbols). The open symbols represent the result of a linear accumulation calculation. (a) Test on a thin foil with $\Delta\varepsilon_1/2 < \Delta\varepsilon_{th}/2$; (b) test on a bulk sample at $\Delta\varepsilon_{pl,1}/2 > \Delta\varepsilon_{pl,th}/2$.

In contrast, in fatigue tests where both amplitudes of the loading block lay above the threshold value $\Delta\varepsilon_{pl,th}/2$ martensite formation was no longer independent of the loading history. In the test on the bulk sample (Fig. 3b) with $\Delta\varepsilon_{pl,1}/2 = 0.4\%$ and $\Delta\varepsilon_{pl,2}/2 = 0.5\%$ more α'-martensite was formed than was expected on the basis of a linear accumulation calculation taking the behaviours in the corresponding constant amplitude tests into account.

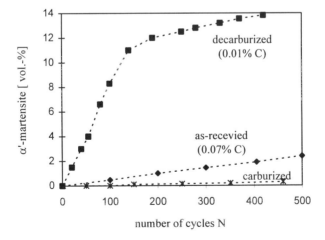

Fig. 4. Volume fraction of the α'-martensite in thin foils of AISI 304 as a function of the number of cycles for different carbon contents.

A systematic variation of the foil thickness showed that the rate and amount of martensite formed during cycling decreases drastically, if the sample thickness reaches the dimensions of the grain size (30μm in this case). This is in accord with results obtained in an earlier study where specific dislocation arrangements were found to trigger the martensite formation [5]. These dislocation configurations are less likely in surface grains since dislocations can leave the grains at the surface.

In order to modify the stability of the austenitic phase, the concentration of carbon in AISI 304 foils was changed. For this purpose foils were annealed at 950°C in a methane/hydrogen atmosphere and in pure hydrogen, respectively. Figure 4 demonstrates that the amount of martensite formed in decarburized thin foils increases drastically as compared to the as-received material. If the carbon content is raised so that carbides form a continuous film along the grain boundaries, the transformation rate becomes very low. These results correspond to the idea of an austenite-stabilizing effect of carbon.

SUMMARY

The effect of sample geometry, austenite stability and changing amplitude on the cyclic deformation-induced austenite/martensite transformation in stainless steels of type AISI 304 and AISI 304L at room temperature was studied. The main results can be summarized as follows.

- In agreement with earlier work [5] a certain plastic strain amplitude must be exceeded to trigger the formation of martensite. This threshold value was found to be $\Delta\varepsilon_{pl,th}/2 \approx 0,3\%$. If this value is exceeded the martensite volume fraction increases with the number of cycles. The rate of transformation depends strongly on the plastic strain amplitude.
- The threshold value is not changed with the martensite content. Consequently, the cycles in a block loading test with amplitudes below the threshold value do not affect martensite formation. However, in those tests where several amplitudes of the loading block are above the threshold the transformation rate exceeds the prediction based on linear summation.
- Fatigue tests on thin foils with different carbon content showed that the stability of the austenitic phase depends strongly on the carbon concentration. The amount of martensite formed in decarburized thin foils is drastically increased as compared to the as-received material, and is extremely low if the carbon content is raised so that carbides form a continuous film along the grain boundaries.

REFERENCES

1. Vilay Shrinivas, Varma, S.K. and Murr, L.E. (1995) *Metall. Mater. Trans.* **26A**, 662
2. Maxwell, P.G., Goldberg, A. and Shyne, J.C. (1974) *Metall. Trans.* **5**, 1305
3. Lecroisey, F. and Pineau, A. (1972) *Metall. Trans.* **3**, 387
4. Ganesh Sundara Raman, S. and Padmanabhan, K.A. (1994) *Mater. Sci. Techn.* **10**, 614
5. Bayerlein, M., Christ, H.-J. and Mughrabi, H. (1989) *Mater. Sci. Eng.* **A114**, L11
6. Tjong, S.C. and Ho, N.J. (1988) *Mater. Sci. Eng.* **A102**, 125
7. Suzuki, T., Kojima, H., Suzuki, K., Hashimoto, T., Koike, S. and Ichihara, M. (1976) *Acta Metall.* **10**, 353
8. Baudry, G. and Pineau, A. (1977) *Mater. Sci. Eng.* **28**, 229

Low Cycle Fatigue and Elasto-Plastic Behaviour of Materials
K-T. Rie and P.D. Portella (Editors)
327

CORRELATION BETWEEN GRAIN BOUNDARY STRUCTURE AND CYCLIC PLASTIC STRAIN BELOW YIELD STRESS

LUIZ CARLOS ROLIM LOPES
Universidade Federal do Rio Grande do Sul, Escola de Engenharia, Centro de Tecnologia,
Av. Bento Gonçalves, 9.500, C.P. 15.021, Porto Alegre, 91.501-907, RS, Brazil
CHRIS BOYD THOMSON
VALERIE RANDLE
University of Wales Swansea, Department of Materials Engineering, Singleton Park, Swansea,
SA2 8PP - Wales, UK

ABSTRACT

The influence of grain orientation and grain boundary structure on the onset of plastic strain in an austenitic stainless steel submitted to fatigue process under a stress amplitude bellow yield strength has been studied. The electron back-scatter diffraction technique and the coincidence site lattice model were used to measure grain texture and grain boundary characterization. The results reveal that crystallographic orientation of grains by itself has no effect on the persistent slip bands nucleation. Bands nucleation is affected by grain boundary structure. Random boundaries are preferred sites for nucleation. Twin boundaries also nucleate bands in a lesser degree. Few boundaries with Σ value ranging from 5 to 29 were involved in plastic deformation. Low angle boundaries are resistant to intergranular plasticity. These results are discussed in terms of stress concentration at grain boundaries.

KEYWORDS

Fatigue, microplasticity, grain boundary structure, crystallographic orientation, microtexture.

INTRODUCTION

The role of grain boundaries (GB's) in the deformation process was studied by different authors [1-5]. Experimental works showed that the onset of plastic deformation is a highly localized process, taking place at GB and in regions closely adjoining them [2-5]. Different orientations of grains and incompatibility effects resulting from elastic anisotropy of crystals cause stress concentration and create a GB affected zone. Meyers and Ashworth calculated stress concentration at GB and found that local stress at GB can be almost three times higher than the applied stress [6].

These effects have a particular consequence for the fatigue behavior of polycrystals. Cyclic straining below the yield strength produces microplastic strain which can modify the initial strength of the material [7-11]. Persistent slip bands (PSB's) are activated to accommodate the intergranular stresses [11].

In this work, an austenitic stainless steel has been fatigued below yield strength and the effects of grain orientation (microtexture) and GB structure on the microplastic activity and PSB's nucleation have been studied. The electron back-scatter diffraction (EBSD) technique associated to scanning electron microscope (SEM) and the coincidence site lattice (CSL) model have been used.

EXPERIMENTAL PROCEDURES

The material used was a sample of a round bar of 316L steel in the hot rolled and solution treated condition, with a mean grain size of 48μm. A fatigue specimen with parallel flat surfaces has been specially designed for metallographic analysis after fatigue test in a rotating beam fatigue testing device. The fatigue test has been conducted at a constant stress amplitude of 232 MPa, corresponding to about 80% of the yield strength of the material. Fatigue specimen has received a standard metallographic preparation, including a final polish with colloidal silica. The fatigued specimen has been analysed in a Jeol 6100 SEM equipped with an EBSD system and crystallographic orientation of grains, measured [12]. The structure of grain boundaries taking part in plastic strain accommodation has been characterized by the CSL model [13-16]. The Brandon's criterion of deviation from exact CSL misorientation has been used [17]. Microtexture experiments have also been performed on the material in the as-received conditions to determine texture and grain boundary character distribution (GBCD) resulting from the material's processing.

RESULTS AND DISCUSSION

Surface Observations of Fatigued Material

Cyclic loading produced PSB's belonging to a single slip system. About 21% of the material surface was covered by PSB's. The analysis of 100 grains in the fatigued material has provided data for 77 boundaries. 36 boundaries lay between grains both containing PSB's and 41 boundaries lay between grains for which only one reveals PSB's. The results for these two different behaviours have been classified in two categories: **PSB-PSB** and **PSB-non PSB** boundaries.

Microtexture Experiments

The analyses of the as-received material have shown no preferred orientation. The number of grains analysed in this material were 306 and 229 for longitudinal and transverse samples respectively.

Figures 1 and 2 show the 111 pole figures corresponding to the records of 68 grains with PSB's and 32 grains without PSB's respectively of the fatigued material. The figures show that no preferred orientation is observed for none of the these two sets of data. Therefore, the crystallographic orientation of grains with respect to stress axis itself has no effect on the PSB's nucleation process, bellow the yield strength.

Figure 3 shows the GBCD data of the 77 boundaries analysed in the fatigued material. For comparison, the GBCD data of the as-received material are also presented. The proportion of CSL boundaries involved with plastic strain in the fatigued process is lower than the CSL existent in the material. On the other hand, the proportion of random (R) boundaries ($\Sigma \geq 31$) involved with plastic strain is significantly higher than the proportion of R boundaries present in the material. This indicates that R boundaries have higher tendency to nucleate PSB's. The proportion of CSL boundaries ranging from $\Sigma=5$ to 29 which take part in the strain accommodation process is less then 10%. Less than 3% of low angle boundaries (LAB) have nucleated bands. Twin boundaries ($\Sigma=3$) and R boundaries are the most important interfaces in the PSB's nucleation process. Llanes and Laird, using transmission electron microscopy, also observed that annealing twin boundaries were preferred nucleation sites for PSB's in polycrystalline copper submitted to intermediate strain amplitude [18].

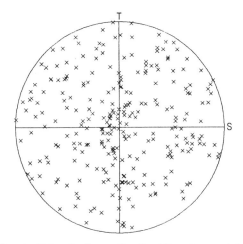

Fig. 1. 111 pole figure for the 68 grains containing PSB's in the fatigued material. S indicates stress axis direction.

GBCD data for PSB-PSB and PSB-non PSB categories of boundaries are presented separately in fig.4. The most significant differences between these two categories are observed in the proportion of twin boundaries (TB) and CSL boundaries with $11 \leq \Sigma \leq 29$. TB of PSB-PSB type are present in a significantly lower proportion than the existing proportion of TB in the material. This difference was investigated in terms of mean deviation from exact $\Sigma=3$ misorientation. This is expressed as the normalized deviation v/v_m [17]. Lower v/v_m values are an indication of more stable, i.e. lower energy, boundary configurations. The results are shown in table 1. They lead to the conclusion that the more stable TB's are more resistant to intergranular plasticity.

Since crystallographic orientation of grains alone is not a determining factor for PSB's nucleation on primary slip system at the onset of plastic strain, the nucleation process at GB must be controlled by stress concentration due to intergranular stresses. The stress concentration factor seems to be related to GB structure, once EBSD experiments indicates that PSB's nucleation is affected by GB boundary structure. R boundaries are preferred nucleation sites for primary PSB's.

330

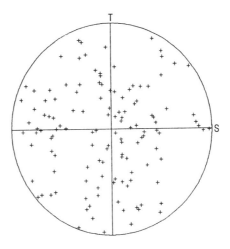

Fig. 2. 111 pole figure for 32 grains without PSB's of
the fatigued material. S indicates stress axis direction.

The principal physical feature which characterizes the structure of a GB is its excess free volume and associated elastic strain fields. So, one can assume that a GB with a higher Σ value have a higher concentration factor. This is supported by the fact that few LAB and few CSL boundaries with low Σ value have nucleated PSB's. The exceptions are TB's. TB's must have lower elastic strain energy associated. However, they are also responsible for primary PSB's nucleation below yield strength. One factor that also must affect GB behavior is the GB plane orientation with respect to stress axis. The experimental analysis of GB plane orientation is very difficult to perform. Its effect should be considered with aid of calculation methods. Peralta and coworkers calculated stress in the region next to a TB and found that it reaches to a maximum when the tensile axis is parallel to a <111> direction [19]. Twin plane in FCC crystals is a {111} high density plane. So, it could be able to emit dislocations and nucleate PSB's, when conveniently oriented with respect to stress axis.

TB separating PSB-non PSB grains have lower deviation from the exact $\Sigma=3$ misorientation and must have lower stress concentration associated. So, it is stronger, i.e. more resistant to plastic strain than TB of PSB-PSB type. In a polycrystal, a local system formed by grains limited by stable TB's and CSL interfaces with low Σ are more resistant to intergranular plasticity.

However grain orientation does not affect the behavior of GB's, the present results encourage further research on the effect of texture on cyclic behavior of polycrystalline materials, because texture affect grain misorientation and, in some degree, the frequency of CSL boundaries [20]. It possible affects also GB plane orientation. Research oriented to determine relationships between GB structure and cyclic properties as cyclic stress-strain curve, fatigue life or crack nucleation mechanisms are required.

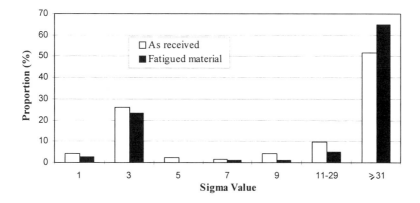

Fig.3 Grain boundary character distribution data corresponding to 100 grains analysed in the fatigued material. Data for the material in the as-received condition is also shown.

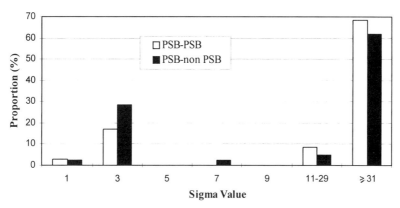

Fig. 4. Grain boundary character distribution data for PSB-PSB and PSB-non PSB boundary categories.

Table 1. Mean deviation from exact $\Sigma=3$ misorientation for both PSB-PSB and PSB-non PSB boundaries.

Grain Boundary Category	Mean deviation v/v_m
PSB-PSB	0.838 ± 0.143
PSB-non PSB	0.529 ± 0.183

332

CONCLUSIONS

The microtexture experiments performed on a 316L steel submitted to cyclic loading below yield strength lead to the following conclusions:
1. Orientation of the grains with respect to stress axis alone is not a determining factor for PSB's nucleation in the primary slip system, in a microplastic process. PSB's nucleation is controlled by stress concentration at boundaries which is affected by the grain boundary structure;
2. Random boundaries, with a proportion of 65%, are preferred nucleation sites of PSB's. Twin boundaries are also sites for nucleation with a proportion of about 24%. Twin boundaries with misorientation close to exact $\Sigma=3$ condition are more resistant to plastic strain. The proportion of CSL boundaries with $\Sigma=5$ to 29 which nucleates PSB's is less then 10%. The proportion of low angle boundaries is less than 3%;
3. These results encourage further research on the effect of texture on cyclic behaviour of polycrystals, since texture may affect grain misorientation and the frequency of CSL boundaries.

ACKNOWLEDGMENTS

One of the authors (L.C.R.L) is grateful to the Rio Grande do Sul State Research Council (FAPERGS), Brazilian Research Council (CNPq) and The Royal Society for sponsoring this work. C.B.T. acknowledges the financial contribution of the EPSRC.

REFERENCES

1. Hirth, J. P., (1972). Metall. Trans., **3**, 3047.
2. Shen, Z., Wagoner R. H., Clark, A. T., (1989). Acta Metall., **36**, 3231.
3. Worthington, P. J., Smith, E., (1965). Acta Metall., **12**, 1277.
4. Carrington, W. E., McLean, D., (1965). Acta Metall., **13**, 439.
5. Tandon, K. N., Tangri, K., (1975). Metall. Trans. A, **6**, 809.
6. Meyers, M. A., Ashworth, E., (1982). Phil. Mag., **46**, 737.
7. Klesnil, M., Lukas, P., Rys, P., (1965). J. Iron Steel Inst. **203**, 47.
8. Klesnil, M., Lukas, P., (1967). J. Iron Steel Inst., **205**, 746.
9. Abel, A., Muir, H. (1973). Acta Metall., **21**, 93.
10. Abel, A. Muir H., (1973). Acta Metall., **21**, 99.
11. Rolim Lopes, L. C., Charlier, J., (1993). Mater. Scie. & Engn. A, **169**, 67.
12. Randle, V., (1992). Microtexture Determination and its Applications, The Institute of Materials, London.
13. Bollmann, W., (1982). Crystal Lattice, Intefaces, Matrices - An Extention of Crystallography, Published by the Author, Geneva.
14. Grimmer, H., Bollmann, W., Warrington, D.H., (1974). Acta Cryst. **30A**, 197.
15. Brandon, D.G., Ralph, B., Ranganathan, S., Wald, M.S., (1964). Acta Metall., **12**, 813.
16. Warrington, D. H., (1980). In:Grain Boundary Structure and Kinetics, 1-12, American Society for Metals, Ohio.
17. Brandon, D. G., (1966). Acta Metall., **14**, 1479.
18. Llanes, L., Laird, C., (1992). Mater. Scie. & Engn. A, **157**, 21.
19. Peralta, P., Llanes, L., Bassani, J., Laird, C., (1994). Phil. Mag. A, **70** , 219.
20. Randle, V., Ralph, B., Dingley, D., (1988). Acta Metall., **36**, 267.

Low Cycle Fatigue and Elasto-Plastic Behaviour of Materials
K-T. Rie and P.D. Portella (Editors)
333

SIGNIFICANCE OF MICROSTRUCTURE AND PORE MORPHOLOGY ON THE INITIATION AND GROWTH OF FATIGUE CRACKS IN SINTERED STEEL

G. BIALLAS** and A. PIOTROWSKI*

Department of Materials Science, University of Essen, 45117 Essen, Germany
** *Institute of Materials Research, German Aerospace Center, 51170 Cologne, Germany*

ABSTRACT

The influence of sintering temperature and of a higher content of alloying elements on micro-structure, pore morphology and fatigue crack initiation and growth of sintered steels is discussed. Strain, temperature and electrical resistance measurements are combined with scanning electron microscopy investigations to characterise the fatigue behaviour. It is found that the pores become larger and more regular with a higher sintering temperature, resulting in a notably higher amount of plastic deformation necessary for crack initiation, although the deformation processes are still concentrated in the pore regions. A higher content of alloying elements leads to smaller and rougher pores but, nevertheless, to a more homogeneous deformation state and a higher cyclic plasticity. This is caused by the microstructure of the metallic matrix which makes possible the induction of crack closure effects. Therefore, non-pore-oriented slip bands and crack paths could be observed.

KEYWORDS

Sintered steel, microstructure, cyclic deformation behaviour, crack initiation and growth

INTRODUCTION

The fatigue behaviour of sintered steels is determined by the mixed influence of their porosity and microstructure. Depending on the pore morphology, the chemical composition and the homogeneity of the microstructure either microplastic deformation or early crack initiation and growth dominate the fatigue processes. In this paper, strain, temperature and electrical resistance measurements are combined with extensive electron microscopy investigations to study the material's reaction on cyclic loading.

MATERIALS

The materials investigated were a mixed alloyed Fe 1.2Cu sintered at a temperature of 1120°C or 1280°C and a partially pre-alloyed Fe 4Ni 1.5Cu 0.5Mo sintered at 1120°C (sintering time 30 minutes, sintering atmosphere 70 % N_2 and 30 % H_2). For all materials the sintered density was 7 g/cm^3 which results in a porosity of about 11%. The yield ($R_{p0.2}$) and the ultimate tensile strength (R_m) as well as the Young's modulus (E) and the fracture strain (ε_f) increase with increasing sintering temperature (see Table 1). A higher content of alloying elements leads to a marked increase in the tensile strength, whereas the Young's modulus and especially the fracture strain are only slightly affected. The endurance limits σ_{el} were determined for different failure probabilities P. The increase in the endurance limit with increasing sintering temperature is clearly lower than the increase caused by a higher content of alloying elements.

Table 1. Mechanical properties

Material, Sintering Temperature	E kN/mm^2	$R_{p0.2}$ N/mm^2	R_m N/mm^2	ε_f %	$\sigma_{el, P=5\%}$ N/mm^2	$\sigma_{el, P=95\%}$ N/mm^2
Fe 4Ni 1.5Cu 0.5Mo, 1120°C	151	235	380	6.8	109	125.5
Fe 1.2Cu, 1120°C	147	185	255	6.7	94	104
Fe 1.2Cu, 1280°C	156.5	220	310	10.6	99	110

EXPERIMENTAL METHODS AND PROCEDURES

To characterise the pore morphology, the pore area as well as three different shape factors for bulkiness, elongation and roughness of the pores were determined. The value of each shape factor ranges from 0 to 1, in which a higher value corresponds to a more regular pore morphology. It is important to take into account stereometrical and statistical requirements. Frequency distributions are generally accepted for estimating the change in pore area and in the shape factors. Medians (values at a cumulative frequency of 50%) are highly suitable for quantifying the distribution characteristics [1-3].

To obtain a more detailed description of the fatigue behaviour strain, temperature and electrical resistance measurements as well as scanning electron microscopy investigations were performed simultaneously. While plastic strain amplitude and temperature change caused by heat dissipation of the deformation work are sensitive to microplastic deformation processes, electrical resistance is mainly influenced by the effective cross-section, which depends on crack initiation and growth. The additional use of temperature and electrical resistance measurement permits a more comprehensive and more sensitive description of the fatigue behaviour as could be shown in previous investigations [1,3-5].

All fatigue tests were performed in a tension-compression mode under stress control without mean stress (R = -1) at room temperature and a frequency of about 35 Hz. Characteristic values of plastic strain amplitude, temperature and electrical resistance change at the half of fatigue life (N = N_f/2) were used to evaluate Coffin-Manson (CM, cyclic ductility exponent c'), cyclic stress-strain (CSS, cyclic hardening exponent n'), cyclic stress-temperature (cyclic temperature hardening exponent t') and cyclic stress-electrical resistance curves (cyclic electrical resistance exponent r'). The difference between n' and the energy equivalent exponent t'/(1-t') is suitable for estimating the inhomogeneity of the deformation state (n' - t'/(1-t') $\rightarrow$ 0 for an ideal ductile material), and the temperature-electrical resistance ratio $f_{TR} = \Delta T(N_f/2) / \Delta R(N_f/2)$ can be used to approximate the amount of plastic deformation necessary for crack initiation [1,3,5].

EXPERIMENTAL RESULTS AND DISCUSSION

Microstructure and Porosity

Micrographs of the materials investigated are shown in Fig. 1. Along the pores (black areas) and grain boundaries, regions of higher copper concentration are visible for Fe 1.2Cu sintered

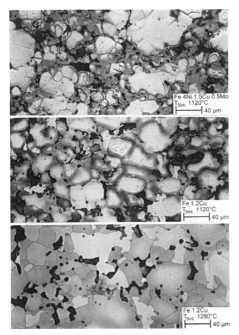

Fig. 1. Micrographs

at 1120°C, whereas the concentration is nearly equalised for the material sintered at 1280°C. An increase in the content of diffusion-bonded alloying elements leads to a more pronounced solid-solution hardening effect, especially in the near pore regions. For Fe 4Ni 1.5Cu 0.5Mo this has been proved by microhardness measurements and complementary EDS X-ray analyses [1-3], but it is also indicated by the strongly etched zones surrounding the pores in the upper picture of Fig. 1. In addition to the temporary liquid phase sintering, caused by copper, the higher sintering temperature enhances the sintering processes. Thus, the pores of Fe 1.2 Cu become larger and more regular at a higher sintering temperature (see medians in Table 2). In comparison to Fe 1.2Cu sintered at 1120°C, Fe 4Ni 1.5Cu 0.5Mo shows smaller and distinctly rougher pores, although the bulkiness is nearly unchanged and the pores are somewhat less elongated. An increase in pore area is equivalent to an increasing interpore distance, because the total amount of porosity is identical for all materials investigated. A more regular pore shape brings about a smaller notch effect which leads to a more homogeneous stress distribution in the bulk of the material during cycling. Therefore, the pore morphology of Fe 1.2Cu sintered at 1280°C is most favourable of the present materials.

Table 2. Medians of pore area and shape factors

Material	Pore area	Bulkiness	Elongation	Roughness
Fe 4Ni 1.5Cu 0.5Mo, 1120°C	6.4 μm^2	0.487	0.349	0.731
Fe 1.2Cu, 1120°C	8.0 μm^2	0.483	0.335	0.817
Fe 1.2Cu, 1280°C	11.9 μm^2	0.533	0.372	0.830

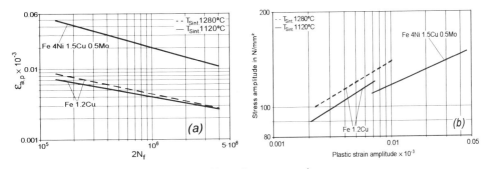

Fig. 2. (a) Coffin-Manson diagram and (b) cyclic stress-strain curves

336

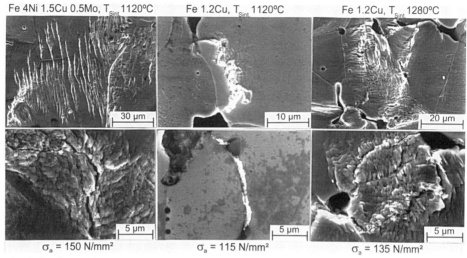

Fe 4Ni 1.5Cu 0.5Mo, T$_{Sint}$ 1120°C   Fe 1.2Cu, T$_{Sint}$ 1120°C   Fe 1.2Cu, T$_{Sint}$ 1280°C

| 30 µm | 10 µm | 20 µm |
| 5 µm | 5 µm | 5 µm |

σ_a = 150 N/mm² σ_a = 115 N/mm² σ_a = 135 N/mm²

Fig. 3. Surface deformation features, $N = N_f \approx 10^{-}$

Cyclic Deformation Behaviour

CM curves show a slight increase in cyclic plasticity with increasing sintering temperature but a strong increase for a higher content of alloying elements if specimens are loaded to the same number of cycles to failure (see Fig. 2a). For the higher sintering temperature CSS curves even reveal a lower cyclic plasticity for a certain stress amplitude, whereas a higher content of alloying elements is still connected with an increase in plastic deformation (see Fig. 2b). These tendencies seem to be contradictory to the change in pore morphology but are in accordance with the extent and density of typical slip features (see Fig. 3, upper line). For Fe 1.2Cu sintered at 1120°C small extrusions are sufficient for crack initiation, whereas for Fe 4Ni 1.5Cu 0.5Mo and for Fe 1.2Cu sintered at 1280°C crack initiation preferentially occurs in regions with a high density of slip bands (see Fig. 3, lower line). This observation coincides excellently with the f_{TR}-values given in Table 3. The higher sintering temperature leads to larger and rounder pores, resulting in a considerably lower crack sensitivity (f_{TR} ↑) but in less homogeneous cyclic deformation processes (n'- t'/(1-t') ↑). A higher content of alloying elements leads to smaller and more irregular pores, but this is connected with a reduced crack sensitivity (f_{TR} ↑) and an even more homogeneous deformation state (n'- t'/(1-t') ↓, cf. Table 3).

Table 3. Cyclic exponents and f_{TR}-factor

Material	n'	t'	r'	$n' - \dfrac{t'}{1-t'}$	c'	$\overline{f_{TR}}$
Fe 4Ni 1.5Cu 0.5Mo, 1120°C	0.163	0.129	0.129	0.014	-0.450	0.97 K/%
Fe 1.2Cu, 1120°C	0.236	0.137	0.113	0.078	-0.293	0.17 K/%
Fe 1.2Cu, 1280°C	0.224	0.121	0.120	0.087	-0.340	1.53 K/%

Initiation and Growth of Fatigue Cracks

The contradictions described in the previous section can be resolved by a detailed examination of the crack initiation and growth. Stress-strain hysteresis loops (see Fig. 4a) include first

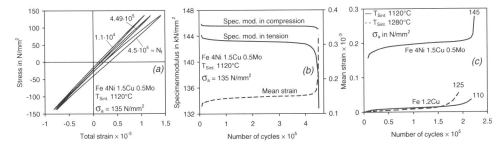

Fig. 4. (a) stress-strain hysteresis loops, (b) change of specimen moduli and (c) mean strain

information about crack behaviour. For all materials investigated the hysteresis loops are shifted to positive mean strain (see Fig. 4b) and a change in slope of the loops is noticeable since the shifts of the tension and compression arrest point are not equal. Thus, a distinct difference between the specimen moduli in tension and in the compression phase develops during the first cycles and scarcely changes for the greatest part of further cycling. This difference is caused by the opening of fatigue cracks which have been initiated near the pores. For both states of Fe 1.2Cu the cracks grow straight to the next pore which is exemplarily shown for the lower sintering temperature in the upper line of Fig. 5. For Fe 4Ni 1.5Cu 0.5Mo, numerous slip features form near the crack tip and cause a retarded crack propagation so that the crack does not reach the next pore (see Fig. 5, lower line).

EDS X-ray analyses of Fe 4Ni 1.5Cu 0.5Mo, combined with microhardness measurements, have revealed that the metallic matrix areas along the pores are markedly solid-solution hardened by a high nickel and copper concentration, whereas the ferritic areas inside the grains are low alloyed [1-3]. Obviously, such microstructure retards the growth of short cracks. Highly deformable areas inside the grains, which make possible the induction of crack closure effects, are combined with high strength areas near the pores to reduce the effect of internal notches. Therefore, a high density of short cracks can be tolerated, which is an explanation for the clearly higher level of mean strain observed for this material (see Fig. 4c).

Fe 1.2Cu, $T_{Sint.}$ 1120°C, σ_a = 115 N/mm^2, N_f ≈ 2.1·10^5, |——| 5μm

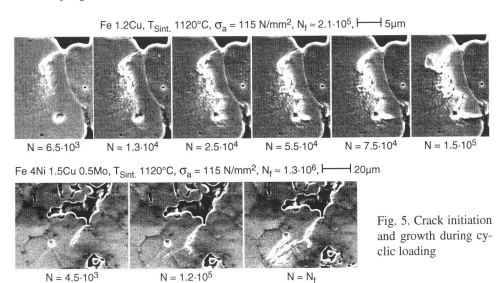

| N = 6.5·10^3 | N = 1.3·10^4 | N = 2.5·10^4 | N = 5.5·10^4 | N = 7.5·10^4 | N = 1.5·10^5 |

Fe 4Ni 1.5Cu 0.5Mo, $T_{Sint.}$ 1120°C, σ_a = 115 N/mm^2, N_f ≈ 1.3·10^6, |——| 20μm

| N = 4.5·10^3 | N = 1.2·10^5 | N = N_f |

Fig. 5. Crack initiation and growth during cyclic loading

338

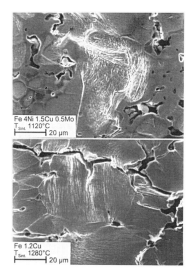

Fe 4Ni 1.5Cu 0.5Mo
T_Sint. 1120°C
|————| 20 µm

Fe 1.2Cu
T_Sint. 1280°C
|————| 20 µm

Fig. 6. Propagation of the final crack, loading conditions as in Fig. 3

Furthermore, the development of deformation features is strongly influenced by the microstructure of the metallic matrix. Fe 4Ni 1.5Cu 0.5Mo shows homogeneously distributed slip bands which are not pore-orientated, i.e. a high amount of the material is really stressed, whereas the deformation processes concentrate in the pore regions for Fe 1.2Cu (cf. Fig. 3). Thus, the more irregular pore structure is of low significance for Fe 4Ni 1.5Cu 0.5Mo, but the change in pore morphology caused by a higher sintering temperature is essential for the fatigue behaviour of Fe 1.2Cu. The interaction of pores and matrix material is impressively illustrated by the propagation of the final crack (see Fig. 6). For Fe 4Ni 1.5Cu 0.5Mo slip bands have formed in the low alloyed ferritic area inside the grain and the crack propagates through this highly deformed area and not along the pores. For Fe 1.2Cu the crack links up neighbouring pores even if it propagates parallel to the horizontal loading direction.

CONCLUSIONS

(1) The pores of Fe 1.2 Cu become larger and more regular at higher sintering temperature, while a higher content of alloying elements leads to smaller interpore distances and distinctly rougher pores.

(2) Nevertheless, Fe 1.2 Cu sintered at the higher temperature shows a lower cyclic plasticity compared to the NiCuMo-alloyed steel. This corresponds to less homogeneously distributed slip features of smaller extent and density.

(3) For all materials investigated fatigue cracks are initiated near the pores during the first cycles. For Fe 1.2Cu the cracks grow straight to the next pore, whereas the crack propagation is remarkably retarded for Fe 4Ni 1.5Cu 0.5Mo, because highly deformable areas inside the grains are combined with high strength areas near the pores.

(4) For the NiCuMo-alloyed steel, slip bands and crack paths are not pore-orientated. Therefore, the more irregular pore structure is of low significance. The deformation processes concentrate in the pore regions for Fe 1.2Cu. Consequently, the change in pore morphology caused by a higher sintering temperature is essential for the fatigue behaviour.

REFERENCES

[1] Biallas, G. (1997) *Ph.D.-Thesis,* VDI Fortschritt-Berichte Reihe 5 Nr. 495 Düsseldorf
[2] Biallas, G., Piotrowski, A., Eifler, D. (1995). In: *Sonderbände der Praktischen Metallographie*, M. Kurz, M. Pohl (Ed.), DGM Informationsgesellschaft Oberursel, pp. 359-362
[3] Piotrowski, A., Biallas, G.,(1997) In: *Proceedings of the 1997 European Conference on Advances in Structural PM Component Production*, EPMA Shrewsbury UK, pp. 107-114
[4] Piotrowski, A., Eifler, D. (1995) *Mat.-wiss. u. Werkstofftech.* 26, pp. 121-127
[5] Biallas, G., Piotrowski, A., Eifler, D. (1995) *Fat. Fract. Eng. Mat. Struct.* 18, pp. 605-615

Low Cycle Fatigue and Elasto-Plastic Behaviour of Materials
K-T. Rie and P.D. Portella (Editors)
339

EFFECTS OF MATRIX-STRUCTURES ON LOW CYCLE FATIGUE PROPERTIES IN DUCTILE CAST IRONS

YASUO OCHI

Department of Mechanical & Control Engineering,
University of Electro-Communications,
1-5-1, Chofugaoka, Cofu-city, Tokyo 182, Japan

MASANOBU KUBOTA

Department of Mechanical Engineering, Kyushu University,
6-10-1, Hakozaki, Higashi-ku, Fukuoka, 812, Japan

ABSTRACT

Total-strain controlled low cycle fatigue tests were carried out at room temperature on three kinds of ductile cast irons having different matrix structures. Fatigue deformation properties such as a stress amplitude, a plastic strain range and a cyclic stress-strain curves were measured during the fatigue process, and the relationship between the strain amplitude and the fatigue life was investigated on the three kinds of DI materials. The propagation behaviors of the surface cracks itiniated from the small drilled notch and the fractographical examinations of fracture surface were also performed in detail regarding the effects of matrix structures and graphite distributions. The surface crack propagation rate da/dN was evaluated in the J integral range ΔJ, and the results for DI materials were compared with the results for other metals.

KEYWORDS

Low cycle fatigue, fatigue deformation, fatigue life, surface crack propagation, J integral range, fracture surface, matrix-structure, ductile cast iron.

INTRODUCTION

Recently, ductile cast irons have been widely using for elements of mechanical structures as mainly automobile parts because of those forte such as high strength, high ductility, high toughness and excellent castability. With expansion of further use, the demands for fatigue strength properties and strength reliability are increasing [1]. It will be especially need for the investigation for low cycle fatigue properties in cases of mechanical parts for use in severe conditions as higher cyclic straining or stressing. Several researches about the low cycle fatigue of the DI materials have been done such as the effects of microstructure for fatigue life and fracture mechanism by S. Harada et al [2,3,4], the effects of prestrain for fatigue life by A. Onuki et al [5] and the comparison of fatigue strength for as-casting and worked materials by M.S.Starkey [6].

In this study, total-strain controlled low cycle fatigue tests were carried out at room temperature on three kinds of ductile cast irons. The fatigue deformation properties such as a stress amplitude, a plastic strain range and c yclic stress-strain curves were measured, and the relationship between the strain amplitude and the fatigue life, the surface crack propagation behaviors , and also tha fractgraphical examination were investigated as regarding the matrix-structures and the distribution of nodular graphites.

EXPERIMENTAL PROCEDURES

The materials used in this study have three different matrix structures ; a ferritic matrix ductile cast iron FDI (100% ferritic matrix), a pearlite-ferritic matrix one P-FDI (nearly same area fractions 40% of pearlitic and ferritic matrix) and a pearlitic matrix one PDI having 80% pearlitic matrix). Figure 1 shows the surface structures of the and Tables 1, 2 show the mechanical properties and the surface microstructural characterics of three kinds of DI materials. The strength increases and the ductility ecreases with increasing the area fraction of pearlitic matrix F_p.

The fatigue tests were carried out by using electro-hydraulic servo controlled testing machine. The tests were performed with total-strain range $\Delta \varepsilon_t$ controlled condition, and the range of $\Delta \varepsilon_t$ of $0.5 \sim 2.0\%$. The specimens used were round-bar type of 7mm in diameter and have a small drilled notch (0.3mm ϕ $\times$ 0.3mm depth) on the center of specimen surface.

Fatigue deformation properties such as a stress amplitude, a plastic strain and a cyclic-strain curves were measured, and the relationship between the strain amplitude and the fatigue life was investigated on the three DI materials. And the propagation behaviors of the surface cracks initiated from the small notch were observed in detail regarding the surface structures of matrix and graphite distributions at the surface. The effects of graphite distribution and the matrix structures on the crack propagation behaviors into the depth direction were also investigated from the SEM observations of the farcture surfaces.

RESULTS AND DISCUSSIONS

Figure 2 shows cyclic work hardening curves of FDI at ranges of $\Delta \varepsilon_t = 0.5 \sim 2.0\%$ as an example. The work hardening increased with increasing of $\Delta \varepsilon_t$. The fatigue life N_f was defined as the number of cycles when the peak stress falled at 5% from the maximum tensile stress. Figure 3 shows the cyclic stress-strain curves of the three DI materials and other metals. In the figure, $\Delta \sigma /2$ and $\Delta \varepsilon_p/2$ show the values of the number of cycles at the half lives $N_f/2$. The curves of the three DI materials showed nearly position with the aliminum alloys. The cyclic stress-strain curves were expressed as the following equation.

$$(\Delta \sigma /2) = k' (\Delta \varepsilon_p)^{n'} \tag{1}$$

where, k' is a cyclic hardening factor and n' is cyclic hardening exponent. The values of K' and n' are shown in the Table 2.

Figure 4 shows the relation between the each strain ranges (total: $\Delta \varepsilon_t$, plastic: $\Delta \varepsilon_p$, and elastic : $\Delta \varepsilon_e$) and the N_f of the FDI as an example. Figure 5 shows the relation between the $\Delta \varepsilon_t$ and the N_f of the three DI materials. From the figure, the relations between them were expressed as Manson-Coffin's relations as the follows,

$$\Delta \varepsilon_p \cdot N_f^{\alpha p} = C_p \tag{2}$$

$$\Delta \varepsilon_e \cdot N_f^{\alpha e} = C_e \tag{3}$$

$$\Delta \varepsilon_t = C_p \cdot N_f^{\alpha p} + C_e \cdot N_f^{\alpha e} \qquad (4)$$

Where, αe, αp, Ce, and Cp are materials constants, and the values of the constans are shown in Table 3.

Figure 6 shows the surface crack propagation curves of P-FDI of the strain range $\Delta \varepsilon_t$ =0.52～1.2% as an example. From the figure, the cracks propagated more quickly with increasing of $\Delta \varepsilon_t$, and the final fracture occured at the crack length reached to 3～4mm. Figure 7 shows the detailed observation results the effects of nodular graphites and microstructures on the surface crack propagation behaviors of P-FDI. From the results, the crack propagated discontinuously by the effects of nodular graphites and ferrite-pearlitic matrix distributions.

Figure 8 shows an example of the fractographical examinations of P-FDI at the final crack length $2a_f$=2156μm. Figure (a) is the macroscopic observation arroud the notch and the other (b)～(e) are the maginified observations. At near the fracture surface of early crack growth from the notch, contact traces between ferrites matrix arround the nodular graphite and laminar structures of pearlites matrix were observed as shown in the figures (b) and (c). Figure (d) is near at the fatigue crack tip, and the dimple fracture originating the graphites and the flat cleavage fracture were observed partially. Figure (e) is the final fravcture surface, and the dimple fracture occured on all surface. From the measurment of the area fraction of graphites F_{gf} of fracture surface as shown in Fig.9, the F_{gf} increased with propagating into the depth direction and reached to 40～60%. This values of F_{gf} were 3～5 times of the values of the surface. This means that the cracks propagated into the depth direction preferentially through the graphites.

Figue 10 shows the relation between t he crack propagation rate da/dN and the J in-tegral range ΔJ of P-FDI as an example. The values of ΔJ was evaluated by the method by T.Hoside et al [7]. The relations were expressed as the following equation.

$$da/dN = C_J \cdot \Delta J^{m_J} \qquad (5)$$

Where, C_J and m_J are materials constants, and the values for three DI materilas are shown in Table 5. Figure 11 shows the comparison of the da/dN-ΔJ relations with other metals. The crack propagation rate of the DI materials were in the position be-tween tose of Ti alloy and aluminum alloys, and were about same position as the low carbon steel and the stainless steel.

CONCLUSION

The results obtained in this study were summarized as the followings.
(1) The cyclic stress-strain curves were expressed as the equation $(\Delta \sigma /2)=k'(\Delta \varepsilon p)n'$. k'and n' are material constants and the values of them increased with increasing the area fraction of perlitic matrix Fp.
(2) The relationships between the each strain range (an elastic, a plastic and a total strain range) and the fatigue life were evaluated as Manson-Coffin' relation, re-spectively and the fatigue life at the constant strain increased with increasing the Fp.
(3) Cracks propagated in discontinuity at the surface because of the effects of matrix structures graphites existing at the surface and at just under the surface. In FDI materials, some cracks propagated with passing through many graphites near surface.
(4) From the observation of fracture surface, the area fraction of graphite F_{gf} in-creased with propagating into the depth direction and the cracks propagated into the depth direction preferentially through the graphites.

(5) The relationships between the crack propagation rate da/dN and the J integral range ΔJ in the three kinds of DI materials were evaluated as the straight lines respectively as the equation, $da/dN = C_J \, \Delta J^{\, m_J}$. C_J. In the three DI materials, the FDI showed the highest crack propagation rate.

REFERENCES

1. Nishitani, H. et al. (1992) ,Seminar Textbook, "Strength Evaluation and Advanced Techlology of Ductile Cast Irons", JSME MMD.
2. Harada, S. et al. (1989). Trans. JSME. 55A, 393.
3. Harada, S. et al. (1990). J. Soc. Mater. Sci. Japan, 39, 1133.
4. Harada, S. et al. (1992). Trans. JSME. 58A, 1306.
5. Onuki, A. et al. (1986). Pre-print of JSME meeting. No.860-3, 183.
6. Stakey, M.S. and Irving, P.E. (1982). Int. J. Fatigue. 4, 129.
7. Hoshide, T. and Tanaka, K. (1982). Trans. JSME. 48A, 1102.

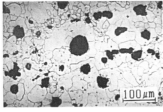

(a) FDI

(b) P-FDI

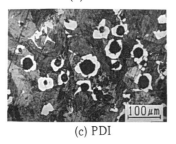

(c) PDI

Fig.1. Surface microstructures of DI materials.

Table 1. Mechanical properties of three DI materials.

Materials	Yield strength σ_Y [MPa]	Tensile strength σ_B [MPa]	Elongation ϕ [%]	Reduction of area R.A. [%]
FDI	294	424	25.0	20.3
P-FDI	305	647	9.75	7.45
PDI	335	805	5.68	4.75

Table 2. Characteristics of surface microstructures.

Materials		FDI	P-FDI	PDI
Average ferrite grain size D [μm]		27.7	—	—
Average SG size d_g [μm]		21.1	21.3	19.8
Ratio of nodularity h_g [%]	Area method	67.0	79.3	60.4
	Shape factor method	73.3	86.4	61.9
Area fraction of pearlite F_p [%]		0	39.7	80.2
Area fraction of graphite F_{gs} [%]		12.1	13.3	9.80
Graphite number n_g [/mm^2]		286	315	265

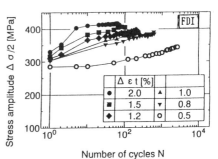

Fig.2. Cyclic work hardening curves of FDI.

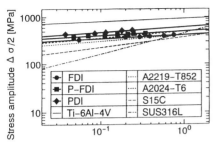

Fig.3. Cyclic stress-strain curves of three DI materials and other metals.

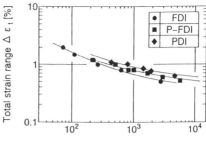

Fig.5. Relations between $\Delta \varepsilon_t$ and the N_f in three DI materials.

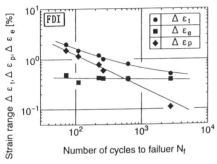

Fig.4. Relation between $\Delta \varepsilon_t$, $\Delta \varepsilon_p$, $\Delta \varepsilon_e$ and the N_f of FDI.

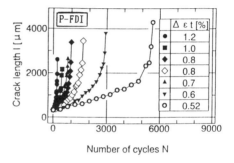

Fig.6. Surface crack propagation curves of P-FDI.

Table 3. Materials constants of cyclic stress-strain curves.

Materials	Coefficient of cyclic work hardening,k' [MPa]	Exponent of cyclic work hardening, n'
FDI	420	0.0772
P-FDI	500	0.0875
PDI	611	0.1079

Table 4. Materials constants of strain ranges - N_f relations.

Materials	α_p	α_e	C_p	C_e
FDI	0.712	0.0231	0.335	0.00464
P-FDI	0.638	0.0411	0.251	0.00613
PDI	0.792	0.0591	0.658	0.00932

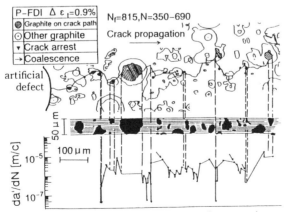

Fig.7. Effects of microstructures on crack propagation behaviors on surface of P-FDI.

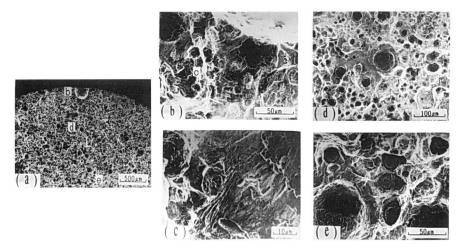

Fig.8. SEM observation of fracture surface of P-FDI.

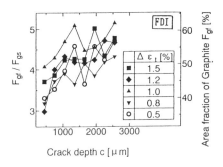

Fig.9. Relation between the area fraction of graphites F_{gf} and the crack depth.

Table 5. Materials constants of da/dN - ΔJ relations.

Materials	C_J	m_J
FDI	1.78×10^{-3}	1.56
P-FDI	3.15×10^{-3}	1.77
PDI	2.66×10^{-3}	1.80

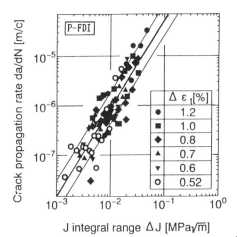

Fig.10. Relation between da/dN and ΔJ of P-FDI.

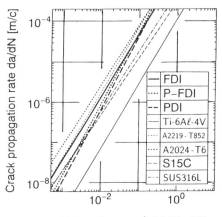

Fig.11. Relation between da/dN and ΔJ of three DI materials and other metals.

Low Cycle Fatigue and Elasto-Plastic Behaviour of Materials
K-T. Rie and P.D. Portella (Editors)
345

EFFECT OF PREDEFORMATION ON THE HIGH-TEMPERATURE LOW CYCLE FATIGUE BEHAVIOUR OF POLYCRYSTALLINE NI-BASE SUPERALLOYS

K. SCHÖLER and H.-J. CHRIST
Institut für Werkstofftechnik,
Universität-GH-Siegen, 57068 Siegen, Fed. Rep. Germany

ABSTRACT

The influence of a monotonic predeformation on cyclic deformation behaviour and cyclic life at elevated temperature was studied on polycrystalline Ni-base superalloys in three characteristic precipitation conditions (precipitation-free, peak-aged and overaged). The dependence on prehistory increased with decreasing test temperature, decreasing amplitude and increasing hardening effect of precipitates. Predeformation caused predamage and led to a reduction of cyclic life under most conditions. However, in certain cases a small predeformation was found to act beneficial increasing cyclic life. In order to describe the effects arising from predeformation, a new predeformation parameter was developed and applied. This predeformation parameter was calculated from stress and strain data resulting from low cycle fatigue tests with and without predeformation. It could be shown that its value correlates with the resulting life.

KEYWORDS

Monotonic predeformation, low cycle fatigue, predeformation parameter, Ni-base superalloy.

INTRODUCTION

In most cases engineering materials undergo a (thermo-)mechanical treatment prior to their application in service, either in order to optimize their mechanical properties or just during the production process. Consequently, the influence of this prehistory on the subsequent cyclic deformation behaviour is of particular interest. Unfortunately, the prehistory dependence is a complex function of several parameters.

It has been shown that the effect of predeformation depends strongly on the dislocation slip behaviour [1]. Materials exhibiting wavy slip such as copper behave history-independent if the plastic strain amplitude is sufficiently high [2-4], whereas planar-slip materials such as α-brass are strongly affected by a predeformation [2,5].

For materials with planar-slip behaviour a monotonic predeformation leads to a reduction of cyclic life and shifts the cyclic stress-strain curve to higher stresses. The microstructure reflects this strong prehistory dependence that manifests itself in the macroscopic mechanical behaviour [1]. Especially at low amplitudes, a predeformation causes a stronger activity of secondary slip and the spacing of the glide planes of the primary slip system is reduced.

The objective of the present study was to improve the knowledge on the effects of predeformation on the cyclic deformation behaviour, the microstructure and the cyclic life, taking especially the influence of temperature, loading amplitude and precipitation condition into account. Furthermore, a parameter is introduced in order to quantify and assess the effect of predeformation. The study was performed on commercially available Ni-base superalloys.

EXPERIMENTAL DETAILS

In Tab. 1 the chemical compositions of the two alloys used in this study are listed. In order to study the influence of precipitation hardening on the cyclic deformation behaviour and the prehistory dependence, the γ'-hardenable alloy Nimonic 105 was heat treated to a peak-aged and an overaged condition. The precipitation-free Nimonic 75 was used as reference material representing the behaviour of the γ matrix of Nimonic 105 in reasonable approximation.

Table 1. Compositions of the alloys under investigation according to producer (weight %).

	Ni	Co	Cr	Ti	Al	Mo	Fe	C	others
Nimonic 105	47-58	18-22	14-16	0.9-1.5	4.5-4.9	4.5-5.5	< 1	< 0.12	< 2.4
Nimonic 75	70-82	-	18-21	0.2-0.6	-	-	< 5	< 0.15	< 2.5

The peak-aged condition showed a mean γ' diameter of 73 nm and a γ' volume fraction of 43 %. This precipitation condition corresponds to maximum hardness at room temperature. The overaged condition is characterised by particle by-passing processes which occur in Nimonic 105 at RT for γ' sizes greater than 600 nm [6]. In the overaged condition studied in this work three main γ' sizes were found. The largest type of γ' particles showed a mean size of 665 nm and the corresponding γ' volume fraction was 38 %.

The cyclic deformation tests were carried out in the temperature range from RT to 800°C under true plastic strain control with superimposed control of the plastic strain value at zero stress. This so-called bimodal control keeps the plastic strain amplitude (at zero stress) constant in spite of changes in Young's modulus. Furthermore, by means of a triangular function generator signal a nearly constant plastic strain rate $\dot{\epsilon}_{pl}$ results. A value of $\dot{\epsilon}_{pl}=5\cdot10^{-4}\,s^{-1}$ was used in the tests reported in this paper. Monotonic predeformations were carried out at RT under total strain control up to given total tensile strains ranging from 1 % to 8 %.

RESULTS AND DISCUSSION

Deformation behaviour without predeformation
In order to determine the effects arising from a predeformation, first the behaviour without predeformation had to be characterized. The influence of precipitation state and temperature on the cyclic deformation behaviour and the corresponding microstructure was studied in tests on purely heat-treated samples by comparing cyclic deformation curves and corresponding dislocation arrangements.

The cyclic deformation curves of the precipitation-free Nimonic 75 exhibited pronounced cyclic saturation at all temperatures. The cyclic saturation state is in contrast to the planar dislocation slip character observed in Nimonic 75 but can be attributed to the microstructure which did not show major changes.

At RT Nimonic 105 in the peak-aged condition showed a weak cyclic softening that is caused by the shearing of precipitates, the formation of γ'-free deformation bands [7-9] and an early microcrack formation. This cyclic softening can be seen in Fig. 1. The cyclic deformation curves of the peak-aged condition are represented for various temperatures. Because of the great differences in cyclic life, the stress amplitude $\Delta\sigma/2$ is plotted versus the logarithm of the cumulative plastic strain $\varepsilon_{pl, cum}$.

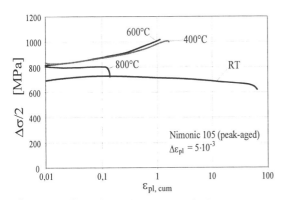

Fig. 1: Cyclic deformation curves of Nimonic 105 in the peak-aged condition for different temperatures.

At 400°C and 600°C a significant cyclic hardening continued until failure. This can be attributed to dynamic strain ageing processes which reduce the mobility of the dislocations and consequently slow down the rate in which a steady state condition is established. At 400°C and 600°C dynamic strain ageing was found for all precipitation conditions studied, its effect being most marked in the peak-aged condition. At 800°C cyclic saturation prevailed during cyclic life. The dislocation/particle interaction changed from particle-shearing to the Orowan process and the dislocation slip character became more wavy.

In the overaged condition the main dislocation/precipitation interaction mechanism was particle by-passing independent of the temperature considered. Hence, an approximate cyclic saturation was found at all temperatures studied.

It is noteworthy that the crack initiation site depended strongly on temperature. As can be seen in Fig. 2, crack initiation at low temperatures takes place transgranularly at slip bands, whereas at high temperatures cracks form at grain boundaries. This behaviour was found to be independent of the precipitation state. The type of crack initiation strongly effects the fatigue life and its dependence on temperature (Fig. 1).

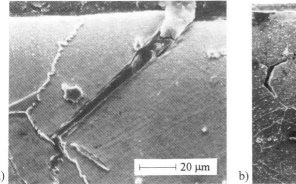
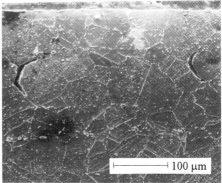

Fig. 2: Scanning electron micrographs of cross sections of cyclically deformed Nimonic 75 ($\Delta\varepsilon_{pl}=5\cdot10^{-4}$) **a)** RT, **b)** 800°C.

Deformation behaviour after predeformation

The prehistory dependence of Ni-base superalloys is mainly a function of temperature, precipitation state and plastic strain range $\Delta\varepsilon_{pl}$. Isothermal LCF tests after RT predeformation showed that the dependence of the cyclic stress-strain behaviour on the prior tensile deformation be-

348

comes more evident with decreasing temperatures, increasing γ' hardening and decreasing plastic strain amplitude. Figures 3a and b show extreme cases as examples. The course of the stress amplitude $\Delta\sigma/2$ and the mean stress σ_m versus the cumulative plastic strain are represented for Nimonic 105 in the peak-aged condition fatigued at RT with $\Delta\varepsilon_{pl}=5\cdot10^{-4}$ (Fig. 3a) and for the precipitation-free Nimonic 75 fatigued at 800°C with $\Delta\varepsilon_{pl}=5\cdot10^{-3}$ (Fig. 3b). In both cases the predeformed condition is compared with that without predeformation. Under the conditions corresponding to Fig. 3a, high mean stresses, increased stress amplitudes and a strong reduction of life are caused by the predeformation. By contrast, Fig. 3b documents a very small effect of a predeformation on the fatigue behaviour.

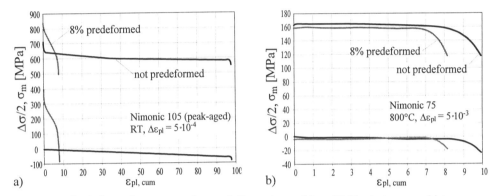

Fig. 3: Cyclic deformation curves after predeformations of 0 and 8 %, **a)** strong prehistory dependence and **b)** weak prehistory dependence.

Even if there is no pronounced influence of predeformation on the (macroscopic) mechanical behaviour, there might be predeformation-induced changes in the microstructure. This is due to slip band formation during monotonic prestraining. It was found that these predeformation-induced slip bands remained active, even under fatigue conditions where usually no slip bands are formed. In Fig. 4a a dislocation arrangement is shown that is typical of the overaged condition (not predeformed) fatigued at 800°C with $\Delta\varepsilon_{pl}=5\cdot10^{-4}$. The dislocation arrangement which corresponds to identical test conditions but applied after a predeformation of 2 % is depicted in Fig. 4b. Without predeformation Orowan-loops and homogeneously distributed dislocations could be observed. Predeformation resulted in slip bands, in addition to the dislocation

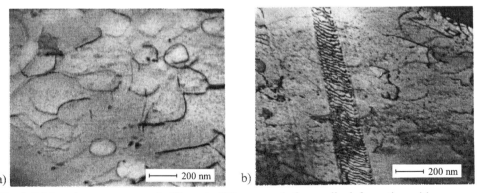

Fig. 4: Dislocation arrangement in Nimonic 105 overaged after cyclic deformation with $\Delta\varepsilon_{pl}=5\cdot10^{-4}$ at 800°C **a)** not predeformed, **b)** 2% predeformed.

arrangement which is characteristic of the plastic strain amplitude applied (Fig. 4b).

Predeformation parameter and cyclic life

In a first attempt to assess the effect of predeformation on cyclic life it is assumed that the number of cycles to failure N_f of the Ni-base superalloys studied is determined mainly by mean stress, stress amplitude and total strain amplitude. Thus, a damage parameter such as P_{SWT} after Smith, Watson and Topper [10] should be suitable to describe the influence of monotonic pre-straining on the cyclic life taking characteristic data of the cyclic stress-strain behaviour into account. P_{SWT} can easily be calculated using Eq. (1) for a given plastic strain amplitude and temperature from the resulting stress amplitude $\Delta\sigma/2$, the mean stress σ_m, the total strain amplitude ε_t and the Young's-modulus E.

Based on P_{SWT} a new predeformation parameter S_{P_SWT} was defined according to Eq. (2). The calculation of S_{P_SWT} requires the values of the damage parameter P_{SWT} for the prede-formed and the not predeformed condition. For the sake of simplicity, the corresponding values were determined at half cyclic life ($N_f/2$).

$$P_{SWT} = \sqrt{(\Delta\sigma/2 + \sigma_m)\cdot\varepsilon_t\cdot E} \quad (1) \qquad S_{P\_SWT} = 1 - \frac{P_{SWT\,(not\ predeformed)}}{P_{SWT\,(predeformed)}} \quad (2)$$

In Fig. 5 the ratio of cyclic life without and with predeformation is plotted versus the prede-formation parameter S_{P_SWT}. A relatively narrow band results describing the position of the data points from various tests. This behaviour shows, that cyclic life is a function of the cyclic deformation parameters contained in P_{SWT} and that P_{SWT} is basically suitable to define a prede-formation parameter. Figure 5 documents, that for one precipitation condition test results ob-tained after different degrees of prede-formations, at different temperatures and at different plastic strain amplitudes fall into a common data band. That means, S_{P_SWT} can be used to quantify the prehistory dependence in a broad field of test parameters. However, it should be noted that the degree of pre-deformation is limited to moderate val-ues, because very large predeformations cause too strong damage and therefore strong reduction of life that cannot be described reasonably using S_{P_SWT}.

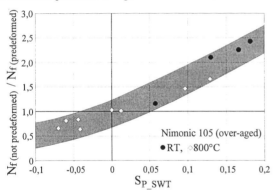

Fig. 5: Normalized cyclic life versus the predefor-mation parameter of various fatigue tests under dif-ferent conditions.

It is remarkable that there are nega-tive values of S_{P_SWT} in the precipita-tion-hardened alloy. Negative values of S_{P_SWT} give rise to an increase in life. A second look at Fig. 5 reveals that a predeformation is only beneficial and increases the life if the temperature of fatigue is high. Furthermore, the plastic strain amplitude and the degree of predeformation has to be low. These test parameters lead to a reduction of the stress amplitude, a very small initial hardening stage and an increase of the cyclic life compared with the corresponding not predeformed specimen (Fig. 6). An ex-planation of this behaviour can be given on the basis of the findings regarding microstructural changes (Fig. 4) and crack initiation site (Fig. 2). The precipitation-hardened alloy, especially in the overaged condition, exhibits no or only few slip bands after high-temperature fatigue at small amplitudes, whereas slip bands are extensively produced by the predeformation process.

350

These predeformation-induced slip bands lead to a more homogeneous distribution of plastic deformation during high-temperature fatigue, reducing the initial hardening as well as the cyclic saturation stress. Therefore the forces acting against the grain boundaries are lowered. Since at high temperatures the crack initiation takes place primarily at the grain boundaries, the relieve of the grain boundaries as a consequence of predeformation gives rise to an extension of cyclic life.

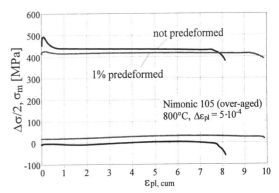

Fig. 6: Cyclic deformation curves of Nimonic 105 (overaged) at 800°C and $\Delta\varepsilon_{pl}=5\cdot10^{-4}$ with and without predeformation.

CONCLUSIONS

(1) Isothermal LCF tests at RT and elevated temperatures on polycrystalline Ni-base superalloys after tensile prestraining at RT showed that the dependence on the prior load history increases with decreasing temperatures, increasing γ' hardening and decreasing plastic strain amplitude.

(2) Fatigue life was found to be determined mainly by the stress amplitude and the mean stress. Consequently corresponding damage parameters can be used to assess cyclic life independent of the degree of predeformation. This holds at least up to moderate degrees of predeformations.

(3) Based on the damage parameter P_{SWT} a new predeformation parameter was introduced. It could be shown that this parameter can be used to quantify the effect of predeformation on cyclic life, independent of the test temperature, the plastic strain amplitude and the degree of predeformation, as long as the predeformation does not exceed a critical level.

(4) It could be shown that the fatigue life at high temperatures and small plastic strain amplitudes can be increased as a result of a small predeformation. This behaviour can be attributed to predeformation-induced slip bands which remain active during subsequent cyclic loading.

REFERENCES

1   Hoffmann, G., Öttinger, O. and Christ H.-J., (1992). In: Proceedings of The Third International Conference on Low-Cycle Fatigue and Elasto-Plastic Behaviour of Materials, K.-T. Rie (Ed.). Elsevier Applied Science, London, pp. 106-111.
2   Feltner, C. E. and Laird, C., (1967) Acta met., **15**, 1621.
3   Feltner, C. E. and Laird, C., (1967) Acta met., **15**, 1633.
4   Feltner, C. E. and Laird, C., (1968) Trans. AIME, **242**, 1253.
5   Lukáš, P. and Klesnil, M., (1973) Mater. Sci. Eng., **11**, 345.
6   Reppich, B., Schepp, P. and Wehner, G., (1982) Acta metall., **30**, 95.
7   Sundararaman, M., Chen, W. and Wahi, R. P., (1989) Scripta Met., **23**, 1795.
8   Sundararaman, M., Chen, W., Singh, V. and Wahi R. P., (1990) Acta metall. mater., **38**, 1813.
9   Sundararaman, M., Chen, W., Wahi, R. P., Wiedemann, A., Wagner, W. and Petry W., (1991) Acta metall. mater., **40**, 1023.
10   Smith, K. N., Watson, P. and Topper, T. H., (1970) J. of Mater., JMLSA, **5**, 767.

Low Cycle Fatigue and Elasto-Plastic Behaviour of Materials
K-T. Rie and P.D. Portella (Editors)

ESTIMATION OF THE CRITICAL STRESS FOR γ' SHEARING IN A SINGLE CRYSTAL SUPERALLOY FROM LCF DATA

J. AUERSWALD, D. MUKHERJI, W. CHEN and R. P. WAHI

Hahn-Meitner-Institut Berlin GmbH, Glienicker Str. 100, 14109 Berlin

ABSTRACT

The deformation behaviour of the single crystal superalloy SC16 has been studied under low cycle fatigue (LCF) loading at 1223 K, 1123 K and 298 K with a strain amplitude range of 0.4% to 1.0% and a constant strain rate of 10^{-3} s^{-1}. Microstructural investigation by transmission electron microscopy of the deformed specimens revealed that the deformation mechanism depends on strain amplitude as well as on temperature of deformation. At large strain amplitudes at the elevated temperatures, the γ' precipitates are sheared by the matrix dislocations leading to stacking fault (SF) formation within the precipitates, while at small strain amplitudes deformation is confined within the γ matrix. At room temperature, the planar faults in γ' precipitates are complex. The deformation under cyclic loading is compared with that under monotonic loading. The threshold stresses for shearing of γ' precipitates at different temperatures are estimated.

KEYWORDS

Nickel-base superalloy, single crystal, deformation, low cycle fatigue, microstructure, TEM, stacking faults, threshold stress.

INTRODUCTION

While a considerable number of studies have been conducted in the area of the mechanisms of fracture, crack initiation, crack propagation and fatigue life prediction in nickel-base superalloys with high volume fraction of γ' precipitates, the evolution of the microstructure under LCF condition is relatively less studied. In a general overview, the influence of microstructure on the material fatigue is discussed by Christ and Mughrabi [1]. The material response to low cycle fatigue (LCF) loading is generally influenced by the material parameters like crystallographic structure, the stacking fault energy and the testing parameters [2-4]. The situation in single crystal nickel-base superalloys appears to be more complex since some additional material parameters may affect the evolution of the microstructure. Some of these parameters e.g. volume fraction and morphology of the γ' precipitates, misfit stresses between the precipitate and the matrix, dislocation networks at the γ/γ' interfaces, single crystal orientation, distance from the fracture surface and rate of deformation etc. have been investigated [5-7]. Low cycle fatigue induced microstructural changes have been studied mainly to explain the observed softening or hardening [8,9] and the tension-compression stress anisotropy [10-11] during the cyclic deformation. Different aspects of the high temperature deformation behaviour of a single crystal alloy SC16 (Ni- 16 Cr - 3 Mo - 3.5 Ti - 3.5 Al - 3.5 Ta ; in wt. %) are being investigated in an ongoing research activity on

352

superalloys at the Hahn-Meitner-Institut and the Technical University in Berlin e.g. [12-14]. microstructural investigations on SC16 so far are mostly after monotonic loading. The characterisation of the deformation microstructure in this alloy after cyclic loading is lacking. Only recently, Jiao et al [11] studied the deformation microstructure to explain the tension-compression asymmetry in the SC16 alloy with a bimodal distribution of γ' precipitates at elevated temperatures. In the present study we examine the microstructural evolution of SC16 containing unimodal distribution of γ' precipitates after cyclic loading conditions and compare its deformation behaviour with that under monotonic loading.

EXPERIMENTAL PROCEDURES

The single crystal bars of the superalloy SC16 were approximately [001] oriented with a maximum of 6° deviation in the rod axis from the crystallographic direction. They were heat treated (1523K/3h/air cool + 1373K/1h/slow cool at 15K/min to 1123 K + 1123K/24h/air cool) to produce a monomodal distribution of cuboidal γ'-precipitates with an average edge length of 450 nm and a volume fraction of about 40% (Fig. 1). LCF tests were carried out on a MTS servo hydraulic machine operating in a total axial strain controlled mode. The fully reversed (R = ε_{min} / ε_{max} = -1) tests were conducted at a constant strain rate of 10^{-3} s^{-1} at 1223 K, 1123 K and 298 K. The strain amplitude was varied between 0.4% and 1.0% at each test temperature. The tensile tests at a constant strain rate of 10^{-3} s^{-1} were also conducted at 1223 K, 1123 K and 298 K until failure. At 1223 K two tests were interrupted after 0.2 and 2% strains respectively, to study the microstructural evolution with increasing deformation under monotonic loading. The microstructures of the deformed specimens were examined by transmission electron microscopy (TEM).

RESULTS

At elevated temperatures and smaller strain amplitudes an initial softening is observed followed by a stable stress region until failure, while at larger strain amplitudes a stable stress response over the entire life is observed (Fig. 2a). Fig. 2b shows the mechanical behaviour at room temperature. At room temperature and the highest strain amplitude of 1%, a very strong hardening with increasing deformation is visible. At the strain amplitude of 0.8%, hardening is comparatively less. The stress response becomes somewhat stable at the lower strain amplitudes of 0.4 and 0.6% (fig. 2b). The stress versus strain plots at different

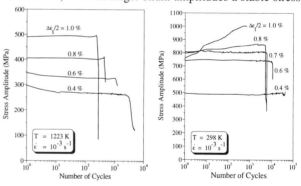

Fig. 2: Cyclic stress response at (a) 1223 K and (b) 298 K

temperatures under monotonic tensile loading at a constant strain rate of 10^{-3} s^{-1} are shown in Fig. 3. After an initial hardening until approximately 2% strain there is a continuous softening until fracture in samples tested at the elevated temperatures. The tensile behaviour at room temperature is quite different. It shows continuous hardening after a sharp yield point. The ultimate tensile stress values under the monotonic loading are generally much higher than the stress amplitude at half life in the LCF tests.

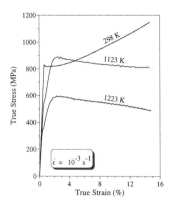

Fig. 3: Stress response in monotonic tensile tests at different temperatures.

The microstructures after LCF deformation at 1223 K with low strain amplitudes (0.4 and 0.6%) show that plastic deformation is confined to the γ matrix phase only. Dislocations density in g channels is however small, while their networks at the γ/γ' interfaces are quite evident (Fig. 4a). Dislocation density in the networks increases with increasing strain amplitude. On deformation at higher strain amplitudes (0.8 and 1.0%) stacking faults are observed within the γ' precipitates (Fig. 4b) indicating shearing of the precipitates. In addition to the stacking faults, the microstructure contained dense dislocation network at the γ/γ' interfaces. Fig. 5 shows TEM micrographs obtained from interrupted tensile tests at 1223 K. No stacking faults are observed up to a strain of 0.2% (Fig. 5a). The microstructure contains dislocation network at the γ/γ' interfaces. On the contrary the microstructure of the specimen interrupted after 2.0% strain clearly shows presence of stacking faults (Fig. 5b). Thus at 1223 K, increasing strain amplitude in LCF

deformation shows similar changes in deformation mechanisms as increasing strain in the tensile deformation. No stacking faults in γ' precipitates were observed in LCF tested specimens at 1123 K for the entire range of strain amplitude (0.4 - 1.0%) covered in the present study. However, stacking faults similar to those found in the LCF and tensile tested specimens at 1223 K were also observed in a failed (strain = 28.5%) tensile sample at 1123 K. For both the monotonic and the cyclic tests at 1123 K and 1223 K where

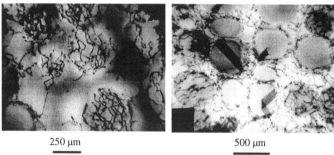

250 µm   500 µm

Fig. 4: Microstructure after cyclic deformation at 1223 K (a) small strain amplitude = 0.6%, (b) large strain amplitude = 0.8%.

the specimens were deformed at the strain rate of 10^{-3} s^{-1}, no appreciable change in γ' morphology (e.g. rafting) was observed.

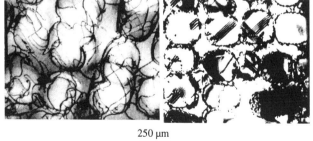

250 µm

Fig. 5: Microstructure after monotonic tensile deformation at 1223 K, (a) after 0.2% strain, (b) after 2% strain.

Under fatigue loading at room temperature, when strain amplitudes were small, plastic deformation was found to be quite inhomogeneous (i.e. areas with and without dislocation

354

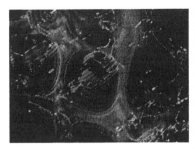

250 μm

Fig. 6: Microstructure after cyclic deformation at 298 K.

activity were observed). The proportion of the deformed regions to undeformed regions increased with increasing strain amplitude. Within the deformed region plastic deformation was mainly localised in microscopic slip bands. Macroscopic slip bands were also observed on the surface of the LCF specimens. In contrast to the observation of single or isolated stacking faults within the γ' precipitates after high temperature deformation, a high density of planar faults were observed in all LCF specimens tested at room temperature (Fig. 6). Their frequency increased with increasing strain amplitude. The faults have a complex image contrast, often showing overlap. Thus they are quite different in appearance from those found after deformations at the elevated temperatures. The partials bounding the faults also have irregular line configurations and are quite frequently located entirely within the γ' precipitates, in some cases appearing as faulted loops.

DISCUSSION

In the following we discuss the microscopic deformation mechanisms and their dependence on the evolving stress response during cyclic and monotonic loading: The plastic deformation is quite inhomogeneous at room temperature but becomes more homogeneous with increasing temperature. At the elevated temperature of 1223 K the deformation mechanism under cyclic loading depends on the imposed strain amplitude. At low strain amplitudes where the stress is also low (Table -1) deformation is confined within the matrix γ channels. The dislocations glide in the matrix channels between the γ' precipitates and are arrested at the γ/γ' interfaces where they interact to form networks [13]. The dislocations under these test conditions overcome the precipitate barrier presumably by climb processes [15]. At higher strain amplitudes the corresponding stresses are also higher (Table -1) and the matrix dislocations are now able to shear the γ' precipitates leading to stacking fault formation within the precipitates. A similar change in the deformation mechanism from dislocation climb to particle shear is also observed in the constant strain rate tensile tests with increasing strain from 0.2 to 2%. In this case there is a corresponding increase in the stress value from 188 MPa at 0.2% strain to about 585 MPa at 2% strain. Load cycling or stress reversals apparently do not cause any observable difference in the deformation substructure as compared to that observed under monotonic loading at this temperature.

The stacking faults observed in SC16 at 1223 K (and also at 1123 K in the fractured tensile test specimen) very closely resemble those found in other nickel-base super alloys at elevated temperatures e.g. IN738LC [16] and PWA 1480 [17]. It can be assumed that the mechanism leading to high-temperature stacking fault formation within the γ' precipitates in all of these alloys is similar. Various models have been proposed in literature to describe the mechanisms of stacking fault formation in γ' phase [18,19]. These models essentially describe the different possibilities for the dislocations interaction and dissociation resulting in various combinations of partials. One of them [19] suggests SF formation requires thermal activation of shockley partials. It is expected that the shearing of the γ' precipitate by the matrix dislocations occurs when a certain threshold shear stress is exceeded. The threshold stress for γ' shearing leading to the formation of anti phase boundary (APB) is related to the APB energy [20]. Similarly the threshold stress for γ' shearing should be related to SF energy. The SF energy as well as the shear modulus of γ and γ' phases are temperature dependent [21,22], therefore, the critical shear stress necessary for the precipitate shearing should also depend on the test temperature. The dependence of the threshold stress on temperature will further be affected by facts such as whether SF formation mechanism involves thermal activation or not.

From the axial stress amplitudes in the LCF tests we obtained the resolved shear stress on the primary slip system. The results are shown in Table -1. The resolved shear stress on the primary slip system in the tensile tests at 0.2 and 2% strains have values of 84 and 260 MPa, respectively. From the microstructural observations and the data in Table - 1 we conclude that the critical shear stress for γ' shearing leading to stacking fault formation at 1223 K can therefore be stated to lie between 144 MPa and 174 MPa. A similar consideration leads to an estimate of the critical shear stress for stacking fault formation at 1123 K to be between 282 and 361 MPa. Lowering of the threshold stress with increasing temperature suggests that the mechanism of stacking fault formation at the elevated temperatures is thermally assisted.

In the present study the low temperature faults were found after LCF deformation at all investigated strain amplitudes. The critical external shear stress for shearing of γ' precipitates in SC16 at 298 K and a strain rate of 10^{-3} s^{-1} is therefore less than 214 MPa (Table - 1). This is the maximum shear stress on the primary slip system corresponding to the smallest investigated axial strain amplitude of 0.4%. The threshold stress for particle shearing at room temperature is therefore smaller than 214 MPa and is also lower than that at 1123 K. This observation suggests that the mechanisms of planar fault formation at low and high temperatures are different. A similar suggestion is made by Milligan and Antolovich [25].

REFERENCE

1. H. -J. Christ, H. Mughrabi: Microstructure and Fatigue. In: K.-T. Rie (Ed.) et al: Low Cycle Fatigue and Elasto-Plastic Behaviour of Materials, Elsevier Science Publishers Ltd., London 1992, pp. 56-69.
2. T. Magnin, C. Ramade, J. Lepinoux, L. P. Kubin: The Sensitivity of Low Cycle Fatigue Properties to the Crystallographic Structure in Single and Polycrystals. In: H. Kitagawa (Ed.) et al.: Fatigue 90, Vol. 1, MCE Publications Ltd., 1990, pp. 225-229.
3. J. Y. Wang, Z. H. Ma, Y. X. Tan, D. W. Guo: Constant Strain Fatigue of Stainless Steel and its Corresponding Substructure Variation. In: H. Kitagawa (Ed.) et al.: Fatigue 90, Vol. 1, MCE Publications Ltd., 1990, pp. 173-178.
4. J. -B. Vogt, T. Magnin, J. Foct: Factors Influencing Planar Slip During Fatigue in fcc Stainless Steels. In: H. Kitagawa (Ed.) et al.: Fatigue 90, Vol. 1, MCE Publications Ltd., 1990, pp. 87-92.
5. D. L. Anton, *Acta metall.* **32**, 1669 (1984).
6. T. P. Gabb and G. Welsch, *Acta metall.* **37**, 2507 (1989).
7. F. Jiao, J. Zhu, R. P. Wahi, H. Chen, W. Chen, H. Wever: Low Cycle Fatigue Behaviour of IN738LC at 1223 K. In: K.-T. Rie (Ed.) et al: Low Cycle Fatigue and Elasto-Plastic Behaviour of Materials, Elsevier Science Publishers Ltd., London 1992, pp. 298-303.

356

8. U. Glatzel and M. Feller-Kniepmeier, *Scripta metall.* **25**, 1845 (1991).
9. B. Decamps, V. Brien and A. J. Morton, *Scripta metall.* **31**, 793 (1994).
10. L. M. Hsiung and N. S. Stoloff, *Acta metall. mater.* **40**, 2993 (1992).
11. F. Jiao, D. Bettge, W. Österle and J. Ziebs, *Acta metall. mater.* **44**, 3933 (1996).
12. H. Gabrisch, T. Kuttner, D. Mukherji, W. Chen, R. P. Wahi, H. Wever: Microstructural Evolution in the Superalloy SC16 under Monotonic Loading. In: Materials for Advanced Power Engineering, Part II, Kluwer Academic Publishers 1994, The Netherlands, pp. 1119-1124.
13. H. Gabrisch, D. Mukherji and R. P. Wahi, *Phil. Mag. A.* **74,** 229 (1996).
14. D. Mukherji, H. Gabrisch, W. Chen, H. J. Fecht and R. P. Wahi, *Acta mater.* **45**, 3143 (1996).
15. T. M. Pollock and A. S. Argon, *Acta metall. mater.* **40**, 1 (1992).
16. D. Mukherji, F. Jiao, W. Chen, R. P. Wahi, *Acta metall. mater.* **39**, 1515 (1991).
17. W. Milligan, S. Antolovich, *Metall. Transc.* **22 A**, 2309 (1990).
18. B. H. Kear, A. F. Giamei, J. M. Silcock and R. K. Ham, *Scripta metall.* **2**, 287 (1986).
19. M. Condat and B. Decamps, *Scripta metall.* **21**, 607 (1987).
20. E. Nembach and G. Neite, *Progr.Mater. Sci.* **29**, 177 (1985).
21. L. E. Murr: Interfacial Phenomena in Metals and Alloys, Addison-Wesley Publishing 1975, London, pp.138-162.
22. Simmons and H. Wang: Single Crystal Elastic Constants and Calculated Aggregate Propperties, MIT Press 1971, Cambridge, pp. 56-59.

PHASE INSTABILITY IN TITANIUM ALLOY GTM -900 DURING LOWCYCLE FATIGUE AT ELEVATED TEMPERATURE

P.N.SINGH, B.K.SINGH and V. SINGH
Centre of Advanced Study, Department of Metallurgical Engineering
Institute of Technology, Banaras Hindu University, Varanasi - 221005, India.

ABSTRACT

Low cycle Fatigue behaviour of an $\alpha+\beta$ type high temperature titanium alloy, GTM-900, was studied in total strain control mode at ambient temperature and 773K in air, following solution treatment in the $\alpha+\beta$ phase field, air-cooling and stabilization treatment at 823K for 6 hours. Stress response curves revealed that there was cyclic softening, increasing with strain amplitude, at ambient temperature. However, there was tendency for stabilization and subsequent hardening, increasing with decrease in strain amplitude at 773K. Also the tendency for cyclic hardening was found to increase with decrease in strain rate and increase in tensile hold period at 773K. The unusual behaviour of cyclic hardening observed at the elevated temperature was analysed to be associated with oxygen and nitrogen enrichment and consequent precipitation of Ti_3Al phase. Also the initial microstructure consisting of equiaxed primary α and transformed β was completely modified into fibrous structure, on LCF testing at 773K, in particular with introduction of tensile hold period.

KEY WORDS

Titanium alloy, low cycle fatigue, cyclic hardening/softening, Ti_3Al, $\alpha+\beta$ alloy, fatigue life.

INTRODUCTION

Titanium alloys are considered highly attractive material for Aerospace industry because of their high specific strength. Titanium alloy GTM-900 is an $\alpha+\beta$ type allloy, designed for high temperature application as compressor disc in jet engines. Its β transus temperature is 1243K [1]. It develops a wide variety of microstructures depending upon the solution treatment and the subsequent rate of cooling. [2]. Occurrence of dynamic strain aging has been observed during tensile deformation at elevated temperatures [3]. Limited investigations have been made on low cycle fatigue (LCF) behaviour of this alloy at room temperature (RT). The present work is concerned with LCF behaviour of the alloy GTM-900 at ambient temperature and at 773K, in the $\alpha+\beta$ solution treated condition. An attempt has been made to analyse the cyclic stress response observed at RT and 773K.

EXPERIMENTAL

The alloy GTM-900 was procured from M/S MIDHANI, Hyderabad in the form of rods of 30 mm diameter, in hot rolled and mill-annealed condition. Its chemical composition is given in Table 1. Cylindrical blanks of 120 mm length and 12.5 mm diameter were vacuum (10^{-3} torr) sealed in silica tube with titanium getter.

Table 1. Chemical composition of the alloy GTM-900 (wt%)

Al	Zr	Mo	Si	Fe	C	O	N	H	Ti
6.6	1.81	3.1	0.3	0.055	0.013	0.077	0.0053	0.0025	Balance

The sealed blanks were solution treated in the $\alpha+\beta$ phase field at 1233 K for 1 hour and cooled in air. The solution treated blanks were subjected to stabilization treatment at 823K for 6 hours. Cylindrical LCF specimens with threaded ends of 30 mm length and 12 mm diameter, gauge length and gauge diameter of 14 mm and 5.5 mm, respectively and shoulder radii of 25 mm were machined from the heat treated blanks. The gauge section of the LCF specimen was mechanically polished. Fully reversed (R = -1) LCF tests were conducted under total strain control mode at RT and at 773K, mounting extensometer in the gauge section, using servohydraulic machine (MTS-810). Specimens for optical metallography were prepared by mechanical polishing and etching by a solution of 10HF + 5HNO₃ + 85H₂O (volume %). The phase instability and precipitation of the Ti₃Al phase in the LCF tested samples at 773K, was examined by transmission electron microscopy. Thin slices were sectioned from the gauge section of the LCF tested samples. Thin foils for TEM were prepared by electrolytic thinning of 3 mm diameter disc, using a twin-jet polisher. Electrothinning was carried out in an electrolyte containing 59 methonal, 35 n- butanol and 6 perchloric acid (volume %) at 12.5 volts. The temperature of the electrolyte was maintained at 223K.

RESULTS AND DISCUSSION

The microstructure of the alloy GTM 900, in the $\alpha+\beta$ solution treated (1033K-1h), air cooled and stabilised (823K-6h) condition is shown by the scanning electron micrograph in Fig. 1. As expected the microstructure, consists of two phases, the primary α (dark) phase and the transformed β (light) phase. The cyclic stress response curves for the LCF tests carried out at RT and

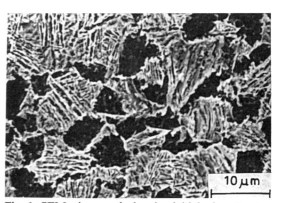

Fig. 1. SEM micrograph showing initial microstructure of the alloy GTM 900 in the $\alpha+\beta$ solution treated, air cooled and stabilized condition.

773K are shown in Fig. 2a. and 2b respectively. It may be seen that there is continuous cyclic softening at RT and the degree of softening increases with increase in strain amplitude (Fig. 2a). On the other hand the stress-response for the 773K test is quite different. There is mild softening during the initial 100 cycles, followed by stabilization and subsequent hardening till the failure of the specimen. It may also be seen that the degree of hardening increases with decrease in the strain amplitude. The variation of fatigue life with plastic strain amplitude at RT and 773K is shown by Coffin-Manson plot in Fig. 3. It is obvious that there is dual slope behaviour both at RT as well as 773K and fatigue life is higher at the elevated temperature. The dual slope behaviour was found to be associated with the difference in the mode of deformation at low and high strain amplitudes [3,4]. The higher fatigue life at 773K than that at RT may be attributed to higher ductility at 773K (Table 2).

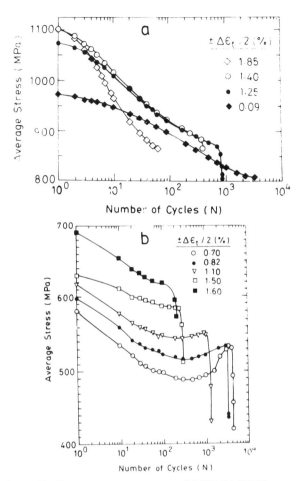

Fig. 2. Cyclic stress response curves (a) RT (b) 773K

Table 2. Tensile properties of the alloy GTM-900 in ($\alpha+\beta$) treated condition, at RT and 773K

Test Temperature	σ_{YS} (MPa)	σ_{UTS} (MPa)	Elongation (%)
RT	1033.2	1168.9	16.23
773K	657.4	815.7	23.20

Cyclic hardening or softening in metals and alloys has been observed to depend on the ratio of tensile strength to 0.2% offset yield strength. When this ratio is >1.4, cyclic hardening is observed and when it is <1.2, softening occurs [5]. The forecast becomes difficult for ratios between 1.2 and 1.4. The observed cyclic softening at RT is in accordance with the

360

above criterion based on tensile properties (Table 2). The cyclic stress response of initial mild softening followed by stable behaviour and finally cyclic hardening at 773K in the present investigation is similar to that observed by Bania and Antolovich [6] in the $\alpha+\beta$ alloy : Ti-5522 Si (Ti-5Al-5Sn-2Zr-2Mo-0.24Si) during LCF at elevated temperature forms 700K to 810K. Bania and Antolovich [6] attribute the initial cyclic softening in the alloy Ti 5522Si to planarity of slip in the primary α phase and the observed cyclic hardening to general work hardening of the material during the later stage of cycling, once the planar deformation of the primary

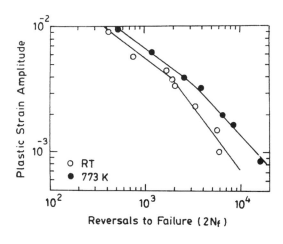

Fig. 3. Coffin-Manson plot showing dual slope behaviour.

α phase by planar slip is saturated. It is difficult to understand the observed behaviour of cyclic stress response observed in the alloy GTM-900, the material of present investigation, either at RT or at 773K on the basis of the mechanism proposed by Bania and Antolorich [6]. The continuous cyclic softening, increasing with strain amplitude, in the alloy GTM 900 at RT, may be attributed to high planarity of slip in the primary α phase and subsequent initiation of cracks in the planar slip bands and their propagation.

In contrast to expected increase in cyclic hardening with increasing strain amplitude, based on the concept of work hardening [6], there is opposite behaviour in the cyclic stress response of the alloy GTM 900 at 773K. The degree of hardening increases with decrease in the strain amplitude. Since the LCF tests also at 773K were conducted in air it becomes essential to analyse the observed cyclic stress response at 773K in terms of environmental influence because testing in air leads to enrichment of solid solution strengtheners like oxygen and nitrogen [7]. Oxygen and nitrogen are known to increase the aluminium equivalent in titanium alloys. Nitrogen is twice

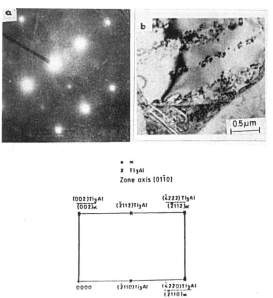

Fig. 4. Diffraction pattern and BFTEM micrograph showing Ti$_3$Al precipitates in the alloy GTM-900 in $\alpha+\beta$ solution treated condition, tested in LCF at 773K ($\Delta\epsilon_t/2 : \pm 1\%$).

as effective in increasing the aluminium equivalent as oxygen [8]. Precipitation of Ti₃Al phase occurs, once the equivalent concentration of aluminium exceeds the solubility limit. TEM examination of the specimens, tested in LCF at 773K, revealed phase instability and precipitation of Ti₃Al phase. (Fig. 4). The observed behaviour of increasing cyclic hardening with decrease in strain amplitude may thus be understood in terms of higher concentrations of oxygen and nitrogen from the atmospheric air due to longer periods of exposure of the specimens, at decreasing strain amplitudes.

Also marked change was observed in the microstructure of the material due to LCF testing at 773K. The initial microstructure consisting of equiaxed primary α and transformed β was modified to fibrous structure [Fig. 5]. The process of the such microstructural modification is not fully understood and the investigations are in progress.

CONCLUSION

Titanium alloy GTM-900, exhibits continuous softening during LCF at room temperature, increasingly with increase in strain amplitude. However, at 773K there is distinct tendency for stabilization and subsequent hardening in the later stage, increasing with decrease in strain amplitude. Cyclic hardening at 773K is essentially due to solid solution strengthening and precipitation of Ti₃Al phase due to increase in the aluminium equivalent because of enrichment of oxygen and nitrogen from air.

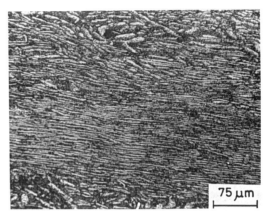

Fig. 5. Optical micrographs showing microstructural modification of the alloy GTM-900, α+β solution treated and tested in LCF at 773K ($\Delta\epsilon_t/2 : \pm 1\%$, tensile hold period : 90 seconds per cycle).

REFERENCES

1. Banerjee, D., Mukherjee D., Saha R.L. and Bose K. (1983) Metall. Trans. 14A, 413.
2. Mukherjee, D., Banerjee, D. And Saha R.L. (1982) Trans. Ind. Inst. Metals 35(6), 583.
3. Singh, P.N. (1996). Ph.D. Thesis, Banaras Hindu University, India.
4. Singh, V., Sundararaman, M., Chen, W. And Wahi, R.P. (1991) Metall. Trans. 22, 499.
5. Smith R.W., Hirschberg M.H. and Manson S.S. (1963) NASA TND-1574.
6. Bania, P.J. and Antolovich, S.D. (1984). In : *Proceedings of the fifth International Conference on Titanium Science and Technology*, pp. 2305-2312, G. Lutjering, U. Zwicker and W. Bunk (Eds.). DGM, FRG.
7. Satyanarayana, K. And Singh V. (1997). In : *Proceedings of the Fifth International Conference on Recent Advances in Metallurgical Processes*. pp. 1155-1160. D.H. Sastry, E.S. Dwarkadasa, G.N.K. Iyengar and S. Subramanian (Eds.), New Age Int. (P) Limited Publishers, Bangalore.
8. Polmear, I.J. (1981). *Light Alloys : Metallurgy of the Light Metals*. Edward Arnold, London.

Low Cycle Fatigue and Elasto-Plastic Behaviour of Materials
K-T. Rie and P.D. Portella (Editors)

STRUCTURAL ASPECTS OF SURFACED ELEMENTS FATIGUE LIFE

P. ADAMIEC and J. DZIUBIŃSKI

*Faculty of Material Science, Metallurgy and Transport, Silesian Technical University,
PL-40 019 Katowice, Poland*

ABSTRACT

Surfaced weld metals structures and their structural models have been presented. Analysis of these structural models behaviour in fatigue loading conditions has been carried out. These considerations conducted to foresights of surfaced elements fatigue life. These foresights have been verified by fatigue tests of surfaced elements with surface layer features different structure. The results of metallographic investigations showed that there is a close connection between the structural constitution and mechanical properties of surfaced weld metals and the carbon equivalent. Fatigue tests of surfaced elements with a surface layer of diverse structure corresponding to the models assumed for these considerations have confirmed the foresight based on an analysis of the behaviour of surfaced weld metals structural models concerning the surfaced elements fatigue.

KEYWORDS

Fatigue life, surfacing, microstructure of surfaced weld metal.

INTRODUCTION

While in use, machine parts are being geometrically worn, which brings about shortening of their working life. Worn out elements can be regenerated by means of surfacing where, depending on the needs, weld metals of very diverse properties and structures are used. In case of exploitation of surfaced machine parts fatigue wear occurs very often which is the result of character of loading of such parts. Basic factors which influence the fatigue life of surfaced machine parts are: conditions of surfacing, defects in surfaced weld metals, internal stresses in the element and the surfaced weld metal microstructure. The influence of the first three of the above mentioned factors has been presented in papers [1,2] whereas determining the influence of the fourth factor is the subject of this paper.

MODELS OF SURFACED WELD METALS STRUCTURE

Structures of surfaced layer are an after-effect of epitaxial growth of crystals and structural transformations resulting from continuous cooling of a surfaced weld metal. Anisotropy of structure which is a result of crystal overgrowth perpendicularly to the direction of heat flow is distinctly visible in C-Mn surfaced weld metals structures where granular proeutectoid ferrite (primary ferrite - PF) of hardness of about 130 HV is observed on primary boundaries of austenite. In order to simplify the terms, other structural constituents such as fine-grained ferrite (acicular ferrite - AF), lamellar forms of ferrite (ferrite with aligned martensite-austenite-carbides - AC) and small amounts of pearlite (up to 5%), are called bainitic ferrite (FB). Models of structures are presented in Fig.1.

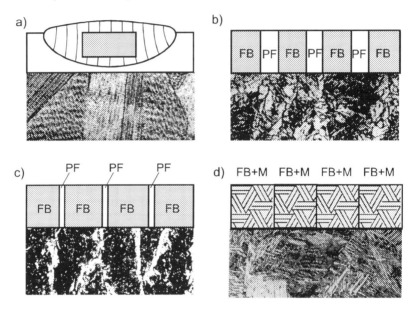

Fig. 1. Models of surfaced layers structures. a) epitaxial growth of primary crystals visible, b) C-Mn surfaced weld metal structure 0.10%C and 1.5% Mn, c) C-Mn surfaced weld metal structure 0.15% C and 1.7% Mn, d) C-Mn-Mo surfaced weld metal structure 0.10%C; 1.1% Mn, and 0.5% Mo. PF - proeutectoid ferrite, FB - bainitic ferrite, B- bainite, M - martensite

In order to evaluate mechanical properties of the surface layer structure the considerations concerning composites properties can be used [3]. These considerations lead to a conclusion that mechanical properties of model structures of two constituents can be described by means of an expression:

$$H = f_{PF} \cdot H(PF) + f_{FB} \cdot H(FB) \qquad (1)$$

$$\sigma_e = f_{PF} \cdot \sigma_e(PF) + f_{FB} \cdot \sigma_e(FB) \qquad (2)$$

$$\sigma_m = f_{PF} \cdot \sigma_m(PF) + f_{FB} \cdot \sigma_m(FB) \qquad (3)$$

where: H - surfaced weld metal hardness, f_{PF} - proeutectoid ferrite fraction, f_{FB} - bainitic ferrite fraction, H(PF) - proeutectoid ferrite hardness, H(FB) - bainitic ferrite hardness; $\sigma_e(PF)$, $\sigma_e(FB)$ - yield point of proeutectoid ferrite and bainitic ferrite respectively; $\sigma_m(PF)$, $\sigma_m(FB)$ - ultimate tensile stress of proeutectoid ferrite and bainitic ferrite respectively.

Assuming that the bainitic ferrite fraction is proportional to the carbon equivalent $C_E=C+Mn/6$ where C and Mn - elements fractions in surfaced weld metal, it can be written:

$$H = H(PF) + A \cdot C_E \qquad (4)$$

$$\sigma_e = \sigma_e(PF) + B \cdot C_E \qquad (5)$$

$$\sigma_m = \sigma_m(PF) + D \cdot C_E \qquad (6)$$

where: A, B and D - constants taking into account $H(FB) - H(PF)$, $\sigma_e(FB) - \sigma_e(PF)$ and $\sigma_m(FB) - \sigma_m(PF)$.

Similarly, an equation describing the true tensile stress σ'_f depending on C_E can be written:

$$\sigma'_f = \sigma'_f(PF) + E \, C_E \qquad (7)$$

where: E - constant taking into account $\sigma'_f(FB) - \sigma'_f(PF)$.

Taking advantage of the results in paper [4] these expressions can be presented in a form of diagrams (Fig.2).

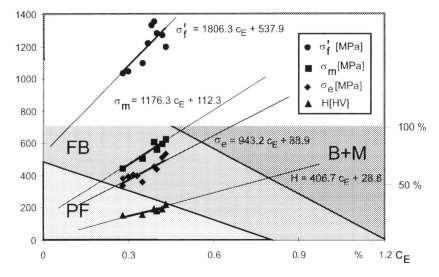

Fig. 2. Influence of the equivalent $C_E = C + Mn/6$ on strength properties of C-Mn weld metals and its structural constitution.

Deformations of surface layer during fatigue loading bring about slip bands, the number of which grows along with the growth of number of cycles. This happens in areas of proeutectoid ferrite of distinctly lower hardness and strength compared to average strength of a surfaced weld metal. Slip bands create geometric forms on proeutectoid ferrite grains boundaries. A scheme of mechanism of slip bands creation as well as cracks initiation and propagation in areas of these slip bands are presented in paper [5]. It is mentioned there that the considerations connected with cracks in slip bands are also important in case of fatigue cracking propagation at high stress amplitudes (Fig. 3).

366

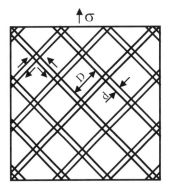 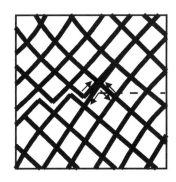

Fig.3. A model of slip bands creation and growth of cracking in them.

SPECIMENS PREPARATION AND MATERIALS USED FOR INVESTIGATIONS

In order to assess the fatigue strength of elements with surfaced weld metals of different structures, fatigue tests have been carried out in which clepsydric specimens loaded with a rotating bending moment were used. For surfacing tests of 45 steel (0.44% C, 0.66% Mn, 0.2% Si) C-Mn weld metals of ferritictic structures were used which simulated regeneration layers as well as C-Mn-Cr-Mo, C-Mn-Mo and C-Mn-Si-Cr weld metals of bainitic-ferritic and bainitic-martensitic structures which simulated hardening layers (Table 1).

Table 1. Chemical constitution of weld metals, structure and hardness of surfaced layer

No.	Weld metal type	Chemical constitution %					Structure	Hardness HV	C_E %
		C	Mn	Si	Cr	other			
1	steel 45	0.43	0.6	0.25	-	-	P+F	240	0.53
2	C-Mn	0.1	1.5	0.90	-	-	F(PF+FB)	171	0.35
3	C-Mn-Cr-Mo	0.07	0.7	0.30	0.9	0.6%Mo	B+F	189	0.49
4	C-Mn-Mo	0.10	1.1	0.70	-	0.5%Mo	B+F	223	0.38
5	C-Mn-Si-Cr	0.35	1.0	1.2	1.2	-	B+M	280	0.75

Remark: $C_E = C + Mn/6 + (Cr + Mo)/5$, P+F - ferritic-pearlitic structure of steel, F - ferritic structure of surfaced weld metal in which proeutectoid ferrite PF and bainitic ferrite FB are distinguished, B+F - bainitic-ferritic structure of surfaced weld metal, B+M - bainitic-martensitic structure of surfaced weld metal.

INFLUENCE OF SURFACED WELD METAL STRUCTURE ON ELEMENTS FATIGUE LIFE

Surfaced specimens were used for metallographic investigations. Their transverse microsections were ground, polished and after etching they were used for the structure quantitative assessment. The assessment was carried out by means of point method for 500 measuring points and its results are presented in Fig. 2.

While using structure models (Fig. 1) for a surfaced elements fatigue life analysis it was assumed that the initiation of cracking appears in the slip bands which are created in soft structure constituents. Development of these slip bands can be restrained by hard structure constituents. Additionally it was assumed that the period of initiation of macrocracks 0.1÷0.5mm long lasts 0.5÷0.95 of the cycles total number until the element's failure [6]. Structural models of surfaced weld metals of a diverse carbon equivalent $C_E = C + Mn/6 + (Cr+ Mo)/5$ are presented in Fig. 1. The model in Fig. 1b represents relatively soft (170HV) regenerating surfaced weld metals C-Mn where $C_E = 0.35\%$ in which cyclic deformations causing slips and creation of slip bands will be situated in proeutectoid ferrite (PF). Simultaneously an obstruction in slip bands development will be mainly harder fine-grained ferrite boundaries (AF) (in Fig. 1b FB). Appearance of a lamellar ferrite phase will be conducive to creation of numerous slip bands. The model in Fig. 1c characterizes harder surfaced weld metals (223 HV) where $C_E = 0.38\%$. The growth of carbon equivalent C_E leads to a reduction of column crystals size as well as a reduction of proeutectoid ferrite (PF) fraction at the cost of an increase in bainitic ferrite (FB) volume fraction. Bainitic structures limit the slip bands creation and development and that is mainly because of carbides which restrain their development. The model in Fig. 1d is typical of harder (280 HV) surfaced weld metals where $C_E = 0.75\%$. Further growth of carbon equivalent C_c leads to a bainitic-martensitic structure (B + M) in which slips can only create and develop along the boundaries of former austenite and along the boundaries of bigger martensite needles [7].

The above considerations can be used for a quantitative assessment of the surfaced weld metal structure on elements fatigue life, where a transformed form of Morrow formula is used [7]:

$$N_f = \frac{1}{2}\sqrt[b]{\frac{\sigma_a}{\sigma'_f}} \qquad (8)$$

where: N_f - number of cycles till test piece failure, σ_a - stress amplitude, σ'_f - true tensile stress calculated from the formula (7), b - fatigue life exponent, -0.12 [7] is assumed.

Assuming $\sigma_a = 375$ MPa and calculating σ'_f according to the formula (7) N_f has been determined according to the formula (8) and plotted onto the diagram (Fig. 4).

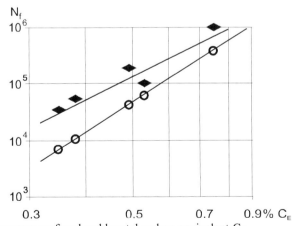

Fig. 4. Fatigue life versus surfaced weld metal carbon equivalent C_E
♦ - experiment, o - calculations

For a verification of models fatigue tests of rods of 18 mm diameter made of steel 45 were carried out. Clepsydric samples were made of these rods (the diameter of measuring part - 10mm) for material and surfaced elements examinations. The samples, after longitudinal alternate surfacing, were mechanically worked to fit the measuring part size of $\varnothing10$. The quality of obtained surfaced welds was estimated visually as well as by means of penetration examinations. Fatigue diagrams in the limited fatigue strength area as well as fatigue strength at rotational bending were determined and the results are presented in Fig. 4.

FINAL REMARKS AND CONCLUSIONS

The analysis of surfaced welds structural models behaviour (Fig. 1) in fatigue load conditions shows that at a low carbon equivalent C_E (Fig. 1b) slips and appearing slip bands will be situated in proeutectoid ferrite. In these areas the microcracks initiation will also take place. Thus, the presence of proeutectoid ferrite in the surfaced weld metal structure is unfavourable, since the fatigue life decreases along with an increase of this constituent.

Along with an increase of alloying additions in the surfaced weld grows the value of C_E, which is connected with a decrease and disappearance of proeutectoid ferrite PF (Fig. 1c) and slip bands are situated on the former austenite boundaries. Further increase of contents constituting the structure of surfaced weld metal, such as bainitic ferrite, bainite and martensite, leads to restraining the process of slip bands creation and development, which in consequence brings about longer life of surfaced elements for surface layers with a structure where bainitic-martensitic phases predominate.

The results of metallographic investigations showed that there is a close connection between the structural constitution and mechanical properties of surfaced weld metals and the carbon equivalent (Fig. 2).

Fatigue tests of surfaced elements with a surface layer of diverse structure corresponding to the models assumed for these considerations have confirmed the foresight based on an analysis of the behaviour of surfaced weld metals structural models concerning the surfaced elements fatigue life (Fig. 4).

REFERENCES

1. Adamiec, P. and Dziubiński, J. (1995). *Cracking and durability of surfaced machine parts* (in Polish). Wyd. Politechniki Śl., Gliwice.
2. Dziubiński, J., Adamiec, P. and Okrajni, J. (1996). *Archive of Mechanical Engineering*, **43**, 153.
3. Somov, A. and Tikhonovsky, M. (1975) *Eutectic Composites* (in Russian). Metallurgia, Moscow.
4. Adamiec, P. (1984). *ZN Politechniki Śl., Seria Mechanika*, **80**, 7.
5. Hornbogen, E. (1975). *Z. Metallkde*, **66**, 511.
6. Kocańda, S. and Szala, J. (1985). *Bases of Fatigue Calculations* (in Polish), Państw. Wyd. Nauk., Warsaw.
7. Kocańda, S.: *Fatigue Cracks of Metals* (in Polish), WNT, Warsaw 1985.

Low Cycle Fatigue and Elasto-Plastic Behaviour of Materials
K-T. Rie and P.D. Portella (Editors)
369

INVESTIGATION OF STRUCTURAL CHANGES OF STRUCTURAL STEEL (IN THIN FOIL) AT LOW CYCLE DEFORMATIONS

T.ETERASHVILI

Republican Center of Structural Investigations, Georgian Technical University, 380075, Tbilisi Georgia

ABSTRACT

By electron microscopic investigations the structural changes in steel structure were detected after cyclic deformation. The value of stresses of the twin relaxed at generation was calculated. It is shown that at cyclic deformation fatigue cracks may generate in the places of packet joints and slip band crossings.

KEYWORDS

Cyclic deformation, twinning, reorientation, stress, slip.

Thin foils of structural steels were investigated before and after cyclic deformation by electron transmission microscopic method. Panoramic images and microdiffraction pictures were taken; then the same foil was strained up to 200 cycles, the earlier snapped places were found and again investigated.

 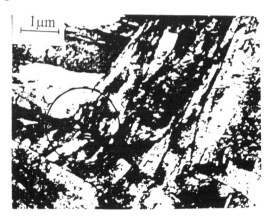

Fig.1. Steel 10X2Г3M structure.   Fig.2 The same after 200 cycles
     before deformation

Comparison of the pictures of the same places before and after deformation showed the change of internal structure. The change is expressed in break-down of initial latch structure of martensite packets. The slip begins along latches on interlatch boundaries. Consequently, the boundaries are widened and gradually pass to slip bands (Fig.1,2).

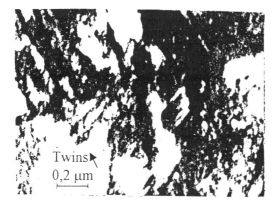

Fig. 5. Deformation twins after cyclic deformation section from Fig.2.

For twin genesis the formation of packing fault is necessary. The value of stresses for packet fault formation is determined by formula:

$$\sigma = F/b \qquad (1)$$

where b is Burgers vector, and F is stress. Besides stresses it is necessary to induce the dislocation to move and to create dislocation loop. In loop center with diameter L stress

$$\sigma = \mu b/2L \qquad (2)$$

is acting [4].

Along dislocation pile-up the number of loops near the peak is equal to n=h/d and twin flatness is determined by expression:

$$h/L \approx 2\sigma/S\mu \qquad (3)$$

where S=b/d is the value of shift at twin formation and for metals body-centered cube in plane (211) it equals ≈0,71. Thus, the final formula for stress value determination is

$$\sigma = \mu Sh/2L \qquad (4)$$

where μ is the modulus of shift elasticity (for the investigated steel $\mu=10 \times 10^{10}$ N/M^2), h is the thickness of the twin, S is shift value and L is the length of the twin (Fig.6).

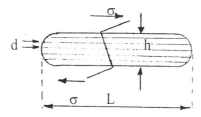

Fig. 6. Scheme of a twin

Crystallogeometrical analysis and stereographical investigations show that slip occurs along habit latch planes (110) <111>$_M$ [1]. If crystal boundaries are at an angle to slip direction, latch boundaries are crossed by slip bands. In this case azimuthal reorientation increases up to 12^0 and horizontal reorientation increases up to 6^0 and substructure of latches and packets changes (Fig.3,4).

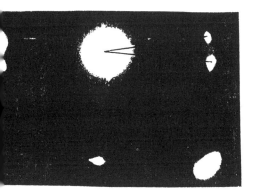 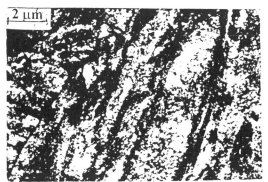

Fig.3. Azimutal reorientation in planes (110)$_M$ [001].

Fig.4. Dislocation structure and separation of carbide phase. The same after 200 cycles

After cyclic deformation the disappearance of residual austenite localized on latch boundaries is also observed. It is supposed that at deformation austenite changes into martensite [2].But because of its small volume (1÷3%), it is difficult to detect.

The separation of carbide phase of two directions observed in wide latches at cyclic deformation are broken down by moving dislocations and loose their elongated shape. On dislocations there can be observed chaotic formations in the form of separate particles (Fig.4). Dislocation structure of martensite is also changed. Before cyclic deformations the dislocations and their distribution are well detected in latches. After deformation they are difficult to detect, as their density increases approximately for order 2.

After deformation in places of jointing of two-three packets deformation twins are generated. Stereographical and microdiffraction investigations show that twinning happens on (211) <$\bar{4}9\bar{1}$>$_M$ (Fig. 5.) [3].

It should be mentioned that besides formation of new twins, intensive growth of the already existing small twins is also observed. Of course, when slip bands and deformation twins are generated internal stresses are relaxed.

Let's try to estimate the value of relaxed stresses at twin formation.

372

The value of relaxed stresses determined by formula (4) gives the values from 3000 MGPa to 6000 MGPa. In case of cyclic deformation the stresses on the microsection are relaxed on the average of 4000 MGPa on generation of each twin.

Thus, as our test investigations have proved at low cyclic deformation the essential changes of the initial structure take place. They consist in formation of new and growth-up of the already existing twins, in the change of residual austenite, in breaking up of the initial structure, in creation of slip bands, etc. In this case external stresses are relaxed, as it was shown on example of twins.

More advantageous in energetic sense is growing up of the already existing twins. This is conditioned by the fact that twin boundaries on slipping become the barriers for dislocations. In this case twinning dislocations which will provide twin growth can freely be formed.

It should be noted that the possibility of such dislocation mechanism is known long ago [5]. But the doubt was expressed about the reality of the proposed scheme of dislocation cleavage. This was caused by energetic disadvantage of cleavage, as it needed great external stresses. Besides, direct experimental proof was lacking.

The carried out tests proved that in our case we directly observed the formation of new twins, as well as the growth of the already existing ones. This undoubtedly proves the correctness of the proposed mechanisms [5] and experimentally verifies their realization at small deformations.

It should be stated that after twin formation is stopped and slip systems are strengthened due to the difficulties of relaxation stresses, the microcracks begin to form just in these places. Crack formation will also favor stress relaxation until total break up occurs.

REFERENCES

1. Этерашвили Т.В. 1978. Развитие пластической деформации пакетного мартенсита. ФММ, т. 76, вып. 4, стр. 772-790.
2. Eterashvili T. 1980. Electron microscopic study of the residual austenite behavior at plastic deformation of packet martensite. 7-th Europian Congress on electron microscopy, Hague, p. 1885-1886.
3. Этерашвили Т.В. 1989. Кристаллография двойникования пакетного мартенсита и внутренние напряжения ФММ, т. 67, вып. 3, стр. 516-524.
4. Фридель Ж. 1970. Дислокации. Москва, Изд. "Мир", стр. 502.
5. Хирт Дж., Лоте И. 1972. Теория Дислокации, Москва "Атомиздат", стр. 489.

Low Cycle Fatigue and Elasto-Plastic Behaviour of Materials
K-T. Rie and P.D. Portella (Editors)

ELASTO-PLASTIC TRANSITION IN TEMPCORE REINFORCING STEEL

H. ZHENG and A. ABEL

Department of Civil Engineering, The University of Sydney
NSW 2006, Australia

ABSTRACT

An elasto-plastic transition before reaching macro yield has been observed in TEMPCORE reinforcing steel in static loading, a feature which is microstructure dependent. As the material can be considered as a composite the observed phenomena could not be explained successfully by the rule of mixture or by a model developed by Elbert based on elastic-plastic theory. In this paper, the interaction between the soft core and the surrounding hard case materials is considered and a model is developed. Based on dislocation theory, this model explains the elasto-plastic transition of TEMPCORE reinforcing steels together with the microstructure dependence.

KEYWORDS

Tempcore, elasto-plastic deformation, microstructure, preyield microstrain.

INTRODUCTION

TEMPCORE reinforcing steel bars are produced with a special technology so that the final bar product is a composite although the chemical composition of the steel remains constant across the bar diameter[1-8]. In this technology the hot bar exiting the last rolling mill is quenched with water so that the surface transforms into a martensitic structure. During the subsequent cooling stage, the trapped heat dissipates from the core through the case of the bar and by doing so the hardened layer is tempered while the core remains soft.

TEMPCORE reinforcing steel is described in the literature as having tempered martensite in the hardened outer layer and fine ferrite-pearlite in the core section leading to a marked yield point and large Lüders strain. In the present investigation, coarse Widmanstätten ferrite plus pearlite structure was often found in the core area, and testing indicated substantial "preyield" before the attendance of Lüders yield. This preyield or elasto-plastic transition phenomenon has been investigated in order to understand the strengthening mechanism of TEMPCORE reinforcing steel.

TEST DETAILS AND THE ELASTO-PLASTIC DEFORMATION

Test Method

Tests were conducted in a 2000 kN Dartec machine with strain rates of 10^{-6} / S and 10^{-5} / S according to Australian Standard AS 1391-1991. Strain measurement was completed by using a 50 mm gauge length extensometer and electrical strain gages. A computer equipped with software "spectran" was used for data acquisition.

Test Results

Two 36 mm diameter bars were tested with the following microstructures: BarF with fine ferrite (grain size 9.4 μm) plus pearlite and BarC with coarse Widmanstätten ferrite (grain size 45 μm) containing some pseudo-eutectoid colonies. The etched cross sections of the two bars are basically the same, Fig. 1, and the corresponding microstructures in the core areas are shown in Fig. 2 while mechanical properties of the case and core materials are given in Table 1. The case materials of both bars yield continuously and are stronger than the corresponding core materials which yield in Lüders style.

The incipient yield parts of the stress-strain curves are given in Fig. 3. Distinctive differences in the yield behavior are witnessed: BarF exhibits a marked yield point, while BarC preyields about 1000 micro strain starting from 310 MPa before Lüders yield. The same phenomena were observed in static compression. Previous tests on bars with other diameters also showed this difference in the yielding of bars with coarse and fine microstructure.

MODELING

The tests revealed an elasto-plastic transition region on bars with coarse microstructure followed by Lüders type yielding. A "developed yield" section has been reported by Kugushin et al[9] starting at 80% of yield strength in a quenched and tempered angle section (basically TEMPCORE process), and this was attributed to the variety of mechanical properties across the cross section. According to Kugushin et al, analogous phenomenon was also observed by Mulin et al, on in-line interrupted quenching and self-tempering reinforcing bars (similar to TEMPCORE).

To understand this elasto-plastic transition behavior the bars are treated as consisting of a hardened layer and soft core composite material. The preyield of BarC starts from 310 MPa a value which coincides with the the yield strength of the core material. Thus the preyield is related to the yield of core material. Theoretically, bars with fine microstructure bars also have an elasto-plastic transition but it is too small to detected with the applied equipment.

Using the rule of mixture and model developed by Elbert[10], a departure from elastic behavior can be predicted immediately after the average stress exceeds the yield strength of the core material, Fig. 3. The two models successfully predicted similar results on BarC, but both failed to predict the behavior of BarF as the effect of microstructure is not incorporated into the approaches.

The failure of the two models above must relate to the assumption that the core material yields in Lüders style as in the case when tested individually. On further examination of the test results it became evident that the core material did not yield in Lüders style during the elasto-plastic transition, as the strain measured in different parts of the bar were homogenous. The preyield is essentially a micro plastic effect and therefore there is no dislocation multiplication taking place at that stage. During this process more microplasticity will take place with increasing stress as more dislocation sources are activated and piled up, and thus the core material is strengthened. This strengthening mechanism is similar to the film effects observed in other metals[11] where the coated film inhibits the passage of dislocations out to the surface of the crystal.

If one considers the TEMPCORE reinforcing steel bar as a concentric cylinder composed of only hard case and soft core material, the core material yields initially with a certain hardening rate imposed by dislocation pile ups with the restraining effect provided by the case material which is still elastic (Fig. 4). Using the theory of Brown[12] the microstrain preyield can be incorporated in describing the yield of core member as:

$$\varepsilon_{pl}^{core} = \rho D^3 (\sigma^{core} - \sigma_0^0)^2 / 2G\sigma_0^0 = C(\sigma^{core} - \sigma_0^0)^2$$

where ρ is the density of dislocation sources (sources/cm^3), D is the grain diameter, σ^{core} is the applied stress on the core material, σ_0^0 is the stress to activate the first dislocation source in the core and G is the shear modules. The term $\rho D^3 / 2G\sigma_0^0$ can be expressed as a constant for a given material and it will be simply C.

The stress of the microyielded core member is then written as: $\sigma^{core} = E(\varepsilon - \varepsilon_{pl}^{core})$, where ε is the strain of the bar under an average bar stress, σ^{bar}

According to the preceding assumption: yielding commences with the activation of a dislocation source, that is $\sigma_0^0 = \sigma_0^{core}$, the above expression can be written as:

$$\sigma^{core} = \sigma^{case} - EC(\sigma^{core} - \sigma_0^{core})^2$$

With the condition that $\sigma^{core} - \sigma_0^{core} > 0$, the solution is:

$$\sigma^{core} = \frac{\sqrt{4EC(\sigma^{case} - \sigma_0^{core}) + 1} - 1}{2EC} + \sigma_0^{core}$$

If the fraction of core and case material in the cross section is A and B respectively, according to the rule of mixture:
$$A\sigma^{core} + B\sigma^{case} = \sigma^{bar}$$
thus

$$\sigma^{bar} = A\{\frac{\sqrt{4EC(\sigma^{case} - \sigma_0^{core}) + 1} - 1}{2EC} + \sigma_0^{core}\} + B\sigma^{case}$$

When the case is still in the elastic state, that is σ^{case} = E ε, from the above equation the stress strain curve of the bar can be calculated.

DISCUSSIONS
The above equation shows that the stress strain plot of the bar is not only a function of the strength of the core material but it also depends on parameter C in which the grain size and the density of dislocation sources of the core material are included. Taking the yield strength of the core material as 300 MPa and A = 0.6 for example, the calculated stress strain curves for

various C parameters will be different as shown in Fig. 5. It is clearly shown that when $C \leq 10^{-8}$ there is very small amount of preyield up to 50 MPa beyond the yield strength of the core, while for $C \geq 10^{-7}$ large preyield occurs.

Suppose each grain in the core of the fine structure (grain size 9.4 μm) contains one dislocation source, then the density of dislocation sources will be 0.925 X $10^{15}/m^3$, and thus the C parameter will be of the order of 10^{-8}. If the dislocation source density with the 45μm grains is the same to that of fine microstructure, the C parameter will be of the order of 10^{-6}. This explains that bars with fine microstructure exhibit no micro plasticity while the bars with coarse microstructure exhibit large amount of plasticity in the elasto-plastic transition. Since the appropriate values of dislocation sources are as yet not well defined for fine ferrite and coarse Widmanstätten ferrite, the use of the model requires a number of assumptions.

CONCLUSIONS
1. TEMPCORE reinforcing steel with coarse Widmanstätten ferrite in the core area exhibits large elasto-plastic transition before Lüders yield, while bars with fine microstructure remain essentially elastic until large scale yielding takes place.
2. The elasto-plastic transition of the TEMPCORE reinforcing steel bar is related to the yielding of the core material.
3. A model based on dislocation source density and grain size has been proposed which can predict the observed performance.

REFERENCE
1. Ammerling, W. J. (1984). "Bar and rod mills - influence of design parameters on the economics and properties of rolled products." Metallurgical plant and technology, No. 1, 46-57.
2. Defourny, J., and Bragard, A. (1977). "TEMPCORE-process, the solution to rebar welding problems." C. R. M., No. 50, 21-26.
3. Economopoulos, M. (1981). "The use of drastic and mild water cooling techniques for controlling bar and rod properties." rod and bar production in the 1980's, Metal society, London, 76-80.
4. Economopoulos, M., Respen, Y., Lessel, G., Steffes, G. (1975). "Application of the TEMPCORE process to the fabrication of high yield strength concrete-reinforcing bars." C. R. M., No. 45, 3-19.
5. Killmore, C. R., Barrett, J. F., and Williams, J. G. (1985). "Mechanical properties of high strength reinforcing bar steels accelerated cooled by "TEMPCORE" process." Proceedings of a symposium on accelerated cooling of steel, Pittsburgh, Pennsylvania, August 19-21, 541-558.
6. Simon, P. (1990). "Optimization of TEMPCORE installations for rebars." Metallurgical plant and technology." No. 2, 61-69.
7. Simon, P., Economopoulos, M., and Nilles, P. (1984a). "TEMPCORE: a new process for the production of high quality reinforcing bars." Iron and steel engineer, March, 53-57.
8. Simon, P., Economopoulos, M., and Nilles, P. (1984b). "TEMPCORE, an economical process for the production of high quality rebars." Metallurgical plant and technology, No. 3, 80-93.
9. Kugushin, A. A., et al. (1986). "Improving strength and subzero brittle fracture resistance in angle sections by quenching and self-tempering from rolling heat." Steel in USSR, Vol. 16 September, 442 - 446.

10. Ebert, J., et al. (1968). "The stress strain behaviour of concentric composite cylinders." Journal of composite materials, Vol. 2, No. 4, P458-479.
11. Gilman, J. J. (1955). "The role of thin surface films in the deformation of metal monocrystals." ASTM STP 171, 3-11.
12. Brown, N., and Lukens, K. F. (1961). "Microstrain in polycrystalline metals." ACTA Metallurgica, Vol. 9, 106-111.

Table 1. Mechanical properties of core and case materials

Material	Yield Strengths MPa	UTS MPa	Yield Style	ψ %	δ_5 %	$\delta_{Lüders}$ %
Case 16	$\sigma_{0.02} = 465$ $\sigma_{0.2} = 495$	573	Continue	–	15	–
Case 36	$\sigma_{0.02} = 540$ $\sigma_{0.2} = 607$	680	Continue	-	13	-
Core16	Up 380 Low 354	481	Lüders	67	31	3.25
Core36	Up 308 Low 297	476	Lüders	71	28	0.73

Fig. 1. Cross section of TEMPCORE reinforcing steel showing the domains of various metallurgical phases

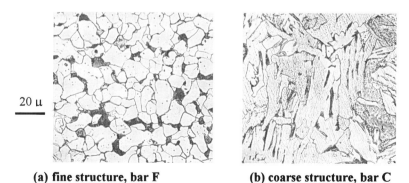

20 µ

(a) fine structure, bar F   **(b) coarse structure, bar C**

Fig. 2. Microstructures in the core areas of the tested bars

378

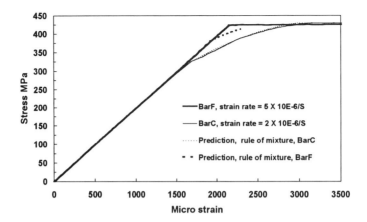

Fig. 3. Test results and predictions using the rule of mixture

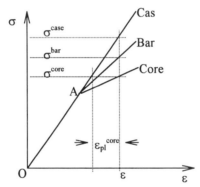

Fig. 4. Schematic illustration of stress strain in the case and core areas of the composite bar

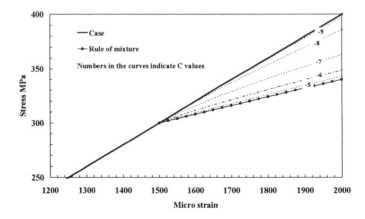

Fig. 5. Calculated stress strain curves for different C parameters

Low Cycle Fatigue and Elasto-Plastic Behaviour of Materials
K-T. Rie and P.D. Portella (Editors)
379

MICROSTRUCTURAL ASPECTS OF STRAIN LOCALIZATION DURING LOW-CYCLE FATIGUE IN MODEL AL-LI ALLOYS

M.LEWANDOWSKA, J.MIZERA, J.W.WYRZYKOWSKI
Faculty of Materials Science and Engineering, Warsaw University of Technology
Narbutta 85, 02-524 Warsaw, Poland

ABSTRACT

In this work the microstructural aspects of the strain localization during low-cycle fatigue in model Al-Li alloys were investigated by TEM. The study has shown that plastic deformation in AL-Li alloys is inhomogeneous. The heterogeneities having the form of slip bands (microbands) are present in the alloys that contain δ' precipitates. They lie in the slip planes {111} through several grains and form characteristic steps on the intersecting grain boundaries or dislocation bands, which indicates that they are regions of localized shear. Within these bands the δ' particles have evidently been sheared. For small precipitates their dissolution within slip bands was observed. An increase of the δ' precipitates size and/or the grain size makes the strain localization more pronounced; this has a detrimental effect on the fatigue life. The presence of the T_1 precipitates (in addition to δ') causes the plastic deformation to become more homogeneous (no slip bands were found) and in consequence the number of cycles to failure to increase 2-3 times compared with that needed for the alloy with δ' precipitates, of the same size, alone.

KEYWORDS
Al-Li alloys, low-cycle fatigue, strain localization, δ' precipitates, microstructure

INTRODUCTION

In many cases, plastic deformation can be considered to be homogeneous. This approach permits e.g. strength calculations to be performed. However, many observations of the dislocation structures formed during plastic deformation have shown that the dislocation distribution may be spatially inhomogeneous even though the macroscopic deformation is homogeneous. The development of the microstructure inhomogeneities during plastic deformation has been studied in detail in many materials, such as e.g. ferritic and austenitic stainless steel, copper and aluminium alloys [1-6]. These studies were concerned with heavy plastic deformations, primarily due to rolling. However, various other straining techniques, for example cyclic compression/tension procedure (low-cycle fatigue), also give inhomogeneus plastic deformation even in homogeneous materials such as copper or single-phase aluminium alloys [7,8].

On the other hand the Al-Li alloys show a strong tendency to strain localization even in monotonic straining. This is caused by two different mechanisms: (i) cutting of shearable δ' precipitates and (ii) preferential deformation in softer precipitate-free zones (PFZ's) [9]. The spherical δ' particles uniformly distributed in the matrix are present in the Al-Li alloys. These precipitates have an ordered L1$_2$ structure and are sheared relatively easy by the moving dislocations. The δ' precipitate free zones form along some grain boundaries due to solute or vacancy depletion. The material within a PFZ is considerably softer than the precipitate hardened matrix. As a result, plastic deformation may be concentrated in this region.

The strain localization does not improve the strength but reduces the fracture toughness and fatigue life. The aim of the present work was to study, by TEM, the microstructural aspects of the localization of plastic deformation during low-cycle fatigue in model Al-Li alloys.

EXPERIMENTAL

The chemical compositions of the model Al-Li alloys examined are given in table 1. The alloys were extruded at a high temperature and then heat treated: solutionized at 550°C for 1 hour, water quenched and aged. The ageing conditions were: (i) natural ageing, (ii) ageing at 200°C for 2 hours, (iii) ageing at 200°C for 24 hours.

Table 1. Chemical composition of the investigated alloys (in weight %).

alloy	Li	Cu	Zr	Fe	Si	Al
Al-Li	2.3	-	-	0.024	0.0058	balance
Al-Li-Zr	2.2	-	0.1	-	0.0058	balance
Al-Li-Cu-Zr	2.2	1.2	0.1	-	0.0058	balance

A metallographic study has shown that the Al-Li alloy possesses a recrystallized microstructure with an average grain size of 480 μm, whereas the two other alloys are unrecrystallized because of the Zr addition which is known as a recrystallization inhibitor [10].

The precipitate state was examined by TEM. After precipitation hardening, δ' precipitates coherent and homogeneously distributed within the matrix are present in all the materials (fig.1). Their average size depends on the ageing conditions. After natural ageing, the δ' precipitates are very small with a mean diameter of 2 nm, after ageing at 200°C for 2 hours their sizes are below 10 nm and after ageing at 200°C for 24 hours their average size is approx. 20 nm. In the alloys containing a Zr addition, composite δ'/β' particles were found. They are visible in fig. 1 as the dark spheroids within some δ' precipitates. In the Al-Li-Cu-Zr alloy aged at 200°C for 24 hours, in addition to δ', the T$_1$ precipitates were observed (fig.2). They have the shape of plates and are oriented along the {111} planes of the matrix.

The specimens were subjected to the low-cycle fatigue test at a total strain control. Thin foils for TEM examination of the dislocation structure and localisation of plastic deformation were prepared from fatigued samples and observed using a Philips electron microscope operated at 100kV.

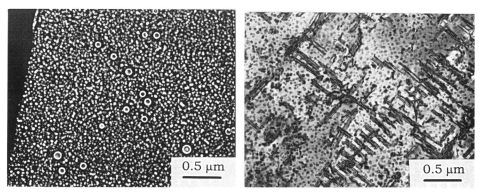

Fig.1. δ' precipitates in the AL-Li-Zr alloy aged at 200°C for 24 hours.

Fig.2. δ' and T_1 precipitates in the Al-Li-Cu-Zr alloy aged at 200°C for 24 hours.

RESULTS AND DISCUSSION

A characteristic feature of the low-cycle fatigue deformation of the Al-Li alloys is its inhomogeneity. On the one hand, in TEM examinations we observe regions with an increased density of dislocations in the form that resembles dislocation walls. Examples of such dislocation accumulations are shown in Fig.3. These walls are arranged along the easy slip planes {111}. In addition to the dislocation lines, numerous dislocation loops can be seen. No shearing of the δ' precipitates was observed here.

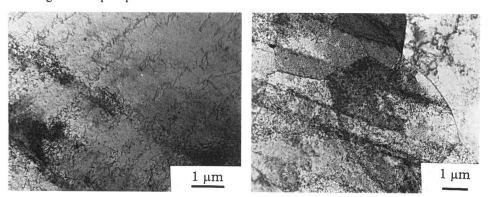

Fig.3. Dislocation walls in the Al-Li alloy (a)and Al-Li-Zr alloy (b) formed during cyclic deformation (both alloys were aged at 200°C for 24 hours)

On the other hand, the characteristic dislocation structures formed during low-cycle fatigue are bands (Fig.4). In the literature concerning the low-cycle fatigue of the Al-Li alloys these bands are known as slip bands [11,12], but the investigators who focused their attention on the plastic strain localization effect call them the microbands or shear bands [4-6]. The bands formed in the alloys examined in the present experiment are parallel to one another and lie in the {111} planes of the matrix. In small grains (the Al-Li-Zr and Al-Li-Cu-Zr alloys), they run

382

through several grains without changing their direction. Usually, we observed several such bands within a single grain (subgrain). At the intersections of the bands with the grain boundaries, the boundaries show characteristic steps which gives evidence that intensive shearing takes place within the bands.

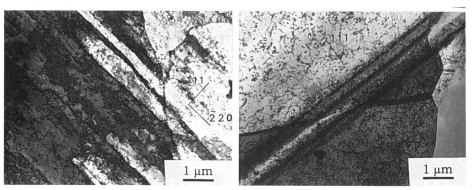

Fig.4. Slip bands formed during cyclic straining in the Al-Li-Zr alloys aged at 200°C for 2 (a) and 24 (b) hours.

Dark field examinations have shown that the δ' precipitates within the slip bands are sheared by dislocations. We observed this shearing effect when the δ' precipitates were relatively large (e.g. in the Al-Li or Al-Li-Zr alloys aged at 200°C for 24 hours), Fig.5a; after having been sheared, the precipitates changed their shapes from spherical into elongated and their effective diameter decreased. TEM observations have shown that small δ' precipitates may be dissolved as a result of being repeatedly crossed by dislocations. This is well illustrated in Fig. 5b. When a single dislocation of the matrix enters the ordered lattice of a precipitate, an antiphase boundary forms in it [13]. In small precipitates, the energy of this boundary elevates the energy of the whole particle highly enough to make it unstable and to cause it to dissolve. This effect was observed in the binary Al-Li alloy [14].

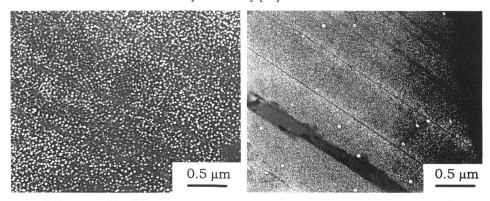

Fig.5. Behaviour of the δ' precipitates within the slip bands:(a) Al-Li alloy aged at 200°C for 24 hours and (b) Al-Li-Zr alloy aged at 200°C for 2 hours.

In the Al-Li alloys containing relatively large precipitates, plastic deformation may also be localized in the precipitate-free zones. In the present experiment, such zones were observed in the samples aged at 200°C for 24 hours. They formed along certain grain boundaries of large misorientation angle.

In the Al-Li alloys, the formation of slip bands is associated with the ordered structure of δ' precipitates and the nature of their interaction with dislocations [13]. The precipitates have an ordered $L1_2$ structure, are coherent with the matrix and the mismatch is negligibly small. This ordering is the major mechanism of strengthening, since the energy of the antiphase boundary, which forms when a matrix dislocation enters the precipitate lattice, is proportional to the strengthening (and so is the particle size). Thanks to this, the subsequent dislocations entering the precipitate lattice can move in the same slip planes (in order to restore the ordering) more easily. Moreover, the effective diameter of a sheared precipitate is reduced, which additionally weakens locally the matrix. As a result, plastic deformation within the bands needs lower stresses to be induced and, thus, tends to localize there.

The plastic strain localization within slip bands is an important effect affecting the fatigue behaviour of the Al-Li alloys, in particular their fatigue life. This was supported by mechanical examinations of the Al-Li-Zr alloy which have shown that: (i) as the mean size of the δ' precipitates increases from 2 to 20nm, the number of cycles to failure decreases by an order of magnitude, and (ii) the cyclic hardening more pronounced, the smaller the size of δ' precipitates. This can be clearly seen in Fig.6, which shows the cyclic hardening curves determined for some of

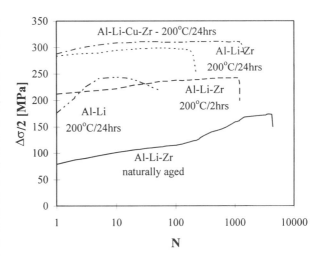

Fig. 6. Cyclic hardening curves for the investigated Al-Li alloys.

The investigated Al-Li alloys. The above observations suggest that with increasing size of the δ' precipitates, the role played by plastic strain localization within the bands increases. When the δ' precipitates are large enough (20nm), the local softening of the material due to the strain localization may predominate over the hardening due to the increased dislocation density and in the cyclic hardening curve we observe softening (see the hardening curves for the Al-Li and Al-Li-Zr alloys aged at 200°C for 24 hours).

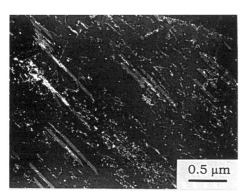

Fig. 7. Dislocation loops formed during low-cycle fatigue in the Al-Li-Cu-Zr alloy.

Another adverse factor that affects the plastic strain localization and the fatigue resistance of the Al-Li alloys is their grain size. The number of cycles to failure is 4 times smaller

383

in the Al-Li alloy (with a grain size of 48um)than it is in the Al-Li-Zr alloy (non-recrystallized) at the same size of δ' precipitates. This is so because of the different slip lengths. Large slip lengths can encourage strain localization and the associated stress concentration at the end of the deformation process [9].

One of the attempts to homogenise slip is an addition of partially coherent or incoherent precipitates, as is the case with the Al-Li-Cu-Zr alloy aged at 200°C for 24 hours in the present study. This alloy contains T_1 precipitates in addition to δ'. In this alloy no slip bands were found but only numerous dislocation loops (fig.7). In consequence it shows the fatigue life 2-3 times greater than that of the alloy containing δ' precipitates of the same average size alone.

CONCLUSIONS

The cyclic plastic deformation in the investigated model Al-Li alloys occurs in an inhomogeneous way. The heterogeneities observed by TEM look like microbands in many materials subjected to large plastic deformation. They run through several grains and form characteristic steps on the intersecting grain boundaries. Their formation results from shearing of the δ' precipitates and they are regions of localized shear.

An increase of the δ' precipitate size and the grain size leads to enhance the strain localization which has a detrimental effect on the fatigue life of the alloy.

T_1 precipitates present in the AL-Li-Cu-Zr alloy cause the slip distribution to become more homogeneous. No slip bands (microbands) were observed in this alloy which is advantageus since improves the fatigue life.

ACKNOWLEDGEMENT
This work was supported by the Polish Committee of Scientific Research (grant No 7 T08A 048 10). The FIAT Corp. is also gratefully acknowledged for financial support for M.L.

REFERENCES

[1]   D.A.Hughes (1993): Acta metall. 41, 1421
[2]   J.J.Gracio (1994): Mat.Sci.Eng. A174, 111
[3]   C.Laird, Z.Wang, B.-T.Ma, H.-F.Chai (1989): Mat.Sci.Eng. A113, 245
[4]   B.Bay, N.Hansen, D.Kuhlmann-Wisdorf (1992): Mat.Sci.Eng. A158, 139
[5]   A.Korbel, J.D.Embury, M.Hatherly, P.L.Martin, H.W.Erbsloh (1986): Acta metall. 34, 1999
[6]   M.Gasperini, C.Pinna, W.Swiatnicki (1996): Acta metall. 44, 4195
[7]   C.Laird, P.Charsley, H.Mughrabi (1986): Mater.Sci.Eng. 81, 433
[8]   J.H.Driver, P.Rieux (1984): Mat.Sci.Eng. 68, 3543
[9]   C.P.Blankenship, E.A.Starke (1991): Proc. of 6th Int. Conf. Al-Li alloys, (ed. M. Peters, P.-J.Winkler), Ga-Pa Germany 1991, 187
[10] P.L.Malkin, B.Ralph (1984): J.Mater.Sci. 19, 3835
[11] D.Khhireddine, R,Rahouadj, M.Clavel (1989): Acta metall. 37, 191
[12] T.S.Srivatsan, E.J.Coyne (1989): Mater.Sci.Tech. 5, 548
[13] J.Lendvai, W.Wuderlich, H.J.Gudladt (1993): Phil.Mag. 67A, 99
[14] Y.Brechet, F.Louchet, C.Marchionni, J.-L.Verger-Gaugry (1987): Phil.Mag. 38, 353

Chapter 5

INFLUENCE OF ENVIRONMENTAL CONDITIONS AND SURFACE TREATMENTS

Low Cycle Fatigue and Elasto-Plastic Behaviour of Materials
K-T. Rie and P.D. Portella (Editors)
387

HYDROGEN EFFECTS
ON THE SATURATION STRESS AND THE ABILITY TO FRACTURE
OF NICKEL SINGLE CRYSTALS FATIGUED UNDER CATHODIC CHARGING

N. RENAUDOT, K. WOLSKI and T. MAGNIN
Ecole des Mines, Centre SMS, URA CNRS 1884
158, Cours Fauriel 42023 Saint Etienne cedex 2, France

ABSTRACT

Fatigue tests were done on monocrystalline Ni oriented for single slip to form a well known structure, i.e. Persistent Slip Bands (PSB). This structure enables the localisation in the same area of (i) an important plasticity and (ii) a hydrogen supply by cathodic charging (hydrogen enter the material via the intrusions formed where PSB emerge at the specimen surface). In these conditions, a macroscopic decrease of the saturation shear stress was observed after few weeks of cathodic charging. Fracture surface observations on cathodically charged specimen showed a cleavage-like fracture along the PSB, where hydrogen-dislocation interactions take place ; i.e. it shows the enhanced ability to microscopic fracture of nickel with hydrogen on $(1\overline{1}1)$ slip planes. Finally, the macroscopic softening was explained in terms of hydrogen-dislocation interactions and a quantitative approach of the softening was developed in the case of a PSB structure.

KEYWORDS

Fatigue, corrosion fatigue, persistent slip bands, hydrogen, stress corrosion cracking

INTRODUCTION

Among the mechanisms of Stress Corrosion Cracking (SCC), one can consider two different types : (i) « the Hydrogen Enhanced Localised Plasticity (HELP) model [1,2] », (ii) « the Hydrogen Induced Decohesion (HID) model » [3]. The aim of this research was to obtain direct evidences of these microscopic features.

Hydrogen-dislocation interactions ahead of a crack tip in SCC are supposed to induce a local softening, i.e. a local decrease of the shear stress [2]. The main goal of this study was to reproduce these interactions at a macroscopic level. In order to have a noticeable effect of hydrogen on the macroscopic shear stress, fatigue tests were performed on monocrystalline Ni oriented for single slip. Moreover, the applied plastic strain was chosen in the stress plateau of the cyclic stress-strain

388

curve of Ni [4], for which a well known dislocation structure (i.e. Persistent Slip Bands (PSB)) is formed. With this structure, the whole plastic deformation is localised in the PSB. Finally, the formation of extrusions-intrusions on the material surface where PSB emerge favours a local hydrogen absorption in the PSB via the intrusions. Therefore, plastic strain and hydrogen absorption are localised in the same area, the PSB. This is supposed to induce significant hydrogen-dislocation interactions with measurable variation of the macroscopic shear stress. In these conditions, the evolution of the saturation stress was analysed before, during and after the cathodic charging and the fracture surfaces, with or without previous charging, were observed.

EXPERIMENTAL PROCEDURE

The fatigue tests were done on Ni single crystals with a [153] tensile axis, for which there is only one activated slip system (Schmid factor : 0.49). The square section smooth specimens (4 x 4 mm) are tested in H_2SO_4 (0.5 N), under a cathodic charging to promote hydrogen reduction, at an applied plastic strain of $\pm 5^.10^{-4}$, to form about 5% of PSB in the material, and at a total strain rate of 10^{-3} s^{-1}.

Figure 1 shows the successive steps of the two representative tests presented in this paper. For both specimens, the first step consist in the material cyclic hardening until saturation (in about 6000 cycles), corresponding to the formation of PSB in the bulk and beginning of the extrusion-intrusion growth on the specimen surfaces. Then cycling is conducted in a discontinuous way, i.e. 15 cycles every 12 hours (at $\dot{\varepsilon}_t = 10^{-3}$ s^{-1}), to see the evolution of the saturation stress over few weeks. For the specimen n°1, discontinuous cycling is first conducted in air over 125 hours, then in H_2SO_4 at a cathodic current (between -200 mA/cm² and -700 mA/cm²) over 525 hours. For the specimen n°2, the discontinuous cycling is directly conducted in H_2SO_4 at a cathodic potential over 325 hours (- 1000 mV/SCE, i.e. a current of -150 mA/cm² at the beginning of the cathodic charging, slowly decreasing to -700 mA/cm² at the end), then in air over 250 hours. Finally, continuous cycling is used on both specimens, at I_c = -700 mA/cm² for the specimen n°1, and in air for the specimen n°2, until the final fracture.

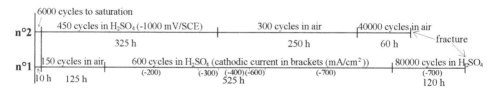

Fig.1. Description of the experiments on specimens n°1 and n°2

RESULTS

The evolution of the saturation stress during the cathodic charging for both specimens is shown in fig. 2. In both cases, a significant softening starts after a cathodic charging of one week and reaches respectively 1.5% of the saturation stress in 2 weeks (specimen n°2) and 3% in 3 weeks (specimen n°1). This softening is not present at the beginning of the cathodic charging, either because of a too small cathodic current (-150 mA/cm² or -200 mA/cm²) or probably because of a very small hydrogen absorption and diffusion rates. The very slow hydrogen diffusion rate in Ni ($6^.10^{-10}$ cm²/s at room temperature [5]) is the reason why discontinuous cycling was used. It allowed a significant time for hydrogen absorption and diffusion in the specimens. The 525 hours of cathodic charging for specimen n°1 corresponds to a diffusion over 80% of the whole section $(4 (D_{Ht})^{0.5} = 1.3$ mm$)$.

No softening was noticed when discontinuous cycling was done in air (fig.1). The stability of the saturation stress in air indicates that the softening obtained during the cathodic charging can only be related to hydrogen absorption and diffusion in the specimens. Table 1 shows a hydrogen content (measured by thermo desorption technique) significantly higher for specimens n°1 and n°2 than for the 3 others cycled without cathodic charging. These results clearly show that, in the cathodically charged specimens, hydrogen has significantly diffused and is responsible for the macroscopic softening observed. The observed softening is attributed to hydrogen-dislocation interactions (see next paragraph).

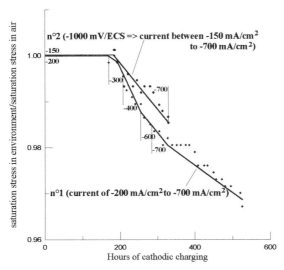

Fig. 2. Evolution of the saturation stress (normalised by the saturation stress in air) as a function of the cathodic charging time.

Table 1

Specimen	cathodic charging time (in hours)	Global hydrogen content measured (in atomic ppm)	estimated local hydrogen content in PSB walls * (in atomic %)
n°1	525	215 ± 5	7.2
n°2	325	160 ± 5	5.3
n°3	0	35 ± 5	-
n°4	0	30 ± 5	-
n°5	0	40 ± 5	-

* see next paragraph.

The second very interesting result is the fracture surface obtained for the Ni monocrystal. As it can be observed on the fig. 3.a and 3.b for specimen n°1 and specimen n° 3 respectively (specimen n°3 is a non charged specimen), both fracture surfaces are brittle. However, a closer observation shows that specimen n°1 have a very important cleavage-like fracture on the $(1\overline{1}1)$ slip planes (fig.3.c and 3.e), along the PSB with steps between the different PSB (see fig. 3.f for crystallography), whereas specimen n°3 fracture surface has no such structures (see fig. 3.d) despite an approximate macroscopic fracture along a $(1\overline{1}1)$ plane. The fracture surface of specimen n°1 shows that hydrogen has favoured the fracture around the PSB, either by a local decrease of the cohesive

390

energy, or by a more intense plasticity in the PSB.

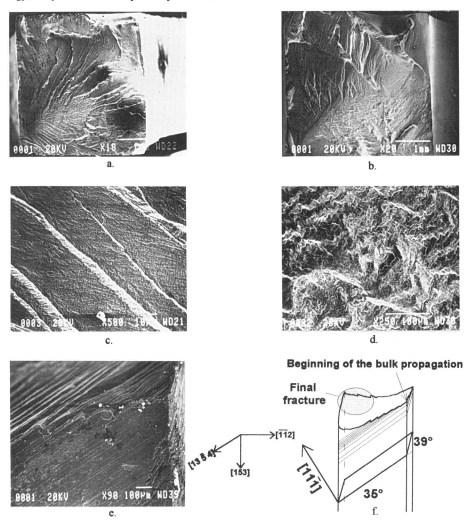

Fig. 3. a. specimen n°1 fracture surface ; b. non charged specimen fracture surface ; c. detail of a. ; d. detail of b. ; e. beginning of the bulk propagation for specimen n°1 (side view) ; f. Schematic representation of the fracture for specimen n°1.

DISCUSSION OF THE SOFTENING EFFECT

First of all, it is important to know if the observed softening is due to a decrease of the internal shear stress or to the thermal shear stress. This is why, during the discontinuous cycling, the saturation stress was measured for two different strain rates : $\dot{\varepsilon}_t = 10^{-3}$ s^{-1} (strain rate of the continuous cycling) and $\dot{\varepsilon}_t = 5\,10^{-6}$ s^{-1} (the lowest strain rate for which we can assume that the thermal stress is equal to zero). The stress difference between these two strain rates is considered to be the thermal stress at $\dot{\varepsilon}_t = 10^{-3}$ s^{-1}. This difference is about 2.5% of the total stress and keeps constant with hydrogen charging. It can be concluded that the softening is due to the internal stress decrease.

The hydrogen effect on the internal stress can be explained by interactions between hydrogen and dislocations. It has been shown that (i) hydrogen segregates on edge dislocations in the tensile zone (in an asymmetric way) and (ii) hydrogen decreases hydrostatic stress fields of edge dislocations causing a decrease of the long distance interactions between dislocations, and consequently a decrease of the internal stress [6].

The link between global hydrogen content and modification of the shear stress will be established in three points :

1 - Estimation of the local hydrogen concentration on edge dislocation dipoles in the PSB walls

2 - Applying the Sofronis approach [7], evaluation of the modification by hydrogen of the edge dislocation energy per unit length

3 - Assuming that the edge dislocation energy per unit length is proportional to the internal shear stress in PSB walls, estimation of the macroscopic softening.

Knowing that the volume of the walls constitutes about 10 % of PSB [8, 9] which occupies about 3% of the whole volume, and assuming that the whole detected hydrogen comes exclusively from the edge dipoles in the PSB walls where hydrogen is supposed to be trapped, one can easily estimate the local hydrogen content in the walls, which is : $(0.0215\%)/(0.03 \times 0.1) = 7.2\%$ for specimen n°1 and $(0.0160\%)/(0.03 \times 0.1) = 5.3\%$ for specimen n°2 (see table 1). This hypothesis certainly induces an overestimation of the local hydrogen content in the PSB walls for two reasons : first, measured hydrogen may also come from the veins in the matrix, and second, during testing, hydrogen is probably not confined in the walls and may be present in the channels.

Sofronis [7] links up the hydrogen content around edge dislocations and their energy per unit length, by a modification of the Poisson ratio ν. The Poisson ratio modified by hydrogen ν_H is :

$$\nu_H = (\nu - \xi\lambda^2 E)/(1 + \xi\lambda^2 E) \qquad (1)$$

$\xi = C_0\Omega/9kT$ with C_0, the atomic fraction of H in Ni $(= n_H/n_{Ni})$
Ω, the atomic volume of Ni
k, the Boltzmann's constant
and T, the temperature.

$\lambda = V^*/N_a\Omega$ with V^*, the atomic volume variation due to hydrogen, i.e. 2 cm³/mol
and N_a, the Avogadro's number.

With T = 298 K, $V_{Ni} = N_a\Omega = 6.596 \ 10^{-6}$ m³/mol, E = 219.2 GPa, $\nu_{Ni} = 0.306$, R = kN_a = 8.3144 J·mol⁻¹·K⁻¹, it leads, for nickel, to a decrease of ν from $\nu = 0.306$ without hydrogen to $\nu_H = -0.0862$ for specimen n°1 and to $\nu_H = -0.00755$ for specimen n°2.

Knowing the energy per unit length is $\mu b^2/2\pi(1 - \nu)$ for edge dislocations and with the assumption that the internal stress in the walls is proportional to the energy per unit length of edge dislocation dipoles in the PSB walls, one can deduce the internal stress decrease in the walls for specimen n°1 and n°2 :

specimen n°1 : 1 - [(1 - (-0.0862)) / (1 - 0.306)] = 36%
specimen n°2 : 1 - [(1 - (-0.00755)) / (1 - 0.306)] = 31%

392

Finally, the wall contribution in the whole PSB internal stress must be estimated. It is assumed [10] that the internal stress in the PSB can be expressed as for a composite :

$$\tau_i = \tau_{i,w}\, \mathbf{f_w} + \tau_{i,c}\, \mathbf{f_c} \qquad\qquad (2)$$

where $\tau_{i,w}$ is the internal stress in PSB walls, $\tau_{i,c}$, the internal stress in the PSB channels and f_w and f_c, the volume fraction of walls and channels respectively.

Assuming the $\tau_{i,w}/\tau_{i,c}$ ratio is about $(d_c/d_w)^{0.5} = (1.2/0.15)^{0.5} \approx 3$ (where d_c is the width of channels and d_w that of walls) [9], the « internal stress with hydrogen/internal stress without hydrogen », or directly the « saturation stress with hydrogen/saturation stress without hydrogen » ratio can be expressed by :

$$\boldsymbol{\sigma}_{sat,H}/\boldsymbol{\sigma}_{sat} = \tfrac{3}{4} + \tfrac{1}{4}\,(\tau_{i,w,H}/\tau_{i,w}) \qquad\qquad (3)$$

where ¾ concerns the contribution of channels which is assumed to be constant. In these conditions, the softening is overestimated ($\sigma_{sat,H}/\sigma_{sat} = 0.91$ as compared to the experimental value of 0.97 for specimen n°1 and 0.923 as compared to 0.985 for specimen n°2). This difference is attributed to the overestimation of the local hydrogen content in the PSB walls.

CONCLUSION

The present study performed on Ni single crystals points out for the first time that hydrogen pick-up in a FCC metal can induce a significant decrease of the saturation stress in corrosion fatigue under cathodic charging (3% decrease of saturation stress as compared to air). This macroscopic effect indicates the possibility of a microscopic softening ahead of a stress corrosion crack tip, one of the critical steps in the Corrosion Enhanced Plasticity Model. Secondly, the cleavage-like fracture observed along the PSB, i.e. where hydrogen-dislocation interactions take place, shows the enhanced ability to microscopic fracture of nickel with hydrogen on $(1\bar{1}1)$ slip planes, either by a local decrease of the cohesive energy, or by a more intense plasticity in the PSB. Finally, the quantitative approach based on hydrogen-dislocation interactions in PSB has been proposed to explain the softening. However, a better estimation of the local hydrogen content in the PSB remains necessary in order to achieve reliable quantification.

[1]   Birnbaum H. K. and Sofronis P. (1994) *Material Science and Engineering* **A176** 191
[2]   Magnin T. (1995) *Advances in Corrosion-Deformation interactions*, Materials Science Forum, Vol 202, Trans Tech Publications
[3]   Troiano A. R., *Trans. ASM 52* (1960) 54
[4]   Magnin. T. (1991) *Mémoires et Etudes Scientifiques, Revue de Métallurgie*, **88** (1) 33
[5]   Brass A. M. and Chanfreau A. (1996) *Acta Mater.*, **44** (9) 3823
[6]   Chateau J.-P., Delafosse D. and Magnin T. (1998), *Annales de Chimie*, to be published
[7]   Sofronis P. (1995) *J. Mech. Phys. solids*, **43** 1385
[8]   Esmann U., Gösele U. and Mughrabi H. (1981) *Phil. Mag.*, A 44 (2) 405
[9]   Tippelt B., Bretschneider J. and Hähner P. (1997) *Phys. Stat. Sol. (a)* **163** (1)
[10]   Esmann U. and Differt K. (1996) *Materials Science and Engineering* **A208** 56

Low Cycle Fatigue and Elasto-Plastic Behaviour of Materials
K-T. Rie and P.D. Portella (Editors)

EVALUATION OF STRESS CORROSION RESISTANCE AND CORROSION FATIGUE FRACTURE BEHAVIOR OF ULTRA-HIGH-STRENGTH P/M AL-ZN-MG ALLOY

K. MINOSHIMA*, M. OKADA† and K. KOMAI*

* Department of Mechanical Engineering, Graduate School of Engineering, Kyoto University,
Kyoto 606-8501, Japan
** JR Tokai Co., Ltd., Tokyo 103-8288, Japan

ABSTRACT

Quasi-static tensile tests in air and slow strain rate tests in a 3.5% NaCl solution were conducted in a ultra-high-strength P/M Al-Zn-Mg alloy fabricated through powder metallurgy. Attention is also paid to fatigue strength and fatigue crack growth behavior in air and in a 3.5% NaCl solution. The alloy has extremely high strength over 800 MPa. However, elongation at break remains small, at about 1.3%. The SSRT strength in a 3.5% NaCl solution decreases slightly at a very low strain rate, that is smaller than those observed in aluminum alloys sensitive to stress corrosion. This means that the crack initiation resistance to stress corrosion is superior. However, the corrosion fatigue strength becomes lower than that conducted in air, because pitting corrosion on a sample surface acts as a stress concentrator. Fatigue crack growth resistance of the alloy is inferior to conventional Al-Zn-Mg alloys fabricated by ingot metallurgy, because the fatigue fracture toughness, or ductility, of the alloy is inferior, and intergranular cracking promotes crack growth. However, no influence of 3.5% NaCl solution on corrosion fatigue crack growth is observed.

KEYWORDS

Corrosion fatigue, stress corrosion cracking, fatigue strength, fatigue crack propagation, fractography, powder metallurgy, Al-Zn-Mg alloy

INTRODUCTION

Due to further progress in science and technology, new high-strength materials have become especially important; among these are metallic materials manufactured by powder metallurgy (P/M). Recently, remarkable improvement in mechanical properties of P/M Al-Zn-Mg-Cu alloys has been achieved [1] by controlling the morphology and distribution of metastable precipitates acting as pinning centers for preventing dislocation motion. However, little is known about its failure mechanism. Besides, in order to facilitate the application of these materials to machines and structures, the mechanical properties, including fatigue in various environments, have to be investigated [2]. In this investigation, quasi-static tensile tests, slow strain rate tests (SSRTs) in a 3.5% NaCl solution, and fatigue strength and fatigue crack growth behavior in air and in a 3.5% NaCl solution were investigated in a ultra-high-strength P/M Al-Zn-Mg-Cu-Mn-Ag alloy, and the fracture mechanisms are discussed.

EXPERIMENTAL PROCEDURES

The material tested was a high-strength Al-9Zn-3Mg-1.5Cu-4Mn-0.5Zr-0.04Ag alloy, fabricated by powder metallurgy [1]. The powders prepared by air atomizing were pressed into a rodlike shape by a cold isostatic pressing, and extruded to an extrusion ratio of 20. The material was then machined to a specimen and heat-treated to a T6 condition. Specimens used were smooth round specimens machined in a LL orientation, with dimensions of 8 mm in diameter and 40 mm in gage length. To investigate crack growth behavior, compact tension type (CT) specimens were machined in T-L and L-T orientations, with dimensions of 7.5 mm in thickness and 28 mm in width. Four lots were used to prepare the specimens. Round smooth specimens were machined from a rod (Lots 1, 2 and 3) in 40 mm in diameter, whereas CT specimens from a square rod (Lot 4) of 40 mm on a side.

Tensile tests were conducted in air at a displacement rate of 2mm/min by using a computer controlled tensile testing machine. Slow strain rate tests were conducted at a displacement rate of 0.05 and 0.005 mm/min in a 3.5% NaCl solution at 25°C. Axial fatigue tests at a stress ratio R of 0.1 were conducted on a smooth specimen in air and in a 3.5% NaCl solution at 25°C: sinusoidal waveform at a frequency f of 1 - 10 Hz was used in air, whereas in a 3.5% NaCl solution, a triangular wave at f = 0.1 Hz was selected.

Fatigue crack growth rates in air were evaluated at R of 0.1 for T-L and L-T specimens, using sinusoidal waveform at f = 1- 5 Hz. For L-T specimens, the influence of 3.5% NaCl solution was studied, using a triangular waveform at f = 0.1 Hz. The testing machine employed was a computer controlled electro-hydraulic fatigue testing machine. Crack opening stress intensity was measured by an elastic unloading compliance technique with a strain gage adhered to the back surface of a CT specimen. Specimen and fracture surfaces were examined with a high-resolution, field-emission type scanning electron microscope (S-4500 by Hitachi, Ltd.)

EXPERIMENTAL RESULTS AND DISCUSSIONS

Quasi-Static Tensile Fracture Behavior in Air

The mechanical properties of the alloy obtained in air were: tensile strength 808 MPa, elastic modulus 80.5 GPa and elongation at break 1.31%. The tensile strength was extremely high compared with ingot-metallurgy (I/M) 7XXX series Al alloys (σ_B = 570 - 595 MPa: 7075-T6, 7175-T66) [3]. However, the elongation at break was smaller than I/M 7XXX series Al alloys (about 11% for 7075-T6 and 7175-T6) [3]. The crack was initiated at an inside defect, a void or an inclusion. The macroscopic fracture surface was normal to the loading direction, and little shear lip could be seen. However, the microscopic fracture surface was dominated by ductile fracture characterized by dimples (Fig. 1). The strengthening mechanisms of the alloy [1] are 1) small grain, less than 1 μm, 2) metastable η' phase in a matrix, and 3) intermetallic compound ($Al_{20}(Cu,Zn)_2Mn_3$) that acts as a fiber reinforcement. The size of dimples was about 0.5 - 1 μm, and this suggested the strengthening mechanism due to small grain was well achieved. The other mechanism that could be obtained from the fracture surface is intermetallic compound, which is shown by arrows in Fig. 1. The intermetallic is about few micrometers in length and the width was about few hundreds nanometers. The longitudinal direction of the intermetallic was parallel to the extruding direction, and they contributed to fiber reinforcement for the present P/M Al alloy.

Slow-Strain Rate Tests

Figure 2 illustrates the SSRT strength in a 3.5% NaCl solution as a function of displacement rate. The fracture strengths in air are also plotted in the figure. In a 3.5% NaCl solution, a crack was initiated at the sample surface. Figure 3 shows SEM photographs of a crack initiation site: a crack was initiated at a pitting corrosion site (Fig. 3(a)), which will be refereed to as "pitting corrosion". In some cases, corrosion severely occurred, and the size of the corrosion defect became larger than pitting corrosion: this severe corrosion is presented in Fig. 3(b). In Fig. 2, symbols with a superscript, "*", indicate the crack initiation site was such a large corrosion defect. The SSRT strength at 0.05 mm/min ($1.5 - 2.0 \times 10^{-6}$ 1/s) was the same as that in air, when the crack initiation site was pitting corrosion. However, at 0.005 mm/min ($1.5 - 2.0 \times 10^{-7}$ 1/s), the

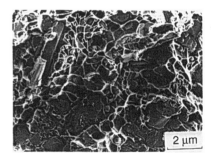

Fig. 1 Fracture surface of a quasi-static tensile test, imaged with SEM. Arrow shows examples of an intermetallic compound that contributes to fiber reinforcement.

395

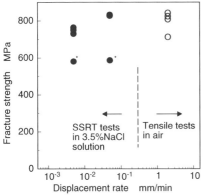

Fig. 2 Fracture strength as a function of strain rate in a 3.5% NaCl solution and in air.

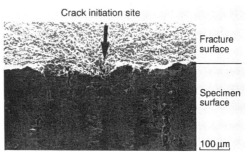

(a) Pitting corrosion type (Displacement rate: 0.005mm/min, $\sigma_B = 732$MPa)

(b) Large corrosion defect type (Displacement rate: 0.05mm/min, $\sigma_B = 587$MPa)
Fig. 3 Crack initiation site of an SSRT in a 3.5% NaCl solution.

strength decreased from air data. When a material is sensitive to stress corrosion cracking (SCC), SSRT strength decreased with decreasing a loading rate, and a minimum strength exists at a certain loading rate. Such a loading rate was reported to be 10^{-5} to 10^{-6} 1/s [4]: this is higher than the loading rate of 0.005 mm/min ($1.5 - 2 \times 10^{-7}$ 1/s) where the SSRT strength of the alloy decreased from air data.

SSRTs evaluate the sensitivity of SCC, in particular, the strength of crack initiation and passive films formed on a surface. When the crack initiation site was pitting corrosion, the strength slightly decreased at a lower loading rate of 0.005 mm/min. This indicates that the alloy had a superior SCC strength to I/M Al alloys. When an Al alloy has very small grains such as this P/M alloy, large precipitates in a grain do not exist, and therefore, less damage of passive films is achieved, which in turn cause an increase in SCC strength. When the crack initiation site was a large corrosion defect, the SSRT strength decreased from that conducted in air at both 0.05 and 0.005 mm/min, because a large defect increased the stress concentration, thereby decreasing the SSRT strength.

Fatigue Strength of Smooth Round Specimens

Figure 4 shows S-N curves in air and in a 3.5% NaCl solution. The symbols with "F" indicate that a foreign phase could be seen in the fracture surface, which will be discussed later in detail. The fracture surface was normal to the axial loading direction, with little shear lip on the periphery of the sample. A fatigue crack in air initiated at an inside inclusion near specimen surface. Electron probe microanalysis showed that inclusions could be classified into three types: oxide inclusion, inclusion with foreign metals (Fe, Zn), and silica (SiO_2). Inclusions with foreign metals or silica are considered to be brought about during manufacturing. Oxide inclusions are the most important and inherent problems for P/M metallurgy. These results show that an amount of inclusions in the material should be decreased to improve tensile and fatigue strengths of the present alloy.

A corrosion fatigue crack was initiated at either an inclusion or a corrosion pit. A superscript "*" is added to the symbol, when the initiation site was an inclusion. When the corrosion fatigue life was extremely short at 300MPa ($N_f = 5.83 \times 10^2$), the crack was initiated at an extremely large corrosion induced defect of 500 to 600 μm. This large corrosion defect caused a large decrease in corrosion fatigue life from other specimens. Except this, the fatigue strength in a 3.5% NaCl solution was even smaller than that in air.

As is discussed later, little influence of the corrosion environment on fatigue crack growth in the alloy was observed. Therefore, a decrease in corrosion fatigue strength was due to promotion of crack initiation by pitting corrosion. In some cases, a large corrosion defect was formed on a specimen surface, shown in Fig. 3(b), resulting in a further decrease in SSRT and corrosion fatigue strength. A reason why such large corrosion defects occurred may be due to some segregation of constituents: further investigation is required.

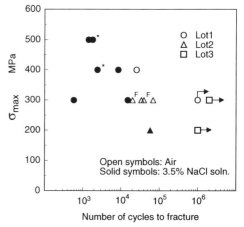

Fig. 4   S-N curves under uniaxial tension-tension fatigue loading.

Fatigue Crack Growth in Air

Figure 5 shows a macroscopic fatigue fracture surface of a CT specimen conducted in air: three different types of fracture morphology could be seen. These will be refereed to as Type 1, Type 2, and Type 3, respectively. The microscopic characteristic features of each fracture surface are summarized as follows: Type 1 fracture surface looked dark, and the fracture surface was completely different from the other types. In the case of L-T specimens, a crack tended to be retarded when the crack reached Type 1 fracture surface. Type 2 fracture surface involves some typical fracture surface morphology which looked as if the mating surface had contacted each other. As for Type 3 fracture surface at higher stress range, characteristic fracture morphology of this P/M alloy, small grains and intermetallic compound acting as a reinforcement, could be seen. Type 1 and Type 2 fracture surfaces were in particular observed in Lot 4, that was machined into CT specimens. A ratio of each fracture surface morphology depended on a specimen, and therefore, crack growth rate varied in sample by sample. Hence, in this investigation, crack growth rates obtained by a CT specimen, of which a Type 3 fracture surface dominated over the fracture surface, are plotted.

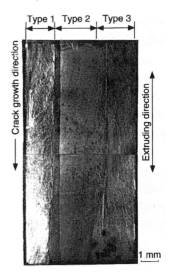

Fig. 5   Macroscopic fatigued fracture surface of T-L orientation (CT specimen).

Figure 6 illustrates the fatigue crack growth rate of T-L specimens in air as a function of ΔK. The crack growth rates of an I/M Al-Zn-Mg alloy, 7075-T6 and a P/M Al-Zn-Mg alloy, 7090-T6 [5] are also plotted for comparison. At lower stress intensity, there was little difference in fatigue crack growth rate. At an intermediate ΔK, crack growth rate of T-L2 was accelerated. At higher stress intensity, acceleration of crack growth was observed in both T-L1 and T-L2 specimens. We must note that crack growth rate at higher stress intensity of the present alloy was the highest, and decreased in the order of 7090-T6 and 7075-T6.

The crack growth rates as a function of effective stress intensity factor, ΔK_{eff}, are shown in Fig. 7. In terms of ΔK_{eff}, no difference in crack growth rate between the T-L1 and T-L2 specimens was attained: the small difference in the ratio of foreign phase could be compensated, using effective stress intensity. Secondly, the crack growth rate in the present P/M Al alloy was higher than 7090-T6 P/M alloy. One reason is that the alloy was very brittle, and had a low fracture toughness. At lower stress intensity, crack grew with transgranular failure, that looked brittle. With an increase in stress intensity, an amount of intergranular cracking increased. At higher stress intensity, crack grew with dimples: an increase in fatigue crack growth rate may be explained by presence of intergranular cracking as well as dimples due to low fracture toughness.

Figure 8 illustrates the influence of crack plane orientation on fatigue crack growth rate of the present Al alloy, as a function of ΔK_{eff}. At lower stress intensity factor, there was little difference in crack growth rate between T-L and L-T crack plane orientation. However, with an increase in stress intensity factor, da/dN of T-L orientation was accelerated and was faster than that of L-T orientation, and the fatigue fracture toughness of L-T orientation was higher than T-L orientation. A most noticeable difference of fracture surface between T-L and L-T specimens is that a pull-out or fracture of intermetallic compound existed in L-T specimens. As was discussed before, intermetallic compound acted as a fiber reinforcement, and a reason why the fatigue crack growth rate in an L-T specimen was smaller than that of a T-L specimen can be partly explained by prevention of crack growth by intermetallic compound aligned in the longitudinal direction.

Corrosion Fatigue Crack Growth Behavior

Just after a fatigue pre-cracked specimen was exposed to a 3.5% NaCl solution, a crack grew for a while. However, with further stress cycles of about 10^4 (30 hours at 0.1 Hz), the crack growth rate became smaller, and finally it was retarded. During this unsteady process, a crack opening stress intensity increased with stress cycles. The fracture surface where the crack was retarded was covered with thick corrosion products. These mean that a unsteady crack growth, or crack retardation was due to corrosion-product induced crack closure [6].

Because of this crack retardation, an applied load was increased step by step, and finally the crack grew in a steady manner above ΔK_{eff} of 3 MPa•m$^{1/2}$. Figure 9 shows the corrosion fatigue crack growth rate under steady state as a function of ΔK_{eff}. The figure also illustrates the crack growth rates obtained before a decrease in growth rate and its consecutive crack retardation, which are plotted with a superscript "*". The fatigue crack growth rate in a 3.5% NaCl solution was the same as that in air: the present P/M Al alloy was insensitive to NaCl solution as far as the present experiments are concerned. Under a steady state, the stress intensity was relatively large, and dimples could be seen and the fracture surface was similar to that conducted in air. This is consistent with that no influence of 3.5% NaCl solution on fatigue crack growth was observed. Of course, further investigation into corrosion fatigue crack growth behavior at lower crack growth rate is required.

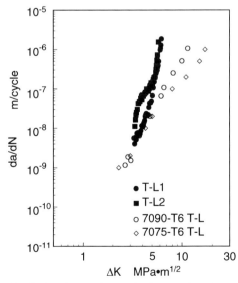

Fig. 6 Crack growth rate, da/dN, as a function of ΔK for T-L crack plane orientation.

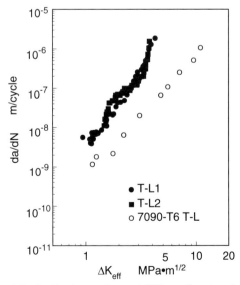

Fig. 7 Crack growth rate, da/dN, as a function of ΔK_{eff} for T-L crack plane orientation.

398

CONCLUSIONS

1. The developed P/M Al alloy has extremely high-tensile strength over 800 MPa. However, elongation at break was about 1 %, and a specimen fractured in a brittle manner.
2. SSRT strength decreases at lower displacement rate. However, the rate is smaller than that of other I/M Al alloys sensitive to SCC. This means that the alloy has superior resistance to initiation of SCC.
3. The fatigue crack growth rate of the alloy is higher than those of I/M Al-Zn-Mg alloys. A reason is that the alloy has brittle nature. Another reason is that the fatigue fracture is associated with some amounts of intergranular cracking.
4. The fatigue crack growth rate in L-T orientation is lower than that in T-L orientation, because of an intermetallic compound $(Al_{20}(Cu,Zn)_2Mn_3)$ aligned in the longitudinal direction.
5. Quasi-static tensile and fatigue crack is initiated at an inclusion: decreasing inclusions in the material is important for further improvement of mechanical properties including fatigue strength.
6. The influence of 3.5% NaCl solution on fatigue crack growth is negligible. However, in NaCl solution, a crack is initiated at pitting corrosion on the surface, resulting in lower corrosion fatigue strength than that conducted in air.

REFERENCES

1. Osamura, K., Kubota, O., Promstit, P. Okuda, H., Ochiai, S., Fujii, K., Kusui, J., Yokote, T. and Kubo, K. (1995) *Metallurg. Trans.* **26A**, 1597.
2. Komai, K. and Minoshima, K. (1989) *JSME International Journal*, Series I, **32**, 1.
3. *Metals Handbook, Vol.2: Properties and Selection: Nonferrous Alloys and Pure Metals*, ASM, Metals Park, Ohio, USA, 1979, p.62
4. e.g., Ugiansky, G.M. and Payer, J.H. (Eds.) (1979) *'Stress Corrosion Cracking–The Slow Strain-Rate Technique'*, *ASTM STP 665*.
5. Minakawa, K., Levan, G. and McEvily, A.J. (1986) *Metallurg. Trans.* **17A**, 1787.
6. Endo, K. Komai, K. and Shikida, S. (1984) In: *ASTM STP 801*, 81; Endo, K. Komai, K. and Ohnishi, K. (1968) *J. Mater. Sci., Japan* **17**, 160.

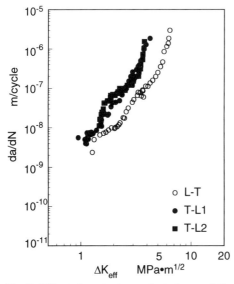

Fig. 8 Effect of crack plane orientation on fatigue crack growth rate in laboratory air as a function of ΔK_{eff}.

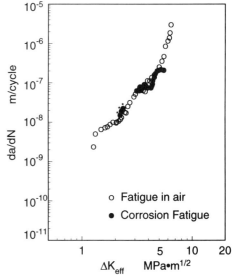

Fig. 9 Corrosion fatigue crack growth behavior of L-T specimens in a 3.5% NaCl solution.

Low Cycle Fatigue and Elasto-Plastic Behaviour of Materials
K-T. Rie and P.D. Portella (Editors)
© 1998 Elsevier Science Ltd. All rights reserved.

FATIGUE OF CEMENTED CARBIDES WITH DIFFERENT SURFACE MODIFICATIONS UNDER CYCLIC LOADS

P. SCHLUND[a], S. KURSAWE[a], H.-G. SOCKEL[a], P. KINDERMANN[a],
U. SCHLEINKOFER[b], W. HEINRICH[c] AND K. GÖRTING[c]

[a] Universität Erlangen-Nürnberg, Institut für Werkstoffwissenschaften, Lehrstuhl I,
Martensstr. 5, 91058 Erlangen, F.R. Germany

[b] Kennametal Inc., Latrobe, PA 15650, U.S.A.

[c] Kennametal Hertel AG, Eckersdorferstr. 10, 95490 Mistelgau, F.R. Germany

ABSTRACT

In the technical application as cutting tools, hard metals are exposed to wear and to complex mechanical loads, which include also thermomechanical and cyclic loads. In the past, the authors have shown that these materials exhibit a strong lifetime-limiting fatigue under cyclic loads.

Investigations of the mechanical behaviour under different loading conditions, i.e. static, monotonically increasing and cyclic loads, and of the microstructural processes have revealed that the processes in the binder phase ligaments and the crack tips influence the subcritical crack growth from the surface which is responsible for the fatigue behaviour observed.

Therefore, in addition, the influences of surface modifications such as CVD- and PVD-coatings and gradient surface layers, which were produced by heat treatments, on the fatigue behaviour of hard metals have been investigated. These studies included mechanical testing under the loading conditions mentioned above and also microstructural investigations of the damage processes.

It could be shown that CVD coating systems can cause a strong reduction of the lifetimes under cyclic loads, while gradient surface layers lead to longer lifetimes. The lifetime of PVD-coated hard metals is influenced by compressive stresses in the coating. The responsible processes for these changes in the fatigue behaviour were investigated by TEM and SEM. The results obtained on the damage processes are discussed in detail.

KEYWORDS

cemented carbides, hard metals, fatigue, subcritical crack growth, damage processes in fatigue.

INTRODUCTION

Cemented carbides are metal ceramic composite materials which consist of a hard ceramic phase and a ductile metallic binder phase. During the technical application as cutting tools, these materials are exposed to static, monotonically increasing and cyclic loads due to the cutting forces, the machine vibrations and the interrupted cutting. Furthermore, the inserts are exposed to thermomechanical loads and to wear due to friction [1].

Earlier works of the authors have proved that these materials show a lifetime-limiting fatigue [2] and have revealed the microstructural processes which are responsible for this behaviour [3].

The observed damage process, subcritical crack growth from the surface, motivated the presented studies on the influence of surface modifications such as CVD- and PVD-coatings and gradient surface layers produced by heat treatment on the mechanical behaviour of cemented carbides under cyclic loads.

METHODS OF INVESTIGATION AND EXPERIMENTAL DETAILS

The mechanical behaviour of a hard metal (P4M) with a TiN-Ti(C,N)-TiN-HTCVD- and a TiN-PVD-coating and a graded hard metal (RW4), all provided by Kennametal Hertel, were tested at room temperature. The results of the tests were compared with those of the uncoated and ungraded materials.

The chemical compositions of the tested materials are given in Table 1.

Table 1. Chemical composition of the hard metals P4M and RW4 in weight percent

Components	P4M	RW4
WC	86.5	83.0
Co	6.0	8.0
TiC	2.5	-
(Ta,Nb)C	5.0	-
(W,Ta,Ti,Nb)(C,N)	-	9.0

The microstructures of the materials are shown in Fig. 1 for the CVD-coated (a) and the graded hard metal (b).

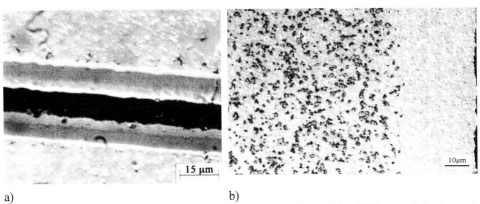

a)   b)

Fig. 1. Optical micrographs of the near-surface cross sections of the CVD-coated hard metal P4M (a) and the graded hard metal RW4 (b).

Figure 1a shows a CVD-coating (TiN/TiC,N/TiN) on the hard metal, and Fig. 1b a cobalt-enriched gradient layer at the sample surface.

All mechanical tests were carried out in a cantilever bending apparatus based on a principle of a test method developed by Fett *et al.* [4]. The apparatus can be modified in order to apply static, monotonically increasing, or cyclic loads to bar-shaped hard metal specimens. All tests under cyclic loads were performed with a stress ratio $R=\sigma_{max}/\sigma_{min}=-1$ at a frequency of 2 Hz for the graded hard metal and of 22 Hz for the coated an uncoated hard metal P4M.

RESULTS

The results on the inert strengths, i.e. bending strengths without the influence of subcritical crack growth, under monotonically increasing loads are presented in Weibull plots in Fig. 2.

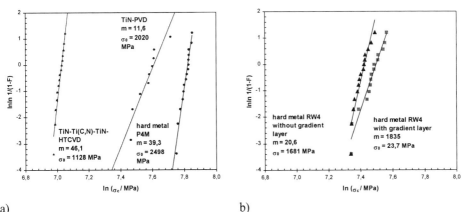

a)   b)

Fig. 2. Weibull plots of the inert strengths for P4M and RW4

The mean inert strength σ_0 of the TiN-Ti(C,N)-TiN-HTCVD coated hard metal, in comparison to the hard metal substrate P4M, is significantly lowered, while the Weibull modulus m which describes the scatter of the data is increased. This means that a strong homogenous damaging of the hard metal P4M substrate has occurred during the HTCVD coating process. Also the PVD-coated hard metal has suffered a moderate reduction of the mean inert strength σ_0, and additionally the Weibull modulus m has decreased. The decrease of the Weibull modulus m can be explained by the directed mass transport during the coating process. If the position of the specimen in the PVD reactor is not optimized, the initial defects in the substrate are not homogeneously healed. The latter gives rise to a broadened defect size distribution causing an increased scattering of the measured inert strength data for the PVD-coated hard metals.

The results of the measurements of the life times under cyclic loads are given in the Wöhler plots in Fig. 3.

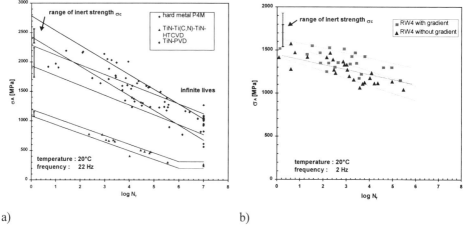

a)   b)

Fig. 3. Wöhler plots for coated and uncoated P4M (a) and graded and ungraded RW4 (b). Stress amplitude σ_A versus N_f, the number of cycles until fracture.

Both, the uncoated and coated hard metal, show a reduction of the endurable stress amplitude σ_A with increasing number of cycles N_f. This indicates a strong fatigue effect [2]. The lifetime behaviour of the TiN-Ti(C,N)-TiN-HTCVD-coated material has suffered a strong deterioration which also affects the technical application as cutting tools. The scattering of the lifetimes is very low which is consistent with the results for the inert strength measurements. The lifetime of the TiN-PVD-coated hard metal in comparison to the hard metal substrate is lowered for high stress amplitudes σ_A. In the range of 10^6 cycles, the endurable stress amplitudes for the uncoated and the TiN-PVD-coated hard metal are nearly similar. The slope of the measured lifetimes versus log N_f points to an improved lifetime behaviour of the TiN-PVD-coated hard metal for $N_f > 10^6$ compared to the substrate. The reasons for this behaviour are compressive residual stresses in the PVD coating hindering the crack formation and the crack growth in the coating at low stress amplitudes σ_A. In the TiN-PVD layer, single delaminations (see Fig. 4a) can be observed. These are caused by adhesion flaws, associated with compressive stresses in the layer. Due to different thermal expansion coefficients of the coating and of the substrate in the TiN-Ti(C,N)-TiN-HTCVD coating, cracks have formed (see Fig. 4b) during the cooling phase of the HTCVD process. In the TiN-Ti(C,N)-TiN-HTCVD coating, residual tensile stresses with a magnitude of 400 MPa were measured by the $\sin^2\psi$ method. At the beginning of the HTCVD process, a decarburisation of the substrate surface has occurred and caused the formation of brittle η-phases. These η-phases with their high density give rise to the formation of pores and a decrease of the binder content at the substrate surface. This leads to an embrittlement of the substrate surface which facilitates the propagation of cracks starting from the coating into the substrate (see Fig. 5).

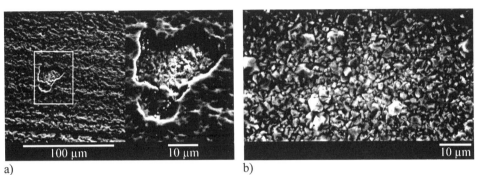

a)   b)

Fig. 4. Delaminations in the TiN-PVD layer in the initial state (a), cracks on the TiN-Ti(C,N)-TiN-HTCVD coating caused by different thermal expansion coefficients of the hard metal substrate and the coating (b). SEM micrographs.

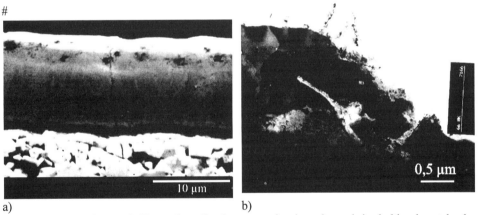

a)   b)

Fig. 5. Ti(C,N)-TiN-HTCVD coating: Crack propagation into the embritteled hard metal substrate (a, SEM micrograph), η-phase formation at the hard metal substrate (b, TEM micrograph).

The graded hard metal shows a higher inert strength σ_c compared to the substrate material as well as a slightly increased Weibull modulus m. Additionally, the graded material exhibits increased lifetimes under cyclic loads. These effects can be explained by the increased content of cobalt in the gradient surface layer. In homogenous hard metals, a higher fracture toughness and a higher bending strength with increasing content of the Co-binder phase have been observed [5]. The tougher surface layer also hinders the propagation of cracks from the specimen surface into the substrate under cyclic loads and therefore decreases the fatigue sensitivity of the graded hard metal. The occurrence of different damage mechanisms under static and cyclic loading conditions was evidenced by the results of the testing under static loads. Under static stresses of 85 % of the inert strength, all tested samples showed infinite lifetimes.

CONCLUSION

The investigation of hard metals with CVD- and PVD-coatings and gradient surface layers led to the following main results:

1. The deposition of PVD coatings on hard metals is connected with an unavoidable decrease of the inert strength σ_c, while the lifetime behaviour of PVD-coated hard metals is controlled by the residual compressive stresses in the coating.
2. The deterioration of the TiN-Ti(C,N)-TiN coated hard metal with respect to the inert strength and the lifetime behaviour is caused by an embrittlement of the substrate surface and an additional loading of the hard metal substrate by residual stresses in the coating.
3. Hard metals with gradient surface layers show an improved mechanical behaviour under monotonically increasing and cyclic loads.

Based on this progress in understanding, further important improvements can be achieved for hard metals as cutting tools by combining gradient surface layers and coatings with appropriate morphologies.

ACKNOWLEDGEMENTS

The authors want to thank Prof. Munz and Dr. Fett at the University of Karlsruhe for the support in the experimental area and the methodology, the Deutsche Forschungsgemeinschaft for the financial support, and Prof. Mughrabi at the University of Erlangen-Nürnberg for the critical reading of the manuscript.

REFERENCES

1. Schedler, W. (1988). *Hartmetall für den Praktiker.* VDI Verlag, Düsseldorf.
2. Schleinkofer, U., Sockel, H.-G., Görting, K. and Heinrich, W. (1996) *Mater. Sci. Eng.* **A209**, 313.
3. Schleinkofer, U., Sockel, H.-G., Görting, K. and Heinrich, W. (1996) *Mater. Sci. Eng.* **A209**, 103.
4 . Fett, T., Martin G., Munz, D. and Thun G. (1991) *Journal of Materials Science* **26**, 3320.
5. Amman, E. and Hinnüber, J. (1951) *Stahl und Eisen* **71**, 1081.

Low Cycle Fatigue and Elasto-Plastic Behaviour of Materials
K-T. Rie and P.D. Portella (Editors)

DAMAGE ACCUMULATION AND CRACK GROWTH IN METALS UNDER HYDROGEN EMBRITTLEMENT

V.ASTAFIEV and L.SHIRYAEVA
Department of Mathematics and Mechanics, Samara State University,
443011 Samara, Russia

ABSTRACT

The governing equations for elastic-plastic material describing the damage accumulation under HE conditions are presented. Damage evolution equation generalised Kachanov's equation in creep is proposed. The change of failure surface mode during HE conditions is described. The determination of material constants from available experimental data is proposed.

KEY WORDS:

Damage accumulation, cracking, hydrogen embrittlement, threshold stress, delayed fracture.

INTRODUCTION

The process of hydrogen embrittlement (HE) in metals is not completely understood yet despite the intensive experimental investigation which has been done all over the world. HE is manifested in time degradation of mechanical properties such as elongation to failure, yield and tensile strength, fracture toughness, etc. HE may also change the mode of fracture from ductile transgranular mode to brittle intergranular one.

HE in metals is usually caused by ions of hydrogen generated during the corrosion reaction in wet H_2S containing environments. It is assumed that hydrogen enters the metal continuously and interacts with defects of microstructure. This defect-hydrogen interaction results in trapping of hydrogen. The traps can be a single solute atom, carbide particles, grain boundaries, internal voids, microcracks and other types of single atom or multi-atom defects. Plastic deformation plays an important role in HE through the interaction between dislocations and hydrogen. Even if applied stress level is less than elastic limit, local yielding can occur at sharp pits, which were developed during exposure to corrosive media.

The exact description of all features of HE process is represented as a meaningless task. Instead of trying to reproduce all fine details of this process it is supposed to be more reasonable to introduce some internal variable (damage parameter) ω reflecting only the main features of damage accumulation. This approach in the creep of metals has been done by Kachanov and Rabotnov [1,2]. The objective of this paper is to generalise the Kachanov-Rabotnov's idea for HE conditions and to analyse from this point of view some features of failure both in uniaxial and multiaxial cases.

1. CONSTITUTIVE EQUATIONS

Experimental facts of HE taken into consideration are follows:
- there is some critical hydrogen concentration below which no hydrogen-induced cracking occurs;
- degradation of mechanical properties under HE increases with the increasing of hydrogen concentration;
- degradation of mechanical properties under HE is determined by the quantity of dissolved hydrogen and doesn't depend on the manner of hydrogen penetration in metal;
- there is the threshold stress - the maximum applied stress below which no failure occurs;
- mechanical properties of material charged with hydrogen under HE can be restored during relaxation without stress as well as after removing the sources of hydrogen embrittlement.

Based on the above-mentioned experimental observations the following damage evolution equation has been proposed [3]:

$$d\omega/dt = A\,sign(\omega_* - \omega)|\omega_* - \omega|^m \qquad (1.1)$$

where A, m are material parameters, $\omega_*(\sigma_{ij})$ - the ultimate value for damage accumulated, which depends on stress state, environment, temperature, etc. The simplest approximation of dependence $\omega_*(\sigma_{ij})$ has been proposed in the following form:

$$\omega_* = \begin{cases} \alpha\sigma_0 + \beta, \ \sigma_0 > 0 \\ \beta, \ \sigma_0 \le 0 \end{cases} \qquad (1.2)$$

where $\alpha > 0, \beta \ge 0$ are material parameters, $\sigma_0 = \sigma_{ii}$.

Material parameter α reflects the influence of stress level on damage accumulation process. Its value depends on hydrogen ion concentration, environment, temperature, and microstructure of material. The value of parameter α is connected with the adsorption of hydrogen at some critical sites within the material such as microvoids, microcracks and other defects in presence of stress. The hydrogen concentration in these critical sites can exceed the average hydrogen concentration in the bulk metal that can promote the development of hydrogen embrittlement in microvolumes.

The value of parameter β coincides with the ultimate value of damage accumulated in material charged with hydrogen without load. This value is connected with such sources of hydrogen embrittlement as particles, impurities, dislocations, alloying elements, etc. Due to accumulated hydrogen these sources begin to transform with time at first to bubbles and then to three-dimensional hydrogen traps.

Let us postulate the yield surface under HE conditions in the following form:

$$\sigma_e = \sigma_*(\omega) \qquad (1.3)$$

where $\sigma_e = \sqrt{3/2 s_{ij} s_{ij}}$ is the effective stress, $s_{ij} = \sigma_{ij} - (1/3)\sigma_0 \delta_{ij}$ is the deviator of stress tensor, $\sigma_*(\omega)$ is the yield strength, changeable under the action of hydrogen environment during the deformation process. The constitutive equations should be completed by some fracture criterion. For elastic-perfectly-plastic theory it is offered to use the deformation type criterion

$$\varepsilon^P_{max} = \varepsilon_*(\omega) \qquad (1.4)$$

where ε^P_{max} is the maximum plastic strain and $\varepsilon_*(\omega)$ is the ultimate plastic strain, changeable under the action of hydrogen environment during the deformation process.

Simple approximations for $\sigma_*(\omega)$ and $\varepsilon_*(\omega)$ vs ω dependencies can be written in the following manner:

$$\sigma_*(\omega) = \sigma_*^0 (1 - k_1 \omega) \qquad (1.5)$$

$$\varepsilon_*(\omega) = \varepsilon_*^0 (1 - k_2 \omega) \qquad (1.6)$$

where σ_*^0 and ε_*^0 - are the yield strength of specimen, ultimate plastic strain of specimen under air conditions, k_1, k_2 are the material parameters. Taking into consideration the dimensionless character of damage parameter ω one of parameters k_1 or k_2 can be equated to 1, i.e. $k_2 = 1$ and $k_1 = k$.

2. DETERMINATION OF HE - PARAMETERS

To obtain the values of material parameters β and k it is necessary to consider the results of tensile tests for specimens charged with hydrogen without stress. In this case $\omega_* = \beta$ and from (1.3) and (1.4) it is followed:

$$\beta = 1 - \varepsilon_*(0) / \varepsilon_*^0 \qquad (2.1)$$

$$k\beta = 1 - \sigma_*(0) / \sigma_*^0 \qquad (2.2)$$

where $\varepsilon_*(0)$ and $\sigma_*(0)$ are ultimate plastic tensile strain and yield strength. Thus, the material parameters k and β reflect the change of strength and strain characteristics of specimens charged with hydrogen without stress in comparison with these characteristics of specimens in air conditions.

HE is examined usually by uniaxial tension load tests (UTLT) or slow strain rate tests (SSRT) in hydrogen environments. The UTLT are performed on cylindrical specimens using dead-weight-type constant tester. The results of UTLT include the experimental data for the threshold stress σ_{th}.

From theoretical point of view the failure under UTLS doesn't occur if $\sigma \leq \min_{t>0} \left\{ \sigma_*^0 (1 - k\omega(t)) \right\}$. Hence, the threshold stress σ_{th} can be written in the following form:

$$\sigma_{th} = \frac{\sigma_*^0 (1 - k\beta)}{1 + k\alpha\sigma_*^0} \qquad (2.3)$$

The value of material parameter α can be obtained from (2.3) with already known parameters k and β as follows:

$$k\alpha\sigma_*^0 = \frac{\sigma_*(0)}{\sigma_{th}} - 1 \qquad (2.4)$$

The SSRT are performed on cylindrical specimens using a tensile machine where specimens are loaded slowly to fracture at the constant strain rate. $\dot{\varepsilon}_0$. The fracture of specimen occurs in the time t_f when the fracture criterion (1.4) takes place $\varepsilon^p = \dot{\varepsilon}_0 t_f - \sigma(t_f)/E = \varepsilon_*^0(1 - \omega(t_f))$. It can be shown that for $\alpha E \dot{\varepsilon}_0 / A \ll 1$

$$\omega_f = \omega(t_f) \approx \omega_* = (\alpha\sigma_*^0 + \beta)/(1 + k\alpha\sigma_*^0) \qquad (2.5)$$

Hence, for the ultimate plastic strain ε_f accumulated in SSRT and for ultimate stress σ_f we have the following condition:

$$\varepsilon_f = \varepsilon_*^0(1 - \omega_f) \approx \varepsilon_*^0(1 - \omega_*)$$
$$\sigma_f = \sigma_*^0(1 - k\omega_f) \approx \sigma_*^0(1 - k\omega_*) \qquad (2.6)$$

The value of parameter α can be estimated from (2.5) using the already known parameters k, β as follows:

$$k\alpha\sigma_*^0 = \frac{\sigma_*(0)}{\sigma_f} - 1 \qquad (2.7)$$

The values of material parameters A and m can be found by means of the least squares method from experimental data for tensile stress vs time to failure (UTLS) or for tensile stress vs strain (SSRT) relations.

Unfortunately, the complete set of experimental data (the tensile tests for specimens charged with hydrogen without stress and UTLT, SSRT) which allow to determinate all HE material parameters α, β, A, m and k is not available. As a rule, the experimental data on the change of mechanical characteristics of specimens, charged with hydrogen without stress are not available and it is impossible to estimate the values of material parameters β and k.

When the experimental data both on UTLT and SSRT are known, one can estimate two parameters k and α (or β). Kaneko [4] investigated the influence of microstructure of AISI 4130 steel on the value of threshold stress σ_{th} and the value of the ductility loss I. The ductility loss was determined as $I = 1 - L/L_0$, where L_0 is the elongation of specimen in the air tensile test and L is the same in SSRT and can compare with ω_*. Using equation (2.3) the expressions for the σ_{th} and ω_* can be rewritten in the form

$$\begin{cases} \alpha\sigma_*^0 r + \beta = \omega_* \\ k\omega_* + r = 1 \end{cases}$$

Hence, the parameter k can be estimated by experimental values $\omega_* = I$ and $r = \sigma_{th}/\sigma_*^0$, but the parameter α can't be found without the knowledge of material parameter β. The results of calculation for the parameter k that correspond to experimental data of Kaneko [4] are represented in table 1.

The hydrogen influence on the change of mechanical properties of some Russian grade steel (38HNZMFA steel) charged under the high hydrogen pressure without external stress was investigated by Korchagin [5]. In this case the value of the parameter β can be determined by means of (2.1). For this steel quenched and tempered to various strength levels and charged up to average hydrogen concentration 5-6 $cm^3/100$ g, the values of parameter β are represented in table 2.

Table 1.

σ_*^0	σ_{th}	r	ω_*	k	$\alpha\sigma_*^0$
MPa	MPa				
650	530	0,815	0,89	0,21	1,092-1,243 β
700	560	0,8	0,895	0,22	1,119-1,250 β
750	600	0,8	0,91	0,22	1,138-1,250 β
800	600	0,75	0,96	0,26	1,280-1,333 β
850	500	0,588	0,97	0,42	1,649-1,701 β
900	400	0,444	0,98	0,57	2,207-2,252 β

Table 2.

C_H	σ_*	ε_*	β
	MPa		
-	1620	0,1	
5,7	1420	0,0	1
-	730	0,185	
5,4	810	0,16	0,14
-	640	0,18	
5,2	660	0,17	0,06

3. MULTIAXIAL FAILURE CONDITIONS

The governing equations (1.1)-(1.6) allow to describe HE process both in uniaxial and multiaxial stress state. In the case of multiaxial stress state under the constant stress tensor σ_{ij} the threshold stress corresponds to the threshold surface, i.e. some surface in the stress space within which no failure occurs, $\sigma_e \leq \min_{t \geq 0}\left\{\sigma_*^0(1 - k\omega(t))\right\}$. Hence, the equation for the threshold surface is

$$\sigma_e = \begin{cases} \sigma_*^0(1 - k(\alpha\sigma_0 + \beta)), & \sigma_0 > 0 \\ \sigma_*^0(1 - k\beta), & \sigma_0 \leq 0 \end{cases} \tag{3.1}$$

Thus, the failure must occur only for the stress state out of the threshold surface (3.1). For example, the failure occurs under any shear stress if the first invariant of stress tensor σ_0 satisfies the following condition $\sigma_0 > (1 - k\beta)/k\alpha$. For the plane stress conditions the equation (3.1) can be written as follows

$$\begin{cases} \sqrt{\xi_1^2 - \xi_1\xi_2 + \xi_2^2} + \alpha k\sigma_*^0(\xi_1 + \xi_2) = 1, & \xi_1 + \xi_2 > 0 \\ \xi_1^2 - \xi_1\xi_2 + \xi_2^2 = 1, & \xi_1 + \xi_2 \leq 0 \end{cases} \tag{3.2}$$

where $\xi_1 = \sigma_1/\sigma_*^0(1 - k\beta)$, $\xi_2 = \sigma_2/\sigma_*^0(1 - k\beta)$ are the normalised principal stresses. The curve (3.2) is the cross of the surface (3.1) by the plane $\xi_3 = 0$ and consists of conjunction of the half-ellipse under the condition $\xi_1 + \xi_2 \leq 0$ with the half-ellipse ($0 < 2k\alpha\sigma_*^0 < 1$), the half-parabola ($2k\alpha\sigma_*^0 = 1$) or the half-hyperbola ($2k\alpha\sigma_*^0 > 1$) under the condition $\xi_1 + \xi_2 > 0$. In the ultimate case ($2k\alpha\sigma_*^0 = \infty$) this curve is degenerated to the half-ellipse, limited by the straight line $\xi_1 + \xi_2 = 0$. The schematic form of surface (3.1) is represented in Fig.1 and the form of the curve (3.2) for various values of $2k\alpha\sigma_*^0$ is represented in Fig.2. Thus, the value of

410

$2k\alpha\sigma_*^0$ reflects the influence of the first invariant of stress tensor σ_0 on the damage character in metals under HE conditions. The threshold surface changes weakly if $2k\alpha\sigma_*^0 \ll 1$ and keeps the elliptical form, which is typical of ductile transgranular failure. If $2k\alpha\sigma_*^0 \geq 1$ then the principal change of curve form (3.2) from elliptical to parabolic and hyperbolic occurs. In this case the value of σ_0 plays the prevailing role which is typical of brittle intergranular failure.

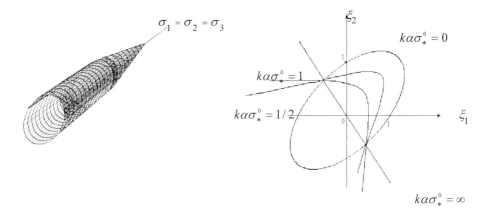

Fig. 1. The schematic form of threshold surface in the stress space.

Fig. 2. The cross of threshold surface by the plane $\xi_3 = 0$.

ACKNOWLEDGMENTS

Authors are grateful to Russian Foundation of Basic Research (grand 96-01-01064) for financial support.

REFERENCES

1. Rabotnov, Yu.N. (1969). *Creep Problems in Structural Members.* North Holland, Amsterdam.
2. Kachanov, L.M. (1958). *Izv. AN SSSR. Otd. Tekn. Nauk* **8**, 26 (in Russian).
3. Astafiev, V.I. and Shiryaeva, L.K. (1997). *Izv. RAN. MTT* **3**, 115 (in Russian).
4. Kaneko, T., Okada, Y. and Ikeda, A. (1989). *Corrosion* **1**, 2.
5. Korchagin, A.P. and Uraido B.F. (1976). *Fiz.-Him. Mehanika Mater.* **2**, 113 (in Russian).

Low Cycle Fatigue and Elasto-Plastic Behaviour of Materials
K-T. Rie and P.D. Portella (Editors)
411

ENVIRONMENTALLY ASSISTED LOW-CYCLE FATIGUE CRACK INITIATION AND GROWTH

V.V.BOLOTIN, V.M.KOVEKH and A.A.SHIPKOV
Institute of Mechanical Engineering Research
Russian Academy of Sciences
M.Kharitonyevsky per. 4, 101830 Moscow, Centre, Russia

ABSTRACT

Models of low-cycle fatigue crack initiation and growth in an active environment are developed based on the synthesis of fracture mechanics and mechanics of damage accumulation. Environmentally affected damage, such as corrosion, ageing and embrittlement, and purely mechanical damage as well their interaction are taken into account during all the stages of crack growth beginning from the near-threshold initiation stage until the final failure. Special attention is given to the problem of the active agent transport to the moving crack tip under cyclic loading. The competition between various damage mechanisms during the crack propagation is demonstrated.

KEYWORDS

Crack growth initiation, fatigue crack growth rate, corrosion fatigue, environmentally assisted fatigue, stress corrosion cracking, damage accumulation.

INTRODUCTION

Prediction of fatigue crack initiation and growth is the subject of many studies, both experimental and theoretical ones. A recent comprehensive survey is given in the two-vo-lume handbook edited by Carpinteri [1]. The problem has been discussed from several viewpoints, material science, mechanics of solids, and as well as special engineering domains. Material science and metallurgical aspects are widely discussed in the book by Kocanda [2]. An analytical approach to modeling fatigue and related phenomena covering all the stages, beginning from the crack initiation till the final fracture, was suggested by Bo-lotin [3]. A systematic presentation of the theory can be found in the books by Bolotin [4, 5]. Flexibility and adaptability of the theory allows its application to a wide range of material properties, load conditions, body and cracks geometry. In particular, the influence of environmental effects can be included into consideration. A preliminary discussion of modeling of corrosion fatigue and stress corrosion cracking was given in [4]. A detailed numerical analysis of crack growth under the combination of cyclic loading and corrosion damage was recently given by Bolotin and Shipkov [6]. The material was modelled as linear elastic in all the bulk of the body except the process zone near the crack tips. The load level was taken in [6] corresponding to the high-cycle fatigue model when plastic straining effects are neglected. In this paper the environmentally affected fatigue is considered for such load levels when plastic strains cannot be neglected. However, the contained yielding is assumed when the thin plastic zone model may be used to cover all the plastic effects. Some applications of this model to conventional low-cycle fatigue are given in [4,5]. A detailed analysis of low-cycle fatigue crack growth can be found in [6,7]. In this paper the above mentioned approach is extended upon the environmentally assisted fatigue.

GENERAL THEORY

The theory of fatigue crack growth is based on an extended version of fracture mechanics named in [3,4] analytical fracture mechanics. The general idea is to consider crack para-meters as a kind of generalized coordinates of mechanical system named in [3] Griffithian coordinates or, briefly, G-coordinates. Together with common Lagrangian coordinates (L-coordinates), they describe the current state of a cracked body under load and environmental actions. In contrast to L-coordinates, G-coordinates almost always are subjected to unilateral constraints because fatigue cracks in solids are, as a rule, "incurable".

Let consider quasistatic processes only and let assume that the first part of the problem, namely, the evaluation of L-coordinates, is already solved. The principle of virtual work for the system cracked body – loading or cracked body – loading device requires that in equilibrium $\delta_G W \le 0$. Here $\delta_G W$ is the virtual work performed on G-coordinates, i.e. calculated for two neighbouring states of the body which are equilibrium states in the ordina-ry sense but differ in crack dimensions. Depending on the sign of virtual work, the state of the body may be stable or unstable with respect to the further crack propagation. The cracks do not propagate at $\delta_G W < 0$ for all G-coordinates that corresponds to the subequilibrium state with respect to G-coordinates. If $\delta_G W > 0$ even for one of G-coordinates, the state of the system is unstable. The case $\delta_G W = 0$ is named equilibrium state. It is of most importance in the theory of fatigue. A crack propagates in a continuous way with respect to a certain G-coordinate at $\delta_G W = 0$, $\delta_G(\delta_G W) < 0$ where the second variation is taken with respect to this coordinate. For the sake of brevity, consider hereafter a body containing an one-parameter crack with G-coordinate a (the general case is discussed in [4,5]). The virtual work can be presented in the form $\delta_G W = G\delta a - \Gamma\delta a$ where δa is the variation of G-coordinate, G and Γ are driving and resistance generalized forces. In terms of generalized forces the subequilibrium state takes place at $G < \Gamma$, and the stable equilibrium state at $G = \Gamma$, $\partial G / \partial a < \partial \Gamma / \partial a$. The state of the system becomes unstable at $G > \Gamma$, or at $G = \Gamma$, $\partial G / \partial a > \partial \Gamma / \partial a$. The latter case corresponds to an unstable equilibrium state.

Additional variables ought to be included to take into account damage. The simplest way is to use the models of continuum damage mechanics. The current damage field $\omega(x,t)$ (scalar or tensorial) is a result of loading history $s(t)$ and crack growth history $a(t)$ and can be presented as a hereditary functional of $s(t)$ and $a(t)$. The tip damage variable $\psi(t)$ is of special importance. This variable may be identified with $\omega(x,t)$ at the crack tip, i.e. $\psi(t) = \omega[a(t), t]$. In case of fatigue the generalized forces depend not only on $s(t)$, $a(t)$ (which takes place in fracture mechanics), but on $\psi(t)$. In particular, the condition of crack nonpropagation is as follows:

$$\max_{t_{N-1} \le t \le t_N} \{G[s(t), a(t), \psi(t)] - \Gamma[s(t), a(t), \psi(t)]\} < 0 \qquad (1)$$

Here $[t_{N-1}, t_N]$ is the time segment corresponding to N-th cycle when cyclic loading is considered. In slowly varying loading, Eq. (1) is applied just for the current time t. The typical situation of fatigue crack growth is that when conditions $\delta_G W = 0$, $\delta_G (\delta_G W) < 0$ are satisfied within each cycle of loading. Then

$$\max_{t_{N-1} \le t \le t_N} \{G[s(t), a(t), \psi(t)] - \Gamma[s(t), a(t), \psi(t)]\} = 0 \qquad (2)$$

THIN PLASTIC ZONE MODEL FOR CORROSION FATIGUE

In the presence of localized plastic straining, let us use the thin plastic zone model. Consider a mode I crack in a wide plate of ideal elastoplastic material. Under monotonous lo-ading, the length of the plastic zone λ and the crack tip opening displacement δ depend on the crack length a, applied stress σ_∞, and a certain characteristic stress σ_0 similar to the yield stress in tension. When the loading process is cyclic, and the ideal Bauschinger effect is assumed, we must replace σ_∞ with the stress range $\Delta \sigma_\infty$, and the yield stress σ_0 with $2\sigma_0$. The stress distribution under monotonous and cyclic loading is illustrated in Fig. 1,a. There $\Delta \sigma$ is the range of the tensile stress on the crack prolongation. The generalized driving force G in this model may be taken according to [4] with the correction factor Y^2 corresponding to an edge crack. The generalized resistance force Γ can be presented as $\Gamma = \gamma_0 f(\psi_f, \psi_c)$. Here γ_0 is the fracture work for the nondamaged material related to the unit area of the newborn crack (without doubling the crack surfaces); $f(\psi_f, \psi_c)$ is a function of two damage measures at the crack tip. The measure $0 \le \psi_f \le 1$ is associated with purely mechanical damage, the measure $0 \le \psi_c \le 1$ with the corrosion damage. When $\psi_f = \psi_c = 1$, we have $\Gamma = \gamma_0$. Function $f(\psi_f, \psi_c)$, as a rule, is diminishing with respect to both variables. The damage measures ahead of the crack, i.e. at $x > a$ are denoted $\omega_f(x, t)$ and $\omega_c(x, t)$, respectively. Thus, $\psi_f(t) = \omega_f[a(t), t]$, $\psi_c(t) = \omega_c[a(t), t]$. The mechanical damage is accumulating both near the tip and in the far field. As to the corrosion damage, it is natural to introduce the corresponding process zone with length λ_c. For simplicity, assume that $\omega_c(x, t) = \psi_c(t)(1 - \xi / \lambda_c)$ at $0 < \xi < \lambda_c$, and $\omega_c(x, t) = 0$ at $\xi > \lambda_c$. Here $\xi = x - a$ (Fig. 1,a).

414

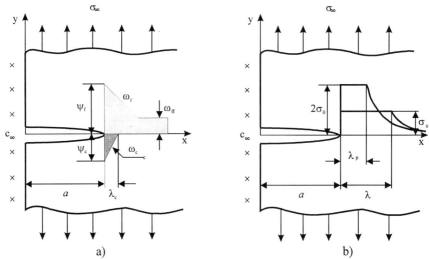

a)   b)

Fig. 1. Edge mode I crack: damage distribution (a); stress distribution (b).

The simplest equations of damage accumulation under sustained (or slowly varying) actions are power-threshold equation used primarily in [3]. Here we use two equations of this type. The first equation describes the accumulation of mechanical damage at $x \geq a$. This equation contains the tensile stress $\sigma(x,t)$ at $y = 0$ and three material parameters, σ_d, σ_{th} and m_f. The second equation describes the accumulation of corrosion damage at $x = a$ when the content of the active agent at the tip is c_t. Material parameters c_d, c_{th} and m_d characterize resistance to corrosion damage. To evaluate the tip content c_t, one need to consider the transport of the agent from the crack mouth to its tip. In general, the latter problem belongs to hydrodynamics. However, one must take into account the diffusion and mixing processes within the crack which tip is moving ahead. In addition, cracks are "breathing" when cyclic loading is concerned. The boundary condition at the crack mouth is usually simple, say $c(0,t) = c_\infty = $ const (see Fig.1,b). A phenomenological model was suggested in [6] based on the first order ordinary differential equation with respect to $c_t(t)$. This model will be used in the later analysis.

NUMERICAL EXPERIMENTATION AND DISCUSSION OF RESULTS

The presented theory is illustrated hereafter by numerical examples. A mode I edge crack is considered under the tension with applied stresses $\sigma_\infty = \sigma_{\infty,m} + \sigma_{\infty,a} \cos 2\pi f t$. Thus, the stress range is $\Delta\sigma_\infty = 2\sigma_{\infty,a}$, and the load frequency measured in Hz's is f. The material is assumed linearly elastic with Young's modulus $E = 200$ GPa, Poisson's ratio $v = 0,3$ everywhere except the thin plastic zone where the tensile yield limit $\sigma_0 = 500$ MPa is assumed. The generalized resistance force is taken as $\Gamma = \gamma_0(1 - \psi_f - \psi_c)$. This means that the additiviness of damage of mechanical and corrosion origin is assumed. The specific fracture work for the nondamaged material is $\gamma_0 = 10$ kJ/m^2, material parameters in the equation of mechanical damage are: $\sigma_f = 5$ GPa, $\sigma_{th} = 250$ MPa, $m_f = 4$. The part of the problem concerning corrosion is not so clear. To avoid too far-going assumptions, the normalized agent content c_∞ / c_d is used as a control parameter and characteristic time $\tau_d = 10^3$ s, characteristic crack length $a_\infty = 1$ mm, and characteristic frequency $f_\infty = 1$ Hz are assumed in computations.

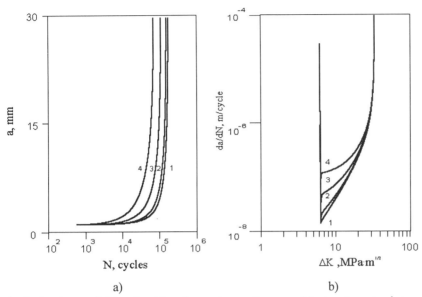

Fig. 2. Crack growth histories (a) and growth rate diagrams (b) at various c_∞ / c_d.

The numerical results in Fig. 2 are plotted for $\Delta\sigma_\infty = 100$ MPa, $R = 0.2$, $f = 1$ Hz and the initial conditions $a_0 = 1$ mm, $c_t(0) = c_\infty$. Lines 1, 2, 3, and 4 correspond to the active agent content at the crack mouth $c_\infty / c_d = 0.25, 0.5, 0.75$ and 1.0. Crack growth histories are shown in Fig. 2,a, and the crack growth rate da / dN as a function of the range ΔK of the stress intensity factor in Fig. 2,b. When the agent content increases, the total fatigue life is subjected to shortening. As to Fig. 2,b, in the case of low content the diagram is very similar to standard fatigue crack growth rate diagrams except for the "tail up" feature which is a result of crack tip jumps when the initially corroded process zone is ruptured. The acceleration of crack growth is observed when the content increases. At high content, a kind of "plateau" occurs which is in agreement with experimental observations [1, 2]. All lines in Fig. 2,b are merging with growing ΔK. These effects can be easily understood when we consider the comparative contribution of mechanical and corrosion damage into the total damage and, as a result, into the generalized resistance force. Figure 3,a is drawn for $c_\infty / c_d = 0.25$, and Fig. 3.b for $c_\infty / c_d = 1$. In the first case, the mechanical damage measure ψ_f dominates from the beginning. In the second case, the situation is more complicated. In the beginning we observe intensive corrosion damage. This explains the presence of "plateau" in Fig. 2.b. During the further crack growth the contribution of corrosion damage diminishes, and when the crack is propagating rapidly, almost all damage is of mechanical nature.

416

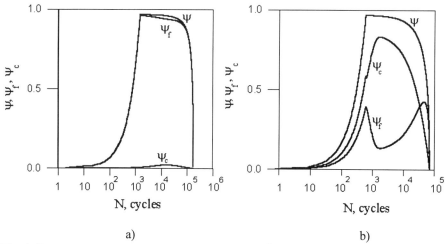

a)   b)

.Fig. 3. Damages measures at moving crack tip at c_∞ / c_d = 0.25 (a) and c_∞ / c_d = 1 (b).

ACKNOWLEDGMENT

This study was partially supported by the Russian Foundation for Basic Research (grant 96-01-01488).

REFERENCES

1. Carpinteri, A. (1994) Handbook on Fatigue Crack Propagation in Metallic Structures, Elsevier, Amsterdam.
2. Kocanda, S. (1985) Fatigue Fracture of Metals. Scientific Engineering Publishers, Warsaw (in Polish).
3. Bolotin, V.V. (1983) Izv. AN SSSR, Mekh. Tverd. Tela (MTT), 4, pp. 153-160 (in Russian).
4. Bolotin, V.V. (1989) Prediction of Service Life for Machines and Structures. ASME Press, New York.
5. Bolotin, V.V. (1996) Stability Problems in Fracture Mechanics. John Wiley, New York.
6. Bolotin, V.V. and Shipkov, A.A. (1998) Prikl. Matem. Mekh. (PMM), 62, pp. 91-102 (in Russian).
7. Bolotin, V.V. and Lebedev, V.N. (1996) Int. J. Solids and Structures, 33, 9, pp. 1229-1242.
8. Bolotin, V.V. and Kovekh, V.M. (1996) Prikl. Matem. Mekh. (PMM), 60, 1029-1038 (in Russian).

Low Cycle Fatigue and Elasto-Plastic Behaviour of Materials
K-T. Rie and P.D. Portella (Editors)
417

ELASTO-PLASTIC TRANSITION OF CARBON TOOL STEELS SURFACE TREATED BY LASER

SANDA MARIA LEVCOVICI [a], ION CRUDU [a],
BOGDAN GEORGE LEVCOVICI [b], DAN TEODOR LEVCOVICI [c].
[a] University "Dunarea de Jos" of Galatzi, 47 Domneasca, 6200 Galatzi, Romania.
[b] S.C."GALFINBAND"-S.A., 2 bis Smârdan, 6200 Galatzi, Romania.
[c] S.C."ICPPAM"S.A., 2 Smârdan, 6200 Galatzi, Romania.

ABSTRACT

The selection of the hardening procedure is a significant way of increasing the load carrying capacity of contact surfaces and reducing the sticking susceptibility. In this paper the plastic deformation transition of 1%C steel surface layers, laser strengthened by hardening, alloying and reinforcing with tungsten carbide particles was investigated compared to the standard treatment condition - volume quenching and low tempering. The quasi-static load carrying capacity and the friction coeffcients were determined using an instalation having sphere-plane point contact with sliding dry friction. The results have been correlated to the traces profile diagrams, microstructure, hardening depth and surface hardness. The load carrying capacity increased with the hardness and depth of laser processed layers. The composite microstructures resulted from alloying and reinforcing caused the drastic reduction of the friction coefficients and gripping susceptibility.

KEYWORDS

Elastic-plastic transition, tool steel, laser, load carrying capacity, friction coefficient.

INTRODUCTION

The need to increase the reliability of some highly stressed tribological systems of carbon tool steels led to the promotion of local surface hardening technologies by laser beam such as hardening, alloying, reinforcing with hard particles. The ultrarapid thermal cycle in the Ac_1 - solidus temperature range provides the martensitic hardening. Ultrafine martensite, over-saturated by partial carbide dissolving, chemically non-homogenous and with a developed sub-structure followed by an increased amount of residual austenite is formed [1]. By symultaneous melting of a suitable added material and steel substrate composite or/and alloyed surface layers, hardened from the liquid phase can be obtained [2]. The microstructure of these layers contains non-equilibrium solidification structures, oriented, finished, dispersed quasi-eutectoids and hard reinforcing particles [3]. The microstructural features of these layers result in hardness, wear resistance and higher thermal stability of the standard heat treatment [4, 5, 6]. The limits of these surface hardening procedures are given relatively low depths and the presence under the hardened layer of a tempered zone with low hardness which can cause the hard layer spalling in operation.

This paper aims at characterising the laser alloyed and hardened layers [7, 8, 9], under limit conditions of transition to the plastic deformation and determination of gripping tendency by the evolution of friction coefficients in the plastic range.

METHODS

The 1%C steel specimens have been subjected to the preliminary heat treatment of volumetric quenching and tempering at the hardness HV_{49}=8000MPa (load 49N). Single bends of laser hardening, alloying and reinforcing of pastes have been traced on the plane specimen surface using continuous-wave CO_2 laser GT 1200 W (Romania). Hardening was performed under the conditions below: radiation power P=900W; displacement rate v=10mm/s; beam equivalent radius r_e=1.89mm, ZnO absorber.

For alloying and reinforcing two formulae of pastes resulted from the mechanical mixture of powders with hydroxiethilcellulosic binder were used: W1(77%WC+3%Co+10%Si); W2(73%WC+12%Co+5%Mo+10% Si). The predeposited paste layers with 0.15mm in thickness were processed under the conditions: P=900W; r_e=1.41mm; v_1=2.15mm/s (W1), v_2=1.54mm/s (W2) respectively. The dimensions of the hardened (h_c - depth, l_c - width) respectively alloyed (h_a - depth, l_a - width) layers and the surface hardness for the four structurally distinct states studied in elastic-plastic transition are shown in table 1.

Table 1. Experimental data for the four structurally distinct states

Speci-men	Treatment conditions	Dimensions (mm)				HV_{49} (MPa)	$Rp_{0,2}$ (MPa)
		h_a	h_c	l_a	l_c		
1	Volume quenching + tempering at 175°C one hour	-	-	-	-	8000	2800
2	Laser hardening + subcooling at -60°C, 30 min. + tempering at 175°C one hour	-	0.51	-	2.50	9790	3426
3	W1 alloying + subcooling at -60°C, 30 min.	0.17	0.78	1.97	2.59	11120	3892
4	W2 alloying + subcooling at -60°C, 30 min.	0.29	1.08	2.17	3.49	10960	3836

The microhardness variation with the laser alloyed and hardened layer depth is shown in fig. 1.

In order to study the plastic deformation transition an instalation with sphere-plane point contact was used with sliding dry friction (fig. 2), [10]. The ⌀10mm spheric penetrator is made of bearing steel with HV_{49}=7800 MPa. The load increases gradually with the sliding of the spheric penetrator on the specimen inclined plane. The stress conditions were: vertical pressing force N=50-1850N, penetrator displacement rate v=0.6mm/s. The plastic deformation evolution in the specimen plane half-spacing with the normal force variation characterized by recording the Q force of advance resistance and the profile diagrams of traced marks. The quasi-static load carrying capacity was defined as the maximum tension where the relative plastic deformation δ_p/R=0.001.

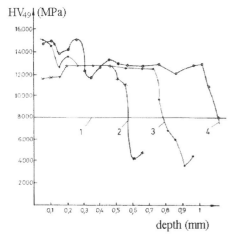

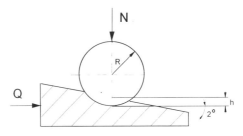

N

R

Q

2°

h

Fig. 2: Surface layer stress conditions.

Fig. 1: $HV_{0.98}$ microhardness variation with the laser processed layer depth: 1 - volume quenching; 2 - laser hardening; 3 - W1 alloying; 4 - W2 alloying.

For hardened steels the elastic-plastic transition is continuous and gradual and the tension state at the beginning of plastic deformation (δ_p=0.1-10μm) is expressed, with an acceptable approximation by Hertz equations [11]:

$$\sigma_{K_{max}} = 0.388 \sqrt[3]{\frac{F_n E^2}{R^2}} \tag{1}$$

The yield point of the surface layer is expressed as a function of the surface hardness by the equation:

$$Rp_{0.2} = 0.35\ HV_{49}\ (MPa) \tag{2}$$

The friction coefficient f was estimated as a function of the measured values of plastic deformation δ_p, normal F_n and tangential F_t forces of contact by equation [10]:

$$\frac{\delta_p}{R} = 1 - \frac{fF_t + F_n}{\sqrt{\left(F_t^2 + F_n^2\right)\left(1 + f^2\right)}} \tag{3}$$

The plastic deformation δ_p of the plane semispace was given by the maximum depth of the trace profile in axial section.

RESULTS AND DISCUSSIONS

Quasi-static Load Carrying Capacity

The analysis of the traced mark profile diagrams (fig. 3) shows that the order of entering the plastic deformation of the layers is volume quenching, laser hardening, W2 alloying and W1 alloying, in agreement with the hardness, yield point respectively of the material. The plastic deformation is maximum in volume hardened layers and minimum in the ones alloyed with W1 and W2 formulae.

The profile diagrams transversal to the traces show the location of plastic deformation near the penetrator at volume hardened steels, emphasizing the sharp overheight with stress increase. The laser hardening reduces the overheight but the surface relief character is maintained. The material overheight is low and extended towards the contact neighbouring zones in the alloyed layers.

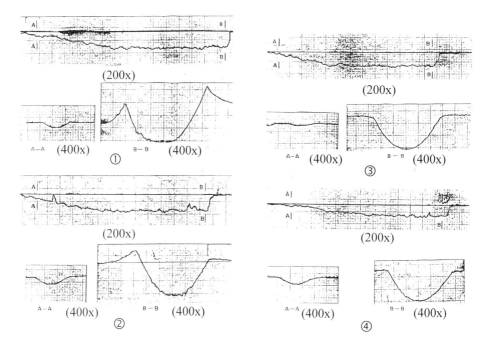

Fig. 3: Axial and cross profilogram trace for various states of the material: volume quenching ①, laser hardening ②, W1 alloying ③, W2 alloying ④.

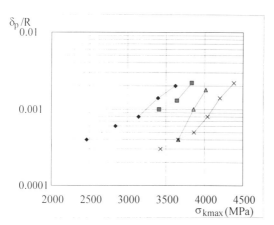

Fig. 4: The variation σ_{Kmax} as a function δ_p/R when entering the plastic deformation: volume quenching (◊), laser hardening (□), W1 alloying (△), W2 alloying (×).

The variation σ_{Kmax} as a function δ_p/R when entering the plastic deformation (δ_p=0.1-10 μm) in fig. 4 is shown. The experimental data processing using the EXCEL computing program resulted in the expression of relative plastic deformation:

$$\frac{\delta_p}{R} = K_1\sigma_{max}^{K_2} \qquad (4)$$

where: K_1, K_2 - constants specific for the specimen structural conditions.

The values of these constants and the size of the load carrying capacity determined by equation (4) are shown in table 2. The quasi-static load carrying capacity increases with the hardness, yield point, respectively, and with the hardened layer depth.

The maximum tangential tension of plastic deformation initiation under contact is displaced at higher pressing forces and depths.

The increase of the load carrying capacity by laser processing varies between 8.7 and 28.8% related to the reference structural condition.

Table 2. The values of K_1, K_2 and the size of the load carrying capacity.

| Specimen | Material constants | | r^2 | Load carrying capacity |
	K_1	K_2		σ_{Kmax} (MPa)
1	3.18×10^{-18}	4.142	0.950	3162
2	2.35×10^{-27}	6.681	0.932	3439
3	1.93×10^{-56}	14.089	0.991	3877
4	2.38×10^{-33}	8.206	0.939	4073

NOTE: r^2 - correlation coefficient of experimental data.

Friction Coefficient

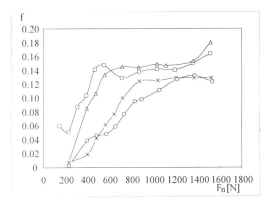

Fig. 6 - The evolution of the friction coefficient with the size of contact normal force for 1%C steel: volume quenching (□), laser hardening (Δ), W1 alloying (×), W2 alloying (o).

The evolution of the friction coefficient with the size of contact normal force (fig. 6) points out the increase of the friction coefficient during the pressing of the plane semispace roughnesses, followed by a stabilization in the volume plastic deformation field. The normal force of entering the plastic deformation increases with the yield point ot the surface layer. The hardened layers have maximum friction coefficient.

The occurrence of the plastic deformation collar around the penetrator and the material accumulation in front of it increase the advance resistance and the friction surface. The oxide film is damaged and the adhesion susceptibility of the contact materials increases.

Until entering the plastic deformation (~ 550N) the friction is heavier with the volume quenching, after which it reaches higher values with laser hardness. It can be correlated to the stronger carbide dissolution and the increased amount of residual austenite when laser quenching. Over 1200N in both conditions the friction coeffcient increases strougly with a gripping tendency.

The lowest friction coeffcients occur in the alloyed layers with maximum load carrying capacity where the plastic deformation component of the friction coefficients is lower. The composite structure where the dispersed WC, M_6C - type carbides are present, has an adhesion inhibiting feature, with no occurrence of gripping tendency in the stress range. The W2 formula alloying, providing the maximum thickness of hardened layer, shows the maximum load carrying capacity and the lowest friction coefficients.

CONCLUSIONS

The laser hardening and alloying of 1%C steel are efficient ways of hardening and displacement of elastic-plastic transition at higher contact tensions. The quasi-static load carrying capacity increases with the hardness and depth of the hardened layers.

The laser hardening increases the load carrying capacity by 8.7% compared to the volume quenching, reduces the friction coefficient until entering the plastic deformation without changing the gripping tendency.

The obtaining of composite structures by alloying with W1 and W2 formulae increases the load carrying capacity by 22.6% respectively 28.8%, reduces drastically the friction coeffcient and the gripping tendency.

REFERENCES

1. Burakov V.A., Kanapenas R.M. (1987). *Lazernaia tehnologia II,* Vilnius, pp. 12.
2. Barton G., Bergmann H.W., Mordike B.L., Gros N. (1988) *Laser Materials Processing II, SPIE* **vol. 455**, pp. 113.
3. Porter D.A., Easterling K.E. (1992). *Phase Transformations in Metals and Alloys.* Chapman & Hall, London.
4. Munteanu V., Levcovici D.T., Paraschiv M.M., Levcovici S.M. (1997). In: *Surface Engineering vol. 13,* **1**, The Institute of Materials, London, pp. 75.
5. Sallamand P., Pelletier J.M., Vannes A.B. (1994). In: *Proceedings of the Surface Modification Technologies VIII,* The Institute of Materials, Nice, pp. 287.
6. Pelletier J.M., Fouguet F., Dezert D., Robin M., Vannes A.B. (1992). In: *Proceedings of the Laser Treatment of Materiales,* Göttingen.
7. Levcovici S.M., Oprea F., Levcovici D.T., Paraschiv M. (1994). In: *Proceedings of the Surface Modification Technologies VIII,* The Institute of Materials, Nice.
8. Levcovici B.G., Levcovici S.M.., Levcovici D.T. (1995). In: *Proceedings of the 4th European Conference on Advanced Materials and Processes, EUROMAT '95,* Padua/Venice, vol. D, pp. 381.
9. Levcovici S.M., Levcovici D.T., Gologan V., Farkaş L. (1997). In: *Proceedings of the 5th European Conference on Advanced Materials and Processes, EUROMAT '97,* Maastricht, pp. 131.
10. Krageliskii I.V. (1962) *Trenie i iznos.* MAŞGIZ, Moskva.
11. Klaprodt T. (1981) *Antriebstechnick nr. 6,* pp. 271.

Low Cycle Fatigue and Elasto-Plastic Behaviour of Materials
K-T. Rie and P.D. Portella (Editors)
© 1998 Elsevier Science Ltd. All rights reserved.

EFFECT OF THE NEAR-SURFACE LAYERS AND SURFACE MODIFICATION ON BEHAVIOUR OF BCC METALS UNDER FATIGUE.

V.F. TERENT'EV, A.G. KOLMAKOV and G.V. VSTOVSKY
Institute of metallurgy of Russian Academy of Sciences,
Leninsky prospect 49, Moscow, 117911, Russia

ABSTRACT

It is proposed to consider the near-surface layers as a subsystem of an overall system of deformed material. A capacity to control the material properties due to the modification of its near-surface layers is considered.

KEYWORDS

Influence of surface state on mechanical properties, surface modification, self-organization, BCC metals, fatigue crack initiation, multifractal analysis

PHENOMENOLOGICAL BACKGROUND

Under mechanical loading, material manifests itself as a thermodynamic open system. The properties of such a system are defined by the processes of self-organization of dissipative multifractal structures. In the near-surface layers of deformed materials the processes of self-organization are faster as compared with an internal volume of the material [1-9].

For the BBC metals (Mo, Mo-alloys and steels) are under fatigue loading, there is a relation between the surface effects and period of fatigue cracks initiation. The latter occurs in several stages: 1) cyclic micro-flow; 2) cyclic flow (non-homogeneous deformation); 3) cyclic strengthening (ends with beginning of submicrocracks generation); 4) surface micro-cracks initiation (ends with the initiation of micro-cracks with the depth of the order of grain size) (Fig.1) [11].

An appearance of the physical fatigue limit can be explained by a phenomenological model based on the idea of dislocation barrier in the stronger near-surface layer (with the depth of the order of grain size) formed at the early

424

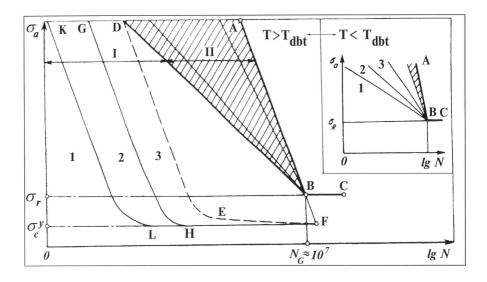

Fig. 1. Periods and stages of cyclic deformation : I - the period of the initiation of cracks, II = the period of the expansion of cracks; 1 - the stage of cyclic micro-flow, 2 - the stage of cyclic flow, 3 - the stage of cyclic strengthening; ABC - the fatigue curve, DB - the curve of fatigue damaging, DEF - the line of fatigue submicrocracks initiation, GH - the end line of cyclic flow stage, KL - the line of cyclic flow stage beginning, BF - the line of |appearance of stable structure of the material with defects at $\sigma < \sigma_r$; σ_r is a fatigue limit, σ_c^y is a cyclic yield stress, N_G is the basic cycle number of test; T - the temperature, T_{dbt} - the temperature of a ductile-brittle transition.

stages of cyclic loading. When the stresses become close to the fatigue limit, there exist two stages of strengthened near-surface layer formation: 1) the predominant flow of this layer at the stages of cyclic micro-flow and flow; 2) strain hardening in combination with the effect of the dynamic deformation aging of the near-surface layer with the appearance of barrier effect (even in the case of submicrocracks appearance) in presence of the overall plastic flow of the metal up to the basic number of loading cycles. When the cycle number achieves its basic value, an advanced dislocation cellular structure forms in the metal, but in the near-surface layer, in addition to cellular and band dislocation structures, there are the fatigue bands of gliding with a system of non-propagating sub- and micro-cracks that are hold by the grain boundaries and by relaxation processes at crack tips For the greater bases of tests ($N>10^7...10^8$) and, respectively, lower amplitudes, the initiation of fatigue cracks occurs under surface layer. The change in the mechanical properties of near-surface layer (due to the faster growth of dislocation density in this layer and the processes of dynamic deformation aging), barrier effect of this layer (due to obstruction to the exit of dislocations onto the surface) and formation of damage zones (the plastic zones at the tips of non-propagating submickrocracks) influence the behavior of BCC metals at the fatigue limit. When the temperature of fatigue test T is below the temperature of ductile-brittle transition T_{dbt}, $T<T_{dbt}$, formation of the near-surface layer and its effect on resistance to break are the same as for $T>T_{dbt}$. However, at $T<T_{dbt}$ the processes of cyclic deformation accumulation and micro-crack initiation are more localized in the near-surface layer (the structural changes are insignificant in the most of the material), and the near-surface layers are more plastic as compared with the internal volume (the surface dislocation sources are active merely in these layers).

Taking into account the data above, it is proposed to consider the near-surface layers as the subsystem of the overall system of deformed material. In such an approach the subsystem of the internal volumes of material is responsible for the intrinsic properties of the system, and the subsystem of the near-surface layers is responsible for an exchange by energy, substance and information with an environment and determines the external properties of system under given conditions. Hence there is a capacity to control the properties of material by the modification of its near-surface layers.

EXPERIMENTAL RESULTS

An approach described was used in the investigations of changes of the state of the near-surface layers of the number of BCC metals and alloys under mechanical loading:

- the PVD-deposition of the Re-coatings (of thickness $h=8...10$ μm), Ni-coatings ($h=4$ μm) and Mo-47%Re-alloy-coatings ($h=0.5$ μm) on sheeted (thickness $h=1$ μm) and wire-formed (diameter $d=0.2$ and 0.8 mm) specimens of technically pure Mo with subsequent diffusion annealing;
- the ion implantation ($D=10^{16}..10^{17}$ cm^{-2}, $E=20...80$ keV) of the specimens of 30ХГСНА ($0.29C-0.94Mn-1.02Si-1.09Cr-1.61Ni-0.18Cu-0.05Mo$ in weight %) and 12X2H4AШ ($0.12C-0.5Mn-1.4Cr-3.5Ni$ in weight %) steels ($d_{min}=1.52$ mm) with the depth of the intrusion of C, N and B ions down to 40 nm (Fig.2);
- the change of geometrical structure of surface micro-defects by the surface treatment of Mo ($d=0.2...1$ mm) and Mo-0.1HfC-0.1HfN-0.03C alloy ($d=0.3$ mm) wires. In these investigations the traditional methods of investigations were accomplished by the use of an original method of multifractal parametrization of structures. A very good correlation of mechanical properties with a uniformity index of defect structure was obtained (correlation coefficient $>0,9$).

The surface modification led to the increase in fatigue limit: in the first case - by 5...25%, in the second case - by 18...31% and in the third case - by 90%. The relative effective depth of modified surface layer (the relation of the depth of modified layer to the specimen diameter or thickness) was $0.001...0.002$. Modification of the greater depth was not accompanied by the noticeable change in the mechanical properties.

This work was supported by RFFI in the context of grant No 96-15-98243 for the state support of the leading scientific schools of Russian Federation.

426

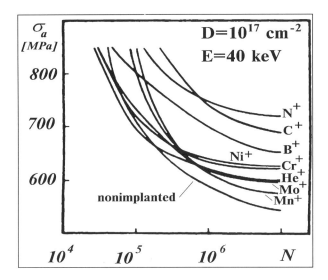

Fig. 2. Fatigue curves of the specimens of 30ХГСНА steel after ion implantation.

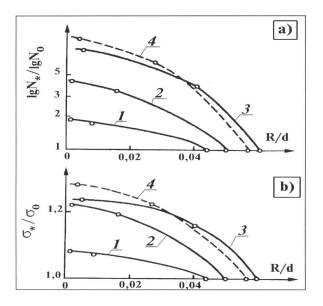

Fig. 3. The dependence of relative change of fatigue life N_*/N_0 at $\sigma_a=1000$ MPa (a) and the relative change of bounded fatigue limit σ_*/σ_0 on basis $N=10^5$ (b) on the relative maximum size of surface micro-defects R/d of wires: 1 - Mo (d=1 mm), 2 - Mo (d=0.5 mm), 3 - Mo (d=0.2 mm), 4 - Mo-0.1HfC-0.1HfN-0.03C alloy (d=0.3 mm); the figures marked with 0 and * are for the specimens before and after the treatment, respectively.

REFERENCES

1. Kramer, I. R. (1986) Adv.Mech. and Phys.Surface **3**, 109.
2. Terent'ev, V. F. (1969) Russian Physics (Doklady) **185**, 83.
3. Terent'ev, V. F. (1969) Russian Physics (Doklady) **185**, 324
4. Kolmakov, A. G., Geminov, V. N., Vstovsky, G. V. et al. (1995) Surface and Coatings Technology **72**, 43.
5. Kolmakov, A. G., Rybakova, L. M. and Terent'ev, V. F. (1994) Phys. and Chem. Mater. Treatment **2**, 76.
6. Vstovsky, G. V., Kolmakov, A. G. and Terent'ev, V. F. (1994) In: *Proceedings of Conference «EUROMAT 94 TOPICAL»*, Hungary, Balatonszeplak, Vol.IV, pp. 1187-1191.
7. Kolmakov, A.G., Geminov, V.N. and Terent'ev, V.F. (1994) In: *Proceedings of Conference «EUROMAT 94 TOPICAL»*, Vol.IV, pp. 1164-1168.
8. Kolmakov, A.G., Geminov, V.N. and Terent'ev, V.F. (1996) Phys. and Chem. Mater. Treatment **1**, 87.
9. Journal of Offshore and Polar Engineering, 1997, 7, (1), 44-47.
10. Kolmakov, A.G. (1997) Phys. and Chem. Mater. Treatment **3**, 49.
11. Terent'ev, V. F.. (1996) Russian Metallurgy (Metally) **6**, 14.

Chapter 6

ADVANCED MATERIALS

CYCLIC STRESS-STRAIN AND FATIGUE BEHAVIOUR OF PARTICULATE-REINFORCED AL-MATRIX COMPOSITES

O. HARTMANN, H. BIERMANN AND H. MUGHRABI

Institut für Werkstoffwissenschaften, Lehrstuhl I, Universität Erlangen-Nürnberg,
Martensstr. 5, D-91058 Erlangen, Fed. Rep. Germany

ABSTRACT

The low-cycle fatigue behaviour of the Al-composite AA6061 reinforced with Al_2O_3-particles of different volume fractions f_p (15 and 20 vol. %) has been investigated. Cyclic deformation tests were carried out on material in T6-condition at temperatures between $T = -30\,°C$ and $T = 150\,°C$ in total-strain control mode. Both cyclic hardening and cyclic softening were observed, dependent on the strain amplitude and test temperature, respectively. Cyclic hardening was obtained for all total-strain ranges $\Delta\epsilon_t$ ($\Delta\epsilon_t = 0.004$ up to 0.02) at temperatures below and at room temperature. At higher temperatures, cyclic hardening was only observed for $\Delta\epsilon_t \geq 0.014$. The composite with the higher volume fraction of reinforcement phase exhibited higher cyclic strengths but lower fatigue lives. The number of cycles to failure, N_f, could be related to the applied strain amplitude by the laws of Manson-Coffin and Basquin. In the evaluation of the plastic-strain hysteresis loops, the observed non-linear elastic behaviour was taken into account by extension of the linear Hooke's law by a second-order term. The variation of non-linearity during fatigue was found to be a sensitive indicator of damage.

KEYWORDS

Al-based MMC, $AA6061/Al_2O_3p$, low-cycle fatigue, fatigue life, non-linear elastic behaviour, precipitation hardening

INTRODUCTION

There have been increased efforts in developing reinforced light metal alloys in the last decade. These metal-matrix composites (MMCs) are now used increasingly in different applications because of their higher strength and their improved strength-to-weight ratio. As the fatigue behaviour of materials often differs from that under monotonic loading, it is important to know the influence of loading parameters on the fatigue behaviour. In MMC-materials with a precipitation-hardened matrix, it is of interest to study the decrease of the cyclic strength at higher temperatures at which ageing and recovery processes are accelerated. Another important factor is the influence of the residual stresses which are induced by the production processes and thermal treatments and which can change during

432

mechanical straining. Thus, the main objectives of the present study were to investigate in detail the cyclic stress-strain and fatigue behaviour and its dependence on temperature and mechanical loading for specimens with different volume fractions of reinforcement.

MATERIAL AND EXPERIMENTAL PROCEDURE

In this study, the low-cycle fatigue behaviour of the composite AA6061-Al$_2$O$_3$-15p-T6 has been investigated in the temperature range between $T = -30\,^\circ$C and $T = 150\,^\circ$C and that of the composite AA6061-Al$_2$O$_3$-20p-T6 at room temperature. The aluminium-matrix AA6061 is hardened by Mg$_2$Si-precipitates (β") and reinforced by Al$_2$O$_3$-particles with a volume fraction of 15 % and 20 %, respectively. The mean size of the mostly rectangular particles is about 15 μm. Figure 1 shows the microstructure of the 20 % particulate-reinforced MMC. The particles are aligned in extrusion direction (horizontal) and distributed nearly uniformly, except for a few zones showing agglomeration. During the production process, the alumina particles and magnesium of the AA6061-matrix react to the spinel-phase MgAl$_2$O$_4$ [1–3] which leads to good cohesion between the matrix and the alumina particles. A scanning electron microscopic (SEM) image of the spinel is shown in fig. 2.

The material was aged to peak-hardness by T6 heat treatment. The ageing treatment was controlled by hardness tests. The ageing behaviour of the two investigated MMCs is similar, therefore both materials were heat-treated in the same way to achieve T6-condition ($560\,^\circ$C/30 min/H$_2$O + $165\,^\circ$C/16 h/air). After mechanical grinding and polishing the specimens, isothermal, total-strain controlled fatigue tests were carried out with total-strain amplitudes ranging between $\Delta\epsilon_t/2 = 2 \times 10^{-3}$ and $\Delta\epsilon_t/2 = 10 \times 10^{-3}$ ($R_{\epsilon_t} = -1$). The monotonic stress-strain curves were obtained at the same total-strain rate, $\dot{\epsilon}_t$, as used for the cyclic deformation ($\dot{\epsilon}_t = 2 \times 10^{-3}\,\mathrm{s}^{-1}$).

RESULTS AND DISCUSSION

Figure 3 shows some cyclic deformation curves of the 15 vol. %-reinforced MMC deformed at different total-strain amplitudes ($\Delta\epsilon_t/2 = 3 \times 10^{-3}$, 5×10^{-3} and 8×10^{-3}) and at two temperatures ($T = -30\,^\circ$C and $T = 150\,^\circ$C) in a semi-logarithmic plot of the stress

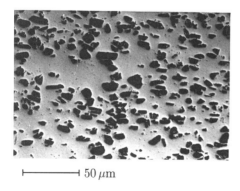

$\longmapsto$ 50 μm

Fig. 1: Optical micrograph of the composite AA6061-Al$_2$O$_3$-20p.

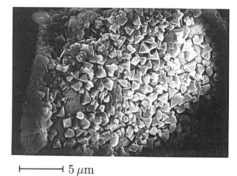

$\longmapsto$ 5 μm

Fig. 2: Spinel MgAl$_2$O$_4$ on the surface of an Al$_2$O$_3$-particle.

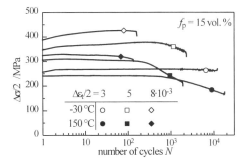
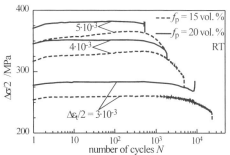

Fig. 3: Cyclic deformation curves at different temperatures for several total-strain
amplitudes.

Fig. 4: Cyclic deformation curves for different volume fractions of reinforcement and
total-strain amplitudes.

amplitudes $\Delta\sigma/2$ as a function of the number of cycles N. The symbols on the curves
at half the number of cycles to failure describe the test parameters. All curves show initial cyclic hardening and a subsequent continuous cyclic softening until fracture. The
amount of initial cyclic hardening increases with the applied total-strain amplitude and
decreases with increasing temperature. The cyclic softening becomes more pronounced
and occurs earlier with increasing temperature and/or increasing total-strain amplitude.
At $\Delta\epsilon_t/2 = 3 \times 10^{-3}$ and $T = -30\,°\mathrm{C}$ nearly no cyclic softening appears. In contrast, the
curve at the same total-strain amplitude but at $T = 150\,°\mathrm{C}$ shows more pronounced cyclic
softening. Hence, one can suppose that the microstructure is not stable at the higher temperature. A possible explanation for this behaviour is that repeated cutting of the Mg_2Si
precipitates during cyclic deformation leads to dissolution at higher temperatures, because
the size of the precipitates becomes smaller than the critical nucleus size. The results from
the fatigue test at the total-strain amplitude of $\Delta\epsilon_t/2 = 5 \times 10^{-3}$ and $T = 150\,°\mathrm{C}$ corro-
borate this assumption. One can recognize a decrease of the cyclic stress amplitude which
is more rapid and occurs earlier than in comparable tests at $\Delta\epsilon_t/2 = 3 \times 10^{-3}$. Because of
the higher deformation the precipitates will be cut more often, hence the cyclic softening
appears earlier and is more pronounced. It is assumed that the observed cyclic softening at
$\Delta\epsilon_t/2 = 5 \times 10^{-3}$ and $T = -30\,°\mathrm{C}$ is due to failure of the reinforcing particles or decohesion
in addition to precipitation cutting. In fig. 4 cyclic deformation curves of MMCs with diffe-
rent volume fractions of reinforcement f_p are compared at several total-strain amplitudes.
In the case of specimens with higher f_p (20 vol. %) higher stress amplitudes are required at
the same strain amplitudes, cyclic softening is less pronounced and fatigue lives are shorter.

Figure 5 shows cyclic stress-strain (css) curves for the temperatures $T = -30\,°\mathrm{C}$ and
$T = 150\,°\mathrm{C}$ which were obtained by plotting the values of the stress amplitude at half
the number of cycles to failure as "saturation" stress amplitudes $\Delta\sigma_s/2$ versus the plastic-
strain amplitude $\Delta\epsilon_{pl}/2$. The plot includes also the monotonic stress-strain curves. Comparing the css curves and the monotonic stress-strain curves, the investigated MMC with
$f_p = 15$ vol. % shows cyclic hardening for all strain amplitudes below and at room temperature. At higher temperatures, the fatigue behaviour is different. Cyclic hardening
occurs only at high strain amplitudes and therefore short test times. For small amplitudes and long test durations, cyclic softening is observed. Comparing room temperature
css-curves of MMCs with different volume fractions of reinforcement shows that the MMC
with $f_p = 20$ vol. % softens cyclically at small strain amplitudes (see fig. 6). The composite

434

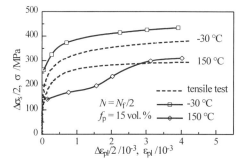
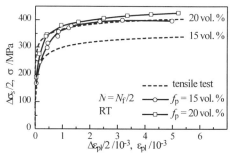

Fig. 5: Cyclic and monotonic stress-strain curves at different temperatures.

Fig. 6: Cyclic and monotonic stress-strain curves for different volume fractions of reinforcement.

with $f_p = 15\,\text{vol.}\,\%$ hardens cyclically over the entire investigated range of cyclic deformation. These results differ from those of Perng *et al.* [4] and Llorca [5], who obtained cyclic softening at small strain amplitudes even at room temperature. The increase of cyclic strength (compared to monotonic strength) is more pronounced for the 15 vol. %-MMC.

The relationships between the applied strain amplitudes $\Delta\epsilon_t/2$ and $\Delta\epsilon_{pl}/2$ and the elastic strain amplitude $\Delta\epsilon_{el}/2$ on the obtained number of cycles to failure, $2N_f$, are represented in fig. 7. Here, the results of room temperature fatigue tests on the 15 vol. % particulate reinforced MMC (solid lines) and on the MMC with a fraction of particles of 20 vol. % (dashed lines and symbols) are compared in a double-logarithmic plot. The strain amplitudes $\Delta\epsilon_{pl}/2$ and $\Delta\epsilon_{el}/2$ can be represented by straight lines in a good approximation. For reasons of clarity in the case of the 15 vol. %-MMC, the actual data points are not shown. The slopes and the ordinate-values at $N_f = 0.5$ of these lines yield the coefficients (ϵ_f, σ_f/E) and exponents (c, b) of the equations of Manson-Coffin and Basquin:

$$\text{Manson-Coffin:} \quad \frac{\Delta\epsilon_{pl}}{2} = \epsilon_f \cdot (2N_f)^c \qquad \text{Basquin:} \quad \frac{\Delta\epsilon_{el}}{2} = \frac{\sigma_f}{E} \cdot (2N_f)^b.$$

The parameters of these equations for fatigue tests at room temperature are given in table 1. The MMC with the higher volume fraction of reinforcement exhibits shorter fatigue lives in all performed fatigue tests. This is in good agreement with results of other groups [4, 6, 7]. SEM-investigations of the samples show a high density of broken particles around the crack tip. It is assumed that stress concentrations near particles lead to particle fracture and failure of the specimens.

Table 1: Parameters of the Coffin-Manson and Basquin equations for room temperature.

f_p in vol. %	c	ϵ_f	b	$\frac{\sigma_f}{E}\,/10^{-3}$
15	-0.71	0.35	-0.14	12.0
20	-0.70	0.15	-0.12	8.7

Another objective of the present work is the examination of the non-linear elastic behaviour of the composites. Unusual sickle-shaped hysteresis loops are obtained, when the plastic strain is calculated according to the linear Hooke relation:.

$$\epsilon_{pl}(E) = \epsilon_t - \epsilon_{el} = \epsilon_t - \frac{\sigma}{E} \tag{1}$$

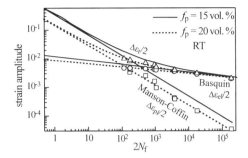

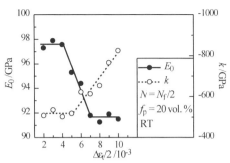

Fig. 7: Relationship between applied strain amplitude and twice the number of cycles to failure $2N_f$ for different volume fractions of reinforcement.

Fig. 8: Values of E_0 and k obtained from calculations of plastic-strain hysteresis loops as a function of the applied total-strain amplitude.

The non-linear elastic behaviour was taken into account by extension of Hooke's law by a second-order term (see eqn. 2), as in an earlier work [8]. This leads to eqn. 3 for the plastic strain as a function of the two parameters E_0 and k.

$$\sigma = E_0\epsilon_{el} + k\epsilon_{el}^2 \tag{2}$$

$$\epsilon_{pl}(E_0, k) = \epsilon_t - \frac{2\sigma}{E_0 + \sqrt{E_0^2 + 4k\sigma}} \tag{3}$$

k represents the degree of deviation from the loading/unloading linear-elastic branches of the plastic-strain hysteresis loop calculated with eqn. 1. Applying eqn. 3, the originally bent hysteresis loops (σ vs. $\epsilon_{pl}(E)$) could be converted into hysteresis loops of the usual shape (σ vs. $\epsilon_p(E_0, k)$), for which both elastic branches of the hysteresis loops are parallel. The extension of Hooke's law was originally related to the asymmetrical shape of the potential energy vs. distance-curve [8, 9]. The stress-dependence of the elastic modulus E of Al-based MMCs has also been reported in other studies [10, 11]. For pure Al and Al-alloys, k could be calculated from the elastic constants of third order [9] and varies from $\sim$-350 GPa (Al) to $\sim$-500 GPa (Al-Si alloys). It is emphasized that k is negative. In the case of damaged particulate-reinforced alloys, it is supposed that the non-linear "elastic" behaviour stems partially from the reduction of the compliance by particle/matrix failure (fracture, decohesion, voids) during mechanical loading and can hence be regarded as a sensitive indicator of damage. Particle reinforcement enhances E_0 from $\sim$ 68 GPa to $\sim$ 95 GPa ($f_p = 20$ vol. %). A closer examination of the Young's modulus in tension and compression as a function of the number of cycles shows that both parameters decrease during cyclic deformation, but the decrease of the (smaller) Young's modulus in tension is more rapid than in compression. The value of E_0 is also a function of the applied total-strain amplitude. Figure 8 shows the values of E_0 and k used for the calculation of different plastic-strain hysteresis loops as a function of the total-strain amplitude. For all applied $\Delta\epsilon_t$, E_0 varies between 91.3 GPa and 97.9 GPa and k is in the range from -502 GPa to -825 GPa. The curve of E_0 vs. $\Delta\epsilon_t/2$ can be divided into three regions. For $\Delta\epsilon_t/2 \leq 0.004$, E_0 is nearly constant. As $\Delta\epsilon_t$ is increased, the E_0-values decrease and then merge into another plateau for $\Delta\epsilon_t/2 \geq 0.007$. On the other hand, $|k|$ is also nearly constant up to $\Delta\epsilon_t/2 \leq 0.005$ and increases for higher strain amplitudes. It is hence inferred that particle fracture occurs during cyclic deformation at strain amplitudes which exceed a given amount ($\Delta\epsilon_t/2 = 0.004$). This is in agreement with results from other studies. Elomari

436

et al. [12] found that the fraction of broken particles increases with prestrain and that large particles have a greater tendency to fracture than smaller ones. Lloyd [13] associated a decrease of the Young's modulus with an increase of the fraction of broken particles. This behaviour is more pronounced with increasing volume fraction of reinforcement [14].

CONCLUSIONS

1. The results of fatigue tests between $T = -30\,°C$ and $T = 150\,°C$ on 6061 T6 composites reinforced with 15 and 20 vol. % Al_2O_3-particles showed cyclic hardening and subsequent cyclic softening, depending on test temperature and strain amplitude. The cyclic hardening is more pronounced with decreasing temperature and/or increasing total-strain amplitude, while cycling softening is enhanced at higher temperatures.
2. Specimens with a higher volume fraction of particles have higher strengths and lower fatigue lives.
3. The investigated MMCs exhibit a non-linear elastic behaviour. Taking into account the extension of the linear Hooke's law by a term of second order ($\sigma = E_0\epsilon_{el} + k\epsilon_{el}^2$), the originally sickle-shaped hysteresis loops assume normal shapes. The two parameters E_0 and $|k|$ decrease and increase, respectively, with the increase of the applied total-strain amplitude. It is concluded that the variations of the values of E_0 and k are sensitive indicators of particle/matrix damage.

ACKNOWLEDGEMENTS

This research was financially supported by Deutsche Forschungsgemeinschaft (DFG) within the Gerhard-Hess Programme. The authors are grateful to Leichtmetall-Kompetenzzentrum, Ranshofen, Austria, for providing the material.

REFERENCES

1. Suéry, M. and L'Esperance, G. (1993). *Key Eng. Mat.* **79/80**, 33.
2. McLeod, A.D. and Gabryel, C.M. (1992). *Met. Trans. A* **23A**, 1279.
3. Berek, H., Zywitzki, O., Degischer, H.P. and Leitner, H. (1994). *Z. Metallkd.* **85**, 131.
4. Perng, C.C., Hwang, J.R. and Doong, J.L. (1993). *Comp. Sci. Technol.* **49**, 225.
5. Llorca, J. (1994). *Scripta metall. mater.* **30**, 755.
6. Hall, J.N., Jones, J.W. and Sachdev, A.K. (1994). *Mat. Sci. Eng.* **A183**, 69.
7. Srivatsan, T.S. (1995). *Int. J. Fatigue* **17**, 183.
8. Sommer, C., Christ, H.J. and Mughrabi, H. (1991). *Acta metall. mater.* **39**, 1177.
9. Wasserbäch, W. (1991). *phys. stat. sol. (a)* **164**, 121.
10. Biermann, H., Beyer, G. and Mughrabi, H. (1996). In: *Verbundwerkstoffe und Werkstoffverbunde*, pp. 197-200, Ziegler, G. (Ed.). DGM Informationsgesellschaft, Oberursel.
11. Maier, H.J., Rausch, K. and Christ, H.-J. (1996). In: *Proc. 6th Int. Fatigue Congress (Fatigue '96)*, pp. 1469-1474, Lütjering, G. and Nowack, H. (Eds.). Elsevier Science Ltd., Oxford.
12. Elomari, S. and Boukhili, R. (1995). *J. Mat. Sci.* **37**, 3037.
13. Lloyd, D.J. (1991). *Acta metall. mater.* **39**, 59.
14. Hunt, W.H., Brockenbrough, J.R. and Magnusen, P.E. (1991). *Scripta Metall.* **25**, 15.

Low Cycle Fatigue and Elasto-Plastic Behaviour of Materials
K-T. Rie and P.D. Portella (Editors)

INFLUENCE OF LOCAL DAMAGE ON THE CREEP BEHAVIOUR
OF PARTICLE REINFORCED METAL MATRIX COMPOSITES

C. BROECKMANN and R. PANDORF

Institute for Materials, Ruhr-University Bochum, 44780 Bochum, Germany

ABSTRACT

The creep behaviour of a particle reinforced metal matrix composite is numerically modelled by use of the FE-method. In order to describe the influence of ceramic particles on the overall creep curve, idealised unit cell models, representing a periodic arrangement of particles are generated. The particles are assumed to react linear elastically. Particle cleavage is taken into account. For the matrix an additive viscosity hardening law coupled with an isotropic damage variable is used. Numerical studies depicting the influence of the particle's aspect ratio and the evolution of local damage on the global behaviour are presented.

KEYWORDS

Finite element method, creep, local damage, metal matrix composite

INTRODUCTION

The use of particle reinforced metal matrix composites (PMMC's) results from their excellent stiffness-to-weight ratio, superior wear resistance as well as the reduced thermal expansion. In technical applications, MMC's are often exposed to elevated temperatures. Therefore, the creep behaviour of these materials is of particular interest. Results of creep tests with particle reinforced aluminium alloys were published by [1-7]. In tensile tests a pronounced region of secondary creep with a constant strainrate was not observed. In contrast to this, creep tests under uniaxial pressure show distinct secondary creep. Thus, an influence of stress controlled damage on the global strain rate was supposed. Wether or not particles retard creep straining can not be answered uniformly. At low temperatures and high stresses an increasing volume fraction of particles leads to an increasing minimum strainrate. This relation is reversed for high temperatures and low stresses. It can be concluded that the creep behaviour of PMMC's strongly depends on the combination of the external parameters stress σ and temperature T.

The aluminium alloy 6061 reinforced with 22 vol% Al_2O_3-particles was tested under uniaxial constant stress at a temperature of T=300 °C. The creep curves are characterized by a very short

438

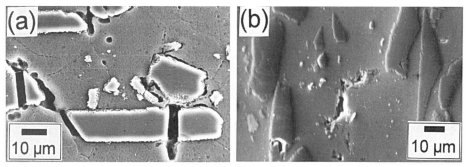

Fig. 1: Local damage in particle reinforced AA 6061 after creep loading at T = 300 °C:
a) broken Al_2O_3-particle, b) voids in the metal matrix

primary creep stage and a large tertiary creep regime. Longitudinal sections of crept specimens were prepared after interupting the tests in the teriary creep stage. Two different damage mechanisms were found in the material: damage due to particle cleavage (fig. 1a) and matrix damage by void initiation and growth (fig.1b). Most of the cleavage cracks are oriented perpendicular to the direction of macroscopic loading. Thus, it is assumed that particle fracture is governened by a critical normal stress within the particle. In order to investigate the evolution of particle

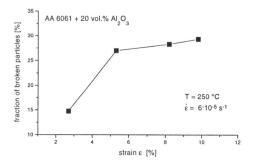

Fig. 2: Fraction of broken particles as a function of the macroscopic strain

cracking strain controlled tests with a constant strain rate of $\dot{\varepsilon} = 6 \cdot 10^{-5}$ s^{-1} were carried out. Using scanning electron microscopy the fraction of broken particles was determined in longitudinal sections. Figure 2 shows the evolution of carbide cleavage as a function of the global strain.

NUMERICAL MODEL

The creep behaviour was numerically simulated on a microscopic scale using the finite element program CRACKAN [8]. A multiphase model was used for the simulation of the microstructure. The carbides are assumed to react linear elastically. They may fracture due to a critical local maximum principal stress. The metal matrix was modelled with a viscoplastic material law taking into account isotropic strain hardening [9] and damage. The total strain rate is divided into an elastic and a plastic part

$$\dot{\varepsilon} = \dot{\varepsilon}_e + \dot{\varepsilon}_p \tag{1}$$

The stress is connected with the elastic strain by Hooke's law. The viscoplastic strain rate depends on the state of stress, the total plastic strain and an isotropic damage variable ω:

$$\dot{\varepsilon}_p = \frac{3}{2} \left\langle \frac{\sigma_{eq} - \sigma_i(\varepsilon_p)}{K_a (1 - \omega(\sigma_{eq}))} \right\rangle^{N_a} \frac{s}{\sigma_{eq}} \quad . \tag{2}$$

σ_{eq} denotes the effective von Mises stress and s is the deviatoric part of the stress. The internal stress σ_i depends on the total plastic strain:

$$\sigma_i(\varepsilon_p) = \sigma_s + Q_1 \varepsilon_{p,\,eq} + Q_2 [1 - e^{-b\varepsilon_{p,\,eq}}] \quad . \tag{3}$$

$\varepsilon_{p,eq}$ is the equivalent plastic strain, defined by

$$\varepsilon_{p,\,eq} = \frac{2}{3} \sqrt{\varepsilon_{p,\,ij}\, \varepsilon_{p,\,ij}} \quad . \tag{4}$$

The parameters K_a, N_a, σ_s, Q_1, Q_2 and b have to be determined experimentally. It is assumed that $\omega = 0$ for the undamaged material and $\omega = 1$ at time to fracture t_f. An evolution equation for the damage parameter ω was developed by Kachanov (see also [10]):

$$\dot{\omega} = \left(\frac{\sigma_{eq}}{B_0 (1 - \omega)} \right)^r \quad . \tag{5}$$

B_0 and r are material parameters which have to be determined by creep tests. All parameters used in the numerical studies are given in tab. 1. They were determined by relaxation, hardening and creep tests with pure AA 6061 matrix material at a testing temperature of T=300 °C.

Table 1: Parameters for the numerical studies

E_p [GPa]	v_p	E_M [GPa]	v_M	N_a	K_a [MPa]	σ_s [MPa]	Q_1 [MPa]	Q_2 [MPa]	b	B_0 [MPa]	r
400.0	0.22	51.3	0.422	4.55	1712.0	3.0	657.0	12.2	614	1026	4.7

NUMERICAL STUDIES

Two numerical studies are presented in this article. The first one investigates the influence of a particle's aspect ratio on the creep behaviour of the undamaged composite and the second deals with the effect of local damage on the global creep deformation. Both studies were calculated under the assumption of plane strain. In order to represent a periodic arrangement of particles a unit cell technique was used.

Figure 3 shows the FE-model for the first study. Spacial periodicity is guaranteed by the use of kinematic boundary conditions where the borders of the structure remain straight and parallel

440

during deformation. A volume fraction of f_p=16 vol.% Al_2O_3-particles was modelled. The aspect ratio h/b of the particle was varied between 0.25 and 4.0. The structure was loaded by a constant nominal stress σ_N. In this study damage effects were not taken into account. The resulting creep curves are plotted in fig. 4. It is known from previous numerical investigations that the creep rate is decreased due to the presence of plastically undeformable particles [11]. The present result shows that the profit in creep resistance is the higher the more the aspect ratio deviates from 1. This result seems to be rather independent of the orientation of the particle. An aspect ratio of 4 leads to a creep curve close to that obtained by an aspect ratio of 0.25. It will be seen later on that the stress within the particle strongly depends on its orientation with respect

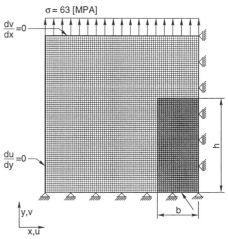

Fig. 3 FE-model, arrow indicates the point for the plot of the particle stress

to the direction of global loading. Thus, the reason for the decrease of creep strain in case of very slender particles is not the stress bearing capacity of the particles, but the impede in strain near the interface in the matrix. In fig. 5 the normal stress σ_{yy} in the centre of the particle is plotted as a function of time. The initial local stress depends on the aspect ratio. For particles with an aspect ratio h/b$\leq$1.0 this stress is approximately as high as the macroscopic stress σ_N. Due to the so called fibre effect the normal stress in the centre of the particle increases with an increasing aspect ratio. A distinctive increase of stress with time is only monitored for high aspect ratios, while for aspect ratios $\leq$1.0 the particle's stress relaxes with time. As stated before, cleavage of particles is controlled by a local normal stress criterion. The numerical study points out that high stresses may only occur in particles with high aspect ratios. This observation is verificated by experimental investigations. Most of the broken particles have aspect ratios higher than 1.0 (see fig. 1a).

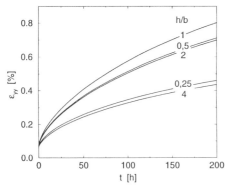

Fig. 4: Creep curves for particles with different aspect ratios h/b

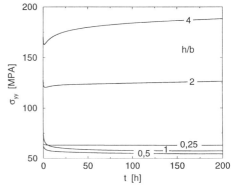

Fig. 5: Normal stress σ_{yy} in the center of the particle for different aspect ratios h/b

In the second study the influence of damage was investigated. A microstructure containing 19 vol.% particles each with an aspect ratio h/b=10.0 was modelled (fig. 6). It is known that the local stress in a particle is influenced by dislocation reactions at the interface. Continuum

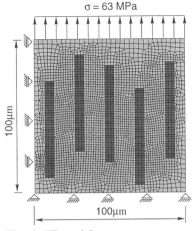

σ = 63 MPa

100µm

100µm

Fig. 6: FE-model

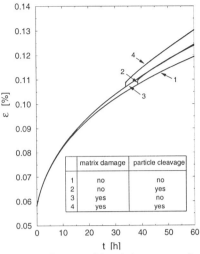

	matrix damage	particle cleavage
1	no	no
2	no	yes
3	yes	no
4	yes	yes

Fig. 7: Influence of local damage on the creep curve ε=ε(t)

mechanical methods generally underestimate the stress concentration in the particles [12]. Therefore, a relative low critical normal stress for particle cleavage was chosen ($\sigma_{cl,c} = 250$ MPa). In order to investigate the influence of matrix damage and particle cleavage on the overall creep behaviour all possible combinations of these two mechanisms were modelled. The resulting creep curves are plotted in fig. 7. Consideration of matrix damage leads to an increase of the global strain. In case of particle cleavage an immediate increase of global strain, followed by subsequent deformation with a higher strainrate is observed. The occurance of particle failure is shifted towards shorter times by the consideration of matrix damage. The distribution of the damage parameter ω after 35 h is presented in fig. 8. Damage starts at the end of particles, where the stresses in the matrix are relatively high. Damage bands develop under 45° to the macroscopic loading direction, giving rise to assume that damage is connected with deformation bands, which were also found to spread in this direction [13]. In case of particle cleavage regions of high damage develope between cracked particles and the ends of adjacent particles.

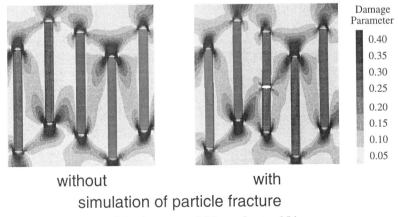

without   with
simulation of particle fracture

Damage Parameter
0.40
0.35
0.30
0.25
0.20
0.15
0.10
0.05

Fig. 8: Distribution of the damage variable ω after t = 35 h

CONCLUSIONS

- The creep behaviour of particle reinforced aluminium alloy 6061 was modelled. The particles were assumed to react linear elastically, while the matrix was described by a viscoplastic material law taking into account isotropic hardening and damage. Particle cleavage was explicitly simulated.
- Unit cell models were developed, representing periodic arrangements of particles.
- A variation of the particle's aspect ratio shows that the creep resistance of the composite is the higher the more the aspect ratio derivates from 1.
- Stress concentrations in the particle occur only at aspect ratios higher than 1.
- Particle cleavage leads to an immediate increase of macroscopic strain.
- Matrix damage is concentrated at the ends of elongated particles and near microcracks caused by broken particles.

REFERENCES

[1] Bhagat, R.B.; Amateau, M.F.; House, M.B.; Meinert, K.C.; Nisson, P.; J. Comp. Mat. 26 (1992) 1578-1593
[2] Luster, J.W.; Thumann, M.; Baumann, R.; Mat. Sci. Techn. 9 (1993) 853-862
[3] Pandey, A.B.; Mishra, R.S.; Mahajan, Y.R.; Scr. Met. 24 (1990) 1565-1570
[4] Krajewski, P.E.; Allison, J.E.; Jones, J.W.; Met. Trans. 24A (1993) 2731-2741
[5] Furukawa, M.; Wang, J.; Horita, Z.; Nemoto, M.; Ma, Y.; Langdon, T.G.; Met. Trans. 26A (1995) 633-639
[6] Nieh, T.G.; Xia, K.; Langdon, T.G.; J. Eng. Mat. Techn. 110 (1988) 77-82
[7] Park, K.T.; Lavernia, E.J.; Mohamed, F.A.; Acta Met. 38 (1990) 2149-2159
[8] Broeckmann, C.; VDI Fortschr.-Ber., Reihe 18, Nr. 169, Düsseldorf, 1995
[9] Lemaitre, L.; Chaboche, J.-L.: Mechanics of Solid Materials; Cambridge University Press, Cambridge, 1990
[10] Leckie, F.A.; Hayhurst, D.R.; Acta Met. 25 (1977) 1059-1070
[11] Berns, H.; Broeckmann, C.; Weichert, D.; Proc. 11th Irish Materials Forum Conference, University College Galway, Ireland, 1995, 163-169
[12] Cleveringa, H.H.M.; van der Giessen, E.; Needleman, A; Acta Mater. 45 (1997) 8, 3163-3179
[13] Broeckmann, C.; Hinz, H.; Pandorf, R.; Proc. Localized Damage 98, Bologna, Italy, in press

Low Cycle Fatigue and Elasto-Plastic Behaviour of Materials
K-T. Rie and P.D. Portella (Editors)
443

THERMAL AND THERMAL-MECHANICAL FATIGUE OF DISPERSION STRENGTHENED Al-ALLOYS

M. BECK, I. KRÖPFL, K.-H. LANG and O. VÖHRINGER
Institute of Material Science and Engineering I
University of Karlsruhe, Kaiserstr. 12, D-76128 Karlsruhe

ABSTRACT

Due to the thermodynamically stable dispersoids, dispersion strengthened aluminium alloys are expected to have better material properties at service temperatures higher than 250° C, than agehardened aluminium alloys. Especially they should have much better high temperature strength and creep resistance than conventional aluminium alloys. Therefore, dispersion strengthened aluminium alloys are suitable for components underlying high mechanical, as well as high thermal loadings. To design such components suitable strength properties determined under thermal fatigue and thermal mechanical fatigue conditions are required.

Two dispersion strengthened Al-Alloys, one with 1,5 vol-% Al_2O_3, the other with 12 vol-% Al_4C_3 and 1,5 vol-% Al_2O_3 were investigated in the as-received state under pure thermal fatigue without and with a additional external constant load for maximum temperatures between 300 and 500° C. Further on the material behaviour under thermal-mechanical fatigue of both as-received materials was studied by suppressing completely the thermal expansion during the temperature cycles. The different maximum temperatures varied between 300 and 600° C.

KEYWORDS

aluminium alloys, dispersion strengthening, thermal fatigue, thermal-mechanical fatigue, high temperature properties

INTRODUCTION

Dispersion strengthened aluminium alloys produced by reaction milling and consolidated by a final extrusion step at 550° C are distinguished from conventional aluminium alloys by much better high temperature strength and creep resistance up to 500° C [1]. The processing route leads to a extremly fine grained microstructure, which is well stabilized by thermodynamically stable oxides (Al_2O_3) and carbides (Al_4C_3) also generated during processing [2, 3, 4]. In this manner the material properties are preserved for a long time at service temperatures. Therefore, this group of metallic materials is suitable for many applications for example as components in combustion engines, which are exposed to both high temperatures and high mechanical load, with the goal to raise the efficiency. Several investigations with isothermal cyclic loading at elevated temperatures were already done [5]. But extensive information about the deformation behaviour under thermal fatigue and thermal-mechanical fatigue is still missing.

MATERIALS

The investigations were carried out with two dispersion strengthened aluminium base materials: Al 99,5 with 1,5 vol-% Al_2O_3 (marked as Al-1,5 Al_2O_3) and aluminium Al 99,5 with 12 vol-% Al_4C_3 and 1,5 vol-% Al_2O_3 (marked as Al-12 Al_4C_3/1,5 Al_2O_3). The chemical composition of both materials is shown in table 1.

Table 1. Marking, chemical composition and volume contents of dispersoids of the investigated materials

marking	O [wt-%]	C [wt-%]	Al [wt-%]	Al_2O_3 [vol-%]	Al_4C_3 [vol-%]
Al-1,5 Al_2O_3	1,0	0,5	rest	1,5	-
Al-12 Al_4C_3/1,5 Al_2O_3	1,0	3,0	rest	1,5	12

The materials were produced by reaction milling by PEAK, Germany. Permant breaking and subsequent rewelding of powder particles during the milling process, presents permanently new surfaces to the milling atmosphere so that alumina (Al_2O_3) is formed. At the same time additives or alloying elements are very finely dispersed in the granulates. The chemical reaction of added carbon leads to the formation of aluminium carbide (Al_4C_3). After cold compaction, heating of milled granulates results in crystallization of hitherto amorphous alumina. A final hot extrusion step at $480°$ C $\leq$ T $\leq 550°$ C completes the production of nearly 100 % dense materials. In the material Al-1,5 Al_2O_3 the dispersoids Al_2O_3 are mainly located at the grain boundaries. Only few dispersoids are inside the grains. The mean diameter of the rounded dispersoids is abaout 100 nm. The average grain size is 6,4 μm in extrusion direction and 2,3 μm in transverse direction. The average dislocation density inside the grains is about $2*10^9$ cm². In the material Al-12 Al_4C_3/1,5 Al_2O_3 the carbides Al_4C_3 and oxides Al_2O_3 are positioned exclusively at grain boundaries. The Al_4C_3-dispersoids have a cylindrical or platelike shape with lengths between 30 and 200 nm. Inside the grains there are only few dislocations, whereas near to many grain boundaries the dislocation density is much higher. The average grain size is about 0,5 μm. The oxides and carbides located at the grain boundaries hinder the grain growth. Both materials were investigated in the as-received state.

EXPERIMENTAL PROCEDURE

The thermal fatigue (TF) and thermal-mechanical (TMF) fatigue tests were performed with a servohydraulic testing machine with a maximum capacity of 100 kN. The specimens were heated with an induction heating system. For both kinds of tests (TF and TMF) solid round specimens with a gauge length of 10 mm and a gauge diameter of 7 mm were used.

The thermal fatigue investigations were carried out without, as well as with an additional superimposed constant tensile or compressive external force. The specimens were heated up without any external load to $T_{min} = 50°$ C. After the temperature compensation the constant tensile or compressive force was superimposed and the temperature was cycled between $T_{min} = 50°$ C and T_{max}, with a constant heating and cooling rate of $10°$ C/s up to an ultimate number of temperature cycles of 100. As maximum temperatures 300, 400 and 500° C were choosen. The plastic deformation ε_p of the specimens was measured before each temperature cycle.

For the thermal-mechanical Out-of-Phase fatigue tests the specimens were also heated under load control to $T_{min} = 50°$ C. After the temperature compensation the testing machine was switched to strain control and the temperature was cycled between T_{min} and T_{max} suppressing simultaneously the thermal expansion of the specimen completely, so that $\varepsilon_t = \varepsilon^{th}(T_{min})$. The TMF tests were stopped after failure of the specimen or at an ultimate number of cycles of 10^4. The constant heating and colling rate was here also 10° C/s. These Out-of-Phase TMF tests were carried out for the maximum temperatures of 300, 400, 500, 550 and 600° C.

RESULTS

Thermal Fatigue without additional external Loading

The figures 1a) and b) present the development of the plastic deformation ε_p of both investigated materials Al-1,5 Al_2O_3 and Al-12 $Al_4C_3/1,5$ Al_2O_3 under pure thermal fatigue without an additional external force for maximum temperatures of 300, 400 and 500° C versus the number of temperature cycles. For all investigated maximum temperatures negative plastic deformations occur for Al-1,5 Al_2O_3. They increase with increasing number of temperature cycles for all T_{max}. The higher the maximum temperature the more pronounced is the shortening of the specimens. Also for Al-12 $Al_4C_3/1,5$ Al_2O_3 the negative plastic deformations increase continuously during the whole test for $T_{max} = 400$ and 500° C. For $T_{max} = 300°$ C the plastic deformation saturates after the third temperature cycle.

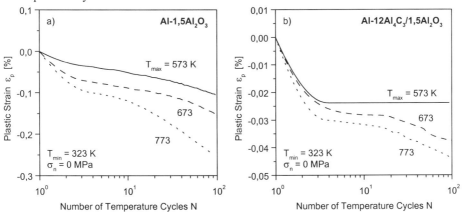

Fig.1 Plastic strain ε_p versus number of temperature cycles N for different maximum temperatures without additional external force a) for Al-1,5 Al_2O_3 and b) for Al-12 $Al_4C_3/1,5$ Al_2O_3.

Thermal Fatigue with an additional external Loading

In the figures 2a) and b) the development of ε_p versus the number of temperature cycles is shown for Al-1,5 Al_2O_3 at thermal fatigue load with several additional constant external tensile and compressive loads for the maximum temperatures of 300 and 400° C . For both investigated T_{max} the absolute values of ε_p under tensile load are smaller than under compressive load with the same absolute value. Fig 2a) shows, that at $T_{max} = 300°$ C the specimen shortens during the temperature

446

cycling although an external tensile stress of 10 MPa is acting. Also at $T_{max} = 400°$ C the specimen shortens under a tensile stress of 10 MPa between the third and the 50th temperature cycle, (Fig. 2b)). As shown in Fig. 2c) and 2d), specimens made of Al-12 $Al_4C_3/1,5$ Al_2O_3 develop also smaller plastic deformations under tensile than under compressive stresses with the same absolute value at both maximum temperatures investigated. A tensile stress of 90 MPa leads to fracture of the specimen after 30 temperature cycles, probably due to the damaging of the interface between dispersoids and aluminium. For both materials the difference between the absolute value of the plastic deformation under compressive stress and the absolute value of the plastic deformation under tensile stress developing during the TF-test is the larger the higher the absolute value of the superimposed external force is.

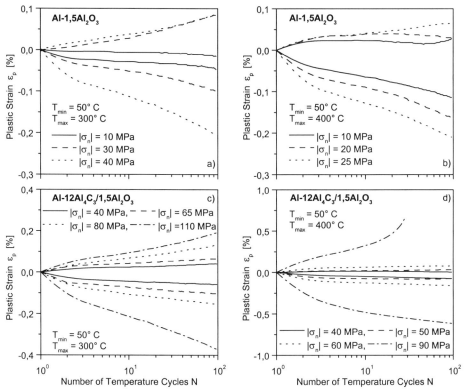

Fig. 2. Plastic strain ε_p versus number of temperature cycles N at different maximum temperatures and different absolute values of additionally superimposed external forces a) and b) for Al-1,5 Al_2O_3 and c) and d) for Al-12 $Al_4C_3/1,5$ Al_2O_3.

Thermal-mechanical fatigue

Results from thermal-mechanical Out-of-Phase TMF-tests with different maximum temperatures are presented in Fig. 3 of Al-1,5 Al_2O_3. Figure 3a) shows, that the maximum (σ_{max}), mean (σ_m) and minimum (σ_{min}) stresses appear at all investigated maximum temperatures in one scatter. The absolute value of σ_{min} decreases during the first three cycles and positive mean stresses are built up at all T_{max}-values. The maximum stress decreases slightly with the number of cycles till macro crack initiation. The $\varepsilon_{a,p}$-curves in fig. 3b) reveal hardening at all T_{max}-values during the first three cycles. In the remaining life time the material softens slightly. For Al-12 $Al_4C_3/1,5\ Al_2O_3$ the maximum, mean and minimum stresses at $T_{max} > 300°\ C$ form a relativly wide scatter band. Within the scatter bands the respective values seem to be ordered in accordance with T_{max}. The maximum and the mean stress at $T_{max} = 300°\ C$ are smaller than those for the other maximum temperatures, Fig. 3c). Tensile mean stresses develop at all T_{max} because the absolute values of the minimum stresses are significantly lower than the corresponding maximum stresses. Cyclic hardening occures also for Al-12 $Al_4C_3/1,5\ Al_2O_3$ during the first three cycles. For the rest of the life $\varepsilon_{a,p}$ remain nearly constant for all maximum temperatures, Fig 3d).

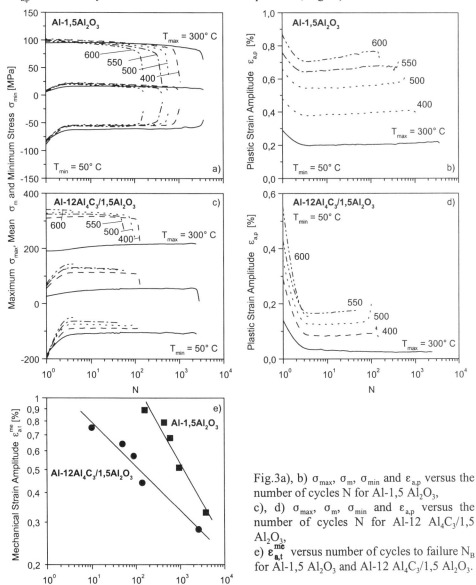

Fig.3a), b) σ_{max}, σ_m, σ_{min} and $\varepsilon_{a,p}$ versus the number of cycles N for Al-1,5 Al_2O_3,
c), d) σ_{max}, σ_m, σ_{min} and $\varepsilon_{a,p}$ versus the number of cycles N for Al-12 $Al_4C_3/1,5$ Al_2O_3,
e) $\varepsilon_{a,t}^{me}$ versus number of cycles to failure N_B for Al-1,5 Al_2O_3 and Al-12 $Al_4C_3/1,5\ Al_2O_3$.

The comparison of $\varepsilon_{a,p}$ appearing at both materials shows that Al-1,5 Al_2O_3 is more ductile than Al-12 Al_4C_3/1,5 Al_2O_3 at all T_{max}. Fig 3e) proves that at all T_{max} investigated the life-time under Out-of-Phase TMF load of Al-1,5 Al_2O_3 is higher than the life-time of Al-12 Al_4C_3/1,5 Al_2O_3 despite of the higher total mechanical strain amplitudes $\varepsilon_{a,t}^{me}$ induced at Al-1,5 Al_2O_3.

DISCUSSION

Specimens of both investigated materials Al-1,5 Al_2O_3 and Al-12 Al_4C_3/1,5 Al_2O_3 shorten under pure thermal fatigue without superimposed external stresses at all maximum temperatures. This unexpected material behaviour is caused by dislocation networks, twist and tilt boundaries, which are arranged at the dispersoids and the grain boundaries in the as-received state [5, 6]. They originate from heavy plastic deformation during hot extrusion and could not be completely relaxed by dynamic recrystallisation during cooling down after the hot extrusion. The internal stress fields in these dislocation structures are not fully saturated, therefore long range stress fields are induced which cause so called backstresses. Due to their nature of origin they operate in compressive direction. Therefore, the backstresses cause directed dislocation movement during cyclic thermal loading. Another effect results from the much smaller coefficients of thermal expansion of the Al_2O_3 and Al_4C_3 dispersoids compared to that one of the aluminium matrix. During temperature cycling mis-fit strains and consequently mis-fit stresses at the interfaces of oxides/carbides and aluminium are induced and the oxides and carbides act as dislocation sources [7]. Due to the internal backstresses the effective stress acting on the dislocation is under tensile load smaller than under compressive load with the same absolute value. So the plastic deformation under additional superimposed external compressive load is higher than under a tensile load of the same absolute value. The influence of these backstresses on the material behaviour under Out-of-Phase TMF load can not be seen clearly in the development of the σ_m-N-curves, because the plastic deformations in the first cycles are probably too high.

CONCLUSIONS

Due to backstresses shortening of the specimens occurs during thermal cycling without an additional external force for both investigated materials. Additional superimposed external forces lead to higher absolute values of the plastic deformation under external compressive force than under tensile forces of the same absolute value. TMF Out-of-Phase load leads to higher life-times of Al-1,5 Al_2O_3 than for Al-12 Al_4C_3/1,5 Al_2O_3 for all investigated maximum temperatures.

REFERENCES

1. El-Magd, E., Nicolini, G. (1996) Metallwissenschaft und Technik **6**, p. 404.
2. Arnhold, V., Hummert, K. (1989). In: *New Materials by Mechanical Alloying Technique*, E.Arzt, L. Schultz, (Eds.), DGM Verlag, 263.
3. Hawk, J.A., Mirchandani, P.K., Benn, R.C. and Wilsdorf, H.G.F. (1988). In: *Dispersion Strengthened Aluminum Alloys*, Kim, Y.-W., Griffith, W. M. (Eds). The Minerals, Metals & Miners Society, Warrendale, pp. 517-538.
4. Jangg, G., Kuttner F., Korb, G. (1975), *Aluminium*, **51**, p. 641
5. Kröpfl, I. (1996) *Verformungs- und Bruchverhalten von oxid- sowie karbid- und oxidverfestigten Aluminiumwerkstoffen*. Dissertation, Universität Karlsruhe.
6. Kröpfl, I., Vöhringer, O. (1996). *In: Proc.Int.Conf. Aluminium Alloys, ICAA 5*, Transtec.Publ., Zürich, pp.199-204.
7. Wilsdorf, H.G.F. (1988). In: *Dispersion Strengthened Aluminum Alloys*, Kim, Y.-W., Griffith, W.M. (Eds). The Minerals, Metals & Miners Society, Warrendale, pp. 3-30.

Low Cycle Fatigue and Elasto-Plastic Behaviour of Materials
K-T. Rie and P.D. Portella (Editors)
449

IN-CYCLE STRAIN EVOLUTION IN A SHORT-FIBER REINFORCED ALUMINUM-ALLOY DURING THERMAL CYCLING CREEP

A. FLAIG, A. WANNER* and E. ARZT,
*Institut für Metallkunde, Universität Stuttgart and
Max-Planck-Institut für Metallforschung, D-70174 Stuttgart*

ABSTRACT

Thermal cycling creep experiments were performed on a short-fiber reinforced aluminum matrix composite. In addition to the assessment of average creep rates, a close examination of the strain evolution within the thermal cycles was conducted. This was done with the aid of appropriate methods to compensate for the thermal strain, similar to those used in thermo-mechanical fatigue testing. Striking differences of creep rates between heating and cooling segments of the thermal cycles and even negative strain rates were discovered. The findings are discussed in terms of the evolution of the residual stresses on thermal cycling and the creep behavior of the matrix under these complex cyclic loading conditions.

INTRODUCTION

It is well known that thermal cycling during creep loading can have significant effects on the creep behavior of metal matrix composites (MMCs). In most cases, a strong increase of the average creep rate compared to isothermal experiments is found, as well as a reduced stress sensitivity [1]. Experimental findings have usually been presented and modeled in terms of strain increments per thermal cycle only. To our knowledge, the strain evolution within the individual thermal cycles has so far been addressed only by Furness and Clyne [2], who used the technique of scanning laser extensometry to study the creep strain evolution of short-fiber reinforced aluminum. They compared their results with predictions from a model based on the elastic inclusion method by Eshelby.

In the present work the thermal cycling creep properties of a short-fiber reinforced Al-Si alloy are investigated with special attention to the plastic strain evolution within thermal cycles.

EXPERIMENTAL

The composite material examined consists of an age-hardening aluminum cast alloy with the composition AlSi12Cu1Mg1Ni1 which is reinforced by 15 vol.% of Al_2O_3-fibers (Saffil). It is

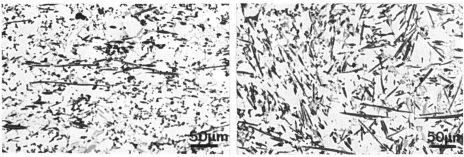

Fig. 1. Typical microstructure of the fiber-reinforced aluminum alloy. Note the preferential orientation of the fibers perpendicular to the loading direction (random-planar).
    a) longitudinal section of a tested sample   b) cross section of a tested sample.
    (loading direction vertical)

* currently on leave to Dept. of Mater. Sci. & Eng., Northwestern University, Evanston, IL, USA

manufactured via high pressure squeeze-casting by the company MAHLE GmbH, Stuttgart, Germany. After casting, the material was subjected to a T6 heat treatment. The cylindrical fibers have an average diameter of 3 μm and an average length of 200 μm with a wide length distribution [3]. They have a slightly preferential orientation in one particular plane (see Fig. 1 a) and b)). Metallography also shows strong local variations in fiber content with limited regions of fiber clustering.

Cylindrical button head specimens with a parallel gage length of 15 mm and a diameter of 7 mm were cut from 20 mm thick plates so that the plane of preferential fiber orientation is perpendicular to the length axis. The specimens were diamond ground and successively polished on the gage length. Afterwards 5 type K thermocouples were spot-welded to different positions along the gage length. The specimens were mounted in a uniaxial electro-mechanical testing machine with water-cooled grips. Bending strains due to misalignment and gripping were found to be below 5% of the axial strain. Specimen heating was accomplished by an induction coil and a 5 kW RF-heater. One of the thermocouples was used for closed-loop temperature control. By careful design of the induction coil, the thermal gradients along the gage length were minimized so that the maximum temperature difference within the gage length during a typical test procedure could be maintained reproducibly below 4 K. Strain measurement was performed using a high-temperature side-contact extensometer with gage length 12.9 mm. Axial strains could be measured up to ±10% with a resolution of better than $2 \cdot 10^{-5}$, being limited mainly by the resolution of the digital data acquisition.

The testing procedure was as follows: After the specimen was mounted in the testing machine under zero load control, it was heated up to 350 °C and kept at this temperature for at least 30 minutes to overage the matrix alloy. Then the thermal expansion behavior of the material was measured by subjecting the sample to typically 15 to 20 thermal cycles between 350 and 150 °C until a stable strain-temperature hysteresis was reached. The nominal heating and cooling rates during these cycles were 50 K/min and the cycle period was 10 minutes

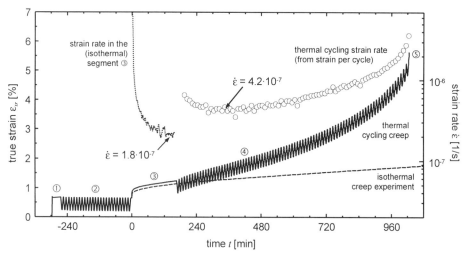

Fig. 2. Thermal cycling creep at 40 MPa and 350↔150 °C. Phases of the experiment are: ① over-aging, ② thermal cycles under zero-load control, ③ isothermal creep (40 MPa, 350 °C) ④ loaded thermal cycling, ⑤ fracture. Also shown are the strain rate in the isothermal segment and the average thermal cycling strain rate (from the strain increment per cycle) as well as the strain evolution for a conventional creep test (40 MPa, 350 °C).

including dwell times of 1 minute at the maximum and minimum temperatures (see also Fig. 5). After this initial load-free cycling stage, an isothermal creep segment under external load was introduced in order to establish a defined condition of stationary creep in the composite. Temperature was held constant at 350 °C and the load was increased to the desired value and held constant for the rest of the experiment. After a certain amount of isothermal creep strain had accumulated, thermal cycling was resumed until specimen fracture occurred. Load, temperature distribution, and strain were measured continuously throughout the experiment. Control of the experiment, data acquisition and storing was accomplished using self-written software. Tests were performed with different loads from 20 to 50 MPa and isothermal pre-strains of 0.3% for loads below 30 MPa and 0.6% for loads above 30 MPa.

RESULTS AND DISCUSSION

Figure 2 shows the typical strain and strain rate evolution for a thermal cycling creep experiment. One can clearly see that with the start of loaded thermal cycling the strain rate immediately shifts to higher values. It then undergoes a transition stage of decreasing strain rate until an almost stationary rate is reached. Towards the end of the experiment, the strain rate increases until final fracture occurs. The results of this study are summarized in table 1. For comparison, isothermal creep data measured by Bidlingmaier et al. [4] on the same material using a conventional creep machine are also listed. Fig. 3 shows the minimum creep rates under isothermal and thermal cycling conditions as a function of temperature.

Table 1: results from thermal cycling creep and isothermal creep experiments

thermal cycling creep experiments ($150 \leftrightarrow 350$ °C)						
tensile stress	σ [MPa]	23,7	26,9	30,4	40,7	51,6
creep strain (isothermal segment)	ε_{itc} [%]	0,3	0,3	0,56	0,6	0,6
total fracture strain	ε_t [%]	4,2	4,3	4,1	5,0	3,3
time to fracture	t_t [h]	123	81	57	17	6,7
thermal cycling creep rate	$\dot{\varepsilon}_{tcc}$ [1/s]	$2,8 \cdot 10^{-8}$	$4 \cdot 10^{-8}$	$2,6 \cdot 10^{-7}$	$4,0 \cdot 10^{-7}$	$7,5 \cdot 10^{-7}$
for comparison: isothermal creep experiments at 350 °C [4]						
tensile stress	σ [MPa]			30,6	39,9	50,0
total fracture strain	ε_t [%]			1,53	1,7	1,12
time to fracture	t_t [h]			104	25,7	5,9
min. strain rate	$\dot{\varepsilon}$ [1/s]			$2,2 \cdot 10^{-8}$	$1,1 \cdot 10^{-7}$	$2,7 \cdot 10^{-7}$

It can be seen that the creep rates and the fracture strain are increased considerably by thermal cycling. At the same time, the time to fracture is reduced. For the samples with only 0.3% pre-strain a reduction in the stress exponent is observed. They also show significantly lower thermal cycling creep rates.

In addition to the conventional assessment of thermal cycling creep behavior based on strain per cycle measurements, a close examination of the strain evolution within the thermal cycles was conducted. The strain data from at least 5 subsequent stable load-free cycles were averaged over time after smoothing and spline-interpolation to obtain the thermal strain (see Fig. 4) as a function of time in the cycle. The thermal expansion curve in Fig. 4 shows a small hysteresis where the high-temperature expansion coefficient is higher on heating than on cooling. Comparison with experimental results obtained on similar material [5] and with analytical [6] or micromechanical [7] models for aligned short-fiber composites shows that

452

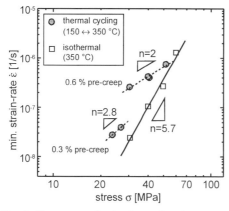

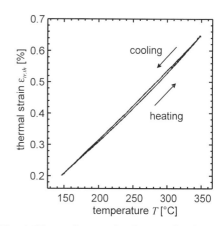

Fig. 3. Creep rates for isothermal and thermal cycling creep

Fig. 4. Thermal expansion hysteresis of the material under no external load

such a "positive" hysteresis is characteristic for the thermal expansion behavior perpendicular to a preferential fiber orientation. This is in good agreement with the metallographic findings of the fibers being preferentially oriented perpendicular to the loading direction.

The averaged reference cycle was subtracted from the strain evolution in the loaded thermal cycles, yielding a "thermally compensated" strain that contains only the excess strain due to loading. The maximum deviation of strain in the stable unloaded cycles from the reference cycle was always smaller than $4 \cdot 10^{-5}$ at a total strain amplitude of about $5 \cdot 10^{-3}$. This means that the precision of the thermally compensated strain according to this compensation method is $4 \cdot 10^{-5}$ or better.

Examples of cyclic strain evolutions are given in Fig. 5 a) for the first and second cycle of a specimen loaded with 40 MPa and in Fig. 5 b) for cycles 83 and 84 of the same experiment. In the first cycle with the beginning of the cooling segment, the thermally compensated strain suddenly increases with a high rate compared to the isothermal creep rate. When the minimum temperature of the cycle is reached, this increase stops but it is even stronger towards the end

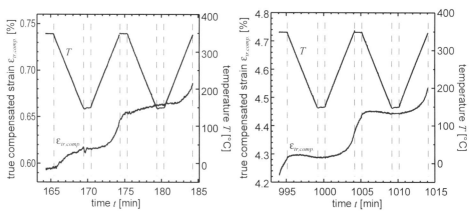

Fig. 5.   Evolution of thermally compensated strain at 40 MPa for
  a) first and second cycle   b) 83rd and 84th cycle

of the heating segment. In this first cycle a remarkable difference is seen in the strain rates at identical temperatures in the cooling and heating half-cycles. In the second cycle, the strain rate stays relatively low for the duration of the cooling half-cycle but is again high when the maximum temperature at the end of the cycle is reached. The thermal asymmetry of strain rates becomes more and more pronounced with further cycles. In Fig. 5 b) even negative strain rates can be observed in the cooling half-cycles of cycle 83 and 84. It was found that negative strain rates occurred more readily the lower the load and therefore the lower the average strain per cycle was. The amount of pre-strain seems to have no effect on the strain evolution in the subsequent thermal cycling creep experiment. It should be noted that after the end of isothermal creep there is a remarkable transition regime of at least 2 thermal cycles until the form of the creep cycles stabilizes.

The phenomenon of thermal cycling creep of MMCs is usually attributed to the permanent regeneration of internal stresses caused by the difference in thermal expansion coefficient of matrix and reinforcement. The complex interplay between reinforcement geometry, thermal stresses, and the multitude of possible stress relaxation mechanisms that are usually strongly dependent on both temperature and stress has made analytical approaches to the thermal cycling behavior extremely difficult.

In the following discussion, we focus mainly on the in-cycle strain evolution and try to set up a very simple scheme to explain the effects, based on the following restrictive assumptions. The fibers are treated as long fibers with a strong fiber/matrix interface. The local thermal stresses in the matrix are assumed to be uniaxial in the direction of the fiber. Radial and hoop stresses and fiber end effects are neglected. Because of the preferential fiber orientation, the thermal stresses are expected to be mainly in-plane stresses perpendicular to the loading direction. The matrix carries global tensile stresses at room temperature and the cycling between 150 and 350 °C causes no global plastic yielding in the matrix. On heating to 350 °C the (thermal) matrix stress state reverts to compressive stresses while at 150 °C tensile stresses prevail. On the basis of these assumptions we can conclude the following for the loaded thermal cycles:

At the end of the heating half-cycle, the thermal matrix stresses in the fiber direction become compressive. As these thermal stresses act perpendicular to the tensile matrix stress caused by the external load, high deviatoric stresses in the matrix can be expected. At the same time the temperature is high enough to allow matrix relaxation by creep or local plastic yielding. This will cause high composite strain rates until the assisting in-plane stresses are relaxed to low values and the creep rate reduces to the for isothermal conditions. As the temperature decreases, the in-plane stresses will immediately become tensile, reducing the deviatoric stress and thus slowing down plastic and creep deformation. This agrees well with the results shown in Fig. 5, where the high strain rates always concentrate at the end of the heating cycles. Negative creep rates may occur, if the in-plane stresses become more tensile than the matrix stress in the loading direction and so creep and local plasticity are likely to cause the composite to extend in the fiber plane and contract perpendicular to it. This is expected and experimentally found for the low-temperature cooling part of the cycle (see Fig. 5b).

This view is consistent with some of the predictions of the model by Furness and Clyne [2], who used Eshelby methods to calculate the average deviatoric stresses in the matrix as a driving force for creep. They investigated an aluminum matrix composite with 20% short fibers aligned in the loading direction. For low loads, they found negative in-cycle strain rates both experimentally and predicted by their model. In their model this happens because the transverse matrix stresses (perpendicular to the fiber orientation and loading direction) are considered and these exceed the axial matrix stresses for certain conditions. This argument conforms to the one we made for our composite, although for the random planar fiber distribution it is not required to explicitly take into account the thermal stresses perpendicular to the fibers to explain negative strain rates.

The two main causes for the high strain-rate asymmetry according to our concept can be summarized as a) the assistance of thermal stresses for matrix creep under favorable fiber orientation on heating and b) the relaxation of these stresses at the end of the heating half-cycle to low values.

Some phenomena still remain to be clarified e.g. the differences between the very first and the following thermal cycles. These cannot be explained on basis of the simple model set up above. However, this transition regime is a strong indication that the distribution of the internal stresses between matrix and reinforcement during isothermal creep is very different compared to the stresses under stabilized thermal cycling creep because it takes more than one cycle to achieve this steady state.

CONCLUSIONS

The thermal cycling creep behavior of a fiber-reinforced material has been investigated with respect to the in-cycle strain evolution. The strain-rate asymmetry found could be explained with a simple scheme: 1) thermal stresses on heating assist the external stresses under favorable fiber orientation to cause accelerated matrix creep and 2) the relaxation of these stresses at the end of the heating half-cycle prevents the effect from occuring on the cooling half-cycle. For a detailed analysis of all these effects, analytical or micromechanical calculations will be necessary. These should at first address the possible influence of plastic yielding of the matrix and of radial and hoop stresses around the fibers. With this work and further work in progress, we aim to make a contribution to the understanding of metal-matrix composites under conditions close to those in applications.

ACKNOWLEDGMENT

The authors wish to thank the company MAHLE GmbH, Stuttgart, Germany, for provision of the composite material. Part of this work was supported by the Deutsche Forschungsgemeinschaft (DFG), project SFB 381/A1.

REFERENCES

1. Derby B. In: Hansen N, et al. eds., 12th Risø International Symposium on Materials Science: Metal Matrix Composites - Processing, Microstructure and Properties. Risø National Laboratory Roskilde, Denmark, 1991, 31-49
2. Furness JAG, Clyne TW. In: Hansen N, et al. eds., 12th Risø International Symposium on Materials Science: Metal Matrix Composites - Processing, Microstructure and Properties. Risø National Laboratory Roskilde, Denmark, 1991, 349-354
3. Bär J., Dissertation, Universität Stuttgart, 1992
4. Bidlingmaier T, Vogt D, Wanner A. In: Verbundwerkstoffe und Werkstoffverbunde, Bayreuth, DGM Informationsgesellschaft mbH, 1995
5. Neite G, Mielke S. Mater. Sci. Eng. A, 1991, 148(1), 85-92
6. Wakashima K, Tsukamoto H. Mater. Sci. Eng. A, 1991, 146(1-2), 291-316.
7. Weissenbek E, Rammerstorfer F.G., Acta metall. mater., 1993, 41(10), 2833-2843.

Low Cycle Fatigue and Elasto-Plastic Behaviour of Materials
K-T. Rie and P.D. Portella (Editors)
455

INFLUENCE OF Al$_2$O$_3$ FIBER REINFORCEMENT ON THE THERMAL-MECHANICAL FATIGUE BEHAVIOUR OF THE CAST ALUMINIUM ALLOY AlSi10Mg

T. BECK, K.-H. LANG, D. LÖHE
Institute of Materials Science and Engineering I
University of Karlsruhe, Kaiserstrasse 12, D-76131 Karlsruhe

ABSTRACT

Thermal-mechanical fatigue (TMF) tests under total strain constraint were carried out on the cast Aluminium alloy AlSi10Mg - T6, unreinforced and with 15% Al$_2$O$_3$ (Saffil) fiber reinforcement, respectively. The number of cycles to fracture is increased markedly by the fiber reinforcement, especially for maximum cycle temperatures above 250 °C. Surprisingly, during TMF cycling higher stress amplitudes are induced in the unreinforced alloy than in the fiber reinforced one. The cyclic softening behaviour observed at both materials is more pronounced for the unreinforced alloy. In tensile tests at 20 °C, the unreinforced material revealed a higher 0,2% proof stress than the fiber reinforced one. These results are explained by a magnesium enrichment in the fiber-matrix interface of the compound material which was proved by EDX analyses. According to this, the Mg content of the compound's matrix is lowered and the hardenability of the fiber reinforced material is reduced below the one of the unreinforced alloy.

KEYWORDS

Thermal-mechanical fatigue, cyclic deformation behaviour, tensile properties, aluminium alloys, fiber reinforcement

INTRODUCTION

Cast aluminium alloys are applied broadly in car engines due to their high specific strength, even at elevated temperatures, and their good castability. Thermal fatigue cycles induced in several engine parts by repeated start-stop processes must be endured up to 10000 times during the lifetime of an engine. Recently, however, the development of efficiency optimised pistons and cylinder heads causes service temperature and load peaks which cannot be endured by conventional aluminium alloys any more. One possibility of using aluminium alloys in spite of these service conditions, is reinforcement by Al$_2$O$_3$ short fibers especially in these parts of a component which are subjected to the highest thermal and mechanical loadings. Due to the fiber reinforcement, a completely new material is produced, whose properties must be characterised carefully under near service loading conditions for providing a reliable engineering database.

MATERIAL AND TESTING SPECIMENS

All tests were carried out on the cast aluminium alloy AlSi10Mg unreinforced as well as reinforced with 15 Vol.-% Al_2O_3 (Saffil) short fibers in random planar orientation. The average length of the fibers was 300 μm, their average diameter 3 μm. The unreinforced raw material was produced by conventional die casting, the compound was manufactured by a squeeze casting process using preforms with a SiO_2 binder. After T6 heat treatment of these raw materials, solid cylindrical specimens were machined with a gauge length of 10 mm and a diameter of 7 mm in the gauge length [1]. The chemical composition of the unreinforced as well as of the matrix alloy of the compound is given in Table 1.

Table 1: Chemical composition of the tested cast aluminium alloys [Ma.-%]

AlSi10Mg	Al	Si	Mg	Fe	Zn, Ni, Ti	others
unreinforced	balance	9,53	0,41	0,13	< 0,1	< 0,01
15 % Saffil	balance	9,38	0,34	0,13	< 0,1	< 0,01

EXPERIMENTAL DETAILS

The thermal-mechanical fatigue (TMF) tests were carried out on a servohydraulic testing machine with 100 kN maximum load. The specimens were heated by an inductive system. The cooling by thermal conduction to the water cooled grips could be forced additionally by a controlled air jet to the surface of the specimen. The TMF tests were performed in total strain control. The total strain ε_t is composed of two parts: the thermal strain ε^{th} and the total mechanical strain ε_t^{me}. The total strain ε_t was held constant to the thermal strain value at the mean temperature of the TMF cycles $\varepsilon^{th}(T_m)$. This results in a total strain amplitude equal to zero, a mechanical total strain amplitude equal to the thermal strain amplitude given by $\varepsilon_a^{th} = \frac{1}{2}[\varepsilon_{th}(T_{max})-\varepsilon_{th}(T_{min})]$ and a strain ratio $R_\varepsilon = \varepsilon_{t,min}^{me}/\varepsilon_{t,max}^{me} = -1$. For all tests, the minimum temperature T_{min} of the triangle shaped temperature cycles was 50 °C. The maximum temperature T_{max} was varied between 200 °C and 350 °C for the unreinforced alloy and between 250 °C and 400 °C for the fiber reinforced material, respectively. The heating and cooling speed, respectively, was 10 °C/s and the ultimate number of cycles generally was 10^4.

Tensile tests were performed using the same testing facilities as for the TMF experiments. Both materials were tested in the T6 state and after solution annealing at 525 °C for 3 hours with subsequent slow cooling to room temperature (heat treatment state O), respectively. All tensile tests were done in total strain control with a total strain rate $d\varepsilon_t/dt = 5\cdot10^{-2}$ s^{-1} at a temperature of 20 °C.

Vickers HV 0,025 microhardness measurements were done at the the unreinforced alloy and the α-Al matrix of the fiber reinforced material, respectively. In order to estimate the Mg distribution throughout the microstructure, qualitative EDX line analyses were carried out on both materials, additionally. The fracture surfaces were investigated by SEM after TMF and tensile testing.

RESULTS AND DISCUSSION

The lifetime behaviour of the cast Al-alloy AlSi10Mg, unreinforced and with 15% Saffil fiber reinforcement, respectively, is shown in Fig. 1. The maximum cycle temperature T_{max} is plotted versus the number of cycles to fracture N_f. It can be seen that fiber reinforcement causes a large increase of lifetime in the whole investigated temperature range. This increase is the more pronounced, the higher the maximum temperature is. Furthermore, the highest maximum temperature for which the geometrical shape of the specimen remains stable throughout the test, is increased by the fiber reinforcement from T_{max}=300°C for the unreinforced alloy to T_{max}=400°C for the compound material.

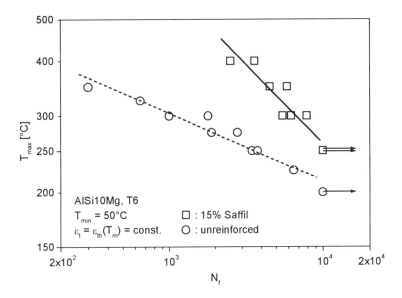

Fig. 1: Maximum cycle temperature versus lifetime for the unreinforced and the fiber reinforced material, respectively

As an example for the cyclic deformation behaviour of both investigated materials, Fig. 2 shows the stress amplitude resulting from TMF loading versus the number of cycles for a maximum cycle temperature of 300 °C.

Surprisingly, there are higher stress amplitudes induced in the unreinforced alloy than in the fiber reinforced material, while one would have expected, that TMF loading induces higher stress amplitudes in the fiber reinforced material than in the unreinforced alloy. Furthermore, the cyclic softening, which takes place at both materials, is more pronounced for the unreinforced alloy than for the fiber reinforced one. For this reason, progressive fiber detachment, which could result in pronounced cyclic softening during TMF loading, obviously does not occur. According to this, SEM investigations of the compound's fracture surface after the TMF tests showed very good fiber matrix bonding. Fiber pull-out cold not be found even after testing with a maximum cycle temperature of 400 °C.

458

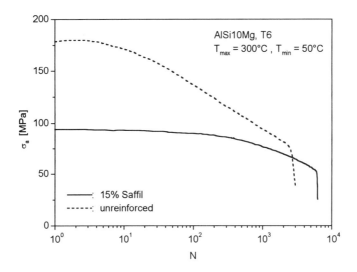

Fig. 2: Cyclic deformation behaviour of the unreinforced and the reinforced alloy, respectively

In order to explain this surprising TMF behavior, tensile tests were carried out at T=20 °C on both materials after T6 and O heat treatment, respectively. The results are shown in Fig. 3 as nominal stress versus total strain curves.

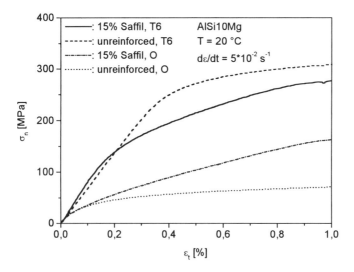

Fig. 3: Tensile stress-strain behaviour of both investigated materials after T6 and O heat treatment, respectively

Comparing the stress-strain curves for both materials in the T6 state, it firstly can be seen, that, as expected, Young's modulus is enlarged by the fiber reinforcement. This is another evidence for good fiber matrix bonding in the compound material. Secondly, in accordance with the cyclic deformation behaviour under TMF loading as shown in Fig. 2, the 0,2 % proof stress, for example, is reduced by the fiber reinforcement. On the other hand it can be seen, that in the O state the 0,2 % proof stress is enlarged by the reinforcement. So the conclusion can be drawn that the reduction of the material's strength by fiber reinforcement observed in TMF and tensile tests after T6 treatment, must be caused by a reduction of the strength of the matrix. According to that, the results of microhardness measurements in the α-Al matrix of the unreinforced and the fiber reinforced material, which are given in Table 2, show that the decrease of the matrix strength due to fiber reinforcement is much more pronounced in the T6 than in the O state.

Table 2: Microhardness values HV 0,025 in the α-Al matrix of the investigated materials, T6
        heat treated

Material	HV 0,025, T6 state	HV 0,025, O state
AlSi10Mg, unreinforced	113	55
AlSi10Mg + 15 % Saffil	80	45

Additionally, the Mg content of both alloys was measured qualitatively by EDX line analysis. Fig.4 shows a scan over the α-Al matrix and a fiber of the compound material.

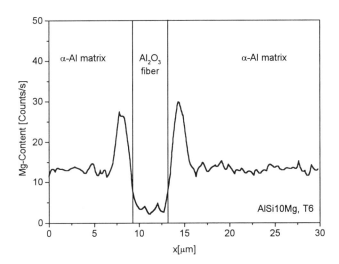

Fig. 4: Mg content measured by EDX in the microstructure of the unreinforced and the fiber
        reinforced alloy, respectively

A marked enrichment of Mg near the fiber-matrix interface of the compound can be seen. This effect is also reported by other authors and may be caused by Mg adsorption at the fiber surface

460

[2] or by reactions between the SiO_2 content in the Saffil fibers [3] or the SiO_2 preform binder [4] with the Mg content of the alloy during squeeze casting and/or heat treatment. The Mg enrichment near the fibers leads to lower Mg contents within the rest of the α-Al matrix. Because an Mg content of at least 0,3 Ma.-% is required for providing full hardenability of Al-Si-Mg type cast aluminium alloys [5], and because the overall Mg content of the compound is 0,34 Ma.-%, the matrix hardenability is reduced markedly by the fiber reinforcement. By this fact, the lower strength of the fiber reinforced alloy compared with the unreinforced one can be explained consistently.

SUMMARY AND CONCLUSIONS

The experimental results presented above show, that fiber reinforcement of a well known metal matrix can lead to unexpected properties of the compound. This is a consequence of the modification of the matrix properties due to the fiber reinforcement. So an optimisation of the matrix alloy may be necessary for the development of suitable metal matrix composites.
An increase of the Mg content in the matrix alloy of the compound would increase the tensile properties of the composite compared with those of the unreinforced matrix in all heat treatment states. This could be shown in a previous work [6] for the piston alloy AlSi12CuMgNi with a Mg content of 1,08 Ma.-%. However, it should be kept in mind, that the lifetime behaviour under TMF conditions possibly will get worse by increasing the hardenability of the matrix due to higher stress amplitudes induced by TMF loading and embrittlement of the material. For those reasons, AlSi10Mg with a higher Mg content will be investigated additionally in the current research project.

ACKNOWLEDGEMENTS

The support of the investigation by the Forschungsvereinigung Verbrennungskraftmaschinen (FVV), Frankfurt, Germany, is gratefully acknowleged.

REFERENCES

1:      B. Flaig: Isothermes und thermisch-mechanisches Ermüdungsverhalten von GK-AlSi10Mg wa, GK-AlSi12CuMgNi und GK-AlSi6Cu4, Diss. Univ. Karlsruhe, 1995

2:      T. W. Clyne, M. G. Bader, G. R. Cappleman, P. A. Hubert: The Use of a δ-Alumina Fibre for Metal-Matrix Composites, J. Mat. Sci. **20**, 85-95

3:      H. J. Dudek, S. Wang, R. Borath: Grenzflächenreaktionen in Al_2O_3-Al, DLR Institut für Werkstofforschung, Statusbericht 1996, 47-49

4:      W. Henning, G. Neite, E. Schmid: Einfluß der Wärmebehandlung auf die Faserstabilität in verstärkten Al-Si-Leichtmetall-Kolbenlegierungen, Metall **48**, 451-454

5:      W. Schneider, F. J. Feikus: Wärmebehandlung von Aluminium-Gußlegierungen für das Vakuum-Druckgießen, Teil 6, Giesserei **84,** 30-42

6:      T. Beck, K.-H. Lang, D. Löhe: Quasistatisches und zyklisches Verformungsverhalten der Al_2O_3-faserverstärkten Kolbenlegierung AlSi12CuMgNi, in: Verbundwerkstoffe und Werkstoffverbunde, DGM-Informationsgesellschaft, Frankfurt/Main, 1997

Low Cycle Fatigue and Elasto-Plastic Behaviour of Materials
K-T. Rie and P.D. Portella (Editors)
461

ISOTHERMAL LOW CYCLE FATIGUE OF TITANIUM MATRIX COMPOSITES

H. ASSLER, J. HEMPTENMACHER and K.-H. TRAUTMANN
Institute of Materials Research, German Aerospace Center (DLR)
D-51147 Cologne, Germany

ABSTRACT

Ceramic fibre reinforcement has been suggested as a means to improve the resistance of a titanium alloy to mechanical fatigue. Low cycle fatigue tests at room and elevated (600°C) temperature were conducted on specimens, fabricated by DLR, a part of them heat treated at 700°C/2000h. In comparison with results described in the literature the tested material showed an increase of stress of nearly 30% at the same fatigue life. An investigation of metallographically polished specimens gave a first insight in the damage mechanisms. Defects at the surface of the specimens and the brittle reaction zone at the fibre/matrix interface acted as points of crack initiation. The crack propagation mainly depended on fibre bridging mechanisms, stress level in the matrix, microstructure of the matrix and properties of the fibre/matrix interface.

KEYWORDS

titanium matrix composite, IMI834, SiC-fibre, crack initiation, crack propagation

INTRODUCTION

The today's development of aircraft turbine engines aims at e.g. a reduction of pollutant emission and an increase of efficiency resulting in a reduction of energy consumption. A possibility to reach this aim is to make use of advanced materials. Most important features of materials in use for aircraft turbine components are elevated temperature strength, high stiffness and low specific weight.

Monolithic titanium alloy is an adequate material for the compressor stages of aero-engine gas turbines. The stiffness and strength improvements realised by fibre reinforcement could significantly extend the range of application. Therefore metal matrix composites, especially SiC-fibre reinforced titanium alloys, are under development to fulfil the above-mentioned requirements [1].

The German Aerospace Center (DLR) has developed a special method of processing, which can be divided into two steps (Figure 1). At first SiC-fibres (SCS-6) are coated with titanium matrix (e.g. Ti-IMI834) by magnetron sputtering, i.e. a single fibre composite is produced. After that bundles of coated fibres are encapsulated in a container of the desired component shape (here, a tube with an outer diameter of 10mm and an inner diameter of 3.2mm) and hot isostatic pressed (HIPed) to obtain the consolidated component. Most important advantages of this type of processing are a variable fibre volume fraction V_f between 0.2 and 0.6, a nearly ideal fibre distribution and a matrix with a small grain size [2].

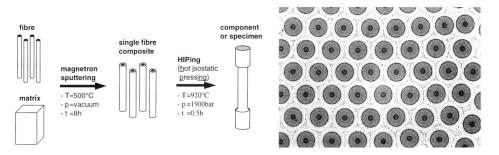

Fig. 1. Processing of titanium matrix composites Fig. 2. SCS-6/IMI834 composite

A comprehensive investigation of unidirectional SiC-fibre reinforced titanium alloys includes the determination of their low cycle fatigue properties. For mechanical characterisation isothermal fatigue tests were performed in air at room temperature and at 600°C. The specimens were tested in as-processed condition and after a thermal treatment at 700°C/2000h.

MATERIAL DETAILS

The SCS-6 fibre (SiC-fibre, Textron Specialty Materials) is coated with a thin layer of pure pyrocarbon followed by a SiC/C mixture (in the following designated as carbon layer) to protect the SiC filament from matrix interaction. IMI834, used in this investigation, is a near-alpha titanium alloy of medium strength and temperature capability up to about 600°C combined with good fatigue resistance.

As described above 10mm diameter cylinders of which the inner 3.1mm is reinforced with SCS-6 fibres with a fibre volume fraction of V_f=0.5 were fabricated (Figure 2). Half of the cylinders were heat treated at 700°C/2000h in air. Tensile tests at room temperature have shown that the heat treatment does not degrade the ultimate tensile strength significantly due to stability of the fibre/matrix interface [3]. Fatigue specimens, having a 3.5mm gage diameter and 6mm gauge length, were machined from the cylinders. Due to the outer unreinforced layer of titanium with a thickness of about 0.2mm the fibre volume fraction over the total specimen cross section measures V_f=0.39. The fatigue specimens were mechanically polished to minimise the influence of surface defects on fatigue properties.

EXPERIMENTAL DETAILS

For mechanical characterisation of the SiC/Ti composite with fibres in 0° direction low cycle fatigue tests were carried out. All fatigue testing of the composite specimens were conducted at 0.25Hz in a load-controlled, tension-tension mode and at a stress ratio of R=0.1 (R=minimum stress/maximum stress) using a servohydraulic axial fatigue test facility which could be operated by a servo-controlled hydraulic test system. Both, tests at room temperature and at 600°C (beyond operating temperature) were performed. At elevated temperatures the specimens and the grips were resistance heated. The specimens had a thread on both ends and were gripped in the female screw such that the load was introduced not only via the threads but also by clamping stresses, as the female screw was cut into four sections with a slot in between enabling a mechanical clamping. Strain measurements were performed using an extensometer with ceramic arms. The stress/strain response was periodically recorded during fatigue life [4].

RESULTS AND DISCUSSION

Low cycle fatigue tests on reinforced titanium alloy specimens (0.39-SCS-6/IMI834) were performed in air at room temperature at a stress ratio of R=0.1 and at a frequency of 0.25Hz. Mechanical tests were followed by first microstructural characterisation to identify damage and failure mechanisms. First results are presented in Figure 3 as applied maximum external stress versus number of cycles to failure (S-N curve).

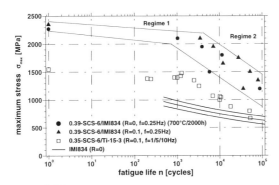

Fig. 3   Room temperature S-N-curve

The low cycle fatigue response of TMCs can generally be divided into two regimes: Regime 1 is a high-stress low-life regime, during which failure is catastrophic and probably arises because the stress in the fibre approaches its breaking strength. Regime 2 which includes both crack initiation and matrix crack propagation including fibre-bridging will be discussed later in detail. The fatigue behaviour of TMC composites is observed to be superior when compared to the fatigue behaviour of bulk material. The principal reason for this improvement is that the fibres stiffen the composite significantly so that for a given fatigue stress level, the composite matrix will be subjected to a much smaller strain than in case of unreinforced matrix. Because of the reduced matrix strain, the composite should display superior fatigue behaviour. Typically, the ceramic fibre reinforcement does not fatigue; thus, it is the matrix and/or the fibre/matrix-interface that determines the composite fatigue life.

Obviously the SCS-6/IMI834 specimens, produced and tested by the DLR, show an increase of maximum stress of about 40% in comparison with results described by Gabb et al. [5] and others. The lower results are caused by the failure mechanisms in composites produced by foil-fibre-foil technique. These are triggered by cut fibres at the outer surface of a rectangular specimen and molybdenum wires in the composite used to weave SiC fibres into mats (crack initiation).

Similar to the results of tensile tests, the described heat treatment of 700°C/2000h seems to have no significant influence on the low cycle fatigue properties of SCS6/IMI834 composites.

464

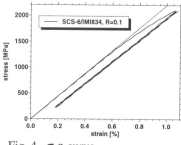

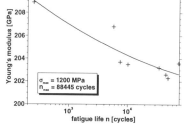

Fig. 4. σ-ε-curve   Fig. 5. Young's modulus

Figure 4 shows a typical stress-strain behaviour for the SCS-6/IMI834 composite during the first two load cycles with a maximum stress of 2100MPa. Up to a strain of about 0.6% the composite deforms elastically. With increasing stress a knee occurs due to plastic deformation of the matrix. From tensile tests it is known that the yield strain of IMI834 is at a higher level. This decrease of strain level is caused by residual stresses which are induced after cooling down from the manufacturing temperature due to the different thermal expansion coefficient of fibre and matrix ($\alpha_f < \alpha_m$). Yielding of the titanium alloy begins at about 0.8% strain, whereas in the composite matrix yielding is already observed at a strain of about 0.6% strain. From this it can be concluded that the residual matrix stress is equivalent to about 0.2% strain. This means a residual tensile stress of about 200 MPa is stored in the matrix.

In Figure 5 the Young's modulus which is estimated with the aid of the recorded σ-ε curves at different numbers of cycles is plotted. During the whole fatigue life a slight reduction of the stiffness is found.

Crack initiation and propagation

Two different crack initiation sites are found to influence the fatigue life. Cracks start at defects on the surface like grooves or notches caused by machining or handling. Sometimes the contact surface between specimen and extensometer is the point where a surface crack starts. The crack mainly propagates in the unreinforced matrix because the crack resistance of the titanium is lower then that of the composite. The „low-stress" (σ_{max}=1350MPa) fatigue specimen shown in Figure 6.a exhibits a corresponding flat area of crack growth in the unreinforced matrix and larger area indicating tensile overload. The fatigue crack in the surface layer of unreinforced titanium surrounds ¾ of the reinforced section. The lower the maximum stress is the higher the number of detectable surface cracks is.

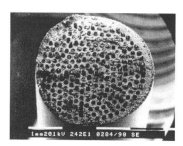

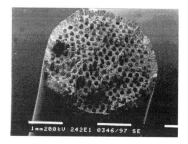

Fig. 6: (a) „Low-stress" fracture surface   (b.) „High-stress" fracture surface

In addition to that the reaction-zone acts as an important internal crack initiation site at room temperature. This brittle reaction zone mainly consists of TiC that results from an interaction of carbon (fibre coating) and titanium (matrix) at high temperatures during processing (consolidation) of the composite. Figure 7.a is a micrograph of a fatigued specimen showing fatigue crack initiation in the reaction zone layer. Several cracks spread into the matrix material, however due to low bond strength between SiC-fibre and its outer C-layer the crack also deflects in the corresponding interface (Figure 7.b). These cracks have a strong influence on the fracture surface at high maximum stresses. In contrast to the flat „low-stress" fracture surface, Figure 6.b represents a rugged „high-stress" surface in which no significant decrease of the pull-out length is found. In addition to that shear failure under roughly 45° of the unreinforced titanium surface layer indicates the ductility of the matrix.

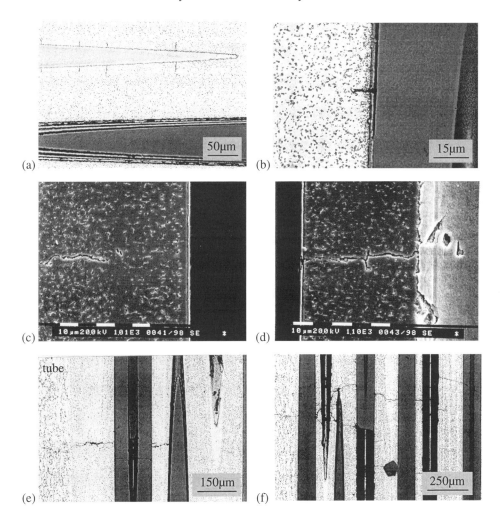

Fig. 7. Metallographically polished specimens are examined optically and using electron microscopy (longitudinal section)

466

During crack propagation a very interesting effect occurs which is linked to the method of processing described above. In spite of hot isostatic pressing above 900°C the outer borders of the single fibre composites are still visible after consolidation (see Figure 2, honeycomb structure). These borders form microstructural irregularities which have a positive influence on the resistance against crack propagation. Figure 7.c and 7.d show that the cracks stop or are delayed at these irregularities. Figure 7.e makes clear that cracks, visible at the surface, not only result from defects at the surface of the specimen but also from internal defects in the reaction zone between fibre and matrix. A crack starts near the fibre surface, propagates through the fine sputtered matrix material and after that through the coarse matrix material of the tube. This tube is the container which encloses the single fibre composites during hot isostatic pressing. As represented in Figure 7.f extensive matrix cracking, combined with fibre bridging, can be detected after a high number of cycles. These effects explain the reduction in modulus which is shown in Figure 5.

SUMMARY AND CONCLUSIONS

A comprehensive investigation of the low cycle fatigue behaviour of titanium matrix composites was performed. Load controlled axial fatigue tests on cylindrical specimens with and without heat treatment (700°C/2000h) were performed and evaluated.
The composite with as well as without heat treatment shows a significantly better fatigue resistance better than that of both unreinforced matrix and other titanium matrix composites, described in the literature. At high maximum stresses the titanium alloy deforms plastically, indicating that tensile residual stresses are induced after processing. These stresses are found to decrease fatigue resistance.
Fractographic analyses exhibit areas of crack growth in the matrix and larger areas indicating tensile overload. The crack regions are relatively featureless and a middle degree of fibre pull-out suggests of fibre bridging. Metallographic examination of fatigue specimens revealed that fatigue cracks initiate either at the specimen surface and/or in the brittle reaction zone at the fibre/matrix interface. Extensive matrix cracking is observed combined with fibre bridging explaining the reduction of modulus. Some of the factors which increase fatigue resistance are crack deflection at the fibre/matrix interface and micromechanical irregularities (texture as a result of the fabrication process) coupled with the process of fabrication. In conclusion the damage mechanism is a complex interaction of different effects.
In the near future the experiments at elevated temperature (600°C) will be completed including a detailed metallographic investigation. The latter is necessary because the knowledge of failure mechanisms support the development of improved TMCs by the modification of the architecture and processing routes.

REFERENCES

1. Ward-Close, C.M. and Loander, C. (1995) *Recent Advanced in Titanium MMCs*, pp. 19-32
2. Leucht, R. and Dudek, H.J. (1994) *Mat. Sci. Engn.*, A 188, pp. 201-210
3. Dudek, H.J., Borath, R., Leucht, R., Hemptenmacher, J. and Kaysser, W.A. (1997) *Proceeding of Interface Phenomena in Composite Materials*, Eger, Hungary
4. Schulte, K., Trautmann, K.-H. and Leucht, R. (1994) *ASTM STP 1184*, pp. 32-47
5. Gabb, T.P., Gayda, J., Lerch, B.A. and Halford, G.R. (1991) *Scr. Met.*, Vol. 25, pp. 2879-2884

Low Cycle Fatigue and Elasto-Plastic Behaviour of Materials
K-T. Rie and P.D. Portella (Editors)
467

MECHANICS AND MECHANISMS OF THERMAL FATIGUE DAMAGE IN SCS-6/Ti-24Al-11Nb COMPOSITE.

M.OKAZAKI
Department of Mechanical Engineering, Nagaoka University of Technology,
Tomioka, Nagaoka 940-21, JAPAN.
H.NAKATANI
Kawasaki Heavy Industries Ltd.,
Kakamigahara, Gifu 504, JAPAN.

ABSTRACT

Thermal fatigue tests of an unidirectional SiC fiber reinforced Ti-24Al-11Nb matrix composite, SCS-6/Ti-24Al-11Nb, have been carried out without external load in air. Based on the systematic investigations on the crack initiation and the change in density during the thermal cycling, as well as on the chemical reaction near the matrix/fiber interface by EPMA and SEM, and on the interface shear strength employing the push-out tests, special attentions are paid to understand mechanisms and mechanics of thermal fatigue failure, and how the thermal fatigue life should be assessed.

KEY WORDS

Thermal fatigue, SCS-6/Ti-24Al-11Nb, C coating, push-out test, interfacial shear strength.

INTRODUCTION

Silicon-carbide fiber reinforced titanium alloy composites have received considerable interests as the materials for aerospace structural applications, because they possess many attractive characteristics: higher specific strength at high temperature and higher specific stiffness, compared with traditional monolithic materials [1-7]. Many studies have been carried out on fracture under static loading [2], fatigue [3] and creep [4]. On applying this kind of composite to actual components in engineering field at high temperatures, on the other hand, special consideration should be paid to thermal stress which is induced by mismatches in elastic moduli and thermal expansion coefficient [1]. Thus, information on thermal fatigue strength is inevitable [5,6], however, there remain many problems to be clarified. For example, no clear answer on how thermal fatigue life should be assessed in this kind of composite and on what the mechanisms is, has been established yet.

In this work thermal fatigue tests of an unidirectional SiC fiber reinforced Ti-24Al-11Nb matrix composite, SCS-6/Ti-24Al-11Nb, were carried out without external load in air, under various thermal cycle conditions. Investigating systematically on the crack initiation and the change in density during the thermal cycling, on the chemical reaction near the matrix/fiber interface by EPMA and SEM, and on the interface shear strength by push-out tests, discussions are made on mechanisms and mechanics of thermal fatigue failure, and how the thermal fatigue life should be assessed.

468

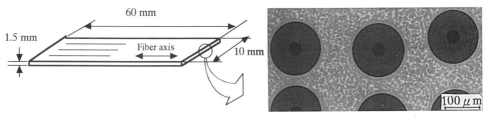

Fig. 1 Geometry of specimen and the metallograph of composite material tested.

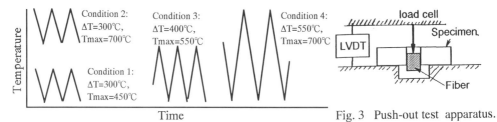

Fig. 2 Thermal cycle conditions given.

Fig. 3 Push-out test apparatus.

EXPERIMENTAL PROCEDURE

The material specified is a Ti-24Al-11Nb matrix composite reinforced unidirectionally with continuous SiC fiber (; SCS-6 Textron Corp., Lowell, MA), where the SCS-6 fiber is a β-SiC fiber of 140 μ m diameter with a carbon rich graded silicon carbide coating on the surface. The 7-plies composite sheet of 180x40x1.5 mm has been fabricated at Kawasaki Heavy Industries Co. Ltd. (Kakamigahara, Gifu, Japan), by hot pressing alternate layers of thin-foils of Ti-24Al-11Nb alloy and green tapes of the SCS-6 fibers at 1050℃ for 2 hrs. in vacuum. The volume fraction of the fibers in the composite is about 33 %.

From the composite sheet, the plate specimens of 60x10x1.5 mm (Fig. 1) were taken out by electro-discharge machining, where the SiC fibers align parallel to the longitudinal (60 mm) direction, and some of them appear on the specimen side and bottom surfaces. The thermal cycles schematically illustrated in Fig.2 were repeated to the specimen in a frequency of about 1/3 cpm. without external load in air, by using a test apparatus which consists of an induction heating system for heating and an air compressor for cooling. The temperature distribution in the specimen was less than 7 degree. All the heating-cooling waveforms shown in Fig.2 were almost triangle. During the thermal cycling the replica of the specimen surface was taken to monitor the damage evolution process and the crack nucleation. The change of fiber/matrix interface was investigated by SEM and EPMA, with specific attention being paid to the oxidation and reaction. The interfacial shear strength after thermal cycling was also measured, by employing the push-out test (Fig. 3) .

RESULTS AND DISCUSSIONS.

Thermal Fatigue Cracks

The following visible crackings were found by repeating the thermal cycle: one is the longitudinal cracking along the fiber/matrix interface parallel to the fiber axis; the second is the transverse, or circumferential cracking in the matrix normal to the fiber axis; and the third is the fiber breakage (Fig. 4). As the results of these crackings which release the fibers some of fibers were found protruded out from the composite.

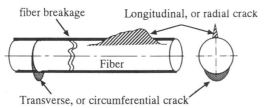

Fig. 4 Typical cracks nucleated during thermal cycling.

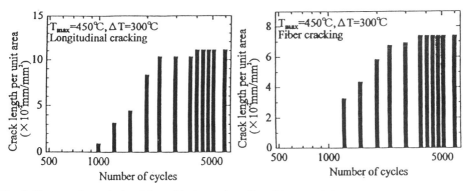

Fig. 5 Change of crack densities with thermal cycling (T_{max}=450℃, ΔT=300℃.)

The crack length per unit area, or line density, of these types of cracks was measured from the replica of the specimen surface. Some of the typical results are given in Fig.5. From these the following characteristics have been found out:

(a) The line densities of all types of cracks increased with the number of thermal cycles, and finally they saturated. The final density in each condition is summarized in Fig. 6.

(b) The higher the temperature range, ΔT, the earlier all the types of cracks initiated.

(c) The line density of the longitudinal cracks was higher than any other types of cracks in all conditions. The longitudinal cracks generally initiated earliest.

(d) The change of the fiber cracking

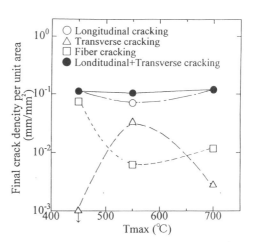

Fig. 6 Summary of crack densities saturated.

behaved in a similar manner to that of the longitudinal cracking. The dependence of thermal cycle condition in the saturated crack density also resembles in the both types of cracks (Fig. 6). On the the hand, the transverse and longitudinal cracks seemed to be nucleated in a confliction manner each other: when the density of the latter was high, that of the former became low, and vise versa. In fact, the sum of the transverse and longitudinal crack densities reveals insensitive to the thermal cycle condition (Fig.6).

The crack density is one of candidate parameters to define the "thermal fatigue damage" quantitatively. In this work the thermal fatigue failure life was defined by the number of cycles at which the sum of densities of transverse and longitudinal cracks reached the final, or

470

saturated value (Fig. 6), and the damage ratio on the way to the final failure was defined by the crack density normalized by the final value. According to the above definition the thermal fatigue life is correlated in Fig. 7 in terms of temperature range, ΔT, which corresponds to the thermal stress range induced by thermal cycling (hence, Fig.7 is equivalent to S-N curve in normal fatigue failure). A good linear relationship can be found on log-log coordinates in Fig. 7. This figure must be also interesting to evaluate the remaining life, because it is expressed as the function of damage ratio.

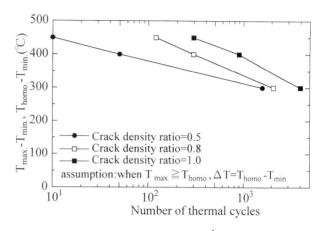

Fig. 7 Life assessment curve proposed.

Interface Damage .

By means of EPMA and SEM the matrix/fiber interface was observed and analyzed in the sample taken out from the midsection of the composite which had been exposed to thermal cycling. Before applying the thermal cycling, as shown in Fig. 8(a), a carbon rich coating of a few micrometers in thickness, which had been originally performed to achieve full strength by healing surface flaws of the fiber and to protect from the reaction with the matrix during the consolidation process of the composite [7], can be seen on the fiber surface. Even after applying the thermal cycles of $N=6000$, $T_{max}=450°C$ and $\Delta T=300°C$, the C-rich coating still remains on the fiber surface (see Fig. 8(b)). It is worth noting in Fig. 8(b)) that the intensity of

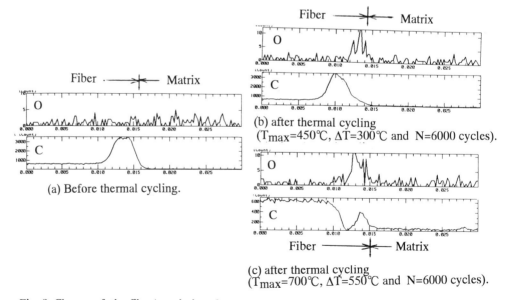

(a) Before thermal cycling.

(b) after thermal cycling ($T_{max}=450°C$, $\Delta T=300°C$ and $N=6000$ cycles).

(c) after thermal cycling ($T_{max}=700°C$, $\Delta T=550°C$ and $N=6000$ cycles).

Fig. 8 Change of the fiber/matrix interface.

O increased at the interface, which means the formation of oxide. These results suggest that oxidation would begin prior to the depletion of C coating during the thermal cycling. Furthermore, when the thermal cycles of N=6000, T_{max}=700°C and ΔT=550°C was given, the C-rich coating on the fiber surface has been completely depleted, and the oxidation at the interface got more remarkable (Fig. 8(c)). These observations indicate not only mechanical factors, such as thermal stress induced by thermal cycling, but chemical factors; oxidation and depletion of carbon coating, play important roles in thermal fatigue failure in this composite.

The interfacial shear strength, τ_C, above which fiber is protruded in push-out test, is generally expressed by

$$\tau_C = \mu\sigma_r + \tau_r \qquad (1)$$

where μ, σ_r and τ_r are frictional coefficient between fiber and matrix, the residual stress in radial direction of fiber, and the bonding strength of reaction zone, respectively. While the term of τ_r is generally negligible in this composite, the μ in Eq.(1) is significantly affected by the roughness between fiber and matrix.

The τ_C of the composite subjected to thermal cycle is shown in Fig. 9 on the Weibull diagram, in comparison with that of the material exposed to the isothermal aging in air, where these were evaluated in the sample taken from the midsection of the material by the push-out test. It is found from Fig. 9 that when the material was isothermally exposed at 400°C in which the carbon rich coating still remained on the fiber surface, the τ_C is almost comparable to that of the virgin material. However, when the exposure temperature was 700°C at which the carbon coating disappeared, the reduction in τ_C is very significant. What is important in this reduction is that the τ_C's not only on relatively higher level but on lower level are reduced comparably (i.e., the distribution of the τ_C shifts to left hand side along the abscissa). The reduction in σ_r in Eq.(1) resulting from the carbon coating depletion must be responsible for this reduction. When the material was experienced the thermal cycling, on the other hand, the τ_{cr} is also lowered (Fig.9). The feature in this case is that the τ_C on relatively lower level is significantly decreased, but not on higher level (i.e., on the Weibull diagram the slope get gentle). It is worthy to remind that once cracks are nucleated during the thermal cycling, the thermal stress which combines the fiber with the matrix is released, resulting in the decrease of σ_r in Eq.(1). Hence, not only the oxidation and the depletion of carbon coating on the fiber surface but the cracking play essential roles in the change of τ_C shown in Fig. 9.

From these aspects the thermal fatigue damage in air

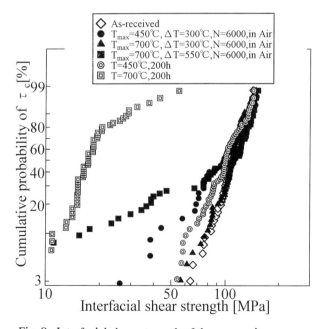

Fig. 9 Interfacial shear strength of the composite.

472

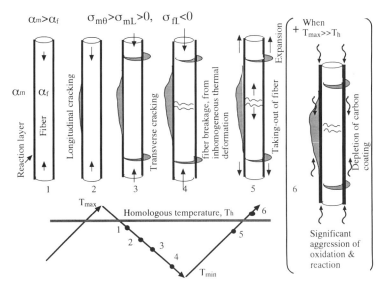

Fig. 10 Thermal fatigue damage process proposed.

is found to be evolved according to the mechanisms and the process as shown in Fig.10.

CONCLUSIONS

The main conclusions reached are summarized as follows:
(1) Thermal cycling introduces the longitudinal and transverse cracking in the matrix, as well as the the fiber breakage. In general the density of the longitudinal cracking was higher than any other types of cracking. These types of cracking were found to increase in density rapidly at the beginning of thermal cycling. They evolved interacting with each other, and finally saturated in density, depending on the condition of thermal cycle.
(2) The interfacial shear strength of the composite, τ_c, was significantly changed by applying the thermal cycle in air. Not only the cracking induced by the thermal stress cycling but also the oxidation and the depletion of carbon coating on the fiber surface played essential roles in the change of τ_c.
(3) The evolution process and the mechanisms of thermal fatigue damage in air were clarified and proposed.

ACKNOWLEDGEMENT:Author (M. Okazaki) expresses his great gratitude to the financial support by the Grant-in-Aid for sceintific reserach by Japan Ministry of Education (No.08555023). The authors would like to thank Mr. K. Hayashi and Mr. S. Asahi, Graduate students, for their assistance on the experiments.

REFERENCES

1.   Budianski, B., Hutchinson, J. and Evans, A.G. (1986)*J. Mech. Physc. Solids*, **34**, 167.
2.   Chiu, H. and Yang,J.M. (1995)*Acta Met.* , **43**, 2581.
3.   Davidson, D.L. (1991)*Metall. Trans.-A* , **23-A**, 865-879.
4.   Eggleston, M.R. and Ritter, A.M. (1995)*Metall. Trans.-A*, **26-A**, 2733.
5.   Russ, S.M. (1990) *Metall. Trans.*, **21-A**, 1595.
6.   Neu, R.W. and Nicholas, T. (1994)*J. Comp. Tech. & Research*, **16**, 214 .
7.   Baumann, S.F. ,Brindley, P.K. and Smith, S.D(1990) *Metall. Trans.*, **21-A**, 1559.

Low Cycle Fatigue and Elasto-Plastic Behaviour of Materials
K-T. Rie and P.D. Portella (Editors)
473

INVESTIGATIONS ON THE THERMOMECHANICAL FATIGUE BEHAVIOUR OF PARTICLE REINFORCED ALUMINIUM ALLOYS APPLYING A GLEEBLE 1500

P. PRADER and H.P.DEGISCHER

Department of Materials Science and Testing, Vienna University of Technology
1040 Vienna, Austria

ABSTRACT

Experiments on the thermomechanical fatigue behaviour of a 6061 aluminium metal matrix composite, reinforced by 10 vol.-% Al_2O_3-particles were performed with a GLEEBLE 1500 apparatus. Relatively fast thermal cycles were provided by optimizing the specimen geometry with respect to low dynamic thermal gradients. To analyse the development of surface failures, some experiments were interrupted and the specimens were investigated by means of scanning electron microscopy.

KEYWORDS

Metal matrix composite, particle reinforced aluminium, thermomechanical fatigue, thermo-cyclic strain, Gleeble apparatus

INTRODUCTION

Combining a high strength ceramic phase with a matrix of a ductile light metal to form metal matrix composites (MMC) offers attractive choices for use in combustion engines and brake components because of improved strength, stiffness, dimensional stability, fatigue and wear resistance compared with unreinforced alloys.

In components that are mechanically loaded and suffer additional temperature cycles during service, the large mismatch in the thermal expansion coefficient of the matrix and reinforcement lead to the generation of misfit strains and large internal elastic stresses during temperature changes, producing significant plastic deformations locally. Internal stresses can be regenerated at each thermal cycle (from ambient temperature up to 300°C) and the development of excessive plastic strains lead to local damage. The understanding of the composite behaviour under such thermomechanical fatigue conditions is of significant importance for practical applications and determines the lifetime of the component.

474

There are different experimental methods to test thermomechanical fatigue [1,2,3]. Frequently arising problems are dynamic thermal gradients, precise temperature control and long total testing times because of slow heating and cooling rates. This paper describes experiments on the thermomechanical fatigue behaviour of particle reinforced aluminium alloys by adaptation of a Gleeble 1500 apparatus [4, 5].

EXPERIMENTAL CONDITIONS

The GLEEBLE 1500 testing device combines full resistance heating thermal capabilities with hydraulic servo-mechanical testing performance. The temperature is controlled by a spot welded thermo-couple and experiments are performed by computer control, the test results are displayed in real time. Water cooled grips and additional pressurized air cooling provide an efficient cooling rate necessary for fast thermal cycles. Among others, the advantage of this apparatus is the realization of high heating rates (up to 10000 K/s, dependent on specimen material and geometry) without loss in accuracy.

Much effort was focused to realize fast thermal cycles without dynamic thermal gradients which can cause barreling of the specimens during thermomechanical fatigue [1]. Barreling is a ratchetting phenomenon promoting material flow and localized straining during cycling. On a cylindrical specimen this behaviour is evidenced as circumferential bulging.

For that purpose specimens of particle reinforced aluminium were made in various geometries and subjected to thermal cycles. The axial temperature profiles within the measuring range were monitored by thermo-couples spot-welded onto the specimens in the length-wise direction. The aim was the development of a specimen geometry with dynamic thermal gradients in the measuring range less than ±5 K during thermal cycling.

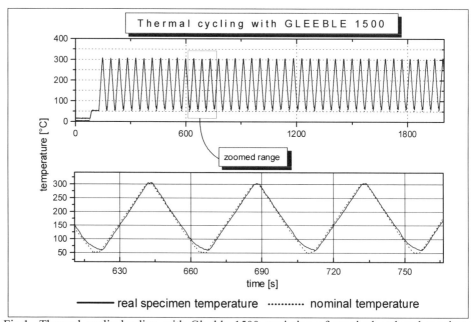

Fig.1. Thermal cyclic loading with Gleeble 1500, variation of nominal and real specimen temperature

475

The cyclic thermal loading consisted of heating from 50°C to 300°C and cooling back to 50°C in 45 seconds. The minimum and maximum temperatures were hold constant for 2 seconds. Figure 1 shows the variation of nominal and real specimen temperature. Reaching 90°C during cooling sequence additional pressurized air cooling provided a negligible difference between nominal and real specimen temperature.

Analysing the temperature profile within the measuring range during thermal cycling between 50° and 300°C an optimized specimen geometry was developed (Fig.2). This specimen geometry provides a homogeneous temperature distribution within the measuring range during the whole cycle with a negligible gradient of about ±2 K (Fig.3)

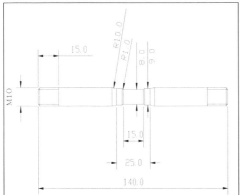
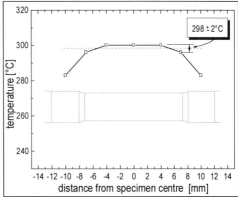

Fig.2: Optimized specimen for thermo-mechanical cycling with GLEEBLE 1500

Fig.3: Temperature distribution in the measuring range of specimens made of 6061+10 vol.-% Al$_2$O$_3$

The thermal-cyclic behaviour of a 6061 aluminium metal matrix composite, reinforced with 10 vol.-% Al$_2$O$_3$ particles by stir-casting (Duralcan) [6] and hot extruded, was analysed (Table 1). Specimens of this material were subjected to thermal cycles between 50°C and 300°C and an initial stress of 20 MPa. To investigate the development of surface defects the specimens were demounted after several thermal cycles ,analysed by scanning electron microscopy (SEM) and subjected to further cycles afterwards. The surface of the specimens was polished with diamond paste to remove machining grooves. The extension of the specimens was measured by means of a high temperature extensometer.

Table 1. Nominal composition of the Duralcan aluminium matrix composite

matrix	Fe	Si	Cu	Mn	Mg	Ti	Ni
AlMg1SiCu / 6061	0,20	0,69	0,28	0,009	0,93	0,10	--
+10 vol.-% Al$_2$O$_3$	particles of about 8 µm in diameter						

RESULTS

A typical plot of extension against time of the 6061+10% Al$_2$O$_3$ for thermal cyclic loading with ΔT=250 K and a constant load corresponding to a stress of σ = 20 MPa is shown in figure 4.

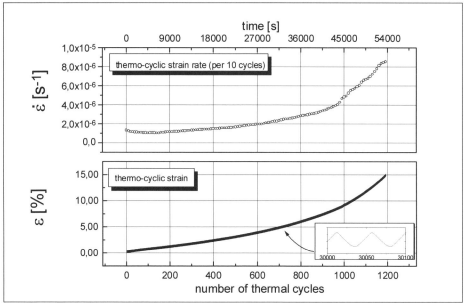

Fig.4. Thermal cycling of 6061+10% Al$_2$O$_3$: thermo-cyclic strain and thermo-cyclic strain rate dependent on number of cycles

Similar to conventional creep the composite shows a slightly decreasing thermo-cyclic strain rate during the first 200 cycles. The strain rate in the nearly constant range is between $\dot{\varepsilon} = 1 \cdot 10^{-6}$ and $\dot{\varepsilon} = 4 \cdot 10^{-6}$ [s^{-1}]. Flaig et al [7] found a strain rate of $\dot{\varepsilon} = 4.2 \cdot 10^{-7}$ [s^{-1}] (150°C$\leftrightarrow$350°C, 40 MPa) for a Saffil-fibre reinforced cast aluminium alloy, which indicates that particle reinforced aluminium has a less creep resistance when thermally cycled.

Figure 5, a-f, show the results of the SEM-investigations on a thermally cycled specimen at different stages of exposure. The composite consists of Al$_2$O$_3$-particles, which are partially fractured by machining or hot extrusion, embedded in the aluminium matrix (Fig.5a). The SEM investigations were focused on this area, especially on the relatively coarse and fractured particle in the centre of figure 5a. After 300 cycles, the second branch of the crack has been opened and at the sharp edge of the particle a crack is initiated in the matrix (Fig.5c, arrow). The subsequent cycles cause the further propagation of this crack into the aluminium matrix (Fig.5 d-e). It can be assumed, that the misfit created by the different thermal expansions of the matrix and the reinforcement is responsible for the development of that crack at the narrowest gap between two particles [8]. In addition after 1100 cycles the two branches of the crack in the particle are widely opened and crack initiation takes place from the interface region into the aluminium matrix (Fig.5f, arrows).

During thermomechanical cycling the particulate fragments of a ruptured particle are moved apart and after 1100 cycles a large cavity is formed between the two parts (Fig.5e, below arrow).

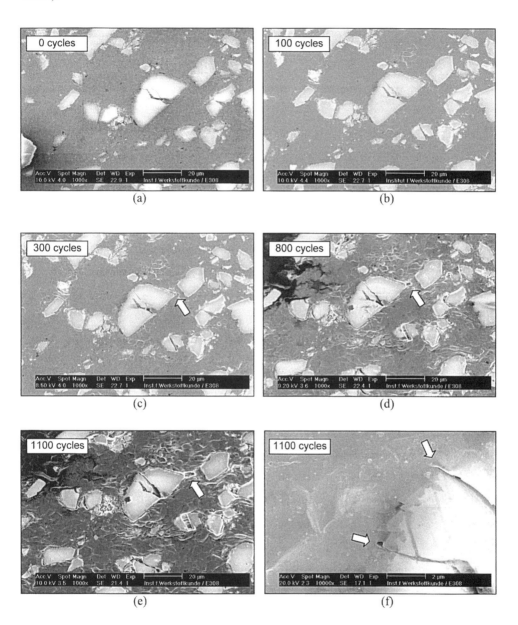

Fig.5: 6061+10% Al$_2$O$_3$ composite after various thermal cycles: before exposure(a), after 100 (b), 300 (c), 800 (d) and 1100 (e) cycles. Crack initiation in the matrix caused by particle cracks after 1100 cycles (f).

478

SUMMARY

The facility of a Gleeble 1500 apparatus for thermomechanical fatigue investigations on aluminium matrix composites is demonstrated. With the developed specimen geometry relatively fast thermal cycles sequences of about 45 seconds are possible by providing a negligible dynamic thermal gradient of ± 2 K in the measuring range. The thermal cyclic creep behaviour of a 6061 aluminium alloy reinforced by 10 vol.-% Al_2O_3-particles was analysed (ΔT=250 K and $\sigma = 20$ MPa). The development of surface failure mechanisms was monitored during the experiment. It was found that cracks and edges of particles can initiate cracks in the aluminium matrix. In addition, ruptured particles form cavities during fatigue loading.

ACKNOWLEDGEMENT

The authors acknowledge the financial support by the University Fund of the Vienna Municipality.

REFERENCES

1. Castelli, M.G., Ellis, J.R. (1991). In: *Thermomechanical Fatigue Behaviour of Materials*, H.Sehitoglu (Ed.). ASTM STP 1186, Philadelphia, pp 195-211
2. Bressers, J., Rèmy, L., Steen, M., Vallès, J.L. (Eds., 1995): *Proceedings of the Symposium "Fatigue under Thermal and Mechanical Loading: Mechanisms, Mechanics and Modelling"*, Kluwer Academic Publishers, Dordrecht
3. Shakesheff, A.J., Pitcher, P.D. (1996). In: *Proceedings ICAA5*, J.H. Driver et al. (Eds.), Materials Science Forum Vols.217-222, Transtec Publications, Switzerland, pp.1133-1138
4. Ferguson, H. (1993): *Advanced Materials & Progresses* **4/93**, pp.33-36
5. Lasday, S.B. (1996): *Industrial Heating* **1/96**, pp.23-29
6. Degischer, H.P., Kaufmann, H., Leitner, H. (1992): *VDI Berichte* **Nr.965.1**, VDI Verlag GmbH, Düsseldorf, pp. 179-188
7. Flaig, A., Wanner, A., Arzt, E. (1997) In: *Proceedings of the conference "Verbund-werkstoffe und Werkstoffverbunde"*, K.Friedrich (Ed.). DGM Informationsgesellschaft, Frankfurt, pp.557-562
8. Pettermann, H., (1997). *Derivation and Finite Element Implementation of Constitutive Material Laws for Multiphase Composites Based on Mori-Tanaka Approaches.* Doctoral Thesis, Vienna University of Technology, Austria

Low Cycle Fatigue and Elasto-Plastic Behaviour of Materials
K-T. Rie and P.D. Portella (Editors)
479

OFF-AXIS SHEAR BEHAVIOR OF SELECTED MMC's

P.N. MICHELIS, P.D. NICOLAOU, and P.G. KORNEZOS
IMMG S.A.
22 Askiton Str., Pendeli, GR 15236, Greece

ABSTRACT

Off-axis shear tests on unidirectional and cross-woven MMC's were conducted using a prototype shear testing machine, in order to determine important material constitutive parameters such as shear modulus, strength, and strain to failure. The variation of these parameters as a function of the orientation angle θ was evaluated. Specifically, it was found that the shear modulus increases with θ, while the reverse occurs for the other two parameters. The experimental results were compared to predictions of existing models; macromechanical and micromechanical disciplines were used to justify deviations between the predictions and the data. Cyclic tests showed internal damage accumulation in the composite, which increases with the number of cycles. Finally, it was shown that cross-woven composites possess higher shear strength, but their strains to failure are much lower than those of their unidirectional counterparts. The sensitivity of the material behavior to the stress sign disables existing models to predict accurately the experimental results obtained. However, the latter may lead to the appropriate direction for future modeling efforts.

KEYWORDS

Metal matrix composites, off-axis shear behavior, monotonic loading, cyclic loading.

INTRODUCTION

For the past few years, the determination of the mechanical response of metal matrix composites under shear loading conditions has attracted the attention of the composite community. In order to obtain accurate and reliable data a number of issues which arise from the anisotropic nature of these materials is required to be considered. Ideally, for quantitative shear measurements, the test method should provide a region of pure, uniform shear stress. The test method should also be relatively simple to conduct, employ small, easily fabricated specimens, and be capable of measuring both shear strength and shear modulus[1-3]. Since the failure behavior varies with the type and direction of loading and it is also intimately related to the properties of the individual composite constituents (i.e. fiber, matrix, and their interfaces) it is of main interest to determine of the material response under off-axis shear loading, i.e. when the principal directions of the material do not coincide with the testing machine shear plane.

If elastic behavior is assumed, then the variation of the shear modulus with respect to the orientation angle θ can be predicted from the following equation[4]:

$$\frac{1}{G_{xy}} = \frac{1}{G_{12}}\cos^2 2\theta + \left(\frac{1+\nu_{12}}{E_{11}} + \frac{1}{E_{22}}\right)\sin^2 2\theta \qquad (1)$$

480

where G_{xy}, G_{12} are the shear moduli at an angle θ and at $\theta = 0°$ respectively, E_{11}, E_{22} are the composite elastic moduli in the longitudinal and transverse direction, while ν_{12} is the Poisson ratio.

Several micromechanical and macromechanical models have been developed in order to predict the shear strength of a composite as a function of the orientation angle; macromechanical models have been proved to be more accurate than micromechanical. Almost all of them are based on assumptions of homogeneity and linear stress-strain behavior to failure. The Tsai-Hill model/criterion[5] is one of the commonly used for the prediction the shear strength versus θ since it is relatively simple and also it allows for interaction among the stress components:

$$\frac{1}{\tau_{xy}^2} = \frac{2\sin^2 2\theta}{\sigma_1^2} + \frac{\sin^2 2\theta}{\sigma_2^2} + \frac{\cos^2 2\theta}{\tau_{12}^2} \qquad (2)$$

where σ_1, σ_2, τ_{12} are the longitudinal, transverse, and shear strength respectively at $\theta = 0°$, and τ_{xy} is the shear strength at an angle θ.

To this end, a prototype shear testing apparatus was utilized to investigate the shear constitutive parameters as a function of fiber orientation, fiber architecture (unidirectional or cross-woven) and loading mode (monotonic or cyclic). A number of experimental data were obtained and the results were compared to predictions of the above Eqs. (1) and (2).

MATERIALS AND TESTING PROCEDURES

Materials

The composite materials used in this research effort comprised either an Al-0.15Mg or Al-6Zn-1Mg matrix reinforced with unidirectional or cross-woven Altex SN fibers. The fiber volume fraction was ~50% and they were manufactured by pressure assisted investment casting by GI, University of Aachen. The specimens were removed from prismatic samples in such a way that the fibers formed different inclinations to the testing machine shearing plane. The deformation was measured by a small, rectangular (~2 x 2 mm) three-element strain rosette, directly bonded on the uniform shear strain region of the specimen.

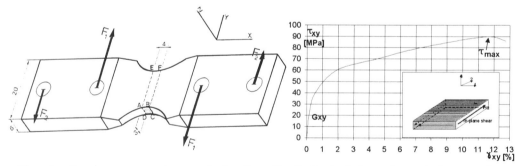

Fig. 1: Specimen geometry used in the experiments; the boundary loading conditions of the shear device is also shown.

Fig. 2. Typical shear stress-strain curve for a Al-0.15Mg matrix-Altex SN fiber UD composite for an orientation angle $\theta = 45°$.

Shear Testing Machine

A prototype shear testing machine was designed and validated, which develops a uniform shear field at the central zone ABCD of a suitably configured specimen (Fig. 1). The shear stress calculation is based on the boundary forces without resolting to any constitutive relation for the material. Shear strain can be easily measured by bonding strain rosettes or positioning suitable extensometer on the one or both sides of the specimen central zones (ABFE). Extensive 3-D finite element analysis revealed that in the central zone all the normal stress reach small level values in comparizon to shear stresses. The shear stress distribution, within the elastic domain, remains the same when the material becomes anisotropic. This also applies for orthotropic materials and particularly for transversely anisotropic when the fibres change direction.

RESULTS AND DISCUSSION

The shear failures were reflected on the form of stress-strain curve discontinuity and/or audible sound emission. The crack always initiated from the test section and there was no damage from the loading fixtures. Figure 2 represents a typical shear stress-shear strain curve for a monotonically loading test specimen, with a rotation of the fiber angle θ equal to 45 degrees. A number of important shear parameters such as shear modulus, ultimate shear strength, shear strain to failure, and stress-strain behavior were determined. The results from a large number of tests are summarized and discussed in the subsequent sections. Unless otherwise stated, the results correspond to UD composites deformed under monotonic loads.

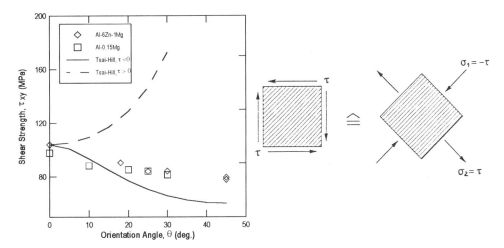

Fig. 3. Variation of the shear strength with the orientation angle θ for the two matrices of UD composites.

Fig. 4. Schematic of decomposition of a negative shear stress applied on a UD composite at an orientation angle of 45°.

Shear Strength.

The shear strength of the composite versus the orientation angle θ is shown in Fig. 3. The individual data points denote measurements for two different matrix materials, while the continuous lines represent predictions of the Tsai-Hill criterion. The values of the material parameters (determined from tension, compression, and shear testing) used in Eq. 2 for a negative shear stress were $\sigma_1 = 490$ MPa, $\sigma_2 = 61$ MPa, $\tau_{12} = 104$ MPa, while for a positive shear stress they were $\sigma_1 = 609$ MPa, $\sigma_2 = 391$ MPa, and $\tau_{12} = 104$ MPa. Note that for positive τ, the shear strength increases with θ. On the other hand the measurements and the predictions, for negative τ, have the same trend, i.e. τ_{xy} decreases as θ increases. This indicates that a negative shear stress is applied on the composite specimen. A first order, qualitative explanation is provided in Fig. 4, which shows a composite loaded at an angle $\theta = 45°$ with a negative shear stress τ. The negative shear stress is equivalent to the

482

combination of a compressive stress in the longitudinal direction of the composite and a tensile in the transverse. This stress state corresponds to the weakest situation of the composite, the tensile stress/load is carried totally by the matrix. Therefore, the shear strength is minimum at the 45° orientation angle. On the other hand, as θ gets closer to zero, the high strength fibers are supporting a part of the tensile stresses; the maximum support occurs at $\theta = 0°$, where τ_{xy} (i.e. τ_{12}) has its maximum value. Notice that the reverse holds when the applied shear stress is positive.

From the quantitative standpoint, the Tsai-Hill (T-H) criterion does not predict the experimental results quite accurately for reasons which include the positive strain hardening exhibited by the composite (Fig. 2), the highly non-linear stress-strain behavior, the high failure strains, and mainly the positive-negative shear stress interactions at θ's other than 0 and 45° which lead to changes of the deformation mode. Non of the above is accounted on the T-H criterion, since the latter is based on simple measurements and combinations among longitudinal, transverse and shear strengths. Further deviations between predictions and data arise from the omission of important micromechanical issues such as stress distributions, and the behavior of the fiber/matrix interface. As it is widely accepted, the strength of the latter plays a quite important role at least in cases of transverse tensile stresses[6]. The above lead to the conclusion that accurate models describing the composite behavior have to be developed, with the results of this effort pointing to the appropriate directions.

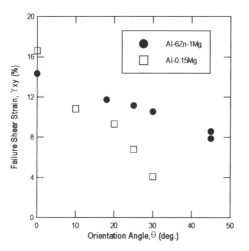

Fig. 5. Failure shear strain versus the orientation angle θ.

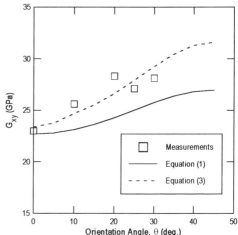

Fig. 6. Shear modulus versus the orientation angle θ for a Al-0.15Mg matrix UD composite.

Shear Strain at Failure.

The shear strain at the peak stress (which essentially corresponds to the failure strain) as a function of the orientation angle θ is shown in Fig. 5. It is seen that the shear ductility decreases rapidly as θ increases. This behavior may be also explained by utilizing arguments concerned with the fiber/matrix interface and the direction of the principal tensile stresses with respect to the fiber direction. Specifically, since the interface is, usually, the weakest "constituent" of a composite, it is expected that failure initiates and proceeds through them. Since a negative shear stress at 45° corresponds to the weakest orientation of the composite and its interface, it is expected that crack will nucleate and propagate quite early during shearing process and the strains to failure will be the lowest.

On the other hand, as θ gets closer to zero, the perpendicular to the fibers compressive stresses increase, the apparent interfacial strength increases, hence failure is delayed to higher strains.

Shear Modulus

Figure 6 show the variation of the shear modulus Gxy as a function of the angle θ. In agreement with earlier experimental and theoretical analyses, G_{xy} increases with θ, and it acquires its maximum value at $\theta= 45°$. The minimum values of G_{xy} are reached at $\theta=0°$ and $\theta=90°$. In addition, Fig. 6 includes two continuous lines. The solid line represents predictions of G_{xy} versus θ based on Eq. (1). It is seen that Eq. (1) underestimates the magnitude of G_{xy} for a given θ. This deviation can be attributed to following reasons which are related to the material, the testing procedure, and the nature of the model:
  (i) deviations between the magnitude of the elastic constants (i.e. the input to Eq. (1) $E_{11}= 101$ GPa, $E_{22}= 35$ GPa, $v_{12}= 0.27$) which have been determined from tension and compression tests to the elastic constants of the material tested in shear. These deviations arise from variations of the fiber volume fraction, processing defects, statistical variations of the fiber strength, batch to batch variability of the mean fiber strength, interfacial conditions among others.
  (ii) the stress field developed in the composite specimen, which is not perfectly uniform during the shear testing.
  (iii) the linear stress-strain behavior which is inherently assumed for the derivation of Eq. (1).
  To this end, an empirical constant (F), that accounts for the individual and/or combined action of the effects mentioned above was determined and introduced into Eq. (1) in order to more accurately predict the materials' shear modulus. Specifically, Eq. (1) was slightly modified to the following form:

$$\frac{1}{G_{xy}} = \frac{1}{G_{12}}\cos^2 2\theta + F\left(\frac{1+v_{12}}{E_{11}} + \frac{1}{E_{22}}\right)\sin^2 2\theta \qquad (3)$$

The dash line in Fig. 6 corresponds to predictions from Equation (3). It is seen that, within the limits of the experimental error, the actual variation of the shear modulus with respect to the orientation angle is predicted quite well.

Table 1. Constitutive material parameters for the "12", "21", and "23" directions of UD Al-6Zn-1Mg matrix composites.

Direction	12	21	23
τ (MPa)	103.9	129.5	99.3
γ (%)	14.3	4.51	0.89
G (MPa)	17.2	22.2	57.2

Table 2. Shear test results for Al-6Zn-1Mg reinforced with woven fabric Altex SN fibers.

θ (°)	τ_{xy} (MPa)	γ_{xy} (%)	G_{xy} (GPa)
0	158.4	0.97	22.2
0	161.7	0.93	24.9
25	149.4	1.09	20.0
30	147.6	1.05	21.6

"21" and "23" directions

The anisotropic nature of these MMC's can be also observed in Table 1 where shear strength, strain to failure, and modulus are presented for the "21" and "23" directions; for comparison the "12" direction results are also shown. It is seen that there is a 2-fold increase in the shear modulus in the "23" direction, while the strains to failure have dropped to less than 1%. The "21" direction has the highest shear strength, while the

484

magnitude of its shear modulus and shear strain to failure are intermediate to those of the other two directions.

Cyclic Tests.

A limited number of cyclic tests have been conducted, at a frequency of 0.05 Hz, and at an R ratio ($R=\tau_{max}/\tau_{min}$) equal to -1. Although the stress level used at the tests represented less than 40% of the materials' peak strength, permanent deformation was observed. In particular, for a Al-6Zn-1Mg matrix composite with fibers inclined at $\theta= 0^o$, it was found that the internal damage was rather limited for the first 200 cycles ($\gamma_{12}^P < 0.1\%$); however damage subsequently accumulates quite rapidly reaching a $\gamma_{12}^P= 0.6\%$ at the 500th cycle. It should be noted that for $\theta= 0^o$, the change of the sign of the applied shear stress does not affect the local stress state inside the composite, and damage is governed from the material properties. On the other hand, as θ deviates from zero, tensile and compressive stresses are alternating at the fiber/matrix interface due to the stress sign change and it is expected the amount of damage accumulation to increase.

Woven Composites.

Shear test results for Al-6Zn-1Mg matrix reinforced with woven fabric Altex fibers are shown in Table 2. The first observation is that the woven fabric fiber reinforcement improve substantially the shear strength; the strain to failure decreases dramatically. The variation of the angle θ indicates a change of the constitutive parameters of the composite. However, a mathematical description of the composite constitutive behavior will not be attempted at this point because of (i) the small number of experiments performed to date, and (ii) the difference in the quality of the specimens received and tested.

CONCLUSIONS
A prototype shear testing apparatus was fabricated and utilized to determine the shear behavior of MMC's. Off-axis shear test data were developed and revealed the following trends:
1. The variation of the composite constitutive parameters such as shear strength, shear strain to failure, and shear modulus versus the orientation angle θ was determined, showing a decrease of the first two and an increase of the latter at θ increases.
2. The results were compared to model predictions. The disagreement between the two was justified using arguments concerned with the models' nature, and the composite macromechanics and micromechanics.
3. Cyclic shear loading at stress levels less than 40% of the yield point causes internal damage; the rate of damage accumulation increases with the number of cycles.
4. In comparison with their unidirectional counterparts, cross-woven fiber composites exhibited an increase in the shear strength and a dramatic decrease in the shear strain to failure, indicating that their utilization is more appropriate under shear loads.

REFERENCES
1.   Michelis, P.N., and Dafalias, Y.F., (1994). In: *Proc. of the Int. Symp. On Advanced Materials for Lightweigh Structures*, ESTEC, Noordwijk, 219.
2.   Broughton, W.R., Kumosa, M. and Hull, D., (1990) *Comp. Sci. Tech.* 38,299.
3.   Anand, K., Gupta, V. and Dartford, D. (1994) *Acta Matall. Mater.* 42, 797.
4.   Daniels, I.M., and Ishai, O. (1994) *Engineering Mechanics of Composite Materials*. Oxford University Press, Oxford.
5.   Azzi, V.D., and Tsai, S.W. (1965) *Exp. Mech.* 5, 283.
6.   Taya, M. and Arsenault, R.J. (1989) *Metal Matrix Composites*. Pergamon Press, New York.

ACKNOWLEDGMENTS
The main part of the presented work was partially supported by the DGXII of the CEU under the contract number BRPR95-0126, within the framework of the Brite-EuRam Programme.

Low Cycle Fatigue and Elasto-Plastic Behaviour of Materials
K-T. Rie and P.D. Portella (Editors)

SOME EXPERIMENTAL DATA OF LOW-CYCLE IMPACT FATIGUE FOR SEVERAL METALLIC MATERIALS

PINGSHENG YANG

Department of Materials Science and Engineering , Nanchang University
Nanchang 330047, P . R . China
RONGRU GU , YONGZE WANG AND YUNFU RAO
Nuclear Industry General Co. 720 Factory
Nanchang 330101 , P. R . China

ABSTRACT

This is the companion volume of the paper [1], in this paper some experimental data of LCIF for several metallic materials are presented in the form of curve and table . There are 1 4 figures which show the cyclic hardening(softening) behaviours , fatigue life and so on .

KEYWORDS

Low-cycle impact fatigue , experimental data .

INTRODUCTION

The investigation on low-cycle impact fatigue (LCIF) is significant in theory and practice . A push-pull impact fatigue testing apparatus has been designed by the authors . By using the apparatus the LCIF tests have been made of a series of metallic materials such as low carbon steel , austenitic stainless steel , medium carbon alloy steel , aluminum , duralumin , brass , and SiC/Al composites . The results show that there are not only similarities but also dissimilarities in cyclic deformation and fracture behaviour between LCIF and LCF of materials ; these are the strain rate effects . Different strain rate effects occur in each material ; therefore LCIF is an independent behaviour of materials , and it can neither be replaced by LCF nor be estimated by impact test .

In this paper some experimental data of LCIF for several metallic materials are presented in the form of curve and table . The experimental method , results and discussions have been presented in the companion volume [1] .

EXPERIMENTAL DATA

Table 1 Heat treatment and mechanical properties of the materials

Material	Heat treatment	0.2% proof stress (MPa)	Tensile strength (MPa)	Elongation (%)
0.1%C steel	920℃ annealing in vacuum	234	413	
1Cr18Ni9Ti steel	1050℃ water quenching	247	663	
40Cr steel	850℃ quenching , 600℃ tempered	800	900	
A₁ aluminum	240℃ annealing	82	85	
Ly12 duralumin	Quenched and aged	424	600	25.0
H62 brass	580℃ annealing	150	370	
SiCp/Ly12 composite with 12%V_f SiC	Quenched and aged	325	457	2.7
SiCp/Ly12 composite with 15%V_f SiC	Quenched and aged	334	464	2.5
SiCp/Ly12 composite with 20%V_f SiC	Quenched and aged	338	470	2.0

Low Carbon Steel

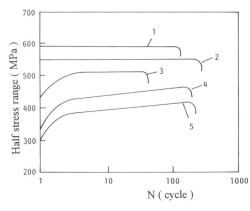

Fig . 1 Cyclic hardening curves of 0.1%C steel : 1, LCIF, $\Delta\varepsilon_T/2 = 0.009$; 2 , LCIF, $\Delta\varepsilon_T/2 = 0.007$; 3 , LCF , $\Delta\varepsilon_T/2 = 0.045$; 4 , LCF , $\Delta\varepsilon_T/2 = 0.023$; 5 , LCF , $\Delta\varepsilon_T/2 = 0.017$

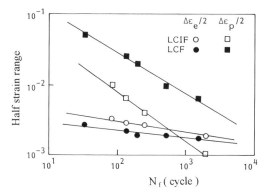

Fig . 2 Relationship between half strain range $\Delta\varepsilon_e/2$, $\Delta\varepsilon_p/2$ and cycles to failure N_f for 0.1%C steel

Austenitic Stainless Steel

Fig . 3 Cyclic hardening curves of 1Cr18Ni9Ti steel : 1, LCIF, $\Delta\varepsilon_T/2 = 0.017$; 2 , LCIF, $\Delta\varepsilon_T/2 = 0.012$; 3 , LCIF , $\Delta\varepsilon_T/2 = 0.009$; 4 , LCIF , $\Delta\varepsilon_T/2 = 0.006$; 5 , LCF , $\Delta\varepsilon_T/2 = 0.025$; 6 , LCF , $\Delta\varepsilon_T/2 = 0.011$

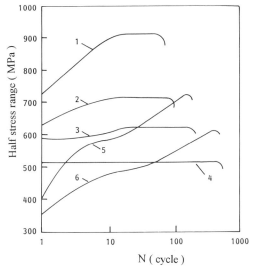

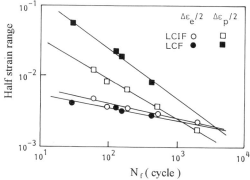

Fig . 4 Relationship between half strain range $\Delta\varepsilon_e/2$, $\Delta\varepsilon_p/2$ and cycles to failure N_f for 1Cr18Ni9Ti steel

488

Medium Carbon Alloy Steel

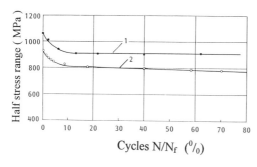

Fig . 5 Cyclic softening curves of 40Cr steel : 1, LCIF, $\Delta \varepsilon_{t}/2 = 0.010$, $N_{f} = 110$; 2 , LCF , $\Delta \varepsilon_{t}/2 = 0.013$, $N_{f} = 197$

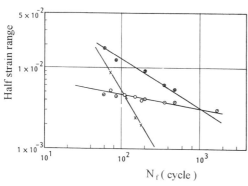

Fig . 6 Relationship between strain range and cycles to failure N_f for 40Cr steel : ○ , LCIF , $\Delta \varepsilon_{e}/2$; ✕ , LCIF , $\Delta \varepsilon_{p}/2$; ⊙ , LCF , $\Delta \varepsilon_{e}/2$; ● , LCF , $\Delta \varepsilon_{p}/2$

Aluminum

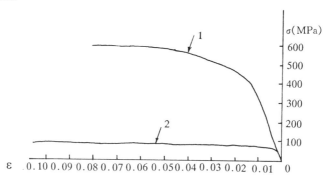

Fig . 7 Tensile stress-strain curves : 1 , Ly12 duralumin ; 2 , A1 aluminum

489

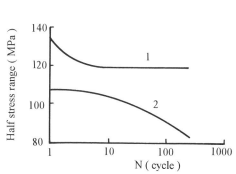

Fig . 8 Cyclic softening curves of A l aluminum :
1 , LCIF , $\Delta\varepsilon_T/2 = 0.0055$, $N_f = 272$; 2 , LCF ,
$\Delta\varepsilon_T/2 = 0.0097$, $N_f = 280$

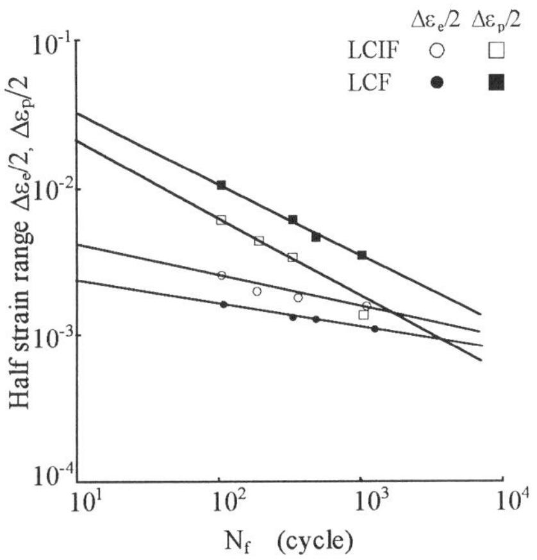

Fig . 9   Relationship between half strain range
$\Delta\varepsilon_e/2$, $\Delta\varepsilon_p/2$ and cycles to failure N_f for A l
aluminum

Duralumin

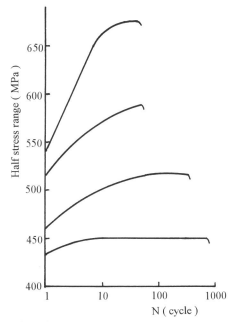

Fig . 10 Cyclic hardening curves of Ly12
duralumin : 1, LCIF, $\Delta\varepsilon_T/2 = 0.0118$, $N_f = 42$;
2 , LCF , $\Delta\varepsilon_T/2 = 0.0145$, $N_f = 45$; 3 , LCF ,
$\Delta\varepsilon_T/2 = 0.0091$, $N_f = 296$; 4 , LCF , $\Delta\varepsilon_T/2$
$= 0.0076$, $N_f = 596$

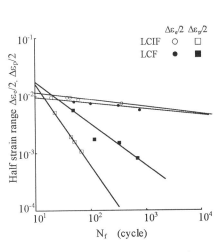

Fig . 11 Relationship between half strain range
$\Delta\varepsilon_e/2$, $\Delta\varepsilon_p/2$ and cycles to failure N_f for Ly12
duralumin

490

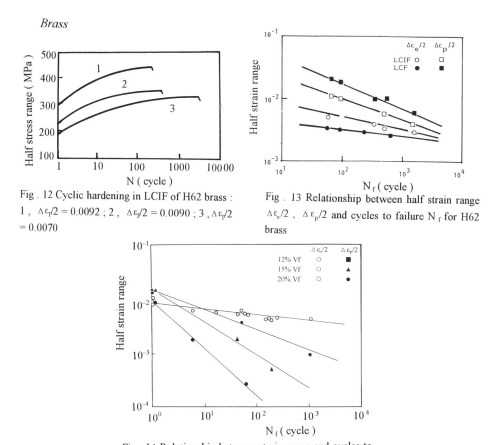

Fig . 12 Cyclic hardening in LCIF of H62 brass :
1 , $\Delta\varepsilon_T/2 = 0.0092$; 2 , $\Delta\varepsilon_T/2 = 0.0090$; 3 , $\Delta\varepsilon_T/2$ = 0.0070

Fig . 13 Relationship between half strain range $\Delta\varepsilon_e/2$, $\Delta\varepsilon_p/2$ and cycles to failure N_f for H62 brass

Fig . 14 Relationship between strain range and cycles to failure N_f in LCIF of the $SiC_p/Ly12$ composite ;

ACKNOWLEDGMENT

The research was supported by the National Natural Science Foundation of China under Subcontract 58971075 and 59371038 .

REFERENCES

1. Yang , P . S . , GU , R . R . , Wang , Y . Z . and Rao , Y . F . (1998) In : this proceedings .
2. Yang , P. S . and Zhou , H . J. (1994) Int . J. Fatigue 16 , 567 .
3. Yang , P. S ., Liao , X . N ., Zhu , J. H . and Zhou , H . J. (1994) Int . J. Fatigue 16 , 327 .
4. Yang , P. S ., Liu , Y. and Zhang , M . (1996) Journal of Nanchang University (Natural Science) 20 , 156 .
5. Yang , P. S . and Zhang , M . (1996) In : Proceedings of the 6th International Fatigue Congress , G . Lütjering and H . Nowack (Eds) . Pergamon , Oxford , pp . 197-202 .
6. Yang , P. S ., Zhang , M ., Xu , F. and Gong , S . P. (1997) Materials Science and Engineering A236 , 127 .

Chapter 7

BEHAVIOUR OF SHORT AND LARGE CRACKS

Low Cycle Fatigue and Elasto-Plastic Behaviour of Materials
K-T. Rie and P.D. Portella (Editors)

CYCLIC STRAIN LOCALISATION, CRACK NUCLEATION
AND SHORT CRACK GROWTH

J. POLÁK

Institute of Physics of Materials, Academy of Sciences,
Žižkova 22, 616 62 Brno, Czech Republic

ABSTRACT

Physical manifestation of the fatigue damage evolution in crystalline materials is reviewed. Basic stages of the fatigue process are demarcated and the attention is concentrated to cyclic strain localisation, crack nucleation and short crack growth. Cyclic strain localisation is an substantial feature of the fatigue damage. It results in the formation of the surface relief, crack nucleation and short crack growth. The experimental results and current theories and models of crack initiation and short crack growth are described and discussed.

KEYWORDS

Fatigue damage, cyclic strain localisation, crack nucleation, short crack growth.

1. INTRODUCTION

The study of fatigue damage in materials started shortly after the phenomenon of fatigue had been recognised at the end of the last century. Two basic lines were followed. The early investigations [1] (see also [2-6]) opened the study of the physical manifestation of the effect of variable stress acting on the material and of the mechanisms leading eventually to the fatigue failure. Fatigue damage could be identified by the changes of a number of physical and mechanical properties of materials subjected to alternating loading.

Second line of studies attempted to express quantitatively the degree of damage using specific quantity *D*, called "damage". This quantity is zero before loading and achieves unity at fracture. In constant amplitude loading it is changed during the fatigue life in a characteristic way, which depends on the type of loading and stress or strain amplitude [7]. The important application of this approach is the possibility to yield fatigue life predictions for more complex load histories using e. g. the life data in constant amplitude loading. Generally, fatigue damage is summed up until unity is reached.

Both these approaches have been intensively developed and vast experimental basis has been collected on the physical manifestation of the fatigue damage and mechanisms leading to crack

494

initiation and crack growth. Especially in recent years, modern experimental techniques allowed to study fine details of the internal structure evolution and surface relief evolution, which precede crack nucleation and growth. Simultaneously, numerous damage accumulation procedures were proposed and applied to various practical situations [7]. Comparison with experimental results allowed the verification of various hypotheses. Due to immense variability of loading conditions, variety of materials and manifold external conditions that influence the fatigue life it is not surprising that different hypothesis proved to have only limited validity.

Therefore, both approaches are presently combined in order to learn maximum information on the mechanisms of fatigue damage and to include this knowledge into the damage accumulation procedures allowing better life predictions in a number of specific and industry important applications. In this review the information on the physical manifestation of the fatigue damage in crystalline materials is presented with emphasis on the cyclic slip localisation and its role in crack nucleation and early crack growth. The experimental data on short crack growth are confronted with existing theories and descriptions of short crack growth.

2. DAMAGE EVOLUTION IN FATIGUE

Fatigue damage in materials can be traced by changes of numerous physical and mechanical properties as e. g. electrical resistivity, density, elasticity modulus, hardness, etc. Changes of these properties are the result of the internal structure of the cyclically deformed material. Therefore, closer attention is paid to the modification of the internal structure and to the changes of these properties that are directly lied to the damage and to the development of the crack and final failure of a structural member.

The most important stages and sub-stages of the fatigue process are depicted on the time basis in Fig. 1. From the engineering point of view only two major stages are present, i. e. the macrocrack initiation and macrocrack growth followed by sudden fracture. However, since

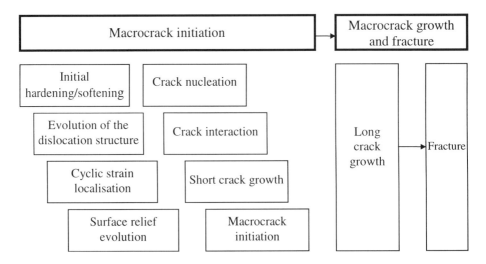

Fig. 1. Evolution of the fatigue damage in crystalline materials.

the macrocrack initiation can occupy important fraction of the total fatigue life, it is reasonable to look for the important sub-stages or identify the mechanisms important for damage evolution. The elastoplastic cyclic straining modifies the internal dislocation structure and as a result the cyclic stress-strain response is changed appreciably. The material cyclically hardens or softens. The tendency to develop low-energy dislocation structures leads to inhomogeneous cyclic deformation. Cyclic strain is localised to specific volumes called "persistent slip bands" (PSBs). Localisation of the cyclic strain results in the formation of the surface relief, preferably at sites where PSBs emerge on the crystal surface. The high local amplitude in PSBs and notch-like geometry of intrusions result in crack nucleation. In case of high applied plastic strain numerous cracks are nucleated and interact mutually. The linkages of cracks contribute to crack growth and dominant crack formation. Propagation of a dominant crack leads to sudden failure.

3. CYCLIC STRAIN LOCALISATION

The early observations of the fatigue damage in polycrystalline Swedish iron using optical microscope by Ewing and Humfrey [1] revealed localised slip in surface grains and formation of the pronounced surface relief. Ewing and Humfrey noticed "a heaving-up of the surface of the crystal in the neighbourhood of each slip band, the markings being decidedly above the level of the other parts of the crystal". The localisation of the cyclic plastic strain in the bands of localised slip (PSBs) [8] is a general and very important feature of cyclic straining. In low amplitude cyclic straining, low density of PSBs is formed. Within one grain either none or only one slip system is activated. In high amplitude straining, high density of PSBs with two or more slip systems in individual grains arises. The cyclic slip localisation can be detected by observation of the surface relief since extrusions and intrusions are formed at emerging slip bands and play an important role in crack nucleation. Figure 2 shows characteristic surface relief of a fatigued copper single crystal [9]. Both individual PSBs and persistent slip macrobands are present and are separated by an originally smooth specimen surface.

Cyclic slip localisation can be also detected by measurement of the cyclic stress-strain response. Stress amplitude characteristic for localised volumes has in each material characteristic value and therefore the cyclic stress-strain curve of a copper single crystal has a

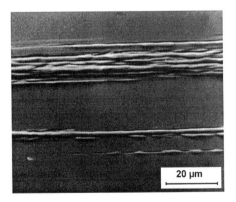
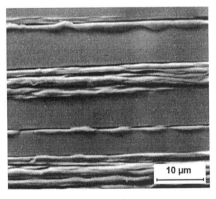

Fig. 2. Surface of copper single crystal fatigued to saturation of the stress amplitude.

496

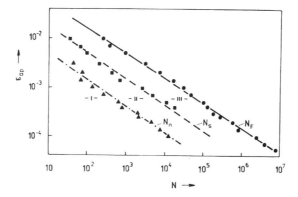

Fig. 3. Fatigue mechanism map in polycrystalline copper.

plateau extending over two orders of magnitude of the plastic strain [10]. When the strain is localised, the hardening within the loop decreases considerably and the hysteresis loop becomes more rectangular. The loop shape parameter, V_H, introduced by Mughrabi [10], thus increases. Loop shape parameter is equal to the ratio of the hysteresis loop area and the area of the parallelogram with the basis $2\varepsilon_{ap}$ and the height $2\sigma_a$. Cyclic strain localisation affects the loops shape parameter in single crystals [11,12] as well in polycrystals [13,14]. Using the position of local minima and maxima of the loop shape parameter, the number of cycles to PSB nucleation, N_n, the number of cycles to reach the saturated density of PSBs, N_s, and the number of cycles to fracture, N_f, can be plotted vs. the plastic strain amplitude. The fatigue mechanism map can be constructed and is shown in Fig. 3.

In this map the domain of homogeneous cyclic plastic straining (domain I) is separated from the domain in which PSBs are formed (domain II) and the domain in which cyclic plastic strain is concentrated in the network of PSBs and crack nucleate and grow (domain III). Domain III occupies the most of the fatigue life since surface relief is built, the high local plastic strain amplitude in PSBs results in multiple crack initiation, crack interaction, short crack growth, formation and propagation of the dominant crack.

4. FATIGUE CRACK NUCLEATION

4.1 Surface Relief Formation

All experimental observations of the crack nucleation since the early findings of Ewing and Humfrey [1] revealed that crack nucleation is related to the formation of the surface relief at the emerging PSBs. The detailed models of the surface relief formation were not possible before the discovery of dislocations and their role in the plastic deformation [15]. Theoretical models were developed that tried to model dislocation motion during cyclic loading that could produce the surface relief in agreement with experimental observations. The early models of surface relief formation and crack nucleation based on the specific motion of individual dislocations were reviewed by Thompson and Wadsworth [16] and by Laird and Duquette [17]. All of these early models were found unsatisfactory.

The experimental observations of the fatigue crack initiation in high strain fatigue loading by Kim and Laird [18] showed the role of high angle boundaries. Crack is formed where the dominant slip system of a grain is directed to the intersection of grain boundary with the specimen surface. The step with sharp root radius promotes the strain irreversibility and leads to crack nucleation. Cheng and Laird [19] studied the PSB offsets in fatigued copper single crystals using interference microscope. They have found that preferred sites of crack nucleation were the narrow PSBs with the highest strain offsets. They proposed the random slip model for the formation of the surface relief and crack nucleation. The crack nucleation is related to the probability of finding a critical offset produced in one cycle.

Fatigue crack initiation model proposed by Tanaka and Mura [20] and refined by Liu et al. [21] and Mura and Nakasone [22] does not take into account the actual dislocation structure of a PSB but models it as a plate-like distribution of dislocation vacancy type dipoles. The accumulation of dipoles during cycling increases the internal tensile stress. The crack initiates suddenly when fracture criterion is satisfied. The model predicts the initiation life is inversely proportional to the grain size. Fatigue crack initiation at grain boundaries is considered in the model proposed by Mughrabi [23]. Grain boundary cracks can develop in the region where PSBs impinge on grain boundaries. This model predicts the initiation life is inversely proportional to the cube of the grain size.

The real dislocation structure of PSBs in copper single and polycrystals consisting of alternating thin dipolar walls and wide dislocation-poor channels was used by Brown et al. [24] and Essmann et al. [25] to create models of surface relief formation. The well-known EMG (Essmann, Mughrabi, Gössele) model considers cyclic slip localisation in PSBs, slip irreversibility and point defect generation producing a static extrusion. The random slip superimposed on the extrusion [26] predicts a relief, which is close to some experimental observations.

Detailed observations of the surface of fatigued copper single crystal oriented for single slip [27] simultaneously with the dislocation structure of the typical PSB [28] led Polák et al. [27] to propose a relation between the surface relief and dislocation arrangement in PSB. Figure 4 shows schematically the lamella of a PSB and alternating tongue-like extrusions and intrusions on the surface of the crystal. If d is the spacing between the walls and α the angle between the intersection of the primary slip plane with the surface and the primary Burges vector the spacing s between the tongue-like extrusions is

$$s = \frac{d}{\cos\alpha} \qquad (1)$$

The continuous production and annihilation of dislocations in the PSB and simultaneous production of point defects proposed by Essmann et al. [25] was refined by Polák [29] by considerations of the point defect migration resulting in mass transfer. Polák's model predicts formation of the surface relief in close agreement with experimental observations and shows how the irreversibility of slip in the tip of an intrusion can result in microcrack nucleation. The production rates of point defects in cyclic straining of copper polycrystals [30] and single crystals [31] were quantitatively evaluated using electrical resistivity measurements. Several mechanisms of point defect production in a PSB due to interaction of mobile dislocations were discussed [32].

An individual PSB, which carries high cyclic plastic deformation (about 1%) in a quasi-elastic matrix, is considered. Schematic structure of a PSB and dislocation interactions within PSB in a single crystal or in a grain of a polycrystal are shown in Fig. 5. A slab of PSB consists of thin parallel dipolar walls separated by thick dislocation poor channels. The surrounding matrix structure consists of loop patches or veins separated by irregular channels. High plastic strain

498

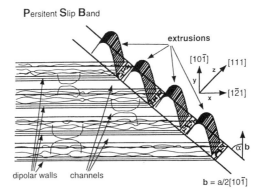

Fig. 4. Relation between the internal structure and surface relief in fatigued crystal.

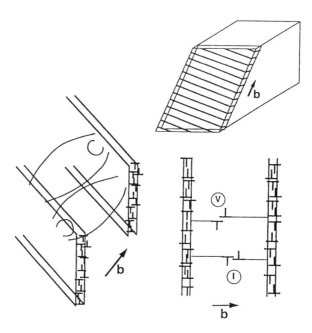

Fig. 5. Schematics of dislocation motion and point defect production in a PSB.

amplitude within the PSB is accommodated by bowing of dislocation segments from the walls and their extension until edge segments reach the neighbouring wall. Screw segments carry most of the strain until they are annihilated by cross-slip with opposite screw dislocation moving on the parallel plane. The point defects are formed either by the interaction of two edge dislocations moving on the neighbouring planes (Fig. 5) or by dragging jogs on screw dislocations. Since activation energy of vacancy formation is much smaller than the activation energy of the formation of an interstitial, mostly vacancy type defects are produced in cyclic straining. The rate of vacancy production in room temperature cyclic straining of copper cycled with plastic strain amplitude of 1% is evaluated from resistivity measurements to about 2.5×10^{-6} per cycle [31].

If vacancies or vacancy type defect (e. g. di-vacancies) are mobile at ambient temperature, they will migrate to sinks. The excess vacancy concentration will thus depend on the sink density. Edge dislocations or edge dipoles represent a perfect sink for vacancies. Since their density is much higher in the walls than in the channels, inhomogeneous excess vacancy concentration arises. Figure 6 shows a section through a PSB in yz plane. The profiles of the excess vacancy concentration in two directions are shown too. The one along the PSB results in periodic variation of the vacancy concentration that reaches minimum at the walls and maximum in the middle of the channels. Vacancies within the PSB thus systematically migrate from the channels to the walls. The arrows in Fig. 6 show the direction of vacancy flux.

In a section cut in the centre of the channel normal to the Burgers vector the maximum vacancy concentration is reached in the centre of the band. It decreases in direction of the matrix since no point defects are generated in the matrix but matrix structure is a perfect sink for migrating vacancies. In a quasi-steady situation, the vacancy gradient results in a systematic flux of vacancies according to Fick's law. The arrows in Fig. 6 show the direction of vacancy flux out of the PSB to the matrix.

Let us consider the consequences of the steady flux of vacancies within the PSB (Fig. 6). The steady flow of vacancies in one direction results in transport of atoms in the opposite direction. The mass will be thus accumulated in the channels and decline in the walls. The accrual and the loss of the mass in y-direction compensate mutually. The width of the PSB in z-direction is small and, moreover, even when mass is accumulated here, it can expand freely. The surplus of mass will produce compressive stresses only in x-direction. Since PSB is continuously traversed by dislocations and thus deformed plastically, these stresses will be continually relaxed and strain offsets in x-direction will arise. The strain offsets produce extrusion at the site where the channel intersects the surface and intrusion at the site where the wall intersects the surface (Fig. 4). The periodicity of extrusion-intrusion pairs can be derived from the periodicity of walls and channels (see eq. (1)).

Section through the PSB (plane x=0)

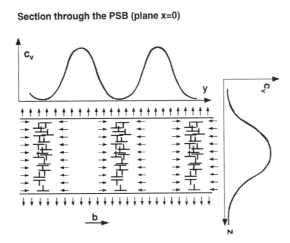

Fig. 6. Schematic section through PSB in yz plane and vacancy profiles in y- and z-directions.

500

The flux of vacancies out of the PSB to the matrix and the flow of the atoms in the opposite direction has different effect on surface relief formation than the migration of vacancies within the band. In this case, the mass accumulates in the whole PSB and since free expansion is possible only in z-direction, the whole PSB is extruded both in x- and y-directions. As a result a ribbon-like extrusion is produced. The extrusion of the whole PSB is compensated by the two intrusions at the interface between the PSB and the matrix.

Since vacancies in cyclic straining migrate both within the band and from the band to the matrix, both processes pass simultaneously and resulting relief should show both ribbon- and tongue-like extrusions and intrusions. The combination of both processes at a later stage of development results in ribbon/and/tongue-like extrusions whose height changes periodically and is accompanied by two intrusions at the interface with the matrix. It corresponds reasonably to the actual observations of the surface relief (see Fig. 2).

The same reasoning can be applied to persistent slip macroband [29] and a net protrusion connected with the macroband is predicted. This prediction is in agreement with experimental observations [27].

Repetto and Ortiz [33] adopted the mechanisms of dislocation interaction and point defect migration from PSB to the matrix and performed finite element modelling of extrusion and intrusion formation. Reasonable agreement with the experimental observations of the shape of ribbon-like extrusions and accompanying intrusions was obtained.

4.2 Models of Crack Initiation

Formation of the surface relief on the originally flat surface due to localised cyclic plastic straining represents the first step in the nucleation of a fatigue crack. Sharp intrusions represent an effective stress raiser since the radius of the intrusion is very small. The sharp corner technique, developed by Basinski and Basinski [34], and the sectioning technique, applied by Hunsche and Neumann [35], allow to estimate the notch radius to 0.1 μm. The stress and strain concentration in the tip of an intrusion is thus comparable with that of a crack of the same dimension. The high stress concentration in the tip of an intrusion gives rise to local slip-unslip mechanism along the primary plane. The irreversible slip-unslip mechanism has been already

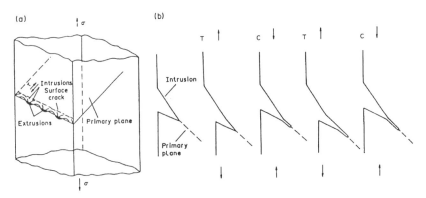

Fig. 7. Schematics of crack initiation; a) from the intrusions, b) the slip-unslip mechanism.

considered by Wood [36]. The more specific model taking into account the line of alternating intrusions and extrusions at the emerging PSBs has been proposed by Polák and Liškutín [37]. Figure 7a shows schematically the environmentally assisted nucleation of surface cracks from a row of intrusions that link to create a shallow surface crack. Figure 7b shows schematically the slip-unslip mechanism starting from the tip of an intrusion under the action of corrosive environment, which prevents the rewelding of newly created surfaces. The mechanism shown in Fig. 7b results in the formation of the shallow microcrack within PSB.

In numerous studies of microcrack initiation the row of tongue-like cracks or later shallow cracks are observed early in the fatigue life [38,39]. Only later the crack grows in the interior of the grain or crosses the grain boundary.

5. SHORT CRACK GROWTH

The growth of initiated cracks represents an important and often decisive period in the fatigue life of a component or of a laboratory specimen. The description of the fatigue crack growth has been facilitated by application of fracture mechanics [40]. The fracture mechanics gives a good description of stress and strain fields at the tip of a crack in an isotropic material with a plastic zone appreciably smaller than the crack size or the dimension of a loaded body. Since structural materials consist of a large number of small crystallites with anisotropy properties, it is natural that fracture mechanics approach breaks down if the crack has dimension close to the dimension of a structural unit, e. g. grain size.

The early observations of the anomalies in crack growth rates of short cracks in comparison with long cracks have been made by Pearson [41] in aluminium alloys. Since then many experimental studies of short crack growth and their analysis have been performed. Several reviews [42- 47] and conferences [48-50] were devoted to this subject. Suresh and Ritchie [43] in an early review of short crack growth distinguish "physically small cracks" (smaller than 1mm), "mechanically small cracks" (smaller than the plastic zone size), "microstructurally small cracks" (smaller or comparable to the size of the structural unit of the material) and "chemically small cracks" (cracks whose growth is substantially influenced by environment). McClung et al. [47] demarcate "small" cracks and "short" cracks and consider specific criteria concerning the size of a small crack relative to the size of the structural unit. Generally, however, "small" and "short" cracks are considered as equivalent expressions.

Considering the definition of a small crack, any natural crack nucleated within a grain of the material or in a single crystal is small (or short). Two basic stages of crack growth can be usually distinguished [51]. In "stage I" the crack follows the primary slip plane in which the microcrack has been nucleated. It grows later in "stage II" in a plane perpendicular to the stress axis. In single slip oriented single crystal the crack follows primary slip plane until secondary slip system becomes active [52]. Only limited number of short crack growth studies in single crystals has been performed. Blochwitz and Heinrich [53] adopted the sectioning technique at various stages of the fatigue life to measure the microcrack densities and their depths. They obtained the crack growth rates proportional to the crack depth for small crack depths and proportional to the square of the crack depth for larger crack depths.

Majority of experimental data on short crack growth in structural materials was compared with the master curve for long cracks, i. e. the plot of the growth rate vs. the stress intensity range [43]. The stress intensity of short cracks was calculated in the same way as that of long cracks though sound physical interpretation is missing. In this plot the crack growth rates of short cracks are higher than those of long cracks, have higher scatter and the crack grows at stress intensities below the threshold value for long cracks.

Early analysis of crack growth in stage II by Tomkins [54] proposed the crack growth rate to be proportional to the crack length. Considerable attention to the study of short crack behaviour was paid by K. Miller and his collaborators [45,46]. He proposes to use microstructural fracture mechanics (MFM) in the early stage of crack growth, elastic-plastic fracture mechanics (EPFM) in intermediate range and linear elastic fracture mechanics (LEFM) for long cracks and small stresses. The crack growth equation for individual domains were derived from the work of Hobson et al. [55] in the form

$$\frac{da}{dN} = A\Delta\gamma^{\alpha}(d-a) \quad \text{for MFM} \tag{2}$$

$$\frac{da}{dN} = B\Delta\gamma^{\beta}a - D \quad \text{for EPFM} \tag{3}$$

$$\frac{da}{dN} = C\left(\Delta\gamma\sqrt{\pi a}\right)^{n} \quad \text{for LEFM} \tag{4}$$

where a is the crack length , d the grain size and $A, B\ C, D, \alpha, \beta$, and n are constants; D represents the threshold condition for EPFM cracks. Similar classification was used by McDowell [56] who uses the same three regimes of crack growth in case of multiaxial loading. Such three crack growth regimes were found by Buirette et al. [57] in the study of short crack growth in the welds of bainitic steel.

Short crack growth kinetics of natural cracks has been studied extensively in low carbon steel by Nishitani et al. [44] and Murakami et al. [58] using smooth specimens or specimens with a small hole. The cracks initiated in a slip band and grew at constant stress amplitude. The growth rate of small cracks could be described using simple law

$$\frac{da}{dN} = H\sigma_a^{h}a \tag{5}$$

where H and h are constants, h is approximately equal to 8. The integration of equation (5) yields the logarithm of the crack growth rate proportional to the number of cycles.

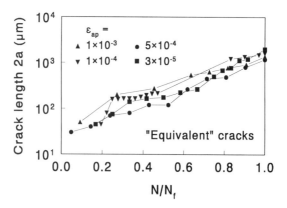

Fig. 8. Growth of "equivalent" cracks in constant plastic strain amplitude loading of 316L steel for different plastic strain amplitudes.

Experimental studies of short crack growth in constant plastic strain amplitude regime of several structural materials [59-61] revealed that initial crack growth rate is either constant [60] or is proportional to the crack length [61]. This is in agreement with original equation proposed by Tomkins [54] and also with equation (5) provided cyclic stress-strain response is stabilised. The initial crack growth rate was found to be power law function of plastic strain amplitude. Taking into account the form of the cyclic stress-strain curve this dependence is equivalent to eq. (5). Consideration of two contributions to the crack growth rate in constant amplitude loading allowed the description of the growth of the crack from initiation to the size of a macroscopic crack [59].

Recent study of crack growth rates in variable amplitude block loading of 316L steel [61] shows that logarithm of crack length is proportional to the number of loading blocks. This compares well with the dependence of the crack length on the number of cycles in constant plastic strain amplitude loading (Fig. 8). Simple logarithmic law in constant amplitude loading allows the prediction of crack growth rates in variable amplitude loading provided the crack increment due to all closed hysteresis loops in a block can be summed up independently. The study of short crack growth in variable amplitude loading thus shows why the simple linear cumulative Palmgren-Miner rule works in majority of life predictions.

REFERENCES

1. Ewing, J. A. and Humfrey, J. C. W. (1902) *Philos. Trans. R. Soc. London* **A 200**, 241.
2. Kocanda, S. (1978) *Fatigue Failure of Metals*, Sijthoff and Nordhoff, Amsterdam.
3. Klesnil, M. and Lukáš, P. (1993) *Fatigue of Metallic Materials*, Elsevier, Amsterdam.
4. Polák, J. (1991) *Cyclic Plasticity and Low Cycle Fatigue Life*, Elsevier, Amsterdam.
5. Suresh, S. (1991) *Fatigue of Materials*, Cambridge University Press, Cambridge.
6. Laird, C. (1996) In: *Physical Metallurgy*, R. W. Cahn and P. Haasen (Eds.) 4th edition, Elsevier, Amsterdam, pp. 2293-2397.
7. Fatemi, A. and Yang, L. (1998) *Int. J. Fatigue* **20**, 9.
8. Thompson, N., Wadsworth, N. and Louat N. (1956) *Philos. Mag.* **1**, 113.
9. Polák, J., Lepisto, T. and Kettunen, P. (1985) *Mater. Sci. Engng* **74**, 85.
10. Mughrabi, H. (1978) *Mater. Sci. Engng* **33**, 207.
11. Polák, J., Obrtlík, K. and Helešic J. (1988) *Mater. Sci. Engng* **A 101**, 7.
12. Wittmer, D. E., Farrington, G. C. and Laird, C. (1987) *Acta Metall.* **35**, 1895 and 1911.
13. Polák, J., Obrtlík, K. and Helešic J. (1991) *Mater. Sci. Engng* **A 132**, 67.
14. Polák, J., Obrtlík, K., Hájek, M. and Vašek, A. (1992) *Mater. Sci. Engng* **A 151**, 19.
15. Nabarro, F. R. N., Basinski, Z. S. and Holt, D. B. (1964) *Adv. Physics* **50**, 193.
16. Thompson, N. and Wadsworth, N. (1958) *Adv. Physics* **7**, 72.
17. Laird, C. and Duquette, D. J. (1972) In: *Corrosion Fatigue*, A. J. McEvily, O. Devereux and R. J. Staehle (Eds.) NACE-2, pp. 88-117.
18. Kim, W. H. and Laird, C. (1978) *Acta Metall.* **26**, 777 and 789.
19. Cheng, A. S. and Laird, C. (1981) *Mater. Sci. Engng* **51**, 55 and 111.
20. Tanaka, K. and Mura, T. (1981) *Jour. Appl. Mech.* **48**, 97.
21. Lin, M. R., Fine, M. E. and Mura, T. (1986) *Acta Metall.* **34**, 619.
22. Mura, T. and Nakasone, Y. (1990) *Jour. Appl. Mech.* **57**, 1.
23. Mughrabi H. (1983) In: *Defects, Fracture and Fatigue*, G. C. Sih and J. W. Provan (Eds.) Martinus Nijhoff, Hague, pp. 139-146.
24. Brown, L. M. (1981) In: *Proc. Int. Conf. Dislocation Modelling of Physical Systems*, M. F. Ashby et al. (Eds.), Pergamon Press, Oxford, pp. 51-75.

25. Essmann, U. Gössele, U. and Mughrabi, H. (1981) *Philos. Mag.* **44**, 405.

26. Differt, K., Essmann, U. and Mughrabi, H. (1986) *Philos. Mag.* **54**, 237.

27. Polák, J., Lepistö, T. and Kettunen P. (1985) *Mater. Sci. Engng* **74**, 85.

28. Laird, C., Charsley, P. and Mughrabi, H. (1986) *Mater. Sci. Engng* **81**, 457.

29. Polák, J. (1987) *Mater Sci. Engng* **92**, 71.

30. Polák, J. (1970) *Scripta Metall.* **4**, 761.

31. Polák, J. *Mater. Sci. Engng* **89**, 35.

32. Polák, J. (1994) In: *Dislocations 93*, J. Rabier, A. George, Y. Bréchet and J. Kubin, (Eds,) Scitec Publ., pp. 405-410.

33. Repetto, E. A. and Ortiz, M. (1997) *Acta Mater.* **45**, 2577.

34. Basinski, Z. S. and Basinski, S. (1985) *Acta Metall.* **33**, 1307 and 1319.

35. Hunsche, A. and Neumann, P. (1986) Acta Metall. **34**, 207.

36. Wood, W. A. (1958) *Philos. Mag.* **3**, 692.

37. Polák, J. and Liškutín, P. (1990) *Fatigue Fract. Engng Mater. Struct.* **13**, 119.

38. Kwon, I. B., Fine, M. E. and. Weertmann, J. (1989) *Acta Metall.* **37**, 2937.

39. Basinski, Z. S. and Basinski S. (1992) *Progress in Material Science* **36**, 89.

40. Paris, P. C. and Erdogan, F. (1961) *J. Basic Engng* **85**, 528.

41. Pearson, S. (1975) *Engng Fract. Mech.* **7**, 235.

42. Hudak, Jr., S. J. (1981) *J. Engng Mater. Technol.* **103**, 26.

43. Suresh, S. and Ritchie, R. O. (1984) *Int. Metall. Rev.* **29**, 445.

44. Nishitani, H. (1985) In: *Current Research on Fatigue Cracks*, Soc. Mater. Science, Japan, pp. 1-21.

45. Miller, K. J. (1993) *Mater. Sci. Technol.* **9**, 453.

46. Miller, K. J. (1997) In: *Fatigue and Fracture Mechanics*, R. S. Piacsik, J. C. Newman and N. E. Dowling (Eds.) ASTM STP 1296, ASTM, Philadelphia, pp. 267-285.

47. McClung. R. C., Chan, K. S. Hudak, Jr., S. J. and Davidson, D. L. (1996) In: *ASM Handbook Volume 19, Fatigue and Fracture*, ASM, pp. 153-158.

48. Ritchie, R. O. and Lankford, J. (Eds.) (1986) *Small Fatigue Cracks*, The Metalurgical Society.

49. Miller, K. J. and los Rios, E. R. (Eds.) (1986) *The Behaviour of Short Fatigue Cracks*, ESIS Publ. 1, MEP, London.

50. Miller, K. J. and los Rios, E. R. (Eds.) (1992) *Short Fatigue Cracks*, ESIS Publ. 13, MEP, London.

51. Forsyth, P. J. E. (1963) *Acta Metall.* **11**, 703.

52. Cheng, A. S. and Laird, C. (1983) *Mater. Sci. Engng* **60**, 177.

53. Blochwitz, C. and Heinrich, D. (1988) In: *Basic Mechanisms In Fatigue of Metals*, P. Lukáš and J. Polák (Eds.) Elsevier, Amsterdam, pp. 316-322.

54. Tomkins, B. (1968) *Philos. Mag.* **18**, 1041.

55. Hobson, P. D., Brown, M. V. and de los Rios, E. R. (1986) In: *The Behaviour of Short Fatigue Cracks*, K. J. Miller, E. R. de los Rios (Eds.), IME, London, pp. 412-459.

56. McDowell, D. L. (1997) *Int. J. Fatigue* **19**, Suppl. 1, S127.

57. Buirette, C., Degallaix, G. and Menigault, J. (1998) - this Conference.

58. Murakami, Y., Harada, S., Endo, T., Tani-ishi, H. and Fukushima, Y. (1983) Engng Fract. Mech. **18**, 909.

59. Polák, J., Obrtlík, K. and Vašek, A. (1997) *Mater. Sci. Engng* **A 234-236**, 970.

60. Obrtlík, K., Polák, J. and Vašek, A. (1997) *Int. J. Fatigue*, **19**, 471.

61. Polák, J., Obrtlík, K. and Vašek, A. (1998) In: *Fatigue Design 98* - accepted for publication.

Low Cycle Fatigue and Elasto-Plastic Behaviour of Materials
K-T. Rie and P.D. Portella (Editors)
505

AN ANALYSIS OF THE GERBER PARABOLIC RELATIONSHIP BASED UPON SMALL FATIGUE CRACK GROWTH BEHAVIOR

A. J. McEvily* and S. Ishihara**

*Metallurgy Dept., University of Connecticut, Storrs, CT 06269, USA
**Department of Mechanical Engineering, Toyama University, Toyama 930, JAPAN

ABSTRACT

Over a century ago Gerber developed an empirical equation to account for the influence of mean stress on the fatigue strength. Since that time his relation as well as empirical relations developed by Goodman, Haigh, Soderberg and others have been widely used by design engineers in estimating the effect of a mean stress on fatigue lifetimes. However, there appears to be a lack of fundamental understanding as to why the fatigue strength is dependent upon the mean stress. The purpose of the present paper is to show that the fatigue lifetime calculated on the basis of an elastic-plastic analysis of short crack growth behavior can provide this understanding.

KEYWORDS

Gerber relation, mean stress, small fatigue crack growth, fatigue lifetime.

INTRODUCTION

In 1874 Gerber [1] empirically accounted for the dependence of the endurance limit on the mean stress through the following parabolic relationship:

$$\sigma = \sigma_{R=-1}\left\{1-\left(\frac{\sigma_m}{\sigma_{UTS}}\right)^2\right\} \qquad (1)$$

where σ is the stress amplitude corresponding to the fatigue strength at a given mean stress level, σ_m; $\sigma_{R=-1}$ is the fatigue strength at $R=-1$, and σ_{UTS} is the tensile ultimate strength. (Note that because of the symmetrical nature of this equation about zero mean stress, the relation is only valid for positive mean stresses.) Over the years much data has been accumulated which is in general agreement with Eq. 1, yet very little work has been done in trying to explain the physical basis underlying the relationship.

In 1991 McEvily, Eifler and Macherauch [2] provided an initial rationale for the relationship on the basis of the growth of short fatigue cracks using the following constitutive equation:

$$\frac{da}{dN} = A\left(\Delta K_{eff} - \Delta K_{effth}\right)^2 \tag{2}$$

where a is the crack length, N is the number of cycles, A is a material constant of dimensions $(MPa)^{-2}$, ΔK_{eff} is the effective range of the stress intensity factor, and ΔK_{effth} is the effective range of the stress intensity factor at the threshold level. The stress intensity factor was defined as

$$\Delta K = \left\{\sqrt{2\pi r_e} + Y\sqrt{\frac{\pi}{2}a\left(\sec\frac{\pi}{2}\frac{\sigma_{max}}{\sigma_y} + 1\right)}\right\}\Delta\sigma \tag{3}$$

where r_e is a material constant which relates ΔK_{effth} to the endurance limit; Y is a function of geometry, and for a semi-circular surface crack has a value of 0.65; σ_{max} is the maximum stress in a loading cycle; and σ_y is the yield strength. However because the maximum cyclic stress often exceeded the yield stress at high mean stress levels, σ_{UTS} was used in place of σ_y. The term involving the secant stems from the Dugdale's analysis [3] of the plastic zone size at a crack tip, and is included to incorporate Irwin's suggestion [4] that the actual crack length be increased by one-half of the plastic zone size to take into account elastic-plastic behavior. In the analysis of McEvily, Eifler and Macherauch, only the endurance limit as a function of mean stress was determined, which involved setting the right hand side of Eq. 3 equal to ΔK_{effth}, with a taken to be equal to r_e. Since σ_{max} is equal to $2\sigma_m/(1+R)$, a value for $\Delta\sigma$ at each R level or each mean stress level, σ_m, could then be calculated. Note that as σ_{max} increases, $\Delta\sigma$ must decrease to keep the value of ΔK_{eff} constant. The results of such calculations were in good agreement with the form Gerber's parabolic relationship, but were not compared to experimental data.

The present paper takes this method further by including consideration of some experimental data for the fatigue strength as well as for finite fatigue lifetimes.

ANALYSIS

The approach taken to the calculation of the finite fatigue lifetime is based upon the assumption that the fatigue lifetime is determined by the number of cycles required to grow a crack from a total length $2r_e$ to a total length sufficient to cause failure (herein taken to be 10-mm). Since we are dealing with small cracks, the effect of crack closure development on the rate of fatigue crack growth was taken into account in the calculations. The following expression was used to describe the magnitude of crack closure developed in the wake of a newly formed crack [5]:

$$K_{op} = \left(1 - e^{-ka}\right)K_{op\,max} \tag{4}$$

where K_{op} is the crack tip opening level, a function of the crack length a, K_{opmax} is the crack tip opening level for a macroscopic crack, and k is a material parameter related to the rate of crack closure development. With the inclusion of the crack closure term, Eq. 2 becomes:

$$\frac{da}{dN} = A\left\{\left[\sqrt{2\pi r_e} + Y\sqrt{\frac{\pi}{2}a\left(\sec\frac{\pi}{2}\frac{\sigma_{max}}{\sigma_{UTS}}+1\right)}\right]\Delta\sigma - \left(1-e^{-ka}\right)\left(K_{op\,max}-K_{min}\right)-\Delta K_{effth}\right\}^2 \qquad (5)$$

An example of the use of Eq. 5 for the aluminum alloy 2024-T4 will be given. Experimental results [6] for this material were chosen for comparison purposes because of its Gerber-like behavior and wide range of experimental results. The tensile strength of this alloy in extruded bar form was 580 MPa, and as in ref. [2], this value was used in place of the yield strength in Eq. 2. The values assigned to the constants in Eq. 5 were: $A=1.5\times10^{-9}$ (MPa$\sqrt{m}$)$^{-2}$, $Y=0.65$, $k=10,000$ m^{-1}, and $\Delta K_{effth}=1.4$ MPa$\sqrt{m}$, K_{opmax}, a function of R, was assumed to depend upon R as follows: for $R=0$ and higher, $K_{opmax}=2.0$ MPa$\sqrt{m}$; for $R=-1$, $K_{opmax}=1.6$ MPa$\sqrt{m}$; and for $R=-2$, $K_{opmax}=1.2$ MPa$\sqrt{m}$. The fatigue strength, a function of R, was taken to correspond to a fatigue lifetime of 5×10^8 cycles. The calculated fatigue lifetime was determined by numerically integrating Eq. 5 in 10 μm increments at the shorter crack lengths and in larger increments as the total crack length at the surface approached the 10-mm final length. The following equation was used in the evaluation of r_e:

$$\left[\sqrt{2\pi r_e} + 0.65\sqrt{\frac{\pi}{2}r_e\left(\sec\frac{\pi}{2}\frac{\sigma_{max}}{580}+1\right)}\right]\Delta\sigma_e = \Delta K_{effth} = 1.4\,MPa\sqrt{m} \qquad (6)$$

where $\Delta\sigma_e$ also a function of R, is the experimental value of the fatigue strength at 5×10^8 cycles. An average value of 3.2 μm was determined for r_e on the basis of Eq. 6.

RESULTS OF ANALYSIS

Figure 1 compares the calculated and experimental values of the stress amplitude as a function of mean stress for a fatigue lifetime of 5×10^8 cycles. It is seen that the calculations agree reasonably well with the experimental results with each other, and also with the trend predicted by the Gerber relation.

Figure 2 shows the calculated σ_{max}-log N relationships for R values ranging from $R=-2$ to $R=0.5$. In accord with observed behavior, it is noted that these curves become increasingly more flat as the R value increases. Also the curves tend to be less separated at negative R values, in part because the compressive portion of a loading cycle becomes less important once closure has developed.

Figure 3 is a plot of maximum stress as a function of minimum stress for four different lifetimes. Agreement between calculations and experiment is generally good except for a fatigue life time of 10^5 cycles where the calculated maximum stress values are less than the experimental values.

CONCLUDING REMARKS

It is noted that at small mean stress values above and below zero, the allowable stress amplitude for a given lifetime does not change significantly. This behavior is in accord with the experimental results of Pippan [7] who showed that for small cracks grown under low constant

508

ΔK conditions at low mean stress levels, i. e., in the linear elastic fracture mechanics (LEFM) range, the initial rate of fatigue crack growth was independent of the mean stress, and changed only as crack closure developed. The present results show that at higher mean stress levels the monotonic plastic zone size becomes large with respect to the crack length and the resultant elastic-plastic behavior necessitates a decrease in stress amplitude with increasing mean stress to maintain a given rate of fatigue crack growth.

The effect of crack closure was found to be most significant at low R values and for stress amplitudes just above the fatigue limit, where K_{opmax} values just slightly above the values used could lead to a predicted non-propagation of the fatigue cracks. At high stress amplitudes and high R values there was little, if any, effect of crack closure on the calculated results.

It is also noted that in contrast to the Gerber relation, the present approach can be used to predict lifetimes at R values less than -1.

Finally, it is concluded that fatigue lifetime predictions based upon the growth of small cracks lead to results which are in reasonably good agreement with experimental findings.

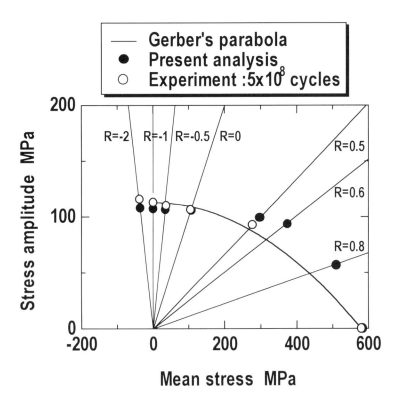

Fig. 1   A comparison of the experimental and calculated fatigue strengths for 2024-T4 aluminum alloy at 5×10^8 cycles with those expected on the basis of the Gerber relation, Eq.1.

ACKNOWLEDGEMENT

The authors express their appreciation to NASA (Research Grant No. NAG 1 1382) and to Toyama University for support of this study. Particular thanks are due to Mr. Koji Kaya for assistance with the numerical calculations.

REFERENCES

1.   Gerber, H. (1874) *Z. Bayerischen Architecktin und Ingenieur-Vereins*, **6**, 101-110.
2.   McEvily, A. J., Eifler, D., and Macherauch, E., (1991) *Engng. Fract. Mechs.*, **40**, 571-584.
3.   Dugdale, D. S. (1960) *J. Mechs. and Physics of Solids*, **8**, 100-108.
4.   Irwin, G. R. (1960) in *First Symp. Naval Struct. Mech.*, Pergamon Press, 557-594.
5.   McEvily, A. J. and Minakawa, K. (1984) *Scripta Met*, **18**, 71-76.
6.   Howell, F. M. and Miller, J. L. (1955) *Proc. ASTM*, **55**, 955-968.
7.   Pippan, R. (1987) *Fatigue Fract. Eng. Mater. Struct.*, **9**, 319-328.

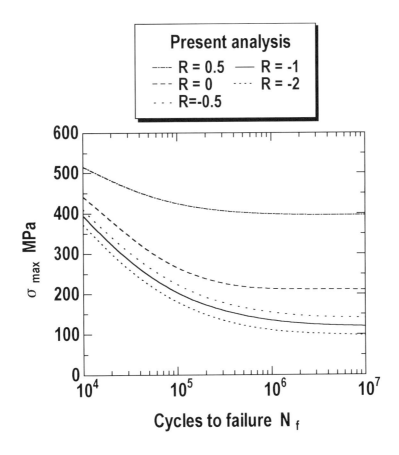

Fig. 2   Calculated σ_{max}-log N curves for 2024-T4 aluminum alloy at R levels ranging from -2 to +0.5 under axial load.

510

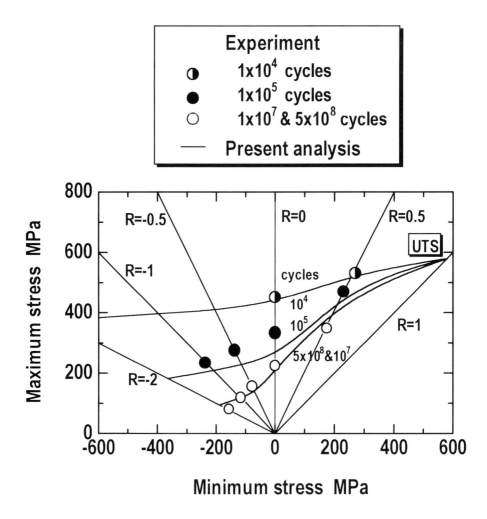

Fig. 3 A comparison of the predicted and the experimental values of the fatigue strength of 2024-T4 aluminum alloy at several fatigue lifetimes and several R values.

Low Cycle Fatigue and Elasto-Plastic Behaviour of Materials
K-T. Rie and P.D. Portella (Editors)
511

A CONSIDERATION OF SCATTER IN SMALL FATIGUE CRACK GROWTH

R. L. CARLSON*and M. D. HALLIDAY**
*School of Aerospace Engineering, Georgia Institute of Technology, Atlanta, Georgia, U.S.A.
**IRC in Materials for High Performance Applications and School of Metallurgy and Materials,
University of Birmingham, Birmingham, U.K.

ABSTRACT

Test results for the growth of small corner cracks in aluminum 6061-T651 and small surface cracks in aluminum 2024-T351 are presented. When the crack front intersects a small number of grains, there is considerable scatter. A transition to reproducible growth for the corner cracks began when the number of grains intersected by the crack front became about six. A measure of the scatter of the corner and surface crack data is obtained by the application of statistical analyses which provide confidence intervals for crack length during the growth histories. .

KEYWORDS

Small fatigue cracks, microstructure, crack path morphology, scatter, statistical representation

INTRODUCTION

It is well known that, because of data scatter, the use of fatigue S-N diagrams for predicting lives of components and structures requires a statistical representation [1]. In contrast, a deterministic approach using fracture mechanics parameters [2] has been very successful in predicting the growth of long fatigue cracks. The small fatigue crack growth regime refers to crack growth between crack initiation and long crack growth. The spread in growth rates encountered in this regime is in part responsible for the scatter in low cycle fatigue lives [3].

Both deterministic [4, 5] and statistical [6, 7] methods of analysis have been proposed to describe growth in this transition regime. Statistical representations are based on the implicit recognition that growth in the small crack regime is a statistical phenomenon; i.e. data scatter should be an anticipated feature of the growth process.

Deterministic analytical methods have usually interpreted the anomolous growth behavior of short/small cracks as being a consequence of a reduction in closure obstruction. An effective crack tip stress intensity factor range, ΔK_{eff} , has been introduced to account for these differences in crack closure [4, 5]. Although an effective ΔK can be adjusted to correlate test data, the validity of this approach can be questioned [8], and it does not incorporate the inherent statistical features of the growth behavior.

The objectives of the present paper are to present experimental evidence of scatter in small fatigue crack growth, to describe sources of scatter, and to demonstrate how a statistical analysis of crack growth test data can be used to describe the transitional behavior between small and long crack growth.

EXPERIMENTAL

Small crack growth tests were conducted using two types of specimen: one containing a corner crack; one with surface or 'thumbnail' cracks. The corner crack test specimens were of aluminum alloy 6061-T651, and were machined from 16 mm diameter bar stock. The average transverse grain size was 200 microns. The longitudinal grains were elongated and varied widely about an average size of about 350 microns. The 0.2 per cent offset yield stress was 283 MPa and the ultimate strength was 293 MPa. The specimens were loaded in three point bending with a load ratio of $R = 0.2$ [9]. They had square cross sections which were oriented to produce a maximum tensile stress at a mid-point corner. At this position, a small notch with a 60 degree included angle and a depth of about 150 microns was cut using a digitally controlled sliting saw. The faces adjacent to the notch were prepared by grinding with a range of abrasive papers, and polishing media to a 1 micron diamond finish.

The 'thumbnail' crack specimen was obtained from 10 mm thick plate of 2024-T351 aluminum alloy. The grains were pancake shaped with mean linear intercept grain dimensions of 590 x 160 x 40 microns in the longitudinal, transverse and short transverse directions, respectively. Longitudinal values for the 0.2 per cent yield stress and the ultimate strength were 375 MPa and 478 MPa, respectively. The specimens were machined for fatigue loading at $R = -1$ in push-pull along the longitudinal direction [10]. They were circular in cross-section with a reduced centre section of 5 mm diameter, and were screw threaded at the ends. A flat surface, normal to the transverse direction, was prepared on the reduced section by removing material to a depth of 0.6 mm, initially by spark erosion. The flat was ground and polished, in similar fashion to the other specimens to a 1/4 micron diamond finish, and was then etched by swabbing with Kellers's reagent to reveal microstructural features.

The fatigue tests were carried out on Instron servo hydraulic testing machines under sinusoidal loading at 10Hz for the bend specimens, and at 0.5 Hz for the push-pull specimen. A telemicroscope system was used to monitor crack growth in the bend specimens. Details of the experimental system are presented in a previous publication [11]. A nominal maximum stress of 0.9 of the yield stress was used to initiate small cracks from the notches. After initiation, this was reduced to 0.8 of the yield strength for continued crack growth. Crack length measurements, which included the notch depths, were commenced when the crack lengths were about 400 microns.

A stress range of ± 0.7 of the yield stress was used to initiate and grow small 'thumbnail' cracks in the push-pull specimen. Cracks self initiate in this material at intermetallic inclusions [12]. Crack growth was measured using intermittant cellulose acetate replication.

RESULTS

Crack length/cycles data for three corner crack growth tests are presented in Fig. 1. The data are for growth on one face for each specimen. The results shown as upper and lower bounds and the top curves are discussed in the DISCUSSION section. Test results for 'thumbnail', surface crack growth are presented in Fig. 2. The data are for three cracks which were initiated and grew from three intermetallic particles in the same test specimen. Measurements were made for growth to the left hand side (lhs) and to the right hand side (rhs) of each 'thumbnail' crack, so a total of six data points are shown for each crack growth interval. The indicated upper and lower bounds are discussed in the DISCUSSION section.

DISCUSSION

An examination of the results presented in Figs. 1 and 2 reveals a common feature; i. e. , there is considerable scatter in the crack growth histories, even for cracks growing simultaneously in the same specimen. A second feature is the growth-arrest behavior which is exhibited most clearly in Fig. 1. In Fig. 1 the number of grains intersected by the crack front ranges from about three for a crack length of 400 microns to about six for a crack length of about 800 microns [11]. For a length of about 2000 microns the crack front intersects about sixteen grains. Note that there is a substantial reduction in scatter for crack lengths greater than about 800 microns. The abrupt decrease in scatter observed at about 100×10^3 load cycles may be attributed to an observation that for some tests two separate, unequal corner cracks ultimately coallesced, and subsequent growth was more uniform [13].

For the surface crack results of Fig. 2 the growth has occurred within a grain or a grain and a half on either side of the crack initiating intermetallic particle. The unequal growth on the two sides beyond the particle is clear from the data in Fig. 2. Note that often in surface crack growth tests, the total surface crack length is measured, and post-examinations of the fracture are concerned with the shape of the crack surface. It is clear from these results, however, that a crack surface can translate as well as expand. In the micrograph of Fig. 3 for site cr3, rhs growth is to the right, and lhs is to the left. Note that the direction of crack advance changes abruptly at the grain boundaries which pass through the cracked particle, and also at a boundary to the left of the particle. Individual grain orientations also have a key influence on both crack path direction and on crack growth rate. Corner cracks have also been observed to exhibit nonplanar crack surface growth [13].

Scatter of small crack growth data has been considered by Torng and McClung [7], who have proposed the use of a generalized Paris type relation. In their statistical representation crack growth rate has been chosen as the random variable. It is assumed that the use of the LEFM stress intensity factor is valid for small cracks, and that the 'anomolous' behavior of small cracks is a result of a reduction in plasticity- induced crack closure. Because of the oscillatory character of the crack growth histories, the use of an analytical procedure for computing growth rates is required. Often the rates so determined are presented on plots of growth rate, da/dN, versus the range of the stress intensity factor, ΔK, as clusters of unconnected points, rather than curves.

The primary objective of the analysis used here was to examine the statistical features of the growth behavior. The measured crack length, however, rather than a derived growth rate, was chosen as the random variable. Although the amount of data available cannot be considered to serve as a basis for a rigorous statistical analysis, the statistical features of the growth behavior can be examined by the use of the Student t distribution. The equation used to estimate the confidence intervals is

$$\mu = X_{sm} \pm (t\,S)\,m^{-0.5}, \tag{1}$$

where μ provides upper and lower confidence bounds, and t has a value which depends on the level of confidence to be evaluated. X_{sm} is the sample mean, S is the variance and m is the sample size, which here is the number of crack length readings for a given number of load cycles. A 95 per cent confidence level has been chosen for the results which are presented here. For the data of Fig. 1, m = 3, and t = 4.303. For the data of Fig. 2, m = 6, and t = 1.05.

An examination of Fig. 1 indicates that although the initial crack lengths were about the same, the growth values soon exhibited considerable divergence. This is reflected in the statistical bounds which have been connected by line segments. In this range the number of grains intersected by the crack front ranges from about three to six. Beyond 100×10^3 cycles, however, the range between the upper and lower bounds has been reduced. At about 75×10^3 cycles, the variation in the deviation from the sample mean is about $\pm$ 50 per cent. At 250×10^3 cycles the deviation is about $\pm$ 6 per cent. It could be inferred from these data that fairly reproducible growth begins to occur when the number of grains intersected by the crack front reaches a value of six, which corresponds to a crack length of about 800 microns. It should be noted, however, that since barriers to crack advance may be expected to be more effective for lower load intensities than those applied here, the transition to reproducible growth then may occur at a longer crack length than that observed here; i. e., the transition crack length may be dependent on both grain size and applied load.

Confidence limits for surface crack growth are shown in Fig. 2. Note that the data points for the rhs of specimen cr1 are above the upper limits. Also, some of the data points for the rhs of specimen cr3 are below the lower limits. These data are, then, in the 5 per cent of the population that are outside the confidence limits. It can be inferred from the data that the rate of crack growth is of the order of 10^{-9} m/ cycle. This indicates that growth rate within single grains was of this order within the resolution of the data.

The importance of microstructural features is emphasized by focusing on growth when a small number of grains are intersected by the crack front. When corner and surface cracks of the same length are present, then, on average, the crack front/grain interaction will be different and this could result in differences in growth behavior. If, for instance, the surface crack front intersects four grains, the corner crack would intersect two grains. For the corner crack both grains would have uncontrained surfaces, whereas the surface crack would have two interior, constrained grains, and two grains with unconstrained surfaces. It is reasonable to expect that the micromechanics of the fatigue processes would differ for these two cases. For short, edge through cracks the differences would be greater. It has been observed that the stress intensity factor cannot differentiate between these cases [13].

CONCLUSIONS

Test results for small surface and corner cracks exhibit considerable scatter when the number of grains intersected by the crack front is small. The fracture surface of 'thumbnail' cracks can deviate significantly from the idealized assumption of planar growth, and the crack surface can translate relative to its origin, as well as expand within this scale of growth. Differences in growth behavior between small surface cracks, small corner cracks and short edge, through cracks are not incorporated in the stress intensity factor. For the applied loading conditions corner crack growth reproducibility appears to begin when the crack front intersects about 6 grains. Statistical analyses of the crack length growth data can be used to provide a measure of scatter in terms of crack length confidence intervals.

ACKNOWLEDGEMENTS

The first author acknowledges the benefit of discussions with Prof. G. A. Kardomateas and Dr. D. L. Steadman. The second author would like to thank the Engineering and Physical Sciences Research Council for the provision of research facilities.

REFERENCES

1.   Pook, L. P. (1989). *The Role of Crack Growth in Metal Fatigue*, The Metals Society, UK.
2.   Paris, P. C. and Erdogan, F. (1963). *Trans. ASME, Journal of Basic Engineering* **85**, pp. 528-534.
3.   Miller, K. J., Mohamed, H. J. and de los Rios, E. R. (1986). In: *The Behavior of Short Fatigue Cracks*, K. J. Miller and E. R. delos Rios (Eds). pp. 491-511
4   Liaw, P. K. and Logsdon, W. A. (1985). *Eng. Fract. Mechanics* **22**, pp. 115-121.
5.   Edwards, P. R. and Newman, J. C. Jr. (1990). In: *Short Crack Growth in Various Aircraft Materials,* AGARD REPORT 767
6.   Cox, B. N. and Morris, W. L. (1988). *Engineering Fracture Mechanics* **31**, pp. 591-610.
7.   Torng, T. Y. and McClung, R. C. (1994) *Proc. 35th SDM Conference*, AIAA, pp. 1514-1524.
8.   Halliday, M. D., Poole, P. and Bowen, P. (1995). *Fatigue and Fracture of Engineering Materials and Structures* **18**, pp. 717-729.
9.   Steadman, D. L. (1997). *Growth-arrest Behavior of Small Cracks*, PhD Thesis, School of Aerospace Engineering, Georgia Institute of Technology, Atlanta, Ga., USA
10.   Halliday, M. D. (1994). Unpublished research; IRC in Materials for High Performance Applications, University of Birmingham, UK.
11.   Carlson, R. L., Dancila, D. S. and Kardomateas, G. A. (1996). In: *Proc. Sixth International Fatigue Congress*, G. Lütjering and H. Nowack (Eds). pp. 289-294.
12.   Halliday, M. D. and Beevers, C. J. (1992). In: *Aluminum Alloys-Their Physical and Mechanical Properties*, L. Arnberg, E. Nes and N. Ryum (Eds). The Norwegian Institute ofTechnology / SINTEF, Vol. 3, pp. 526-532.
13.   Carlson, R. L., Steadman, D. L., Dancila, D. S., and Kardomateas, G. A. (1997). *Int. J. Fatigue* **19**, pp. 119-125.

516

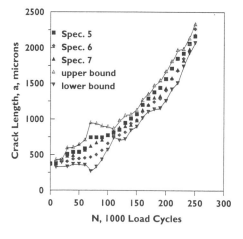
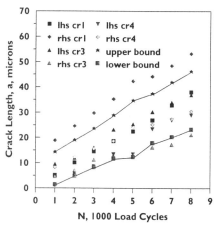

Fig. 1. Corner crack length vs loading cycles for aluminum 6061T-651. R = 0.2.

Fig. 2 Thumbnail crack length vs loading cycles for aluminum 2024-T351. R = -1.

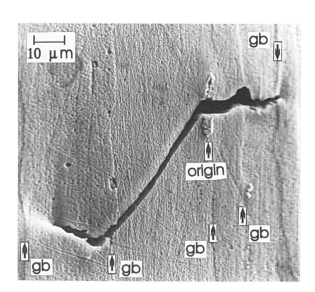

loading direction ↑

Fig. 3. Initiation and growth of the thumbnail crack at site crs 3.

PROPAGATION OF SHORT FATIGUE CRACKS IN AUSTEMPERED DUCTILE IRON UNDER CYCLIC BLOCK LOADING

M. SCHLEICHER, H. BOMAS, P. MAYR

Stiftung Institut für Werkstofftechnik, Badgasteiner Straße 3, D-28359 Bremen, Germany

ABSTRACT

Low cycle fatigue tests were carried out at two ductile cast irons: a ferritic-pearlitic and an austempered ductile iron. The growth characteristics of short surface cracks were investigated under different block loading programmes. The lengths of the cracks leading to failure and the crack densities were determined in dependence of the cycle number. In general the growth rates under short block loading are higher than under long block loading. It was found that the failure causing cracks consisted of up to 10 independent short cracks that had linked. The crack densities show some differences of the crack nucleation rates between the short and the long block loading condition. In general the reduced lifetime under the short block loading is caused by an interaction of higher pure growth rates and an increased propagation due to the absorption of more other cracks.

KEYWORDS

short fatigue cracks, block loading effect, ductile cast iron, Miner sum, crack densities

INTRODUCTION

Austempered ductile iron (ADI) is an iron with excellent mechanical properties under static load because of its exceptional combination of high strength and ductility and so it has the potential to replace forged components in order to reduce costs. So far, there has been a lack in systematic studies of ADI under cyclic loading. To understand the behaviour of ADI in the fatigue test, it is necessary to investigate the growth characteristics of short fatigue cracks, which was done in an extensive study. In general, short cracks play a dominant role in the LCF-region. Usually materials show a so called short crack effect, which means that short cracks are able to grow at levels of ΔK that are lower than the threshold value ΔK_{th} for long cracks. Another characteristic feature of short cracks is, that they can be influenced by the microstructure of a material, grain boundaries for example. As a result of this the growth rates of short cracks often show large scatter. There are several articles giving a survey of the short crack problem, see for example [1, 2].

518

RESULTS

The investigated materials were the nodular cast iron GGG-60 and the ADI version
GGG-90 B, which was produced by heat treating of GGG-60 (870°C,1h → 365°C, 2h) in order
to achieve a bainitic-austenitic matrix with a portion of roughly 35 % retained austenite. Table 1
shows the chemical composition, mechanical properties are given in table 2 [3].

Table 1: Chemical composition, wt%

C	Si	Mn	S	Ni	Cu	Mg
3,8	2,55	0,62	0,002	0,023	0,33	0,026

Besides the determination of the nodule count per area (108 ± 17 mm^2), a statistical examination
of the graphite nodules diameter and the distance to the neighbouring nodule was carried out.
The mean diameter of the graphite nodules is 32 ± 10 µm. 10 % of the graphite nodules have a
distance < 5 µm to their neighbour. Stress controlled fatigue tests were conducted with
cylindrical smooth specimens with a diameter of 8 mm in a servo-hydraulic fatigue testing
machine at a frequency of 3 Hz. Constant amplitude experiments were carried out to determine
the stress levels that lead to N_f = 10000 (High level) and N_f = 150000 (Low level). Both R = -1
and R = 0 stress ratios were investigated. In order to examine the influence of the load history
on N_f block loading experiments were carried out. Figure 1 gives an idea of the experimental
programme, table 3 shows details of the applied loading conditions.

Table 2: Mechanical properties [3]

	$R_{p0,2}$ [MPa]	R_m [MPa]	A_5 [%]	Hardness
GGG-60	415 ± 7	733 ± 16	8 ± 2	237 ± 9 HB 2,5 30
GGG-90 B	794 ± 7	1062 ± 12	6 ± 1	350 ± 14 HV 1

Fatigue process was successively observed on the specimen surface by the aid of a plastic
replication technique. It was possible to conserve the major part of the specimen surface at
different numbers of load cycles. After the specimen was broken, the crack that led to failure
could be traced back. In addition to that, crack densities in dependence of the cycle number
were determined. Crack coalescence occurred even at low stresses and dominated the growth
rates of longer cracks. It was found that the failure causing cracks consisted of up to 10
independent short cracks that had linked during the fatigue process. The majority of the cracks
start from graphite nodules and microshrinkage pores, which was found by other workers too
[4, 5].

Figure 2 shows the crack growth rates of the dominant cracks, that caused failure. These
propagation rates display a large scatter, which is typical for short cracks. Because of relative
high surface crack densities in the investigated materials, the crack propagation rates are
influenced by crack coalescence and other interactions between neighbouring cracks, causing
large scatter. If the propagation rates under **HLHL1** and **HLHL2** load are compared, in spite of
the scatter, in the case of **HLHL2** higher values especially for crack lengths < 200 µm can be
recognized.

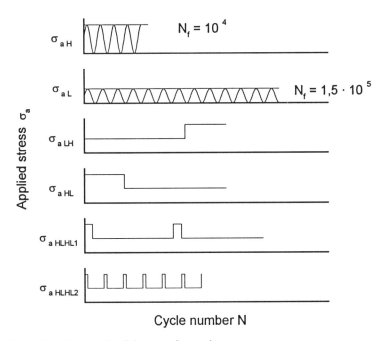

Fig. 1: Schematic sketch of the experimental programme

Table 3: Detailed experimental programme

		H	L	HL	LH	HLHL1	HLHL2
GGG-60 **R = 0**	σ_{min} [MPa]	0	0	0	0	0	0
	σ_a [MPa]	300	215	300→215	215→300	300→215	300→215
	N_H	N_f	0	6300	N_f	500	5
	N_L	0	N_f	N_f	74500	6000	60
GGG-60 **R = -1**	σ_{min} [MPa	-420	-320	-420→-320	-320→-420	-420→-320	-420→-320
	σ_a [MPa]	420	320	420→320	320→420	420→320	420→320
	N_H	N_f	0	4950	N_f	500	5
	N_L	0	N_f	N_f	77500	8000	80
GGG-90 B **R = 0**	σ_{min} [MPa	0	0	0	0	0	0
	σ_a [MPa]	450	280	450→280	280→450	450→280	450→280
	N_H	N_f	0	5175	N_f	500	5
	N_L	0	N_f	N_f	68250	6500	65
GGG-90 B **R = -1**	σ_{min} [MPa	-640	-410	-640→-410	-410→-640	-640→-410	-640→-410
	σ_a [MPa]	640	410	640→410	410→640	640→410	640→410
	N_H	N_f	0	5825	N_f	500	5
	N_L	0	N_f	N_f	75200	6500	65

520

Table 4 shows the Miner sum D [6, 7] under block loading conditions. If the block loading conditions **HLHL1** and **HLHL2** are compared under the assumption of a linear damage accumulation rule, the values of N_f and D would remain constant. The experiment however shows that the Miner sums under the short block loading condition (**HLHL2**) are between 31 and 50 % smaller than under the long block loading condition (**HLHL1**), which is in accordance to the behaviour of many materials. The smaller Miner sums under the short block loading condition **HLHL2** are consistent with higher crack propagation rates compared to the long block loading condition **HLHL1**.

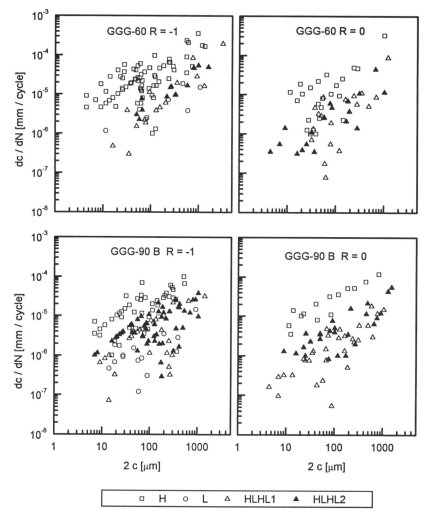

Fig. 2. Surface propagation rates of cracks leading to failure under different load conditions

So the smaller Miner sums under **HLHL2** can be explained with higher crack propagation rates. One reason could be different crack densities under the loading conditions **HLHL1** and **HLHL2**. It is well known that the crack density has an influence on the propagation rate [8]. In general, one can separate pure growth and propagation due to the absorption of other cracks located in the growth plane of the dominant crack. It is obvious that a higher crack density leads to a higher propagation rate due to crack coalescence. Extensive investigations of the crack densities in dependence of the number of load cycles show some references to differences of the crack nucleation rates between the loading conditions **HLHL1** and **HLHL2** but only at GGG-60 R = 0 and GGG-90 B R = -1 the crack densities at a certain number of load cycles are significant higher.

Differences in the pure crack growth rates of long cracks in steel and aluminium alloys were often investigated. A change of the stress amplitude from low to high leads to an accelerated growth rate, whereas a change from high to low leads to a retarded growth rate. Often these effects are explained with differences in the crack closure behaviour [9, 10, 11]. Results of crack closure measurements of short fatigue cracks in ferritic-pearlitic and austempered ductile iron could not be found in the literature.

Table 4. Miner sums D in the block loading tests

	D(HLHL1)		D(HLHL2)	
Material	R = -1	R = 0	R = -1	R = 0
GGG-60	$0,62 \pm 0,09$	$1,04 \pm 0,14$	$0,42 \pm 0,09$	$0,63 \pm 0,04$
GGG-90 B	$1,44 \pm 0,29$	$1,61 \pm 0,13$	$0,72 \pm 0,11$	$1,13 \pm 0,09$

SUMMARY

The block length has a strong influence on the lifetime in the fatigue test of specimens made of ductile cast irons. A reduced block length leads to a decreasing number of cycles to failure. There is a close connection between the growth behaviour of short cracks and the number of cycles to failure. The propagation rates of the failure causing cracks under short block loading are higher than under long block loading. It was found that the failure causing cracks consisted of up to 10 independent short cracks that had linked. Because of this the crack densities were investigated. There are some differences of the crack nucleation rates between the short and the long block loading condition. In general the reduced lifetime under the short block loading is caused by an interaction of higher pure growth rates and an increased propagation due to the absorption of more other cracks. Further work will be necessary to clarify the observed effects of acceleration and retardation on the crack propagation rates.

ACKNOWLEDGEMENTS

This work was funded by the „Arbeitsgemeinschaft industrieller Forschungsvereinigungen (AiF)" and the Bundesministerium für Wirtschaft (AiF-number 9673B).

REFERENCES

1. Miller, K. J. (1993). Materials science perspective of metal fatigue resistance. *Materials Science and Technology* 9, pp. 453-462

2. Schijve, J. (1996). Predictions on fatigue life and crack growth as an engineering problem. A state of the art survey. In: Lütjering, G.; Nowack, H.: *Fatigue '96, Proceedings of the sixth international fatigue congress*, Berlin, pp. 1149-1164

3. Böschen, R. (1995). *Schwingfestigkeitsverhalten von randschichtgehärtetem bainitisch-austenitischen Gußeisen mit Kugelgraphit.* VDI-Fortschrittbericht Reihe 18 Nr.168, Düsseldorf

4. Lin, C.-K.; Lai, P.-K.; Shih, T.-S. (1996). Influence of microstructure on the fatigue properties of austempered ductile irons-I. High-cycle fatigue. *International Journal of Fatigue* 18, pp. 297-307

5. Lin, C.-K.; Hung, T.-P. (1996). Influence of microstructure on the fatigue properties of austempered ductile irons-II. Low-cycle fatigue. *International Journal of Fatigue* 18, pp. 309-320

6. Palmgren, A. (1924). Die Lebensdauer von Kugellagern. *VDI-Zeitschrift* 68, pp. 339-341

7. Miner, M. A. (1945). Cumulative damage in fatigue. *Trans. ASME, Journal of Applied Mechanics* 12, pp. A159-A164

8. Bomas, H.; Linkewitz, T.; Mayr, P. (1996). Short cracks and lifetime of a normalized carbon steel under cyclic block loading. *Fatigue and Fracture of Engineering Materials and Structures* 19, pp. 299-307

9. Buschermöhle, H.; Vormwald, M.; Memhard, D. (1997). Einfluß von Überlasten auf die Rißfortschrittslebensdauer. *DVM-Bericht 800*, pp. 93-115

10. Davidson, D. L. (1992). The experimental mechanics of microcracks. In: Larsen, J. M.; Allison, J. E.: *Small-Crack Test Methods*, ASTM STP 1149, Philadelphia, pp. 81-91

11. Dabayeh, A. A.; Topper, T. H. (1995). Changes in crack-opening stress after underloads in 2024-T351 aluminium alloy. *International Journal of Fatigue* 17, pp. 261-269

Low Cycle Fatigue and Elasto-Plastic Behaviour of Materials
K-T. Rie and P.D. Portella (Editors)
523

SMALL CRACK PROPAGATION DURING LOW CYCLE FATIGUE TESTS AT ELEVATED TEMPERATURE

Thomas Hansson, Torsten Koch, Peter Georgsson, Jan Börjesson
Dept. of Materials Testing, Volvo Aero Corporation, 461 81 Trollhättan, Sweden

ABSTRACT

A procedure to indirectly study the degree of damage and the propagation of cracks under standard low cycle fatigue (LCF) tests was investigated. A damage parameter, D, derived from the maximum and minimum stress for each cycle, was introduced. Based on a correlation between D and the reduction of effective cross-section under the tensile part of the hysteresis loops an apparent crack depth, a, was introduced. The apparent crack depth was defined as the depth of a surface layer with a dense population of small cracks that would result in a reduction of the maximum stress at maximum strain equal to that observed during the tests. This definition has a number of limitations but the information from this exercise may provide useful information about the crack initiation and crack growth process under LCF conditions. The feasibility of this procedure are discussed by analysing test data from a nickel base superalloys, Haynes 230, at elevated temperature.

KEYWORDS

Small crack propagation, low cycle fatigue, elevated temperature

INTRODUCTION

Low cycle fatigue is considered to be one of the major life limiting factors in aircraft engine components [1]. Romanoski et al. [2] concluded, that the low cycle fatigue life is found to be dominated by propagation of micro-cracks to a critical size and that crack initiation occurs very early in the fatigue life, probably even during the first cycle. Analytical treatments of small crack growth often attempt to derive predictions of small crack growth rates from large crack growth data [3]. This type of predictions are dangerous since small cracks may propagate orders of magnitude faster than what can be predicted using crack propagation data for long cracks [3]. Small cracks may also grow at stress intensity ranges, ΔK, well below the threshold value, ΔK_{th}, for long cracks.

Several experimental techniques are available for measuring the size of small fatigue cracks during testing. Examples of available techniques are replication, optical microscopy, potential drop, ultrasonic measurements, laser interferometry, and scanning electron microscopy [3,4]. These techniques however, require sophisticated instruments and/or are very time consuming, especially for investigations at elevated temperatures.

The approach for the present work was to define a damage parameter and to estimate a measure of the small cracks and their propagation rates under low cycle fatigue conditions by using data from standard low cycle fatigue tests. This has been done by studying the relation between the maximum and minimum stresses for the hysteresis loops as a function of the cycle number.

DESCRIPTION OF METHOD AND RESULTS

Damage Parameter, D

A simple procedure to determine the degree of damage and crack growth from a standard LCF tests, for a loading ratio $R_\varepsilon = -1$, has been developed. The procedure is based on the assumption that an increase in the difference between the absolute value of the minimum stress, $|\sigma_{min}|$ and the maximum stress, σ_{max}, provide a measure of the damage that has accumulated in the specimen. The change of the ratio between the maximum and minimum stress during a fully reversed cycle is then associated with the occurrence of a propagating damage, i.e. a reduction of the effective cross-section due to initiation and propagation of cracks. The reduction of the effective cross-section due to cracks reduces the maximum tensile stress whereas the reduction of the absolute value of maximum compressive stress will be less pronounced. Using this assumption a damage parameter D can be deduced as:

$$D = \frac{\left| \sigma_{min} \right| - \sigma_{max}}{\left| \sigma_{min} \right|} \tag{1}$$

The above procedure is applied on data obtained from LCF investigations of a Ni-base superalloy, Haynes 230. The data used was obtained from high temperature LCF tests at a total strain range of 1%.

In Fig. 1, D is plotted as a function of cycles, N, for a LCF test performed at 800K at 1% total strain range. The parameter D has been calculated using stress levels compensated by the change in diameter during loading and also for regulation errors:

$$\sigma_c = \sigma_{meas} + \Delta\varepsilon \cdot d\sigma / d\varepsilon \tag{2}$$

where

$$\Delta\varepsilon_{max} = \left| \varepsilon_{t\arg et} \right| - \varepsilon_{max} \tag{3}$$

and

$$\Delta\varepsilon_{min} = \left| \varepsilon_{t\arg et} \right| - \varepsilon_{min} \tag{4}$$

The value σ_c is the compensated stress level, σ_{meas} is the measured stress level and $d\sigma/d\varepsilon$ is derived from the slope of the $\sigma - \varepsilon$ - curve at the relevant strain level.

2

It is interesting to note that even during the first cycles the average of the introduced damage parameter, D, is larger than zero. This indicates that cracks initiate and grow very early in the fatigue life which is consistent with the findings in Ref. [2].

The scatter of D can be reduced by applying a running average of the data with an appropriate window and to delete half the window at the beginning and at the end. In order to reduce the number of data points the values of D were rounded off to three decimals and the average cycle number for each value of D was calculated. The effect of the smoothing procedure can also be seen in Fig. 1.

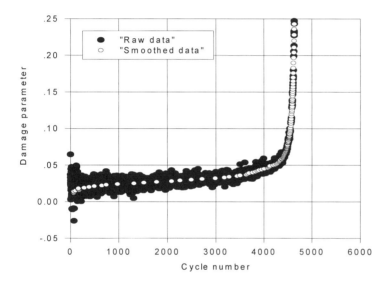

Fig. 1. Damage parameter D as a function of cycles N for Haynes 230 at 800K and a total strain range of 1%.

Crack Length, a, and Crack Growth Rate, da/dN

The next step is now to estimate a crack length, a, from D. This is a critical step which may be done in a number of ways. In this work we have assumed that during the first part of the fatigue life a dense population of small cracks initiate and grow. It is assumed that these cracks are randomly distributed over the surface and completely release the stresses in a surface layer with a depth equal to the depth of the crack. It is also assumed that these cracks reduce the effective area of the specimen in tension (but not in compression). Of course, this assumption is not completely correct, but by using this simple approach trends in the crack growth behaviour may be revealed.

During compression the effective cross-section is approximately equal to A_{nom} while the effective cross-section at maximum tensile load is $A_{nom} - A_{crack}$, where A_{nom} is the nominal area of the cross section and A_{crack} is equal to the area of the surface layer containing cracks. Since the tests were performed in strain control the measured stresses should be directly proportional to the effective cross-section which gives the following relation :

$$\frac{\sigma_{max}}{\left|\sigma_{min}\right|} = \frac{A_{nom} - A_{crack}}{A_{nom}} \tag{5}$$

The effective area of the cracks when the cracks are small compared to the diameter of the specimen may be estimated as:

$$A_{crack} = a \cdot \pi \cdot d \tag{6}$$

where a is the depth of the cracked zone, and d is the specimen diameter. By using Equations (1), (5) and (6) the crack depth, a, may be expressed in terms of D as:

$$a = \frac{d}{4} \cdot D \tag{7}$$

As can bee seen in Fig. 2, the cracks grow rapidly to a depth of about 30 microns for the Haynes 230 material at 1% total strain range. By using the data shown in Fig. 2 the crack growth rate versus crack depth can be derived (see Fig. 3). After the initial growth the propagation rate decreases and remains more or less constant until the stage when the crack propagation rate increases very rapidly with increasing cycle number (see Figs. 2 and 3). The fatigue life is mainly determined by the duration of the regime where the crack propagation rate is small. It should be kept in mind though, that the presented data are based on a number of assumptions and should not be treated a an absolute measure of the crack depth. A reasonable explanation for the observed behaviour may be that the crack propagation rate decreases when the crack front approaches a grain boundary. The mean grain size of the tested material was 50-65 microns.

The crack propagation rate for specimens tested under the same condition exhibit similar crack propagation. The above described crack propagation behaviour is consistent with the findings in other studies [3, 5].

RECOMMENDATIONS FOR FURTHER WORK

The described procedure have been demonstrated on a number of other superalloys tested at elevated temperatures and strain ranges [6] but further work is needed to verify a more general significance.

Further work should be concentrated on the development of a more thorough understanding of the physical meaning of the introduced damage parameter. This would provide the means for a better estimation of the crack length from the damage parameter. The results should also be verified by microscopy studies of specimens interrupted at different stages before failure.

The effect of a population of surface cracks on the maximun stress level is not studied in detail in this work and it is here assumed that the cracked zone is stress free. This is not a correct assumtion, but is made for simplicity. A more accurate measure of the crack depth may be obtained if this problem is studied by means of FE methods.

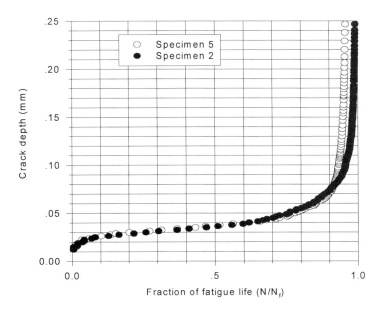

Fig 2. Crack depth as a function of fraction of the fatigue life of Haynes 230 at 800K

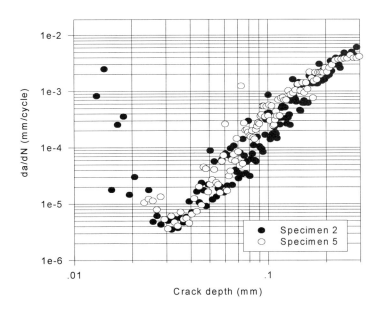

Fig. 3.   Crack propagation rate as a function of crack depth for Haynes 230 at 800K

5

CONCLUSIONS

It was shown that a measure of the damage during low cycle fatigue testing at elevated temperature could be obtained by simply using the relation between the maximum and minimum stress during each cycle for specimens tested at R_ε =-1. The scatter of the data was reduced by averaging procedures and the crack depths corresponding to the values of the damage parameter were estimated by using a number of assumptions. The main assumptions were that the cross-section during the tensile part of the fatigue cycle was reduced by the area of the surface layer containing a dense population of cracks and that this layer is not capable of carrying any load.

It was shown that the cracks propagated fast during the first cycles and reached a steady state regime with low crack propagation rates before a very rapid increase in the propagation rate towards the end of the tests. Most of the fatigue life is clearly spent in the steady state regime with low crack propagation rates.

ACKNOWLEDGEMENTS

Discussions with Dr. F. Kandil, National Physical Laboratory, Teddington, UK, are greatly acknowledged.

REFERENCES

1. Dai, Y., Marchand, N.J. and Hongoh, M. (1992). In: *Proceedings of the third International Conference on Low Cycle Fatigue and Elasto-Plastic Behaviour of Materials*, pp. 594-600, Elsevier, Berlin.
2. Romanoski, G.R., Antolovich, S.D. and Pelloux, R.M. (1988). In: *Low Cycle Fatigue ASTM STP 942*, pp. 456-469, H.D. Solomon, G.R. Halford, L.R. Kaisand and B.N. Leis (Eds.) ASTM, Philadelphia.
3. McClung, R.C., Chan, K.S., Hudak, S. J. and Davidson, D.L. (1996). In *ASM Handbook Vol. 19, Fatigue and Fracture*, pp. 153-158, S.R. Lampman, G.M. Davidson, F. Reidenbach, R.L. Boring, A. Hammel, S.D Henry and W.W. Scott (Eds). ASM
4. Davidsson, D. L. (1996), In: *Fatigue '96, Proceedings of the 6th International Fatigue Congress*, pp. 1925-1935, G. Lütjering and H. Novack (Eds). Pergamon, Oxford.
5. Miller, K. J. , In: *Fatigue '96, Proceedings of the 6th International Fatigue Congress*, pp. 253-264, G. Lütjering and H. Novack (Eds). Pergamon, Oxford
6. Hansson, T. , Manuscript in preparation

Low Cycle Fatigue and Elasto-Plastic Behaviour of Materials
K-T. Rie and P.D. Portella (Editors)
529

FATIGUE LIFE CURVES OF MATERIALS AND THE GROWTH OF SHORT CRACKS

J. POLÁK*, **, K. OBRTLÍK* and A. VAŠEK*
*Institute of Physics of Materials, Academy of Sciences,
Žižkova 22, 616 62 Brno, Czech Republic
** Laboratoire de Mécanique de Lille (URA CNRS 1441), Ecole Centrale de Lille,
BP 48, Cité Scientifique, 59651 Villeneuve d'Ascq Cedex, France

ABSTRACT

Fatigue life curves and cyclic stress-strain curves on smooth cylindrical specimens of several engineering materials were measured and on similar specimens the crack initiation and short crack growth were studied. The dependence of the short crack growth rates on the crack length and plastic strain amplitude was measured and analysed. The integration of the short crack growth law yields the fatigue life. The relation between the fatigue life and initial rate of short cracks is discussed and the possibility to evaluate the fatigue life from the crack growth rate is proposed.

KEYWORDS

Fatigue life, life curves, short cracks, crack initiation, crack growth.

INTRODUCTION

Fatigue life curves are traditionally used to characterise the fatigue resistance of materials, structural parts and structures. The fatigue life curves in terms of stress have been introduced already in the last century by Wöhler, those in terms of strain or plastic strain in the fifties by Manson and Coffin (see e.g. [1]). They represent integral fatigue characteristics of materials.

The study of the fatigue crack propagation in cracked bodies and the application of fracture mechanics in the sixties allowed obtaining the crack growth rates, which represent differential fatigue characteristic. The integration of the crack propagation law in the domain of long crack growth law yields the perfect agreement with the fatigue life of a cracked body.

The study of short cracks in smooth bodies revealed that their growth determines the fatigue life. The crack growth rates of short cracks thus can represent a differential fatigue characteristic of a material. Several relations between the life curves and the kinetics of short crack growth have been suggested [1,2,3] but they lack experimental verification and a detailed analysis. The aim of the present paper is to report the life curves and the crack growth rates of short fatigue cracks in several engineering materials and to discuss their relation.

EXPERIMENTAL

Strain controlled, constant plastic strain amplitude loading has been applied in computer controlled electrohydraulic testing system to the smooth cylindrical specimens of copper, 316L stainless steel, duplex stainless steel and aluminium-lithium alloys. Manson-Coffin curves in a wide interval of fatigue lives were obtained using one type of smooth cylindrical specimen and one type of fatigue testing system. Similar cylindrical specimens, containing shallow surface notch, were subjected to constant plastic strain amplitude or constant stress amplitude loading and the fatigue damage mechanisms and kinetics of short crack growth were studied. Further experimental details can be found elsewhere [4,5].

RESULTS

The cyclic stress strain curves (CSSCs) and the Manson-Coffin curves of the four materials studied are shown in Figs. 1 and 2. All CSSCs can be represented reasonably well by a power law in the form

$$\sigma_a = K_b \varepsilon_{ap}^{n_b} \tag{1}$$

where K_b and n_b are parameters, and the Manson-Coffin curves by a law

$$\varepsilon_{ap} = \varepsilon'_f \left(2N_f\right)^c \tag{2}$$

where ε'_f and c are parameters. The parameters in the eqs. (1) and (2) were evaluated using least-square procedure and are shown in Table 1. The derived Wöhler curves can be evaluated, too, and the parameters of the Basquin relation

$$\sigma_{ap} = \sigma'_f \left(2N_f\right)^b \tag{3}$$

σ'_f and b were calculated and are also given in Table 1.

Table 1. Parameters of the cyclic stress-strain curves and life curves.

Material	K_b(MPa)	n_b	ε_f'	c	σ_f'(MPa)	b
Cu polycrystal	648	0.240	0.243	- 0.505	461	- 0.121
316L steel	581	0.118	1.738	- 0.647	620	- 0.076
duplex steel	1096	0.127	0.972	- 0.642	1092	- 0.081
Al-Li alloy	734	0.077	1.130	- 1.090	741	- 0.083

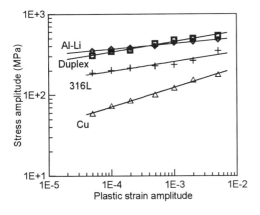

Fig.1. Cyclic stress-strain curves.

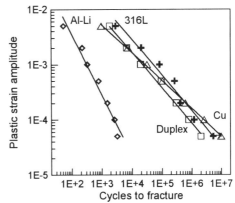

Fig. 2. Manson-Coffin curves.

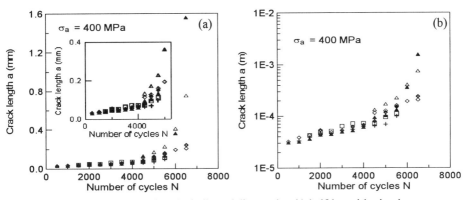

Fig. 3. Growth of short cracks in Al-Li alloy, a) linear plot, b) half-logarithmic plot.

The growth of short cracks was studied in all four materials under constant plastic strain amplitude loading. The plastic stress-strain response in the domain of crack initiation and growth for all materials studied is stabilised and thus the loading conditions are very close to constant stress amplitude loading.

Since the kinetics of short crack growth has been already reported for polycrystalline copper [4] and for 316L steel [5], the experimental data on short crack growth will be given only for duplex steel and Al-Li alloy. Figure 3 shows the growth of short cracks at one stress amplitude in Al-Li alloy. Only the crack growth rates on the S-side (the side perpendicular to the short direction of the extruded bar) are shown here since principal cracks always started on the S-side. Several cracks are initiated early in the fatigue life, usually from non-metallic inclusions. The length of the initiated crack is thus only slightly larger than the size of an inclusion (~ 60 μm in Fig. 3). Initially the crack length of all cracks increases approximately linearly with the number of cycles, which suggests constant initial crack growth rate. The cracks interact and some of them link with smaller cracks. Most cracks stop growing and only one of them becomes a dominant crack. The straight line can also approximate very well the experimental data in the initial part of the crack growth in the half-logarithmic plot (Fig. 3b). This linear dependence, however, corresponds to the proportionality of the crack growth rate to the crack length.

532

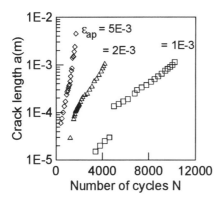 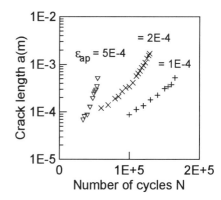

Fig. 4. Growth of short cracks in duplex steel.

Only dominant cracks that were initiated and grew under constant plastic strain amplitude loading in a duplex steel are shown in Fig. 4. At higher stress amplitudes multiple cracks were initiated starting from persistent slip bands. At lower stress amplitudes the cracks were initiated from the small surface defects that were present after electrolytic polishing of the surface in places where non-metallic inclusions were close to the surface. Usually one large crack was present in the area of observation, which later became a dominant crack. The crack growth curve can be reasonably well approximated by a straight line in half-logarithmic coordinates.

ANALYSIS

The aim of our analysis is to show quantitative relation between the fatigue life of smooth bodies and the short crack growth rate. Based on the experimental observations of several materials two laws describing the growth of short cracks in smooth bodies from their initiation (initial crack length a_i) up to the appearance of a crack of a critical dimension (final crack length a_f) were proposed [5,6]. Both of them are based on the assumption that the crack growth rate $v = da/dN$ results from two major contributions which are additive, i. e.

$$v = v_g + v_p \qquad (4)$$

where v_g is the crack generation rate and v_p the crack propagation rate. The crack propagation rate can be derived from fracture mechanics considerations and is proportional to the power of stress intensity amplitude. This term influences only the final stage of the fatigue life and thus only approximate evaluation of this contribution is adequate. Since v_p is proportional to K_a^4 and K_a is proportional to $a^{1/2}$, the dependence of v_p on the crack length is

$$v_p = k_p \, a^2/a_i \qquad (5)$$

where k_p is the crack growth coefficient and a_i the initial crack length.

In order to evaluate the fatigue life from the early crack growth rate the dependence of the crack generation rate on the crack length must be known. There are at least three important obstacles which make the experimental evaluation of this dependence difficult:
(i) In the study of crack initiation and growth usually only surface crack length can be measured. The surface crack length does not characterise completely the crack topology. In stabilised

situation the crack shape tends to be close to a semicircle and the crack diameter a is calculated as the half of the surface length. However, in the early growth of a crack, when the crack is shallow, or in case of the crack linkage, this approximation is not satisfactory.

(ii) The crack growth rates of individual cracks fluctuate since the crack rate slows down when the crack approaches grain or phase boundary.

(iii) In multiple initiation and crack interaction the dominant crack need not be the same crack for all the fatigue life. In this case the concept of an "equivalent crack" [5] proved to be useful.

The analysis of experimental data on duplex steel and Al-Li alloys and previous data on 316L steel [5] and copper [4] revealed that both the crack growth law according to which v_g is independent of the crack length and the $v_g \sim a$ law describes reasonably well the experimental data. It was shown earlier [5] that in the first case, i. e. if

$$v_g = g_i a_i \tag{6}$$

where g is constant. The integration of the crack growth equation (4) with (5) and (6) leads to

$$\frac{a}{a_f} = \frac{a_t}{a_f} \tan\left[\arctan\left(\frac{a_i}{a_t}\right) + \frac{N-N_i}{N_f-N_i}\left(\arctan\left(\frac{a_f}{a_t}\right) - \arctan\left(\frac{a_i}{a_t}\right) \right) \right] \tag{7}$$

where a_t is the transitional crack length defined as a crack length for which $v_g = v_p$, i. e. $a_t = a_i(g/k_p)^{1/2}$. The number of cycles necessary to propagate a crack from the initial size (a_i) to the final size (a_f) is

$$N_f - N_i = \frac{a_t}{g a_i}\left[\arctan\left(\frac{a_f}{a_t}\right) - \arctan\left(\frac{a_i}{a_t}\right) \right] . \tag{8}$$

In case that

$$v_g = k_g a \tag{9}$$

The integration of crack growth equation (4) with (5) and (9) leads to

$$\frac{a}{a_f} = \frac{\dfrac{a_t}{a_f}}{\left(1+\dfrac{a_t}{a_f}\right)^{\frac{N-N_i}{N_f-N_i}}\left(1+\dfrac{a_t}{a_i}\right)^{1-\frac{N-N_i}{N_f-N_i}} - 1} \tag{10}$$

where a_t is again the transitional crack length defined as a crack length for which $v_g = v_p$. Here it is $a_t = a_i k_g/k_p$. The number of cycles to propagate a crack from initiation (a_i) to the end of fatigue life (a_f) is

$$N_f - N_i = \frac{1}{k_g}\left(\ln\left(1+\frac{a_t}{a_i}\right) - \ln\left(1+\frac{a_t}{a_f}\right) \right) . \tag{11}$$

The relative crack length a/a_f is plotted vs. relative number of cycles spent in the growth of a crack $n_r = (N-N_i)/(N_f-N_i)$ in Fig. 5 for two ratios of the initial crack length to the transitional crack length a_i/a_t. In both cases, in an appreciable interval of the fatigue life, the plot of the logarithm of the relative crack length vs. relative number of cycles is linear. This explains the

534

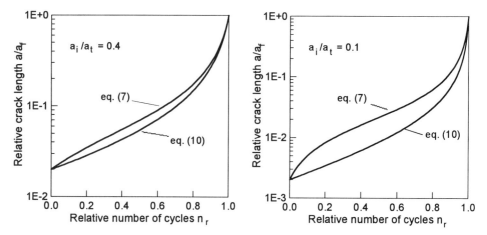

Fig. 5 Plot of the relative crack length vs. relative number of cycles for two values of a_i/a_t.

difficulty in distinguishing between these two cases on the basis of experimental data that have an inevitable scatter.

Equation (8) or eq. (11) gives the relation between the fatigue life and the crack growth rate when the crack reaches a dimension for which both contributions to the crack growth rate are equal. This gives us the possibility to predict fatigue life of a specimen on the basis of the early crack growth rate. If the number of cycles to crack initiation can be neglected, the fatigue life of a smooth body is primarily determined by the crack growth rate $(d \ln a/dN)_{a=a_i}$ in both cases. The proportionality constant is either $(arctg(a_t/a_i)-arctg\ (a_t/a_j))$ or $(\ln(1+a_t/a_i)-\ln(1+a_t/a_t))$. Therefore, the fatigue life can be evaluated from the initial crack growth rate. On the other hand, this relation gives the possibility to obtain the stress dependence of the rate of short cracks from the fatigue life curve of a material found on smooth specimens without a necessity to perform complicated measurements of the kinetics of short crack growth.

The authors appreciate the support of the present research by the Grant No. A2041704/1997 of the Grant Agency of the Academy of Sciences and by the Grant No. 106/97/1034 of the Grant Agency of the Czech Republic. They are also obliged to Dr. Jan-Olof Nilsson from Sandvik Steel for the provision of duplex steel.

REFERENCES:

1. Polák, J. (1991). *Cyclic Plasticity and Low Cycle Fatigue Life of Metals.* Academia-Elsevier, Praha-Amsterodam..
2. Miller, K. J. and de los Rios, E. R. (Eds.) (1986). *The Behaviour of Short Fatigue Cracks.* Institution of Mechanical Engineers, London.
3. McDowel, D. L. (1996) *Int. J. Fracture,* **80**, 103.
4. Polák, J. and Liškutín, P. (1990) *Fatigue Fract. Engng Mater. Struct.* **13**, 119.
5. Obrtlík, K., Polák, J., Hájek M. and Vašek A. (1997) *Int. J. Fatigue,* **19**, 471.
6. Polák, J., Obrtlík, K. and Vašek, A. (1997) *Mater. Sci. Engng,* **A234-236**, 970.

Low Cycle Fatigue and Elasto-Plastic Behaviour of Materials
K-T. Rie and P.D. Portella (Editors)
535

STEREOLOGICAL ANALYSIS OF FATIGUE SHORT CRACK
PROPAGATION IN ZIRCALOY-4

J. STOLARZ * and K.J. KURZYDLOWSKI **
* Ecole des Mines de Saint-Etienne, Centre SMS, URA CNRS 1884
F-42023 Saint-Etienne Cedex 2, France
** Politechnika Warszawska, Instytut Inzynierii Materialowej
PL-00-661 Warszawa, Poland

ABSTRACT

Short crack propagation in low cycle fatigue in Zircaloy-4 has been investigated. Efforts were focused on crack propagation in the bulk. Experiments were carried out on smooth specimens under symmetrical push-pull loading at imposed plastic strain amplitude under vacuum (LCF) and in iodised methyl alcohol (LCCF). Short crack populations were first investigated at surface using direct observations. In the second stage, short crack propagation in the bulk was studied using stereological methods on vertical sections of samples tested until 75% N_F. A distribution of crack depths could thus be determined and correlated with that of surface lengths. The mean shape factor of short cracks is slightly higher in tests carried out under vacuum than in iodised CH_3OH. The difference is explained in terms of crack nucleation and propagation modes. The results obtained indicate clearly that short crack propagation is not limited to the surface layer of the material and that in case of coalescence the parameter which controls the kinetics of damage is crack depth rathet than crack extension at surface.

KEYWORDS

Low cycle fatigue, zircaloy, short crack propagation, shape factor of cracks, multicracking.

INTRODUCTION

The damage evolution in metallic materials under cyclic loading generally evolves in three stages: (i) modification of dislocation structures leading directly or indirectly to the short crack nucleation (microscopic scale), (ii) short crack growth, nucleation of new short cracks, interactions between cracks (mesoscopic scale), (iii) fatal crack propagation and failure (macroscopic scale). The intermediate stage in which the damage evolution is strongly microstructure dependent extends in most cases over at least 80% of the fatigue life N_F. Crack propagation in this stage can be decribed as a succession of "elementary events" which are crack nucleation and crossing of microstructural barriers. The second event is responsible for a strongly discontinuous character of crack propagation at the mesoscale. The ratio between activation energies of crack nucleation and of crossing of barriers determines the ability of a system to develop multiple cracking (multicracking) during cyclic loading as it is generally observed in polycrystalline materials in low cycle regime. Under assumption of a purely superficial character of short cracks, their surface extension was considered to provide the essential part of the driving force for propagation at the mesoscale [1]. Consequently, crack

536

coalescence at surface which occurs very frequently in presence od high surface crack densities (up to 200 per sq.mm) was supposed to accelerate strongly the kinetics of damage. Recently, Stolarz pointed out [2] that if crack propagation takes place simultaneously at surface and in the bulk, the importance of the crack coalescence should be much lower from that postulated in models involving superficial nature of short cracks. Consequently, surface crack density does not seem to be a sufficient criterion for evaluation of the degree of damage.

Some experimental data collected by means of confocal microscopy on several LCF tested c.c. and c.f.c. alloys indicate that short crack propagation at surface and in the bulk take place simultaneously [3]. Indeed, the shape factor of short cracks (between 0,65 and 1,0) remains approximately constant during propagation of mesoscopic cracks with surface lengths between 10 and 1000μm [3]. However, no experimental results on short crack propagation in the bulk in h.c. materials are available. On the other hand, the use of confocal microscopy is strictly limited to the case of planar cracks which is not necessarily true if crack propagation in the bulk is of crystallographic nature.

The aim of this study is to introduce a simple method for the evaluation of short crack depths from vertical sections of multicracked specimens. This method is applied to study the LCF crack propagation in Zircaloy-4 under vacuum and in iodised methyl alcohol.

STEREOLOGICAL ANALYSIS OF SHORT CRACK GROWTH IN THE BULK

On vertical sections of multicracked specimens, fatigue short cracks appear as lines propagating in the bulk from the sample surface (Fig.1c). In low cycle fatigue, surface densities of short crack are generally very high and numerous crack are cut by the testing plane (Fig.1a). In order to estimate the relationship between the surface crack density and the number of cracks per unit length of a vertical testing line (linear crack density), an individual crack is considered (Fig.1b). The probability p of intersection between a vertical testing line and a linear crack segment characterized by the length ℓ and oriented at $\pi/2-\alpha$ versus vertical direction is equal to:

$$p = \frac{\ell \cdot \cos\alpha}{w} \tag{1}$$

For a population of short cracks such as represented in Fig.1a, the number P_L of cracks per unit length of a vertical testing line is given by:

$$P_L = D \cdot \frac{\overline{\ell \cdot \cos\alpha}}{L} \tag{2}$$

where D represents the surface density of short cracks and L the total length of testing lines. For a random distribution of crack orientation at surface, the equation (2) reduces to:

$$P_L = \frac{\sqrt{2} \cdot D \cdot \bar{\ell}}{2L} \tag{3}$$

In which concerns crack extension in the bulk, the depths b observed on vertical sections (Fig.1c) do not represent necessarily maximal ones. If one assumes the semi-eliptical shape of cracks, the maximal crack depths d can be estimated from:

$$E(d) = \frac{4 \cdot b}{\pi} \tag{4}$$

Consequently, the mean crack depth can be calculated from the population of cracks present on vertical sections and it can be compared with the mean surface length giving the average shape factor of short cracks defined as $S=2d/\ell$.

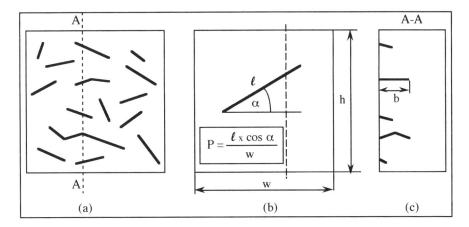

Fig.1. (a) Surface aspect of a multicracked specimen; (b) Probability of intersection of a surface crack by a vertical testing line; (c) Schematic representation of short cracks on a vertical section.

The distribution of crack depths can be estimated using the method of random oriented lines (ROL). According to the basic relationships of stereology, the length of a linear element of microstructure per unit surface (L_A) is related to the number (P_L) of intercepts between randomly oriented testing lines and the element considered using the following relationsip [4]:

$$L_A = \frac{\pi}{2} \cdot P_L \tag{5}$$

The estimation of L_A made at different depths leads to a relationship between the total crack length and the distance from surface. A comparison between this distribution and that of surface crack lengths can thus give an idea about the evolution of the shape factor during crack growth.

EXPERIMENTAL PROCEDURE

Low cycle fatigue tests have been carried out on smooth samples of semi-recrystallized Zircaloy-4 (supplied by FRAGEMA - Lyon, France) containing (in wt%): 1,5%Sn, 0,20%Fe and 0,11% C. The microstructure is composed of equiaxed grains of 20-25 μm diameter. The alloy exhibits a pronounced crystallographic texture. Low cycle fatigue tests have been performed in symmetrical tension-compression under plastic strain control ($\Delta\varepsilon_p/2=2 \cdot 10^{-3}$) and at a constant strain rate ($d\varepsilon/dt=10^{-3}$ s^{-1}) in two environmental conditions: under vacuum ($<10^{-3}$ Pa) and in iodised methyl alcohol (0,1 wt% I_2). The second environment is considered to simulate chemical conditions in pressurized water reactor (PWR) [5]. In the first stage, fatigue life N_F and surface propagation of short crack densities have been investigated. After that,

538

fatigue tests have been carried out in both environments during respective number of cycles corresponding to 75% N_F. Short crack propagation in the bulk was then analysed on vertical sample sections after mechanical polishing and cleaning.

RESULTS

Fatigue Life and Surface Propagation of Short Cracks

Under imposed testing conditions, the fatigue life of Zircaloy-4 is 4000 cycles under vacuum and approximately 220 cycles in iodised CH_3OH.

Under vacuum, first short cracks (ℓ~20μm) nucleate at slip bands at 25% N_F. Between 25% and 95% N_F, numerous short cracks are nucleated. Simultaneously, propagation across grain boundaries leads to the formation of a regular network of cracks over the whole surface (Fig.2).

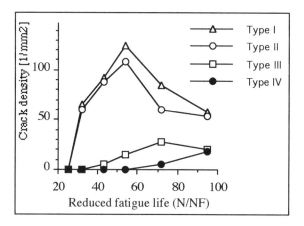

Fig.2. Evolution of surface crack densities in Zircaloy-4 tested under vacuum at $\Delta\varepsilon_p=2 \cdot 10^{-3}$ and at $d\varepsilon/dt=10^{-3}$ s^{-1}, $N_F=4000$ cycles.

Type I cracks are shorter from the mean grain size ($\ell_I \sim 20$μm) while type II ones are formed after crossing of one or two microstructural barriers (grain boundaries). Thieir mean length ℓ_{II} is thus approximately equal to 60μm. At 75% N_F, these two classes of cracks represent over 80% of the total crack population. The limit between type III and type IV cracks has been fixed at 250μm. All cracks observed at the metal surface propagate according to the same transgranular mechanism and their orientation is determined by that of corresponding grains. At this stage, the population of short cracks can be characterized by surface density D of about 170mm^{-2} and by mean surface crack length close to 60μm.

In CH_3OH+I_2, cracks nucleate exclusively at grain boundaries after approximately 10% N_F [6]. Such intergranular cracks propagate very rapidly and reach lengths of 500μm after 35% N_F. At this stage, a network of cracks ($\ell > 500$μm) perpendicular versus sample axis and distant about 100μm one from another, is observed [6]. Further propagation of cracks at surface occurs mainly through coalescence and after 75% N_F cracks reach lengths between 1 and 3mm.

Short Crack Propagation in the Bulk

Cracks present on vertical sections are composed of linear segments 10 to 30µm long. The average direction of all cracks (LCF and LCCF) is perpendicular to the surface even if individual segments have random orientations. Maximal crack penetration is close to 40µm under vacuum and it reaches 250µm in iodised CH_3OH. Crystallographic character of the crack growth is fully conserved in both cases.

Table 1. Summary of experimantal data for crack propagation under vacuum and in the bulk

Environment	Mean surface length (µm)	Surface density (mm^{-2})	Linear density theor.	exper.	Mean depth d (µm)	Shape factor S=2d/ℓ
Vacuum	60	170	7,2	8,0	34	1,13
CH_3OH+I_2	>1000	12	8,5	6,5	154	<0,31

Linear crack densities observed are close to 8,5mm^{-1} (LCF) and 6,5mm^{-1} (LCCF). These densities correspond to theoretical ones calculated from the equations (2) and (3) using respective mean surface crack lengths and densities (Tab.1). Mean crack depths: 27µm in LCF and 123µm in LCCF, are calculated from the data of Fig.3. The estimations of real crack depths (eq.(4)) allow to calculate the mean crack shape factor S defined as the ratio of crack depth d and half surface length ℓ/2 (Tab.1). The shape factor of 1,13 found in LCF is higher from those reported in [3] for individual cracks in cubic alloys (S≤1). In LCCF (intergranular cracking, only long cracks detected) the shape factor (S<0,31) is much smaller than in LCF.

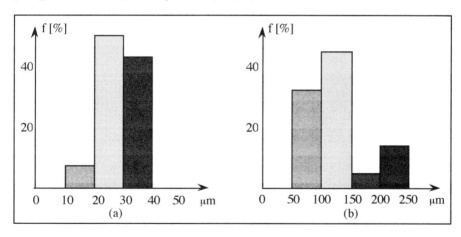

Fig.3. Distribution of LCF crack depths in Zircaloy-4 at 75% N_F at $\Delta\varepsilon_p$=2·10^{-3} and at dε/dt=10^{-3} s^{-1} measured on vertical sections: (a) LCF under vacuum; (b) LCCF in CH_3OH+I_2.

DISCUSSION

A comparison of theoretical and experimental linear crack densities (Tab.1) indicates that almost all short cracks propagte simultaneously at surface and in the bulk.
In LCF tests under vacuum, maximal crack depth observed on vertical sections do not exceed 40µm which correspond to one or two barriers (grain boundaries) crossed while some cracks reach surface lengths higher than 250µm. It seems that the longest surface cracks have not reached depths corresponding to the mean shape factor, at 75% N_F. A possible explanation

involves crack coalescence. Indeed, with high surface densities observed, long cracks forme mainly through coalescence. In such cases, surface propagation is supposed to be stopped until the crack reaches its equilibrium shape factor [2]. Therefore, just after coalescence, shape factors can thus be considerably lower from the ones observed on short cracks

Anyway, the mean shape factor observed in LCF, equal to 1,13, is considerably higher from those found in cubic alloys using confocal microscopy [3]. Since, as we have seen, short cracks in the bulk follow the same crystallographic path as at surface, it is probable that results obtained by confocal microscopy are strongly underestimated.

In LCCF (CH_3OH+I_2), intergranular propagation of individual crack segments is fully conserved in the bulk, at least until the maximal depth observed ($250\mu m$). The shape factor calculated at 75% N_F is considerably lower from that found in LCF under vacuum (Tab.1). It seems that coalescence effects contribue to the overall decrease of the shape factor.in this case. Indeed, the crack network in LCCF is created mainly through coalescence of intergranular crack segments of several hundreds of microns [6]. In such conditions, it cannot be excluded that the depths measured on vertical sections correspond to those of segments of coalescing cracks. On the other hand, the radical difference in crack propagation modes: transgranular in LCF and intergranular in LCCF, can influence the shape factor value as well. Further studies of short cracks in Zircaloy-4 at earlier stages of LCCF damage are supposed to give a solution to this problem.

CONCLUSION

LCF short crack propagation in Zircaloy-4 has been investigated on vertical sample sections.
The propagation of short cracks in the bulk is found to concern the whole population of short cracks. The results obtained provide thus a supplementary argument against the theories based on a purely superficial nature of short cracks in LCF.
The shape factor of short cracks in LCF is much higher from the one measured by confocal microscopy [3]. It seems that this results expresses the risk of underestimation of crack depths when direct observations from the surface are applied which is mainly due to the non-planar crack propagation in the bulk as shown in the present study.
Finally, the effects of crack coalescence are found to play an important role in the damage evolution both in LCF and in LCCF. Nevertheless, the damage acceleration through coalescence appears less pronounced than in LCF models based exclusively on surface cracking in which surface extension provides the main contribution to the driving force for crack propagation.

REFERENCES

1. Bataille, A. and Magnin, T. (1994). *Acta Metall. Mater.* **42**, 3817.
2. Stolarz, J. (1997). *Mat. Sci. Eng.* **A234-236**, 861.
3. Varvani-Farahani, A., Topper, T.H. and Plumtree, A. (1996).
   Fatigue Fract. Eng. Mater. Struct. **19**, 1153.
4. Kurzydlowski, K.J. and Ralph, B. (1995).
   In: *The Quantitative Description of the Microstructure of Materials*. CRC Press.
5. Stolarz, J. and Beloucif, A. (1997). In: *Corrosion-Deformation Interactions (CDI'96)*.
   T. Magnin (Ed.), The Institute of Metals, pp.117-126.
6. Stolarz, J. (1996). In: *Proceedings of the Sixth International Fatigue Congress*,
   G. Lütjering and H. Nowack (Eds.),Pergamon, Vol.I, pp.667-672.

Low Cycle Fatigue and Elasto-Plastic Behaviour of Materials
K-T. Rie and P.D. Portella (Editors)
541

A STOCHASTIC MODEL FOR PLURAL FRACTURE

S. IGNATOVICH and F.F. NINASIVINCHA SOTO
Kyiv International University of Civil Aviation,
KIUCA, 1 Kosmonavt Komarov av., 252601 Kyiv-58, Ukraine

ABSTRACT

Plural destruction of a loaded materials surface is considered at which two main damage processes - nucleation and propagation of small dispersed cracks will be simultaneously realized. Nucleation of cracks was simulated in the form of Poisson point process with constant and dependent from time intensity. The crack propagation with random rate was described by linear function of length increment from time. Various types of crack-growth rate distributions are considered: equil-probability, gamma-distribution, including exponential distribution. Derived in so doing the crack length statistical distributions are similar and in a indicated kind can be described by generalized exponential dependence.

KEY WORDS

Plural fracture, small cracks, crack nucleation, crack propagation, crack length statistical distribution.

INTRODUCTION

General manifestation of strength exhaustion processes (damaging) is the alternation of plural destruction stages [1]. These stages will be realized in hierarchical scales (from microscopic to macroscopic) through the self-similarity mechanism - continuous nucleation and growth of flaws, their coalescence and formation of flaws of higher dimensional level. In this case main factors of damaging are quantity and dimensional parameters of flaws which can be generalized by the appropriate statistical distributions.

According to experimental data for materials with various properties and under a variety of loading kinds the flaw-size distribution is described by negative exponential or hyperbolic functions. This is true for fatigue cracks [2,3], pores under creep [4], cracks under impact loading [5]. Exponential type of crack size distribution is postulated in early studies on the statistical fracture mechanics [6]. The fundamental peculiarity of plural fracture is scaled invariance of flaw-size distributions curves (scaling) [1,7,8]. Some authors ascribe this peculiarity to similarity property of fractal clusters [9], however theoretical substantiation of this manifestation is not elaborated.

If the ultimate state of a machine component is conditioned by formation of crack with inadmissible length l_* and crack length statistical distribution at the fixed time t is given by probability density $f_l(l;t)$, the function of the life (t_f) distribution can be defined as

$$F_t(t_f) = \int_{l_*}^{\infty} f_l(l;t_f)dl \qquad (1)$$

Consequently, information on size non-homogeneity of flaws which is represented as crack length distribution has governing significance for solution of extensive class of topical problems of constructions reliability and materials strength physics.

The objective of work is development of stochastic model for formation of size non-homogeneity of dispersed flaws and reception of theoretically justified small crack length statistical distribution, as well as experimental observations of surface microcracks behaviour under LCF-tests of nickel-base superalloys.

EXPERIMENT

Two nickel-base superalloys of dissimilar ductility: ЭИ698ВД ($\phi = 20\%$) and ЖС6УВИ ($\phi = 8\%$) which are used for making of gas turbine disks and blades of aviation engines have been studied. Were used specimen of 5 mm diameter. Surface of specimen was carefully polished. The loading conditions: interrupted tension to prescribed values of plastic deformation and interrupted LCF-tests with asymmetry coefficient $R = \sigma_{min} / \sigma_{max} = 0$. At intervening points of loading duration the condition of specimen surface was tested by visible dye penetrate testing. With a flaws, specimen were examined under the microscope. Each crack was located at the surface, the crack quantity and length were fixed. Crack samples were performed at the areas with homogeneous dispersing of flaws. Sample sizes were representative and they were varied from 50 to 250 units.

For tension the first microcracks (20...60 μm) on surface of specimens were found at plastic deformation less then 0.2ϕ for plastic alloy and 0.1ϕ for less plastic alloy. As the plastic deformation increases the crack density on surface increases non-linearly (Fig. 1a).

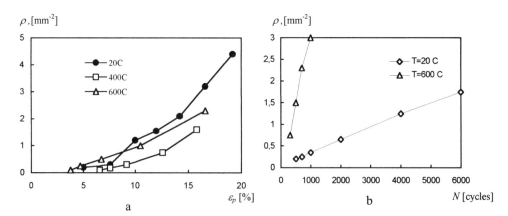

Fig. 1. Evolution of small surface crack density (ρ) under tension (a) and under LCF-tests (b) (σ_{max} = 850 MPa for T = 600°C; σ_{max} = 1150 MPa for T = 20°C) for ЭИ698ВД alloy.

Under low-cycle loading the plural destruction processes proceed intensively. The first small cracks nucleate at the early stage of total life - $(0,02...0,1)N_f$. The surface crack density linearly increase with number of loading cycles (Fig. 1b).

As expected, the considerable non-homogeneity of small crack rate is observed. In each sample for the majority of flaws the relatively small values of rate in the interval of realized quantities are typical. The crack growth rate distribution is close to exponential distribution (Fig. 2).

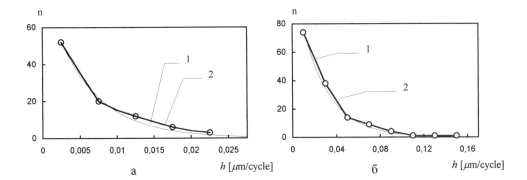

Fig. 2. Distributions of crack quantity over growth rate for LCF-tests of the ЭИ698ВД alloy: $T = 20°C$, $\sigma_{max} = 1150$ MPa, $N = 500$ cycles (a) and $T = 400°C$, $\sigma_{max} = 980$ MPa, $N = 2000$ cycles (b). 1 - design curves for exponential distribution, 2 - empirical distributions.

MODELING OF SIZE NON-HOMOGENEITY

The crack nucleation intensity μ and crack growth rate h are main parameters of model. The cracks nucleation is a Poisson point process. Each crack grows steady (cracks coalescence is not taken into account) and h is random quantity ($h \in [0, \infty)$).

The probability distribution function (PDF) for crack length (l) at the fixed time t is determined as

$$F_l(l;t) = \int_0^t \mu(\tau) F_h\left(\frac{l}{t-\tau}\right) d\tau \tag{2}$$

where $F_h(\cdot)$ is PDF for crack rate.

The formation of the cracks length non-homogeneity was simulated under various variants of cracks nucleation and growth. From (2) the following expressions for crack length probability density were derived

for $\mu = const$:
$$f_l(l;t) = t^{-1} \int_{l/t}^{\infty} h^{-1} f_h(h) dh \tag{3}$$

for $\mu(t) = \beta t$:
$$f_l(l;t) = (2/t)\left[\int_{l/t}^{\infty} h^{-1} f_h(h) dh - (l/t)\int_{l/t}^{\infty} h^{-2} f_h(h) dh\right] \tag{4}$$

where $f_h(h)$ is crack rate PD; β is coefficient.

From formulae (3) and (4) the crack length probability density were determined for various $f_h(h)$:

equi-probability distribution:

$$f_h(h) = 1/h_m; \qquad h \in [0, h_m], \qquad m_h = h_m/2 \qquad (5)$$

gamma-distribution:

$$f_h(h) = \frac{v^k h^{k-1}}{\Gamma(k)} \exp(-vh); \quad h \in [0, \infty); \qquad m_h = k/v \qquad (6)$$

where h_m is the largest value of parameter h; m_h is mathematical expectation of h; v, k are parameters of distribution; $\Gamma(\cdot)$ is gamma-function.

When $k = 1$ from the formula (6) follows exponential distribution of crack growth rate that corresponds to experimental data (Fig. 2). For more complete scope of possible types of h distributions the variants $k = 2$ for gamma-distribution (6) and equi-probability distribution was considered. The crack length statistical distributions were determined by integration of expressions (3) and (4) for various kinds of h distribution which were set by the formulae (5) and (6).

The crack length distributions derived in simulation differ in form of writing (Table 1). For comparability of results these distributions are represented as distribution of normalized parameter $\lambda = l/m_l$, where m_l is mathematical expectation of crack length. The design values of all functions $f_l^{(i)}(\lambda), (i = 1, \ldots, 6)$ submitted in Table 1 are in close agreement and at $\lambda > 1,5$ they can be satisfactorily approximated by generalizing function of a exponential kind. It testifies to similarity of dispersed crack length statistical distributions at fixed t.

Table 1. Crack length non-homogeneity modeling results

Crack rate probability density	Probability density for parameter λ	
	$\mu = const$	$\mu(t) = \beta t$
$f_h(h) = 1/h_m$	$f_l^{(1)}(\lambda) = \frac{1}{4} \ln\left(\frac{4}{\lambda}\right), \ \lambda \in [0, 4)$	$f_l^{(4)}(\lambda) = \frac{1}{3} \ln\left(\frac{6}{e\lambda}\right) + \frac{\lambda}{18}, \qquad \lambda \in (0, 6]$
$f_h(h) = v e^{-vh}$ $(k = 1)$	$f_l^{(2)}(\lambda) = \frac{1}{2} E_1\left(\frac{\lambda}{2}\right)$	$f_l^{(5)}(\lambda) = \frac{2}{3}\left[\left(1 + \frac{\lambda}{3}\right) E_1\left(\frac{\lambda}{3}\right) - \exp\left(-\frac{\lambda}{3}\right)\right]$
$f_h(h) = v^2 h e^{-vh}$ $(k = 2)$	$f_l^{(3)}(\lambda) = \exp(-\lambda)$	$f_l^{(6)}(\lambda) = \frac{4}{3}\left[\exp\left(-\frac{2\lambda}{3}\right) - \frac{2\lambda}{3} E_1\left(\frac{2\lambda}{3}\right)\right]$

$E_1(\cdot)$ - integral-exponential function, $E_1(z) = \int\limits_z^\infty x^{-1} \exp(-x) dx$

Carried out experimental studies of surface destruction of nickel-base superalloys under variety of loading kinds shown satisfactory conformity of theoretical and empirical surface crack length distributions for normalized parameter λ (Fig. 3).

It indicatives of similarity of crack length distribution and it strengthens the hypothesis of damage self-similarity.

$f_l(\lambda)$

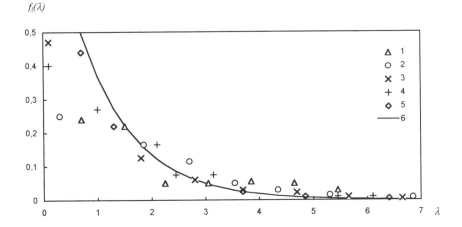

Fig. 3. Crack length distribution for ЭИ698ВД under low-cycle loading: 1 - $T = 20°C$, $\sigma_{max} = 1000$ MPa, $N = 7000$ cycles; 2 - 20°C, $\sigma_{max} = 1100$ MPa, $N = 1500$; 3 - 400°C, $\sigma_{max} = 980$ MPa; $N = 2000$; 4 - 400°C; $\sigma_{max} = 980$ MPa; $N = 4000$; 5 - for ЖС6УВИ under tension: $T = 800°C, \varepsilon_p = 0,35\%$; 6 - approximated by exponential distribution $f_l^{(3)}(\lambda) = \exp(-\lambda)$.

REFERENCE

1. Barenblatt G.I., Botvina L.R. (1986) *Physic-chemical mechanics of materials* **1**, 57.
2. Cheng A.S., Laird C. (1981) *Fatigue Eng. Mater. Structur* **4**, 4, 343.
3. Suh C.M., Yuuki R. and Kitagawa H. (1985) *Fatigue Fract. Eng. Mater. Stractur* **8**, 2, 193.
4. Chen I.W., Argon A.S. (1981) *Acta met.***29**, 7, 1321.
5. Seaman L., Curran D.R. and Shockey D.A. (1976) *J.Applied Physics* **47**, 11, 4814.
6. Fisher J.C., Hollomon J.H. (1947) *Metals Thechnology* **14**, 5, 1.
7. Lebowitz J.L., Marro J. and Kalos M.H. (1982) *Acta met.***30**, 1, 297.
8. Hart E.W., Solomon H.D. (1973) *Acta met.***21**, 3, 295.
9. Mandelbrot B.B. (1982) *The fractal geometry of nature*. Freeman , San-Francisco

Low Cycle Fatigue and Elasto-Plastic Behaviour of Materials
K-T. Rie and P.D. Portella (Editors)
547

FATIGUE CRACK GROWTH AND FATIGUE LIFE ESTIMATION
OF NOTCHED MEMBERS IN A SHORT CRACK RANGE

D. KOCAŃDA, S. KOCAŃDA and H. TOMASZEK
Department of Mechanical Engineering, Military University of Technology,
01-489 Warszawa, Poland

ABSTRACT

The paper presents a simple method for prediction the short fatigue crack growth rate and the lifetime estimation of a notched member in short crack range. In order to model the crack behaviour there was assumed an exponentially decreasing stress concentration factor over the distance ahead of the notch root. The crack propagation law follows the Paris formula. The viability of the method has been verified for the cracks developed at the notches in a medium carbon steel specimens under bending.

KEYWORDS

Notch, short crack, crack growth rate , crack growth model, fatigue life, medium carbon steel.

INTRODUCTION

Theoretical and experimental analysis of the notched components and structures have been continuously developed for sixty years in order to better describe and include the influence of the notch stress field in the fatigue calculations. Simple approaches originated from the thirtieth years and then the Neuber considerations are modified and adapted in the present time as well for resolving the mentioned problem. In the last decade the attempting are focused on an including the short crack range in the notched problem. The references considered that item set rather a little contribution on the fatigue crack growth at the notched components, generally. On the other hand, the importance of microstructurally short crack range in crack growing period and prediction the fatigue life of the components is well known.
The initiation and growth of fatigue cracks at notches covert the formation of surface microcracks in high strain field, crack propagation in a changing stress field to long cracks in an elastic continuum. This wide problem is not readily rationalised by linear elastic fracture mechanics (LEFM). Cracks at small or sharp notches start to grow at a rate faster than predicted by the LEFM model and on the other hand, can lead to formation of non-propagating cracks [1] [2]. Smith and Miller [3] estimated the notch contribution on the growth of a crack. On their approach the notch influences the crack over a distance of approximately $0.13\sqrt{D\rho}$ ahead of the

notch root of D – depth and ρ - radius of the notch root. Convenient expressions for the stress intensity factor of a short crack developing at a notch was derived by Schijve [4].

Yates [5] proposed to use either the conventional geometrical correction or an additional notch correction terms when the stress intensity factor is formulate for cracks at notches in tension. In the analysis of the fatigue at notches is important to be able estimate the extent of notch root plasticity. Hammouda et. al. [6] focused on the behaviour of short cracks within the plastic zone of a notch. Ahmad and Yates [7] developed an elastic-plastic model for predicting the fatigue crack growth rates at notches in the range of short and long cracks. The strain distribution ahead of an elliptical notch originated from the strain energy method was formulated by Glinka [8]. The modified stress concentration factors separately for short and long cracks at notches were given in [9]. In [10] was presented the approach to the description of the crack growth at notches including the effect of crack closure.

The short review of the papers devoted the fatigue crack growth in the notch field indicates that the prediction of crack growth requires further considerations of the factors such as notch contribution to the crack length, modification of the expression for stress intensity factor, new conditions for crack growth taking the distribution of strain ahead of the notch into account. Experimental results associated with fatigue at notches show initially high crack growth rates that decrease and then increase as the crack depth grows away from the notch.

In the present paper the consideration of the notch influence on the short crack depth behaviour is undertaken and then the fatigue lifetime in the range of surface short cracks is estimated.

DESCRIPTION OF THE METHOD

Here was assumed that the notch influences on the crack growth covering the distance of l_o. Within this distance the short crack depth is grown and its length is a ($a \leq l_o$). Beyond this region the crack grows as the long crack of length $l = D + a$ where D is the depth of the notch and a is the length of real fatigue crack. The short and long crack depth growth rates are described by the Paris formula with the power exponent $m = 2$. The elastic stress field at the notch root is characterised by the stress concentration factor K_T derived by Neuber [11]. For tension and bending the factor K_T has the form:

$$K_T = \frac{\sigma_{max}}{\sigma_n} = 1 + 2\sqrt{D/\rho} = 1 + Q \tag{1}$$

and for torsion :
$$K_T = 1 + \sqrt{D/\rho} = 1 + Q \tag{1a}$$

where Q is the parameter of the notch stress field, σ_{max} , σ_n are the peak of stress field and the nominal stress, respectively and ρ is the radius of the tip of the notch.

In the case of torsion the normal stress σ is replaced by the shear stress τ .

According with the earlier assumptions the formula for the stress concentration factor becomes:

$$K_T = 1 + Q \exp\left(-B \cdot a / Q\right) \tag{2}$$

where B is the material constant [m^{-1}] and it expresses the decrease of the stress concentration factor per the crack length unity, crack length ahead of the notch root equals a.

In the range of short cracks the stress intensity factor K_{SC} gives the form :

$$K_{SC} = Y\sigma \cdot (\ 1 + Q \cdot e^{-B \cdot a/Q})\sqrt{\pi \cdot a} \qquad (3)$$

where Y is the conventional geometrical correction factor. Let assume that in the region close the notch root is valid an expandation :

$$e^{-B \cdot a/Q} = 1 - \frac{B \cdot a}{1!Q} + \frac{1}{2!}\left(\frac{B \cdot a}{Q}\right)^2 - \ldots = 1 - \frac{B \cdot a}{Q} \qquad \text{for } B \cdot a/Q < 1 \qquad (4)$$

Substituting this form in Eq. (3) gives:

$$K_{SC} = Y \cdot \sigma \cdot (\ 1 + (Q - B \cdot a)\ )\sqrt{\pi \cdot a} \qquad (5)$$

If crack depth reaches l_o the notch influence on the crack propagation disappears , it means that

$$Q - B\,l_o = 0 \qquad \Rightarrow \qquad l_o = Q/B \qquad (6)$$

Finally, the stress intensity factor becomes:

$$K_{SC} = Y_{SC} \cdot \sigma \cdot (1 + (Q - B \cdot a)\ )\sqrt{\pi \cdot a} \qquad \text{for } a \le l_o \quad \text{i.e. short crack range}$$

$$\qquad (7)$$

$$K_{LC} = Y_{LC} \cdot \sigma \cdot \sqrt{\pi \cdot l} \qquad \text{for } l > l_o \quad \text{i.e. long crack range}$$

The rate of crack growth is described by the Paris formula in whole range of crack propagation according with the assumptions making above:

$$\frac{da}{dN} = C_{SC} \cdot \left[Y_{SC} \cdot \sigma \cdot \sqrt{\pi \cdot a} + Y_{SC} \cdot \sigma \cdot (\ Q - B \cdot a\ )\sqrt{\pi \cdot a} \right]^2 \qquad \text{for } a \le l_o$$

$$\qquad (8)$$

$$\frac{dl}{dN} = C_{LC} \cdot \left[Y_{LC} \cdot \sigma \cdot \sqrt{\pi \cdot l} \ \right]^2 \qquad \text{for } l > l_o$$

Integration of the above equations separately in the range of short ($a_i \le a_d \le l_o$) and long cracks ($l > l_o$) leads to the result :

$$\int_0^{N_{SC}} dN = \int_{a_i}^{a_d} \frac{da}{a \cdot \left[Y_{SC} \cdot \sqrt{C_{SC}} \cdot \sigma \cdot \sqrt{\pi} \cdot (\ (1 + Q) - B \cdot a)\ \right]^2}$$

$$\qquad (9)$$

$$\int_0^{N_{LC}} dN = N_{SC} + \int_{l_o}^{l_d} \frac{dl}{l \cdot \left[\sqrt{C_{LC}} \cdot Y_{LC} \cdot \sigma \cdot \sqrt{\pi} \ \right]^2}$$

$$N_{SC}=\frac{1}{C_{SC}\cdot Y_{SC}^{2}\pi\cdot\sigma^{2}(1+Q)^{2}}\ln\left[\frac{((1+Q)-B\cdot a_{i})\cdot a_{d}}{((1+Q)-B\cdot a_{d})\cdot a_{i}}+\frac{B\cdot(1+Q)\cdot(a_{d}-a_{i})}{((1+Q)-B\cdot a_{d})\cdot((1+Q)-B\cdot a_{i}}\right] \quad (10)$$

$$N_{LC}=N_{SC}+\frac{1}{C_{LC}\cdot Y_{LC}^{2}\cdot\pi\cdot\sigma^{2}}\ln\left[\frac{l_{d}}{l_{o}}\right] \quad (11)$$

where a_d is the length of short crack depth, a_i is the initial crack depth length and l_d is the long crack depth.

Taking into consideration Eq. (10) you can obtain the expression for the expected short crack length a_{av} after N cycles :

$$a_{av}=\frac{a_{i}+\left((1+Q)^{2}-B\cdot a_{i}\cdot(1+Q)\right)\cdot a_{i}\cdot C_{SC}\cdot\pi\cdot\sigma^{2}\cdot N}{1-\left(B^{2}\cdot a_{i}-B\cdot(1+Q)\right)\cdot a_{i}\cdot C_{SC}\cdot\pi\cdot\sigma^{2}\cdot N} \quad (12)$$

Total fatigue life of the notched component is the sum of the lifetime N_{SC} in the short crack range plus the lifetime N_{LC} in the long crack range.

The viability of the discussed method has been verified using experimental data gained for medium carbon steel notched specimens fatigued under rotating bending.

EXPERIMENTAL VERIFICATION

A normalised medium carbon steel of composition (wt. %) : 0,48C, 0,66Mn, 0,24Si, 0,02P, 0.019S, 0,25Cu, 0,24Cr, 0,028Al, 0,03Mo and remainder ferrite was used for the fatigue tests. The mechanical properties of the steel were as follows: yield stress = 425 MPa, ultimate tensile stress = 690 MPa, elongation = 18%, reduction of area = 40% . Notched fatigue tests were performed on 10 mm of diameter round bars with circumferential notches. Three chosen profiles were of semicircular and the notch radius ρ = 3, 4 and 5 mm respectively. The depth of the notches were 1.5 mm or 2.5 mm. The stress concentration factors K_T were: 1.34, 1.28 and 1.24. The fatigue tests were carried out under load controlled rotating bending at 48 Hz of frequency in air and at room temperature. Surface crack growth was monitored by the replication method. The zones of crack initiation and characteristic features of short and long crack depth behaviour in the samples were established on the basis of further analysis of the replicas carried out with the help of TEM microscope. For the micro-observations the replicas were shadowed by platinum or chromium. The applied experimental method has introduced the limitation for the radius of the notches. Sharper notches were escaped in the tests because they made the difficulties at the surface replication.

The analysis of crack advance in the notched specimens derived the plots of crack growth rate against the crack length (Fig. 1a) and against cycle ratio N_i / N_f (Fig. 1b). N_i means current number of cycles and N_f– number of cycles to failure. For reflecting the crack behaviour in the short crack range in the specimens under examination were presented on the diagrams either the curves for the surface cracks a_s or for the cracks depth a_d. The solid lines that join the black marks represent experimental results, whereas the white marks relate to the calculated results obtained on the basis of expressions given above. The courses of the short crack growth rates in the specimens with different notch geometry tested under different stress amplitudes were similar. However, in the same examined specimens you can observe either decrease in surface short crack growth rate for one crack or an increase for another crack. The surface short crack growth has been described by the Paris formula with the power exponent m ≠ 2. The average short crack depth rates have been estimated on the basis of the micro-fractography analysis of

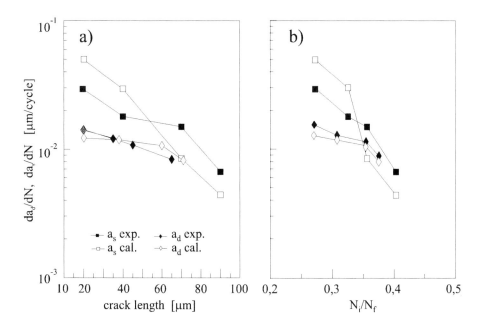

Fig. 1. Short crack growth rate against crack length (a) and cycle ratio N_i / N_f (b); see text for explanation.

the fracture surfaces. For example, in Fig. 1 the plots are referred to the cracks developed in the notched specimen ($K_T = 1,34$) tested under nominal stress amplitude $\sigma_a = 200$ MPa. The calculated fatigue lives N_{SC}^{cal} in the short crack range have been compared with the experimental ones N_{SC}^{exp}. Their values for the specimens with the stress concentration factor $K_T = 1,34$ are given in Table 1.

Table 1. Comparison between experimental and calculated lifetimes in the short crack range

Stress amplitude σ_a [MPa]	N_{SC}^{exp} [cycles]	N_{SC}^{cal} [cycles]
200	$4.20 \cdot 10^4$	$3.87 \cdot 10^4$
220	$3.00 \cdot 10^4$	$2.94 \cdot 10^4$

Agreement between the experimental results and the calculated ones can be recognised as pretty good both in terms of the crack growth description and the estimation of fatigue life in the range of short crack behaviour.

552

SUMMARY

The presented method simplifies the crack growth prediction in the notched components taking into account the short crack behaviour. In the range of short cracks it was assumed the exponential decreasing tendency of the stress concentration factor over the distance ahead of the notch root. The surface crack and the crack depth propagation law follow the Paris formula. The method derives the expressions allow to estimate the fatigue lifetimes in short and long crack range. The viability of the presented approach was confirmed in its description of short crack behaviour for notched specimens made of a medium carbon steel.

Acknowledgements – The research was partly supported by the EU within the COPERNICUS framework No. CIPA-CT-94 0194.

REFERENCES

1.   Tanaka, K.and Nakai, Y. (1983) *Fatigue Fract. Engng Mat. Struct.* **6**, 315.
2.   Yates, J.R. and Brown, M.W. (1987) *Fatigue Fract. Engng. Mater. Struct.* **10**, 187.
3.   Smith, R.A. and Miller, K.J. (1978) *Int. J. Mech. Sci.* **20**, 201.
4.   Schijve, J. (1982) *Fatigue Fract. Engng Mater. Struct.* **5**, 77.
5.   Yates, J.R. (1991) *J. Strain Anal.* **26**, 9.
6.   Hammouda, M.M., Smith, R.A. and Miller, K.J. (1979) *Fatigue Fract. Engng. Mater. Struct.* **2**, 139.
7.   Ahmad, H.Y. and Yates, J.R. (1994) *Fatigue Fract. Engng. Mater. Struct.* **17**, 651.
8.   Glinka, G. (1985) *Engng Fract. Mech.* **22**, 839.
9.   Xu, R.X., Topper, T.H. and Thompson, J.C. (1997) *Fatigue Fract. Engng. Mater. Struct.* **20**, 1351.
10.   Chien, C.H. and Coffin, L.F. (1998) *Fatigue Fract. Engng. Mater. Struct.* **21**, 1.
11.   Neuber, H. (1937). *Kerbspannungslehre. Grundlagen für genaue Spannungsrechnung.*

Low Cycle Fatigue and Elasto-Plastic Behaviour of Materials
K-T. Rie and P.D. Portella (Editors)
553

INITIATION AND PROPAGATION OF CRACKS AT SHARP NOTCHES UNDER CYCLIC LOADING

B. WURM, S. SÄHN and V.B. PHAM

*TU Dresden, Institut für Festkörpermechanik,
Mommsenstr. 13, 01062 Dresden, Germany*

ABSTRACT

Crack initiation and propagation from different sharp notches were observed in fatigue tests under 4-point-bending with two different materials using the replication method. A global equivalent loading parameter was used to describe the behaviour of small and macrocracks. The consideration of the stress and/or strain gradient in a characteristic length $(d^\star)$ is possible but seems to be different in LCF and HCF region.

KEYWORDS

fatigue, notch, crack initiation, crack propagation, small cracks

INTRODUCTION

Fatigue life consists of initiation and propagation of cracks. Therefore it is necessary to detect and to measure them with the required sensitivity.

Material- and manufacture-related defects in components can be simulated as sharp notches. The local stress strain state under loading is highly inhomogeneous and multiaxial. For the evaluation, a suitable equivalent loading parameter and the consideration of the stress and/or strain gradient for instance in a characteristic length $d^\star$ („Ersatzstrukturlänge"by *Neuber* [1]) like

$$\Delta B_{vm} = \frac{1}{d^\star} \int_{(d^\star)} \Delta B_v(r) \, dr \qquad (1)$$

can be used.

ΔB_v and $d^\star$ depend on the type of material and the material state and should be related to the microstructure. A general relation between $d^\star$ and characteristic dimensions of the microstructure like the grain size d could not yet be established, although in literature values for special types of materials can be found (for instance $d^\star = (1...5)d$). Generally the suitability of ΔB_v and the value of $d^\star$ have to be determined experimentally from investigations of strength phenomena like crack initiation lifes of different sharp notched components or propagation of small cracks at different inhomogeneous loadings [2].

MATERIALS

Table 1 and 2 show the chemical composition of the two different materials, which were used for the fatigue tests. In Table 3, there are the tensile properties and the parameters for the cyclic stress strain curve as described by the Ramberg-Osgood law Eq.(2) which were determined from tensile and incremental step tests. The average grain size of the X6CrNiTi18.10 is 58 μm, for the Metasafe 900 10 μm.

Table 1: Chemical composition of Metasafe 900 (in %)

C	Si	Mn	S	P	Cr	Ni	Mo	Cu	Al	V
0.29	0.12	1.78	0.037	0.012	0.29	0.07	0.06	0.19	0.041	0.1

Table 2: Chemical composition of X6CrNiTi18.10 (in %)

C	Si	Mn	S	P	Cr	Ni	Ti
0.044	0.44	1.1	0.008	0.028	17.1	9.03	0.38

Table 3: Material properties for static and cyclic behaviour

	$R_{p0.2}$ [MPa]	R_m [MPa]	A_g %	A_5 %	E [MPa]	K' [MPa]	n'
X6CrNiTi18.10	220.9	632.6	40.4	45.7	200000	2280	0.197
Metasafe 900	632	842	7.6	21.0	206300	1983	0.1924

$$\varepsilon_a = \frac{\sigma_a}{E} + \sqrt[n']{\frac{\sigma_a}{K'}}$$ (2)

SPECIMEN

Different notched specimens as in Fig. 1 were used for the tests. The surfaces, especially the notches, were polished with 1 μm lapping paste to allow an observation under a light microscope at 200 x magnification.

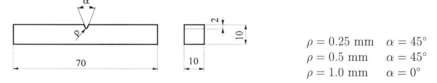

$\rho = 0.25$ mm $\alpha = 45°$
$\rho = 0.5$ mm $\alpha = 45°$
$\rho = 1.0$ mm $\alpha = 0°$

Fig. 1: Specimen

EXPERIMENTAL PROCEDURE

Zero-tension fatigue tests were performed with servohydraulic testing machines (INSTRON, Mayes) under 4-point-bending to determine fatigue lifes in the LCF region. Test frequencies were set dependent on the loading and varied between 0.15 and 20 Hz. In the HCF region a resonance machine was used where the crack growth data is derived from compliance changes.

CRACK LENGTH MEASUREMENT

In the LCF tests, short and long surface cracks were detected and measured with the replication method. It is an indirect optical technique where the trace of the crack is reproduced exactly on a cellulose acetate sheet at given intervals over the test period. The plastic deformation is high and represents a real limit for the identification of initiated small cracks. The X6CrNiTi18.10 has a large plastic range which makes crack detection difficult. In the tests with the Metasafe 900 such problems did not occur. The cracks at the examined surfaces start at different lifetimes in the notch root and grow dependent from each other. The crack fronts are more straight in the region of long cracks and at low strains. Under high loads, phenomenons like crack branching, multiple cracking, crack path changing etc. were observed more frequently than at low loadings (Fig. 2, 3). Their consideration in the value of the crack length has to be defined. Dependent on the loading cracks of 20 μm could be detected and measured exactly. Smaller cracks can be identified but the definition of crack length is uncertain.

In the HCF region crack growth data was evaluated through continuously monitoring the period of resonance vibrations. An increase of the specimen compliance corresponding to a fictitious through-thickness crack of an equivalent length $a_i = 3...10$ μm proved to be reliably measurable.

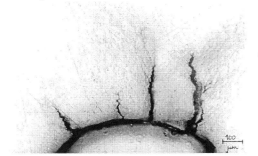

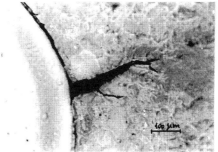

Fig. 2: Metasafe 900, $\rho = 0.5$ mm, $a = 145...387$ μm

Fig. 3: X6CrNiTi18.10, $\rho = 0.5$ mm, $a = 319$ μm

RESULTS AND DISCUSSION

Crack initiation at sharp notches

The experimental results for the LCF tests are shown in Fig. 4 for the Metasafe 900 and in Fig. 5 for the X6CrNiTi18.10 with a defined $a_i = 40$ μm. Separate curves for the different notched specimens can be seen. The X6CrNiTi18.10 is very tough, so there is little difference between the curves for tests with the notch radii $\rho = 0.25$ mm and $\rho = 0.5$ mm. The plastic deformation is high and the ΔB_{vm} (for instance $\Delta \varepsilon_{vm}$) have to be calculated for an elastic-plastic material law numerically or by using an approximation. Using $\Delta \varepsilon_v = \Delta \varepsilon_1$ for plain stress in the ligament and Eq. (1)

$$\Delta \varepsilon_{vm} = \frac{2\sigma_{Na}\alpha_k}{d^\star E}\sqrt{\frac{\rho}{\rho + 2d^\star}}(d^\star(1 - \nu) - \nu(\rho - \sqrt{\rho}\sqrt{\rho + 2d^\star})). \tag{3}$$

It is possible to describe the LCF crack initiation data derived from different sharp notches as follows

$$(\Delta\varepsilon_{vm} - \Delta\varepsilon_D)^n N_i = K_1 \qquad (4)$$

K_1, n and $\Delta\varepsilon_D$ are material constants (Table 4). N_i is the number of cycles until a_i has been initiated. Fig. 6 and 7 show the results for $\Delta\varepsilon_{vm} = f(N_i)$.

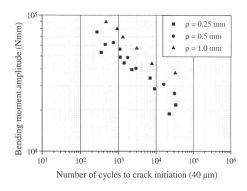

Fig. 4: Metasafe 900, Bending moment versus N_i

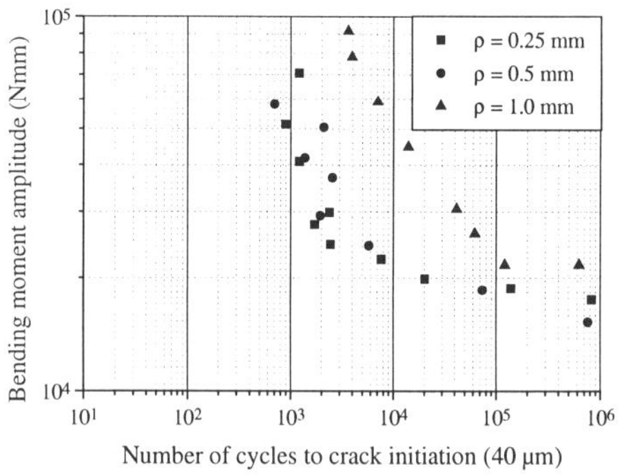

Fig. 5: X6CrNiTi18.10, Bending moment versus N_i

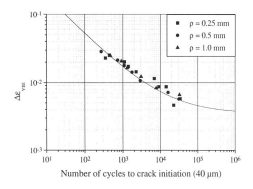
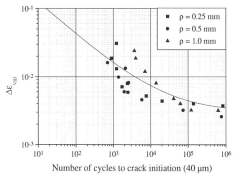

Fig. 6: Metasafe 900, $\Delta\varepsilon_{vm}$ versus N_i

Fig. 7: X6CrNiTi18.10, $\Delta\varepsilon_{vm}$ versus N_i

Table 4: Constants of Eq. (4)

	$d^\star$ (LCF)	$d^\star$ (HCF)	n	K_1	$\Delta\varepsilon_D$
Metasafe	200 μm	17 μm	1.8	0.449	0.35 %
X6CrNiTi18.10	300 μm	108 μm	2.08	0.125	0.3 %

The description of the results for the Metasafe 900 with Eq. (3) and (4) is better than for the X6CrNiTi18.10. It is difficult to consider the large deformation in the notch root correctly.

From former studies in the HCF region $d^\star$ was assumed to be constant for a material. These investigations show that results in the LCF region can be better described with an higher value of $d^\star$. Further examinations are necessary to clarify whether a formulation of $d^\star \neq const.$ is advantageous for life estimation or whether using the $d^\star$ calculated from HCF data is sufficient because crack initiation is more important in this part of life.

Crack propagation

Figure 8 shows crack opening profiles received from the replicas which were taken at mean load.

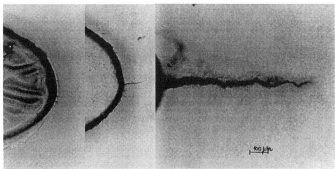

Fig. 8: X6CrNiTi18.10, Sp. 5, $M_{ba} = 22406$ Nmm, $N_1 = 2500$ $a_1 = 10$ μm, $N_2 = 9800$ $a_2 = 183$ μm, $N_3 = 16500$ $a_3 = 833$ μm

The crack growth rate is calculated with the often used secant method:

$$\frac{da}{dN} = \frac{a_{i+1} - a_i}{N_{i+1} - N_i}. \tag{5}$$

Some of the typical crack growth behaviour for the X6CrNiTi18.10 can be seen in Fig. 9 and 10.

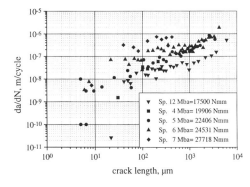

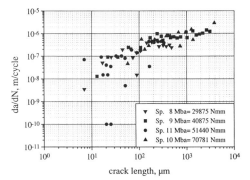

Fig. 9: X6CrNiTi18.10, crack growth rate, $\rho = 0.25$ mm

Fig. 10: X6CrNiTi18.10, crack growth rate, $\rho = 0.25$ mm

In the region of $N_i(a_i \leq 40\mu m) < 10^4$ residual stresses and overloads have only a small influence on crack initiation from sharp notches which is different to crack propagation where crack closure related residual stresses are very important [3]. Residual stresses have the same effect like mean loads and so an effective cyclic stress intensity factor

$$\Delta K_{eff} = f(R)\Delta K \tag{6}$$

has to be used for macrocrack growth under cyclic small scale yielding for instance in

$$\frac{da}{dN} = C_1[\Delta K_{eff}^m - \Delta K_{theff}^m]. \tag{7}$$

C_1 and m are material constants. ΔK_{theff} is the effective threshold intensity factor which should not depend on the load ratio R.

With Eq. (4) ($n = 2$) and a linear damage accumulation hypothesis crack growth can be predicted. For a load ratio $R = constant$ an easy assumption for macrocrack growth [4] is

$$\frac{da}{dN} = C_0 \frac{(\Delta K)^{2.5}}{E^2 \sigma_Y^{0.5} d^{\star 0.25}} \left[1 - \left(\frac{\Delta K_{th}}{\Delta K} \right)^2 \right]. \tag{8}$$

The propagation of small cracks from notches can be an important part of the life. But as shown before for an elastic material behaviour ΔK and ΔK_{eff} are not able to describe the small crack growth. So if a linear-elastic loading parameter after Eq. (1) is used, it is necessary to correct it because of the elastic-plastic material behaviour and because of the crack length dependent crack closure in the region of small cracks. In many cases a description of crack growth with the parameter $\Delta K_{eff}\sqrt{1 + \frac{d_0^{\star}}{2a}}$ is possible

$$\frac{da}{dN} = \bar{C}_1 \left[\left(\Delta K_{eff}\sqrt{1 + \frac{d_0^{\star}}{2a}} \right)^m - \Delta K_{theff}^m \right]. \tag{9}$$

The parameter $d_0^{\star}$ depends on the microsupport approach by *Neuber* ($d^{\star}$), the different crack closure of small and macrocracks and the size of cyclic plastic zone and can be estimated from crack growth data. It depends on the load too, but sometimes a description with $d_0^{\star} = const.$ seems to be sufficient. There have more numerical and experimental investigations to be done for the final evaluation of the presented parameter.

REFERENCES

[1] H. Neuber . *Kerbspannungslehre.* Springer-Verlag Berlin, 1958.

[2] S. Sähn, M. Schaper, V.B. Pham, B. Wurm, T. Husert and E. Winschuh. Damage evolution ahead of sharp notches under cyclic loading. In *Nuclear Engineering and Design.* ELSEVIER, 1996.

[3] S. Sähn, V.B. Pham und T. Hamerník. Einfluß einzelner Überlasten auf die Rißausbreitung an scharfen Kerben bei zyklischer Belastung. In *30. Tagung des DVM-Arbeitskreises Bruchvorgänge,* 1998.

[4] S. Sähn, V.B. Pham, M. Schaper, A. Böhm and T. Hamerník. The evolution of damage at cyclic loading. In *ECF 11,* volume II, pages 1091–1098, 1996.

Acknowledgement: The authors wish to thank the Deutsche Forschungsgemeinschaft for the financial support.

Low Cycle Fatigue and Elasto-Plastic Behaviour of Materials
K-T. Rie and P.D. Portella (Editors)
© 1998 Elsevier Science Ltd. All rights reserved.

TOPOGRAPHY OF THE CRACK NUCLEI AT THE EMERGING PERSISTENT SLIP BAND IN AUSTENITIC 316L STEEL

J. POLÁK*,** and T. KRUML*

*Institute of Physics of Materials, Academy of Sciences of the Czech Republic,
Žižkova 22, 616 62 Brno, Czech Republic
**Laboratoire de Mécanique de Lille (URA CNRS 1441), Ecole Centrale de Lille,
BP 48, Cité Scientifique, 59651 Villeneuve d'Ascq Cedex, France

ABSTRACT

The detailed observations of the surface relief adjacent to the persistent slip bands in polycrystalline 316L stainless steel were performed using high resolution scanning electron microscope. Characteristic features of the extrusions and intrusions emerging on the surface of the grain were obtained by tilting the specimen. The early stages of the relief evolution in the band consist of alternating extrusions and intrusions along the thin band. Well developed band has usually an extrusion in the centre of the band and two parallel intrusions at the interface with the matrix. The crack nuclei arise from the intrusions and link together to form a shallow surface crack. Experimental observations are discussed in terms of the recent models for fatigue crack nucleation.

KEYWORDS

Fatigue crack nucleation, 316L steel, persistent slip bands, extrusions, intrusions.

INTRODUCTION

The nucleation of fatigue cracks in crystalline materials has attracted appreciable attention since the early observations of the cracks on the surface of fatigued Swedish iron by Ewing and Humfrey [1]. Since the nucleation of the cracks was observed in the rough surface relief accompanying the emerging bands of localised cyclic slip, the early theories of fatigue crack nucleation proposed the models for the formation of extrusions and intrusions based on simple dislocation mechanisms (for review see e. g. [2]). The identification of persistent slip bands (PSBs) in fatigued crystalline materials, the mapping of their dislocation structures led to more realistic models of the fatigue crack nucleation. Brown et al. [3] and Essmann et al. [4] proposed the models of extrusion and intrusion formation based on dislocation dipole generation and point defect production.

The relation between the dislocation structure of a PSB and the surface relief in copper single crystal [5] led Polák [6] to propose a model of surface relief formation based on point defect generation and their migration on short distance which results in mass redistribution. Repetto and Ortiz [7] also used point defect generation and migration to obtain the profiles of extrusions and intrusions using the finite element modelling.

The confrontation of the theoretical models with experiments stimulates the improvement of these models. Only few detailed observations of nucleating cracks are available on polycrystalline materials. The general evolution of the surface relief and the density of the PSBs in polycrystalline 316L stainless steel has been reported recently [8].

The subject of the present contribution is to make pertinent observations of the surface topography features in the surface grains in fatigued polycrystalline 316L stainless steel and in combination with the previous results on the internal dislocation structure to discuss the recently proposed models of fatigue crack nucleation.

EXPERIMENTAL

Cylindrical specimens of the 316L austenitic stainless steel (Thyssen) having the average grain size of 110 μm had a shallow notch to facilitate the observations of the surface relief. The notch area was mechanically and electrochemically polished in order to achieve a smooth surface and to reveal grain boundaries. The specimens were cycled with symmetrical strain cycle under constant plastic strain amplitudes to different stages of the fatigue life and the evolution of the characteristic relief in the area of the shallow notch was observed using scanning electron microscope Philips XL 30 FEG. The tilting of the specimens around the axis parallel to the observed slip bands allowed viewing the characteristic features of the extrusions and intrusions. Further experimental details concerning the material and cyclic loading can be found elsewhere [8].

RESULTS

The evolution of the surface relief in cycling with constant plastic strain amplitude is characterised by the appearance of the persistent slip bands that emerge on the surface as prominent extrusions and intrusions. The characteristic relief of the surface grain of a specimen cycled with constant plastic strain amplitude, $\varepsilon_{ap} = 5 \times 10^{-4}$, to approximately one fifth of the fatigue life is shown in Fig. 1. This plastic strain amplitude corresponds to the plateau of the cyclic stress-strain curve [8] and thus to the cyclic strain localisation. Distinct parallel PSBs are equidistantly spaced (average spacing ~ 3.5 μm). The slip bands continue in the neighbouring grain in an other mostly stressed slip system.

The details of the surface topography of individual slip bands are shown at higher magnification in Fig. 2. A narrow band in Fig. 2a is characterised by irregular extrusions and intrusions with a tendency to alternate extrusions and intrusions. Two pronounced extrusions and two intrusions can be identified in the micrograph. The average distance between the extrusions, or intrusions in a particular grain can be, however, only evaluated using numerous PSBs in the early stage of evolution of the surface relief. As cycling progresses, the majority of PSBs is characterised by ribbon-like extrusions and cracks which develop from parallel intrusions at the interface of the band and the matrix. The more developed PSB is shown in Fig. 2b. The ribbon-like extrusion is present along the whole length of the band. The crack like defect can be detected at the interface between the band and the matrix.

Fig. 1 Surface grain of the specimen fatigued with plastic strain amplitude $5x10^{-4}$ for 40 000 cycles.

Fig. 2 Details of the relief of the PSB; (a) at the early stage of evolution, (b) well developed PSB.

562

Both Fig. 1 and Fig. 2 were taken with the primary electron beam perpendicular to the specimen surface and, therefore, they cannot yield the full information on the topography of the extrusion-intrusion along the band. Since the extrusion emanates from the grain in the direction of the primary slip plane we have rotated the specimen around the line of the intersection of the primary slip plane with the surface until the primary electron beam became parallel to the primary slip plane. Figure 3 shows the same segment of the PSB imaged with the beam perpendicular to the specimen surface (Fig. 3a) and with the beam parallel to the active slip plane of the grain (Fig 3b). The angle of rotation was 35°.

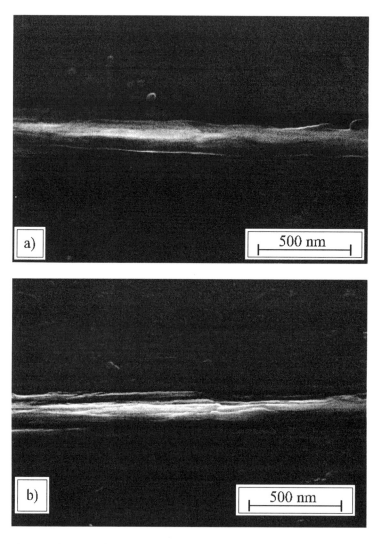

Fig. 3 Surface relief of a well developed segment of a PSB in 316L steel at two inclinations of the specimen: (a) primary beam perpendicular to the specimen surface, (b) primary beam parallel to the primary slip plane of the grain.

Figure 3b shows both interfaces of the PSB and the matrix. The thickness of the PSB emerging on the surface is only about 0.22 μm. It consists of the ribbon-like extrusion in the middle of the band and intrusions or already early cracks at the interface with the matrix, parallel to the extrusion. The extrusion has an irregular surface structure corresponding to the random fine slip. The height of the extrusion can be deduced from Fig. 3a and from the angle of rotation necessary to take the micrograph in Fig. 3b and was evaluated to be 0.14 μm.

Numerous other relief configurations can be documented on the surface grains of fatigued polycrystalline 316L steel. The general features of the relief at the emerging PSBs can be, however, summarised as follows:
surface relief at the PSBs in the embryonic stage of development is characterised by alternating extrusions and intrusions whose height and depth increases with cycling;
mostly ribbon-like extrusions in the centre of the band with two parallel intrusions at the interface with the matrix are formed in well developed PSBs;
the cracks start from the intrusions on any side of the central extrusion and at a later stage can pass from one side to the other.

DISCUSSION

The observations of the relief in the surface grains of the fatigued 316L stainless steel show pronounced localisation of the cyclic strain in a system of nucleated PSBs. This relief can be correlated with the internal dislocation structure studied recently by Kruml et al. [9]. In the foils oriented close to the cross slip plane of the grain they observed the bands of specific dislocation structure parallel to the primary slip plane. The thickness of these bands was typically in the domain 0.25 μm up to 1 μm. They had the structure consisting of thin high density dislocation walls and thick low density dislocation channels which remind the ladder structure. In comparison with the ladder structure of PSBs in copper single or polycrystals, however, the ladders in 316L steel are very irregular with variable wall spacing and the walls inclined to the direction of the band.

The relation between the spacing of the rungs of the ladders and the spacing of the extrusions and intrusions of the embryonic PSBs [5] cannot be thus established. Nevertheless, the periodicity of both the dislocation structure in the band and that of the surface relief of the PSB suggest a strong correlation between them.

The experimental observations of the surface relief formation can be reconciled with two recent models of fatigue crack nucleation [6,7] that are based on point defect generation and short distance migration in cyclic loading. The early stages of the surface relief formation correspond to the redistribution of the mass within the PSB which result in the formation of alternating extrusions and intrusions. The surface relief of the well-developed band corresponds to the redistribution of the mass between the interior of the band and the interface region with the matrix which corresponds to the formation of a central extrusion and two parallel intrusions. Both these features follow from the Polák's model [6] of fatigue crack nucleation. The shape of ribbon-like extrusion is in accord with the simulations of Repetto and Ortiz [7]. The real intrusion is, however, significantly sharper then the grooves obtained on both sides of the extrusion in simulation.

The nucleation of the surface shallow crack results from the linkage of the elementary crack nuclei along the PSB. Since the elementary crack nuclei are in different stages of the evolution when linkage starts the crack passing the whole band can link the elementary cracks from both sides of the extrusion. Therefore, the crack can alternate from one side of the central extrusion to the other side, at least on the surface. Multiple shallow surface cracks are initiated along parallel PSBs within one grain. However, there is a strong competition between the cracks growing in the interior of the crystal. The majority of the shallow parallel cracks stop growing and usually only one of them dominates, penetrates a whole grain and has a chance to propagate into a neighbouring grain.

The authors gratefully acknowledge the support of the present research by the grant No. A2041704/1997 of the Grant Agency of the Academy of Sciences and by the grant No. 106/97/P027 of the Grant Agency of the Czech Republic.

REFERENCES

Ewing, J. A. and Humfrey, J. C. W. (1903) *Philos. Trans. R. Soc. London*, **A 200**, 341.
Laird, C. and Duquette, D. J. (1971). In: *Proc. 2nd Int. Conf. Corrosion Fatigue*, NACE, Houston, pp. 88-123.
Brown, L. M. and Ogin, S. L. (1985). In: *Fundamentals of Deformation Fracture*, B.A. Bilby, K. J. Miller and J. R. Willis (Eds.), Cambridge University Press, Cambridge, pp. 501-528.
Essmann, U., Gösselle, U. and Mughrabi, H. (1981) *Philos. Mag.* **44**, 405.
Polák, J., Lepistö, T. and Kettunen, P. (1985) *Mater. Sci. Engng* **74**, 85.
Polák, J. (1987) *Mater. Sci. Engng* **92**, 71.
Repetto, E. and Ortiz, M. (1997) *Acta Mater.* **45**, 2577.
Obrtlík, K., Kruml, T. and Polák, J. (1994) *Mater. Sci. Engng* **A187**, 1.
Kruml, T., Polák, J., Obrtlík, K. and Degallaix, S. (1997) *Acta Mater.* **45**, 514.

Chapter 8

CRACK INITIATION AND COALESCING; DAMAGE EVOLUTION

INTERACTION BETWEEN LOW CYCLE FATIGUE AND HIGH CYCLE FATIGUE IN 316L STAINLESS STEEL

G. WHEATLEY[1], Y. ESTRIN[1], X.Z. HU[1] AND Y. BRÉCHET[2]

(1) Department of Mechanical and Materials Engineering
The University of Western Australia, Nedlands, WA, 6907, Australia
(2) LTPCM, Institut National Polytechnique de Grenoble
38402 Saint Martin d'Heres, France

ABSTRACT

The effect of low cycle fatigue (LCF) on subsequent high cycle fatigue (HCF) life and the fatigue crack growth rate was investigated. In a tensile specimen, a large number of LCF cycles results in a decrease in the subsequent HCF life. This suggests that damage by LCF causes a decrease in the HCF life. However, following a small number of LCF cycles at a particular LCF waveform, an increase of the HCF life was observed. The effect shows a sensitivity to the particular waveform chosen and the as-received condition of the material.

KEYWORDS

Low/High Cycle Fatigue Interaction, Damage, Fatigue Crack Growth.

INTRODUCTION

This work is part of an investigation of the effect of various types of perturbations on the high cycle fatigue life of 316L steel. One type of perturbation studied so far, is a singular tensile overload which was shown to result in a transient increase in the crack growth rate, yet lead to an overall retardation [1,2]. Damage enhancement was seen to be the cause of transient acceleration with residual compression and strain hardening accounting for the overall retardation. It was suggested that the fatigue damage zone (FDZ) can be thought of as a small volume experiencing LCF. The tests on tensile specimens aim at investigating this idea by imposing LCF pre-strain prior to HCF deformation. This mode of 'perturbation' is of great interest in its own right, as it provides a means of HCF life extension, provided suitable LCF conditions are found [3].

The immediate goal of this study was thus to investigate the effects of LCF on the HCF life and the rate of crack growth in 316L stainless steel and to compare the results with those obtained on a different batch of 316L [2]. A central question in this study was whether low cyclic fatigue preceding high cyclic fatigue deformation can lead to an *increase* of the HCF life found in Ref. 2. The LCF waveforms were varied, as were the number of LCF cycles at each waveform preceding HCF. Identical HCF waveforms were then applied, and the number of cycles to failure was plotted as a function of the LCF cycle number. It was confirmed that an increase in the high cycle fatigue life can be achieved by a suitably chosen regime of LCF pre-straining. The effect contradicts the cumulative damage rules commonly used in fatigue life assessment.

EXPERIMENT

Standard 316L 12.7mm round bar was used in the as-received condition. Two different batches were tested. The chemical composition and mechanical properties are outlined in Tables 1 and 2.

Table 1. Chemical Composition (%) of 316L used as specimens.

	C	Mn	Si	P	S	Ni	Cr	Mo	Cu	N
batch 1	0.022	1.38	0.390	0.037	0.012	11.50	17.49	2.15	0.110	270 ppm
batch 2	0.020	0.61	0.42	0.024	0.026	11.93	17.26	2.05	-	0.050

Table 2. Nominal Mechanical Properties of 316L used as specimens.

	tensile strength (MPa)	yield strength (MPa)	elongation (%)	reduction of area (%)	hardness (HB)
batch 1	687	509	37.4	68.2	207
batch 2	741	632	40	75	217

Tensile specimens were manufactured to the dimensions shown in Fig. 1. The testing part of each specimen was polished to 1μm diamond paste level to assist optical observation.

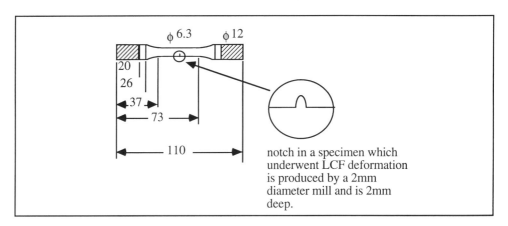

Fig. 1. Specimen dimensions (in mm).

The testing procedure was as follows. A monotonic test was conducted for each material batch in order to determine a suitable strain amplitude for the LCF and HCF waveforms. For the first batch of material, the LCF waveform chosen was strain controlled and ranged from 0.5-1.5% at 5Hz. The resulting LCF life for batch 1 was approximately 5000 cycles (Fig. 2). For the second batch of material, the LCF waveform was also strain controlled but a variety of waveforms were selected (Fig. 3). Following a selected number of LCF cycles, a circular notch was machined into the middle part of the specimen (Fig. 1) to concentrate the HCF at its centre. Following this, HCF was applied. The HCF waveform was load controlled and ranged from 1-9kN at 5Hz. The resulting rate of crack growth, da/dN, and the HCF life were then recorded.

RESULTS

HCF Life

Figure 2 shows the resulting HCF life for batch 1 plotted as a function of the number of LCF cycles which were applied to the specimen prior to HCF testing. Figure 3 shows the resulting HCF life for both batches. Only batch 2 was subjected to differing LCF waveforms. While Fig. 3 summarises all test data, the averages of the data points corresponding to each HCF condition, specified by the number and waveform of the preceding LCF cycles, are presented in Fig. 4.

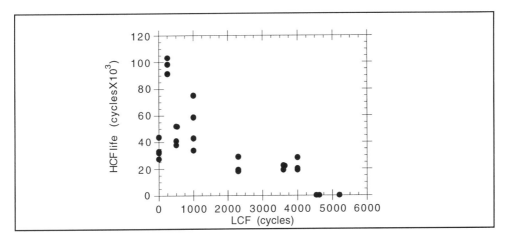

Fig. 2. Effect of number of LCF cycles on HCF life (batch 1).

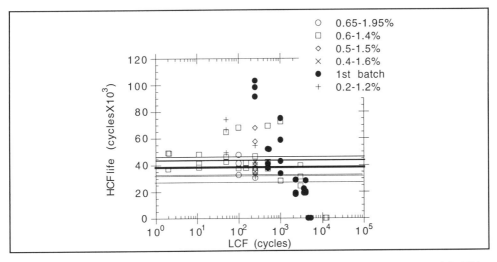

Fig. 3. Effect of number of LCF cycles and their waveform on HCF life (batch 1 and 2). Thin horizontal lines represent HCF life for virgin specimens of batch 1. Bold horizontal lines represent HCF life for virgin specimens of batch 2.

570

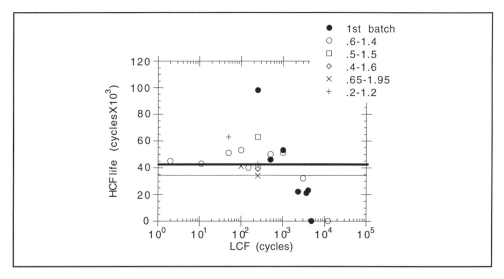

Fig. 4. Effect of number of LCF cycles and waveform on HCF life. Data has been averaged. Thin horizontal line represents average HCF life for virgin specimens of batch 1. Bold horizontal line represents average HCF life for virgin specimens of batch 2.

In spite of the data scatter, Fig. 2 displays the following trend. A specimen subjected only to HCF had a lifetime of about 35,000 cycles. Following 250 cycles of LCF, the HCF life increased to approximately 100,000 cycles. The HCF life almost tripled. When more LCF cycles were applied, the HCF life was decreased. A plateau-like region was observed following about 1,000 cycles of LCF where additional LCF did not result in a further decrease in HCF life. This plateau was approximately 70% of the pure HCF life and 25% of the peak HCF life attained. A precipitous drop of the HCF life from the plateau level to zero was observed for LCF cycles number approaching the pure LCF life.

In Fig. 4, the results for the two different batches are plotted. The pure HCF life of batch 2 was about 43,000 cycles. When batch 2 specimens were subjected to an LCF waveform with an amplitude of 0.4%, a much smaller increase in HCF life was observed as compared to batch 1 with an amplitude of 0.5%. For 2 and 11 LCF cycles, no effect on HCF life was observed, but cycle numbers of up to 1,000 resulted in a small increase in the HCF life. Following an increase in the amplitude of the LCF waveform from 0.4% to 0.5%, ie. identical to that of batch 1 specimens, the HCF life increased moderately again at the 250 cycle mark. However, an increase to 0.6% saw a decrease in the HCF life at 250 cycles. By analysing the standard stress-strain plots for the two batches considered, a similar range of strain was chosen for batch 2 in an attempt to duplicate the LCF waveform used for batch 1. Unlike the batch 1 tests, this did not result in an increase in the HCF life of batch 2 material. Another tact was chosen to duplicate the increase in HCF life by concentrating the LCF waveform around the elbow of the stress-strain plot, where material behaviour shifts from elastic to plastic. An LCF waveform of 0.2-1.2% was selected for batch 2 specimens, and an increase in HCF life was noted at the 50 cycle mark.

HCF Crack Growth

A preliminary analysis of the data on the evolution of the crack length, a, showed that a major change in the a vs. N diagrams under HCF conditions with LCF waveform or with the number of LCF cycles, as compared to the pure HCF specimens, was a parallel displacement of the curves along the N axis, cf. Fig. 5.

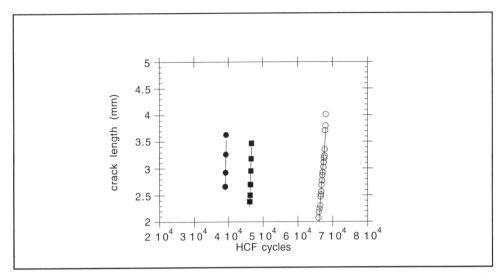

Fig. 5. Crack length vs. HCF cycles plot showing retardation of the crack initiation and growth. Closed symbols correspond to specimens which had not undergone exposure to LCF. Open circles refer to the specimen which has experienced 100 cycles of LCF (0.6-1.4% @ 5Hz).

This suggests that the change in HCF life upon LCF pre-straining is dominated by a change in the number of cycles required for fatigue crack initiation. However, as Fig. 5 shows, the slope of the plot is also changed. This indicates that there is some retardation of the fatigue crack growth with LCF prestraining. Additional studies aiming at clarifying the relative roles of the LCF effects on crack initiation and crack growth are under way.

CONCLUSIONS

It has been shown that high cycle fatigue life is sensitive to prior low cycle fatigue in the tensile strain range. The HCF life can be extended significantly by exposure of material to a small number of LCF cycles prior to HCF deformation. At larger numbers of LCF cycles, HCF life is seen to be insensitive to the duration of LCF exposure. These results suggest that the cumulative damage approach in traditional S-N testing is questionable as 'pre-damaging' the material by LCF initially results in an increase of the HCF life. This conclusion is further supported by the observations of the effects on HCF following a tensile overload reported in Ref. 1.

The as-received condition of the material, the LCF waveform and the number of cycles applied, all influence the subsequent HCF life of a specimen. The results suggest that positioning the LCF waveform in the elastic-plastic transition range is a significant factor in determining an LCF waveform for enhancement of HCF life. Further refining of this method of isolating the ideal LCF waveform and cycle number is in progress. The microstructural changes during LCF and their interaction with the processes at the HCF crack tip are also under investigation. Finally, application of this procedure to other materials is being investigated presently.

These experimental findings reported in this paper may have industrial applications as a way to extend the HCF life of 316L steel by finding an optimum LCF regime preceding service of a structural member.

572

ACKNOWLEDGEMENTS

One of the authors (GW) would like to thank Rio Tinto P/L for their support through an Australian Postgraduate Award (Industry) and the University of Western Australia for their support through a William Lambden Owen Scholarship.

REFERENCES

1.   Wheatley, G., Niefanger, R., Estrin, Y. and Hu, X.Z. (1998) *Key Engineering Materials* **145-149** ,631.
2.   Wheatley, G., Hu, X.Z. and Estrin, Y. *Materials Science and Engineering* (to be published).
3.   Wheatley, G., Estrin, Y., Hu, X.Z. and Bréchet, Y. *Materials Science and Engineering A (Letters)* (submitted).

Low Cycle Fatigue and Elasto-Plastic Behaviour of Materials
K-T. Rie and P.D. Portella (Editors)
© 1998 Elsevier Science Ltd. All rights reserved.

EFFECT OF LOADING WAVE FORM ON THE RESULTS OF LOW CYCLE FATIGUE AND FATIGUE CRACK PROPAGATION TESTS

J. LUKÁCS and GY. NAGY

Department of Mechanical Engineering, University of Miskolc
H-3515 Miskolc-Egyetemváros, Hungary

ABSTRACT

In the course of design and operation of different structural elements initiation and generally propagation of cracks is a concern. Cyclic loading of structures in most cases is not sinusoidal, while material properties used for calculations are obtained from tests carried out with sinusoidal loading wave form. The question is arisen, what is the effect of the loading wave form. In order to assess this effect, low cycle fatigue (LCF) and fatigue crack propagation rate (FCG) tests were carried out on steel base materials having different material structures and mechanical properties. Based on theoretical considerations it was concluded that the effect of loading cycle wave form is significant in the case of both kinds of tests. In the case of greater total strain amplitudes (LCF) and low-strength steels (FCG) this conclusion was confirmed by the performed tests.

KEYWORDS

low cycle fatigue, fatigue crack propagation, loading wave form, total plastic-strain hysteresis energy

INTRODUCTION

In the course of design and operation of a great number of structures, initiation and propagation of cracks should be considered. The most frequent reason is the non-steady loading. This even more applies to pressure vessels and boilers where low cycle fatigue and fatigue crack propagation must be taken into consideration. Loading of these facilities in most cases is stochastical, sometimes it has constant strain amplitude but not necessarily sinusoidal, while majority of tests were carried out with sinusoidal loading wave form. The question is arisen: "how results of low cycle fatigue and fatigue crack propagation rate are modified by the loading wave form?"

The objective of this publication is to introduce the results of tests, which were carried out in order to determine the effect of loading wave form on the results of low cycle fatigue and fatigue crack propagation rate. The presentation represents the changes in the material structure and mechanical properties, which constitute the physical basis of evaluation of tests results.

TESTS AND THEIR RESULTS

Low cycle fatigue tests

Chemical composition and strength properties of steel grade (KL7D) used for the tests are listed in the Table 1.

Table 1. Chemical composition and basic mechanical properties of steel grade KL7D

C	Mn	Si	S	P	Cr	Cu	R_u, MPa	R_y, MPa
0.17	1.31	0.24	0.036	0.020	0.11	0.26	535	390

Specimens were fatigued with triangular, sinusoidal, trapezoidal and rectangular loading wave forms. In order to make the results universal, the controlled total strain amplitudes were $\varepsilon_{a1} = 0.01$ and $\varepsilon_{a2} = 0.005$. Tests were carried out in the air at room temperature, with R = -1 loading ratio by means of MTS type electrohydraulic testing equipment and the maximum tensile force was decreased with 25% while hysteresis curve of each cycle was recorded. Five specimens were tested in the case of all loading wave forms and strain amplitudes.

Fracture cycles (N_t), average and standard deviation of stress amplitudes (σ_{a50}) for each series were determined. The test results are summarised in the Table 2. [1].

Table 2. Effect of loading wave form on the results of low cycle fatigue tests

Shape of the load function	Total strain amplitude	Average of fracture cycle, N_t	Standard deviation	Average of the stress amplitude belonging to the 50% of service life	Standard deviation
sinusoidal	0.010	716	57.75	440	2.05
triangular	0.010	833	89.65	444	1.00
trapezoidal	0.010	843	73.52	452	1.95
rectangular	0.010	799	86.16	447	8.51
sinusoidal	0.005	3592	84	371	2.55
triangular	0.005	3898	367	372	1.3
trapezoidal	0.005	3499	155	371	1.92
rectangular	0.005	3501	184.6	371	10.52

Fatigue crack propagation rate tests

Besides KL7D steel, Armco-iron and X80TM grade HSLA steel (R_y=540 MPa, R_u=625 MPa) were used for fatigue crack propagation rate tests of which chemical compositions are listed in the Table 3.

Table 3. Chemical composition of Armco-iron and X80TM HSLA steel

Material	C	Si	Mn	S	P	Cr	Ni	Mo	Al	Nb	Ti	N
Armco-iron	0.002	0.006	0.06	0.006	0.007	0.02	0.02	0.01	0	–	0.0008	–
X80TM	0.077	0.30	1.84	0.002	0.012	–	–	–	0.036	0.046	0.018	0.0051

TPB specimens were used for the tests (nominal dimensions: W = 20 mm, B = 10 mm). Tests were carried out in the air at room temperature, with R = 0.1 loading ratio by means of MTS type electrohydraulic testing equipment. In the case of all material grade the loading wave forms were triangular, sinusoidal, trapezoidal and rectangular. Propagating crack was observed with video camera in 100x magnification.

Results of tests, i.e. the two constants of the Paris-Erdogan law (C and n) were determined with linear regression, principle of least squares. Fatigue crack propagation rate was calculated with seven point incremental polynomial method [3]. Taking into consideration that there is relationship [4] between the two constants of the Paris-Erdogan law, results are given by means of the exponent (n). Average, standard deviation, standard deviation coefficient of test results are summarised in the Table 4. according to material grades and loading wave forms.

Table 4. Influence of the loading wave form on the results (n) of fatigue crack growth rate tests

Material	Loading wave form	Element number of the sample	Average	Standard deviation	Standard deviation coefficient
Armco-iron	triangular	4	4.497	0.7644	0.170
	sinusoidal	5	3.714	0.5435	0.146
	trapezoidal	5	3.446	1.0521	0.305
	rectangular	4	3.310	0.6619	0.200
KL7D	triangular	6	3.725	0.7009	0.188
	sinusoidal	12	3.335	0.4683	0.140
	trapezoidal	7	3.210	0.5745	0.179
	rectangular	6	3.117	0.2940	0.094
X80TM	triangular	4	2.690	0.1711	0.064
	sinusoidal	13	2.616	0.7499	0.287
	trapezoidal	4	2.652	0.3217	0.121
	rectangular	4	3.167	0.5066	0.160

EVALUATION OF TEST RESULTS

Low cycle fatigue tests

Fracture cycles and stress amplitudes belonging to the 50% of the life cycle obtained from low cycle fatigue tests characterised by different loading wave forms, were considered as two sets. In order to examine their uniformity the Wilcoxon-probe [5] was applied because of the low number of elements and distribution of results was not limited.

Based on statistical calculations the following conclusion can be formulated regarding the effect of loading wave form on the results of low cycle fatigue tests.

- In the case of fatigue test carried out with $\varepsilon_{a1} = 0.01$ total strain amplitude, fracture cycles (N_t) and stress amplitudes belonging to the 50% of the life cycle (σ_{a50}) can be considered different at 95% significance level.

- In the case of test carried out with $\varepsilon_{a1} = 0.005$ total strain amplitude, independently of loading wave forms both fracture cycles and stress amplitudes are identical at 95% significance level.

In order to find reason for this apparent contradiction and physical explanation for the effect of loading wave forms further analysis was carried out. The basic idea was the well known fact that cyclic plastic deformation causes microstructural changes in the material and damage [6] of which extent and number of cycles until fracture proportional to the energy accumulated in the material. Unfortunately, the measurement and calculation of this energy is extremely complicated.

On the contrary, when low cycle fatigue test is being carried out, total plastic-strain hysteresis energy can be determined in a relatively simple way, that is the addition of energy accumulating cycle by cycle. This latter is proportional to the area swept by the hysteresis curves and therefore it can be calculated. On the supposition that in the case of specimens fatigued in similar circumstances, accumulated energy and total plastic-strain hysteresis energy is the same, total plastic-strain hysteresis energy until fracture can be used for characterising the extent of damage. However it is also known that the loading process, the "loading history" i.e. the loading wave form affects the extent of the total plastic-strain hysteresis energy during the plastic deformation [7].

Applying this statement to low cycle fatigue it can be established that shape and area of hysteresis curve, total plastic-strain hysteresis energy until fracture, extent of damage and determinable material properties must be affected by the loading wave form.

Relying on the previous chain of ideas, total plastic-strain hysteresis energy until fracture of specimens was determined and included in the Table 5. In the case of each loading wave form, data related to total strain amplitudes are considered as sets and similarity of work inputs was checked with Wilcoxon-probe again.

Table 5. Effect of the loading wave form on the total plastic-strain hysteresis energy

Shape of the loading function	Total strain amplitude	Average	Standard deviation
		of the total plastic-strain hysteresis energy	
sinusoidal	0.010	8689	667
triangular	0.010	10086	1063
trapezoidal	0.010	10590	991
rectangular	0.010	11077	1438
sinusoidal	0.005	14750	618
triangular	0.005	15467	1254
trapezoidal	0.005	14157	686
rectangular	0.005	15928	2043

Based on the statistical considerations the following conclusions can be formulated.

- In the case of tests carried out with greater strain amplitude ($\varepsilon_a = 0.01$), in agreement with the theoretical considerations amount of total plastic-strain hysteresis energies is different at 95% significance level.

- In the case of less total strain amplitude ($\varepsilon_a = 0.005$) total plastic-strain hysteresis energy until fracture are identical.

- Contradictory results to the theoretical considerations experienced at small strain amplitude are related to the fact that due to the small plastic deformation, less significant effect of loading wave form on the total plastic-strain hysteresis energy is obscured by the high standard deviation of registered properties.

Fatigue crack propagation tests

Based on the test results, considering the data summarised in the Table 4., the following conclusions can be formulated.

- Average of test results carried out with various loading wave forms is different in the case of Armco-iron and KL7D steel, in the case of X80TM HSLA steel it is not. The greater strength the less difference between the greatest and least average value.

- The greater work input per volume per one cycle the less average of exponent (n) of the Paris–Erdogan law in the case of Armco-iron and KL7D steel. Such tendency can not be established in the case of X80TM HSLA steel. The reason for this is the different plastic zone in the crack tip due to the different elastic-plastic behaviour of tested metals.

- There is no unambiguous relationship between the reliability of tests results carried out with different loading wave forms and work input per volume per one cycle in the case of all tested metals.

SUMMARY

Based on theoretical considerations it can be concluded that the shape of loading wave form affects the results of low cycle fatigue and fatigue crack propagation rate tests. This completely agrees with the results of systematic tests, if the low cycle fatigue is carried out with greater strain amplitude ($\varepsilon_{a1} = 0.01$) or if fatigue crack propagation rate was measured on low-strength metals (Armco-iron, KL7D steel).

If low cycle fatigue test is carried out with small total strain amplitude ($\varepsilon_{a2} = 0.005$) tests results can be considered identical, because due to the small plastic deformation, less significant effect of loading wave form on the total plastic-strain hysteresis energy is obscured by the high standard deviation of registered properties.

ACKNOWLEDGEMENT

Authors wish to express their appreciation to the Technische Universität Wien TVFA (Technische Versuchs- und Forschungsanstalt) and the Technical University of Budapest MTAT (Department of Mechanical Technology and Materials Science) for providing the X80TM steel and to the National Scientific Research Found (OTKA) for the financial support (registration numbers of projects: OTKA T 015605 and OTKA T 022020).

REFERENCES

[1] ASTM E 606 (1980). *Constant-Amplitude Low-Cycle Fatigue Testing.*

[2] Paris, P. and Erdogan, F. (1963) *Journal of Basic Engineering, Transactions of the ASME,* p. 528.

[3] ASTM E 647 (1994). *Standard Test Method for Measurement of Fatigue Crack Growth Rates*

[4] Clark, W. G. Jr. and Hudak, S. J. Jr. (1975) *Journal of Testing and Evaluation, JTEVA,* **6**, 454.

[5] Balogh, A., Dukáti, F. and Sallay, L. (1980). *Minőségellenőrzés és megbízhatóság.* Műszaki Könyvkiadó, Budapest.

[6] Bicego, V. and Fossati, M. (1982). In: *Proceedings of the 8th Congress on Material Testing.* OMIKK-TECHNOINFORM, Budapest, pp. 267-271.

[7] Béda, Gy., Kozák, I. and Verhás, J. (1986). *Kontinuummechanika.* Műszaki Könyvkiadó, Budapest.

Low Cycle Fatigue and Elasto-Plastic Behaviour of Materials
K-T. Rie and P.D. Portella (Editors)

RELATION BETWEEN CYCLIC J-INTEGRAL AND CRACK TIP OPENING DISPLACEMENT AT LOW-CYCLE FATIGUE

H. WITTKE[a] and K.-T. RIE[b]

[a] *Institut für Festkörpermechanik, Lehrstuhl für Maschinendynamik und Schwingungslehre, TU Dresden, 01062 Dresden, Germany*

[b] *Institut für Oberflächentechnik und Plasmatechnische Werkstoffentwicklung, TU Braunschweig, Bienroder Weg 53, 38108 Braunschweig, Germany*

ABSTRACT

It is shown that a critical view of different approaches to calculate the effective part of the cyclic J-integral is possible with the aid of experimental data for the crack tip opening displacement. The comparison of calculated and experimental values of the crack tip opening displacement is an aid to find an approach which is physically based.

KEYWORDS

Crack growth, cyclic J-integral, crack tip opening displacement.

INTRODUCTION

Crack growth at low-cycle fatigue is often correlated with cyclic J-integral ΔJ or with its effective part ΔJ_{eff}, respectively. A relation

$$\mathrm{d}a/\mathrm{d}N = C_J \cdot (\Delta J_{eff})^{\alpha} \tag{1}$$

is valid ($\mathrm{d}a/\mathrm{d}N$: crack growth per cycle; C_J, α: constants) for tests with various strain amplitudes $\Delta\epsilon/2$ or plastic strain amplitudes $\Delta\epsilon_p/2$. Surprisingly, even for the same geometries different approaches to calculate ΔJ are given in the literature. It is known that the cyclic J-integral is related to the crack tip opening displacement δ_t. With the aid of experimental data for δ_t a critical view of the different approaches to calculate ΔJ is possible.

580

EXPERIMENTAL DETAILS

Strain controlled LCF-tests at room temperatures were performed at the ferritic steel 2.25Cr-1Mo (10 CrMo 9 10). The strain rate of the tests was $\dot{\epsilon} = 2 \cdot 10^{-3}$ s^{-1}. Specimens with a rectangular cross section and with an edge crack were used. The cross section area was 8.7×5 mm^2 and the gauge length (distance of the tips of the strain extensometer) was 10.5 mm. A crack starter with a depth $a_0 = 0.2$ mm was milled into the specimen along its thin side. Crack growth and the crack tip opening displacement were measured with optical methods: The surface of the wide side of the specimen near the crack tip was recorded by a video system. Crack tip opening displacement in dependence on cycle number N were measured directly on the video screen. As proposed by Tanaka *et al.* [1], the crack tip opening displacement $\Delta\varphi_{250}$ was measured, i.e. the crack opening displacement in a distance of 250 μm from the crack tip. The irregular crack contour on the surface of the specimen was observed in a scanning electron microscope (SEM) after interrupting the test at a given crack depth ($a \approx 1 - 2$ mm). By comparison of the images of the SEM and the video film, crack depth a was measured in dependence on N. The curves a vs. N were transformed into curves da/dN vs. a, the latter were transformed into curves da/dN vs. ΔJ_{eff}. In a similar manner, the curves $\Delta\varphi_{250}$ vs. N were transformed into curves $\Delta\varphi_{250}$ vs. ΔJ_{eff}.

RESULTS AND DISCUSSION

Cyclic J-integral and crack tip opening displacement

In the literature different approaches to calculate ΔJ and ΔJ_{eff} even for the same geometries can be found, e.g. in the case of the edge crack in a flat specimen. Analogous to [2] and by taking crack closure into account by a factor U for the elastic part, ΔJ_{eff} can be calculated as shown in the following equation:

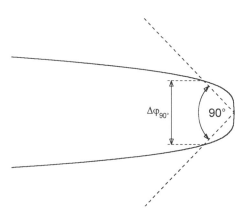

Fig. 1: Crack contour and crack tip opening displacement $\Delta\varphi_{90°}$.

$$\Delta J_{eff} = (7.88 \cdot U^2 \Delta W_e + 4.84 \cdot (1/n')^{1/2} \Delta W_p) \cdot a \qquad (2)$$

The constant n' is the exponent of the cyclic stress-strain curve, ΔW_e and ΔW_p are the elastic and the plastic strain energy density. The relations

$$\Delta W_e = (\Delta\sigma)^2/2E \qquad (3)$$

and

$$\Delta W_p = (\Delta\sigma \cdot \Delta\epsilon_p)/(1-\beta) \qquad (4)$$

are valid, where E is the Young's modulus, $\Delta\sigma$ is the stress range and β is the exponent of the stress-strain hysteresis loop (β is dependent on the strain amplitude).

Crack growth per cycle da/dN correlates well with ΔJ_{eff} calculated according to Eq. (2). But also other approaches for ΔJ_{eff} which can be found in the literature (e.g. [3]) lead to good results. With the aid of experimental data of crack tip opening displacement, the approach of Eq. (2) can be found as a physically based one: Assuming a power law between relative stress and relative plastic strain, a proportional relation between ΔJ_{eff} and $\Delta\varphi_{90°}$ - the crack opening displacement which is two times greater than its corresponding distance to the crack tip, compare Fig. 1 - is valid. The parameter $\Delta\varphi_{90°}$ is more difficult to measure than $\Delta\varphi(x)$ which is the crack opening displacement in a distance x from the crack tip. Therefore, the values of $\Delta\varphi_{90°}$ are calculated in dependence on the exponent n' and the experimental values of $\Delta\varphi(x)$:

$$\Delta\varphi_{90°} = (1/2)^{n'} \cdot (\Delta\varphi(x)) \cdot (\Delta\varphi(x)/x)^{n'} \qquad (5)$$

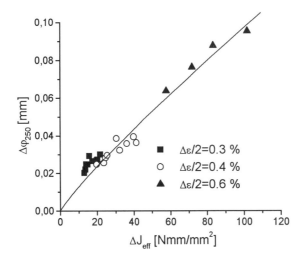

Fig. 2: 2.25Cr-1Mo, crack tip opening displacement $\Delta\varphi_{250}$ vs. ΔJ_{eff}. (ΔJ_{eff} is calculated according to Eq. (2).)

That means that the relation

$$\Delta\varphi_{90°} = 2^{n'} \cdot (\Delta\varphi_{250})^{1+n'} \qquad (6)$$

582

is valid in the case of $x = 250\mu m$ (with $\Delta\varphi_{90°}$ and $\Delta\varphi_{250}$ in mm). The solid line regression curve of Fig. 2 is given by

$$\Delta\varphi_{250} = q_1 \cdot (\Delta J_{eff})^{1/(1+n')} \tag{7}$$

where $q_1 = 1.7 \cdot 10^{-3}$ (with $\Delta\varphi_{250}$ in mm and ΔJ_{eff} in Nmm/mm^2). Using this result, $\Delta\varphi_{90°}$ can be calculated by

$$\Delta\varphi_{90°} = q_2 \cdot \Delta J_{eff} \tag{8}$$

where $q_2 = 7.76 \cdot 10^{-4}$ mm^2/N, see Fig. 3.

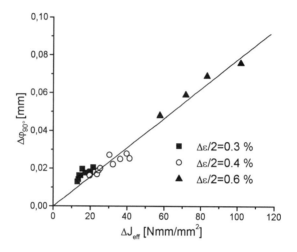

Fig. 3: 2.25Cr-1Mo, crack tip opening displacement $\Delta\varphi_{90°}$ vs. ΔJ_{eff}. (ΔJ_{eff} is calculated according to Eq. (2).)

A theoretical relation between ΔJ_{eff} and $\delta_t = \Delta\varphi_{90°}$ is given by

$$\Delta J_{eff} = (k_c/2) \cdot [2\sigma'_{0.2} - \Delta\sigma(1-U)] \cdot \delta_t \tag{9}$$

where $\sigma'_{0.2}$ is cyclic yield stress, i.e. the stress amplitude for a plastic strain amplitude of $\Delta\epsilon_p/2 = 0.2 \cdot 10^{-2}$. This relation is analogous to the proportional relation between J-integral and δ_t in the monotonic case (see e.g. Robinson [4]). Using the experimental values one gets $2.10 < k_c/2 < 2.13$ which is the realistic order according to [1]. That means that the regression curves in Figs. 2, 3 and 4 are not only fitting curves but curves with a physical meaning.

The parameters used for the calculations are $n' = 0.138$, $U = 0.9$, $\sigma'_{0.2} = 340.6$ MPa and $E = 204$ GPa. The parameters of the solid line regression curve for da/dN vs. ΔJ_{eff} in Fig. 4 are $C_J = 2.75 \cdot 10^{-5}$ and $\alpha = 1.26$ (with da/dN in mm and ΔJ_{eff} in Nmm/mm^2).

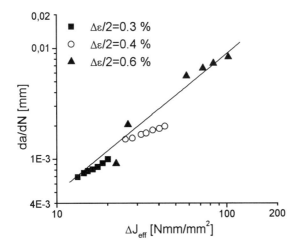

Fig. 4: 2.25Cr-1Mo, crack progress per cycle da/dN vs. ΔJ_{eff}. (For the calculation of ΔJ_{eff}: see Eq. (2).)

Masing and non-Masing behaviour

The ΔJ-concept is valid in the case of Masing behaviour. Many materials show non-Masing behaviour. The physical reasons for the non-Masing behaviour of 2.25Cr-1Mo were discussed in [5]. To take non-Masing approximately into account it was proposed to replace n' by the exponent of a master curve $n^\star$ for the calculation of ΔJ or ΔJ_{eff}, see [5]. It is better physically based to replace n' by the exponent β because the approaches to calculate the J-integral are using simple stress-plastic strain power law relations (compare [6]). The method and the material parameters which are needed to calculate β are given in [7]. It is to mention that the calculation with the aid of n' may be sufficient good, especially when the crack growth rate is small and the hardening/softening rate of the material is high. In any case, β is useful to calculate the exact value of the plastic strain energy density. The values of the parameter β which are used for the calculation differ only a little in comparison with the exponent n': $\beta = 0.17$ for $\Delta\epsilon/2 = 0.3 \cdot 10^{-2}$, $\beta = 0.15$ for $\Delta\epsilon/2 = 0.4 \cdot 10^{-2}$ and $\beta = 0.12$ for $\Delta\epsilon/2 = 0.6 \cdot 10^{-2}$. Therefore, the crack growth parameters C_J and α of Eq. (1) change only slightly when n' in Eq. (2) is replaced by β: $C_J = 3.89 \cdot 10^{-5}$ and $\alpha = 1.16$ (da/dN in mm and ΔJ_{eff} in Nmm/mm^2).

Cyclic J-integral for different geometries

The results of this paper concerning the edge crack in a flat specimen can be compared with the results concerning other geometries. For example, a comparison of the parameters C_J and α of tests with different geometries shows that the equation

$$\Delta J_{eff} = (3.2 \cdot U^2 \Delta W_e + 1.96 \cdot (1/n')^{1/2} \Delta W_p) \cdot a \tag{10}$$

is a good approximation in the case of the semicircular surface crack in contrast to other approaches (for details: see [8]). Equation 10 is important for practical applications because the semicircular surface crack is the typical fatigue crack in turbine blades, pressure vessels and other structures. The above given parameters C_J and α can be used for life prediction. In the case of an elevated temperature $T = 550$ °C the parameters $C_J = 6.92 \cdot 10^{-5}$ and $\alpha = 1.28$ were found for the steel 2.25Cr-1Mo (da/dN in mm and ΔJ_{eff} in Nmm/mm^2), see [6]. The parameters change to $C_J = 1.06 \cdot 10^{-4}$ and $\alpha = 1.14$ when β replaces n' in Eq. (10). It is to mention that the approach of Eq. (10) was used already in [2]. Furthermore, it is to mention that the Eq. (10) differs from Eq. (2) only by a constant factor.

SUMMARY AND CONCLUSION

With the aid of experimental data for the crack tip opening displacement $\Delta\varphi_{250}$ the physical meaning of an approach to calculate ΔJ_{eff} is checked successfully. For future investigation, finite element analysis should be added to calculate crack tip opening displacement, crack closure and ΔJ_{eff} and to compare it with the experimental results. Concerning the experimental results, direct measurements of the crack closure parameter U - the assumption of a constant factor $U = 0.9$ may be a little bit too rough - and also of the crack tip opening displacement $\Delta\varphi_{90°}$ may be useful. Those measurements can be performed by the experimental equipment which is used in this study to measure $\Delta\varphi_{250}$. Additionally, the results of this study should be compared with high-cycle fatigue crack growth data. Such a comparison allows to use known elastic solutions to check different approaches to calculate ΔJ_{eff} also in the case of low-cycle fatigue (see [9]).

REFERENCES

1. Tanaka, K., Hoshide, T. and Sakai, N. (1984) *Engineering Fracture Mechanics*, **19**, 805.
2. Kaisand, C.R. and Mowbray, D.F. (1979) *Journ. of Testing and Evaluation*, **7**, 270.
3. McClung, R.C. and Sehitoglu, H. (1991) *Journ. of Engin. Mat. and Tech.*, **113**, 15.
4. Robinson, J.N. (1976) *Intern. J. of Fracture*, **12**, 723.
5. Rie, K.-T., Wittke, H. and Schubert, R. (1992) In: *Low Cycle Fatigue and Elasto-Plastic Behaviour of Materials - 3*, pp. 514-520, K.-T. Rie (Ed.), Elsevier Applied Science, London and New York.
6. Rie, K.-T. and Wittke, H. (1996) *Fatigue Fract. Engng Mater. Struct.*, **19**, 975.
7. Wittke, H., Olfe, J. and Rie, K.T. (1997) *Int. J. Fatigue*, **19**, 141.
8. Wittke, H. (1997) *Phänomenologische und mikrostrukturell begründete Beschreibung des Verformungsverhaltens und Rißfortschritt im LCF-Bereich.* VDI Verlag, Düsseldorf.
9. Heitmann, H.H., Vehoff, H. and Neumann, P. (1985) In *Advances in Fracture Research (Proceedings of ICF6)*, Vol. 5, Pergamon, pp. 3599-3605.

Low Cycle Fatigue and Elasto-Plastic Behaviour of Materials
K-T. Rie and P.D. Portella (Editors)
585

EFFECT OF STRESS RATIO ON LOCAL FATIGUE DAMAGE ACCUMULATION AROUND NOTCH

Y. IINO

Toyota Technological Institute
468 Nagoya, Japan

ABSTRACT

The stress ratio on fatigue crack initiation from notch of 1mm radius and on fatigue damage accumulation around the notch was studied in the case of a constant maximum cyclic load using notched compact tension specimens of type 304 stainless steel. The damage accumulation was measured by the subsequent recrystallization technique. Accumulated plastic zone with accumulated plastic strain above 0.02 is formed around the notch before fatigue crack initiation at both stress ratio of 0.05 and 0.5. The zone size at the fatigue crack initiation decreases with increasing stress ratio. Highly accumulated plastic zones are also measured.

KEYWORDS

Stress ratio, fatigue crack initiation, notch, fatigue damage accumulation, recrystallization technique.

INTRODUCTION

Fatigue crack initiation from notch has been widely studied from fracture mechanics approach [1,2]. The relation between N_i and a parameter $\Delta K / \sqrt{\rho}$ is usually plotted on log-log scale, where N_i is number of fatigue cycles to crack initiation, ΔK is the stress intensity factor range and ρ is the notch radius. In the experiments, stress ratio R is usually set to be 0.05-0.1. Since the notched components of machine or structure will suffer fatigue with various stress ratio, it will be necessary to examine the effect of R on fatigue crack initiation and also to investigate the fatigue damage accumulation around notch. In the case of $R = -1$, the damage accumulation attending crack initiation in low cycle fatigue was studied by the recrystallization technique [3].

586

In the present experiment, notched compact tension specimens were fatigued to crack initiation under a constant maximum cyclic load at R= 0.05 and 0.5. Fatigue damage accumulation around notch was measured by the recrystallization technique using multiple specimens which were fatigued to various number of cycles before crack initiation.

EXPERIMENTAL PROCEDURE

The material used was the same type 304 stainless steel plate 5.8mm thick as in [3]. Notched compact tension specimen (Fig.1) was machined by electric discharge wire cutting so that the loading direction was parallel with the rolling direction. The notch root was carefully mechanically polished using emery papers of grade number up to 1500. The specimen was fatigued at room temperature using a MTS testing machine under load control of the maximum apparent stress intensity factor $K_{p\text{max}}$ =34.4MPa $\sqrt{m}$ and R= 0.05 and 0.5. During the fatigue test, notch root surface was observed using a stereoscope. The fatigue was continued to various number of cycles. The number of cycles to fatigue crack initiation N_i was defined as the number of cycles after which a $\sim$1mm long crack was visible on the notch root surface. In order to observe the fatigue damage accumulation, the specimens were sectioned at mid-thickness into two parts. The recrystallization technique [4] was used for the damage zone measurement.

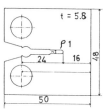

Fig. 1.   Test specimen.

RESULTS AND DISCUSSION

Fatigue crack at notch root surface

R=0.05.   Detection of small fatigue crack by the stereoscope was not easy. So fatigue cycle was continued till the crack grew longer than 1mm. The number of cycle N_i was in range 15000-21000. Number of cycles N of the interrupted specimens before crack initiation was 555, 3000, 6000 and 12000.

R=0.5.   In this case, scatter of N to visible crack formation was larger than at R=0.05; i.e. crack was detected in range 400000-500000, but the fatigue crack in one of the three specimens did not grow even at N=1000000. N of the interrupted specimens was 25000, 210000 and 383000. From the above observations, N_i at $K_{p\text{max}}$ =34.4MPa $\sqrt{m}$ is estimated to be 15000 at R=0.05 and $\sim$400000 at R=0.5. In the case of R= - 1, it was $\sim$550 [3]. The effect of R on N_i was found to be significant.

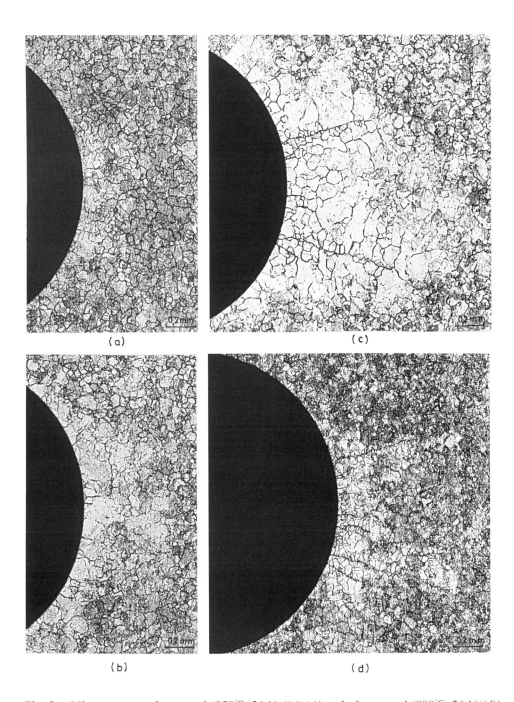

(a)

(c)

(b)

(d)

Fig. 2. Microstructure after anneal (950℃ 24 h) ((a)-(c)) and after anneal (900℃ 24 h)((d)) of the specimen fatigued at stress ratio R=0.05 to number of cycles N=(a) 555, (b) 6000, (c) and (d) 17000. The recrystallized zone shows the accumulated plastic zone with accumulated plastic strain above 0.02((a)-(c)) and above 0.12((d)).

588

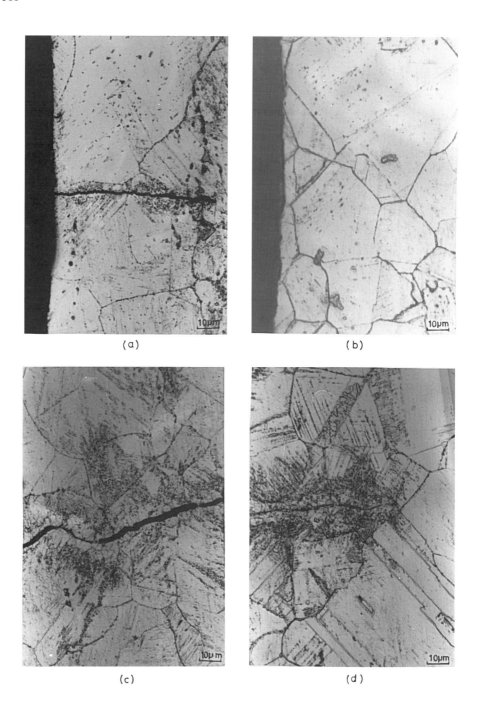

(a)

(b)

(c)

(d)

Fig. 3. Microstructure after anneal (750℃ 24 h) of the specimen fatigued at stress ratio
$R=0.05$((a), (c),(d)) and 0.5((b)). Number of cycles $N=$(a), (c), (d) 17000, (b) 540000.

Accumulated plastic zone

Observation of the accumulated plastic zone APZ was made of the specimens fatigued to
N=555, 6000 and 17000 at R=0.05 and N=25000, 210000 and 540000 at R=0.5. Fig.2 shows
the microstructure after the recrystallization anneal of the specimens fatigued at R=0.05. Some
recrystallized grains can be seen just at the notch tip at N=555 (Fig. 2(a)) and along the notch
tip at N=6000 (Fig. 2(b)). This means that APZ with accumulated plastic strain ε_{ac} above 0.02,
$APZ_{0.02}$ increases with N. At N=17000, some fatigue cracks are seen and the longest is ~1mm
long. The recrystallized zone is formed not only around the notch tip but also around the
fatigue cracks. Recrystallized zone showing APZ with ε_{ac} above 0.12, $APZ_{0.12}$ was not formed
at N=555 and 6000, but can be seen around the notch tip and around the fatigue cracks at
N=17000 (Fig. 2(d)). $APZ_{0.02}$ in the case of R=0.5 is also formed around the notch before
fatigue crack initiation, although the zone size is smaller than at R=0.05, i.e. at N=25000, slight
zone formation and at N=210000 and 540000, smaller zone size than at R=0.05. Fig.3
illustrates the microstructure at the notch tip and fatigue crack after the anneal (750°C 24 h).
No recrystallized region was detected along the notch edge in all the specimens (Fig. 3(a), (b)).
The recrystallized zone which shows APZ with ε_{ac} above 0.5, $APZ_{0.5}$ was detected only along
the fatigue crack as Fig.3(c), (d). Such recrystallized zone is formed also in case of R=0.5,
although smaller than at R=0.05. Thus $APZ_{0.5}$ is found to be formed just at the crack tip and
just near the crack as rake at R=0.05 and 0.5.

Increase of $APZ_{0.02}$ around the notch at R=0.05 with N is shown in Fig.4. $APZ_{0.02}$ at fatigue
crack initiation can be estimated by substituting the $APZ_{0.02}$ around the fatigue crack from the
total $APZ_{0.02}$ around the notch. It is shown in Fig.4 as a dotted line. Fig.5 shows the effect of R
on the zone size of $APZ_{0.02}$ ahead of the notch tip to the zone front in the crack growth direction,

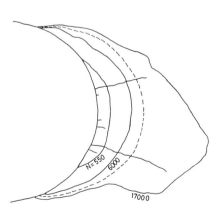

Fig. 4. Increase of the accumulated plastic zone with accumulated plastic strain above 0.02 at
R=0.05 with number of cycles N. The fatigue cracks are in the specimen fatigued to N=17000.
Dotted line is the estimated zone size at fatigue crack initiation.

590

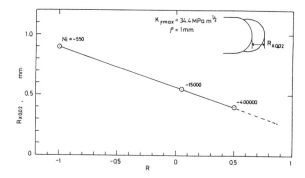

Fig. 5. Relation between stress ratio R and the estimated zone size in crack growth direction of the accumulated plastic zone with accumulated plastic strain above 0.02, $R_{x0.02}$ at fatigue crack initiation. N_i: number of cycles to fatigue crack initiation.

$R_{x0.02}$. It is seen that $R_{x0.02}$ decreases with increasing R. Extrapolated value of $R_{x0.02}$ to $R=1$ is 0.2mm. Since fatigue crack initiation threshold does exist between $R=0.5$ and 1, it could be said that $APZ_{0.02}$ is formed around the notch even at the fatigue limit. Barson and McNicol [2] showed that fatigue crack initiation threshhold under zero-to-tension loads($R=0$) in notched specimens of HY-130 steel is $\Delta K/\sigma_{ys} \sqrt{m} = 0.6$, where σ_{ys} is yield strength. If the relation is used for the present experiment, although the material and stress ratio are different, the calculated value of ΔK is 4.55MPa $\sqrt{m}$. The corresponding stress ratio is 0.87. The estimated $R_{x0.02}$ at the fatigue crack initiation threshold is 0.25mm.

CONCLUSIONS

Accumulated plastic zone with accumulated plastic strain above 0.02 is formed around the notch before fatigue crack initiation at both $R=0.05$ and 0.5. The zone size increases with number of cycles. The zone size at fatigue crack initiation decreases with increasing stress ratio. Accumulated plastic zone with accumulated plastic strain above 0.5 was not detected around the notch tip. However it is formed very adjacent to fatigue crack.

REFERENCES

1. Jack, A. R. and Price, A. T. (1970). *Int. J. Fract. Mech.* **6**, pp.401-09.
2. Barson, J. M. and McNicol, R. C. (1974). In: *Fracture Toughness and Slow Stable Cracking*, ASTM STP-559, ASTM, Philadelphia, pp.183-204.
3. Iino, Y. (1995). *Metall. Mater. Trans. A* **26A**, pp.1419-30.
4. Iino, Y. (1992). *J. Mat. Sci. Lett.* **11**, pp.1253-56.

Low Cycle Fatigue and Elasto-Plastic Behaviour of Materials
K-T. Rie and P.D. Portella (Editors)
591

MODELLING OF THE CREEP FATIGUE CRACK GROWTH BASED ON THE CALCULATION OF THE CAVITY CONFIGURATION IN FRONT OF THE CRACK AND ITS EXPERIMENTAL VERIFICATION

J. OLFE and K.-T. RIE
Institut für Oberflächentechnik und plasmatechnische Werkstoffentwicklung,
Technische Universität Braunschweig, Bienroder Weg 53, 38108 Braunschweig, Germany

ABSTRACT

A model for simulating the creep fatigue crack growth is proposed and has been verified experimentally for the austenitic stainless steel 304L. The simulation is based on the calculation of the local cavity configuration at the crack tip. In this model, the local strain which has been measured in front of the crack is responsible for the inhomogeneous cavity formation. The simulation leads to the same crack growth rates as measured. Furthermore, the calculated cavity density in front of the crack which is responsible for crack growth in this model is similar to the density measured by stereometric metallography.

KEYWORDS

Creep-fatigue, cavity, computer simulation, crack growth, local damage

INTRODUCTION

Austenitic steels often fail under high temperature service conditions due to the formation and growth of grain boundary cavities and fatigue cracks. In the creep fatigue regime the fatigue crack interacts with the grain boundary cavities thus leading to intercrystalline crack growth and failure. The formation of cavities strongly depends on the strain, while the growth of the cavities is governed by the stress. The stress and strain in front of the crack differ considerably from the values in the uncracked material far from the crack tip. As a consequence the formation and growth of the cavities is enhanced in the vicinity of the crack tip compared to the rest of the material. Therefore not only the cavities are influencing the crack path but also the crack influences the cavity configuration in the material in front of the crack tip.
Therefore a new crack growth model for the creep fatigue regime has been developed and experimentally been verified which takes into consideration the local conditions in front of the crack.

EXPERIMENTAL DETAILS

Strain controlled LCF-tests at 600°C were carried out on specimens of the austenitic stainless steel 304L. The strain rate of the tests was 10^{-3} s^{-1}. The strain amplitudes were $\varepsilon_a=0.4\%$ and $\varepsilon_a=0.6\%$. In every tension maximum a hold period of 10 min respectively 60 min was introduced to produce grain boundary cavities, the typical creep damage for this material. Specimens with a rectangular cross section (8.7x5 mm^2) and a gauge length of 10 mm were used. The crack growth and the local strain in front of the crack were measured with a special optical system, which allows video monitoring and photographing of the specimen surface in-situ during the high temperature creep-fatigue test. The crack growth was measured after the test by evaluating the video film. For measuring the local strain the grid method was used. For this purpose the specimens were sprayed with a grating which consists of TiO$_2$ particles with a line distance of 200µm. The grid was photographed at the start of the test and after cycling whenever the local strain should be measured. By evaluating the displacement of the grid crosses by means of a image analysing system the local strain could be calculated. For more details see [1].

Tests were performed with the same parameters and were stopped after different number of cycles to measure the local cavity configuration in dependence on crack length.

After low cycle fatigue testing the specimens were metallographically prepared for analysing the cavity density and size distribution in front of the crack. For this purpose the cavities were photographed by SEM and the cavity density and the distribution of the radii on the polished surface was detected in dependence on the distance to the crack tip. To calculate the corresponding values on the grain boundaries it was supposed that all cavities are on boundaries perpendicular or less than 30° oriented to the load axis and that all grains are of identical size which is the mean value [2,3].

CAVITY CONFIGURATION IN FRONT OF THE CRACK

The results of the stereometric metallography concerning the cavity density for a test with a 10 min hold period in tension are shown in Fig.1. The cavity density in a cyclically deformed material mainly depends on the strain ([2,3], see also equation 1).

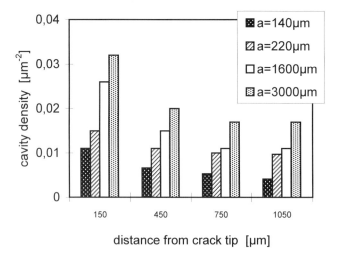

Fig. 1 cavity density on grain boundaries vs. distance to crack tip for different crack lengths a. (304L, 600°C, ε_a=0.6%, tension hold in every cycle: 10 min)

Due to the strain concentration at the crack tip the cavity density for all crack lengths is high at the crack tip and decreases with increasing distance to the crack. Also from Fig.1 the conclusion could be drawn that the longer the crack, the higher is the cavity density in front of the crack tip. Similar results are obtained for tests with 60 min hold period.

CONCEPT OF THE SIMULATION MODEL

A new crack growth model for the creep fatigue regime has been developed which takes into consideration the local conditions in front of the crack. The modelling is divided into 6 steps.

Step 1: Measurement of the local strain

As a first step the local strain in front of the crack has been measured by means of the grid method. The results of the measurements are already published [1,3]. The strain decreases with increasing distance from the crack tip because of the high stress concentration at the crack.

Step 2: Discretization of the strain field

In a second step the region in front of the crack has been divided into several small intervals supposing the strain and stress in every interval to be constant whereas they differ from interval to interval (Fig. 2).

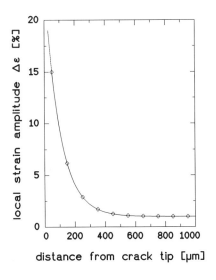 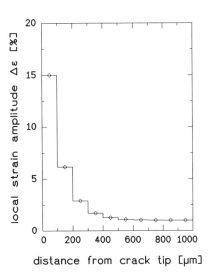

Fig. 2 Discretization of the strain field in front of the crack

Step 3: Setting of the grain boundary orientation and the grain size

In a third step the orientation of the grain boundary on which the crack progresses and the grain size are set by means of a random generator. The mean value of the grain size for simulation is controlled to be the same as the mean value of the material which is used for the experiment.

Step 4: Calculation of cavity formation and growth for every interval

The local cavity density n_{loc} in every interval is calculated from the local strain range $\Delta\varepsilon_{loc}$ by adopting the cavity formation model of the authors which has already been verified in the case of uncracked material and homogeneous strain distribution [2,3]:

$$n_{loc} = p\ \Delta\varepsilon_{loc}\ N^{\kappa} \tag{1}$$

N is the number of cycles, the cavity formation factor p and the exponent κ are constant for the material. The local cavity growth is calculated by means of a cavity growth model [4] which has also been successfully adopted in the case of uncracked material [2]. The cavity growth rate depends particularly on the local stress which is calculated from the strain taking into consideration the grain boundary orientation, and also on the cavity density n_{loc}. For every cycle the cavity density in the interval in front of the crack and the cavity growth during the hold period is calculated. The largest cavities are those which have been formed in the first cycle leading to a corresponding cavity size distribution.

Step 5: Determination of the failure of the interval in front of the crack

When a critical cavity configuration is reached the material in the interval in front of the crack fails and the crack advances to the next interval.
The failure criterion is:

$$\frac{\delta}{2} = \beta\ (\lambda_{loc} - 2\ r_{loc}) \tag{2}$$

δ is the crack tip opening displacement which can be calculated as follows:

$$\delta = a\ (K_1\ \Delta\varepsilon_{el} + K_2\ \Delta\varepsilon_{pl} + K_3\ \varepsilon_{cr}) \tag{3}$$

a is the crack length, $\Delta\varepsilon_{el}$, $\Delta\varepsilon_{pl}$ and ε_{cr} are the elastic, plastic and creep part of the total strain and K_1=4, K_2=6.1 and K_3=8 are constants for the 304L stainless steel.
This criterion with β=1 has already been applied successfully in the case of life prediction and critical crack growth [3,5].

Step 6: Calculation of the crack growth

The calculations are repeated for the next interval (step 3 and 5) taking into consideration that for this interval cavities are already formed and have grown during the previous cycles with a smaller strain and stress amplitude. Step 3 was only operative if the crack tip has reached the end of the simulated grain boundary.

By repeating this procedure the crack growth can be simulated until the material fails provided that the interaction of the crack with the cavities is the dominant damage process.

COMPARISON OF THE MODEL AND THE EXPERIMENT

In the following figure (Fig. 3) an example for the measured crack growth rate in dependence on crack length and the result of a simulation are shown. The good agreement is also obtained for tests with 10 min and 60 min tension hold periods.

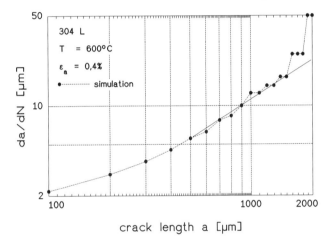

Fig. 3 comparison of simulated (dots) and experimental crack growth rate da/dN
(tension hold in every cycle: 30 min)

A comparison of the measured cavity density and the cavity density as a result of simulation shows that the calculated cavity density is in the right order of magnitude and that the principal increase of the density with decreasing distance to crack tip is a typical feature of the proposed model (Fig.4).

In the experiment also micro cracks in front of the main crack are formed due to cavity coalescence. However, this kind of cracks is not taken into consideration for the simulation. Therefore the measured cavity density is smaller than the results of the simulation.

596

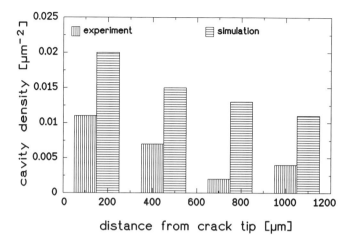

Fig. 4 experimentally detected and simulated cavity density on grain boundaries vs. distance to crack tip for a crack of lengths a=200μm. (304L, 600°C, ε_a=0.6%, tension hold in every cycle: 60 min)

SUMMARY AND CONCLUSION

With the proposed model the crack growth rate can be simulated in the case of creep fatigue if cavitation is the dominant failure process. The orientation of the grain boundaries in front of the crack is taken into consideration.

Furthermore the simulation leads to a cavity configuration which is similar to the experiment. The cavity density is high at the crack tip and decreases with increasing distance to the crack. Differences between simulation and experiment regarding the absolute value can be explained by micro crack formation. Therefore better simulation results can be expected by introducing in the model the formation of micro cracks in front of the crack tip, which will reduce the simulated cavity density.

LITERATURE

[1]   K.-T. Rie, J. Olfe: In-situ measurement of local strain at the crack tip during creep-fatigue, in Proc. of the International Symposium: Local Strain and Temperature Measurements in Non-Uniform Fields at Elevated Temperatures, March 14-15 Berlin 1996

[2]   K.-T. Rie, J. Olfe: Computer Simulation and Experimental Verification of Cavity Formation and Growth during Low Cycle Fatigue, Proc. of the 6th Intern. Fatigue Congress, Fatigue '96, Berlin 1996, Pergamon, 765-770

[3]   J. Olfe: Wechselwirkung zwischen Kriechschädigung und Low Cycle Fatigue und ihre Berücksichtigung bei der Berechnung der Lebensdauer, Dissertation TU Braunschweig, Papierflieger, Clausthal-Zellerfeld, 1996

[4]   Hull, D., Rimmer, D. E.: The Growth of Grain-Boundary Voids Under Stress. Philosophical Magazine 4 (1959) 673 - 687

[5]   K.-T. Rie, J. Olfe,"A Physically Based Model for Predicting LCF Life under Creep Fatigue Interaction", in:Proc. 3nd Int. Conf. on Low Cycle Fatigue and Elasto-Plastic Behaviour of Materials, K.-T. Rie (Hrsg.), Elsevier Applied Science, London/New York (1992), 222-228

Low Cycle Fatigue and Elasto-Plastic Behaviour of Materials
K-T. Rie and P.D. Portella (Editors)
© 1998 Elsevier Science Ltd. All rights reserved.

CRACK PROPAGATION BEHAVIOUR OF THE TITANIUM ALLOY IMI 834 UNDER HIGH TEMPERATURE CREEP-FATIGUE CONDITIONS AND FINITE ELEMENT ANALYSIS

T. KORDISCH and H. NOWACK
Gerhard-Mercator-University of Duisburg, Department of Materials Technology,
Lotharstr.1, D-47048 Duisburg, Germany

ABSTRACT

In the present study the interaction of creep and fatigue at the titanium alloy IMI 834 under cycles with different time dependent and time independent parts was investigated, and, furtheron, how these cycles influenced the propagation behaviour of long cracks. The test temperature was 600 °C.

It was found that crack propagation behaviour was significantly influenced by the type of creep-fatigue cycles applied. A slow increase of load into tension direction within the creep-fatigue cycles had a most detrimental influence on propagation of long cracks.

For an evaluation of the stress and strain distributions at the crack tip under pure fatigue and under creep-fatigue conditions a Finite Element (FE) analysis was performed. The results showed that the stresses, strains and the CTOD-values in the vicinity of the crack tip also develop in a different way depending on the type of creep-fatigue loading applied.

All results of the present study indicated that crack propagation under creep-fatigue loading is a complex interaction of crack tip deformations and environmental influences.

KEYWORDS

Crack propagation, Titanium alloy IMI 834, Creep-fatigue interaction, Finite Element Method

INTRODUCTION

In the present study the creep and fatigue resistant near-α titanium alloy IMI 834 was investigated. This alloy is primarily used for disks and blades in the compressor part of new aeroengines. The alloy investigated exhibited a bi-modal microstructure consisting of primary α-grains at a volume fraction of about 20 % within a lamellar matrix of transformed β-phase. In order to achieve this microstructure the alloy was special thermomechanically treated including ageing at 700 °C for 2h.

598

Under real service conditions of aeroengine components an interaction of fatigue, creep and of the environment occurs. For a safe operation the combined effect of these mechanisms and influences has to be realistically predicted for all stages of fatigue life. The LCF crack initiation life behaviour as it was observed in tests and an evaluation of prediction methods for that life range have already been published [1].

In the present paper the long crack propagation behaviour under different loading conditions for a maximum intended service temperature of 600 °C will be presented.

In order to get more detailed information about the mechanical causes for the observed creep-fatigue crack propagation FE-analyses of the crack tip region were performed. Results will be given.

CRACK PROPAGATION EXPERIMENTS

The crack propagation tests were performed on single edge notched (SEN) specimens (width: 12 mm, thickness: 2.5 mm and length: 46 mm) at R=0 with a maximum applied cross section stress of σ_{max}=250 MPa in laboratory air at 600°C. Crack propagation was measured using the DC potential drop method. Tests with pure fatigue cycles were performed at 2Hz. In the tests with creep-fatigue cycles the cycles exhibited once a slow ramping into tension and/or into compression and once a fast ramping into tension followed by a hold time period (comp. Fig. 1). The cycles with a slow ramping are denoted as slow-fast, slow-slow and fast-slow cycles and the cycles with a hold time as hold time cycles. The fast ramping always occurred whithin 0.25s. The different creep-fatigue cycles exhibited a duration of 90s, 180s or 360s.

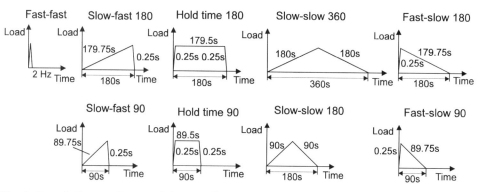

Fig. 1. Pure fatigue and creep-fatigue cycles as applied in the crack propagation tests. The stress ratio was R = $\sigma_{min}/\sigma_{max}$ = 0 and σ_{max} = 250 MPa (600 °C).

RESULTS AND DISCUSSION

Fig. 2 shows that the crack propagation rates in the tests with creep-fatigue cycles were up to a factor of 20 higher than in the tests with pure fatigue (fast-fast) cycles. Although the ramping in the slow-fast and in the slow-slow cycles occurred always at a same velocity (that means that the duration of the slow-slow cycles was twice as large as that for the slow-fast cycles) the crack propagation rates under the slow-fast cycles were slightly higher than under the slow-slow cycles, especially in the small crack regime.

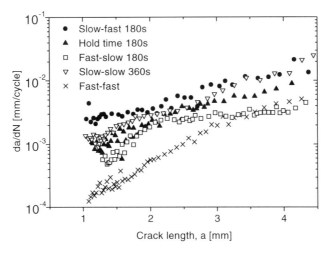

Fig. 2. Experimental crack propagation behaviour under pure fatigue cycles and under creep-fatigue cycles with a different ramping and cycles with hold times (R=0, σ_{max} =250 MPa, 600 °C).

Figure 3 shows the results from the crack propagation tests with slow-fast and with hold time cycles. It can be seen that under slow-fast cycles with a cycle duration of 90s or 180s the cracks propagated up to a factor of 4 faster than under hold time cycles with the same cycle duration. With increasing crack lengths the differences in the crack propagation rates due to the different types of creep-fatigue cycles decreased.

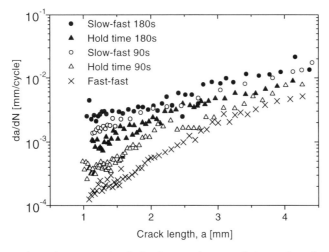

Fig. 3. Experimental crack propagation behaviour under pure fatigue, slow-fast and hold time cycles (R=0, σ_{max} =250 MPa, 600 °C).

Microscopical investigations showed that crack propagation occurred transcrystalline under all types of creep-fatigue cycles. Typical signs of creep damage as cavities could not be observed. Furtheron, the strong dependence of crack propagation on the cycle type cannot simply be explained by creep influences because in the tests with hold time cycles where higher creep strains were expected than in tests with slow-fast cycles the crack propagation rates were lower. In the tests with slow-fast cycles the crack propagation rates were also higher than in the tests with slow-slow cycles. That means that a detrimental influence of a slow ramping into tension direction is not the only reason for the observed behaviour. That means that complex environmental influences exist which depend on the type of loading.

Hydrogen and oxygen may cause an embrittlement of the crack tip region. During the upgoing ramping of the slow-fast cycles deformation permanently accumulates at the crack tip [2]. This can enhance the transport of hydrogen or oxygen to the crack tip region and accelerates crack propagation. Furtheron, during the slow ramping into tension during the cycle a close oxidation layer cannot form at the crack tip because it is permanently ruptured. In the tests with hold time cycles a close oxygen layer can develop.

Other strain-rate sensitive mechanisms may become important, too. When results of other investigations which are reported in the literature [3, 4] are also considered there remain still several questions open regarding the causes for the fast crack propagation under slow-fast cycles. Similar results had also be found for the titanium alloy Ti-1100 at high temperatures in vacuum [4]. This indicates that the deformation at the crack tip may also play an important part - besides the complex environmental influences.

FINITE ELEMENTE ANALYSIS

For an evaluation of the crack tip strains and stresses under different types of creep-fatigue cycles FE-analyses were performed using the FE-code ABAQUS® 5.6 and PATRAN® 6.2 as pre- and post-processor. The SEN-specimen as had been used for the tests was modelled for crack lengths of 1.5, 2.5 and 4 mm using 8-node biquadratic elements (2D-solid, plane stress). The crack tip was modelled using special crack tip elements where one side of the element collapsed in a manner that all three nodes were concentrated at the crack tip with a possibility that they could move independently. The quarter-point technique was also applied which led to a $1/r + 1/\sqrt{r}$ strain singularity at the crack tip. For the calculation the specimen was loaded perpendicular to the crack propagation direction up to the maximum cross section stress of 250 MPa. Once a fast load increase as in the pure fatigue cycles was considered and once loading patterns as in the creep-fatigue cycles with a duration of 89.75s or 179.75s. For the later type of loading patterns the load was once steadily slow increased and once the load was fast increased and hold times of 89.5s or 179.5s at maximum stress were applied in the calculations. The elastic-plastic material law for the FE-calculations was the experimentally determined cyclic stress-strain curve at 600 °C which was piecewise linearised. Creep was considered using the following equation for, $\dot{\varepsilon}_{cr}$ (creep strain rate):

$$\dot{\varepsilon}_{cr} = A \cdot \sigma^n \tag{1}$$

The parameters of the equation were determined from creep tests at 600 °C; n was 5.045 and A=$3.91 \cdot 10^{-19}$. At each calculation step both creep and elastic-plastic material behaviour were simultaneously considered. The step width was maximum 0.2 s.

The FE-results are shown in Fig. 4 and 5. It can be seen that during a hold time at maximum stress a redistribution of the local stresses in the vicinity of the crack tip occurs which becomes stronger as the hold time is increased. At the end of the hold time period of 179.5s the maximum tensile stress dropped below the calculated maximum stress after a slow ramping. This trend was also observed in cyclic FE-calculations assuming a linear kinematic hardening law after one and a half creep-fatigue cycle.

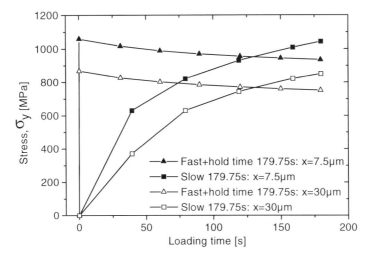

Fig. 4. FE-results for a crack length, a, of 1.5 mm for loading up to a cross section stress, σ_{max}, of 250 MPa (at 600 °C). The stresses into the loading (y-) direction at a distance, x, from the crack tip into the crack propagation direction are shown (the duration of the loading has been indicated).

The calculated crack tip opening displacement CTOD was also by 25 % larger after a hold time than after a slow ramping.

In Fig. 5 the creep strains and the inelastic strains are shown because under creep-fatigue conditions the strains in the vicinity of the crack tip are also of importance.

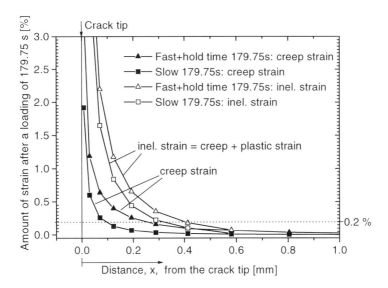

Fig. 5. FE-results of creep and of total inelastic strains into loading direction for a crack length of 1.5 mm as function of the distance, x, from the crack tip (σ_{max} = 250 MPa, loading time 179.75s, 600 °C).

From Fig. 3 and 5 it can be seen that despite of the fact, that the cracks propagated faster under slow-fast cycles in the tests, the calculated creep strains and total inelastic strains (plastic + creep strain) remained smaller than under a hold time cycle. Under hold time loading the inelastic zone which developed became also larger than after a slow ramping or after a fast ramping. In conclusion, the FE-calculations showed that the stresses and strains in the vicinity of the crack tip depended significantly on the type of loading, but could not explain the observed trends in the tests.

Fracture mechanics parameters as the J-Integral and the C(t)-Integral were also evaluated from the above FE-analyses. It turned out that neither the J- nor the C(t)-integral were appropriate parameters to correlate the experimentally observed crack propagation rates under fast-fast, slow-fast and hold-time cycles.

SUMMARY AND CONCLUSIONS

The different types of creep-fatigue cycles investigated significantly influenced the crack propagation behaviour at 600 °C. A slow ramping into tension (slow-fast cycles) had a most detrimental influence on long crack propagation. This has to be considered for damage tolerance design of practical components. A satisfactory microscopical explanation of the observed behaviour could not be found so far. Neither changes in the crack path occurred nor the fracture surface or the other microscopical investigations revealed signs of creep damage.

FE-analyses of the crack tip region for loading patterns which simulated the pure fatigue and creep-fatigue cycles were performed. They showed that the creep deformations did not correlate with the trends in the crack propagation behaviour. However, differences in crack tip blunting and in the redistribution of tensile stresses may lead to differences regarding the environmental influences as the formation of an oxidation layer or the transport of oxygen and hydrogen or regarding the microstructural creep mechanisms [5, 6].

Further microstructural investigations, especially in the crack tip area have to be performed. The utilization of another material model, like the extended Jiang model (which is not yet available in commercial Finite Element codes) may also be advantageous.

ACKNOWLEDGEMENT

The authors thank Prof. Gysler and Prof. Lütjering (TU Hamburg-Harburg) for the thermomechanical treatment of the material and for the fruitful discussion and DFG for financial support.

REFERENCES

1. Kordisch, T. and Nowack, H. (1998) *Fatigue Fract. Engng Mater. Struct.* in print.

2. Rodriguez, P. and Bhanu Sankara Rao, K. (1993) *Progress in Material Science* **37**, 403.

3. Specht, J.U. (1992). In: *Proceedings LCF 3*, pp. 19-24; K.-T. Rie (Ed.). Elsevier Applied Science, London, UK.

4. Foerch, R., Madsen, A. and Ghonem H. (1993) *Metall. Trans. A* **24**, 1321.

5. Gregory, J.K. (1994). In: *Handbook of Fatigue Crack Propagation in Metallic Structures, Vol 1*, pp. 281-322; A. Carpinteri (Ed.). Elsevier Science, Amsterdam, The Netherlands.

6. Evans, W.J. and Bache, M.R. (1995) *Scripta Metall. Mater.* **32**, 1019.

Low Cycle Fatigue and Elasto-Plastic Behaviour of Materials
K-T. Rie and P.D. Portella (Editors)
603

NUMERICAL SIMULATION OF VISCOPLASTIC CRACK CLOSURE

H. ANDERSSON*, C. PERSSON* AND N. JÄRVSTRÅT**
* *Materials Engineering, Lund University, SE - 221 00 LUND, SWEDEN*
** *Volvo Aero Corporation, SE - 461 81 TROLLHÄTTAN, SWEDEN*

ABSTRACT

In this work, numerical simulations of crack closure for a superalloy under out of phase thermo mechanical loading have been carried out. Particular attention is paid to the evolution of crack closure with the number of cycles. When analyzing crack growth with a viscoplastic material model, more than 10 cycles at constant crack length were found to be necessary to the reach steady state crack opening stress. Further, it is shown that imposing a hold time in compression or increasing the maximum temperature has nearly equivalent effects on the evolution of crack closure.

KEYWORDS

High temperature, crack growth, superalloys, numerical simulation, FE, Bodner, hold time, crack opening stress, viscoplasticity

INTRODUCTION

When designing high temperature applications, like gas turbines and jet engines, estimating life is a crucial but difficult task. There are many different approaches for life prediction, but the estimated life often differs from real life by a factor of two or more. This problem is dealt with by introducing safety factors. However, smaller safety factors can be translated into significant weight savings in a component. Clearly seen from the arguments above, there is great need for a better way to calculate life.

Today life can be calculated in many ways, with e.g. strain range partitioning, different modified Coffin-Manson expressions, cumulative models based on Palmgren and Miner's work or (Linear) Elastic Fracture Mechanics. In order to improve the latter calculation, it is common to use ΔK_{eff} instead of ΔK. By doing this, only the part of the cycle when the crack is open is taken into account, or in other words, the crack only grows when it is open. However, all of the methods mentioned above have some drawbacks. In order to give a good prediction, any of the currently used methods must be calibrated for loads very close to actual use. Even then, a life under service conditions might be significantly different than calculated. This means that much of the design today is based on an empirical understanding of previous designs.

One way of using fracture mechanics is to use a modified Paris' law, $da/dN = C(\Delta K_{eff})^m$ and then integrate over the crack length. This can be rewritten as $da/dN = C(\Delta K)^m \cdot U^m$, where U

is a function of stress, temperature, material parameters, crack length, etc. In this work, the parameter U is investigated in a cyclic load with crack growth as a function of cycle parameters e.g. temperature, strain range and compressive hold time.

GEOMETRY AND LOADING

To examine the effects of some load parameters on crack opening stress, a number of simulations have been carried out. A simple geometry was used during the analyses, see Fig. 1.

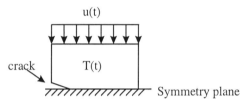

Fig. 1. Geometry used in simulations. The piece was two units wide by one unit high. Load was applied by prescribing displacement of top boundary and the homogeneous temperature as functions of time. The lower edge was next to a rigid surface and which was thus preventing over closure. In this way it was possible to capture contact deformation. One node was fixed in the direction perpendicular to the load to prevent rigid body motion. Crack growth was from the left edge to the right along the rigid surface.

To isolate the effects from crack growth, no notch was incorporated in the model. Since cracks always form within a material, this is no serious shortcoming. This means that the only length scale in the model is the crack itself. Thus, when the plastic zone is large compared to the element size, but small in comparison with the model, the simulation is a good approximation of a crack growing in a homogeneous applied far field strain.

For these numerical simulations, the material model used was developed by S. R. Bodner et al and further refined by numerous other researchers [e.g. 1 - 3]. This constitutive model is a unified visco-plastic (time dependent) model and is thus capable of capturing creep as well elastic and plastic loading. The material properties were derived from tests on B1900Hf by NASA [3]. In the analysis, four node plane stress elements were used to model a thin plate material.

Crack opening in this study was assumed to take place when the node next to the crack tip leaves the rigid surface. Therefore, the movement of this node is of interest. We have studied the simple, but still realistic, loadcase Out of Phase (OP) Thermo Mechanical Fatigue (TMF). Hold time, maximum temperature and minimum strain have been varied in the TMF cycle to find the influence of each parameter.

Case 1: Out of Phase TMF on smooth B1900Hf

For the first case simulated, the load cycle followed the pattern shown in Fig. 2.

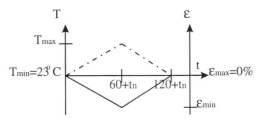

Fig. 2. Load case for TMF simulations with no hold time. The dash dotted line is temperature and the solid line is mechanical strain. t_n is total time after n cycles (120 x n).

Different crack growth schemes were applied, from cycling one full cycle per crack increment to cycling 12 times per crack increment. Comparing the results from these simulations yielded one similarity. With increasing number of cycles per crack increment, the cracks became open for a larger fraction of the cycle. For some load cases, the crack could close for a few cycles and after that it would remain open during the entire load cycle, see Fig. 3.

Figure 3 shows the displacement of the node next to the crack tip as a function of cycle number. For the case with no hold time, $T_{max} = 1000°$ C and $\varepsilon_{min} = -0.5\%$, the crack closed six times and the maximum displacement at cycle 12 was 6.6 x 10^{-4} units. This gives a certain crack opening history that obviously will change when the load changes. In this case, maximum temperature and minimum strain are the independent parameters. A number of simulations similar to those ones in Fig. 4 have been made. They are summarized in table 1.

Table 1. Summary of simulations without hold time. The first number in the table is displacement of the node next to the crack tip in the 12th cycle at crack increment 20 The last number is the number of crack closures. For the relatively low loaded cases crack closure was observed during the 11 full cycles.,

| | $T_{max}(°C)$ | | | | | | | |
$\varepsilon_{min}(\%)$	800	900	925	950	975	1000	1025	1050
-0.4	0.47 \ > 11	0.57 \ > 11		1.0 \ > 11		1.9 \ > 11	2.9 \ 8	4.2 \ 4
-0.45					2.3 \ 10	3.3 \ 10		
-0.50	1.2 \ > 11	1.5 \ > 11		2.7 \ > 11	3.9 \ > 11	6.6 \ 6	10.9 \ 3	17.2 \ 0
-0.55				5.1 \ > 11	7.7 \ 8	11.4 \ 4		
-0.6	3.6 \ > 11	5.0 \ > 11	6.3 \ > 11	8.4 \ 11	11.2 \ 6	16.1 \ 3		

Case 2: Out of Phase TMF on smooth B1900Hf with compressive hold time

Crack length and the number of cycles at each crack length as discussed above are only two of several factors contributing to the crack opening stress. One of the other factors is the hold time, which will be discussed in the following section together with cycle number dependence.

To investigate the effects of hold time, the cycle in Fig. 3 was employed:

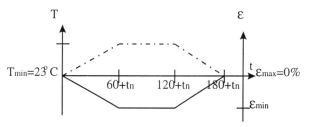

Fig. 3. TMF cycle with compressive hold time.

As expected, the compressive hold time changed the curve a great deal. One possible reason for this is that the mean stress was higher due to compressive creep. This means that the stress state while in compression relaxes and when pulled apart a larger tensile stress will be applied at the crack tip. Because of this the crack will open earlier. To find curves similar to the ones from case 1, the temperature had to be decreased about 50 °C as illustrated in Fig. 4:

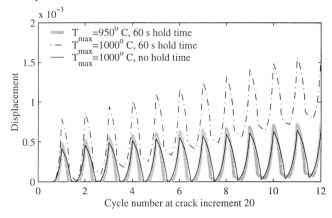

Fig. 4. Displacement of node next to crack tip. $R_\varepsilon = -\infty$. No notch effects. Constant crack length. ε_{min} = -0.5% (mechanical).

A number of simulations with compressive hold time have been conducted and are summarized in table 2.

Table 2. Summary of simulations with 60 s hold time

| $\varepsilon_{min}(\%)$ | $T_{max}(°C)$ | | | | |
	800	900	925	950	1000
-0.50	1.6 \ > 11	3.1 \ > 11		7.4 \ 6	16.3 \ 0
-0.6		8.7 \ > 11	11.6 \ 5		

Comparing table 2 with table 1 it can be seen that crack history with respect to number of closures and maximum displacement of node next to crack tip for this crack increment can be translated to case 1 by decreasing the temperature. Since secondary creep rate follows an Arrhenius' expression, the correlation between hold time at maximum temperature and maximum temperature itself was not surprising.

CRACK OPENING STRESS

One common way of describing crack growth is by using (linear) fracture mechanics. This is usually written in the form $da/dN = C(\Delta K)^m$. If one assumes that cracks grow only when they are open, this can be modified as $da/dN = C(\Delta K_{eff})^m$ where ΔK_{eff} is defined as K_{max} - K_{open}. K_{max} and K_{open} are the stress intensity factors at maximum load and when the crack open respectively. ΔK_{eff} can be rewritten via

$$\frac{\Delta K_{eff}}{\Delta K} = \frac{K_{max} - K_{open}}{K_{max} - K_{min}} = \frac{\sigma_{max}Y\sqrt{\pi a} - \sigma_{open}Y\sqrt{\pi a}}{\sigma_{max}Y\sqrt{\pi a} - \sigma_{min}Y\sqrt{\pi a}} = \frac{\sigma_{max} - \sigma_{open}}{\sigma_{max} - \sigma_{min}} = U$$

where σ_{max} is maximum global stress, σ_{min} is minimum global stress, σ_{open} global stress when the crack opens, a is crack length and Y is a dimensionless constant depending upon crack shape, loading, etc. The resulting U is a factor between 0 and 1 describing how large fraction of the tensile part of the cycle during which the crack is open. U equal to one means that the crack does not close and $U = 0$ that the crack does not open. This means that Paris' law can be rewritten as $da/dN = C(\Delta K)^m \cdot U^m$ where ΔK, C and m can be obtained via standard methods and U must be determined experimentally or calculated. This form of Paris' law has been used with some success in e.g. short crack propagation where the original form predicts the crack growth to be too slow[4]. In this investigation, the factor U has been studied to find dependencies of compressive hold time, maximum temperature and minimum strain. The reason for investigating U is that ΔK is not applicable when the loads are too high. In these cases ΔK_{eff} might be used with some success. Even if it turns out that ΔK_{eff} is not applicable in the present cases, the factor U is believed to be an important factor for crack growth.

For some cases, U at a crack increment is plotted against cycle number, see Fig. 5:

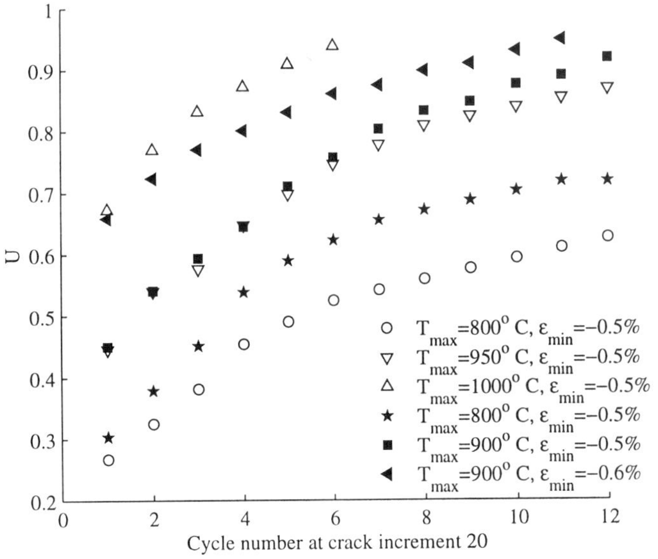

Fig. 5. Crack opening stress vs. cycle number for a stationary crack. The filled markers are with a 1 minute compressive hold time while non-filled markers are without hold time.

For the case without hold time where $T_{max} = 1000°$ C and $\varepsilon_{min} = -0.5\%$ there are not 12 points in the plot, the crack did not close during compression and thereby normalized crack opening stress is meaningless or U could be said equal to unity.

SUMMARY

The simulations without hold time (Case 1) show that the crack do not close when minimum strain range is less than -0.4% while subjected to maximum temperature of 1000° C. This critical strain decreases with increasing temperature. Above this strain range life is very short and the material cannot be used. One plausible reason for this is that the factor U reaches unity as shown.

When a compressive hold time is included, roughly the same crack opening history can be found by decreasing maximum temperature. As secondary creep can be described by an Arrhenius' expression the temperature dependence is obvious. To quantify it, however, is a difficult task which not only requires the activation energy but also some other material constants.

CONCLUSIONS / FURTHER WORK

The simulations indicate that the number of cycles at constant crack length significantly increases the effective part of the load cycle (i.e. when the crack is open). Hold time in the compressive part of cycle also lowers the crack opening stress. A 60 s hold time for all the analyses performed corresponds to about 50° C temperature increase in terms of displacement of first node and number of crack closures in the analyses without hold time.

These simulations will be used to draw a 'map' over U and number of crack closures with respect to mechanical strain and temperature. From this map, experiments will be designed to further investigate and quantify the influence of different parameters on crack closure, and the influence of crack closure on component life.

REFERENCES

1. Rowley, M. A., Thornton, E. A. (1996), *J. of Engineering Materials and Technology* **118**, Transactions of ASME, pp. 19 - 27

2. Kroupa, J. L., Implementation of a Nonisothermal Unified Inelastic-Strain Theory for a Titanium alloy into Abaqus 5.4 - Users guide, WL-TR-95-XXXX, University of Dayton Research Institute

3. Lindholm, U. S, Chan, K .S., Bodner, S. R., Weber, R. M., Walker, K. P., Cassenti, B. N.(1985), *NASA CR-174980*

4. Andersson, T. L.(1995). *Fracture Mechanics Fundamentals and Applications*, 2nd Ed., CRC Press, Boca Raton

Low Cycle Fatigue and Elasto-Plastic Behaviour of Materials
K-T. Rie and P.D. Portella (Editors)
609

INVESTIGATIONS ON INFLUENCE OF THE PLASTIC ZONE AFTER SINGLE-PEAK OVERLOAD ON FATIGUE LIFE OF STRUCTURAL STEEL UNDER TENSION

G.GASIAK and J.GRZELAK

Department of Mechanics and Machine Design, Technical University of Opole
45-233 Opole, Poland

ABSTRACT

The authors proposed a new model of fatigue crack growth in 10HNAP structural steel. The model has been formulated on the basis of analysis of the known models and the experimental data obtained under cyclic tension with single-peak overload. The model proposed includes retardation of the fatigue crack growth, it describes the experimental results very well and allows to determine the structure life. The authors have shown that for low overloads the specimen life N_c reaches its minimum as the stress ratio R increases. For the higher overloads the specimen life N_c monotonically increases as R rises. The effect is connected with formation of a large plastic zone in the crack tip, formation of internal stresses and the slot closing.

KEYWORDS

Tension, fatigue crack, overload, retardation, fatigue life, plastic zone

INTRODUCTION

When the fatigue crack growth takes place, interaction between cycles of different amplitudes plays the most important role [1,2]. A large plastic zone is forming while overloading [3]. In this zone the permanent deformations, being the plastic area, form in the material; the area is surrounded by the material of elastic properties. After unloading, the external material obtains its primary form but the material of the plastic zone remains deformed. When the overload decreases, the external, elastically deformed material shrinks and the plastically deformed material is incompressible, i.e. this zone is too big for the external elastic zone. Thus, the elastic material influences the plastic zone in the crack tip and, as a consequence, compressive stresses, so called internal stresses, occur [4,5]. The internal stresses cause the crack closure. The following cyclic basic loading can cause the fatigue crack growth only when the internal stresses are reduced just enough to open the crack tip again. This effect causes the crack growth retardation directly after single-peak overload [6].

Prediction of the element fatigue life under the given load is the aim of the structural or verifying calculations if there are any slots in the elements. Therefore growth of the slot having the critical length for which rapid decohesion occurs is not permitted. The real models describing the fatigue crack growth rate are necessary for determination of the element life.

THE MODEL TESTS

Quantity of the applied single-peak overload, characterized by the coefficient $k_j = P_{ov} / P_{max}$ (P_{ov} - the maximum load in the overload cycle, P_{max} - the maximum load for the basic cycle) is the most important factor influencing the value of a number of cycles of the fatigue crack growth retardation N_{di}. This number of cycles increases as k_j increases.

Having the experimental data [7], the authors modelled the fatigue life N_c with use of the following relation [8]

$$N_c = \int_{2\ell_0}^{2\ell_{ov}} \frac{(1-R)\sqrt{J_c E} - \Delta K}{C(\Delta K)^m} \, d(2l) + \int_{2\ell_{ov}}^{2\ell_c} \frac{(1-R)\sqrt{J_c E} - \Delta K}{C\left[(1 - C_{di})\Delta K\right]^m} \, d(2l) \, , \tag{1}$$

where

$$C_{di} = \begin{cases} n(R_m^2 - R_e^2) \dfrac{2\ell_i - 2\ell_p}{N_{di}(K_p^2 - K_{max}^2)} & \text{dla } \left[\dfrac{d(2\ell)}{dN}\right]_p < \left[\dfrac{d(2\ell)}{dN}\right]_{nom}, \\[4mm] 0 & \text{dla } \left[\dfrac{d(2\ell)}{dN}\right]_p = \left[\dfrac{d(2\ell)}{dN}\right]_{nom}, \end{cases} \tag{2}$$

C_{di} - parameter of the fatigue crack growth retardation.

The first term of (1) describes the life of specimens starting from the initial crack length $2l_0$ up to application of single-peak overload when the crack length is $2l_{ov}$ and the second term - from the overload to the critical slot length $2l_c$.

THE EXPERIMENTS

The flat specimens made of 10HNAP steel were subjected to variable tension and constant mean loadings $P_m = 50$ kN, 65.5 kN, 75 kN (R = 0.235, 0.379, 0.523) and the constant amplitude of regular load $P_a = 23.5$ kN, 28.1 kN, 31 kN. The single-peak overload coefficient was changed from $k_j = 1.125$ to $k_s = 2.0$. The specimens were tested on a hydraulic pulser under the regular load frequency 14 Hz. The fatigue crack growth was measured with use of travelling microscope with an accuracy of 0.001 mm. In the centre of each specimen there was a hole, 5 mm in diameter. In the hole notches perpendicular to the specimen axis were made (length 1 mm and angle 60°).

THE TEST RESULTS

Fig. 1 shows the test results obtained, namely the specimen lives N_c versus the stress ratio R for eight overload coefficients k_j (k_j = 1.0, 1.125, 1.25, 1.375, 1.5, 1.625, 1.75, 1.875). The curves presented in Fig. 1 were described by the function

$$N_c = b_2 R^2 + b_1 R + b_0 . \qquad (3)$$

Coefficients b_o, b_1 and b_2 of equation (3) are given in Table 1. If there is no overload (Fig. 1, curve 1) or overload is low (k_j = 1.25 - curve 2), the specimen life N_c reaches its minimum as the stress ratio R increases. For the higher k_j, the specimen life N_c monotonically increases together with increase of the stress ratio R.

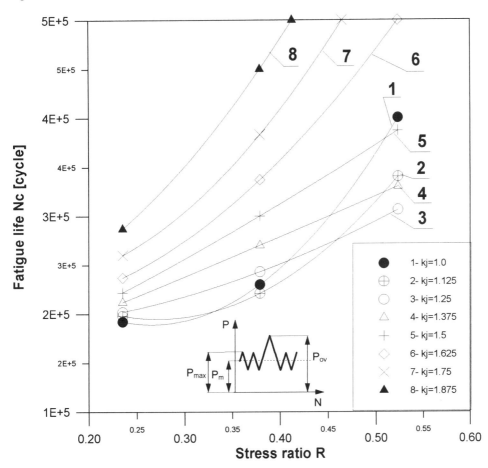

Fig. 1. Influence of the cycle asymmetry ratio R on fatigue life of the specimen N_c for some values of overload k_j; ● ⊕ ○ △ + ◇ × ▲   - the symbols used mean the experimental results.

Table 1. Coefficients of equation (3).

Curve No.	k_j	b_2	b_1	b_0
1	1.000	3,182,870	-1,690,393	413,468
2	1.125	2,314,815	-1,261,574	366,634
3	1.250	530,478	-40,991	182,337
4	1.375	48,225	373,167	121,642
5	1.500	217,013	408,420	114,036
6	1.625	1,519,097	-238,281	209,104
7	1.750	2,201,297	-497,430	255,329
8	1.875	1,902,493	-36,186	190,438

Figure 2 shows distribution of the specimen lives N_c obtained from experiments versus the overload coefficient k_j and the stress ratio R.

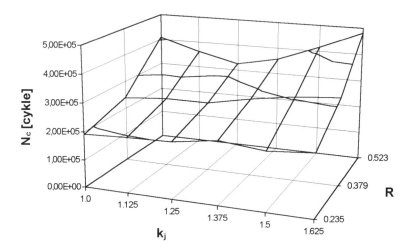

Fig.2 . Distribution of the specimen lives N_c depending on the overload coefficient k_j and the stress ratio R.

CONCLUSIONS

From analysis of the obtained test results the following conclusions can be drawn:
1. For low stress ratios R the single-peak overload causes a small increase of the specimen life. For the high stress ratios and increasing overload, the specimen life at first decreases (reaches its minimum) and next increases.
2. The applied model (1) efficiently describes the experimental data.
3. Equation (3) allows to determine the specimen life for the given values of R.

REFERENCES

1. Gasiak G. and Grzelak J. (1993). In: Zeszyty Naukowe Polit. Świętokrzyskiej. Mechanika 50, Kielce, pp.141-148 (in Polish)
2. Pantelakis S.G., Kermanidis T.B. and Pavlou D.G. (1995). Theoretical and Applied Fracture Mechanics Technology, Vol.22, No.1, pp.35-42
3. Bathias C. (1996). Fatigue Engng Mater.Struct., Vol.19, No 11, pp.1301-1306
4. Daniewicz S.R., Collins J.A. and Houser D.R. (1994). Int. J.Fatigue, 16, 123
5. Gasiak G. and Grzelak J. (1996). In: Proceedings of the ECF11, Mechanisms and Mechanics of Damage and Failure, J.Petit (Ed.). EMAS, Warley UK
6. Gasiak G. and Grzelak J. (1996). Archiwum Budowy Maszyn, Vol.XLIII, Zeszyt 2-3, pp.161-171
7. Gasiak G. and Grzelak J. (1995) In: Zeszyty Naukowe Polit.Świętokrzyskiej. Mechanika 56, Kielce, pp.165-173 (in Polish)
8. Gasiak G. and Grzelak J. (1997). In: Zeszyty Naukowe Polit Świętokrzyskiej. Mechanika 62, Kielce, pp.139-145 (in Polish)

Low Cycle Fatigue and Elasto-Plastic Behaviour of Materials
K-T. Rie and P.D. Portella (Editors)
615

FATIGUE CRACK PROPAGATION IN PONDER HETEROGENEOUS METALLIC MATERIALS

A. A. CHEKALKIN, Yu. V. SOKOLKIN and E. M. YAKUSHINA
Perm State Technical University, 614600, Perm, Russia

ABSTRACT

The cyclic crack growth in the powder heterogeneous materials is investigated by experimental and numerical studies. The fatigue behaviour of the crack opening mode I is tested by cyclic console bending of single edge specimens. The changes of specimens stiffness are measured in connection to the cyclic crack growth . The fatigue crack propagation is computed by macro- and microscale boundary value problems. The macroscale boundary element model of testing specimen is analysed in accordance with fracture mechanics correlation and energy balance in system. The effective strain energy release rate and the effective stress intensity factor is used to analyse the fatigue behaviour of porous inhomogeneous cycling materials. The microscale finite element model of porous sintered metallic materials is simulated the local fracture effects and microstructure interacting mechanisms at the crack tip under high gradient macrostress state in the fatigue crack propagation process. The fatigue crack growth predictions for modes II and III are discussed for powder heterogeneous metallic materials.

KEYWORDS

Fatigue, crack growth, powder material, composite steel.

INTRODUCTION

In safety design of advanced heterogeneous materials and structures, the crack cycling propagation role may be the deciding factor in fatigue fracture, lifetime and durability properties. Significant problem of fracture mechanics was studies by Cherepanov , Atkinson, Craster , N. Gao, M. W. Brown , K . J. Miller[1-3]. The micromechanical approach on the stochastic processes theory base was advanced to fatigue microplastic behaviour of powder porous material in reference [4]. The present item is devoted to localised fatigue fracture of powder sintered metallic materials. The crack propagation in process with micro- and macroscale interaction. The microscale fatigue model allows to predict the fracture toughens for shear and antiplane crack opening modes and to bound changing volume fraction impact effective properties of powder steel.

EXPERIMENTS

Material and specimens

Three different types of chemical composition of powder sintered nickel steel were used to study fatigue crack growth (table 1). The powder rectangle compactant has overall sizes of 95x16x16 mm. Single edge specimen is processed with V-shape cut depth of 4 mm, top cut angle of 60 ^{0} and 65 mm span size among two holes of 8 mm in diameter.

616

Table 1. The chemical composition of powder sintered nickel steel

	C, wt.%	Ni, wt.%	Mo, wt.%	Cr, wt.%		
N1	1.5	1	-	2	remainder	Fe
N2	0.5	9	1	-	remainder	Fe
N3	0.35	0.15	-	-	remainder	Fe

Cycle Test

The crack opening mode I propagation is tested by uncracked ligament length of 12 mm. The change of specimen stiffness during the development of fatigue crack has an effect on the size of moving grip amplitude, which is measured in the process of test in relation to the number of loading cycles . Single edge specimen had been captured by fatigue machine grips and loaded by bending console scheme (SE(B) is test code as [5]). Symmetrical fatigue cycling had been set in motion on stable frequency of 18 Hz with endurance limits up to $5 \cdot 10^6$ cycles. The movable grip amplitudes are proved by the clock type micrometer.

Microstructure

The microscale examination of the powder steel structure was using metallographic optical microscope METAM-PB- 21, computerised microscanning ScanNexIIc equipment and software image processing package. The sintered microstructure contains some porosity and irregular grains of Fe-C phases. The microscanning and image processing procedures are allowed to be wanted the finite element mesh generator parameters or directly using the computerised image by microscale finite element simulation.

MACROSCALE MODEL

Macromoduli

The effective elastic constants of the powder porous material are defined by Christensen [6] as

$$K^* = K \ (1 - (4G + 3K) \ p \ / \ (\ 4G + 3Kp) \ ) \ ,$$
$$G^* = G \ (1 - 15p \ (1-v \ ) \ / \ (7 - 5v \ ) \ ) \tag{1}$$

for isotropic materials

$$G = E \ / \ 2 \ (\ 1 + v \ ) \ , \quad K = E \ / \ 3 \ (\ 1 - 2 \ v \ ) \tag{2}$$

where K. G - bulk and shear modules consecutively, K^*. G^* - effective bulk and shear modules , p - void volume fraction, v - Poisson's ration, E - Young's modules .

Formulation

For the stress-strain analysis of bending single edge specimens is used the two dimensional direct boundary element method by Crouch and Starfield [7]. The contour stress and displacement are computed from following equation

$$
\begin{aligned}
B_{ij}{}^{ss}\,\sigma_j{}^s + B_{ij}{}^{sn}\,\sigma_j{}^n &= A_{ij}{}^{ss}\,u_j{}^s + A_{ij}{}^{sn}\,u_j{}^n\,, \\
B_{ij}{}^{ns}\,\sigma_j{}^s + B_{ij}{}^{nn}\,\sigma_j{}^n &= A_{ij}{}^{ns}\,u_j{}^s + A_{ij}{}^{nn}\,u_j{}^n
\end{aligned}
\tag{3}
$$

where $2N$ contour values are unknowns, other $2N$ values are given from the boundary conditions; A, B is influence matrices; σ^n, σ^s is normal and tangent stress tensor consecutively; u^n, u^s is normal and tangent displacement vector consecutively; $i, j = 1 . . N$, N is the element number. The accordance between the changes of specimen stiffness and growing crack length from edge tip is established by solving boundary value problems.

Fracture Toughness

The strain energy release rate by fatigue crack propagation through the powder steel is described as

$$
G = dU / dl \approx (U(1+\Delta l) - U(1))/\Delta l\,,
\tag{4}
$$

where G - strain energy release rate, Δl - the crack increment in consequence of fatigue cycling.
The work of external loads U was calculated by contour integral

$$
U = {}^1/{}_2 \int_S (\sigma^n u^n + \sigma^s u^s)\, dS\,,
\tag{5}
$$

where S is boundary contour. In linear elasticity G is related to stress intensity factors

$$
G = (K_I{}^2 + K_{II}{}^2)(1-v^2)/E + K_{III}{}^2/2G\,,
\tag{6}
$$

where K_I, K_{II}, K_{III} is stress intensity factor crack modes I, II and III accordingly. The general properties are the effective values and the experimental obtained date allows to compute only G_I and K_I through macroscale model.

MICROSCALE MODEL

Formulation

The microscale stress-strain state of the heterogeneous powder steel near the crack tip is analysed by finite element method. Nodal displacements for Dirichlet s boundary value problem are computed from equilibrium equation

$$
C_{kl} u_k = 0\,,
\tag{7}
$$

618

where C_{kl} is stiffness matrix; $k, l = 1 .. M, M$ - the nodal number. The boundary nodal displacements are given from macroscales solution tailoring through reciprocity theorem[7]. The stiffness matrix is added together the stiffness matrices of ferrite or austenite grains and the void zero matrices. The local displacement state of microscale particle is given as

$$u_{(i)} = L_{(i)} u_k ,$$ (8)

thereafter the microscale stress-strain state is defined through the strain-displacement relationship and the constitutive equations for each grain element.

Damaging

The endurance limits of ferrite and austenite grains are described by Basquin law or Coffin-Manson low respectively as high and low cycle fatigue, the cumulative microdamages in variable amplitude loading are taken into Palmgren-Miner rule. The microstructure stiffness matrix is being decided in line with the chain of grain fractures. The stored elastic energy and external loads work balance in damaging process has main implications for fatigue fracture toughness of powder steels.

EXPERIMENTAL RESULT AND DISCUSSION

The several fracture mechanics parameter that is used to describe fatigue growth of the crack opening mode I is determined from experimental result. The stress intensity factor with respect to loading cycle number will be used to predict the fatigue threshold and stress intensity factor accompanying failure for different moving grip amplitudes. In general, the fatigue threshold for N1 specimens is founded less than $1,0 \cdot 10^7$ Pa√m, for N2 specimens - withing $0,5...1,75 \cdot 10^7$ Pa√m and within $1,0...1,5 \cdot 10^7$ Pa√m for N3 specimens. The stress intensity factor accompanying failure is founded within $2,7...5,0 \cdot 10^7$ Pa√m for N1 specimens; $2,5...4,0 \cdot 10^7$ Pa√m for N2 specimens; $2,8...4.0 \cdot 10^7$ Pa√m for N3 specimens.

The Paris law snows three different stages for material, i.e. fatigue threshold, constant growth and accelerated crack growth. As shown in Fig. 1 the fatigue threshold for N1 is founded within $2,5...7,5 \cdot 10^7$ Pa√m . The fatigue threshold for N2 is $0,5...1,0 \cdot 10^7$ Pa√m (Fig.2) and for N3 is $1,0...1,2 \cdot 10^7$ Pa√m (Fig.3). For all this material it is possible to distinguish the stage of constant rate crack growth. The accelerated crack growth region have scatter of data. The critical value of stress intensity factor is $4.0 \cdot 10^7$ Pa√m for N1, N2 and N3.

Acknowledgements

The authors are indebted to Federal Centre of Powder Metallurgy for test material. This work was supported by Russian Fund of Basic Researches (grant 95-01-00203).

REFERENCES

1.   G. P. Cherepanov (1983) *Fracture Mechanics of Composite Materials*. Nauka, Moscow   [in Russian].
2.   C. Atkinson and R. V. Craster (1994) Theoretical aspects of   fracture mechanics.   Prog. Aerospace Sci. 31, 1-83.
3.   N. Gao, M. W. Brown and K . J. Miller (1995) Crack growth   morphology and microstructural changes in 316 stainless   steel under creep-fatigue cycling. *Fatigue Fract. Engng   Mater. Struct.* **18**, 1407-1422.
4.   A . A . Chekalkin, A .V. Babushkin and Yu . V . Sokolkin (1996)   Fatigue behaviour and cyclic plasticity multiscale modelling of powder porous iron. In: *Fatigue ' 96* (Edited by G . Lutjering and H . Nowack) Pergamon, 1093-1098.
5.   D . Francois (1996) *ESIS TC7D-1-96D Guidelines for Terminology and Nomenclature in the Field of Structural Integrity*. Pre-print from Fatigue Fract. Engng Mater.   Struct. The Great Britain.
6.   R . M . Christensen (1979) *Mechanics of Composite Materials.*   John Wiley & Sons, New York.
7.   S . Crouch and A . Starfield (1983) *Boundary Element Methods in Solid Mechanics.* Georg Allen & UNWIN, London.

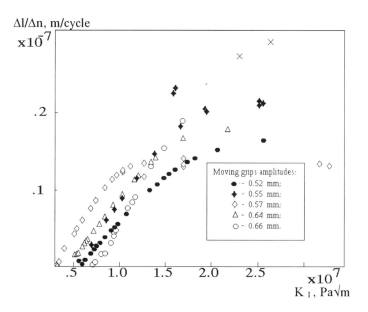

Fig. 1. Fatigue crack rate - SIF curves of powder nickel steel N1.

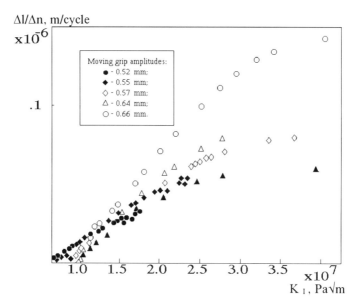

Fig. 2. Fatigue crack rate - SIF curves of powder nickel steel N2.

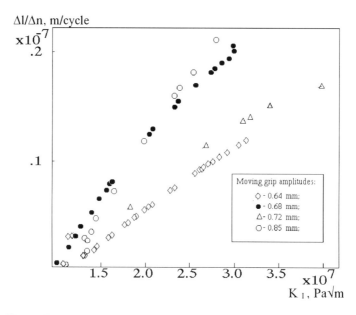

Fig. 3. Fatigue crack rate - SIF curves of powder nickel steel N3.

Low Cycle Fatigue and Elasto-Plastic Behaviour of Materials
K-T. Rie and P.D. Portella (Editors)
621

STABLE DAMAGE ACCUMULATION AND ELASTO-PLASTIC DEFORMATION OF COMPOSITE MATERIALS

V. E. WILDEMANN and A. V. ZAITSEV
Perm State Technical University, 614600, Perm, Russia

ABSTRACT

The primary objective of the research is to describe damage evolution that depends on the loading system properties. Macro-failure of composites can take place in any point of descending branch of the stress-strain diagram in correspondence with the loading system stiffness if the stability conditions are non-carried out. The greater loading system stiffness the greater failure strains. The effects of strain and damage localization, partial or complete unloading under proportional displacement-controlled loading, self-similarity and self-supported damage evolution are found out. The presented results give the possibility of controlling the dissipative processes by optimal design of material structure with the primary object of maximal use of carrying capacity and exception of catastrophic failure of composites.

KEYWORDS

Structural heterogeneity, damage evolution, loading system, stability of inelastic deformation, work-softening behaviour

INTRODUCTION

Research in inelastic behaviour of composites is carried out at the structural or micro-scale and the composite or macro-scale levels and we suppose that the classic mechanics relations and phenomenological equations are valid at these levels. The structural elements can be damaged by different mechanisms. Possible variants of the damage criteria and reduction schemes of material properties after fulfilling of various damage conditions for the anisotropic media are presented in the monograph [1]. Heterogeneous elasto-plastic solids subjected to damage evolution demonstrate stable work-softening behaviour that appears as descending branch in the stress-strain diagrams under loading. This type of deformation could be carried out only for a local domain as a component of the mechanical system with the necessary properties.

LOADING SYSTEM AND STABILITY OF INELASTIC DEFORMATION

Failure of composites is a final non-stable stage of complete multilevel processes, accompanied by accumulation and localization of damaged domains. The observable difference in the character of structural damage evolution of the composites can not always be explained only by the mechanical properties. Special experimental researches confirmed that the resistance to failure also depends on the loading system stiffness. We will name the set of solid, liquid and/or gaseous media, whose deformation takes place as the result of load transmission to a body or to a separate part of it as the loading system. The replacement of one state of material by another due to the replacement of an external load depending on properties of the loading system the characteristics of which in the solution of the boundary value problem for inelastic deformation are taken into account by the boundary conditions of the third kind:

$$\left[d\sigma_{ij}(\mathbf{r})\,n_j(\mathbf{r})+R_{ij}\,du_j(\mathbf{r})\right]\Big|_{\Sigma_S}=dS_i^0(\mathbf{r})\,,\quad \left[du_i(\mathbf{r})+R_{ij}\,d\sigma_{ik}(\mathbf{r})\,n_k(\mathbf{r})\right]\Big|_{\Sigma_U}=du_i^0(\mathbf{r})\,,\quad (1)$$

where $\mathbf{du}$, $\mathbf{dS}$ are the real increments of displacement and external force vectors and $d\sigma$ is that of the stress tensor; $\mathbf{n}$ is a normal vector. Symmetric positively defined 2nd rank tensors $\mathbf{R}(\mathbf{u},\mathbf{r})$ and $\mathbf{Q}(\mathbf{S},\mathbf{r})$ of the loading system stiffness and compliance connect the increment vectors of external forces $\mathbf{dS}^0$ with displacements $\mathbf{du}^0$ by the relations

$$dS_i^0(\mathbf{r})=R_{ij}(\mathbf{u},\mathbf{r})\,du_j^0(\mathbf{r})\,,\quad du_i^0(\mathbf{r})=Q_{ij}(\mathbf{S},\mathbf{r})\,dS_j(\mathbf{r})\,,\quad R_{ik}Q_{kj}=\delta_{ij}\,,\quad \forall\mathbf{r}\in\Sigma\,.$$

The vectors of $\mathbf{dS}^0$ and $\mathbf{du}^0$ are given on the parts $\Sigma_S\subset\Sigma$ and/or $\Sigma_U=\Sigma\setminus\Sigma_S$ of the body surface. The given boundary conditions are mutually reciprocal and so the equation of one type could be used for the whole closed surface Σ only. In the cases of $\mathbf{R}=0$ or $\mathbf{Q}=0$ these boundary conditions correspond to the stress-controlled or displacement-controlled loading.

Using Drucker's approach [2] to total combination of solid and loading systems [1] a sufficient condition of inelastic deformation and damage evolution stability can be put out as:

$$\int_V \widetilde{C}_{ijkl}(\mathbf{r})\,\delta\varepsilon_{ij}(\mathbf{r})\,\delta\varepsilon_{kl}(\mathbf{r})\,dV+\int_\Sigma R_{ij}(\mathbf{r})\,\delta u_i(\mathbf{r})\,\delta u_j(\mathbf{r})\,d\Sigma>0 \qquad (2)$$

for the virtual increments of the strains $\delta\varepsilon$ and displacements δu providing $\mathbf{du}^0=0$. Therefore, equilibrium deformation is possible only in the special conditions defined by the loading system stiffness. Invalidity of the condition (2) corresponds to unstable failure.

INELASTIC DEFORMATION AND WORK-SOFTENING BEHAVIOUR

Interconnected dissipative processes of elasto-plastic deformation of heterogeneous body V with closed surface Σ under quasi-static loading are considered as multi-stage and multi-scale structural damage accumulation and described by the following boundary value problem

$$d\sigma_{ij,j}(\mathbf{r})=0\,,\quad d\varepsilon_{ij}(\mathbf{r})=1/2\left[du_{i,j}(\mathbf{r})+du_{j,i}(\mathbf{r})\right]\,,\quad d\sigma_{ij}(\mathbf{r})=\widetilde{C}_{ijkl}(\mathbf{r})\,d\varepsilon_{kl}(\mathbf{r})\,,\quad \forall\mathbf{r}\in V\,,\quad (3)$$

where $d\varepsilon$ is a real increment of the strain tensor. Inelastic behaviour of isotropic media with evolutionary damaged structure could be described by the tensor-linear constitution equations

$$j^{(1)}=3K(\mathbf{r})[1-\kappa]\,j_e^{(1)}\,,\quad j^{(2)}=2G(\mathbf{r})[1-g]\,j_e^{(2)}\,,\quad \forall\mathbf{r}\in V\,,\qquad (4)$$

where K and G are bulk and elastic moduli; κ and g are independent damage functions the arguments of which are linear and quadratic independent invariants

$$j_\sigma^{(1)}=1/3\sigma_{kk}\,,\quad j_\sigma^{(2)}=\sqrt{\breve{\sigma}_{ij}\breve{\sigma}_{ij}}\,,\quad j_\varepsilon^{(1)}=\varepsilon_{kk}\,,\quad j_\varepsilon^{(2)}=\sqrt{\breve{\varepsilon}_{ij}\breve{\varepsilon}_{ij}}$$

of the stress and/or strain tensors only, where $\breve{\sigma}_{ij}$ and $\breve{\varepsilon}_{ij}$ are the deviator components.

As an example we can consider results of the numerical solution of the boundary value problem (3) with boundary conditions (1) and stability conditions (2) for one of the granular composite (powder composites, ceramics or rock media) representative volume realization that fills a cub unit and contains 3072 homogeneous tetrahedron isotropic structural elements. Weibull distribution parameters of the random strength constants and the values of elastic moduli are given in the article [3]. The authors [4,5] paid attention on a satisfaction of the boundary conditions (1) in the solution by the finite element method.

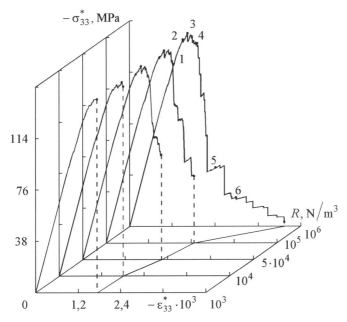

Fig. 1. One axial compression stress-strain diagrams of the granular composite.

The structural damage results in non-linear macroscopic behaviour of the medium even in the case of elastic-brittle properties of components. We will name this type of deformation as pseudo-plastic behaviour. Increasing of the loading system stiffness to stabilise damage evolution and leads to work-softening behaviour. On the ascending branches of the diagrams the value of R_{ij} ($R_{ij} = R\delta_{3i}\delta_{3j}$, where δ_{ij} is Kronecker delta) has no significant influence on the composite behaviour (fig. 1). The curve that plotted at $R = 10^6$ N/m^3 does not differ from the diagram registered in the displacement-controlled regime. We supposed macro-failure of the composite to be the result of loss of deformation stability of the damaged material. The results of the loading system influence on stability of pseudo-plastic deformation confirm the growth of the failure strains while increasing of the loading system stiffness.

MULTI-SCALE DAMAGE EVOLUTION AND LOCALIZATION

At the beginning of loading damage accumulation occurs uniformly in the whole of the representative composite volume and every damage act evolves as a random event. A damaged structural element is a strong local inhomogenisation factor that effects to strain concentration. Further increasing of external forces leads to local and global strain and damage localization. Work-softening behaviour of the composite is accompanied by the uniform accumulation of localized damage centres and partial unloading of the undamaged structural elements. Proceeding of the loading effects to modification of the structural damage mechanisms

624

connected with the beginning of stable formation of a failure cluster and increasing of descending branch angle of inclination. This stage is completed by unstable development and as a consequence by composite macro-failure. It will be noted that localization and failure stages are defined mainly by energy redistribution between the solid and loading system [5].

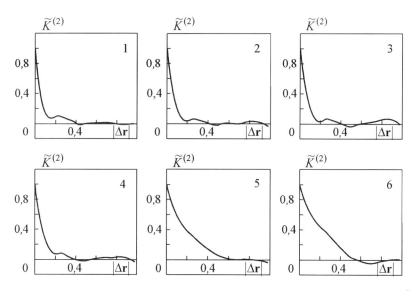

Fig. 2. Normalized correlation functions of the composite damaged structure ($|\Delta r| \cdot 10^3$, m).

Stochastic structure of the damaged composite could be described by the normalized correlation functions $\tilde{K}^{(2)}(|\Delta r|)$. These functions show a character of interaction between the damaged domains under loading (fig. 2). Being obtained for the equilibrium states of the composite marked by points 1–6 on fig. 1, the functions $\tilde{K}^{(2)}(|\Delta r|)$, at the stages of uniform accumulation of the separate damaged elements (fig. 2,1 and fig. 2,2) and localized damage centres (fig. 2,5 and fig. 2,6) differ only by the value $|\Delta r|$. This phenomenon confirms self-similarity of the damage evolution on the different scale levels of the composite.

EFFECT OF LOCAL UNLOADING

The boundary value problem with the special boundary condition $u_i^0 = \varepsilon_{ij}^* r_j$ (where ε^* is macro-strain tensor) corresponding to the macro-homogeneous displacement-controlled loading regime for the laminated composites could be presented as the non-linear simultaneous equations [6]. All the macro-scale values are got by means of homogenization of the structural characteristics. Inelastic behaviour of the layers can be described by the damage functions

$$\kappa = A j_\varepsilon^{(1)}, \quad g = B j_\varepsilon^{(2)} + \left(1 - j_{\varepsilon\,el}^{(2)} / j_\varepsilon^{(2)}\right)\left(1 - G'/G\right) h\left(j_\varepsilon^{(2)} - j_{\varepsilon\,el}^{(2)}\right)$$

with constants $K = 7{,}85 \cdot 10^4\,\text{MPa}$, $G = 2{,}55 \cdot 10^4\,\text{MPa}$, $G' = 7{,}9 \cdot 10^3\,\text{MPa}$, $j_{\varepsilon\,el}^{(2)} = 3{,}3 \cdot 10^{-4}$, $A = 4{,}77$ and $B = 6{,}125$ for the Al layers and $K = 3{,}67 \cdot 10^4\,\text{MPa}$, $G = 1{,}69 \cdot 10^4\,\text{MPa}$, $G' = 3{,}0 \cdot 10^2\,\text{MPa}$, $j_{\varepsilon\,el}^{(2)} = 6{,}3 \cdot 10^{-4}$, $A = 4{,}14$ and $B = 3{,}22$ for the Mg isotropic layers. In the equations given above $h(\circ)$ is the Heaviside function.

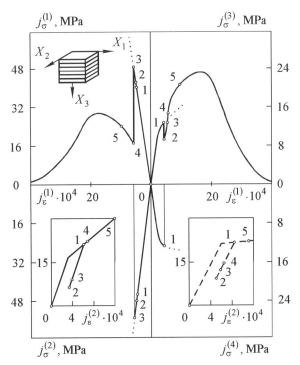

Fig. 3. Stress-strain diagrams in invariant form ($2\varepsilon_{11}^* = 2\varepsilon_{22}^* = \varepsilon_{33}^* = 2\varepsilon_{12}^* = \sqrt{2}\varepsilon_{13}^* = \sqrt{2}\varepsilon_{23}^* \neq 0$) of the laminated composite and local unloading under structural damage conditions; where (··········) are the curves without damage accumulation. The diagrams for the Al (———) and Mg (– – –) layers are presented in $j_\sigma^{(2)} \sim j_\varepsilon^{(2)}$ co-ordinates.

The stress-strain diagrams of the transversal-isotropic on the macro-scale level composite that consists of 14 components (7 groups of Al layers and 7 groups of Mg layers) are presented in fig. 2. The volume fractions of the Al and Mg layers are 0,6 and 0,4. Random strength constants of the layers are also determined by Weibull distribution law

$$F\left(j_\varepsilon^{(\circ)}\right) = 1 - \exp\left\{-\left[\left(j_\varepsilon^{(\circ)} - j_{\varepsilon 0}^{(\circ)}\right)\middle/\left(j_{\varepsilon a}^{(\circ)} - j_{\varepsilon 0}^{(\circ)}\right)\right]^{b(\circ)}\right\}$$

with parameters $j_{\varepsilon 0}^{(1)} = j_{\varepsilon 0}^{(2)} = 6\cdot 10^{-4}$ and $j_{\varepsilon a}^{(1)} = j_{\varepsilon a}^{(2)} = 3\cdot 10^{-3}$ for the Al; $j_{\varepsilon 0}^{(1)} = j_{\varepsilon 0}^{(2)} = 9\cdot 10^{-4}$ and $j_{\varepsilon a}^{(1)} = j_{\varepsilon a}^{(2)} = 4,5\cdot 10^{-3}$ for the Mg layers and $b_{(1)} = 6$, $b_{(2)} = 5$.

According to Pobedrya's theory of anisotropic media plasticity [7], inelastic deformation of damaged transversal-isotropic body could be described by the relations of the four special independent invariants, considering X_3 as the direction of anisotropy:

$$j_\varepsilon^{(1)} = \varepsilon_{11} + \varepsilon_{22}, \quad j_\varepsilon^{(2)} = \varepsilon_{33}, \quad j_\varepsilon^{(3)} = \sqrt{\left(\varepsilon_{11} - \varepsilon_{22}\right)^2 + 4\varepsilon_{12}^2}, \quad j_\varepsilon^{(4)} = \sqrt{\varepsilon_{13}^2 + \varepsilon_{23}^2},$$

$$j_\sigma^{(1)} = \left(\sigma_{11} + \sigma_{22}\right)/2, \quad j_\sigma^{(2)} = \sigma_{33}, \quad j_\sigma^{(3)} = \sqrt{\left(\sigma_{11} - \sigma_{22}\right)^2 + 4\sigma_{12}^2}, \quad j_\sigma^{(4)} = \sqrt{\sigma_{13}^2 + \sigma_{23}^2}.$$

Stress redistribution at the conditions of non-simultaneous transition to plastic deformation of the layers effected by the damage accumulation lead to changing of the mechanical process directions and local unloading of the undamaged structural elements under the condition of monotonous proportional displacement-controlled loading (fig. 3).

SELF-SUPPORTED DAMAGE ACCUMULATION AND LOCAL INSTABILTY

We can specify external (loading system) and internal sources of supplied mechanical energy. The internal energy source is connected with the release of elastic deformation potential energy when the partial unloading of undamaged domains takes place. According to eq.(2) the spontaneous damage propagation without increasing of an external loading is impossible if the supplied external and realized internal mechanical energy is insufficient for damage work.

In the composite structure local domains are detected the non-equilibrium damage of which is not affected neither by the loading system stiffness nor by increasing of loading steps. Similar local unstable discrete energy dissipation appears as separate more or less expanded slopes in the stress-strain diagrams (fig. 1). These slopes are characteristic of the materials having the tendency to self-supported damage accumulation on both work-hardening and work-softening stages on the given loading level. This effect occurs because the damage evolution was defined mainly by an internal source of the supplied mechanical energy. Therefore, we have not succeeded completely in controlling damage character even in displacement-controlled monotonous loading regime on the structural level.

Experimental researches of the sandstone media behaviour under compressive loading demonstrate the smooth stress-strain diagrams without unstable slopes in the conditions of liquid saturation. Otherwise, self-supported damage accumulation takes place at the work-softening stage in the case of gas saturation. If the loading system stiffness is sufficient and the composite structure contains hard incompressible or low-compressible particles then the structural damage process is controlled on both micro- and macro-scale levels.

CONCLUSION

The main mechanisms of non-elastic deformation as a result of the damage evolution of the granular and laminated composites are described. Macro-failure of these media is considered as the final stage of the structural damage nucleation, localization and formation of failure cluster. When the loading system stiffness is sufficient the damage accumulation can occur in the equilibrium regime. The presented model of inelastic deformation with a help of stochastic description allows to research the damage stages of the composite under loading.

REFERENCES

1. Wildemann, V.E., Sokolkin, Yu.V. and Tashkinov, A.A. (1997). *Mechanics of Inelastic Deformation and Faiure of Composites*. Nauka, Moscow [in Russian].
2. Drucker, D.C. (1959) *Trans. ASME. Ser. E. J. Appl. Mech.* **26**, 101.
3. Wildemann, V.E. and Zaitsev, A.V. (1996) *Mechanics of Composite Materials* **32**, 808.
4. Wildemann, V.E., Sokolkin, Yu.V. and Zatsev, A.V. (1997) *Mechanics of Composite Materials* **33**, 329.
5. Wildemann, V.E. and Zaitsev, A.V. (1997). In: *Proceedings of the 5th Int. Conf. on Biaxial-Multiaxial Fatigue & Fracture*, E. Macha and Z. Mróz (Eds). Technical University of Opole Press, Poland. pp. 593–610.
6. Wildemann, V.E., Sokolkin, Yu.V. and Tashkinov, A.A. (1992) *Mechanics of Composite Materials* **28**, 315.
7. Pobedrya, B.E. (1984). *Mechanics of Composite Materials*. Moscow State Universty Press, Moscow [in Russian].

Chapter 9

CONSTITUTIVE EQUATIONS; MODELLING

Low Cycle Fatigue and Elasto-Plastic Behaviour of Materials
K-T. Rie and P.D. Portella (Editors)
629

A MICROSTRUCTURALLY BASED EXPLANATION FOR THE HIGH TEMPERATURE LCF BEHAVIOUR OF THE ODS NICKEL-BASE ALLOY PM 1000

M. HEILMAIER[1] and F. E. H. MÜLLER[2]

[1]*Institut für Festkörper und Werkstofforschung Dresden, D-01171 Dresden*
[2]*PLANSEE AG, A-6600 Reutte/Tirol*

ABSTRACT

The cyclic (LCF) deformation behaviour of the recently developed oxide dispersion strengthened (ODS) nickel-base alloy PM 1000 is described using a microstructural model based on the Haasen-Alexander-Ilschner theory of internal and effective stress. The comparison of the monotonic and the cyclic stress-strain curve at 1123 K and $\dot{\varepsilon} = 10^{-3}s^{-1}$ yields cyclic softening which can be explained by the experimentally verified reduction of dislocation density in the case of cyclic loading.

KEYWORDS

Nickel-base superalloy, low cycle fatigue, dispersion strengthening, microstructural model, cyclic softening

INTRODUCTION

Oxide Dispersion Strengthened (ODS) nickel-base superalloys seem to be attractive candidate materials for high temperature applications in excess of 1273 K because of their favourable combination of good hot corrosion resistance and high strength. The latter is due to the presence of fine and homogeneously distributed oxide particles incoherent and insoluble with the metallic matrix.

Besides their technical importance these materials represent ideal model systems for studying the influence of particles on the deformation behaviour at high temperatures without superimposed changes in the particle microstructure due to the well known thermal stability of these dispersoids [1]. Thus, mechanisms observed in precipitation-strengthened alloys in which cyclic hardening or softening can be promoted by cyclic straining might not be present [2,3]. Consequently, the purpose of this work is to elucidate the influence of stable oxide particles on the high temperature LCF behaviour of a nickel-base superalloy by comparing the cyclic stress-strain (css) curve with its corresponding monotonic counterpart (mss).

EXPERIMENTAL

The recently developed alloy PM 1000 (trademark of Plansee AG, Austria) is essentially a nickel chromium solid solution strengthened by nominal 0.6 weight percent Y_2O_3 dispersoids. It was produced by mechanical alloying and supplied in form of a hot extruded bar with a diameter of 50 mm. A subsequent recrystallisation heat treatment promotes the evolution of a sharp <100>-fibre texture and results in coarse elongated grains with average diameters of $d_l =$

2.6 mm in longitudinal direction and about one tenth perpendicular to that. Hence, the grain aspect ratio yields $GAR = 10$. Fully reversed (R = -1) total strain control loading with a triangular wave shape has been carried out at 1123 K under different total strain amplitudes at a strain rate level of 10^{-3} s^{-1}. Prior to testing, Young's modulus was determined by cycling with a low strain amplitude. For comparison additional monotonic tensile and compressive tests have been performed at the same constant strain rate $\dot{\varepsilon}$. The corresponding particle and dislocation microstructure has been determined by means of transmission electron microscopy (TEM) and quantitative stereology. For further details see [4].

LCF BEHAVIOUR

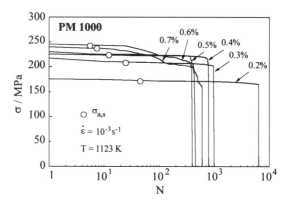

Fig. 1. Cyclic hardening/softening curves of PM 1000 for different applied total strain amplitudes. The open circles mark the saturated stress amplitude.

Cyclic hardening/softening curves of PM 1000 with different total strain amplitudes ranging from 0.2 to 0.7 % are plotted in fig. 1. All curves show cyclic softening at the beginning of cyclic deformation which is, however, only weakly pronounced, followed by an extended region of fairly constant stress amplitudes. As expected, the stress level decreases with reduced strain amplitude leading vice versa to higher cyclic lifetimes.

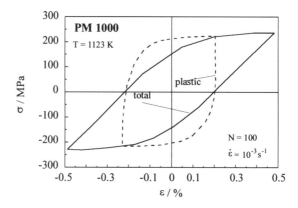

Fig. 2. Stabilized σ-ε-hysteresis loop with a total strain amplitude of $\varepsilon_{a,t} = 0.5$ % (continous curve), dashed: σ-ε_{pl}-hysteresis loop.

A typical hysteresis loop in the saturated state for a total strain amplitude of $\varepsilon_{a,t} = 0.5$ % is depicted in fig. 2 (solid curve). The corresponding σ-ε_{pl}-hysteresis loop has been obtained by substracting the respective elastic strain (dashed curve) calculated via $\varepsilon_{el} = \sigma/E$. For further evaluation and plotting of the css curve in the last section the saturation stress values marked by the open circles in fig. 1 as well as the plastic strain at the point of load reversal (see dashed curve in fig. 2) have been used.

MICROSTRUCTURE

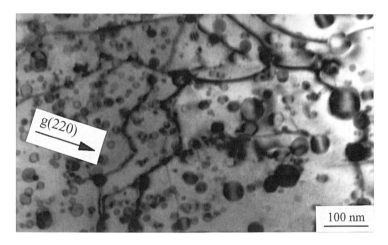

Fig. 3. Bright field TEM micrograph of a fatigued specimen corresponding to fig. 2. The test was interrupted at the point of compressive load reversal.

Figure 3 shows a typical example for the microstructure of a fatigued specimen. It reveals the following characteristics: firstly, a stable Yttria-particle microstructure during mechanical deformation at high temperatures is found revealing an average diameter of $d = 14$ nm, particle spacing of $L = 98$ nm and, hence, a volume fraction of $f = 1$ %. This observation is supported by investigations on the thermal-mechanical stability of oxide particles in MA 754 where also no particle coarsening of Yttria has been detected at 1123 K up to annealing times of 3000 h [1]. Secondly, frequent interfacial pinning [5] has been observed in both cases, but in contrast to monotonic deformation fatigued specimens typically show networks of loosely connected dislocations with a mesh size of 100...200 nm, which corresponds roughly to the planar particle spacing L. However, the evolution of "classical" heterogeneous dislocation structures consisting of subgrain boundaries or cell walls has not been observed during deformation. This can be attributed to the presence of the small incoherent particles with a relatively narrow planar spacing: they cause slip dispersal and deformation becomes homogeneous [6]. Additional TEM investigations on specimens deformed to failure have proven changes in the dislocation microstructure to be negligible. Thus, from similar TEM micrographs such as fig. 3 steady-state dislocation densities ρ_{ss} have been evaluated for different applied strain rates and strain amplitudes, respectively.

Fig. 4 summarizes these data in a plot of the calculated steady-state dislocation spacings $\rho_{ss}^{-0.5}$ versus the applied stress σ. As we will restrict our investigation to only one temperature we omit the commonly used normalization of σ by the shear modulus G in fig. 4. All data follow the inverse stress proportionality [7] with $k_\rho = 1,6$ described in the insert of fig. 3 independently on strain rate and loading conditions. Note that the data obtained from cyclic deformation seem to be limited to $\rho_{ss}^{-0.5}$ values above the grey shaded area for the average particle spacing L and, hence, the formation of characteristic dislocation networks is observed as previously shown in fig. 3. However, this is not fullfilled for the case of monotonic deformation: generally, compared to cyclic deformation one obtains lower $\rho_{ss}^{-0.5}$ values and vice versa higher dislocation densities, compare the corresponding circles and squares in fig. 4. Specifically, at $\dot{\varepsilon} = 10^{-3}$ s^{-1} the dislocation spacings for the monotonic deformation (full squares) are even lower than the average value of L.

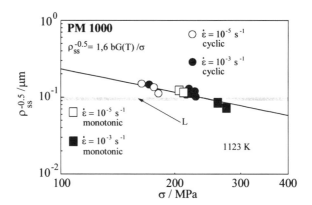

Fig. 4. Inverse stress proportionality of characteristic dislocation spacings after monotonic (squares) and cyclic (circles) deformation. The grey shaded area indicates the average particle spacing L.

MICROSTRUCTURAL MODEL

In this section a microstructural model based on the Haasen-Alexander-Ilschner (HAI-) theory of internal (σ_i) and effective stress (σ_{eff}) [8,9] is presented contributing to the goal of finding a physically based interlink between the observed plastic deformation and the microstructure of the metallic material at high temperature. It is combined with the threshold stress (σ_p-) concept [10] for describing monotonic as well as cyclic stress-strain curves of particle-strengthened alloys like PM 1000.

The approach is based on the fundamental Orowan equation defining $\dot{\varepsilon}$ as the product of mean dislocation density ρ and velocity v. According to [11] one finally obtains for deformation tests at constant true strain rate $\dot{\varepsilon}$:

$$\sigma_{HAI} = \frac{\ln\left[\frac{\dot{\varepsilon} \cdot M}{\rho \cdot b \cdot v_0} + \left(\left(\frac{\dot{\varepsilon} \cdot M}{\rho \cdot b \cdot v_0}\right)^2 + 1\right)^{1/2}\right]}{\beta} + \sigma_\rho + \sigma_p + \sigma_b \quad . \tag{1}$$

The first term on the right side of eq. (1) represents σ_{eff}, which controls dislocation movement. $\beta = 0.08$ MPa^{-1} and $v_0 = 2 \cdot 10^{-9}$ ms^{-1} are the parameters of the sinh-dependence of the dislocation velocity on σ_{eff}. They have been determined from the materials creep response after immediate stress changes at constant (micro-)structure [10]. $M = 2.45$ is the Taylor factor, $b =$

0.254 nm. σ_{eff} is calculated as the difference between the macroscopically measured steady-state stress level σ_{HAI} (open circles in figs. 1 and 4, respectively) and all back stress terms operating under the chosen deformation conditions. These back stresses are determined as follows: the particle hardening contribution σ_p can be evaluated from the yield stress increment at $\varepsilon_{\text{pl}} = 0.02\%$ compared to a single-phase matrix reference material, Nimonic 75, [11] and yields a constant value of $\sigma_p = 91$ MPa for both, monotonic and cyclic, loading conditions. The athermal long-range back stress σ_ρ due to single dislocations can be calculated according to Taylor [12] via $\sigma_\rho = \alpha G b M \sqrt{\rho}$, where $\alpha = 0.33$ is the elastic interaction coefficient and $G = 55.8$ GPa is the shear modulus at 1123 K. Due to the observed homogeneous deformation and dispersed slip (see fig. 3) no back stress resulting from heterogeneous dislocation substructure has to be taken into account [7], hence $\sigma_b = 0$.

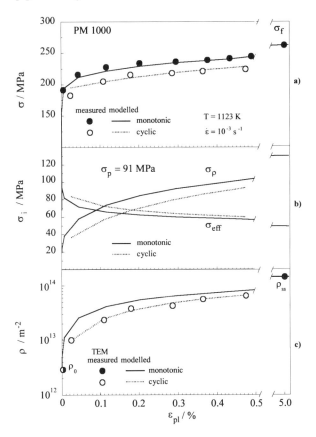

Fig. 5. Evolution of dislocation density and stress components in PM 1000 for the cyclic (dashed lines) and monotonic (continous) stress-strain curve: a) dislocation density, b) stress components and c) measured and modelled stress-strain curves.

To complete the HAI-model an appropriate kinetic law for the dislocation evolution has to be chosen. We follow [7,11] and apply a Johnson-Mehl-Avrami kinetic for the dependence of dislocation spacing $\rho^{-0.5}$ on plastic strain,

$$\rho^{-0,5} = \rho_{ss}^{-0,5} - (\rho_{ss}^{-0,5} - \rho_0^{-0,5}) \cdot \exp(-\varepsilon_{pl} / k)^q , \qquad (2)$$

where k and q are rate constants which control the evolution of the dislocation density of the as-received condition ρ_0 to the steady-state value ρ_{ss} vie the plastic strain ε_{pl}.

COMPARISON OF THE MONOTONIC AND CYCLIC STRESS-STRAIN CURVE

The comparison of the monotonic (full symbols) as well as cyclic (open symbols) stress-strain curve at 1123 K and $\dot{\varepsilon} = 10^{-3}\text{s}^{-1}$ is shown in fig. 5a. The tensile mss curve approaches a dynamic equilibrium of strain hardening and recovery after about 5% of plastic strain revealing steady-state deformation at a constant flow stress level (see σ_f in fig. 5a) until failure occurs at $\varepsilon_{pl} \approx 15$ %. However, as cyclic strain amplitudes are known to be significantly lower we focus our discussion of fig. 5 on the region of initial plastic deformation up to $\varepsilon_{pl} = 0.5$ % (note the broken abscissa), respectively. Obviously, the css curve shows a reduced stress niveau over the entire region of strain investigated. The microstructural origin for that observation is revealed by comparing the corresponding dislocation densities in fig. 5c. The evolution of ρ starts from its initial value value of $\rho_0 = 2.8 \cdot 10^{12}$ m^{-2} (half-filled circle) and approaches steady state ($\rho_{ss} = 1.2 \cdot 10^{14}$ m^{-2}) at $\varepsilon_{pl} = 5$ % in the case of monotonic deformation (full circle in fig. 5c) according to equ. (3). In contrast, the open data points obtained from LCF specimens cyclically saturated with different total strain amplitudes (see open circles in fig. 1) as well as the corresponding dashed curve reveal a retarded kinetic and lower absolute values of ρ. In turn, this reduced dislocation density during cyclic deformation has a twofold effect on the stress components calculated according to equ. (2): on the one hand σ_{eff} is slightly enhanced, however, on the other hand σ_ρ is distinctly lower as for monotonic loading, compare the dashed and continous curves in fig. 5b, respectively. Finally, adding the stress components σ_{eff}, σ_ρ and σ_p (= 91 MPa = const.!) yields the modelled stress-strain response depicted in fig. 5a. Both, the mss as well as the css curve are properly reflected by the model. Hence, the obvious cyclic softening during LCF loading of PM 1000 can be attributed to the experimentally verified difference of dislocation densities (fig. 4).

REFERENCES

1. Heilmaier, M. and Reppich, B. (1996). *Metall. and Mater. Trans.* **27A**, 3861.
2. Brett, S.J. and Doherty, R.D. (1978). *Mater. Sci. Engg.* **32**, 255.
3. Clavel, M. and Pineau, A. (1982). *Mater. Sci. Engg.* **55**, 157.
4. Müller, F.E.H., Heilmaier, M. and Schultz, L. (1997). *Mater. Sci. Engg.* **A234-236**, 509.
5. Schröder, J.H. and Arzt, E. (1985). *Scripta Met.* **19**, 1129.
6. Christ, H.-J. (1989). *Wechselverformung der Metalle,*Springer Verlag, Berlin.
7. Blum, W. (1992). In: *Mater. Sci. and Technology*, R. W. Cahn et al. (eds.), **Vol. 6**, *Plastic Deformation and Fracture of Materials*, H. Mughrabi (ed.), VCH Verlagsgesellschaft, Weinheim, pp. 359.
8. Alexander, H., Haasen, P. (1968). *Sol. State Phys.* **22**, 251.
9. Ilschner, B. (1973). *Hochtemperaturplastizität*, Springer Verlag, Berlin.
10. Brown, L. M., Ham, R. K. (1971). In: *Strengthening Methods in Crystals*, A. Kelly et al. (eds.), Appl. Sci. Publ., London, pp. 9.
11. Heilmaier, M., Wunder, J., Böhm, U. and Reppich, B. (1996). Comp. Mater. Sci. **7**, 159.
12. Taylor, G. (1934). In: *Proc. of the Royal Soc.* **A145**, 362.

Low Cycle Fatigue and Elasto-Plastic Behaviour of Materials
K-T. Rie and P.D. Portella (Editors)
635

A Numerical Calculation on Cyclic Stress-Strain Response in the Vicinity of Crack Tip in Annealed Copper

Kenji HATANAKA*, Takeshi UCHITANI** and Satoshi KAWARAYA***

*Dept. of Mechanical Engineering, Yamaguchi University, Ube, 755 Japan
**Bubcock-Hitachi K.K. Kure Works, Kure, 737 Japan
***Graduate school of Yamaguchi University, Ube, 755 Japan

ABSTRACT

A through-thickness center-cracked specimen of fully annealed copper was cyclically deformed under a load-controlled condition.   Then cyclic stress-strain response in the vicinity of the crack tip and the crack tip opening displacement were calculated using the elastic-plastic finite element method in which the constitutive equation derived on the basis of dislocation dynamics is used.   The distributions of the axial and shear strains along with changes in them and in the crack tip opening displacement during each half of a load cycle were calculated; the results corresponded well with the measurements.   In addition, the calculated stress-strain response showed that the axial strain range and the mean strain decreased to a much greater extent closer to the crack tip as the cyclic loading process progressed.

KEY WORDS

Copper, Constitutive Equation, Load-Controlled Cyclic Deformation Test, Cyclic Stress-Strain Response around Crack Tip, Elastic-Plastic Finite Element Method, Crack Tip Opening Displacement

INTRODUCTION

The cyclic hardening behaviors of annealed copper were successfully described by the constitutive equation which was derived through dislocation dynamics by one of the authors [1].   This equation also expressed well cyclic stress-strain response in the circumferentially notched components of annealed copper, showing its applicability to multiaxial stress state problems [2].   The quantitative description of cyclic stress-strain response and crack opening∕closing behaviors in the vicinity of crack tip are very important for clarifying fatigue crack growth mechanism.   The analysis on such problems has been performed so far [3,4], but cyclic hardening occurring around crack tip has not been taken into account in there.

The present paper aims to calculate numerically the load cycle dependent variations of stress-strain response in the vicinity of crack tip and crack opening displacement, using the constitutive equation proposed earlier by the authors.   Furthermore, the calculated results were compared with the measurements by means of the grid method.

ANALYSIS

The center-cracked specimen with rectangular crosssection, which was employed in the present study, was shown in Fig.1. A quarter of the specimen was modeled for the FEM calculation, which contains crack of $2a_0=2\times2.875$ mm in length including the size of slit introduced as crack precursor; triangular shaped mesh element was employed, where the minimum sized element positioned in the vicinity of crack tip is $15.625\,\mu$m in both the base and height, and total numbers of mesh elements and mesh nodes are 1142 and 659, respectively. The crack extension is not taken into account in the calculation.

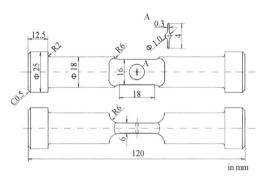

in mm

Fig.1 Configuration and dimensions of test specimen

The coordinate system employed for the FEM calculation is shown in Fig.2. The constitutive equation,

$$\dot{\sigma}_A = E\dot{\varepsilon}_{tA} - Egbv^* \left\{ \rho_{to} + M\varepsilon_{pA}^{\,\exp(-q\varepsilon_{pA})} \right\} \times \exp\left\{ -\left(D + H\varepsilon_{pA}^{\,1/n}\right) \Big/ \left(\sigma_A - \sigma_{iA}\right) \right\} \tag{1}$$

is constituted with respect to the coordinate system $\sigma_A - 0_A - \varepsilon_{pA}$, where the effective stress σ_e is zero

at the origin 0_A. The constitutive equation,

$$\dot{\sigma} = E\dot{\varepsilon}_t - Egbv^* \left\{ \rho_{to} + M(2\varepsilon_p)^{\exp\{-q(2\varepsilon_p)\}} \right\} \times \exp\left[-\left\{ D + H(2\varepsilon_p)^{1/n} \right\} \Big/ (\sigma - \sigma_i) \right] \tag{2}$$

was rebuilt with respect to the coordinate system $\sigma - 0 - \varepsilon_p$ from eq.(1),

where g is the geometrical factor required to transform shear strain into normal one, b the Verger's vector, v^* the terminal velocity of dislocation at infinitively large stress, M the dislocation multiplication factor, ρ_{to} the initial dislocation density of annealed copper, D the characteristic drag stress which is closely related to yield stress, H the constant governing strain hardening rate in large strain regime, n the stress exponent which controls strain hardening rate around yield point and σ_i the internal stress.

The stress σ and plastic strain ε_p are replaced with the equivalent stress σ_{eq} and equivalent plastic strain ε_{eq}^{p} in eqs.(1) and (2). Then these were used for FEM calculations of cyclic deformation around crack tip as the constitutive equations; the cyclic stress-strain response of each mesh-element was expressed by the constitutive equations. The constants settled for the calculations in eqs.(1) and (2) are shown together in the following.

$$E=139[GPa], \quad \rho_{tan\,n}=10^{12}[1/m^2], \quad g=0.5, \quad M=M_0\exp\left(-m_0\Sigma|\varepsilon_p|\right), \quad M_0=5\times10^{14}[1/m^2], \quad m_0=c\varepsilon_{p(\frac{1}{4})}^{n_0-1},$$

$$c=8.58\times10^{-2}, \quad n_0=0.323, \quad b=2.556[\text{Å}], \quad H=3000[MPa], \quad v^*=2270[m/s], \quad n=1.8, \quad D=D_0\eta^{\frac{1}{2}}, \quad D_0=100MPa,$$

$$\eta=\rho_{tN}/\rho_{tan\,n} \quad \text{and} \quad q=-0.5$$

We can calculate cyclic stress-strain response by setting the stress rate which is substituted for $\dot{\sigma}$ in eqs.(1) and (2). The numerical integration of eq.(2) by the Runge-Kutta method gives the equivalent stress-strain response for a first quarter cycles at each mesh-element. The stress-strain curve is approximately expressed by the eighteen straight broken lines and their tangential moduli are determined.

An elastic-plastic FEM calculation is performed based on the coordinate system $\sigma-0-\varepsilon_p$ using the r-min method proposed by Yamada et al. [5], where the above mentioned tangential moduli are employed. Then the stress rate is anew assessed by the stress increment to time-period ratio at each mesh-element during a first quarter cycle. The stress–strain response for a subsequent half cycle is calculated by substituting this stress rate for $\dot{\sigma}$ in eq.(1) on the coordinate system $\sigma_A-0_A-\varepsilon_{pA}$. The elastic-plastic FEM calculation is carried out for every half a cycle, using the calculated stress-strain response as basic material law. The load rate adopted for the calculation is $\dot{P}=\Delta P/\Delta t=1968[N/s]$.

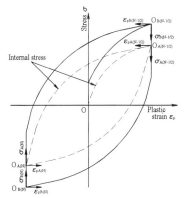

Fig.2 Coordinate systems settled for FEM calculation

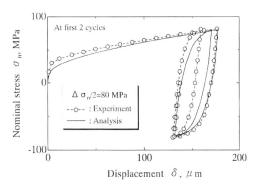

Fig.3 Comparison between cyclic stress-strain responses calculated and measured for the through-thickness center cracked specimen under cyclic loadings

CALCULATION

The cyclic stress-displacement response of the through-thickness cracked specimen calculated under the loading conditions of $\Delta\sigma_n/2 =80$MPa are shown in Fig.3, being compared with the tested one. According to the figure, the calculation describes quite well cyclic hardening behavior which the through-thickness center cracked specimen exhibits.

Figure 4 illustrates cyclic loading pattern and loading points at which strain and crack opening displacement were measured and calculated around the crack tip. Figures 5 (a) to (c) show the normal strain ε_y and the crack opening displacement calculated at loading points $C_{(N=1)}$, $A_{(N=2)}$ and $C_{(N=2)}$ during cyclic loadings at $\Delta\sigma/2 =80$MPa. The followings are shown in the figures: Crack opens greatly and normal strain develops intensively in the direction of $\theta \cong 60°$ ahead of crack tip during the first tensile load stroke from 0 to $C_{(N=1)}$. The crack remains open quite largely at $A_{(N=2)}$, although the ε_y -developed region appreciably shrinks during the following unloading and compressive load stroke from $C_{(N=1)}$ to $A_{(N=2)}$. The crack opening displacement increases and the ε_y - developed region extends during subsequent half a cycle, but these are smaller at $C_{(N=2)}$ than at $C_{(N=1)}$. This might be caused by progress of cyclic hardening around the crack tip. The tested results corresponding to Figs.5(a) to (c) are shown in Figs.6(a) to (c).

Figures 7(a) and (b) show the calculated cyclic stress-strain responses of the mesh-elements situated ahead of the crack tip at $\theta = 0°$ under cyclic loadings at $\Delta\sigma_n/2 =80$MPa. The cyclic strain range and the mean strain decrease to much greater extent at the position nearer to the crack tip in the cyclic loading process in the figures. Furthermore, the negative mean stress is generated in the initial loading cycles and is reduced with progress in the cyclic loading process.

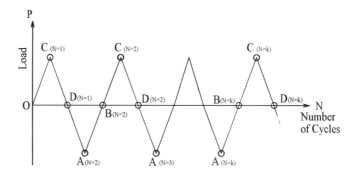

Fig.4   Cyclic loading form and loading points at which strain and
crack opening displacement were measured and calculated

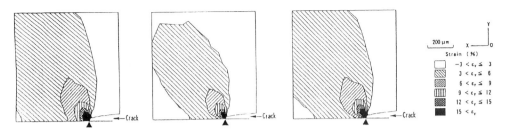

(a) At the loading point C$_{(N=1)}$ (b) At the loading point A$_{(N=2)}$ (c) At the loading point C$_{(N=2)}$

Fig.5 Axial strain and crack profile calculated around crack tip at several loading points shown in Fig.4
under cyclic loadings at $\Delta\sigma_n/2$ =80MPa

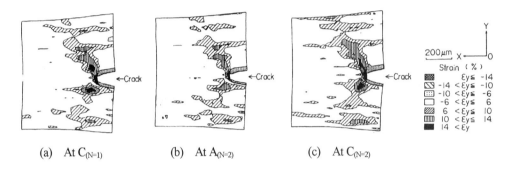

(a) At C$_{(N=1)}$ (b) At A$_{(N=2)}$ (c) At C$_{(N=2)}$

Fig.6 Axial strain and crack profile measured around crack tip under the same loading conditions
as those in Fig.5, where (a) to (c) correspond to (a) to (c) in Fig.5.
The domain where separation of the grid line made strain measurement impossible was represented
by dotted mark.

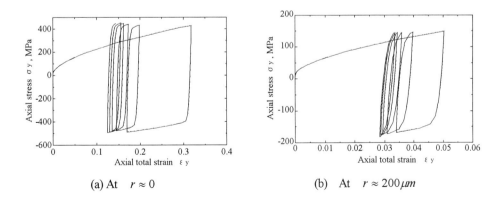

(a) At $r \approx 0$ (b)   At $r \approx 200\,\mu m$

Fig.7 Cyclic axial stress-strain response in crack plane ($\theta = 0°$) ahead of crack tip
during load cycles at $\Delta\sigma_n/2$ =80Mpa

CONCLUSIONS

The cyclic deformation tests were performed for the through-thickness center cracked specimen. Moreover, the calculations were performed for the specimen subjected to cyclic loading, where the constitutive equation proposed earlier by the authors was employed. In addition, the localized strains generated around the crack tip was measured by means of the grid method. The main results obtained are summarized as follows:

(1) The calculated load-displacement response was in quite good agreement with the experimental data.

(2) The calculations showed that the intense normal strain ε_y developed in the direction of the angle inclined at $\theta \cong 60°$ with respect to x-axis and crack opened greatly at a first tensile loading. The strain and crack opening displacement decrease during the following compressive loading, but the positive normal strain is retained and crack remains open at the compressive tip of the cyclic loading wave.

The variation ranges of ε_y and crack opening displacement during load excursion between the tensile and compressive tips in the loading wave decrease with progress in the cyclic loading process. These calculations were in quite good agreement with the measured ones.

(3) The stress-strain hysteresis loops were calculated at several locations along the x-axis ahead of the crack tip under load-controlled condition. The calculation showed that cyclic strain range and the mean strain decreased greater at the location near to the crack tip than at the location distant from the crack tip in the early stage of the cyclic loading process. Moreover, the cyclic stress-strain response attains earlier to the stable state at the latter than at the former.

REFERENCES

1. Hatanaka, K., Fujimitsu, T. and Sumigawa, Y. (1989), Trans. Japan Soc. of Mech. Engrs., Ser. A 55, pp. 1000-1009. (in Japanese)

2. Hatanaka, K., Kakumoto, A. and Ishimoto, Y. (1996), Trans. Japan Soc. of Mech. Engrs., Ser. A 62, pp.1008-1016. (in Japanese)

3. Ohji, K., Ogura, K. and Ohkubo, K., (1976), Trans. Japan Soc. of Mech. Engrs., 42, pp. 643-648. (in Japanese)

4. Shiratori, M., Miyoshi, T. and Miyamoto, H. (1977), Trans. Japan Soc. of Mech. Engrs., 43, pp. 3577-3588. (in Japanese)

5. Yamada, Y. (1972), Plasticity and Viscous Elasticity, Series Monographs on Structure Analysis by Computer, Baihukan No.2-A (in Japanese)

Low Cycle Fatigue and Elasto-Plastic Behaviour of Materials
K-T. Rie and P.D. Portella (Editors)

NUMERICAL AND EXPERIMENTAL INVESTIGATIONS OF TEMPERATURE-RATE TERMS IN MODELLING THE INELASTIC MATERIAL BEHAVIOUR

R. KÜHNER and J. AKTAA

*Forschungszentrum Karlsruhe GmbH, Institut für Materialforschung II,
Postfach 3640, D-76021 Karlsruhe, Germany*

ABSTRACT

To describe inelastic material behaviour under non-isothermal loading conditions on the base of a viscoplastic model it is not sufficient to use only temperature-dependent model parameters. Also the temperature-rate in the evolution equations of the internal variables must be considered. To investigate the influence of the temperature-rate, experiments should be defined in which the model response shows a remarkable dependence on the temperature-rate. Such experiments could be performed on a coupled "two-bar-system" under thermal and complex thermo-mechanical loads, respectively. Using a recently developed testing facility experiments with a coupled "two-bar-system" are realized. This allows different temperature paths to exist at the two bars which can be coupled to produce the same total strain throughout the experiment. At the same time, the sum of the forces of the two bars is controlled as a function of time. First results have shown that temperature-rate terms are necessary to model ratchetting effects correctly. Therefore, approaches taken from the literature were investigated and modified to achieve optimum agreement between modelling and experiment.

KEYWORDS

Modellisation, viscoplasticity, ratchetting, thermo-mechanical loading.

INTRODUCTION

In many areas of technology where components are subjected to varying mechanical and/or varying thermal loadings progressive plastic deformation can occur. This so called ratchetting must be quantitatively predictable because it may lead to failure of component parts. For this purpose constitutive equations are required, being able to describe alternating plastic deformation under cyclic thermo-mechanical loading. Most of the advanced viscoplastic models however are not suitable for predicting ratchetting [1]. In particular those which use the linear kinematic hardening rule by Prager because in these models after a short transition period ratchetting stops [2]. The Chaboche model which uses a nonlinear kinematic hardening law is able to simulate ratchetting effects in principle. But quantitatively there is a significant overestimation of the ratchetting strain. Under cyclic thermal-mechanical loading temperature-rate terms in the equation of kinematic hardening lead to a distinct improvement of the modellisation of the ratchetting behaviour.

TEMPERATURE-RATE TERMS IN THE EVOLUTION EQUATION FOR KINEMATIC HARDENING

The investigations of the present paper base on the Chaboche model [3]. To gain access to ratchetting behaviour we start our considerations with the so-called back stress Ω which was originally introduced to describe kinematic hardening. This back stress can be regarded as a directed inner stress induced by deformation. The Bauschinger effect or, in general, the deformation induced anisotropy appearing under cyclic loading conditions is caused by this directed inner stress. For the Chaboche model the evolution equation for Ω runs:

$$\dot{\Omega} = H\dot{\varepsilon}_{in} - D\Omega|\dot{\varepsilon}_{in}| - R\Omega^m + \frac{\partial H}{\partial T}\frac{\Omega}{H}\dot{T} \tag{1}$$

where $\dot{\varepsilon}_{in}$ and $\dot{T}$ denote the inelastic strain rate and the temperature rate, respectively. H, D, R and m are model parameters depending on material and temperature. The first term corresponds to the linear kinematic hardening rule by Prager [4]. Dynamic and static recovery effects are taken into account by the second and third terms, respectively. For the sake of simplicity, we have only investigated loading conditions with negligible static recovery. Therefore, the temperature-rate independent terms are identical with the non-linear kinematic hardening rule by Armstrong-Frederick [5].

To obtain a better description of the deformation behaviour under nonisothermal loading the evolution equation for Ω was modified by [6] as follows:

$$\dot{\Omega} = H\dot{\varepsilon}_{in} + \frac{\partial H}{\partial T}\frac{\Omega}{H}\dot{T} - \frac{\Omega}{\Omega_s}\left\langle H|\dot{\varepsilon}_{in}| + \frac{\partial H}{\partial T}\frac{\Omega}{H}\dot{T}\right\rangle \tag{2}$$

Here another modification is proposed in order to improve the description of the ratchetting behaviour under thermo-mechanical loading:

$$\dot{\Omega} = H\dot{\varepsilon}_{in} - D\Omega|\dot{\varepsilon}_{in}| + \frac{\partial \Omega_s}{\partial T}\frac{\Omega}{\Omega_s}\dot{T} + \frac{\partial D}{\partial T}(\Omega_s - \Omega)\varepsilon_{in}\dot{T} \tag{3}$$

The first two terms are identical with the non-linear Armstrong-Frederick rule [5]. The other two terms are derived from a non-linear exponential approach for the back stress Ω for monotonous loading. In this terms $\Omega_s = H/D$ is the saturation value for Ω. The fourth term in equation (3) becomes zero when the back stress Ω is saturated. This term is noneffective when the back stress is saturated and takes most effect for little values of back stress and inelastic strain, respectively. Contrary the third term is most effective when the value of the back stress is approximately identical with its saturation value, in other words for large inelastic deformations.

EXPERIMENTAL PROCEDURES

The basic material for the specimens used in the present investigations is a plate of AISI 316L(N), DIN 1.4909 of the Creusot-Marrel (CRM) charge 11477. The thickness of the plate is 40 mm. The parts of alloying elements are 0.02 C, 17.34 Cr, 12.5 Ni, 2.4 Mo, 1.8 Mn, 0.32 Si, 0.12 Cu, 0.080 N, 0.042 Nb+Ta, 0.030 Co, 0.02 P, 0.018 Al, 0.008 Ti, 0.0014 B, 0.0006 S. The state of material as supplied was solution treated (1100°C, vacuum) [7].

To develope temperature-rate terms that allow a satisfactory description of the ratchetting behaviour under cyclic thermal-mechanical loading it is important to have suitable ratchetting experiments. For this purpose suitable experiments on a two-bar-system and on a one-bar-system were defined and performed.

The *one-bar-system* was realised on base of an electromechanical INSTRON 8062 testing machine with digital control. These testing machine was expanded by a heating system with digital temperature control. So nearly any thermal-mechanical loading can be realised. The specimens are heated by direct passage of current and the cooling results from watercooled specimen holders as well as from thermal radiation and convection. The generation of the nominal value as well as the recording of the experimental data are performed synchronous for the testing machine and the heating system by a PC. This PC is equiped with a Microstar DAP-Board. A Ni-Cr-Ni thermocouple element weld on the specimen is used for measurement of temperature. For strain-measuring an extensometer with the gauge length of $l_0 = 6$ mm is directly clamped on the specimen.

The construction of the *two-bar-system* is based on two servohydraulic MTS 810 testing machines. The strain-measuring is realised with two extensometers with the gauge length of $l_0 = 12$ mm that are directly clamped on the specimens. The specimens are inductively heated by a Hüttinger TIG 5/300 high-frequency generator and a watercooled copper spool. The specimens are cooled by watercooled specimen holders, by thermal radiation, convection and if necessary by pressure air. A Ni-Cr-Ni thermocouple element weld on the specimen is used for measurement of temperature. The control of the total system is managed by the four channel digital controller „MTS Test-Star II [8].

EXPERIMENTAL AND NUMERICAL RESULTS

The model parameters used in the present investigation were determined from isothermal tensile and relaxation experiments on the one-bar-system. To investigate the role of temperature-rate terms in modelling the inelastic material behaviour two thermo-mechanical experiments were performed on the one-bar-system (test I) and on the two-bar-system. The loading conditions were chosen so that the experiment shows ratchetting effects without applying a mean stress. Test I was performed under load and temperature control. The load and the temperature histories of test I are shown in figure 1 and the test data are summarized in table 1. The load and the temperature are cycled out of phase while the mean stress is zero. In test II the two specimens of the two-bar-system are coupled so that the total strain of the specimen one is always identical with the strain of the specimen two. At the same time the stress acting on specimen one is of the same amount but of the opposite direction as the stress acting on specimen two so that there is no force from outside acting on the system. The loading of the two-bar-system is caused by the temperature history shown in figure 2. The data of test II are summarized in table 2.

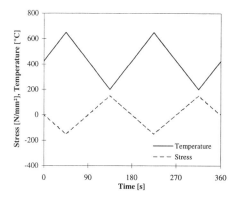

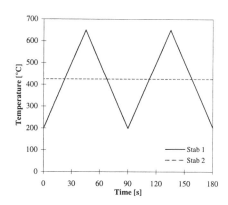

Fig. 1. Load and temperature histories of ratchetting test I.

Fig. 2. Loading history of ratchetting test II.

Table 1. Summary of experimental data for test I

	Tmax [°C]	Tmin [°C]	Tmean [°C]	σmax [N/mm²]	σmin [N/mm²]	σmean [N/mm²]
Test I	650	200	425	150	-150	0

Table 2. Summary of experimental data for test II

	T1max [°C]	T1min [°C]	T1mean [°C]	T2 [°C] (constant)
Test II	650	200	425	425

The experimental and numerical results for ratchetting strain in test I are shown in figure 3. Curve E shows the experimental data whereas the curves A, B and C show the numerical results. Here A is the curve obtained by using equation (1) which was proposed by Chaboche. The result B corresponds to the modellisation using equation (2) and curve C is obtained by using equation (3). The kinematic hardening law by Chaboche overpredicts the ratchetting strain observed in the experiment very much. Equations (2) and (3) produce a better prediction of the ratchetting strain. Curve B shows too less ratchetting at the beginning of the experiment and for a increasing number of cycles it diverges from the experimental curve E. The result C overswings at the beginning of the experiment and then converges for a increasing number of cycles.

The results derived from experiment and calculation for test II are presented in figure 4. In this diagram the curve E shows the experimental data. The curves A, B and C show numerical results. Here A is the curve derived from equation (1) proposed by Chaboche. The result B corresponds to the evolution equation (2) and curve C is obtained by using equation (3). The kinematic hardening law by Chaboche produces much more ratchetting strain than the experiment. Besides, the curve A does not show any tendency of saturation behaviour whereas the experimental result does. The predictions of the ratchetting behaviour obtained using the equations (2) and (3) are much better. Curve B seems quantitatively the closest to the experiment but it shows a qualitatively wrong ratchetting behaviour. Here the ratchetting strain increases while in the experiment the ratchetting strain decreases. The result C shows qualitatively the same evolution as the experiment.

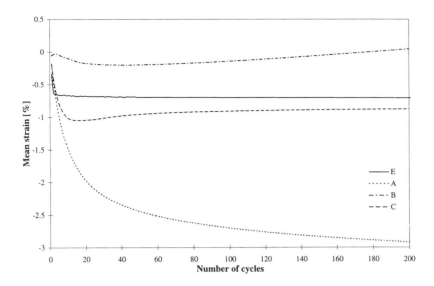

Fig. 3. Uniaxial ratchetting of AISI 316L(N) in test I; experiment and simulations.

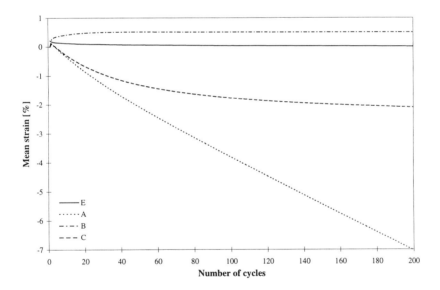

Fig. 4. Uniaxial ratchetting of AISI 316L(N) in test I; experiment and simulations.

CONCLUSIONS

Many advanced constitutive models with nonlinear hardening rules, like the Chaboche model, are not suitable for simulating ratchetting. Particularly the ratchetting produced by thermal or combined thermal-mechanical loading can not be satisfactory predicted by those models because either they have no temperature-rate terms or their temperature-rate terms are too weak. This could be confirmed by applying models known from the literature to describe the ratchetting behaviour of AISI 316L(N) in thermo-mechanical tests. The conditions of the tests performed were selected so that the influence of mean stress on the ratchetting behaviour is excluded in test I and minimized in test II. By a modification of the kinematic hardening rule in the Chaboche model a distinct improvement of the ratchetting behaviour could be obtained. The simulation of the ratchetting in test II is not as well as in test I because among other things mean stress effects are not taken into account by our modifications. This modifications will be the base of our further investigations on the modelling of ratchetting behaviour.

ACKNOWLEDGEMENT

The test on the two-bar-system was performed by Mr. L. Angarita at the Institut für Werkstoffkunde I of the University of Karlsruhe within the framework of a common research project supported by the ministery of science, research and art of Baden-Württemberg.

REFERENCES

1. N. Ohno, Current state of the art in constitutive modelling for ratchetting, *14th International Conference on Structural Mechanics in Reactor Technology (SMiRT)*, Lyon, France, August 17-22, pp. 201-212, 1997.
2. M. Sester, R. Mohrmann, H. Kordisch, Fraunhofer Institut für Werkstoffmechanik, 03.97, T 1/97.
3. J. L.Chaboche, Time-independent constitutive Theories for cyclic plasticity, *International Journal of plasticity*, Vol. 2, No 2, pp. 149-188, 1986.
4. W. Prager, Recent developments in the mathematical theorie of plasticity, *Journal of applied Physics* 20: 235-241, 1949.
5. P. J. Armstrong, C. O. Frederick, A mathematical representation of the multiaxial Bauschinger effect, CEGB report: RD/B/N731. Berkeley Nuclear Laboratories, 1966.
6. R. Sievert, J. Olschewski, Bundesanstalt für Materialforschung und -prüfung (BAM), Modellierung thermisch-mechanischen viskoplastischen Verfestigungsverhaltens und Lebensdauerabschätzung bei mehrachsiger Hochtemperaturermüdung am Beispiel von IN738LC.
7. M. Schirra, S. Heger, Institut für Material- und Festkörperforschung, Kernforschungszentrum Karlsruhe, Zeitstandfestigkeits- und Kriechversuche am EFR-Strukturwerkstoff 316L(N), DIN 1.4909, KFK 4767, 09.1990.
8. G. Pitz, K.-H. Lang, D. Löhe, Institut für Werkstoffkunde I, Universität Karlsruhe (TH), Konzeption und Realisierung eines Prüfsystems für komplexe thermisch-mechanische Belastung.

Low Cycle Fatigue and Elasto-Plastic Behaviour of Materials
K-T. Rie and P.D. Portella (Editors)
© 1998 Elsevier Science Ltd. All rights reserved.

RATCHETING OF AN AUSTENITIC STEEL UNDER MUTLIAXIAL AND THERMOMECHANICAL LAODING: EXPERIMENTS AND MODELING

M. SESTER, R. MOHRMANN, R. BÖSCHEN
Fraunhofer-Institut für Werkstoffmechanik
Wöhlerstr. 11, D-79108 Freiburg, Germany

ABSTRACT

It is now well accepted that classical plasticity models have difficulties in describing the cycle by cycle accumulation of plastic strain under unbalanced cyclic loading called ratcheting [1]. In experiments with non-zero mean stress, the austenitic steel X6 CrNiNb 18-10 showed ratcheting under different kinds of cyclic loading, including biaxial and thermomechanical loading. In all experiments, the deformation increments per cycle quickly decreased with continued cycling. A recent material model according to Jiang [2] was extended with respect to rate dependent plastic flow, temperature dependent cyclic hardening and out-of-phase hardening. The model parameters were adjusted to few standard experiments with constant strain amplitude, constant strain rate, constant mean stress and constant temperature. The transferability of the model parameters was checked by simulating other experiments with varying loads (including hold times), varying temperature and varying mean stress. A good overall agreement of experiments and model predictions was found.

KEYWORDS

cyclic plasticity, kinematic hardening, ratcheting, time dependent plasticity, multiaxial loading, thermomechanical loading, finite elements, local approach

INTRODUCTION

If materials are subjected to cyclic plastic loading with nonzero mean stress, usually there is a cycle by cycle accumulation of plastic strain in the direction of mean stress. This phenomenon is called cyclic creep or ratcheting. It can occur under uniaxial loading, under biaxial loading, under combined mechanical and thermal loading and in other more general situations. Since ratcheting is a progressive deformation accumulating cycle by cycle, it is not easy to predict the development of ratcheting accurately [1].

In cyclic plasticity, the accurate modeling of anisotropic strain hardening is crucial. The modeling of anisotropic strain hardening (also termed kinematic hardening) started with the linear kinematic hardening model of Prager [3]. Among others, Besseling [4], Armstrong and Frederick [5] and Mroz [6] proposed models for representing nonlinear kinematic hardening. A broader discussion is given for example in [1]. It is now well accepted that the above mentioned classical models have difficulties especially in describing the ratcheting [1], [7], [8].

The situation is summarized schematically for uniaxial loading in Fig. 1. Prager's linear kinematic

648

hardening rule gives no ratcheting at all. Under uniaxial loading, the same holds for the Besseling and the Mroz model. The Armstrong and Frederick (AF) model gives ratcheting, however, in most situations, this model overpredicts ratcheting under uniaxial and under general multiaxial conditions. It was suggested by Ohno and Wang that the dynamic recovery term of the AF model is too active for an accurate simulation of ratcheting [1]. In the next section, modifications of the AF model introduced by Chaboche, Ohno/Wang and by Jiang will be discussed.

A model of Jiang [2] was taken as a starting point for extensions regarding rate dependent plastic flow, temperature dependent cyclic hardening and additional hardening under nonproportional loading. In the following, experimental results for the austenitic steel X 6 CrNiNb 18-10, the procedure for determination of model parameters and FE simulations of validation experiments will be presented.

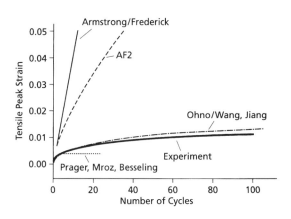

Fig. 1: Capability of models for describing uniaxial ratcheting

MODIFICATIONS OF THE ARMSTRONG-FREDERICK (AF) MODEL

Caboche extended the original AF model by decomposing the kinematic hardening variable or backstress tensor α_{ij} into M components $\alpha_{ij}^{(i)}$ (i = 1,2,...,M). In this model (termed AF2 model)

$$\alpha_{ij} = \sum_{i=1}^{M} \alpha_{ij}^{(i)} \tag{1}$$

$$d\alpha_{ij}^{(i)} = c^{(i)} \left(r^{(i)} d\varepsilon_{ij}^{p} - \alpha_{ij}^{(i)} dp \right), \tag{2}$$

where $d\varepsilon_{ij}^{p}$ is the plastic strain increment, dp is the increment of equivalent plastic strain defined as $dp = \sqrt{\frac{2}{3} d\varepsilon_{ij}^{p} d\varepsilon_{ij}^{p}}$, and $c^{(i)}$ and $r^{(i)}$ are material parameters. In the original AF model, there is only one backstress component (M=1). In order to overcome the excessive ratchetting predicted by the AF and by the AF2 model, Ohno and Wang proposed that the dynamic recovery of $\alpha_{ij}^{(i)}$ is activated in a nonlinear way as $|\alpha_{ij}^{(i)}|$ approaches a critical value $r^{(i)}$. A similar proposal was given by Chaboche [9]. It was shown by Jiang in [2] that the Ohno and Wang model in the form

$$d\alpha_{ij}^{(i)} = c^{(i)} \left(r^{(i)} d\varepsilon_{ij}^{p} - \left(\frac{|\alpha_{ij}^{(i)}|}{r^{(i)}} \right)^{\chi^{(i)}} \alpha_{ij}^{(i)} dp \right), \tag{3}$$

describes experimental results with a rail steel (1070 steel) well, including multiaxial experiments with block loading.

EXPERIMENTAL WORK

Experimental Program

An experimental program was carried out with special emphasis on cyclic loading including variable amplitude and mean stress, temperature dependency, effects of loading rate and multiaxial (including out-of-phase) loading. An overview is given in Tab. 1.

Table 1: Experimental program for the austenitic steel X6 CrNiNb 18-10

Experiments	**Observed Phenomena**
Tensile tests	Stress strain curves, depending on temperature and strain rate
Tensile tests with hold time	Logarithmic creep
Push-pull tests (strain control)	Bauschinger effect, stress strain hysteresis loops, cyclic hardening, damage (microcracking)
Push-pull tests (stress control) with constant mean stress	Ratcheting
Push-pull tests (stress control) with variable mean stress	"Back ratcheting" and other sequence effects
Push-pull tests (stress control) with variable stress and temperature	Sequence and temperature effects
Tension-torsion tests at room temperature (stress control) with proportional and non-proportional loading	Biaxial ratcheting, additional (out of phase) hardening
Biaxial tests with cruciform specimens	Biaxial ratcheting, additional hardening

Experimental Results

In all experiments with nonzero mean stress, the austenitic steel X6 CrNiNb 18-10 showed ratcheting under cyclic loading. However, the deformations per cycle quickly decreased with continued cycling. Between room temperature and 280^0C, significant cold creep was found. In the out-of-phase experiments, the material showed a remarkable amount of additional hardening as compared to uniaxial or proportional loading.

MODELING

Extensions to the Model of Jiang

The material model according to Jiang [2] (based on a proposal of Ohno [1]) was extended with respect to phenomena that were found to be specific to the examined austenitic steel:
- rate dependent plastic flow
- temperature dependent cyclic hardening
- additional hardening under nonproportional loading.

In the Jiang formulation material behaviour is considered time independent. Now, time dependent plastic flow is modeled in the framework of viscoplasticity as given in [9]. Temperature dependent cyclic hardening is described by hardening coefficients $c^{(i)}$ in eq. (3) which are a function of both the accumulated plastic strain and the current temperature. For additional hardening under nonproportional loading a simple approach is used: Viscosity is assumed to depend on a memory variable. This variable has a fading memory of the difference between the current flow direction and an average flow direction. In that way, additional hardening for proportional/nonproportional paths and subsequent softening for nonproportional/proportional paths is described. For details of this extended Jiang model, the reader is referred to [10]. The model was implemented a finite element code in order to enable validation computations of laboratory experiments and structural analyses.

Determination of Material Parameters for the Extended Jiang Model: Procedure and Results

For the determination of material parameters, two approaches were proposed: In the first approach, all material parameters were adjusted *simultaneously* to a subset of experimental data with the help of a numerical optimization routine. In this numerical fit, six isothermal push-pull tests with constant load amplitude and constant mean stress were used as experimental input. In the second approach, different parameters were adjusted *successively* to characteristic experiments. An overview is given in Tab. 2.

Table 2: Proposed parameter adjustment for the extended Jiang model

Parameters	Adjusted to characteristic experiments
Elastic constants	Isothermal tensile tests or push-pull tests
Yield limit	Isothermal tensile tests or push-pull tests
Viscosity, strain rate sensitivity	Tensile tests with hold time
Kinematic hardening parameters $c^{(i)}$, $r^{(i)}$, Eq. (3)	Balanced cyclic tests (strain control)
Exponents, $\chi^{(i)}$, Eq. (3)	Unbalanced cyclic tests (stress control)
Temperature dependent terms in yield limit, viscosity and cyclic hardening	Isothermal push-pull tests at different temperatures
Additional hardening terms	Tension-torsion tests

Jiang showed in [8] that the parameters $c^{(i)}$, $r^{(i)}$, (i=1,2,...M) can be given in closed form if isotropic hardening is neglected and if the stabilized stress strain curves are approximated multilinearly with M+1 sample points. The parameters $\chi^{(i)}$ and the other parameters are adjusted by trial and error such that the simulations fit the experiments. Our experience suggests that the variation of one parameter has a significant effect only on the simulation of its related characteristic experiments and little effect on others.

With the following time consuming FE-simulations in mind, the number of backstress components was restricted to M=5 where Jiang used M=10 [2]. For simplicity, all parameters $\chi^{(i)}$ were assumed to be equal. The result of the fit was $\chi^{(i)} = 10.3$.

Validation of the Extended Jiang Model: Procedure and Results

The transferability of the model parameters was checked by simulating validation experiments with varying loads (including hold times) and varying temperature. Here, results of a push-pull test with block loading and increasing temperature and of a biaxial test are presented.

In the push-pull test, there were 3 blocks with 10 stress cycles followed by a hold time with constant stress and a temperature increase. The applied loading and measured/predicted stress strain curves are given in Fig. 2. and Fig. 3, respectively.

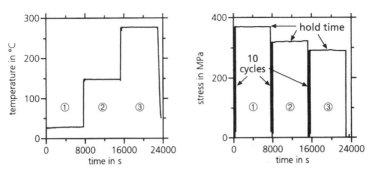

Fig. 2: Push-pull test with block loading: Applied loading

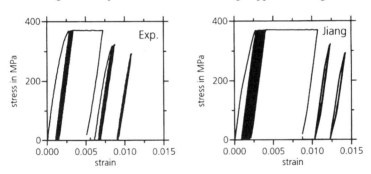

Fig. 3: Push-pull test with block loading: Measured and
predicted stress strain curves

In biaxial experiments with cruciform specimens, a constant force and a cyclic displacement perpendicular to it were applied. Measured and predicted strains in the direction of the applied constant load are given in Fig. 4 for one experiment. Out of the experimental data, only maximum and minimum strain values are shown for some selected cycles.

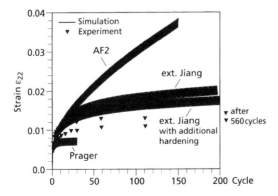

Fig. 4: Biaxial experiment with cruciform specimen: Measured and computed strain values

DISCUSSION

The value of $\chi^{(i)}$ = 10.3 found in the parameter fit is much higher than the $\chi^{(i)}$-values used by Jiang to describe the ratcheting of 1070 steel [2]. On the other hand, Ohno suggests $\chi^{(i)} \rightarrow \infty$ for modeling the ratcheting of 304 and 316 steels at elevated temperatures [1].

In Fig. 2, there is little ratcheting during stress cycling and significant cold creep during the hold time. This is well predicted by the new model. In Fig. 4, there is pronounced ratcheting during the first cycles followed by an almost perfect shakedown. The Prager model gives no ratcheting at all, the AF2 model largely overestimates the increase of plastic strains. The new model is well in line with the experiment.

CONCLUSION

Our simulations once more demonstrate that classical material models offered in commercial FE codes cannot describe material ratcheting accurately. This holds for the Prager model, Mroz model and basic Armstrong-Federick models. On the other hand, modified Armstrong-Federick models with nonlinear recovery terms as proposed by Chaboche or Ohno and used by Jiang and in this work predict ratcheting with reasonable accuracy.

In structural analyses where material ratcheting or mean stress relaxation is of concern, the new model will also give improved estimates of local loading parameters (such as the plastic strain range, stress range, ...) which are needed in local approaches to low cycle fatigue.

REFERENCES

1. Ohno N.: Current State of the Art in Constitutive Modeling for Ratchetting, Transactions of the 14th International Conference on Structural Mechanics in Reactor Technology (SMiRT 14), Lyon, France, August 17-22, 1997
2. Jiang Y., Sehitoglu H.: Modeling of cyclic ratchetting plasticity, Parts I and II. ASME Journal of Applied Mechanics **63**, pp. 720-733, 1996
3. Prager W.: The theory of plasticity: A survey of recent achievements. Proceedings, Institution of Mechanical Engineers, London, Vol. 169, No. 21, pp. 41-57, 1955
4. Besseling J.F.: A theory of elastic, plastic, and creep deformations of an initially isotropic material showing anisotropic strain hardening, creep recovery and secondary creep. *J. App. Mech.* **25**, pp. 529-535, 1958
5. Armstrong P.J., Frederick C.O.: A mathematical representation of the multiaxial Bauschinger effect. Report RD/B/N 731, Central Electricity Generating Board, 1966
6. Mroz Z.: On the description of anisotropic workharding. *J. Mech. Phys. Solids* **15**, pp. 165-175, 1976
7. Lemaitre J., Chaboche J.L.: Mechanics of Solid Materials, Cambridge University Press 1990
8. Jiang Y, Kurath P.: Characteristics of the Armstrong-Frederick type plasticity models, *Int. J. Plast.* **12**, pp. 387-415, 1996
9. Chaboche J.L.: On some modifications of kinematic hardening to improve the description of ratchetting effects. *Int. J. Plast.* **7**, pp. 661-678, 1991
10. Sester M., Mohrmann R., Kordisch H.: Evaluation and assessment of plastic deformations in power plant components due to mechanical and thermocyclic loading. Final Report (in German) Fraunhofer IWM T1/97, Freiburg, 1997

Low Cycle Fatigue and Elasto-Plastic Behaviour of Materials
K-T. Rie and P.D. Portella (Editors)

MATERIAL PARAMETER IDENTIFICATION CONSIDERING SCATTERING AND VERIFICATION BY FULL FIELD EXPERIMENTAL DATA

C. ZORN[*], F. THIELECKE[**], H. FRIEBE[***], K. GALANULIS[****], R. RITTER[***], E. STECK[*]

[*] *TU Braunschweig, Institut für Allgemeine Mechanik und Festigkeitslehre*
PO Box 3329, D-38023 Braunschweig, Germany
[**] *DLR, Institut für Flugmechanik*
PO Box 3267, D-38022 Braunschweig, Germany
[***] *TU Braunschweig, Institut für Meßtechnik und Experimentelle Mechanik*
PO Box 3389, D-38023 Braunschweig, Germany
[****] *GOM mbH, Rebenring 33, D-38106 Braunschweig, Germany*

ABSTRACT

A constitutive model for inelastic behaviour of stainless steel AISI 316 L considering scattering is outlined. For a more realistic description particular material parameters or the material's initial state are modelled as random variables. Identification of the material parameters and implementation of scattering is based on uniaxial tests. An optical full field measurement of real structure leads to an experimental strain distribution which is compared with a corresponding Finite Element Analysis to examine the model's quality.

KEYWORDS

Stainless steel AISI 316 L, inelastic behaviour, scattering phenomenon, Finite Element Analysis, optical deformation measurement

INTRODUCTION

The inelastic behaviour of metalls can be modelled by non-linear ordinary differential equations. The determination of unknown material parameters is based on a Maximum-Likelihood output error method which compares experimental data with numerical simulation. The aim is to reproduce the observed data as close as possible. This demand contrasts with the fact that repeated tests for the same laboratory conditions show significant scattering. Small changes in chemical or physical structure of the material or in initial conditions like preliminary deformation could be reasons for the scattering phenomenon. Based on a statistical analysis this uncertainty can be taken into account by modelling particular material parameters or the material's initial state as random variables. The objective of using constitutive models is to predict the mechanical behaviour of metallic structures. Those problems are usually solved by Finite Element Analysis (FEA). Regarding to a verification of the simulated behaviour, experimental investigations of real structures instead of uniaxial tests are necessary. As an experimental counterpart of FEA an optical full field measurement based on the object grating method is an appropriate way to get strain information over the range of the structure's surface.

UNIAXIAL TESTS

The investigated material is stainless steel AISI 316 L which is used by chemical industry or for high temperature problems. The experimental data base was provided by Kollmann et al. [1]. It includes a total of 192 experiments which consist of creep tests at two different stresses and temperature levels, tension-relaxation and cyclic tests with three different strain rates and at two different temperatures levels as well. In order to provide information on the scattering phenomenon the experiments were repeated up to 12 times at each laboratory condition.

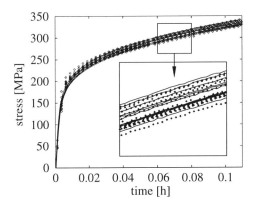

Fig. 1. Observed (dots) and simulated (lines) scattering phenomenon during a tension test

CONSTITUTIVE MODEL

$$\dot{\varepsilon}_{ie} = 2C \exp\left\{-\frac{V_0\sigma_0}{\Re T_0}\right\} \sinh\left\{\frac{V\,(\sigma - \sigma_{kin})}{\Re T}\right\}$$

$$\dot{\sigma}_{kin} = H \exp\left\{\delta\frac{V_0\sigma_0}{\Re T_0}\right\} E \exp\left\{-\delta\frac{V\sigma_{kin}}{\Re T} sign(\sigma_{eff})\right\} \dot{\varepsilon}_{ie} \qquad (1)$$

$$\dot{V} = B\ V_0^{-2}\ (Q-V)V\,|\dot{\varepsilon}_{ie}|$$

The differential equation (1) shows a stochastic model for inelastic behaviour of metallic materials in mean value formulation. It considers micro mechanical mechanisms at low temperature level (see Steck and Gerdes [2,3]). Thielecke [4] introduced scaling factors in equation (1) which are used to normalize the shape terms in order to stabilize the identification of the model parameters C, H, δ, B and Q. The kinetic equation for the inelastic strain rate $\dot{\varepsilon}_{ie}$ describes the dependency on the stress σ and the two inner state variables σ_{kin} and V. The kinematic stress σ_{kin} is directed while the activation volume V is isotropic. For an implementation in a three-dimensional FEA approach the constitutive model (1) has to be formulated as a tensor equation. The elastic behaviour is described by Hooke's law as an additional equation. In the following only the version for strain driven tests at room temperature is described. For further information see Thielecke [4].

TREND MODEL AT ROOM TEMPERATURE

Table 1. Constants of the model

constant	$\Re$	$E_{20°C}$	σ_0	V_0	T_0	$\sigma(0)$	$\sigma_{kin}(0)$	$\varepsilon_{ie}(0)$
unit	$\frac{kJ}{mol \cdot K}$	MPa	MPa	$\frac{kJ}{mol \cdot MPa}$	K	MPa	MPa	-
value	$8.3147 \cdot 10^{-3}$	$2.2 \cdot 10^5$	50.0	0.5	293.0	0.0	0.0	0.0

Table 2. Parameter estimation of the 'trend model'

parameter	C	H	δ	B	Q	$V(0)$
unit	h^{-1}	-	-	$\frac{kJ}{mol \cdot MPa}$	$\frac{kJ}{mol \cdot MPa}$	$\frac{kJ}{mol \cdot MPa}$
value	$7.888 \cdot 10^{-4}$	1.771	$1.1 \cdot 10^{-1}$	$1.5 \cdot 10^1$	$5.2 \cdot 10^{-1}$	$7.0 \cdot 10^{-1}$

The identification of the material parameters is based on uniaxial tests. A basic constitutive model called 'trend model' describes a mean curve which has to be selected within the range of scattering. The parameters are determined by minimizing the output error which describes the deviation of numerical results compared with uniaxial experimental data. In view of the experimentation region the constants of the room temperature model are fixed according to Tab. 1. For tension-relaxation and cyclic tests one joint model could be found. Table 2 shows the resulting parameter estimations.

SCATTERING MODELS

Table 3. Characteristic values of the scattering models

parameter	unit	trend	mean	deviation	correlation
C	h^{-1}	$7.888 \cdot 10^{-4}$	$8.185 \cdot 10^{-4}$	$1.509 \cdot 10^{-4}$	$\left.\begin{array}{l}\\ \\\end{array}\right\} 0.752$
H	-	1.771	1.762	$2.71 \cdot 10^{-1}$	
$V(0)$	$\frac{kJ}{mol \cdot MPa}$	$7.0 \cdot 10^{-1}$	$7.066 \cdot 10^{-1}$	$1.795 \cdot 10^{-2}$	

Investigating the scattering phenomenon various causes of uncertainty can be found. In the following two possible approaches are introduced. Uncertainties caused by changes of the material's chemical or physical structure are modelled by variation of material parameters C and H which prove to be suitable. If there is preliminary deformation the corresponding scattering parameter is the initial value of the state variable V. It is assumed that only one of these two possibilities is responsible for the scattering phenomenon. The statistical behaviour of both influences depends on the scattering of the experimental data. Each curve of a repeated test can be seen as a deviation from the trend and therefore represents a realisation of the material parameters C and H or of the initial value $V(0)$. Identifying each of those realisations one obtains a sample that determines the statistics of the scattering model. Using this scattering model the uncertainties of the uniaxial experimental data could be simulated well (see Fig. 1). The characteristic values of the samples are shown in Tab. 3.

VERIFICATION OF THE MODELS

In view of a verification of the constitutive model whose identification is based on uniaxial experimental data full field measurements on a real structure are necessary. The investigated structure is a notched flat bar shown in Fig. 2. In spite of simple loading conditions this structure leads to complex three-dimensional states of stresses and strains and therefore, it is suited to assess the quality of the constitutive model. The experimental data describes a tension test at room temperature. The global strain rate of $1.35*10^{-5}\,s^{-1}$ is comparable with the conditions of uniaxial tests. Figure 3 shows the specimen in different load stages. For a verification of the scattering model more than one analysed full field measurement is needed.

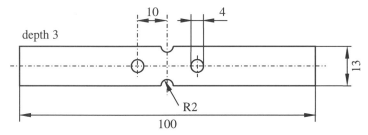

Fig. 2. Notched flat bar

a)   b)

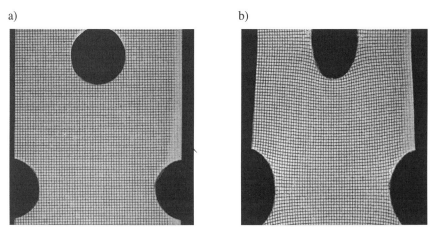

Fig. 3. Initial (a) and deformed (b) pattern

OPTICAL FULL FIELD MEASUREMENTS

A measurement directly on the testing machine in the test field during an on-going test is practical. Also measuring methods without contact and interaction are advantageous. All of these requirements lead to the application of optical full field measuring techniques (Ritter [5,6]) as speckle interferometry and the object grating method described in the following.

The precondition using the object grating method is a structure which is firmly attached to the considered object surface. The structure can be a cross grating (see Fig. 3) or a stochastic pattern. This is recorded by one or two CCD cameras at different load stages of the object. The coordinates of pattern features are determined by the image processing algorithm in the object coordinate system. When the difference of a change of the neighbouring features is related to their initial distance, the average strain in the field is obtained, which is limited by these two features. The results of this technique are the three-dimensional displacement distribution from which the strain values of the object surface can be calculated.

Depending on the used CCD camera, the lateral resolution is approximately 80 by 50 measurement values for a standard CCD camera which corresponds to the same number of strain gauges. Strain distributions from 0,1 % up to more then 100 % can be analyzed with a high measuring resolution.

STRUCTURAL ANALYSIS

A Finite Element model of the notched flat bar in Fig. 2 is developed for the FE program ABAQUS/Standard. Because of the symmetry of geometry and loading conditions it is sufficient to discretize only one eighth of the notched flat bar. A structural analysis which adapts the experimental conditions of the experimental investigations leads to comparable strain field data on the structure's surface. Figure 5 shows the experimental and simulated strain distributions for axial direction.

Both the results of FEA and measurement show a 'butterfly characteristic' near the holes. There is a complex interaction between holes and notches. A region of about 2.5% strain which connects holes with notches is indicated by FEA as well as experimental data. Maximum strain up to 20 % can be observed at the hole boundaries. Shape and level of the 'strain island' in the middle of the bar coincide. In order to categorize the investigated notched flat bar within the range of scattering the global load is printed against time (see Fig. 4). A Monte Carlo Study leads to estimated boundaries for the simulated load curves. Except for the beginning the simulated global structural response covers the experimental data. Hopefully, future investigations will lead to an implementation of the scattering model by stochastic Finite Element Analysis (SFEA) which will also have to be verified by experimental investigation.

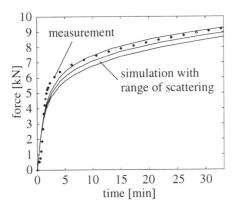

Fig. 4. Experimental and simulated data of structure's global response

658

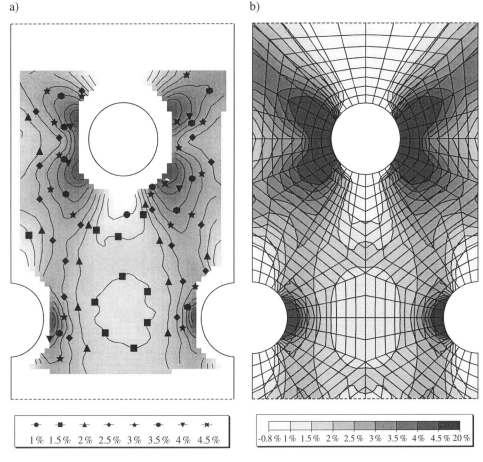

a)   b)

1 % 1.5 % 2 % 2.5 % 3 % 3.5 % 4 % 4.5 %

-0.8 % 1 % 1.5 % 2 % 2.5 % 3 % 3.5 % 4 % 4.5 % 20 %

Fig. 5. Experimental (a) and simulated (b) strain distribution in axial direction (see also [4])

REFERENCES

1.   Lehn, J.; Kollmann, F. G. (1996) *Arbeits- und Ergebnisbericht 1994-1996*, Teilprojekt A4, Sonderforschungsbereich 298
2.   Steck, E.; Gerdes, R. (1997) In: *Acta Mechanica* 120, pp. 1-30
3.   Steck, E. (1990) In: *Res. Mechanica*, pp. 1-19
4.   Thielecke, F. (1998) *Parameteridentifizierung von Simulationsmodellen für das viskoplastische Verhalten von Metallen - Theorie, Numerik Anwendungen*, Braunschweiger Schriften zur Mechanik
5.   Ritter, R. (1996) In: *Local Strain and Temperature Measurement in Non-Uniform Fields at Elevated Temperatures*, pp. 1-12, Woodhead Publishing Limited
6.   Friebe, H.; Galanulis, K.; Winter, D. (1996) In: *Local Strain and Temperature Measurement in Non-Uniform Fields at Elevated Temperatures*, pp. 40-48, Woodhead Publishing Limited

Low Cycle Fatigue and Elasto-Plastic Behaviour of Materials
K-T. Rie and P.D. Portella (Editors)
659

MULTIMODEL ANALYSIS OF THE ELASTO-PLASTIC AND ELASTO-VISCO-PLASTIC DEFORMATION PROCESSES IN MATERIALS AND STRUCTURES

A.S. SEMENOV* and B.E. MELNIKOV**

Dresden Technical University ,Germany.
**St.-Petersburg State Technical University, Russia.*

ABSTRACT

The problem of the universal and reliable description of inelastic behaviour of complex structures under combined variable loading is considered. The strategy of multimodel analysis based on creation and application of the hierarchical sequence of material models is described. Application of different material models is considered both for material element and for complex structures subjected to non-proportional loading.

KEYWORDS

Constitutive modelling, complex loading, finite element analysis.

1. INTRODUCTION

Increasing demands for reliability and durability of structures with simultaneous material economy have stimulated improvement of constitutive equations for description of inelastic deformation processes. This has led to the development of phenomenological modeling of complex phenomena of irreversible deformation including history-dependent and rate-dependent effects. During the last several decades many works have been devoted to the development of plastic and viscoplastic models, in order to better predict the material behavior under combined variable thermomechanical loading.

The results of various viscoplastic models predictions often are considerably differed from the experimental data in some cases of the non-proportional loading. Therefore the problem of choice of the more suitable theory of plasticity is actual for the strength analysis of structures under combined loading. The suggested multimodel approach, which is based on the creation of hierarchical consistency of the models, whose fields of reliable application partially coinciding mutually supplement each other, is most rational. Application of the concrete model must correspond to the complexity of being considered loading processes.

The developed strategy of multimodel analysis consists of the following features:

- creation of the *plastic and viscoplastic models library*, providing solution of the wide spectrum of non-elastic problems,
- determination of the *selection criteria system*, realizing the choice of the simplest variant of theory sufficient for correct problem solution,
- caring out of *multivariant sequential clarifying computations*.

Computation with using of several different theories should be performed in the most responsible cases of strength analysis. Coincidence of the results according different models demonstrates the correctness of computations. Difference between obtaining results for structure demands determination of the most adequate model. This choice can be carried out on the base of comparison between numerical results and experimental data for element of material, subjected to the same history of loading as the most stressed point of structure.

The basic ideas and application of multimodel analysis for the rate-independent material behavior was described in previous works of authors [1,2]. Present analysis is devoted to rate-dependent deformation.

2. LIBRARY OF PLASTIC AND VISCOPLASTC MODELS

The developed library of material models represents generalized data set, including information about limitations on field application, basic experiments for material parameters determination, continuous mathematical model, discrete numerical model, computational algorithm, implementation into finite element program, recommendations about computation strategy.

The set of following criteria has been taken into consideration: conformity to a general principles of physics (thermodynamic laws, principle of fading memory, tensorness of relations and others); experimental verification of models for various classes of loading; possibility of micromechanical interpretation; complexity of determination of the material parameters; algorithmic effectiveness.

2.1 Plastic models

At the present time the developed and implemented into finite element program library of rate-independent (plastic) models includes:

- Plastic flow theories with the various isotropic-kinematic laws of hardening [3-5]. The relations of this "classical" models belong to the first-order linear tensorial equations convenient for computations.
- Structural (rheologic) models theories [6,7]. They possess a clarity of properties, thermodynamic basis, obvious creation and improvement.
- Multisurface theory with one active surface of plastic compliance [8,9]. The model provides high accuracy of the description for the complex paths of passive loading.
- Endochronic theory of plasticity [10-12]. The equations can be applied for a wide class of materials from a metal to a soil. This theory does not use the existence of a yield surface and employs the same equation for the loading and unloading processes.

2.2 Viscoplastic models

The our library of the rate-dependent (viscoplastic) models includes:

- Technical theories of creep (ageing theory [13], flow theory [13], hardening theory [13]). These models are convenient for the primary express analysis and are applicable for weakly variable loading. They are simplest models with least set of the necessary experimental data.
- Elastic/viscoplastic models (Perzyna [14], Chaboche [5,15], Robinson [16], Bodner-Partom [17], Krempl [18], Hart [19]] models at el.). There are most popular in computations models. They demonstrate the viscous effects only after exceedition of static yield limit.
- Viscoelastic/viscoplastic models (Naghdi-Murch theory [20]). They demonstrate the viscous effects always as before as after exceedition of yield limit.
- Elastoviscoplastic models (Endochronic theory [10-12], Nonlinear heredity theory [21,22]). These models don't possess pronounced yield limit and demonstrate simultaneously elastic, viscous and plastic properties. They represent extension of viscoelasticity.

- Structural (rheological, fraction, sublayer) models (Besseling [23], Reiner[24], Gokhfeld-Sadakov [25], Palmov [7], Partom-Keren-Gennusov [26] et al.). They allow to create and easy to modify models with wide spectrum of elastic, viscous and plastic properties in clarified and obvious way.

The detailed reviews of the inelastic constitutive equations have been presented in [27-29].

2.3 Regulation criteria of viscoplastic models

Viscoplastic deformation of materials demonstrates wide range of nonlinear time-dependent effects such as primary and secondary creep, relaxation, rate-dependence of the stress-strain relations, ageing, recovery, restoration, creep delay after partial unloading, cyclic softening and hardening, ratcheting, additional hardening in nonproportional loading, cross hardening, coupling of plasticity and viscoplasticity and et al. Diversity of mentioned effects generates the various viscoplastic models with different underlying conceptions. For the regulation of viscoplastic models can be considered the following criteria and corresponding selected groups of models:

1. Restrictiveness of viscous properties manifestation:
    - elastic/viscoplastic models,
    - viscoelastic/viscoplastic models.

In the first case the viscous effects manifest theyself only in combination with plastic properties after exceeding of yield limit. In the second case the viscous effects are observed as in elastic as in plastic range. It permits to describe more wide phenomena class, but leads to the considerable complication of mathematical description and computation. Schematic representation of the difference between these approaches is illustrated in Fig. 1 by means of rheological models. Elastic/viscoplastic models correspond to the case b) in Fig. 1, when viscous element is interlocked by parallel-connected plastic element .Viscoelastic/viscoplastic models are shown in Fig. 1, a) and c). Use of generalized plastic and viscous elements (marked in Fig. 1 by dashed circles) allows to extend capability of considered models and to display purged peculiarity of strain decomposition. Generalized plastic (viscous) element is arbitrary combinations of arbitrary numbers of perfect plastic (linear viscous) and elastic elements.

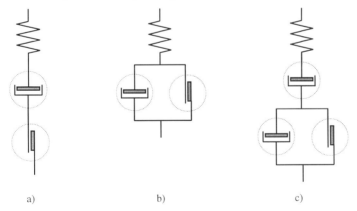

a)   b)   c)

Fig. 1. Distinctive viscoplastic rheological models with using generalized elastic, viscous and plastic elements.

2. Plasticity-viscoplasticity interaction
    - uncoupled models, governed by equation $\varepsilon = \varepsilon^e + \varepsilon^v + \varepsilon^p$ (Fig. 1, a, c),
    - unified models, governed by equation $\varepsilon = \varepsilon^e + \varepsilon^{vp}$ (Fig. 1, b).

The first conception is founded on the existence of different time scale corresponding to different processes. Possibility of such decomposition is defined by variability and duration of loading process, temperature level.

3. Yield conditions (*static yield* conditions, *dynamic yield* conditions, *absent* of yield condition).Description of rate influence on material behavior and loading-unloading conditions can be formulated on the basis of quasistatic or dynamic stress-strain curve.

4. Hardening laws (*perfect* viscoplasticity law, *isotropic* hardening, *kinematic* hardening (linear and nonlinear rules), *anisotropic* hardening.).Description of the complex history of combined loading demands complication of the hardening law.

2.4 Internal state variables approach

The uniform representation of constitutive equations is actual for the creation of inelastic models library with the purpose to simplify program realization and to perform comparative analysis. The thermodynamic approach with internal state variables provides a powerful tool for representing of the constitutive equations of plasticity and viscoplasticity. All considered here models of inelastic material can be written in common quite general mathematical form.

The inelastic strain rate tensor $\dot{\varepsilon}^{vp}$ is assumed as function of the stress tensor σ, a set suitably defined internal state variables $\chi^{(k)}$, k=1,...,n and temperature T and can be defined in form

$$\dot{\varepsilon}^{vp} = \dot{p}\,\mathbf{a}(\sigma, \chi^{(k)}, T) \qquad (1)$$

The internal state variables $\chi^{(k)}$ can be either second-order tensors or scalars. Evolution laws for these internal variables can be represented in the form:

$$\dot{\chi}^{(k)} = \dot{p}\,\mathbf{b}(\sigma, \chi^{(l)}, T) \qquad (2)$$

All members of plastic and viscoplastic models library fit into the frame (1)-(2).

3. SELECTION CRITERIA SYSTEM

The determination of the selection criteria system, based on classification of inelastic theories and their domain of advantageous applicability, is one of the main problem in multi-model analysis. Selection criteria system generates necessary conditions for material model on the basis of information concerning external actions, available experimental data, discrete model of structure. The choice of rational model, which is the simplest among models satisfied necessary conditions, may be corrected by clarifying sequential computational experiments.

Necessity to take into consideration time-dependent effects is defined by the parameters of external actions such as temperature level, duration and rate loading and specific material properties. The viscoplastic analysis succeeds the criteria of plastic analysis [2] and adds some new. Some common viscoplastic criteria has been considered in section 2.3.

The classical examples of plastic criteria are degree of plastic strains development in comparison with elastic strain and proportionality of loading path. The first criterion determines choice between the Prandtl-Reuss and Levy-Mises theories of plastic flow, when the possibility to neglect elastic strain arises. The second criterion defines the conditions of possibility of application of the Hencky-Ilyushin or Prandtl-Reuss theory. The another selection criterion suggested by Ilyushin in [30] are founded on the path curvature of deformation process. Suggested criterion is based on consideration of geometrical regulated levels of plastic deformation analysis. Similarly as in [31] we introduce the following levels:

- Body level **B** considers the body or complex structure as a whole. "Integral" analysis corresponds initial strength problem. In most cases zones of plasticity are local.

- Element level **E** is introduced for separate parts of the structure as fragment of structure, substructure, superelement or individual finite element. "Semi-integral" analysis is carried out in this case. In most of cases zones of plasticity are extensive.
- Point level **P** is the basic level, related to selected points of material continuum or to a model of structure. "Local" analysis is carried out for homogeneous stress state.

Finally, the criterion can be formulated by following manner. Complexity of applying theory of plasticity must correspond to the level of the structure approximation. The levels **P** and **E** with more detail description of the structure geometry and with possibility of extensive zones of plasticity demand the more difficult variants of theory adequate to the loading process. Using of the simple models is sufficiently at the **B** level of the investigation, when the deformation of local zones of plasticity is smoothed by influence of extensive elastic region.

4. RESULTS OF MULTIMODEL COMPUTATIONAL ANALYSIS

Comparison of the results of numerical finite element analysis and experimental data for series investigated constructions corresponding to the first level **B** (frames, pipelines, vapor producing plant, gas generator, vessel of nuclear reactor) says about relative nearness of different theories predictions. However series of computations corresponding to the second level **E** of the structures considerations (fragment of rolling mill, fastenings of vapor producing plant, various fastening knots, socket, circular ring) have shown that the considerable differences of the prediction of stress-strain state by means of different theories of plasticity were displayed for a developed zone of plasticity and complex history of loading. Set of trials according level **P** carried out on tubular specimens of X18H10T steel under a wide range of the combined cyclic loading, including polygonal and circular paths of deformation. In general the results different theories corresponding to the level **P** can be essentially quality differed. Typical example of multimodel computations corresponding to the level **E** for thin circular ring being the part of more complex structure is shown in Fig. 2.

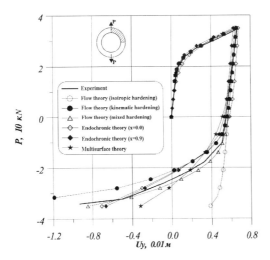

Fig. 2. Different models predictions for circular ring under axial tension-compression.

The levels **P** and **E** with more detail description of the structure geometry and with possibility of extensive zones of plasticity demand the more difficult variants of theory adequate to the loading process. Using of the simple models is sufficiently at the **B** level of the investigation, when the deformation of local zones of plasticity is smoothed by influence of extensive elastic region.

ACKNOWLEDGMENTS

The research described in this publication was made possible in part by Grant No. 96-17-4-32 from the Moscow aviation institute.

REFERENCES

1. Melnikov B.E., Semenov A.S. *Proc. IV Int. Conf. on Computational Plasticity. Fundamentals and Applications (COMPLAS IV)*. Swansea. U.K.: Pineridge Press. 1995. P. 181-189.
2. Melnikov B.E., Semenov A.S *Proc. of 6th Int. Conf. on Comp. in Civil and Build. Eng.* Berlin. 1995. P. 1073-1079.
3. Prager W. *Proc. Inst. Mech. Ing.*, London, **169** (41). 1955. P. 41-57.
4. Kadashevitch Yu.I. *Scientific papers on elast. and plast.*, 6. 1967. P. 25-38.
5. Chaboche J.L and Nouailhas D. *Proc. Int. Conf. Comp. Eng. Sci.*, Atlanta, 1988. P. 19.I.1-4.
6. Iwan W.D. *J. Appl. Mech.* **34**. 1967. P. 612-617.
7. Palmov V.A. *Oscillations of the elasto-plastic bodies*. Moscow: Nauka. 1976. 338 p.
8. Melnikov B.E., Semenov A.S. *Trans. St.-Petersburg State TU*. N 441. 1991. P. 26-31.
9. Melnikov B.E., Izotov I.I., Kuznetsov N.P. *Strength of Materials*, 22, 8. 1990. P. 1122-1127.
10. Valanis K.C. *Arch. Mech. Stasow.* 1971. V. 23, **4**. P. 517-551.
11. Valanis K.C. *Arch. Mech. Stasow.* 1980. V. 32, **2**. P. 171-191.
12. Watanabe O., Atluri S.N. *Int. J. of Plasticity*. V. 2. 1986. P. 37-57.
13. Rabotnov Yu.N. *Creep of construction elements*. Moskow: Nauka. 1966. 752 p.
14. Perzyna P. *Advances in Appl. Mech.* V. 9. 1966.
15. Chaboche J.L., Rousselier G. *J. Press. Ves. Tecn. Trans. ASME*. 1983. V. 105. P. 153-158.
16. Robinson D.N. ONRL-5969. 1979.
17. Bodner S.R., Partom Y. *J. Appl. Mech. Trans. ASME*. 1975. V. 42, **2**. P. 385-389.
18. Yao D., Krempl E. *Int J. of Plasticity*. 1. 1985. P. 259-274.
19. Hart E.W. *Acta Metallurgica*. 1970. V. 18, **6**. P. 599-610.
20. Naghdi P.M., Murch S.A. *J. Appl. Mech.* V. 30. 1963. P. 321-328.
21. Shevchenko Yu. N., Terehov R.G. *Physical equations of thermoviscoplasticty*. Kiev: 1982.
22. Cernocky E.P., Krempl E.A. *Acta Mech.* V. 36. 1980.
23. Besseling J.F. *Trans. ASME*. V. 25. 1958. P. 529.
24. Reiner M. *Rheology*. Berlin: Springer-Verlag. 1958. 223 p.
25. Gokhfeld D.A., Sadakov O.S. *Plasticity and creep in structural members under repeated loading*. 1984.
26. Partom Y., Keren B., Genussov R. *Calculation of viscoplastic and nonlinear viscoelastic response with the phases model*. P. 491-504.
27. Inoue T. *JSME Int.* J. Ser. I .V. 31, **4**. 1988. P.653-663.
28. Kolarov D., Baltov A., Boncheva N. *Mechanics of plastic media*. Moscow: Mir. 1979. 302 p
29. Novozhilov V.V., Kadashevich. Yu.I. *Microstresses in structural materials*. 1990. 223 p.
30. Illyshin A.A. *Continuum mechanics*. 1990. 310 p.
31. Zyczkowski M. *Combined loadings in the theory of plasticity*. Warszawa, 1981. 714 p.

Low Cycle Fatigue and Elasto-Plastic Behaviour of Materials
K-T. Rie and P.D. Portella (Editors)
665

EXPERIMENTAL OBSERVATION OF HEAT GENERATION RULES OF METALLIC MATERIALS UNDER MONOTONIC AND CYCLIC LOADING

X.Y. Tong,   D.Y. Ye   L.J. Yao   Q. Sun
Institute of Aerospace Structure Integrity, Northwestern Polytech. Univ., Xi'An, 710072, P.R. China,

ABSTRACT

For the purpose to understand the heat generating mechanisms and observe the self-heating generation rule to fatigue damage, the thermistor gauge has been used to measure the self-heating generation of three kinds of metallic materials: 8090 Al-Li alloy, Tc4 Alloy and class A 40CrNiMo Alloy, under monotonic and cyclic loading. Meanwhile, the heat generation and its variation in low cycle fatigue under different strain rates and ratios in the materials have been experimentally studied.

KEYWORDS

Thermistor gauge, 8090 Al-Li alloy, heat dissipation energy, self-heating generation, monotonic and cyclic loading, fatigue damage, strain rates and strain ratios.

INTRODUCTION

Irreversible energy dissipation resulting from material deformation can be considered as one of the main parameters to describe the damage in it. In the sixth decades, energy concept on fatigue damage has become one of the main opinions to phenomenalogically explain how fatigue damage is accumulated and been considered as one of the fatigue failure criterions [1,2]. The energy problem on fatigue damage is obscured until the heat dissipation in cyclic loading process can be measured by means of the contact or non-contact thermometer, such as thermistor, thermocouple, thermovison and so on [3]. By means of the thermocouple, Harig, H.[4] measured the temperature on the surface of a specimen and analyzed the heat flow of the metallic specimen at the typical stages in a cyclic loading process. The temperature increment caused by plastic deformation gives a good description of fatigue damage process. Jardon, E.H. and Sandor, B.I.[5,6] used tiny thermocouple welding directly on specimen surface to detect the temperature change associated with plastic deformation as well as to study the plastic deformation at the root of a notched specimen. It is found that the majority of the irreversible plastic strain work in metals is converted into heat. Under adiabatic condition, the spatial distribution of the temperature increment by self-heating is indicative of the spatial distribution of plastic strain work. Gross,T.S.[7] used the thermocouple to measure the temperature rise and relate the plastic strain work done in one cycle to the heat generated in the same cycle. Simultaneously, he studied crack closure effect on temperature rise.

Attermo,R. and Ostberg,G.[8] first used an infrared camera to measure the temperature rise ahead of a fatigue crack tip. Later, Charles,J.A., Appl,F.J. and J.E. Francis[9] used the infrared camera to predict the location of an impending fatigue damage and to monitor the growing fatigue crack as well as to map the temperature field around the region of stress concentration. Using the infrared thermovison system, Huang,Y.[10] studied the changing thermograph picture on the surface of a metallic specimen during high-cycle fatigue. With a similar instrument, Tung,X.[11] described the fatigue damage evolution process by means of observing the thermograph photos and temperature increment in low cycle fatigue under different strain rates and ratios.

In order to study the heat emission mechanisms and its relations to fatigue damage, a contact thermistor gauge has been used to measure the self-heating temperature increment of three kinds of metallic materials: 8090 Al-Li alloy, Tc4 Alloy and class A 40CrNiMo Alloy, under monotonic and cyclic loading. Meanwhile, the heat emission and its variation affected by the loading process in these materials have been analyzed.

EXPERIMENTAL PROCEDURE AND TEST RESULTS

Materials and Specimens

In this paper, 8090 Al-Li alloy, Tc4 alloy and class A 40CrNiMo alloy are used to measure the temperature increment by the self-heating under monotonic and cyclic loading. The thin plate specimen has been designed to perform these tests when the temperature increment was measured directly by an adhesive thermistor gauge at the center of the specimen surface. Meanwhile, the variation of temperature increment under the various loading rates has also been experimentally measured. Fig.1 shows the specimen shape and size.

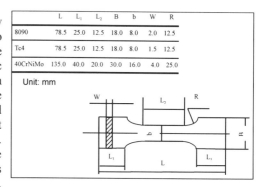

	L	L_1	L_2	B	b	W	R
8090	78.5	25.0	12.5	18.0	8.0	2.0	12.5
Tc4	78.5	25.0	12.5	18.0	8.0	1.5	12.5
40CrNiMo	135.0	40.0	20.0	30.0	16.0	4.0	25.0

Unit: mm

Fig.1 Specimen shape and size

Test Description and Procedure

Three kinds of tests have been arranged to experimentally study self-heating generation rule, i.e.: first at all, the tensile loading test is performed under five various loading rates in the ranges from 1MPaSec^{-1} to 100MPaSec^{-1}. Secondly, the compressive loading test is performed similarly, but its loading rates are in the ranges from 1MPaSec^{-1} to 50MPaSec^{-1}. The last test performance is the one under cyclic loading. The temperature changes by self-heating generation are measured under low-cycle fatigue process in the metallic materials. For class A 40CrNiMo alloy, the temperature increments under different cyclic loading frequencies as well as various loading ratios are measured respectively.

In loading process, the resistance change of the thermistor gauge, resulting from its temperature change, is acquired by a special designated amplifier, then it was converted to voltage signal and amplified, finally through an A/D interface, the analogous signal is transformed into a digital signal. After data processing and analysis, the temperature change can be obtained. In order to obtain a true result, the size of the thermistor gauge is required as small as possible. The gauge size used in the tests is 1mm in length and 1mm in width. To avoid a failure of the gauge in a testing process, the fatigue life of the gauge should be no less than 500,000 cycles under the strain amplitude ε_a=0.3%. Moreover, it must sustain a maximum strain larger than 10 percent in monotonic loading process. While, another important requirement is that the resistance change resulted from the deformation of the gauge's resistant line should not exceed 5 percent of the one resulting from temperature increment.

The temperature measurement system by means of the thermistor gauge can acquire a signal in 10^{-6} second. Its resolution is lower than 10^{-3}K. The real sensitivity of this temperature measurement system is tested about in a range of 0.05~0.1K. So, it is possible to obtain the satisfying data of the temperature increment in a cycle under the low-cycle fatigue test with a loading frequency in the range of 0.05Hz to 2Hz. In order to shorten the data processing hours, the sampling numbers in one cycle are usually chosen in accordance with the sampling theory, i.e.

$$NN > 4 \sim 10 \frac{1}{f}$$

As where: f is loading frequency and NN is the sampling numbers.

EXPERIMENTAL RESULT ANALYSIS AND CONCLUSSION

Relationship of Self-Heating Generation to Temperature Increment

Under a cyclic loading, the irreversible dissipated energy will be caused by the localized or macroscopic plastic deformation and be converted into a hot plane as soon as the start of the loading. The temperature increment by heat emission in a material can indicate sensitively the features at the stages in a fatigue failure process.

Heat Emission and Absorption in Monotonic Loading Process

Temperature Increment Process under Tensile and Compressive Loading

During a tensile loading process, it is initially loaded to a prior chosen maximum value below the elastic limit, to hold it at least for 3~15 minutes, then unloaded to zero stress. This should be repeated once again. Eventually, the specimen will be loaded until rupture. Fig. 2 shows the self-heating generation process of these materials under monotonic tensile and compressive loading. In region of macroscopic elasticity at the initial loading stage, the temperature increment is negative with the increasing tensile loading, it shows that the heat is absorbed by the elastic strain, then changes in a positive value with the increasing load. Opposite, under compressive loading process, the temperature increment increases with the increasing load from the start of loading. At the initial stage, heat absorption occurs in these materials. There exists a valley value after that the heat absorption is changed into heat emission. At the stage of heat absorption, its experimental correlation of heat absorption rate to loading rate are following the form:

$$\Delta \dot{T} = \kappa \dot{\sigma}$$

As where: κ is material constant. Table 1 lists its values of these materials. The temperature increment increases with the increasing of loading. Under different loading rates, the heat absorption and emission change linearly with the increasing loading, the temperature increment rate is proportional to the loading rate. When the plastic yield occurs in the specimen, the heat emits more quickly than before, at the instant of tensile failure, a "heat flash" occurs and its temperature increment rate approaches infinity, as shown in Fig.2a.

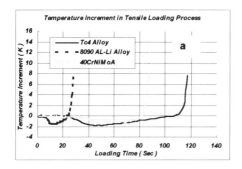 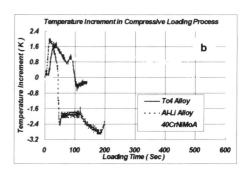

**Figure 2   A Typical Self-Heating Generation Under Monotonic Loading Process
a) Tensile Loading Process   b) Compressive Loading Process**

In monotonic compressive loading process, the performances of heat emission in these materials are contrary to the one under tensile loading. As shown in Fig.2b. For these tested materials, the compressive loading rate is also proportional to heat emission rate, its experimental fitted correlation is similar to the one under the tensile loading, the material constant κ can also be found in Table 1.

Table 1 Coefficient κ of the tested materials

Material Constant	8090	Tc4	40CrNiMo
Compressive κ	18.60	7.50	5.12
Tensile κ	-16.34	-7.35	-4.07

Unit: $MPaK^{-1}$

The saturation value of temperature increment rate can be approximated as a constant because its variation with the loading rates are quite small and of a range in: 0.005K/MPa~ 0.01K/MPa.

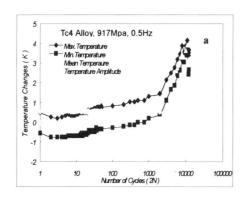 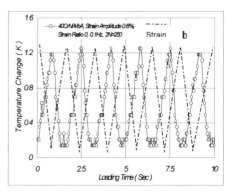

**Figure 3   Temperature Increment By Self-Heating Generation In Cyclic Loading Process
a) Self-Heating Process   b) Fluctuation In A Cycle**

Temperature Increment in Cyclic Loading Process Figure 3a shows a typical temperature increment in cyclic loading process. At the stage of cyclic hardiness or softness, there are dislocation movement and obstruction in the view of the microscopic point. Correspondingly, an obvious temperature increment by heat emission could be observed at this stage. At the stage of fatigue crack initiation and micro-crack growth, the temperature increment will not vary obviously because the temperature increment by heat emission will be balanced by the temperature decrement due to heat exchange. Similarly, it is known the inelastic responses such as plastic strain amplitude as well as plastic strain energy etc are all related steadily to the number of cycle [13]. As a naked eye observed fatigue crack occurs, a short period of temperature decrement is found because of the breakage of equilibrium. The required internal energy is lower than the one produced the heat emission as the increment of the boundary area. Quickly, it changes as the appearance of local plastic yield at the tip of the observed crack, i.e. the energy supplied for heat emission will be lower again than the required internal energy. Therefore, the temperature increases again and become more quickly as the closing to the unsteady crack propagation. At the instant of rupture failure, the temperature increment rate approaches to infinite.

Temperature Increment Fluctuation in a Cycle Contrasting the fluctuation of the load in a cycle to the one of temperature increment by self-heating generation, there is 180 degrees phase difference between them. In the tensile loading branch, it appears heat absorption and tends a valley value at the peak of tensile loading. Similarly, it appears heat emission in the branch of compressive loading and ends at compressive loading valley, as shown in Fig.3b.

Influences on Temperature Increment In cyclic loading process, the maximum and minimum temperature increments are dependent on the loading frequency and cyclic strain ratios, as shown in Fig.4. The amplitude of cyclic temperature increment is independent to the number of cycle and it is independent of strain ratio, as seen in Fig.5, but it is properly effected by the loading rate. At a lower rate, it can also be found to be independent.

The experimental relations of the mean, maximum and minimum temperature increment to fatigue life are shown in Fig 6, it can be found that the relations are effected by loading rate as well as ratio. As shown in Fig.6, the relation of temperature increment to fatigue life is simple. Despite the situation under the cyclic strain ratio $R_\varepsilon = 0$, all the relations under the others are closed to each other. Furthermore, the temperature increment ranges don't change with the fatigue life, i.e. they aren't effected by the loading levels.

Although cyclic ratio has no influence on the temperature increment ranges, it still affects the extent of temperature increment. In initial ten cycles, all the ranges are almost of a same value. After that there exists obvious differences between the ones under different loading rates, the loading rate effects either the its mean values or its ranges.

REFERENCES

G.R. Halford,(1966) *J. Maters*,3
JeDean Morrow, ASTM STP378; 1965, 45
H. Harig and M. Weber Defect, *Crack and Fracture*, 1983
E.H. Jardon, (1985)*Exper. Mech.* ,24
E.H. Jardon and B.I. Sandor, (1978) *J. Test.Evol.*, 325
A.D. Joseph and T.S.Gross, (1985)*Engng. Fract. Mech.*1,63

R.Attermo and G.Ostberg, (1971)*Int. J. Fract. Mech.* 1,122

J.A.Charles, F.J.Appl and J.E.Francis, (1975)*Exper. Mech.* 4,133

Y.Huang, (1984)*Acta Metall.Sinica* 2,120 (in Chinese)

X.Tung, (1991)*Acta Metall.Sinica* 5, 374

X.Tung, ibid; 1992, 5A(4)

D. Zhang and B.I.Sandor (1990)*Exper. Mech.* 1, 68

ACKNOWELDGEMENT

The authors thank National Nature Science Foundation of China (NSFC) and the Fok Ying Tung Education Foundation and Aeronautic Science Foundation of China(ASFC). This work was supported in part by whose auspices.

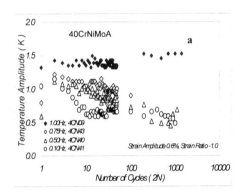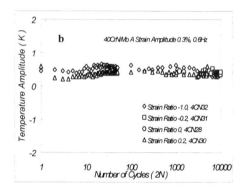

Figure 4   Relations Of The Cyclic Temperature Amplitude To Cycles Strain Ratio And Frequency

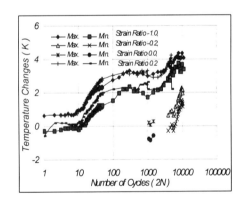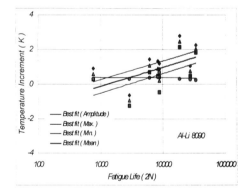

Figure 5 The Relations of Cyclic Maximum And Minimum Temperature Increments To Cyclic Strain Ratios

Figure 6 The Relation Of Temperature Increment To Fatigue Life

Low Cycle Fatigue and Elasto-Plastic Behaviour of Materials
K-T. Rie and P.D. Portella (Editors)
671

COMPUTATIONAL SIMULATION OF SELF-HEATING GENERATION IN CYCLIC LOADING PROCESS

X.Y. Tong,   D.Y. Ye   L.J. Yao
Institute of Aerospace Structure Integrity, Northwestern Polytech. Univ., Xi'An, 710072, P.R. China,

ABSTRACT

A numerical analysis method has been introduced to simulate the self-heating process. Meanwhile, its produced heat energy dissipation has been discussed in terms of the microstructure property. In order to verify the simulation results, the temperature changes in self-heating process and surface temperature fields of the metallic specimen have been measured in cyclic loading process. It is found that the computational simulation could not only recover the temperature fluctuation at a certain moment during cyclic loading process, but also its distribution in the whole calculation interval. It is also concluded that the temperature changes in a cycle obtained respectively by experimental measurement and computational simulation are in good agreement with each other.

KEYWORDS: Heat energy dissipation, Self-heating process, Surface temperature field Cyclic loading, Computational simulation

1. INTRODUCTION

An Irreversible Energy Dissipation in metallic materials can be considered mainly as resulting from material damage during a deforming process (X. Tung, 1991). In continuum media mechanics, materials are usually regarded as homogeneous and isotropic, then the question is suspected that fatigue damage is nucleated in high cycle fatigue, even the cyclic stress amplitude is under the elastic limit. Obviously, it is easy explained with consideration of which materials are, actually, inhomogeneous and anisotropic. The micro-plastic flow is, in fact, random and associated with the microstructure properties, such as grain sizes, orientation, interface defects and residual micro-stress etc. (A. Esin, 1966, J.D. Morrow, 1965, W.L. Morris 1983). Thus, the micro-plastic strain energy could be statistically analyzed in accordance with the microstructure properties. Because of the randomness of the microstructure properties, the mechanical characterizations are also of randomness. With respect of engineering application, the macro-scopic mechanical properties is more attributed. Since the expectation of a microscopic physical quantity can indicate the macroscopic property of materials, it is possible to associate macroscopic physical quantities with microscopic ones.

Although the local micro-plastic flow concept is proposed properly to associate high cycle fatigue with low cycle fatigue, it has no assistant yet to implicit the efficient energy dissipation from total energy dissipated during fatigue damage process. Therefore the presented work should include, firstly, introduction of a Monte-Carlo numerical approxi-mation to connect the macroscopic properties by utilizing the mathematical expectations of microscopic physical quantities. Secondly, the computational simulation method for the temperature distribution and the heat energy dissipation by means of the plastic deforming work, and the heat generation deduced on basis of thermodynamic theories.

2 THE GOVERNING EQUATION OF SLEF-HEATING GENERATION OF MATERIALS

2.1 *Governing equation*

Fatigue damage process can be regarded as an irreversible energy dissipating process. In consideration of Gibbs local equilibrium assumption of non-equilibrium thermodynamics, the energy equilibrium equation in a unit volume can be built up as:

$$\rho C_P \frac{\partial T}{\partial \tau} = \kappa \nabla^2 T + g \tag{1}$$

Where: g is the heat internal generation by the external source. For fatigue problem, it implies the total dissipated energy except the one dissipated to cause damage. Here we assume:

$$g = \eta(D, C)\dot{W}^P(r, t) \tag{2}$$

Then a universal thermodynamic equation on fatigue damage process can be written as below:

$$\rho C_P \frac{\partial T}{\partial \tau} = \kappa \nabla^2 T + \eta(D, C)\dot{W}^P(r, t) \tag{3}$$

Where: κ is thermal conductivity coefficient; ρ is mass density and C_p is specific heat. C is a material constant, D is material damage factor, η is a function correlated with the material constant, $\dot{W}^P$ and is the plastic strain energy rate.

2.2 *Cyclic plastic strain energy*

In macroscopic level, the cyclic plastic strain work can be calculated by Morrow's or Eyllin's formulas (J.D. Morrow, 1965, F. Eyylin 1984) respectively for Masing or Non-Masing property materials.

Masing materials

$$\Delta W^P = \Delta W^P(\bar{r}) = \frac{1-n}{1+n} \Delta\sigma\Delta\varepsilon^P \tag{4}$$

Non-Masing materials

$$\Delta W^P = \frac{1-n''}{1+n''}(\Delta\sigma - \delta\sigma)\Delta\varepsilon^P + \delta\sigma\Delta\varepsilon^P \tag{5}$$

Where $\Delta\sigma$ and $\Delta\varepsilon^P$ are the local cyclic stress and cyclic strain ranges, n' is the cyclic hardness exponent, $\delta\sigma$ is the difference in yield limit between the essential hysteresis loop and normal hysteresis loop, n'' is the cyclic hardness exponent of essential hysteresis loop.

2.3. *Plastic strain energy rate*

The plastic strain energy rate implies the dissipation rate of plastic strain energy, it is defined to be:

$$\dot{W}^P = \sigma \cdot \dot{\varepsilon}^P \tag{6}$$

Where: $\dot{\varepsilon}^P$ is the plastic strain rate. For Masing materials, after reaching the cyclic stability, the cyclic stress strain relation can be described by the Osgood-Ramberg equation, that is:

$$\Delta \varepsilon^t = \Delta \varepsilon^e + \Delta \varepsilon^P = \frac{\Delta \sigma}{E} + 2\left(\frac{\Delta \sigma}{2K'}\right)^{\frac{1}{n'}} \tag{7}$$

It can be deduced form eqn.(6) and eqn.(7) that the plastic strain energy rate can be expressed by:

$$\dot{W}^P = \frac{\sigma \dot{\sigma}}{n'K'}\left(\frac{\Delta \sigma}{2K'}\right)^{\frac{1}{n'}-1} \tag{8}$$

The accumulated plastic strain energy at arbitrary time t in a cycle is:

$$W^P(t) = \int_0^t \dot{W}^P dt \tag{9}$$

3. NUMERICAL SIMULATION OF TEMPERA-TURE FIELD AS SELF-HEATING GENERA-TION IN CYCLIC LOAD-ING PROCESS

Finite difference method and finite element method are utilized in the numerical simulation of self-heating generation in fatigue damage process. For FEM, we directly use the commercial structure analysis software packages, such as ADINA and NASTRON etc. For FDM, according to the *Frank-Nilson* equation, the difference equation of governing equation of can be written as (X. Tung 1993):

$$\rho C_p \frac{T_{i,j}^{n+1} - T_{i,j}^n}{\tau \cdot \Delta m} = \frac{k}{2}\left[\frac{T_{i-1,j}^n - 2T_{i,j}^n + T_{i+1,j}^n}{\Delta x^2} + \frac{T_{i-1,j}^{n+1} - 2T_{i,j}^{n+1} + T_{i+1,j}^{n+1}}{\Delta x^2}\right]$$
$$+ \frac{k}{2}\left[\frac{T_{i,j-1}^n - 2T_{i,j}^n + T_{i,j+1}^n}{\Delta y^2} + \frac{T_{i,j-1}^{n+1} - 2T_{i,j}^{n+1} + T_{i,j+1}^{n+1}}{\Delta y^2}\right] \tag{10}$$
$$+ \dot{w}_{i,j}^n - 2hT_{i,j}^n/B$$

There are three types of boundary conditions. Including the intersection point of the boundary, we totally have six cases for a rectangle smooth plate specimen.

① $i=0, j=0$: $C_5 T_{0,0}^{n+1} - 2C_2 T_{1,0}^{n+1} - 2C_3 T_{0,1}^{n+1} = C_6 T_{0,0}^n + 2C_2 T_{1,0}^{n+1} + 2C_3 T_{0,1}^{n+1} + \dot{w}_{0,0}^n \tag{11}$

As where:

$$C_5 = \frac{\rho C_p}{\tau \cdot \Delta m} + \frac{k}{\Delta x^2} + \frac{k}{\Delta y^2} + \frac{h}{\Delta x} = C_1 + \frac{h}{\Delta x}$$

$$C_6 = \frac{\rho C_p}{\tau \cdot \Delta m} - \frac{k}{\Delta x^2} - \frac{k}{\Delta y^2} - \frac{2h}{B} + \frac{h}{\Delta x} = C_4 - \frac{h}{\Delta x}$$

② $j=0, \ 0<i<r$:

$$-C_2 T_{i-1,0}^{n+1} + C_1 T_{i,0}^{n+1} - C_2 T_{i+1,0}^{n+1} - 2C_3 T_{i,1}^{n+1}$$
$$= C_2 T_{i-1,0}^n + C_4 T_{i,0}^n + C_2 T_{i+1,0}^n + 2C_3 T_{i,1}^n + \dot{w}_{i,0}^n$$

③ $i=r, \ j=0$:

$$C_1 T_{0,0}^{n+1} - 2C_2 T_{1,0}^{n+1} - 2C_3 T_{0,1}^{n+1} = C_4 T_{0,0}^n + 2C_2 T_{1,0}^{n+1}$$
$$+ 2C_3 T_{0,1}^{n+1} + \dot{w}_{0,0}^n$$

674

④ $i=0$, $0<j<s$:

$$-C_3 T_{0,j-1}^{n+1} + C_5 T_{0,j}^{n+1} - 2C_2 T_{1,j}^{n+1} - C_3 T_{0,j+1}^{n+1}$$
$$= C_3 T_{0,j-1}^{n} + C_6 T_{0,j}^{n} + 2C_2 T_{1,j}^{n} + C_3 T_{0,j+1}^{n} + \dot{w}_{0,j}^{n}$$

⑤ $i=r$, $0<j<s$:

$$-C_3 T_{0,j-1}^{n+1} + C_1 T_{0,j}^{n+1} - 2C_2 T_{1,j}^{n+1} - C_3 T_{0,j+1}^{n+1}$$
$$= C_3 T_{0,j-1}^{n} + C_4 T_{0,j}^{n} + 2C_2 T_{1,j}^{n} + C_3 T_{0,j+1}^{n} + \dot{w}_{0,j}^{n}$$

⑥ $j=s$:

$$T_{i,s}^{n} = 0, \quad i = 0, \cdots r$$

The temperature of the specimen equals to the environment at initial moment, so the initial conditions are:

$$T_{i,j}^{0} = 0, \quad i = 0, \cdots, r, \quad j = 0, \cdots s$$

4. DISSCUSSIONS AND CONCLUSIONS

In order to verify the imaging results by the numerical simulation method, the class A 40CrNiMo alloy is used to measure the temperature change, resulted from the self-heating generation under cyclic loading. All the cyclic loading tests are performed in a universal material testing system. The measurement instrument is infrared thermovision.

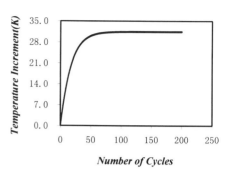

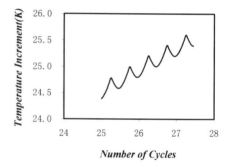

Figure 1 Temperature Increment and Fluctuation in Cyclic Loading Process (a) and in a Cycle (b)

Differing from the tested results, a rapid temperature increasing stage was found in the computational results, as shown in figure 1b. Additionally, the simulating temperature changes is faster than the experimental measured ones because of the cyclic hysteresis energy at initial stage is much less than the ones at the steady stage. The amplitude of the fluctuation is about one centigrade degree and is relevant to the strain amplitude, loading frequency, loading wave, and so on.

Figure 2 shows the computationally simulated surface temperature field of single, double and center notched specimen. It is easy found that the distribution maps are very similar with the experimentally measured ones. Figure 3 shows the comparison between experimental measured 2 dimension and 3 dimension temperature fields and computational simulated ones. Figure 4 shows the comparison of the temperature fields in a cycle obtained by experimental measurement and computational simulation. It is found that the temperature fields and fluctuation are in good agreement with each other, they varies with a range of 1.2 ~1.5. Therefore, it is believed the computational simulation could not only recover the temperature fluctuation at a certain point, but also its distribution in the whole calculation region. Moreover, the possibility is supposed to image the heat dissipate process in fatigue damage process.

Choose continuous time unit to investigate the temperature variation within one load cycle. The average of the temperature increases alone, while temperature fluctuation exists in a cycle, as shown in fig.1. In figure 2a and figure 2b, it is shown that the peak value of the temperature increases as the load cycles increase. At the initial stage of fatigue, the temperature changes rapidly, because the most part of the plastic strain energy is stored in the material and only a little part dissipates into the environment. Some cycles later, the ratio of the dissipation energy increases because of the high temperature difference between the specimen and the environment, and the speed of temperature increment slows down consequently. When the heat dissipation and the heat production due to the plastic strain energy reaches balanced, the temperature becomes nearly unchanged. The temperature increment is very sensitive to the strain amplitude and the loading frequency, since they significantly affect the dissipation rate of the plastic strain energy.

5 REFERENCES

1. Esin,A. and Jones, W.J.D., *Nuclear Engineering and Design*, Vol.4, 1966, pp.292-298
2. Morrow,JoDean ASTM STP378; 1965, 45
3. Morris,W.L. and James, M.R., *Fatigue Mechanisms* ASTM STP811, 1983, 179
4. Ohnami,M. *Plasticity and High Temperature Strength of Materials*, Elsevier Applied Science., London and New York, 1990, 184
5. Tung,X. *Acta Metall.Sinica*; 1991, 4A(5): 374
6. Tung,X. *Acta Metall.Sinica*; 1992, 5A(4)
7. Ye D.Y. *Inter. J. Fatigue*, 1996, 18(2): 105
8. Tung, X. He, J.M. *Proceedings of the First Postdoctoral Academic Congress*, Defense Industry Press , Beijing, 1993, Vol.1, 1
9. Lefebrve D and F. Eyllin, *Inter. J. Fatigue*, 1984 6(1) PP9

ACKNOWELDGEMENT

The authors thank the Fok Ying Tung Education Foundation and National Nature Science Foundation of China (NSFC) and Aeronautic and Astronautic Science Foundation of China(AASFC). This work was supported in part by whose auspices.
They also thank Prof. Dr.-Ing. K.Detert, University GH Siegen, Germany, for his helpful advises and discussions for this manuscript, They also thank Prof. Dr.-Ing. K. Schiffner for the supplement of ADINA software. They also thank Senior Researcher C.Z. He, Director of Rocket Structure and Pipe Line System Division for the supplement of Nastron software.

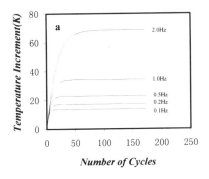
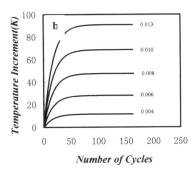

Figure 2   The Peak Value of The Temperature Increases in Cyclic Loading Process

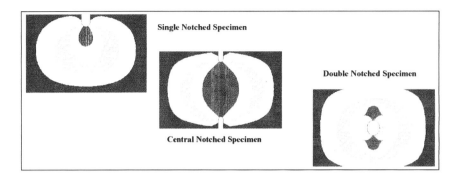

Fig.3 Numerical Simulated Distributions of Different Notched Specimens

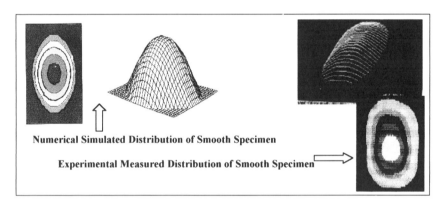

Figure 4 Numerical Simulated and Experimental Measured Distributions of Smooth Specimens

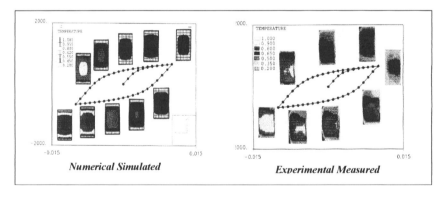

Figure 5 Temperature Fluctuation in a Cycle Comparison by Experimental Measured and Numerical Simulated

Low Cycle Fatigue and Elasto-Plastic Behaviour of Materials
K-T. Rie and P.D. Portella (Editors)
© 1998 Elsevier Science Ltd. All rights reserved.

BP PROGRAM - INTERACTIVE EVALUATION OF
BODNER-PARTOM MODEL MATERIAL CONSTANTS

J.J. VIÑALS
Mechanical Engineer, Structural and Mechanisms Department
Sener, Av. Zugazarte, 56, 48930-Las Arenas, Vizcaya - Spain

ABSTRACT

BP program is a PC computer software developed to calculate the material constants used in the Bodner-Partom model. The Bodner-Partom model belongs to the material specific unified models, where the constitutive equations have not a yield criteria or loading/unloading conditions to be developed. The equations are in the field of multi-dimensional continuous mechanisms which makes them capable of solving problems by analytical and numerical methods. The program described in this paper gives several options for material constant evaluation in order to obtain the best simulation of cyclic viscoplasticity, in line with experimental results via an interactive user procedure.

KEYWORDS

Viscoplastic, Bodner-Partom, material model, unified constitutive equations, constant evaluation.

BRIEF DESCRIPTION OF THE MATERIAL CONSTANTS EVALUATIONS AND RELATED EQUATIONS

The Bodner-Partom material model allows to establish the constitutive equations in such a form that the material constants can be evaluated from test results by separate consideration of basic physical effects. Excluding elastic constants, 13 material constants must be evaluated from experimental data. In most cases, only nine are needed, and can be obtained from tensile and creep tests. Table 1 defines the material constants used in the Bodner-Partom model.

Table 1. Material constants in Bodner-Partom constitutive model

CONSTANTS	DESCRIPTIONS
D_0	Limiting shear strain rate
Z_0	Initial value of the isotropic hardening variable
Z_1	Limiting (maximum) value of Z^I

CONSTANTS	DESCRIPTIONS
Z_2	Fully recovered (minimum) value of Z^I
Z_3	Limiting (maximum) value of Z^D
m_1	Hardening rate coefficient of Z^I
m_2	Hardening rate coefficient of Z^D
n	Kinetic parameter
A_1	Recovery coefficient for Z^I
A_2	Recovery coefficient for Z^D
r_1	Recovery exponent for Z^I
r_2	Recovery exponent for Z^D

Work hardening calculations

The first step is to obtain the work hardening rate from tensile tests (input data: Stress/strain points and elastic modulus), determining the plastic strain, subtracting the elastic component from the total strain. Then, the stress (σ) is defined as a function of the plastic strain (ε^P), considering 3 possible functions (in order to have functions easy for numerical purposes):

1. Polynomial: $\sigma = a_0 + a_1 \varepsilon^P + a_2 (\varepsilon^P)^2 + ... + a_m (\varepsilon^P)^m$

2. Logarithmic: $\sigma = a + b \, l_n (\varepsilon^P)$

3. Ramberg-Osgood: $\sigma = [\varepsilon^P/0.002]^{1/n} \cdot (0.2\% \text{ TYS})$

    n: Ramberg-Osgood parameter
    0.2% TYS: Tensile Yield Stress

The work hardening rate is defined as: $\gamma = d\sigma/dW_P = (1/\sigma) \cdot (d\sigma/d\varepsilon^P)$, so forth:

1. Polynomial: $\gamma = (1/\sigma) \cdot [a_1 + 2a_2 \varepsilon^P + ... + m \, a_m (\varepsilon^P)^{m-1}]$

2. Logarithmic: $\gamma = (1/\sigma) \cdot (b/\varepsilon^P)$

3. Ramberg-Osgood: $\gamma = 1/(n \cdot \varepsilon^P)$

The program allows to choose the best fitting for each case, by means of the plotting capabilities also included in the software developed. In the case of the polynomial approach, the program calculates polynomial functions from grade 2 to 6 (limited in some cases by the number of input data); the user can select the most appropriated curve. Recent studies [1, 2, 3] establish for B1900 + Hf that the cubic polynomial (m=3) is generally sufficient but fourth-order (m=4) polynomial has also been used occasionally.

Hardening coefficients evaluation

The m_1 and m_2 values can be evaluated from the plot of γ versus σ in the uniaxial case. Following Bodner-Partom theory, the mentioned plot should be linear with slope $(m_1 + m_2)$, according to NASA investigation [2], or slope m_2, NASA studies [3], in the range of small plastic strains; this approach (both options) are included in the software developed via interactive menus.

At the larger strains, the relation between γ and σ is also linear with slope m_1. The program plot capabilities permit to select the number of points needed for evaluation of m_1 and m_2, and change the number points in order to improve the fitting. In addition, the program has the option of elimination points in both zones (small and larger plastic zones) in order to avoid strange values that can produce wrong results.

The PC program developed allows to repeat the process in an interactive procedure to re-assess the final values, so many times as user wants.

Final internal calculations

Until now, the calculations are performed for each tested specimen. When all the specimen results are introduced, the recovery constants evaluation is performed, as folows:.

- Evaluation of n and Do in the kinetic equation. Do is usually adopted as 1×10^{-4} sec^{-1}.

- Calculation of $Z_1 + Z_3$.

- Relative contributions of Z_1 and Z_3.

- Evaluation of r_1, r_2, A_1, A_2.

All of these aspects (and also previous contents) are explained in more detail in the user manual [4] developed inside BREU6021.

EXAMPLES

Output result files are shown in Tables 2, 3 and 4, for the 3 (three) approaches (polynomial, logarithmic and Ramberg-Osgood) in the case of 5 tested specimens at 600 °C with different tensile strain-rates.

Table 2. Evaluations of Bodner-Partom Material Constants - Polynomial

Spec.Id.	Temp. (C)	El.Mod. (GPA)	Strain Rate (SEC-1)	M1 (MPA-1)	M2 (MPA-1)	Sat.Str (MPA)	n	Z1 (MPA)	S.O. (MPA)	S.B. (MPA)	Z0=Z2 (MPA)
ten600-2	600.0	91.00	.1000E-01	2.6300	3.4310	555.42	3.118	1026.3	450.0	520.0	767.5
ten600-3	600.0	90.00	.1000E-02	2.1200	2.4820	545.21	3.118	1120.3	475.0	521.0	830.2
ten600-4	600.0	92.00	.1000E-03	2.2800	2.7270	557.89	3.118	1293.3	475.0	517.0	848.0
ten600-5	600.0	86.00	.1000E-04	2.3800	2.2390	535.52	3.118	196.9	450.0	514.0	818.6
ten600-6	600.0	90.00	.1000E-05	2.4560	4.5862	503.91	3.118	1077.0	400.0	419.1	740.0

680

Table 3. Evaluations of Bodner-Partom Material Constants - Logarithmic

Spec.Id	Temp. (C)	El.Mod (GPA)	Strain Rate (SEC-1)	M1 (MPA-1)	M2 (MPA-1)	Sat.Str. (MPA)	n	Z1 (MPA)	S.O. (MPA)	S.B. (MPA)	Z0=Z2 (MPA)
ten600-2	600.0	91.00	.1000E-01	1.1000	2.7030	561.89	3.307	864.1	450.0	519.0	744.4
ten600-3	600.0	90.00	.1000E-02	1.1300	2.7990	554.92	3.307	934.5	475.0	509.0	804.2
ten600-4	600.0	92.00	.1000E-03	1.1400	2.8920	568.08	3.307	977.5	475.0	513.0	820.4
ten600-5	600.0	86.00	.1000E-04	1.1400	2.8080	539.73	3.307	911.6	450.0	499.0	791.1
ten600-6	600.0	90.00	.1000E-05	1.3664	3.6511	515.33	3.307	905.3	400.0	448.4	714.4

Table 4. Evaluations of Bodner-Partom Material Constants - Ramberg-Osgood

Spec.Id.	Temp (C)	El.Mod (GPA)	Strain Rate (SEC-1)	M1 (MPA-1)	M2 (MPA-1)	Sat.Str (MPA)	n	Z1 (MPA)	S.O. (MPA)	S.B. (MPA)	Z0=Z2 (MPA)
ten600-2	600.0	91.00	.1000E-01	1.1000	2.7030	561.89	3.144	1005.8	520.0	519.0	883.0
ten600-3	600.0	90.00	.1000E-02	1.1300	2.9090	560.93	3.144	1025.7	503.0	508.0	875.1
ten600-4	600.0	92.00	.1000E-03	1.1400	2.6150	568.08	3.144	1065.3	509.0	517.0	904.4
ten600-5	600.0	86.00	.1000E-04	1.1400	2.8080	539.73	3.144	1029.2	500.0	499.0	905.1
ten600-6	600.0	90.00	.1000E-05	1.3664	3.6511	515.33	3.144	1025.0	450.0	448.4	828.3

For the last specimen, fig. 1, 2 and 3 show the fitting obtained with logarithmic approach:

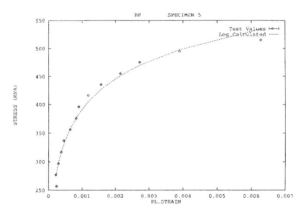

Fig. 1: Stress versus plastic strain; Test and fitting.

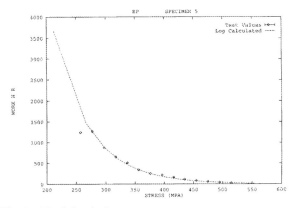

Fig. 2: Work hardening rate versus stress; Test and fitting.

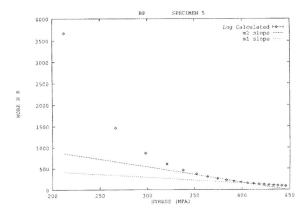

Fig. 3: m_1 and m_2 calculation.

Table 5 shows a comparison between the constants calculated in a recent report [3] for the B19000 + Hf, and the constants obtained with BP program for the same material and test conditions. As it can be seen, the agreement can be considered quite good between both procedures.

Table 5. Result comparison between report [3] and Brite Euram BP program developed

		NASA CR 179522 Bodner-Partom Model Const. for B1900 + Hf		Brite/Euram Program Developed with B1900 + Hf from NASA CR 179522	
		Base Program	Optional Program	Case 1 Polynomial values	Case 2 Logarithmic values
Temp. Ind.	m1	0,27	0,20	1,48	0,71
	m2	1,52	1,85	2,00	2,59
	Z1	3000	29000	60966	24240
	Z3	1150	8500	9351	8634
Temp.Dep.	n	1,03	0,446	0,378	0,451
	Z0=Z2	2400	23000	41411	19058

CONCLUDING REMARKS

The software developed lets flexibility and wide range of tools for material constant evaluation (Bodner-Partom).

The program plot capabilities give a good procedure for user optimization during the process for each material, temperature level and different strain rates.

The viscoplastic simulation performed via BP program and Vpsim program [5], inside project BRITE/EURAM 6021, indicates a good analysis procedure for uniaxial cases and different history loading conditions, with pretty good agreement between experimental data and simulation.

ACKNOWLEDGEMENT

This work was supported by Brite/Euram project 6021 ("Development of an Understanding of Material Properties Under the Combined Influence of Creep, Fatigue and Oxidation") with the co-operation of MTU, ARA, DERA, CNR-ITM, BMW-RR, TU, VAC, SENER, TM and RR.

REFERENCES

1. Lindholm, U.S., Chan, K.S., Bodner, S.R., Weber, R.M., Walker, K.P. and Cassenti B.N. (1984) NASA CR-174718, Constitutive modelling for isotropic materials (Host).

2. Lindholm, U.S., Chan, K.S., Bodner, S.R., Weber, R.M., Walker, K.P. and Cassenti, B.N. (1985) NASA CR-174980, Constitutive modelling for isotropic materials (Host) (Second Annual Status Report).

3. Chan, K.S., Linsholm, U.S., Bodner, S.R., Hill, J.T., Weber, R.M. and Meyer ,T.G. (1986) NASA CR-179522, Constitutive modelling for isotropic materials (Host) (Third Annual Status Report).

4. Viñals, J.J. (1995), BP PROGRAM 2.0, An interactive program for Bodner-Partom model material constants evaluation, BRITE/EURAM 6021, User Manual.

5. Sjöqvist, R. and Järvstrat, N., Volvo Aero Corporation, S-46181 Trollhättan, Sweden. (1996). In: Vpsim-Interactive simulation of cyclic viscoplasticity, volume 2, pp 1333-1338 Fatigue '96 Proceedings of the Sixth International Fatigue Congress, Berlin, G. Lütjering and H. Nowack (Editors). Pergamon.

Low Cycle Fatigue and Elasto-Plastic Behaviour of Materials
K-T. Rie and P.D. Portella (Editors)
683

IMPLEMENTATION OF VISCOPLASTIC CONSTITUTIVE MODEL IN MARC COMPUTER CODE

E. CEARRA and J.J. VIÑALS
Mechanical Engineers, Structural and Mechanisms Department
Sener, Avda. Zugazarte, 56, 48930-Las Arenas Vizcaya, Spain

ABSTRACT

The MARC non-linear finite element computer program [1] has been used for incorporation of the Bodner-Partom model (Viscoplastic behaviour). The main objective is to provide evidence for the applicability of the unified constitutive models to describe the time and temperature dependent response of metals submitted to arbitrary load or deformation histories, and in addition, to demonstrate the practical application that will depend on their adaptability in computational algorithms when incorporated into finite element methods for stress analysis.

The aspects that are explained in the following paragraphs are:

a)   A brief summary of the main concepts of Bodner-Partom model.
b)   Implementation in MARC : Subroutines applied.
c)   Calculations performed and comparison with classical approach and test results.
d)   Main results and conclusions.

The calculations are carried out on test specimens for LCF conditions with and without dwell time at the maximum strain in each hysteresis cycle, for two materials typical in turbine discs for aeronautical applications.

This research is part of a BRITE/EURAM project Programme BREU 6021 " Development of an understanding of materials properties under the combined influence of creep, fatigue and oxidation".

KEYWORDS

Viscoplastic, Bodner-Partom, material model, FEM implementation, constitutive equations.

BODNER-PARTOM MODEL

The Bodner-Partom model is a set of constitutive equations for elastic-viscoplastic material without a yield criterion. The theory is formulated on the basis of internal variables which

depend on the loading history. The model [2] includes isotropic and directional hardening, thermal recovery of hardening and general temperature dependence of plastic flow.
The main features are :

a)   Flow law : Prandtl-Reuss law.
b)   Kinetic equations.
c)   Evolutionary equations.

The material constants, most of them are temperature independent, are obtained from uniaxial tensile tests at different strain rates and temperature, and creep tests.

IMPLEMENTATION IN MARC

The model has been implemented in MARC via proper and user defined subroutines. The internal variables of the model are taken into account using state variables. The first state variable is the isotropic hardening and the second is the directional hardening.

The finite element formulation leads to a set of linear equations which solution is obtained using a direct procedure with the appropriate selection in the SOLVER parameter of the input file. The results are considered pretty accurate.

The material constants are introduced via the USDATA parameter in the USDATA subroutine. The user subroutines are :

- USDATA : Read in material data and store in the common block USDACM.

- FORCDT : Introduction of the load history. In the case analysed (LCF strain control), a forced displacement to a set of nodes.

- UVSCPL : Computation of the inelastic strain increment for an elastic-viscoplastic behaviour.

- HOOKE : Determination of the elastic stress-strain relation for isotropic material.

- FLWRAT : Calculation of the equivalent visco-plastic flow rate.

- ZISOTR : Determination of the isotropic part of the yield stress.

- BETA : Calculation of the beta tensor components (directional hardening).

Subroutine UVSCPL calls internal MARC subroutines [3] for implementation : SCLA, GMPRD, MCPY, ZERO, SMPY, GMAD, ZEROE and GMSUB.

CALCULATIONS AND RESULTS COMPARISON

LCF Strain control simulation has been performed for smooth specimens in a cycle from 0 % of total strain to 1 %, with and without dwell time (120 sec.) at maximum total strain.

Element type 28 (Solid axisymmetric element, cquad of 8 nodes) has been used [4]. The element stiffness is obtained via 9 integration points.

Two cases are analysed. In the first one, the load is introduced as small displacement increments applied in the specimen clamped position, simulating test conditions; in the second analysis, during creep interval, displacements are not introduced.

In the strain-stress analysis (0%-1% not dwell), during the loading process, the inelastic strain begins in a very small area of the radial side near the specimen straight zone. Fig. 1 and 2 show the stress and strains for one cycle in the control element.

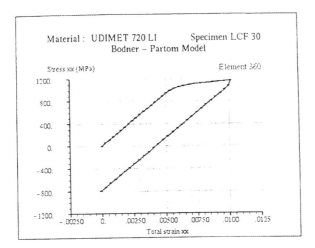

Fig. 1

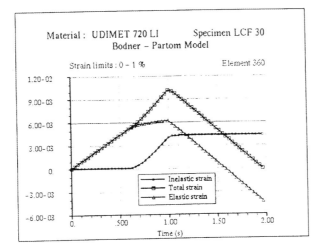

Fig. 2

686

In the case of dwell time (120 sec.) at 1 % of total strain, the loading is similar to the previous one but when creep begins, a stress relaxation occurs during the strain hold and the inelastic strain grows. Total strain increases during all the cycle until reaching t=120 sec. The plastic strain is constant along the moment analysed. Elastic strain decreases during creep influence and creep strain increases during the creep process. Fig. 3 and 4 indicate the control element behaviour in one cycle.

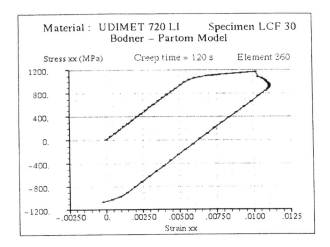

Fig. 3

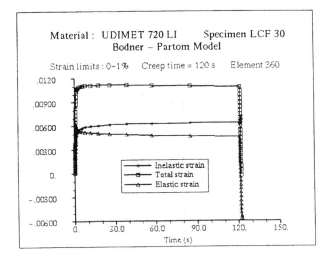

Fig. 4

When during an analysis, the loading is time independent, the inelastic strain that could be produced is plastic strain and when the loading is time dependent, the inelastic strain is creep strain, whole if plastic strain exists, it remains constant.

The results obtained from the implementation has been compared with classical approach (elasto-plastic calculations) and test data for both cases (with and without dwell time) :

a) For both cases, the elasto-plastic analysis and the implementation are considered pretty good. See fig. 5 and 6.

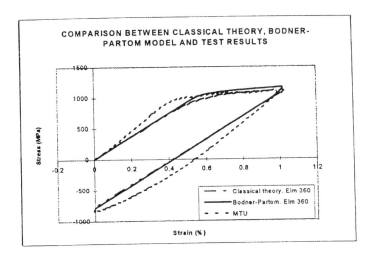

Fig. 5

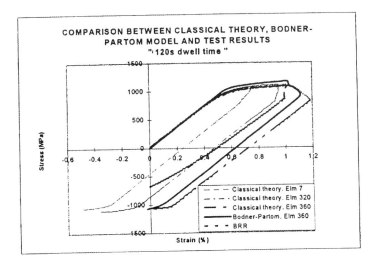

Fig. 6

b) In the comparison of test results and implementation, for both cases, the fitting is also good, but small differences appear due to methodology assumptions and differences between test conditions and material constants data. See also fig. 5 and 6.

MAIN RESULTS AND CONCLUSIONS

The implementation of Bodner-Partom model in MARC allows to simulate any load history without the specific material data for each case, only material constants evaluation are required.

The agreement between theoretical calculations, test results and implementation procedure can be considered good enough.

MARC capabilities and subroutines used have demonstrated that the computer code is a useful tool for non-linear analysis and viscoplastic simulations.

ACKNOWLEDGEMENTS

The work has been supported by Brite/Euram project 6021 with the co-operation of MTU, ARA, DERA, CNR-ITM, BMW-RR, TU, VAC, SENER, TM and RR.

In addition, Mr. Ewout Meijers (MARC - The Netherlands) collaboration and support is grateful.

REFERENCES

1.   MARC User Information (Volume A), Rev. K6.2.

2.   Bodner, S.R. and Parton, Y. (1975), Journal of Applied Mechanics, "Constitutive equations for elastic-viscoplastic strain hardening materials" pp 385-389.

3.   MARC User Subroutines (Volume D), Rev. K6.2.

4.   MARC Element Library (Volume B), Rev. K6.2.

Chapter 10

DESIGN METHODS; LIFE PREDICTION

Low Cycle Fatigue and Elasto-Plastic Behaviour of Materials
K-T. Rie and P.D. Portella (Editors)
691

EXPERIMENTAL INVESTIGATION ON THE CYCLIC BEHAVIOUR OF TRANSFORMED STEEL SHEET

R. MASENDORF and H. ZENNER

Institute for Plant Engineering and Fatigue Analysis, Technical University of Clausthal.
D-38678 Clausthal-Zellerfeld, Germany

ABSTRACT

Components of prestrained sheet metal are frequently subjected to pulsating stress. For estimating the service life of such components on the basis of the local strain concept, data concerning the cyclic behaviour of the material must be available. The sheet samples tend to buckle when subjected to compressive stress. Buckling is prevented by appropriate supporters. Initial results on cyclic behaviour are available for St15 with a thickness of 1 mm. The cyclic behaviour of the material depends on the degree of deformation and deformation ratio; therefore, the material was prestrained by 10, 20, and 30 per cent by tension before taking samples for the strain-controlled tests.

KEYWORDS

LCF, St15, sheet material, prestrained, supporter

INTRODUCTION

The service life of components made of prestrained sheet metal can be predicted on the basis of the local strain concept. For calculations with the use of the local strain concept, the cyclic behaviour of the material must be known. The characteristic data on the cyclic behaviour depend on the degree of deformation and deformation ratio of the sheet. Since no approximate solution is hitherto available for the cyclic characteristics with due consideration of prestraining, tests have been performed with St15 metal sheet. The cyclic behaviour of the material is determined by strain-controlled tensile-compressive tests at constant amplitude and without mean strain. Buckling of the sample when subjected to compressive stress is prevented by means of friction-minimised supporters. Preliminary results are presented.

692

EXPERIMENTAL PROCEDURE

For experimentally determining the cyclic characteristics of the material, strain-controlled tests are performed by means of a servohydraulic loading device. The material samples are subjected to axial stress, figure 1, with an alternating strain amplitude ($R_\varepsilon = -1$).

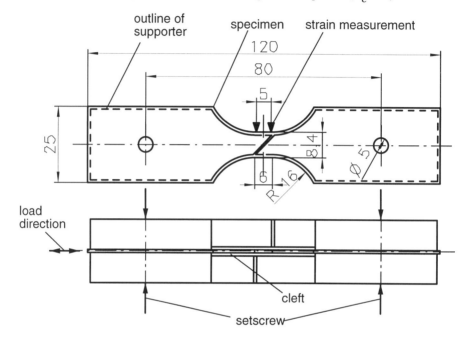

Figure 1: Material sample [1] and clamping with support for strain-controlled single-stage tests

Plastic strain components must be replasticised. Under the compressive forces which thereby occur, the material sample tends to buckle. Buckling is prevented by supporters with an air gap between the sample and supporter in the testing zone, figure 1. The sheet surface is almost completely covered. For the strain measurement, therefore, only the cut edge of the sheet is available, figure 1. Note: Even at a stress amplitude in the pulsating tensile range, plastic strain components occur and must be replasticised.

The strain processes are time-dependent. Consequently, the forces or stresses measured during the tests depend on the frequency and strain amplitude. A sinusoidal strain-time variation was selected for the tests, since the range with a high plastic strain component is traversed at a lower strain rate than range with a low plastic strain component in this case. With a triangular strain-time variation, all tests can be performed at a constant strain rate, of course; however, they must correspond with the strain behaviour of the material with high plastic strain components. With a sinusoidal strain-time variation, the test duration can be shortened. The test frequency is selected in such a way that the difference between the load amplitudes at a reference frequency of 0.1 Hz and at the test frequency is less than 2 per cent, figure 2.

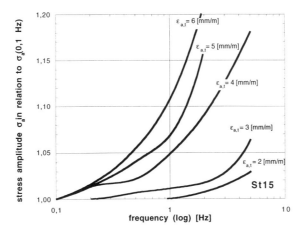

Figure 2: Effect of the frequency on the stress amplitude as a function of the strain amplitude

During the tests the stress amplitude and, at regular intervals, a stress-strain hysteresis loop are recorded. The surface of the sample is almost completely covered by the supporter. Hence, a visual inspection for incipient cracks is not possible. The decrease in load amplitude is therefore employed as criterion for crack initiation. Samples which have been removed shortly before exceeding the stress maximum exhibit crack initiation over a length of 0.5 mm. If the test is continued, these incipient cracks propagate quickly until fracture of the sample. In this case, the maximum of the stress amplitude is assumed as crack initiation point, figure 3.

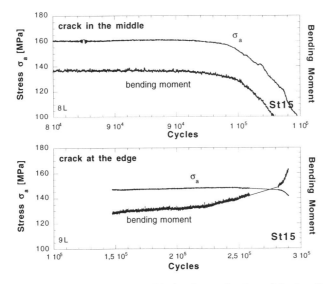

Figure 3: Estimating the position of crack initiation by evaluation of the bending moment

694

Tests are currently in progress for determining the time and position of crack initiation (middle or edge of the sample) from the occurrence of a superimposed bending moment. For this purpose, a load cell is employed for measuring bending moments as well as axial forces. If the sample ruptures in the vicinity of the edges, the bending moment amplitude already increases before the decrease in stress amplitude, figure 3, bottom. In the case of crack initiation in the middle of the sample, both amplitudes decrease synchronously, figure 3, top.

PRELIMINARY EXPERIMENTAL RESULTS

Sheet metal components are frequently subjected to severe deformation during manufacture. For analysing the deformation, samples were taken from prestrained sheet metal strips. Results of the strain-controlled tests include the stress-strain curve, figure 4, and the strain S-N curve, figure 5. For strain amplitudes smaller than 2 mm/m, a considerable increase in service life is evident. In this range, the variation of the strain S-N curve is governed essentially by the elastic strain component, that is, σ_f' and b. The variation of these characteristic values is plotted as a function of the prestraining in figure 6. Moreover, the parameters, K´ and n´, for describing the cyclic stress-strain curve are also indicated in figure 6. In figures 4, 5, and 6, results are presented for the material St15 (ultimate strength R_m = 314 MPa, yield strength $R_{p0.2}$ = 190 MPa) with a thickness of 1 mm.

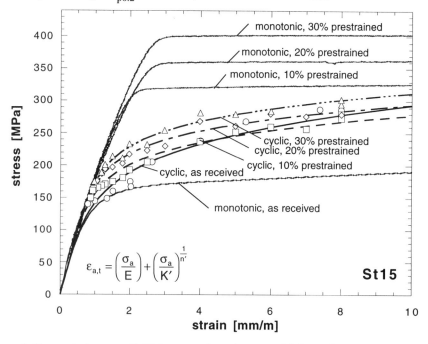

Figure 4: Stress-strain curve, St15 1 mm, various degrees of deformation

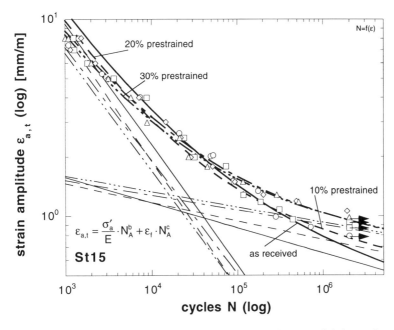

Figure 5: Strain S-N curve [3, 4, 5], St15 1 mm, various degrees of deformation

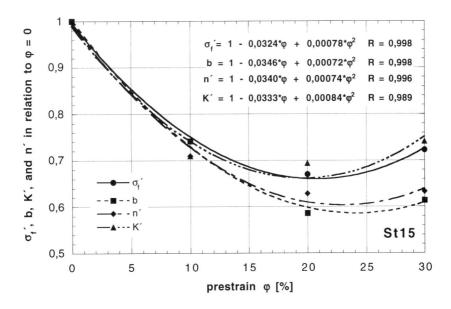

Figure 6: Effect of prestraining on σ_f', b, n´, and K´, referred to the as-received condition

SUMMARY

- In the present article, a possibility is described for clamping sheet metal samples for strain-controlled testing. Buckling under compressive stress is prevented by means of a friction-minimised supporter.
- The samples are subjected to sinusoidal loading at a frequency between 0.1 and 5 Hz.
- Preliminary results are available for determining crack initiation and incipient crack position by means of the associated bending moment.
- The pronounced effect of prestraining on the deformational behaviour and service life has been examined with the use of prestrained samples.

In the future, the effect of prestraining (degree of deformation and deformation ratio) on the cyclic behaviour of various steel and aluminium sheet materials should be investigated. The objective is to extend the uniform material law /2/ by one coefficient, which describes the effect of prestraining.

ACKNOWLEDGEMENTS

The work was supported by Deutsche Forschungsgemeinschaft (SFB 362 "Fertigen in Feinblech", TU Clausthal, Uni Hannover).

REFERENCES

[1] Drewes, E.-J., et al. (1995). Abschlußbericht: Neue Stähle mit hoher statischer, dynamischer und Dauerfestigkeit für den Automobilbau, BMBF-Bericht 03 M 3021

[2] Seeger, T. In: Seminar Betriebsfestigkeit auf der Grundlage örtlicher Beanspruchungen, FG Werkstoffmechanik, TH Darmstadt

[3] Coffin, L. F. (1954). A Study of the Effects of Cyclic Thermal Stresses on a Ductile Metal, Transaction of ASME 76 pp. 931-950

[4] Manson, S. S. (1965). A Complex Subject-Some Simple Approximations. Experimental Mechanics pp. 193-226

[5] Morrow, J. D. (1965). Cyclic Plastic Strain Energy and Fatigue of Metals. In: Internal Friction, Damping and Cyclic Plasticity, ASTM pp. 45-87

Low Cycle Fatigue and Elasto-Plastic Behaviour of Materials
K-T. Rie and P.D. Portella (Editors)
© 1998 Elsevier Science Ltd. All rights reserved.   697

LOW CYCLE FATIGUE BEHAVIOUR OF TWO FERRITE-PEARLITE MICROALLOYED STEELS

V. SUBRAMANYA SARMA[1], K.A.PADMANABHAN[2], G.JAEGER[3] and A.KOETHE[3]
[1] Department of Metallurgical Engineering, Indian Institute of Technology Madras 600 036
[2] Indian Institute of Technology Kanpur 208 016 INDIA
[3] Institut fuer Festkoerper- und Werkstofforschung (IFW) Dresden, GERMANY

ABSTRACT

In this paper the room temperature cyclic stress-strain (CSS) and the low cycle fatigue (LCF) behaviour under total strain control of a medium carbon and a low carbon ferrite-pearlite microalloyed (MA) steel are reported. The experimentally determined life was compared with the life predicted by the modified universal slopes equation and Tomkins' model. The medium carbon MA steel exhibited cyclic softening at lower (total) strain amplitudes (≤0.6%) and cyclic hardening at higher strain amplitudes. The cyclic stress response of the low carbon MA steel in contrast displayed cyclic softening at all the total strain amplitudes employed. Though the life predicted by the modified universal slopes equation for the medium carbon MA steel correlated well with the experiments, it grossly overestimated the life in case of the low carbon steel. On the other hand, life predicted by Tomkins' model correlates rather well with the experiments in case of both the steels. The deleterious influence of the elongated inclusions present in the latter steel was thought to be responsible for the above overestimation.

KEY WORDS

Microalloyed steels, cyclic stress-strain properties, Lüders bands, low cycle fatigue, life prediction

INTRODUCTION

Microalloying followed by controlled processing is a direct and cost effective route to improved strength levels in ferrite-pearlite structures and compares favourably with the conventional approach of normalising or quenching to acicular structures and tempering. The former route has been adopted for low carbon, high strength low alloy (HSLA) steels. More recently, attention has been focused on applying this methodology for the processing of medium and high carbon steels [1-2]. Most engineering components made of microalloyed (MA) steels experience cyclic loading in service, and the cyclic response of many engineering materials is significantly different from their monotonic behaviour. Material behaviour under dynamic loading conditions is characterised by cyclic hardening or softening [3].

As fatigue testing requires considerable time and effort, attempts have been made to estimate fatigue properties from tensile data [4-7]. Park and Song [7] analysed data pertaining to 115 steels and have concluded that of all the proposed methods, the modified universal slopes (MUS) method of Muralitharan and Manson [6] leads to most reliable results. The modified universal slopes equation is given by

$$\Delta\varepsilon_t = 1.17 \left(\frac{\sigma_B}{E}\right)^{0.832} N_f^{-0.09} + 0.0266 \, \varepsilon_f^{0.155} \left(\frac{\sigma_B}{E}\right)^{-0.53} N_f^{-0.56} \tag{1}$$

where $\Delta\varepsilon_t$ is the total strain amplitude, σ_B is the ultimate tensile strength, N_f is the number of cycles to failure, ε_f is the fracture ductility (log (100/100-R.A.), R.A. is reduction in area (%)) and E is Young's modulus. In contrast, assuming that most of the low cycle fatigue (LCF) life is spent in fatigue crack propagation and using a continuum mechanics approach, Tomkins [8] showed that

$$N_f = \frac{\ln\left(\dfrac{a_0}{a_f}\right)}{\sec\left(\dfrac{\pi\sigma}{2\sigma_B}\right)\dfrac{\Delta\varepsilon_p}{2}} \tag{2}$$

a_0 and a_f are respectively the initial crack length (usually taken as 10μm [8]) and the final crack length (depends on fracture toughness K_{IC} for brittle materials [9] and for ductile materials is taken as 2/3 diameter or width of the specimen [8]) and $\Delta\varepsilon_p$ is the plastic strain range. Assuming $\sigma=K'(\Delta\varepsilon_p/2)^{n'}$ (K' is the cyclic strength coefficient and n' is the cyclic strain hardening exponent), the LCF life can be computed from the cyclic stress-strain (CSS) properties. Depending on the shape, size and orientation with respect to crack propagation direction, non-metallic inclusions can have a strong influence on the fracture resistance [10].

In this paper, the room temperature CSS and LCF behaviour of a medium carbon MA forging steel (49MnVS3) and a low carbon MA steel (E-38) are reported. The experimentally determined fatigue life is compared with the life predicted by the modified universal slopes equation and Tomkins' model. The influence of non-metallic inclusions on the LCF life is also discussed.

EXPERIMENTAL

The medium carbon MA steel (49MnVS3) was received in the form of forged bars of 30mm diameter, and the low carbon MA steel (E-38) was received in the form of 7mm thick hot rolled plates. The chemical compositions of the two MA steels are given in Table 1.

Table 1 Chemical composition (wt. %) of the MA steels 49MnVS3 & E-38

Material	C	Si	Mn	P	S	V	Nb	N	Fe
49MnVS3	0.49	0.27	0.98	0.02	0.05	0.09	-	0.035	Balance
E-38	0.09	0.03	0.81	0.02	0.03	0.007	0.005	0.007	Balance

The LCF specimens were prepared according to ASTM Standard E 606-80 and the testing details are given elsewhere [11]. The fractured specimens were observed under a JEOL JSM 6300 scanning electron microscope operating at 25kV.

RESULTS AND DISCUSSION

Cyclic Stress Response

Figure 1 shows the variation of the peak tensile stress with the number of cycles for different total strain amplitudes for 49MnVS3 (Fig. 1a) and E-38 (Fig.1b). In 49MnVS3, cyclic softening at lower total strain amplitudes ($\leq 0.6\%$) and cyclic hardening at higher total strain amplitudes ($> 0.6\%$) were present (Fig. 1a). At total strain amplitudes of 0.5% and 0.6%, mild cyclic hardening followed initial softening (Fig. 1a). At all strain amplitudes a saturation in the cyclic stress response was observed after about 15 cycles (Fig. 1a). In E-38 steel cyclic softening was observed at all the strain amplitudes employed (Fig. 1b). While at a lower total strain amplitude of 0.35% softening all the way till fracture was seen (Fig. 1b), at higher strain amplitudes a more or less stable cyclic stress response followed the initial softening (Fig. 1b). It has been suggested that cyclic softening is the result of the formation and spreading of dislocation sources along the specimen gauge length during the first few cycles. This spreading continues until the dislocation sources have covered the entire gauge volume of the specimen [12-13]. This idea is in general agreement with the explanation for the yield point behaviour in terms of the spreading of Lüders bands along the specimen gauge length. Fig. 2 shows a stereo micrograph of Lüders bands on the surface of a E-38 steel specimen following cyclic straining at $\Delta\varepsilon_t/2 = 0.3\%$ for 10 cycles. Cyclic hardening at higher strain amplitudes seen in steel 49MnVS3 is associated with work hardening (resulting from dislocation-dislocation and dislocation-second phase interactions) being more dominant than the softening resulting from the spread of Luders bands. After the initial softening or hardening, 49MnVS3 and to an extent E-38 at higher total strain amplitudes exhibit a stable stress response. It is possible that at still higher total strain amplitudes in E-38 steel also cyclic hardening would be observed (as seen in steel 49MnVS3). But LCF tests on E-38 steel specimens at total strain amplitudes in excess of 0.6% went out of control very early. (6 unsuccessful tests were performed at a total strain amplitude of 0.8%.) Thus this feature, which would make the cyclic response of the two MA steels very similar except for details, remains to be established unequivocally.

Cyclic Stress-Strain (CSS) and Fatigue Properties

The cyclic-stress strain (CSS) plots for the MA steels 49MnVS3 and E-38 were found to obey the power law (Fig. 3). The stress and the plastic strain amplitude values of the stabilised hysteresis loops or at half-life (when the former was absent) were considered for obtaining the CSS plots. The cyclic strength coefficient K' and the cyclic strain hardening exponent n' for the two steels are presented in Table 2. The fatigue properties of both the steels - i.e., the constants in the Coffin - Manson and Basquin relationships, the fatigue ductility exponent c, fatigue ductility coefficient ε_f', fatigue strength exponent b and fatigue strength coefficient σ_f' are also given Table 2.

Table 2 Room temperature fatigue properties of 49MnVS3 & E-38

Material	σ_{cys} (MPa)	c*	ε_f'* (%)	b**	σ_f'** (MPa)	n'+	K'+ (MPa)
49MnVS3	509	- 0.64	60	-0.09	1152	0.14	1245
E-38	318	- 0.36	4.6	-0.08	645	0.23	1231

$^*\Delta\varepsilon_p/2 = \varepsilon_f' (2N_f)^c$ $^{**}\Delta\sigma/2 = \sigma_f' (2N_f)^b$ $^+\Delta\sigma/2 = K' (\Delta\varepsilon_p/2)^{n'}$

LCF Life Assessment

The predicted (Eqn. 1 - MUS relationship) and the experimental $\Delta\varepsilon_t/2$-N_f plots for the two MA steels are shown in Fig. 4. It could be seen that while the LCF life predicted by the MUS equation correlates well with the actual life in the case of 49MnVS3, it grossly overestimates the life in the case of steel E-38 by a factor of 6 at high strain amplitudes and by a factor of 3 at low strain amplitudes. This result highlights the limitations of this empirical approach and the fact that high tensile ductility does not always enhance the fatigue life. The role of inclusions present in rolled steels is very important in this connection. Under cyclic loading sharp crack-like discontinuities (delaminations) result due to decohesion at the inclusion-matrix interface. (Inclusions introduce localised strain concentration because of the mismatch with the mechanical, physical and chemical properties of the matrix [10]). It is also known that the orientation of the inclusions with respect to the crack propagation direction is important, especially under push-pull type loading [10]. Examination of fracture surfaces revealed extensive inclusion (MnS)- induced delaminations in MA steel E-38 (Fig. 5a). It is believed that the delaminations oriented along the crack propagation direction were responsible for the low cyclic ductility of this steel (see Table 2). In the case of 49MnVS3, where a very good correlation was obtained between the life predicted by the MUS equation and experiments, inclusions being more or less spheroidal/ellipsoidal in shape (Fig. 5b) played no detrimental role [11]. Inclusion-induced delaminations in E-38 steel were therefore thought to be responsible for the difference between the experimental and the predicted lives when the MUS equation was used. In contrast, the lives predicted by Tomkins' model from the CSS parameters K' and n' for 49MnVS3 and E-38 steels plotted in Fig. 6 show very good agreement with the experimental results. (Eqn. 2 was used, $a_0 = 10\mu m$, $a_f = 3mm$ for 49MnVS3, (high strength material; a_f determined from $K_{IC} \approx 50MPa\sqrt{m}$ [14]; $a_0 = 10\mu m$, $a_f = 9.66mm$ ($\approx 2/3$ specimen width) for E-38 .) In E-38 steel, the prediction was somewhat conservative at the lower total strain amplitudes (Fig. 6), although its statistical significance is yet to be established. The better agreement with the experimental results seen in the case of this model is due to the fatigue life being estimated from the CSS parameters which depend on the microstructural state under cyclic loading and, more importantly, are obtained from the experimental results on an empirical basis.

ACKNOWLEDGEMENTS

The authors thank the Volkswagen Stiftung for financial assistance and Drs. J.J. Irani and O.N. Mohanty of Tata Iron and Steel Company, Jamshedpur, India, for supplying the microalloyed steels.

REFERENCES

1.   Llewellyn, D.T. (1996), Ironmaking and Steelmaking **23**, 397
2.   Khalid, F.A. and Edmonds, D.V. (1993), Mater. Sci. Technol. **9**, 384
3.   Krabiel, A. and Reichel, V.J. (1993), steel res. **64**, 425
4.   Manson, S.S. (1965), Exp. Mech. **5**, 193
5.   Ong. J.H. (1993), Int. J. Fatigue **15**, 213
6.   Muralitharan, U. and Manson, S.S. (1987), J. Engng. Mater. Technol. **110**, 55
7.   Park, J.H. and Song, J.H. (1995), Int. J. Fatigue **17**, 365
8.   Tomkins, B. (1968), Phil. Mag. **18**, 1041
9.   Tomkins, B. (1971), Phil. Mag. **21**, 687

10.   Miller, K.J. (1993) Mater. Sci. Technol. **9**, 453
11.   Subramanya Sarma, V. and Padmanabhan, K.A. (1997) Int. J. Fatigue **17**, 135
12.   Roven, H.J. and Nes, E. (1991) Acta Metall. Mater. **39**, 1719
13.   Klesnil, M. and Lucas. P. (1968) JISI **210,** 746
14.   Farsetti, P. and Blarasin, A. (1988) Int. J. Fatigue **10**, 153

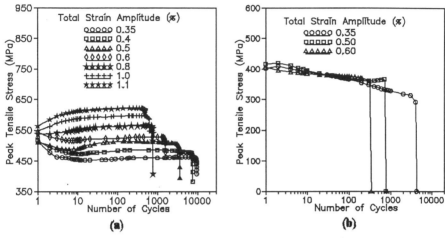

Fig. 1   Cyclic stress response of MA steels (a) 49MnVS3 and (b) E-38

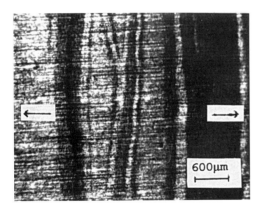

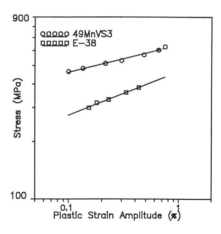

Fig. 2 Stereomicrograph of Lüders bands formed on the specimen surface (loading direction is arrowed). E-38 steel

Fig. 3 Cyclic stress - strain plots. (The ordinate represents the stabilised stress in case of 49MnVS3(same as stress at half-life in this case) and the stress at half-life in case of E-38.)

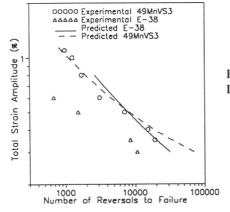

Fig. 4 Experimentally observed and predicted LCF lives based on the MUS equation

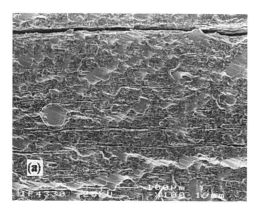

Fig. 5 Fractographs displaying (a) inclusion (MnS)-induced delaminations in E-38 steel, (b) an ellipsoidal MnS inclusion near the surface in 49MnVS3.

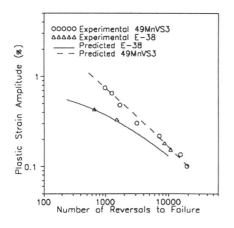

Fig. 6 Experimentally observed and predicted LCF lives based on Tomkins' model

Low Cycle Fatigue and Elasto-Plastic Behaviour of Materials
K-T. Rie and P.D. Portella (Editors)

PROPOSAL OF A HIGH-TEMPERATURE LOW CYCLE FATIGUE
LIFE PREDICTION MODEL

T. OGATA and A. NITTA

Central Research Institute of Electric Power Industry

2-11-1 Iwatokita, Komae, Tokyo, Japan

ABSTRACT

High-temperature low cycle fatigue tests with strain hold time including 100 hours per cycle were carried out on three kinds of ferritic steels and a Type 304 stainless steel, and a new life prediction model was proposed based on the experimental results and theoretical considerations. It was suggested that creep damage during stress relaxation should be divided into the matrix creep damage and the grain boundary creep damage from the comparison between the experimental results and the creep deformation mechanism map. Considering these creep damage mechanisms and interaction between fatigue and creep damage, a nonlinear damage accumulation model was proposed and simplified emprical life prediction equations have been derived. Low cycle fatigue lives of present tests and those in literatures were predicted with high accuracy by the new model.

KEYWORDS

High-Temperature, Creep-Fatigue Damage, Life Prediction, Ferritic Steel, Austenetic Steel.

INTRODUCTION

Most of high-temperature components in power plants such as steam turbines and boilers are subjected to both fatigue damage caused by thermal stress during start-stop cycles and creep damage during steady state operation. Therefore establishment of a creep-fatigue life prediction method is important in design and/or remaining life assessment for safety operation of the components . Presently the linear damage rule (LDR) is widely used to calculate the creep-fatigue damage because of its simplicity. However, life prediction results for creep-fatigue tests by the LDR were unconservative for a Type 304 stainless steel[1] and too conservative for a Mod.9Cr-1Mo steel[2]. Therefore some modifications should be made, when using the LDR, depending on materials. A ductility exhaustion theory (DET) has been applied to the creep-fatigue life prediction for Cr-Mo steels by Miller et. al.[3]. However accuracy of the prediction by the DET depends on the applied strain range relating to the definition of ductility[2] which is still uncertain for the creep-fatigue. In this study, a simple creep-fatigue damage evaluation model is proposed based on theoretical and phenomenological consideration of the long-term creep-fatigue test results on three kinds of ferritic steels and a Type 304 stainless steel, and the applicability of the model is demonstrated by predicting litrerature data .

EXPERIMENTAL PROCEDURE

The materials used in this study are a Cr-Mo-V forging steel (CMV), a modified 12Cr forging steel (12CR), a 2.25Cr-1Mo steel (STPA) and Type 304 stainless steel(304SS). Chemical compositions of these four materials are listed in elsewhere[1,4]. CMV has a temper bainetic structure , 12CR has a temper martensitic structure and STPA has a ferrite-parlite two phase structure. Test specimens were solid bar specimens of 8mm and 10mm in diameter with 10mm gage length, and they were finished with up to no.1000 grade emery paper along the axial direction. Creep-fatigue tests were performed by electro-mechanical driven type high temperature fatigue testing machines, with an induction heating device on three ferritic steels and with an electric furnace on 304SS. Temperature distribution between the gage length was kept within +3°C.

The tests on three ferritic steels were carried out under trapezoidal strain waveform at 600°C with tensile strain hold period of 10 and 60minutes per cycle. One test on CMV had 6 hours hold time. The tests on 304SS were conducted under the same strain waveform with between 1minute and 100 hours tensile strain hold at 500, 550 and 600°C. Tension and compression going strain rate was 0.1%/s. The applied strain range was between 1.0 and 0.5% for all materials.

RESULTS AND DISCUSSION

In order to examine influence of hold time on fatigue life, life reduction ratio , which is defined as the ratio of fatigue life in the hold time tests, N_f to fatigue life in no hold tests, N_{f0}, is plotted against hold time in Fig.1. Although there is some scatter depending on the materials, fatigue life reduced even in 10 min or shorter tensile hold and a considerable life reduction occurred in 1hour or longer tensile hold, in which the life reduction increased with increasing hold time. Transgranular failure occurred and striations were observed on the failure surface in 10 min and shorter tensile hold, while intergranular failure occurred in 1 hour or longer hold where creep cavities and micro cracks on grain boundaries were also observed in the section along the specimen axis of all materials by a scanning electron microscope.

The hold time effect on fatigue life in the creep-fatigue loading may be explained by considering the relation between creep damage during stress relaxation and the failure appearances. Creep deformation and damage mechanism during the relaxation varies with time because of change in

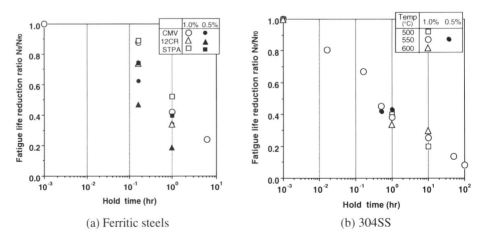

(a) Ferritic steels   (b) 304SS

Fig.1   Influence of strain hold time on low cycle fatigue life at high-temperature

creep strain rate with time. Creep deformation mechanism changes from dislocation motion such as glide and climb in the matrix to diffusion through the matrix and the grain boundaries[4]. In 10 min tension hold, dislocation motion is dominant mechanism during the whole relaxation period[4]. Crack initiation mechanism under the low cycle fatigue without hold time is strain localization at the grain boundary mainly due to irreversible dislocation glide motion. Since dislocation motion resulted in plastic flow could be enhanced by dislocation glide and climb during the tension hold period, crack initiation in the tension hold becomes shorter than that in no hold. A fatigue crack in no hold propagates the matrix by crack sharpening and blunting mechanisms building up striations. The crack propagation rate in 10 min tension hold is also accelerated, because dislocation motion at a crack tip during the hold period makes the crack opening displacement larger than that in no hold. Therefore the fatigue life in 10 min tension hold reduced comparing to that in no hold, in spite of showing transgranular failure. Such creep damage during relaxation resulting from dislocation motion in the matrix is referred as a matrix creep damage.

On the other hand, diffusional flow becomes dominant under 1 hour or longer tensile hold after a certain period of relaxation and creep cavities can initiate and grow on grain boundaries. So that a grain boundary creep damage accumulates during relaxation in addition to the matrix damage due to dislocation motion. In this condition, a crack propagates preferentially at damaged grain boundaries where ductility reduced due to cavity nucleation and growth, and its propagation rate might be accelerated. As a result, intergranular failure caused a greater reduction of fatigue life than transgranular failure observed in 10 min tension hold. Thus the effect of hold time on fatigue life can be explained by considering the creep damage mechanism during relaxation.

CREEP-FATIGUE LIFE PREDICTION MODEL

The important points to be considered for discussing creep-fatigue life prediction model are how creep damage during strain hold should be evaluated and how to formulate interaction between creep and fatigue damages. At first, creep damage during strain hold is considered as follows based on the theoritical background and experimental evidences.

Deformation mechanisms under static creep loading for irons and alloys are classified into a) dislocation glide creep, b) power law creep, c) diffusion creep in relation to stress and temperature[5]. A creep deformation mechanism map for 304SS with a grain size of 50 μm is established baed on the assumption that the above mechanisms occur independently of one another and the fastest mechanism is rate controlling under any selected conditions. The creep deformation mechanism maps represented by normalized stress vs temperature is shown in Fig.2. The material parameters used in drawing the map were referred in [5] and our laboratory creep data. The creep deformation mechanism map for CMV has already been established in the previous study[4]. It should be noted that the creep deformation mechanism map does not indicate a strict division between mechanisms because the deformation mechanism does not change drastically and some mechanisms must be superimposed, but the dominant mechanism for certain condition might be identified. In creep tests at stress levels within the region of dislocation glide or climb, creep deformation and rupture result from matrix deformation showing large elongation and reduction of area. In these deformation mechanisms, creep damage is considered as the matrix creep damage defined above. On the other hand, creep rupture elongation and reduction of area are generally small under low stress levels at higher temperatue where cavities nucleate and grow on grain boundaries. It is considered that creep damage is mainly accumulated on grain boundaries in the region of diffusional flow. Thus the creep deformation mechanism during relaxation process can be illustrated schematically as in Fig.3. Since the creep deformation mechanism varies with hold time, creep damage during relaxation in long hold time should be divided into the matrix creep

706

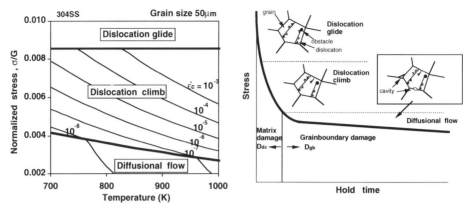

Fig.2 Creep deformation mechanism map for 304SS

Fig.3 Shcematic representation of creep deformation mechanism during relaxation

damage related to dislocation motion and the grain boundary creep damage related to creep cavities on grain boundaries.

Then interaction of creep and fatigue damage is considered as follows. Pure low cycle fatigue damage process is occupied by growth of small cracks whose rate are proportional to crack length. Crack growth rate must be accelerated by propagating through the grain boundary creep damaged field. In order to express such interaction between the crack growth and intergranular damage, not only the creep damage during strain hold at certain cycle but also accumulation of the creep damage until the crack tip approaches to the creep damaged field should be taken into account. In other words, acceleration of the crack growth rate depends on both accumulation of creep damage in the field at the crack tip and creep damage contributing to crack extension during strain hold. Therefore creep-fatigue damage increases nonlinealy under the trapezoidal strain waveform test. Concept of the creep-fatgiue damage evaluation model can be drawn as Fig.4.

Based on above discussion, a nonlinear damage accumulation model (NLDA) is defiend as a damage rate formula by next equations.

$$\left\{ \begin{array}{l} \dfrac{dD_f}{dN} = A^{-1}\{1 + (\alpha_{dc}D_{dc} + \alpha_{gb}D_{gb})\}\Delta\varepsilon_{in}^{-m} \qquad (1) \\[2mm] \dfrac{dD_{dc}}{dt} = \dfrac{\dot{\varepsilon}_{dc}}{\varepsilon_f} \ , \qquad \dfrac{dD_{gb}}{dt} = \dfrac{\dot{\varepsilon}_{gb}}{\varepsilon_r} \qquad (2) \end{array} \right.$$

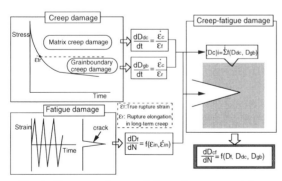

Fig.4 Concept of creep-fatigue damage model

Table 1 Material constans used for life prediction

(a) Ferritic steels

Material	$\dot{\varepsilon}_{tr}$(1/sec)	A	m	α_{dc}	α_{gb}	ε_f	ε_r
CMV	1×10^7	2.40	-1.15	0.38	2.0	1.20	0.03
12CR	1×10^7	3.42	-1.20	0.38	2.0	1.60	0.03
STPA	1×10^7	4.55	-1.20	0.38	2.0	1.60	0.10

(b) 304SS

Temp(℃)	$\dot{\varepsilon}_{tr}$(1/sec)	A	m	α_{dc}	α_{gb}	ε_f	ε_r
500	2×10^8	0.608	-1.7	0.38	6.0	0.22	0.06
550	3.5×10^8	0.204	-1.84	0.38	4.0	0.5	0.1
600	5×10^8	0.127	-1.92	0.38	2.0	0.92	0.1

where D_{cf} is creep-fatigue damage, D_{dc} and D_{gb} are the matrix and grain boundary creep damage, ε_{dc} and ε_{gb} are creep strain rate for D_{dc} and D_{gb}, ε_f and ε_r are creep rupture ductilities corresponding to D_{dc} and D_{gb}, α_{dc} and α_{gb} are matrix and grain boundary damage coefficients determined by creep-fatigue tests, $\Delta\varepsilon_{in}$ is inelastic strain range, A and m are material constants in Manson-Coffin rule. D_{dc} is defined as the ratio of accumulation of ε_{dc} to true rupture strain, ε_f in short-term creep rupture tests where rupture is caused by matrix deformation. D_{gb} is defined as the ratio of accumulation of ε_{gb} to creep rupture elongation, ε_r in long-term creep rupture tests where creep damage progresses by cavity nucleation and growth on grain boundaries. To obtain ε_r, however , long-term creep tests have to be performed. In this study, ε_r for up to 1 hour hold time tests was determined as equal to a constant C of the Monkman-Grant relation (ε_{st} x t_r =C : ε_{st}; steady state creep strain rate, t_r; rupture time) and designated as ε_{r1}. For longer hold time tests, hold time effect on ε_r was considered as $\varepsilon_{r1}(1+\log t_h)^{-1.2}$ where t_h is hold time in hour.

A life prediction equation is obtained by integrating eq.(1) and eq.(2) as follows:

$$N_f = \left[-1 + \left\{ 1 + 2\left(\alpha_{dc}D_{dc} + \alpha_{gb}D_{gb}\right)A\Delta\varepsilon_{in}^{\ m}\right\}^{\frac{1}{2}} \right] \big/ \left(\alpha_{dc}D_{dc} + \alpha_{gb}D_{gb}\right) \qquad (3)$$

The transition strain rates ε_{tr} between D_{dc} and D_{gb} were determained from the creep deformation maps. In spite of the difference of microstructure and mechanical properties between materials, it was found that α_{dc} of 0.38 and α_{gb} of 2.0 were applicable to all materials at 600°C. This means that the present model can be widely applied to both ferritic and austenitic steels. Life prediction results based on mid-life data for all materials are shown in Fig.5 comparing to the actual life. Dotted lines show a factor of 2 on life. Material constants using life prediction by eq.(3) are summarized in Table 1. It is seen that the creep-fatigue life of all materials can be predicted successfully by the eq.(3).

Inelastic strain range, the matrix and grain boundary creep strain in the eq(2) can be obtained from elastic-plastic and creep analysis. But it might be convinient and widely applicable if the creep-fatigue life can be predicted by the NLDA without complicated analysis. In this study, emprical eqations are derived based on the long-term creep-fatigue data for 304SS to estimate ε_{dc} and ε_{gb} during hold period. The relation between amount of ε_{gb} and hold time including our laboratory data is shown in Fig.6. The similar relation was obtained between the ratio of $\Delta\varepsilon_{gb}$ to total creep strain and hold time. Based on these relations, following eqations can be derived.

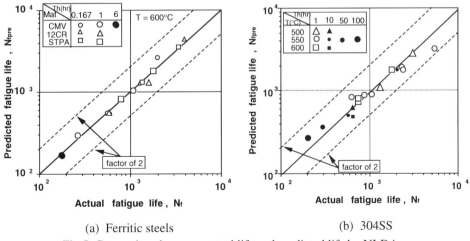

(a) Ferritic steels

(b) 304SS

Fig.5 Comparison between actual life and predicted life by NLDA

708

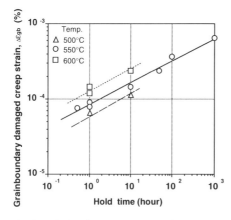

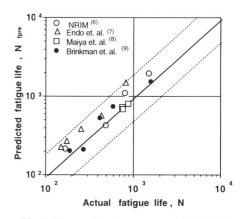

Fig.6 Relation between Δεgb and hold time

Fig.7 Comparison between actal life in the literature and predicted life

$$\Delta\varepsilon_{gb} = 1.876 \times 10^{-6} \Delta\varepsilon \ t_h^{0.3} \exp(6.93 \times 10^{-3} \ T) \tag{4}$$

$$\Delta\varepsilon_{gb}/\Delta\varepsilon_c = \left\{ \Delta\varepsilon - \left(42.88 - 0.15T + 1.256 \times 10^{-4} T^2 \right) \right\}/9.73 + 0.12 \log t_h \tag{5}$$

where Δε is total strain range in % and T is temperature in °C. By using these equations, D_{dc} and D_{gb} can be estimated from Δε at certain temperature and hold time. Then creep-fatigue life can be predicted based on fatigue test data by eq.(3). Life prediction of creep-fatigue tests conducted by other researchers on 304 stainless steel[6-9] was made by the NLDA using above equations. Prediction results is shown in Fig.7. All fatigue lives could be predicted within the factor of 2.

CONCLUSION

A new creep-fatigue life prediction model was proposed based on theoretical consideration and experimental evidences obtained from long-term creep-fatigue tests on three kind of ferritic steels and a Type 304 stainless steel. Simplified emprical life prediction equations were also derived. Applicability of the model for ferritic and austenetic steels was demonstrated by predicting fatigue lives of present tests and those in literatures with high accuracy.

REFERENCES

1. Takahashi, Y., Ogata, T. and Take, K.(1991) Proceedings of the 8th Post SMiRT Seminar, Japan, pp.71-89
2. Ogata, T.(1997) J. Soc. Mater. Sci.Japan, **46**, pp.25-31.
3. Miller, D., Hamm, C. D., and Phillips, J. L.(1982) Matet. Sci. Engng., **53**, pp.233-244.
4. Ogata, T. and Nitta, A.(1994), PVP-**276**, *Determining Material Characterization,* J. C. Spanner, Jr(Ed), Book No. G00844-1994, pp.97-105.
5. Frost, H. J. and Ashby M. F.(1977) *Fundamental Aspects of Structural Alloy Design,* R. I. Jaffe and B. A. Wilcox(Eds). Pelnume, New York, pp.27-65.
6. NRIM FATIGUE DATA SHEET(19785) No.49.
7 Endo, T., Nishida, T., Sakon, T. and Tomioka, M., (1981)MISTUBISIII JUKO GIIIO,**18**,603.
8. Mayer, P. S. (1981) ASME J. Mater. Sci. Engng. , **47**, pp.13-20.
9. Brinkman, C. R. and Korth, G. E.(1973) ASME J. Nucl. Mater. **48**, pp.293-306.

Low Cycle Fatigue and Elasto-Plastic Behaviour of Materials
K-T. Rie and P.D. Portella (Editors)
© 1998 Elsevier Science Ltd. All rights reserved.

FATIGUE PROPAGATION OF SURFACE CRACKS UNDER CYCLIC LOADING

T. BOUKHAROUBA[1], J. GILGERT[2] K., AZOUAOUI[1] and G. PLUVINAGE[3]
*(1) Laboratoire de Mécanique et des Matériaux de l'IGM-USTHB, B.P. 32 El-Alia, 16111
Alger-Algérie*
(2) École Nationale des Ingénieurs de Metz (ENIM), Ile du Saulcy, 57045 Metz-France
(3) Laboratoire de Fiabilité Mécanique Université de Metz, Ile du Saulcy, 57045 Metz-France

ABSTRACT

The use of the local or semi-local stress intensity factor as a fatigue crack propagation criteria
has been experimentally and theoretically analysed. Crack propagation tests on three
dimensions plates and two dimensions normalised specimens have been carried out. The local
stress intensity factor has been determined by a finite element simulation. The analysis of the
m_i and C_i crack propagation parameters shows that these parameters are not constant along the
semi-elliptical crack propagation front, that is, $m_A \neq m_C$ and $C_A \neq C_C$ for any used criteria.

KEYWORDS

Elliptical crack, Shape prediction, Crack propagation parameters, Propagation Criteria.

INTRODUCTION

The problem of the fatigue crack propagation of an embedded semi-elliptical crack in a finite
plate subjected to a cyclic bending is of major importance for the prediction of life duration
and its conditions of failure. The evolution of this crack type is often considered as governed
by approaches based on the hypothesis that the propagation is sensitive to the stress gradient
and then governed by the stress intensity factor. Two approaches are namely used : the local
and the semi- local one.
The local approach [1 to 10] assumes that the propagation is governed by the Paris law and the
local stress intensity factor. Obviously, this approach sets problems on surfaces where the
stress intensity factor is not properly defined.
The semi-local approach [9, 12, 13] consists in taking the average of local values of the stress
intensity factor on a crack front increment ΔS and in applying Paris's law using the semi-local
values $K_{I, A}$ in depth and $K_{I, C}$ on surface.

$$\frac{da}{dN} = C^{*}.\left(\Delta K_{I}\right)^{m^{*}} \tag{1}$$

Taking into account the formula (1) and the two previous approaches for the semi-elliptical form defects, one can write :

at the deepest point   on surface

$$\frac{da}{dN} = C_{1,A}.\left(\Delta K_{loc,A}\right)^{m\,1,A} \qquad\qquad \frac{dc}{dN} = C_{1,C}.\left(\Delta K_{loc,C}\right)^{m\,1,C}$$

$$\tag{2}$$

$$\frac{da}{dN} = C_{2,A}.\left(\overline{\Delta K_{A}}\right)^{m\,2,A} \qquad\qquad \frac{dc}{dN} = C_{2,C}.\left(\overline{\Delta K_{C}}\right)^{m\,2,C}$$

The indices 1 and 2 correspond to the approach type, the A and C letters correspond to the deepest point and on surface respectively, of a semi-elliptical defect.

The set question is to know whether the crack propagation coefficients defined at (2) are the same, i.e., $C_{iA} = C_{iC}$ and $m_{iA} = m_{iC}$, and in the inverse case what is the dispersion percentage. The aim of this work is to examine the crack propagation law in three dimensions which fits better to this hypothesis. To carry out this study, the shape evolution of the semi- elliptical defect has been examined during the crack propagation on 35NCDV12 forged steel plates under three points cyclic bending loads.

CRACK PROPAGATION TESTS

Crack propagation tests have been conducted on three dimension plates and the stress intensity factor has been calculated by finite elements method (10, 11). The calculation has been made based on real measurements of the semi-elliptical crack propagation front for different ratios a/c and a/t.
The used material during these tests was a 35NCDV12 forged steel. The crack has initiated from a semi-elliptical central defect, machined by electro-erosion. The notch tip radius is equal to 0.2mm.

The tests have been conducted under a three point cyclic bending loading of sinusoidal form and in environnement temperature. Twelve tests have been carried out; four for each stress ratio R. Only the results corresponding to the stress ratios R = 0.1 and R = 0.3 were analysed. The used techniques to follow the progression of the fatigue crack are :

    1)- an instrumentation by strain gauges and crack gauges allows the detection of initiating fatigue crack and the following of its progress,
    2)- a method of mechanical marking. Based on failure facies it allows at once :

    - the fatigue crack shape evolution analysis from the initial defect (fig.1)
    - direct measurement of the crack propagation velocity in the thick direction and on surface.

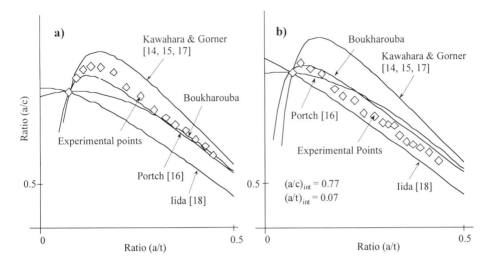

Fig. 1. Crack aspect evolution during crack propagation.

RESULT ANALYSIS

At first, the hypothesis that the crack propagation at the deepest point and on surface is governed by the local value of the stress intensity factor is made. Figures (2) show the local crack propagation velocity curves for a semi-elliptical defect at the deepest point da/dN = f($\Delta K_{A,loc}$) and on surface dc/dN = f($\Delta K_{C,loc}$). The continuous line represents the crack propagation curve determined on a normalised specimen CT, taken as a reference curve.

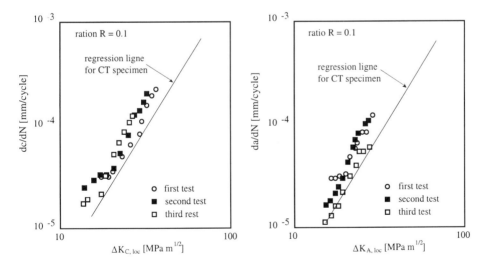

Fig. 2. The local low crack growth rate at the deepest and at surface point.

712

In this case, the stress intensity factor along the crack front is considered to take the average of its local value. The propagation to a crack front point is affected by the K value at the adjacent point. The semi-local value at surface : $\overline{K}_C$, and at the deepest point : $\overline{K}_A$, can be calculated by the equation system given by Cruse and Besuner [19]. Figures (3) give the crack propagation velocity curves using the semi-local stress intensity factor on surface and at the deepest point. The dashed lines and the continuous line have the same signification as previously.

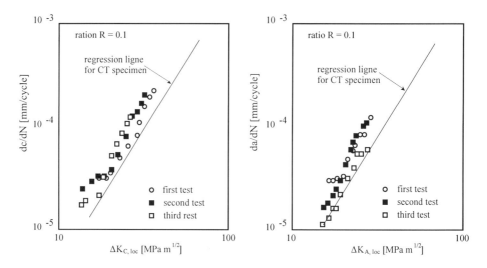

Fig. 3. Average low crack growth rate at the deepest and at surface point.

DISCUSSION

Some rare comparison works of the crack propagation laws of the semi-elliptical cracks have been accomplished, like the ones of Mahmoud and Hosseini [20] and those of Varfolomeyev and al [19]. To our knowledge, there are still different opinions about the sort of the criteria that has to be used (local and semi-local) and about the values of the coefficients C and m that are to be introduced into each law. These coefficients are not forcely the same to the ones belonging to the propagation law taken as reference. A first statistical study on the literature data related to the semi-elliptical crack propagation has been carried out by Mahmoud and Hosseini [20]. These authors conclude that the use of semi-local stress intensity factor amplitude ΔK does not improve the prediction of the a/c ratio. The best solution is the use of local stress intensity factor amplitude for : $m_A = m_C = m^* = 3$ and $C_C = C_A = C^*$. A second study has been carried out by Varfolomeyev and al [19] using five elliptical crack propagation tests. These tests have been conducted on plates under four point bending loading, with a constant load ratio equal to 0.32, in an environment temperature. The plates were made of a 15kh2MPA steel (Re=584 MPa, Rm=700 MPa and A=21%) containing a surface defect of a semi-elliptical shape, the dimensions of which were 220x116x30 mm. The authers conclude that the local stress intensity factor amplitudes $\Delta K_{I, loc}$ do not give good results for $m_A = m_C = m^*$ and $C_C = C_A = C^*$. The best suitable crack propagation law has been investigated based on a statistical test.

A comparison has been made between all the results based on the Student test, that is, to evaluate the difference between C_{Ai}, C_{Ci} and C^* (and between m_{Ai}, m_{Ci} and m^*). The results were represented in terms of Pr% confidence level. The chosen solution was the combination which corresponds to the higher Pr% confidence level according to C_i and m_i. Three values for each series and for each load ratio have been tested. The method brings in two approaches, the first one is based on a finite element analysis of the stress intensity factor at the deepest point and on surface, the second approach is based on a semi-local calculation at the same points [10,11]. The used crack propagation parameters for this test were those determined from the two approaches mentioned above. The test results are presented in the figures (4 and 5). It can be concluded that the semi-local value of the stress intensity factor amplitude does not give good results for m and C. It is clear that the two best results belong to the criteria which use the local stress intensity factor amplitude for a load ratio R = 0.3 in depth and at surface.

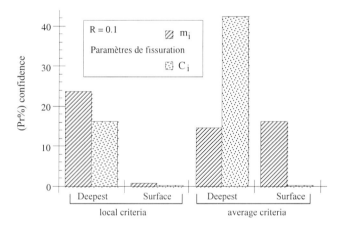

Fig. 4. Student test results, for stress ratio R = 0.1.

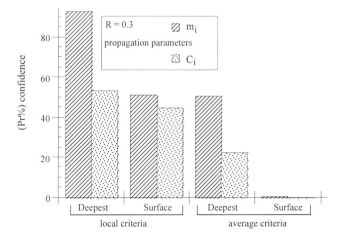

Fig. 5. Student test results, for stress ratio R = 0.3.

714

CONCLUSION

The m and C parameter values of the crack propagation law according to the used criteria type have been examined : m_1 and C_1 for the local criteria, m_2 and C_2 for the semi-local criteria. These coefficients are related to the solution of the semi- elliptical crack at the deepest point (point A) and on surface (point C), for two load ratios R = 0.1 and R = 0.3. The results were compared to m^* and C^* reference values. Figures 4 and 5 show that :

1)- the use of the local stress intensity factor gives the best predictions, but the C and m crack propagation parameters are not constant along the semi-elliptical crack propagation front, i.e., $m_A \neq m_C \neq m^*$ and $C_C \neq C_A \neq C^*$,

2)- the simplifying hypothesis of Mahmoud and Hosseini [20] and Varfolomeyev and al [9] cannot be preserved,

3)- the results obtained with the load ratio R = 0.3 present less dispersion than those at R = 0.1.

These results are a contribution to the study of the semi-elliptical form fatigue crack propagation problems, which is actually an open question.

REFERENCES

1.   Newman, J.C. and I.S. Raju, I.S. (1981) *Eng.Fract.Mech.* **15**, pp. 185-192.
2.   A. Carpinteri, A. (1993) *J. of Strain Analysis* **28**, pp. 117-123.
3.   J.C. Newman, J.C. (1973). In : Sym. *on Fatigue and Fract.,* George Wash. University.
4.   Shah, R.C. and Kobayashi A.S. (1973) *Int.J. of Fracture* **9**, pp. 133-146.
5.   Smith, F.W. and Sorensen D.R. (1976) *Int.J. of Fracture* **12**, pp. 47-57.
6.   Shah, R.C. and Kobayashi, A.S. (1971) *Eng. Fract. Mech.* **3**, pp.71-96.
7.   Heliot, J. (1976). Creusot-Loire, Départ. Nuclaire, DT/D-76. 834, Framatome, France.
8.   Dufresne, J. (1984). In : *Advance in Fracture Research,* **2**, ICF5-Conf. pp. 517-531.
9.   Varfolomeyev, V., Vainshtok, V.A. and Krasowsky A. (1991) *Eng.Fract.Mech.* **40**, pp. 1007-1022.
10. Boukharouba, T., Chehimi, C., Gilgert, J. and Pluvinage, G. (1994). In : *Handbook of Crack Prop. in Met. Struct.,* A. Carpinteri (Ed.). Elsevier Science **1**, pp. 707-732.
11. Boukharouba, T., Gilgert, J. and Pluvinage, G. (1997). In : *Reliability Assessment of Cyclically Loaded Eng. Struct.*, R.A. Smith (Ed.), **39**, pp. 343-375.
12. M.A. Mahmoud, M.A. (1988) *Eng.Fract.Mech.* **31**, pp. 357-369.
13. Mattheck, C., Morawietz, P. and Munz, D. (1983) *Int. J. of Fracture* **23**, pp. 201-212.
14 . Kawahara, M. and Kurihara, M. (1975). In : *J.S.N.A.* Japan, **137**, pp. 86-92.
15. M. Kawahara, M. and Kurihara, M. (1977). In : *Proceedings of the 4th Int. Conf. Fracture*, D. Taplin (Ed). **2**, Waterloo, Canada, pp. 1361-1373.
16. Portch, D.J. (1979). In : Report RD/B/N-4645, Central Electricity Generating Board, Great Britain.
17. Gorner, F., Mattheck, C. and Munz, D. (1983). *Z. Werkstofftech* **14**, pp. 11-18.
18. Iida, K. (1980). In : IIW Doc. XIII-967-80.
19. Cruse, T.Λ. and Besuner, P.M. (1975) *J. of Aircraft* **12**, pp. 369-375.
20. Mahmoud, M.A. and Hosseini, A. (1986) *Eng.Fract.Mech.* **24**, pp. 207-221.

Low Cycle Fatigue and Elasto-Plastic Behaviour of Materials
K-T. Rie and P.D. Portella (Editors)
715

METHODS OF SERVICE DURABILITY ASSESSMENT

C.M. SONSINO
Fraunhofer-Institut für Betriebsfestigkeit (LBF)
Bartningstr. 47, D-64289 Darmstadt / Germany

ABSTRACT

Service durability of a product can be achieved only by careful design taking into account the different aspects of the required mission profile with regard to component strength. In order to include all effects on component strength (like material, manufacturing, local geometry, scattering) on one hand, and to meet competition requirements on the other hand, the design engineer has to apply efficient methods for guarantying service durability. A brief overview over these methods are presented.

Today's developments are focusing on the reduction of time-to-market periods. This fact is putting more emphasis on numerical methods (simulations of dynamic system behaviour, of component properties and local stresses) in order to reduce experimental procedures of optimizing and proof-out.

KEYWORDS

Service durability, component strength, fatigue life, experimental and numerical methods.

INTRODUCTION

The assessment of service durability of components and structures becomes more and more important in different technical areas like vehicle (automotive, railway), aeronautics, transportation, energy production, heavy machinery and maritime technology with regard to the increasing trend towards light-weight constructions and greater demands concerning costs, reduction of time-to-market periods, product safety and liability. In order to fulfil these demands different but interacting methods have to be applied for an effective development and proof-out of the products. In this survey a short overview about the necessary methods for the assessment of service durability, for components expressed also as operating strength or in-service integrity, will be presented.

DEFINITION OF SERVICE DURABILTY

Service durability is nothing else than required product life. It is affected by following loading modes, see Fig. 1: special events/misuse (overloads, impact) and cyclic loads (constant and variable amplitudes and their spectrum) which may interact with each other, i.e. for most of driving and steering automotive components.

716

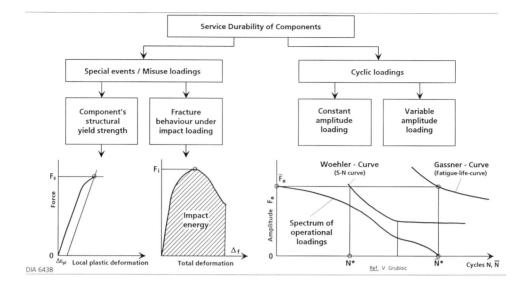

Fig. 1 Loading modes which determine structural durability

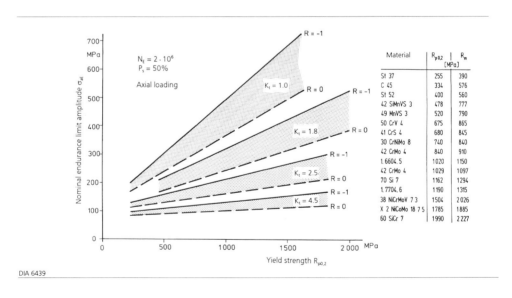

Fig. 2 Endurance limits in dependency of stress concentration K_t

For an aimed selection of needed material data and design concept the design engineer must be aware about the loading mode/s of the product in question, which determine/s the service durability. Thus, with regard to product liability requirements the design engineer has not only to be aware about the expected loading mode of the component but also if it is a safety part which must never fail or a functional part which failure does not affect safety of persons or environment [1, 2].

EFFECTS ON COMPONENT STRENGTH, FATIGUE LIFE AND DIMENSIONS

For performing a required function a component has to reveal a certain strength which depends on material, manufacturing and geometry [3]. While manufacturing (forging, rolling, casting, surface and heat treatments) generally determines the strength level and scattering of mechanical properties the geometry can suppress the influence of material: In case of the casting technology the shaping of components with ribs can increase their impact strength compared to forged components where the technology limits such design possibilities despite superior material properties [4]. Fig. 2 shows that in case of high stress concentrations ($K_t > 2.5$), the fatigue strength is no longer increased by the selection of a high-strength material; this applies not only to steel and cast iron, but also to aluminium [3, 5]. In order to utilize the advantages of a higher strength material, the stress concentration must be reduced by improving the design. Geometry is the major parameter which determines the component strength.

Furthermore, the product life is also determined by the interaction between the strength of the component and its mission profile, i.e. load spectrum. The importance of load spectrum, i.e. variable amplitude loading, on durability and light-weight design was recognized by Ernst Gassner, who formulated 1939 for the first time a procedure for simulating variable amplitude loading, the historical blocked programme sequence with a Gaussian like distribution of loads. [6]. This sequence was frequently used up into the 70ies as a standard until blocked programme tests were substituted by random load sequences applied with modern servo-hydraulical actuators. In the mean time different standardized load spectra for different application areas were developed, mainly for material testing and comparisons [7]. Fig. 3 shows the importance about the knowledge of the load spectrum with regard to fatigue life and structural dimensions. If, for example, the fatigue life for a steering rod is taken to be around 10^8 cycles, then depending on the shape of the spectrum and the maximum permissible stress amplitude, a value that is around 30 to 100 % higher than the corresponding one for the endurance limit is allowable [8]. So, a reduction of cross-sectional sizes and weight can be realized [4]. However, it has to be also assured that an impact load does not lead to a catastrophic failure, even if a crack should be initiated [2].

METHODS FOR SERVICE DURABILITY PROOF-OUT

Generally, the proof-out procedure is carried out by experimental and/or numerical methods. The experimental methods comprise material testing (stress or strain controlled), strain analysis and component testing (uni- or multiaxial), numerical methods the local stress strain analysis by finite or boundry elements, the simulation of dynamic system behaviour with regard to occuring stresses, the simulation of component strength properties and fatigue life calculations.

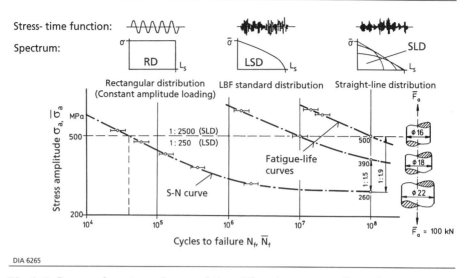

Fig. 3 Influence of spectrum-shape on fatigue life and component dimensions

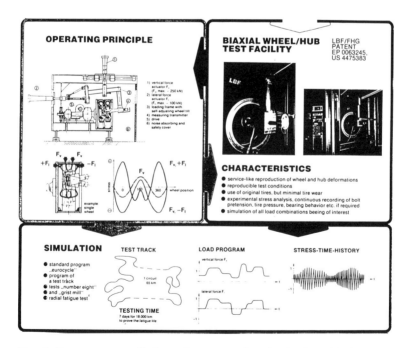

Fig. 4 Computer controlled proof-out tests of truck wheels and hubs

In past years the service durability of structural components was mainly proofed-out by proving ground tests. However, for a systematical product development and cost savings the simulation of service conditions and proof-out was then carried out experimentally in laboratory [9], accompanied by optimization work on particular components. Fig. 4 shows a computer controlled biaxial wheel/hub/bearing test facility for the experimental proof-out of these components. Nowadays, beside the integration of environmental conditions, a major requirement for the performance of laboratory simulation is also the reduction of testing time, e.g. by omission of low, not damaging loads and increased intensity of higher loads, but without exceedence of maximum service loads [2].

However, on one hand the trend for acceleration of product developments and on the other hand the trend for cost savings is focusing more and more on numerical methods, i.e. simulation of dynamic system behaviour, Fig. 5, of component properties and local stresses, in order to reduce experimental procedures of optimizing and proof-out. But presently numerical methods, especially those for fatigue life estimation [10] have not yet reached a satisfactory and reliable standard. Still, the interaction between local stresses /strains, microstructure and crack intraction is not well understood. Therefore, in this area more efforts are necessary.

CONCLUSIONS

While in the past the numerical predimensioning was followed by experimental optimization of particular components and experimental proof-out of a system consisting of different components, the present industrial trend interacts these phases of product development with each other by simultaneous engineering, Fig. 6, in order to reduce time. This procedure can deliver a reliable design only if the numerical assessment considers service experiences and is accompanied by experimental verification. However, more progress with regard to the accuracy of numerical methods is necessary in the future.

REFERENCES

1. Grubisic, V.; Fischer, G. (1997). SAE Int. Congress, Detroit, SAE-Paper No. 970094.
2. Grubisic, V. (1994). Int. Journal of Vehicle Design 15, 1/2, pp. 8-26.
3. Buxbaum, O. (1992). Betriebsfestigkeit, Verlag Stahleisen, Düsseldorf, 2. Edition.
4. Grubisic, V.; Fischer, G.; Sonsino, C.M. (1992). Materialprüfung 34, 3/4, pp. 53-57 and 91-93.
5. Haibach, E. (1989). Betriebsfestigkeit, VDI-Verlag, Düsseldorf.
6. Gassner, E. (1939). Luftwissen 6/2, pp. 61-64.
7. Heuler, P.; Schütz, W. (1998). In: Proc. of the 4th Int. Conf. on Low-Cycle Fatigue and Elasto-plastic Behaviour of Materials, Garmisch-Partenkirchen /Germany.
8. Sonsino, C.M. (1998). In: Proc. of the 6th Conf. on Fracture, Porto/Portugal, pp. 1-44.
9. Naundorf, H.; Wimmer, A. (1998). In: Proc. of the 4th Int. Conf. on Low-Cycle Fatigue and Elasto-plastic Behaviour of Materials, Garmisch-Partenkirchen /Germany.
10. Kotte, K. L.; Zenner, H. (1998). In: Proc. of the 4th Int. Conf. on Low-Cycle Fatigue and Elasto-plastic Behaviour of Materials, Garmisch-Partenkirchen /Germany.

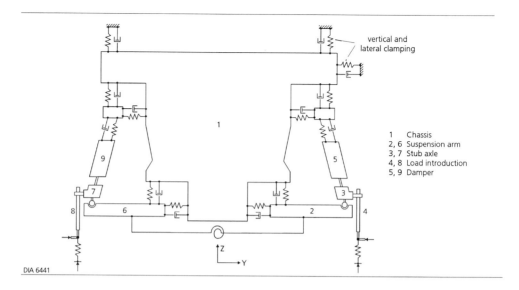

Fig. 5 2D-model for calculating the dynamic system behaviour of a clamped vehicle

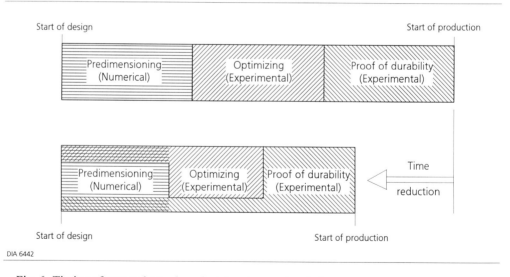

Fig. 6 Timing of past and actual product development periods

Low Cycle Fatigue and Elasto-Plastic Behaviour of Materials
K-T. Rie and P.D. Portella (Editors)

LIFETIME PREDICTION - COMPARISON BETWEEN CALCULATION AND EXPERIMENT ON A LARGE DATA BASE

K. L. KOTTE* and H. ZENNER**
* Institut für Festkörpermechanik, Technische Universität Dresden,
01307 Dresden, Germany
** Institut für Maschinelle Anlagentechnik und Betriebsfestigkeit, Technische Universität
Clausthal, 38678 Clausthal-Zellerfeld

ABSTRACT

Based on a comprehensive and reliable set of test results the accuracy of lifetime prediction has been studied. On the average, the damage sum is situated on the unsafe side with a considerable scatter range.

KEYWORDS

Accuracy of lifetime prediction, comprehensive data base of test results, modification of Miners rule, nominal stress concept, local strain concept, statistical evaluation.

DATA COLLECTION - A NEW EVALUATION OF EXTENSIVE TEST DATA

In most cases, the cost-intensive and time-consuming operating strength tests are not described in sufficient detail in publications and are therefore often not very useful, even for specialists. Consequently, the results of operating strength tests have been collected over a period of almost six years, completely and consistently reevaluated on the basis of the individual tests, and recorded in a computer-aided data base ([1], [2]) - with support from AIF. Special emphasis was placed on the exact description of random loading. In part, the loading was reconstructed with the original test control programs. The Rainflow matrix for loading has thus been successfully reconstructed for all test results, and a turning point sequence has been reconstructed for numerous tests. More than 18 000 test results have been reprocessed in this manner. A general survey of the data base composition is presented in figures 1 to 4. This comprehensive and reliable data base has been employed for testing the usual methods of fatigue life analysis for their accuracy.

RESULTS FOR THE NOMINAL STRESS CONCEPT

All test results in question have been recalculated by the best-known methods (figures 5 and 6, [3], [4], [5]). In addition, calculations with a modified Palmgren-Miner rule [12], [6], [7] have been performed with a fatigue limit which decreases as a function of the damage history. Since all Rainflow matrices are available, the effect of the cycle-counting method has also been successfully separated for the first time from the method of lifetime prediction. If the ratio, $N_{experimental}/N_{calculated}$ is taken, this yields the damage sum which is set equal to unity in the

Miner rule. The statistical evaluation of the multitude of damage sums yields the average value $\overline{D}$ and the scatter range, $T_D = D_{10\%}/D_{90\%}$, as a measure of the scatter. Even in the case where the Miner rule agrees exactly, it has been demonstrated that this scatter is in the range $2 \leq T_D \leq 3.5$, even as a result of the scatter in constant and variable amplitude tests alone. The results of this evaluation are presented in table 1. They clearly indicate that the Miner rule with $\overline{D} \approx 0.3$ is situated on the unsafe side on the average, and that level crossing counting should no longer be applied. As an example of the numerous further studies, only the subdivision into material categories and load types indicated in tables 2 and 3 is presented.

RESULTS FOR THE LOCAL STRAIN CONCEPT

The local strain concept employs only material parameters for the fatigue life analysis of structural components. Hence, its accuracy must be lower than that of the nominal stress concept, which considers the behaviour of structural components. The results of local analyses described in [9] with the damage parameter given by Smith-Watson-Topper [8], are plotted in figure 6. The scatter ranges are decidedly larger than those previously indicated for the nominal stress concept. The causes are the large differences in the average damage sum and especially in the scatter for steel and aluminium. Explanations of this phenomenon on the basis of material physics are presented in [2]. If material parameters are likewise employed with the nominal stress concept, that is, if the component S-N curves are estimated, on the basis of [10], for instance, comparable accuracy results from the local and nominal stress concepts for steel materials (figure 7).

SUMMARY

- The previous assumptions and requirements on the accuracy of fatigue life predictions are not realistic. Even the experimental results of constant amplitude and random fatigue tests employed for comparison result in a scatter range of $2 \leq T_D \leq 3.5$ - even if the damage accumulation hypothesis yields the ideal value of unity in every case.
- The scatter of the fatigue life analysis on the basis of the nominal stress concept is in the range $9.2 \leq T_D \leq 12.7$, in correspondence with the particular Miner modification, even with the favourable Rainflow method. On the average, it is situated on the unsafe side.
- Uncritical application of the local strain concept results in very large uncertainties in the fatigue life prediction, especially for aluminium. In the case of steel, the scatter range of the damage sums is comparable with that of the nominal stress concept with the application of estimated S-N curves.

From the study described in the present excerpts, reliable indications are obtained for the first time for applying the relative Miner rule [11] in the absence of a case history for direct comparison and thus for achieving higher accuracy in the analytical prediction of fatigue life.

ACKNOWLEDGEMENT

This work was carried out under the supervision of the Forschungskuratotium Maschinenbau e.V. (FKM, Frankfurt). Finacial support by the Bundesministerium für Wirtschaft (AiF-Nr. D-81 and AiF-Nr. B9934) is greatfully acknowledged.

723

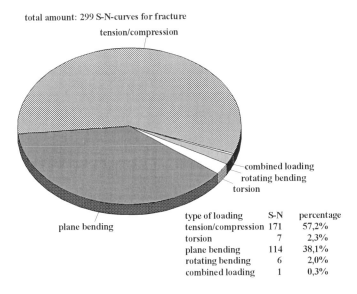

total amount: 299 S-N-curves for fracture
tension/compression

combined loading
rotating bending
torsion

plane bending

type of loading	S-N	percentage
tension/compression	171	57,2%
torsion	7	2,3%
plane bending	114	38,1%
rotating bending	6	2,0%
combined loading	1	0,3%

Fig. 1. Subdivision of S-N-curves in correspondence with the loading type

total amount: 299 S-N-curves for fracture

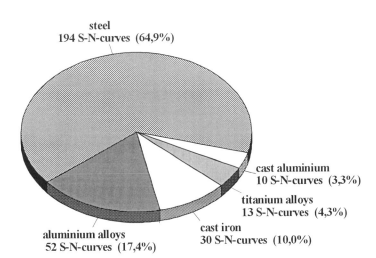

steel
194 S-N-curves (64,9%)

cast aluminium
10 S-N-curves (3,3%)

titanium alloys
13 S-N-curves (4,3%)

cast iron
30 S-N-curves (10,0%)

aluminium alloys
52 S-N-curves (17,4%)

Fig. 2. Subdivision of S-N-curves in correspondence with the material categories

724

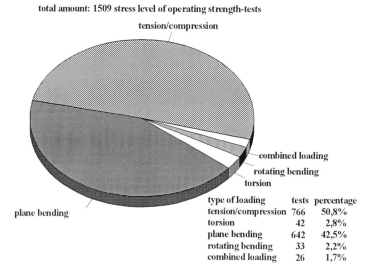

total amount: 1509 stress level of operating strength-tests

tension/compression

combined loading
rotating bending
torsion

plane bending

type of loading	tests	percentage
tension/compression	766	50,8%
torsion	42	2,8%
plane bending	642	42,5%
rotating bending	33	2,2%
combined loading	26	1,7%

Fig. 3. Subdivision of operating strength tests in correspondence with the loading type

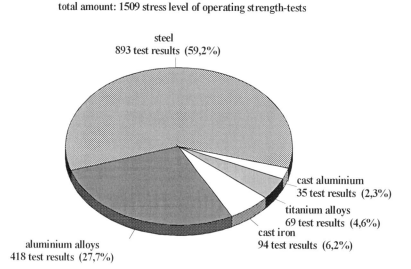

total amount: 1509 stress level of operating strength-tests

steel
893 test results (59,2%)

cast aluminium
35 test results (2,3%)
titanium alloys
69 test results (4,6%)
cast iron
94 test results (6,2%)

aluminium alloys
418 test results (27,7%)

Fig. 4. Subdivision of operating strength tests in correspondence with the material categories

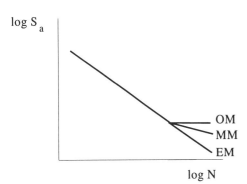

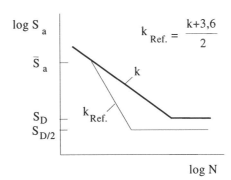

Fig. 5.   Modifications of the Miner Rule:

OM: Miner original
MM: Miner modified by Haibach
EM: Miner elementary

Fig. 6.   Reference S-N-curve refered to the Miner Modification given by Liu/Zenner [5]

Table 1.   Average values and scatter ranges of the damage sums for different Miner modifications and counting methods

| | total | | $\bar{R} = -1$ | | $\bar{R} \neq -1$ | |
| | n = 964 | | n = 587 | | n = 377 | |
	$\bar{D}$	T_D	$\bar{D}$	T_D	$\bar{D}$	T_D
Miner elementary ME						
transform. Amplitude	0,39	12,3	0,40	11,8	0,37	12,3
range pair	0,37	14,7	0,38	13,1	0,35	17,2
level crossing	0,81	24,2	0,69	15,7	1,08	47,6
Miner modified by Haibach						
transform. Amplitude	0,28	12,6	0,30	10,6	0,26	14,8
range pair	0,26	15,1	0,28	11,2	0,23	21,6
level crossing	0,61	26,0	0,53	16,4	0,77	41,5
Miner modified with						
continuous decreasing fatigue						
limit [6, 7]	0,29	12,7	0,30	10,6	0,27	15,3
transform. Amplitude	0,27	15,3	0,29	11,3	0,24	22,2
range pair	0,63	25,9	0,55	16,3	0,80	41,7
level crossing						
Miner-Liu/Zenner	0,75	9,2	0,79	8,6	0,68	9,8
transform. Amplitude	0,70	11,8	0,75	10,9	0,64	12,8
range pair	1,56	18,0	1,35	11,1	2,00	32,0
level crossing						

Fig. 2   Effect of the material on the average value and scatter range of the damage sums

	steel n = 525		aluminium alloys n = 332		cast iron n = 88	
	$\bar{D}$	T_D	$\bar{D}$	T_D	$\bar{D}$	T_D
Miner modified (continuous decreasing fatigue limit)	0,24	10,3	0,34	15,2	0,38	13,8
Miner-Liu/Zenner	0,64	8,7	0,79	8,2	1,25	10,2

Fig. 3   Effect of the loading type on the average value and scatter range of the damage sums

	tension/compression n = 463		plane bending n = 422		torsion n = 21	
	$\bar{D}$	T_D	$\bar{D}$	T_D	$\bar{D}$	T_D
Miner modified (continuous decreasing fatigue limit)	0,25	11,1	0,36	13,5	0,13	3,6
Miner-Liu/Zenner	0,65	8,4	0,90	9,8	0,48	7,2

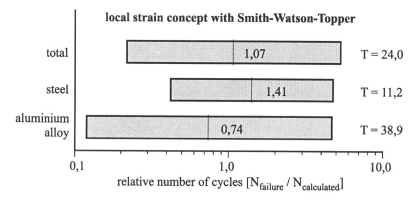

Fig. 7.   Scatter ranges and damage sums with the local strain concept [9]

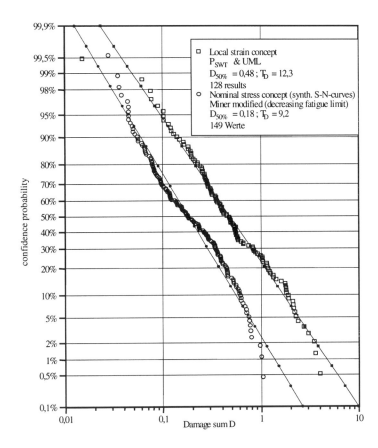

Fig. 8. Comparison of the local strain concept with the nominal stress concept, for estimated initial data on steel materials

REFERENCES

1. Eulitz, K.-G., Esderts, A., Kotte, K. L., Zenner, H. (1994) *Forschungshefte Forschungs-kuratorium Maschinenbau e.V.*, Heft 189, Verbesserung der Lebensdauerabschätzung durch systematische Aufarbeitung und Auswertung vorliegender Versuchsreihen

2. Eulitz, K.-G., Döcke, H., Kotte, K. L., Liu, J., Zenner, H. (1997) *Forschungskuratorium Maschinenbau e.V.*, Heft 227, Lebensdauervorhersage II, Verbesserung der Lebensdauerabschätzung durch systematische Aufarbeitung und Auswertung vorliegender Versuchsdaten, Abschlußbericht

3. Miner, A. M. (1945) *J. of Appl. Mech. Trans ASME*, **12**, pp159-164, Cumulative Damage in Fatigue

4. Corten, H. T., Dolan, T. J. (1956) *Inst. of Mech. Engrs.*, Cumulative Fatigue Damage, London

5. Liu, J., Jung, L., Esderts, A. (1993) *19. Vortrags- und Diskussionsveranstaltung des DVM-Arbeitskreises "Betriebsfestigkeit"*, Vorschläge zur Verbesserung der Lebensdauerabschätzung bei Zufallsbeanspruchung, bei Überlasten und bei zusammengesetzter Beanspruchung, München

6. Gnilke, W. (1980), Lebensdauerberechnung der Maschinenelemente, Berlin

7. Haibach, E. (1989) *VDI-Verlag* Betriebsfestigkeit: Verfahren und Daten zur Bauteilberechnung

8. Smith, K. N., Watson, P., Topper, T. H. (1970) *J. of Materials,* Vol 5, No. 4, pp. 767-78 , A stress - strain function for the fatigue of metals

9. Hickethier, H. Lebensdauerprognose mit Kenngrößen der örtlichen Beanspruchungs-matrix, Abschlußbericht FKM-Forschungshefte,

10. Hänel, B., Wirthgen, G. (1997) *Forschungshefte des Forschungskuratoriums Maschinenbau e.V.*, Heft 183, Festigkeitsnachweis - Rechnerischer Festigkeitsnachweis für Maschinenbauteile, Abschlußbericht

11. Schütz, W., Zenner, H. (1973) *Z. f. Werkstoffmechanik*, 4. Jahrg., Schadensak-kumulationshypothesen zur Lebensdauervorhersage bei schwingender Beanspruchung - Ein kritischer Überblick -, Nr. 1 u. 2

12. Palmgren, A. (1924) *VDI-Zeitschrift* 68, pp 339-341, Die Lebensdauer von Kugellagern

Low Cycle Fatigue and Elasto-Plastic Behaviour of Materials
K-T. Rie and P.D. Portella (Editors)
729

STANDARDIZED LOAD-TIME HISTORIES - STATUS AND TRENDS

P. HEULER[*] and W. SCHÜTZ[**]

[*]AUDI AG, D-85045 Ingolstadt, Germany
[**]IABG, D-85521 Ottobrunn, Germany

ABSTRACT

Standardized load-time histories (SLH) provide advantages to the user in particular in the pre-design stage and for studies of more generic nature. Essential elements for the derivation of a meaningful SLH are discussed, and an overview on SLH presently available and some of their specific features is given followed by an outline of trends and areas for further development.

KEYWORDS

Fatigue, in-service integrity, spectrum loading, usage statistics, load-environment interaction.

INTRODUCTION

Standardized load-time histories (SLH) have been developed and applied for fatigue studies for many years because of various advantages. Perhaps the first SLH proposed and extensively applied was Gassner´s 8 level blocked programme test [1]. It is common understanding that SLH do not refer - in the first instance - to a *specific* design problem, but comprise the *typical* features of the loading environment of a certain class of structures, vehicles etc. Thus SLH are usually applied in the pre-design stage or for studies of more generic nature. These may include an evaluation of how different materials, detail geometries, surface treatments or manufacturing routes affect the fatigue behaviour of specimens and components. SLH are also frequently used in projects to develop or evaluate numerical life prediction models or in round robin pro-grammes with several participating laboratories on experimental or analytical fatigue-related problems. Results generated by use of SLH are readily compared at least with regard to the important aspect of the loading environment of a given test series. Thus a data base is created which allows the evaluation of individual test results or scatterbands and may help to reduce the amount of fatigue testing. SLH spectra may also be used as a reference case for comparison with and assessment of ad-hoc load measurements which necessarily cover a limited time or usage period in most cases.

CHARACTERISATION AND ANALYSIS OF LOADING ENVIRONMENTS

It is generally agreed that the structural load variations with respect to fatigue should be characterized in the time domain since in most cases the range (or amplitude) of a load, stress or strain cycle plus its respective max or mean value can be considered as dominating para-meters. Furthermore, the sequence or arrangement of load cycles of different ranges and means must not be neglected. Analyses in the frequency domain give insight into the frequency con-tent of a load signal which is particularly instructive for flexible structure, but do not deliver the above-mentioned values.

Rainflow counting and synthesis is a widely accepted procedure to reduce the bulk of measured data fatigue-wise and to synthesize test load sequences from measured and/or extrapolated rainflow matrices. This is possible for uniaxial as well as for multiaxial problems [2].

Most structural loading environments can be described as sequences of different modes [3] which may be a particular flight, driving a car on certain road types, a seastate of a given severity etc. These modes of operation contain load cycles of different but typical magnitudes and frequencies. Often distinct patterns of grouped load cycles can be distinguished, called an event, such as braking or cornering of a car, different flight phases or manoeuvres of an aircraft. The respective SLH has to reflect these characteristics in an appropriate manner.

It is this mixture, sequence, *relative* frequency and magnitude of modes, events and scaled load cycles plus time and phase information for multiaxial problems which is "standardized" - generally not the actual load, stress or strain level which has to be assigned to a specific problem. It is preferable - but not always possible - to base a SLH on quantities which are representative for the external loading acting on a structure or vehicle.

BASIS OF GENERATION OF STANDARDIZED LOAD-TIME HISTORIES

It is evident that "load" measurements are most essential, but on top of that some further input is required for the derivation of a meaningful SLH.

Load measurements. Measurements from several similar structures, vehicles etc. should be available, because only then it is possible to identify the characteristic features of the spectrum, the sequence and mixture of events and load variations.

Usage statistics. At best the measurements available cover all relevant modes of operation and environmental conditions (gust levels, road roughness etc), but most probably not in the correct percentage. Therefore, additional information has to be collected and introduced with regard to usage statistics such as the anticipated distribution of driving distance on different types of roads (highway, country road, city road, gravel road), mix of missions planned for a military aircraft, weather statistics (seastates, max wave heights) for certain geographical locations for offshore applications etc.

General definitions. A number of spectrum parameters have to be defined such as the extreme values of loads which may have to be extrapolated from the data pool, truncated for certain reasons [4] or determined from physical limits. Similarly the block size or return period of the SLH has to be agreed which must not be too short in order to avoid unrealistically truncated spectrum shapes (close to a constant amplitude test). This is more or less equivalent to the requirement of creating a spectrum shape where the (calculated) fatigue damage is realistically distributed over large, medium and small cycles of the history. With respect to applicability considerations, it is often necessary (and possible) to omit small "non-damaging" (at least in a relative sense) load variations. "Rules of thumb" have evolved that give some guidance for the selection of omission levels [5].

Non-technical items. Experience shows that the development of a SLH should be a cooperative effort of several contributors in order to consider the various aspects and requirements effectively. The principles followed and the range of application should be clearly documented and published.

EXISTING STANDARDIZED LOAD-TIME HISTORIES

In this section,the SLH developed mainly in Europe and approaches taken in the US are shortly presented. More details are given in some overview papers [3,6,7] and in the respective original papers and reports.

Table 1 lists the SLH generated by European working groups over the last 25 years. With the need for optimum light-weight design, originally the aircraft industry was the main driver for these efforts. The need for development and use of SLH has been pointed already by Gassner

NAME	PURPOSE	STRUCTURAL DETAIL	DATA BASE	BLOCK SIZE [1]	EQUIV USAGE	YEAR /REF.[2]
TWIST	transport aircraft wing	wing root bending moment	meas. on 4 transports, frequ. distr. on 3 aircraft	402.000	4000 flights	1973 [11]
GAUSSIAN	general purpose random sequence	narrow-band to wide-band random		$1 * 10^6$ - $3 * 10^6$	-	1974 [16]
FALSTAFF	fighter aircraft	wing root	load-factor (g) meas. on 4 fighter aircraft	18.000	200 flights	1975 [12]
MiniTWIST	shortened version	as above	as above	62.000	4000 flights	1979 [25]
HELIX, FELIX	helicopt., hinged and fixed rotors	blade bending	operational/stress meas. on 4 helicopters	$2.3 * 10^6$ $2.1 * 10^6$	140 flights	1984 [13]
HELIX/32, FELIX/28	shortened versions	as above	as above	$2.9 * 10^5$ $3.2 * 10^5$	140 flights	1984 [13]
Cold TURBISTAN	tactical aircraft engine discs	bore	rpm and mission data of 5 fighter aircraft	7.700	100 flights	1985 [14]
ENSTAFF	FALSTAFF + temperature	wing root	mission analyses + Europ. meteor. data	18.000	200 flights	1987 [17]
WISPER	wind turbines	blade in-plane and out-of-plane bending	meas. on 9 wind turbines	130.000	2 months	1988 [26]
Hot TURBISTAN	tactical aircraft engine discs	rim	rpm and mission data of 5 fighter aircraft	8.200	100 flights	1989 [15]
WASH I	offshore structures	structural members of oil platforms	meas. on 2 platforms, weather statistics	$5 * 10^5$	1 year	1989 [18]
WAWESTA	steel mill drive		meas./statistics on 6 steel mill drives	28.200	1 month	1990 [27]
CARLOS	car loading standard sequ. (uniaxial)	vertical, lateral, longitud. forces on front suspension parts	meas. on 10 vehicles	84.000 to 136.000	40.000 km	1990 [19]
CARLOS multi	car loading standard (multiaxial)	4-channel load comp.s for front suspension parts	meas. on 12 vehicles	as above	40.000 km	1994 [20]
CARLOS PTM	car power train (manual shift)	power train comp.s e.g. clutch, gear-wheels, shafts,bearings	meas. on 14 vehicles /vehicle variants	24.000 torque cycl., 1 to 6 hrs for revolv. parts		1997 [21]

[1] Cycles

[2] Partly the references do not cite the original report(s), but those available to the public.

Table 1   European standardized load-time histories

and - later on - in the late 60´s and early 70´s [8-10]. Two of the most "successful" SLH are the TWIST [11] and FALSTAFF [12] sequences for transport and fighter aircraft, respectively, which have been and are still being applied for numerous studies on materials, joints and other structural elements. HELIX/FELIX [13] and the two TURBISTAN sequences [14,15], Fig. 1, are further examples from the aerospace field. Because of the extremely high number of cycles of the HELIX/FELIX sequences, shortened versions of these SLH have been included already in the original report where about 85 % of the original sequence has been omitted. To the authors´ knowledge, only the shortened versions have subsequently been applied, for example, within round robin programmes of the American Helicopter Society.

The GAUSSIAN random sequences [16] published in the mid 70´s are unique with respect to the fact that they are not based on particular load or stress measurements, but on general experience mainly regarding automotive chassis components. Using three irregularity factors of I = 0.99, 0.7 and 0.3, narrow-band to medium wide-band to wide-band loading spectra are covered.

732

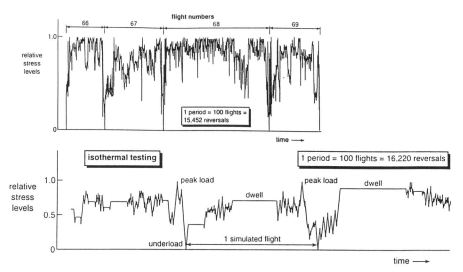

Fig. 1 Segments from the Cold TURBISTAN (top) and Hot TURBISTAN (bottom) SLH

The ENSTAFF [17] and Hot TURBISTAN sequences are specifically mentioned here because of the fact that, for these SLH, in addition to the sequence of (normalized) peaks and troughs further parameters have been standardized. ENSTAFF (Environmental FALSTAFF) combines FALSTAFF with temperature-time profiles in order to allow appropriate testing and quali- fication of composite aircraft structure with its susceptibility to humidity and temperature effects (according to [17] humidity effects shall be taken into account by use of pre-conditioned test articles). The Hot TURBISTAN sequence reflects the complex interaction of mechanical (centrifugal) and temperature(-gradient) induced loading components at critical locations of the rim of engine discs. It is, however, not a "true" thermo-mechanical fatigue (TMF) sequence, since - with a view to experimental aspects - it has been tranformed to an equivalent isothermal sequence with appropriately adjusted hold or dwell times at various stress levels.

As Table 1 indicates, starting in the mid 80´s SLH for non-aerospace applications have been developed. WASH I [18], typical for drag dominated smaller members of platforms operating in the North Sea or similar sites is one outcome of an international working group. It was intended by the working group to additionally provide a framework for the derivation of individual test load sequences where only statistical parameters such as the power spectral density are fixed. This was felt to be more appropriate for inertia dominated big members of platforms. By doing so, however, one of the requirements for SLH, i.e. fixed sequence of (normalized) peaks and troughs, will be disregarded.

With regard to the relevance of fatigue loading for automotive structure, the quite recent SLH activities in the automotive industry appear worth mentioning. It certainly does not mean that the assessment of the in-service integrity was not important before, but indicates the tradition- ally more (ad-hoc) problem-oriented way of working. With continuous pressure for light, but rigid and endurable structure as well as the introduction of advanced materials and production schemes, standardization and collaboration obviously gained attraction also in the automotive industry, resulting in the CARLOS "family" of SLH for front axle suspension parts (both uniaxial and multiaxial) [19,20] as well as the recent SLH for power train components [21].

In the US, activities were mainly centered on the derivation of test load sequences to be used for evaluation and development of fatigue life prediction methodology. The SAE Fatigue and Evaluation Committee took a more pragmatic approach by selecting test load sequences from existing strain measurements which were felt to be typical for the ground vehicle industry. The early sequences ´Suspension´, ´Transmission´ and ´Bracket´ were derived from individual measurements and filtered with an omission level of 20 % of max range [22]. Later, criticism has been expressed with regard to the very short return periods of those sequences, see Table 2, which led to biased life estimates because most of the damage was done by the small number of large cycles.

NAME	PURPOSE	LOAD COMPONENT	BLOCK SIZE [1]	YEAR /REF.
TRANSMISSION	tractor transmission (torque)		854	1977 [22]
SUSPENSION	vehicle suspension (bending moment)		1253	1977 [22]
BRACKET	vehicle mounting bracket (rough road)		2968	1977 [22]
AGRICULTURE TRACTOR	drive axle, several typical operations	bending torsion	31.800 18.700	1989 [23]
LOG SKIDDER	drive axle, operation in forests	bending torsion	$3.1 * 10^5$ $3.1 * 10^5$	1989 [23]

Table 2

Standardized load-time histories from the US

[1] cycles

This has been considerably improved in a further effort [23] where two sequences (´Agriculture Tractor´and ´Log Skidder´) have been selected based on a detailed review of measurements made available from several sources. In particular, care has been taken to achieve a more realistic distribution of damage fractions contributed by medium and smaller cycles. Furthermore, this effort represents the first coordinated approach towards multiaxial spectrum fatigue loading problems. However, only subsets (not the full histories mentioned in Table 2) have subsequently been digitized and applied in some test campaigns [24] as fully biaxial load histories where the phase angle and time relationship between the two channels have been preserved.

DISCUSSION

Reviewing the topic of SLH, several aspects and trends may be worth noting. The earlier SLH were centered on the definition of sequences of peaks and troughs for one load component which consequently meant "uniaxial" testing mainly of specimens and simple structural elements. Similarly these SLH found application for the evaluation of fatigue life prediction methodology with one major stress/strain component such as the local strain concept (uniaxial) or most of the fatigue crack growth models. Increased complexities have been found essential and were introduced later on with regard to problems (a) with more than one load component (multi-channel or multiaxial loading) and (b) where environmentally-assisted or -enhanced fatigue attains increased importance.

As mentioned, the SAE programme [23] was the first to consider sets of multiaxial (biaxial) load histories, but CARLOS multi [20] was the first fully multi-channel SLH adopting four load channels (vertical, lateral and two longitudinal forces). The complex nature and specifics of multi-channel loading problems generally dictate a close connection and dedication of SLH to real structural applications as the front axle suspension system of passenger cars in that particular case.

The aspect of environmentally-assisted or -enhanced fatigue has been mentioned above with regard to ENSTAFF and Hot TURBISTAN. In the automotive industry, the topic has created increased attention because of the incresing use of light alloys (aluminium, magnesium) for chassis and engine/drive train components. Standards are available for accelerated testing ag-

734

ainst pure environmental attack such as corrosive media, UV radiation or dust, but no standard practices or SLH are currently available which give some guidance to, for example, corrosion-fatigue testing. It is expected that this type of SLH would be very useful for many cases, but it is acknowledged that fundamental understanding of these phenomena is poor - it is certainly much less understood than pure fatigue or corrosion mechanisms. Such an extended SLH would be worthwhile for more generic studies on a specimen level; but such an SLH would also have to address factors such as pre-conditioning (for example, gravel-induced pre-damage of automotive coil springs or suspension parts, overloading of rivetted joints of aircraft at low temperatures) in order to allow assessment of the efficiency of surface protection systems.

The more recent developments, in particular the CARLOS activities, are oriented towards real components and structures which certainly make them attractive for real-life qualification purposes. Qualification requirements imposed by OEM´s to suppliers place a heavy burden to those companies, the more so when quite different test scenarios have to be realized for different customers for the same or a similar part. Here extended SLH accepted by many OEM´s (maybe with different design margins) would be advantageous and economic. This tends to impose economic value to SLH, which will limit their free circulation within the fatigue community which is actually true for the newest SLH CARLOS PTM.

REFERENCES

1. Gassner, E. (1939), Deutsche Luftwacht, Ausgabe Luftwissen, Bd. 6.
2. Dressler, K. Carmine R. and Krüger, W. (1992), In: *LCF and El.-Pl.Beh. Mater.*, K.T.Rie (Ed.), Elsevier Science, 325.
3. ten Have, A.A. (1989), In: *Development of Fatigue Loading Spectra*, J.M.Potter and R.T Watanabe (Eds.), ASTM-STP 1006.
4. Crichlow, W. (1973), AGARD-LS-62.
5. Heuler, P. and Seeger, T. (1986), *Int J. Fatigue 8*, 225.
6. Heuler, P. and Schütz, W: (1988), *Mat.-wiss. Werkstofftech. 19*, 282.
7. Schütz, W. (1989), In: *Development of Fatigue Loading Spectra*, J.M.Potter and R.T Watanabe (Eds.), ASTM-STP 1006, 3.
8. Schütz, D. (1969), In: *Review of Investigations on Aeronautical Fatigue in the Federal Republic of Germany*, E.Gassner and O.Buxbaum (Eds.), LBF Darmstadt.
9. Barrois, W. (1972), *Random Load Fatigue*, AGARD-CP-118.
10. Schijve, J. (1972), *Random Load Fatigue*, AGARD-CP-118.
11. Schütz, D., Lowak, H., de Jonge, J.B. and Schijve, J. (1973), LBF-Rep. FB-106, NLR-Rep. TR 73.
12. Branger, J. et al (1976), Common report of F+W Emmen, LBF, NLR, IABG.
13. Edwards, P.R. and Darts, J. (1984), RAE Rep. TR 84084.
14. Bergmann, J.W., Schütz, W., Köbler, H.-G., Schütz, D., Fischersworring, A. and Koschel, W. (1985), IABG-Rep. TF-1934.
15. Bergmann, J.W. and Schütz, W. (1990), IABG-Rep. TF-2809.
16. Haibach, E. Fischer, R., Schütz, W. and Hück, M. (1976), In: *Fatigue Testing and Design*, Proc. Society of Environmental Engineers (SEE), 29.1.
17. Gerharz, J.J. (Ed.) (1987), LBF-Rep. FB-179.
18. Schütz, W., Klätschke, H., Hück, M. and Sonsino, C.M. (1990), *Fatigue Fract. Engng Mater. Struct. 13*, 15.
19. Schütz, D., Klätschke, H., Steinhilber, H. Heuler, P. and Schütz, W. (1990), LBF-Rep. No. FB-191, IABG-Rep. TF-2695.
20. Schütz, D., Klätschke, H. and Heuler, P. (1994), LBF-Rep. FB-201.
21. Klätschke, H. and Schütz, D. (1997), LBF-Rep. 7558 (unpubl.).
22. Tucker, L. and Bussa, S. (1977), In: *The SAE Cumulative Fatigue Damage Test Program*, SAE AE-6, 1.
23 Fash, J W. and Conle, F.A. (1989), In: *Multiaxial Fatigue*, G.L.Leese and D.Socie (Eds.), AE-14, SAE International, 33.
24. Bonnen, J.J.F., Conle, F.A. and Chu, C.C. (1991), SAE Paper 910164.
25. Lowak, H., de Jonge, J.B., Franz, T. and Schütz, D. (1979), LBF-Rep. TF-146, NLR-Rep. MP 79018.
26. ten Have, A.A. (1991), NLR-Rep. CR 91476 L.
27. Brune, M. and Zenner, H. (1990), Rep. ABF40.1, VBFEh, Germany.

IN-SERVICE INTEGRITY IN THE GERMAN AUTOMOTIVE INDUSTRY TODAY AND TOMORROW

A. Wimmer
AUDI AG, 85045 Ingolstadt, Germany

H. Naundorf
TWP, 85598 Baldham, Germany

ABSTRACT

The German automotive industry goes to great lengths to measure the loads placed on their vehicles by customers world-wide. Various counting methods are used to determine load spectra on the basis of load-time histories. Mathematical and experimental simulation methods are applied to determine sectional force and component load spectra. With a sufficient number of measurements, the load scatter can be used to derive reference load spectra which must be tolerated by components and groups of components during the test. Test conditions are as realistic as possible, taking into account component loading in final assembly position and environmental influences. Components are tested until cracks or fractures occur. The mean value and the scatter of load cycles achieved during the tests together with the required load cycles form the basis for strength approval. Road endurance tests covering more than 8 million kilometres provide additional confirmation for statements made by operating strength laboratories.

KEY WORDS

In-service integrity, load spectrum, stress spectrum, operating load simulation test, special events, time-lapse

INTRODUCTION

We take it for granted that none of a car's vital components fails during its service life, irrespective of the weather or geographical location. A self-confident customer will convince himself of such things as acoustics, comfort, vehicle dynamics and other characteristics. But what about in-service integrity, how can he convince himself of something he cannot grasp with his senses?

The employees in the automotive industry are responsible for ensuring that the customers' trust is justified and that the vehicle does not take offence at slight mishaps.

That is why strength tests in the automotive industry must anticipate all stress factors that may be caused by customers world-wide.

IN-SERVICE INTEGRITY IN THE AUTOMOTIVE INDUSTRY

None of a car's safety-critical components must fail during its service life. Vehicle service life differs from carmaker to carmaker, but is always at least 300,000 km, irrespective of driving conditions e.g. sporty driving, mainly city cycle, mainly inferior or unpaved roads and taking into account production scatter for the most unfavourable components.

To ensure this, all automobile manufacturers

- employ a team of specialists responsible for in-service integrity,
- measure all loads to which a vehicle is subjected to by customers in every day use, generally in all climates and on all public roads world-wide,
- determine loads which may occur in borderline cases e.g. when the driver goes over a speed breaker at high speed or hits a kerbstone or similar,
- design components using computer load simulation to determine stress factors and compare these results with modern component measuring methods for subsequent evaluation,
- test components, groups of components and complete vehicles on operating load simulation test rigs until failure to determine the strength limits,
- optimise components mathematically and experimentally together with the design department until all requirements are met,
- conduct road endurance tests to determine whether operating strength targets have been achieved.

METHODS FOR SUITABLE DIMENSIONING OF AUTOMOBILES

Load Determination

In everyday use, automobiles are subjected to loads resulting from a number of recurring driving conditions with rather different effects e.g. acceleration and braking procedures, overtaking manoeuvres, cornering, going over bumps and pot holes etc. The load amplitude is determined by the customer's driving behaviour, road and weather conditions as well as engine type, drive train, running gear, tyres and vehicle load etc. In line with their individual history, the various automobile manufacturers have developed slightly different approaches to vehicle operating strength and the elaboration of reasonable reference values.

What all automobile manufacturers have in common is that they use wheel load cells to determine the forces and torques acting on the car's wheels. The signals are transmitted via telemetrics to the passenger compartment, where they are recorded. However, they use different methods to determine load scatter. Statistics provide us with the percentage of men and women drivers, blue collar and white collar drivers, they tell us how many days a year roads are wet or dry. There are load statistics and statistics about the percentage of urban, rural, motorway and unpaved road usage and the corresponding speed distribution.

Based on this information, the different companies elaborate road tests which provide information about load amplitude and scatter. The immense volume of load time histories is condensed by means of counting methods to obtain load spectra which indicate load amplitude and number of cycles. In addition, damage calculations and statistical methods allow us to classify drivers in high or low strain categories. One of the methods for example selects the most severe driver out of 100 and stipulates that the vehicle must last 300,000 km for this type of driver. The load spectrum for this driver which is derived from a defined measuring course is extrapolated to 300,000 km. This spectrum is then used as a reference for component and vehicle testing, a method employed by all automobile manufacturers.

Not included in this consideration are loads resulting from special events, i.e. forces caused by the driver's carelessness such as going over speed breakers at high speed. Such loads are not covered by the aforementioned methods. It takes many years of studying the market to determine such occasional events and consider them in load determination.

Stress Determination

If the variety of loads and their frequency are known, it is possible to conduct load path studies with the help of dynamic computer models on the basis of multi-element simulation methods. Forces and torques can be determined at any interface and for any component. As all components have already been described according to the finite element method, it is possible to determine the stress distribution over a particular component. On components which are true to drawing, it allows us to closely estimate the amplitude and location of max. stress.

In spite of these excellent achievements, we still require experimental determination of local stress. The reasons for this are manifold, e.g. components no longer correspond to the ideal shape after forging, casting or forming; straight profiles are not straight and even planes not even; weld seam shape differs from the drawing, etc. The most important method for the determination of local stress is the measurement of local strain using strain gauges. Local stress can then be derived from the material characteristics.

Once we know the loads and their scatter, we also know the stresses and their scatter and then we can use the load spectrum to determine the stress spectrum. The task now is to elaborate a standard which a component, a group of components or the entire vehicle has to meet to ensure adequate service life. The procedures used by the different German carmakers are not identical, but very similar because they all had the same master, namely Ernst Gaßner.

Operating Strength of Components

In automotive engineering particularly the frequently recurring small and medium loads acting on a component are of great importance. Failure due to fatigue loading only occurs after a certain operating time and is usually accompanied by only a slight deformation. It is therefore difficult to detect the failure of a component in advance and the consequences can be catastrophical.

Tests are carried out in the laboratory to determine the fracture stability of each component. For this reason the results of load measurements the elaborated requirements are translated into damage-related signals which are used in the development in order to control the test rigs.

There are a number of methods to reduce testing time, but these have to be selected carefully. Usually, small loads which are far below the endurance limit can be omitted. However, this does not hold true for all materials, in the case of aluminium and magnesium one should be careful. We can increase the test frequency i.e. speed up the load-time cycles, but have to be careful not to falsify transmission functions or change material properties through component temperatures due to the higher frequency (e.g. elastomeres). We can also increase the test load by a certain percentage. This method should be rejected, however, if max. physical forces are exceeded because elasticity or local plastification may lead to changes which distort reality. When determining time-lapse methods, an extensive knowledge of material properties is required in addition to a knowledge of the physical limits of vehicle strain. This is not always an easy task, because in the wake of weight reduction, we often come across materials that have not been extensively investigated. In addition, practically all of the car's components are subjected to corrosion, stone chipping and in some cases elevated temperatures. Fatigue tests with these influences are either non-existent or end at $5 * 10^6$ load cycles. However, after 300,000 km a car wheel has already endured $5 * 10^8$ load cycles.

Due to the uncertainties with new materials, components are tested until failure on multi-component test rigs as realistically as possible i.e. with corrosion, chipping, temperature, etc. Operating load simulation tests with 3 or 4 moving axes are used to apply pre-defined loads three-dimensionally on components and groups of components.

Furthermore, real-time simulations based on road test load measurements are carried out on axles, bodies and complete vehicles on highly complex multi-component test rigs with 10, 12 or more servo-hydraulic actuators.

The test volume is supplemented by further simulations extending from special loads to abuse to determine component behaviour under max. admissible strain. The rule here is "leak before fracture", which means that damages to safety-critical parts must be easily detectable so that they can be eliminated before there is a risk of fracture under normal operating conditions.

CO-OPERATION WITH SYSTEM SUPPLIERS

Incorporating the suppliers in the design and development process is a worthwhile endeavour not only with a view to shorter development times. Based on the variety of vehicle concepts and designs in the German automotive industry, we have managed to develop standardised and harmonised test methods such as "Carlos", which enable advance component testing anywhere in the world, provided that the necessary test equipment is available. Through extensive testing the automobile manufacturers ensure that the overall structure with its interacting components meets the set standard.

However, there is still a great need for clarification on issues such as product responsibility, product identification and quality control, to name but a few. The possibilities here are wide-ranging.

PRODUCTION RELEASE

Before a component is released for production, the number of cycles achieved on the test rig is set off against the requirements determined on the basis of measurements. The statistic scatter of stress and the statistic scatter of test rig results play an important part in the definition of required mileage.

It is illusionary to think that we can define mileage with the knowledge of the above scatter values. We can only forecast component service life with a specific failure probability. The failure probability must be set at such a level that no failure occurs during actual operation.

German automobile manufacturers have been applying this method successfully as can be seen from the field results. In spite of the millions of cars on the road, no strength-related failures are known.

Operating load simulation tests are supported by road endurance tests and quality assurance. Road tests are conducted on special test and torture tracks and on public roads. They serve as function tests and provide additional confirmation of test rig results. On series production vehicles, quality assurance conducts material tests in the laboratory, component strength tests on test rigs and road tests on public roads world-wide. Before a vehicle is finally approved for production, the various carmakers have covered between 8 and 10 million test kilometres. This gives operating strength engineers the necessary support and confidence to have done a good job.

OUTLOOK

All future activities aim at:

- elaborating sound statements about representative component stress in all major markets,
- optimising components with the help of computer simulation models based on the elaborated representative component stress histories and, in the long-term facilitating mathematical service life prediction through continuous comparison with experimental results,
- developing time-lapse test methods which facilitate quick and reliable assessment of the total structure according to representative stress values, taking into account material characteristics,
- developing suitable methods for time-lapse simulation of environmental conditions and applying these in overall test programmes,
- facilitating the assignment of development tasks to system suppliers, including testing of these parts according to standardised loading procedures,
- continuing the fruitful exchange of information in the field of in-service integrity within the German automotive industry and institutions working on this subject and extending it to European partners wherever possible.

Special thanks to:

AUDI AG	Dr. Beste
BMW AG	Dr. Stauber
Daimler Benz AG	Mr. Schäfer
Adam Opel AG	Mr. Bräker
Porsche AG	Mr. Käumle

Low Cycle Fatigue and Elasto-Plastic Behaviour of Materials
K-T. Rie and P.D. Portella (Editors)
741

SIMULATION OF LONG TERM CREEP FATIGUE BEHAVIOUR
BY MULTI-STAGE SERVICE TYPE STRAIN CYCLING

A. SCHOLZ, J. GRANACHER and C. BERGER

Institute of Materials Technology, Darmstadt University of Technology,

Grafenstrasse 2, 64283 Darmstadt, Germany

ABSTRACT

The creep fatigue behaviour of two heat resistant steels is investigated by multi-stage service type strain cycling. A loading sequence which is characteristic for a medium load steam power plant contains a coldstart cycle, warmstart cycles and hotstart cycles. To simulate this type of loading three stage service type strain cycling tests were performed up to 15 000 h. Such three stage tests and simpler single stage service type tests lead to similar values of relative creep fatigue life. This is the result of a creep fatigue life analysis based on cyclic deformation and the generalized damage accumulation rule, whereby preloading influences, internal stress and mean stress are taken into account. Only small differences were found from the application of rainflow counting and range mean counting.

KEYWORDS

Creep fatigue, multi-stage strain cycling, service type strain cycles, heat resistant steels, cyclic deformation, life analysis, internal stress, mean stress, preloading, cycle counting.

INTRODUCTION

High temperature components of steam power plants normally operate under complex and partly variable loading conditions [1,2]. At the heated surface of the components temperature transients cause strain cycling with variable thermal (secondary) stresses. In addition, pressure loading and, at rotors, centrifugal loading lead to quasistatic (primary) stresses. As a consequence, strain cycling can be the critical loading condition at the heated surface of large components. In addition, creep occurs due to relaxation in the hold phases of strain cycles. To simulate these creep fatigue conditions, a service type strain cycle was developed [3] (Fig. 1). This cycle provides a compressive strain hold phase 1 simulating start up condition, a zero strain hold phase 2 approximating temperature equibalance during constant loading and a tensile strain hold phase 3 simulating shut down condition. Anisothermal strain cycling is of interest because it is close to service conditions. However, simpler isothermal testing is of interest in the long term region with regard to a design life of power plants of up to 200 000 h or more. As demonstrated earlier [4], creep fatigue life under such conditions can be analyzed with the aid of the generalized damage accumulation rule, if internal stress, mean stress and a damage interaction concept are considered. The current research work is concerned with the extension of this rule to longer test durations and from single stage to three stage strain cycling.

SINGLE STAGE SERVICE TYPE STRAIN CYCLING

Earlier investigations were focussed on single stage service type strain cycling tests according to Fig. 1c. Tests were carried out on two typical heat resistant turbine rotor steels, a bainitic 1Cr-1Mo-0.7Ni-0.3V-steel at 525 °C and a martensitic 12Cr-1Mo-V-steel at 550 °C. The total strain range $\Delta\varepsilon$ is constant and the strain rate at the ramps between the hold times is $\dot{\varepsilon}_r = 6$ %/min. Cylindrical testpieces of 10 mm diameter and 35 mm gauge length were used. The number of cycles to failure N_f is defined by a 1.5 %-drop of the stress range from its linear course. This corresponds to a crack depth of 0.5 to 1 mm.

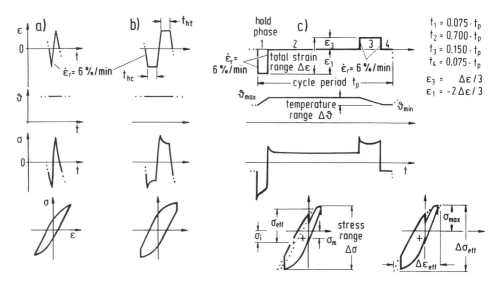

Fig. 1. Different strain cycles: standard cycle without (a) and with (b) hold times as well as service type cycle at a heated surface (c)

For comparison purposes, anisothermal tests (an) with compensated thermal strain [3,4] as well as isothermal tests (iso) at the maximum temperature of the anisothermal cycle were carried out with failure times of up to 8 000 h [4,5]. For long term testing, package type tests (pa) were developed. They comprise short strain cycling packages periodically inserted into creep packages with tension stresses representing the phases 2 and 3 of the full cycle (Fig. 1c). These tests are under continuation and have exceeded 66 000 h on the 1Cr-1Mo-0.7Ni-0.3V-steel and 33 000 h on the 12Cr-1Mo-V-steel. In this region the total strain ranges are below 0.4%, the hysteresis loops show viscoelastic deformation and cyclic softening of the two steels is almost completely disappearing [4,5]. Anisothermal tests as well as isothermal tests and comparable isothermal package type tests gave only insignificantly different numbers of cycles to failure. Thus, for failure times above several 1000 h isothermal package type testing can economically replace anisothermal testing.

CREEP FATIGUE LIFE ANALYSIS

The analysis of creep fatigue life is based on an analysis of cyclic deformation followed by an analysis of creep fatigue damage. As a result the experimental stress strain hysteresis loops with phases of elastoplastic deformation and relaxation can be described on the basis of an effective stress concept governing plasticity and creep. The effective stress σ_{eff} is the

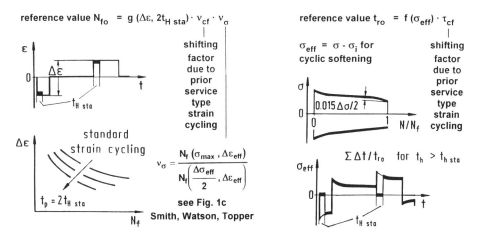

Fig. 2. Analysis and accumulation of creep fatigue damage based on the generalized damage accumulation rule (eq. 1)

difference of the external stress σ and an internal stress σ_i (Fig. 1c). The internal stress σ_i is defined for any point of the measured hysteresis loop by the center of a hypothetical elastoplastic flank curve loop which is inserted in the flank curve loop enveloping the complete loop observed [4,5]. The flank curves are derived from a cyclic or quasistatic yield curve by multiplying the latter by a factor of two. The cyclic yield curve can be experimentally determined by a strain cycle without hold times which is inserted into the service type strain cycle. A special value of the internal stress σ_i is the mean stress σ_m, which is situated in the centre of the flank curve loop.

The analysis of the creep fatigue damage or relative life L is based on the generalized damage accumulation rule

$$\Sigma\, N/N_{fo} + \Sigma\, \Delta\, t/t_{ro} = L \qquad , \tag{1}$$

which combines the Miner rule for fatigue damage and the life fraction rule for creep damage (Fig. 2). A special life analysis method was developed [4,5] which is described in the following. With respect to the accumulation of fatigue damage, the number of cycles to failure N_{fo} in eq. (1) is taken for standard strain cycling (Fig. 1b) with a characteristic hold time $t_{h\,sta}$. This hold time increases with decreasing total strain range. It was empirically determined in the range of 0 up to 1 h. A shifting of the value N_{fo} due to creep fatigue interaction is considered as a function of prior service type strain cycling [4-6]. A corresponding shifting factor v_{cf} depends on the total strain range $\Delta\varepsilon$ of the standard strain cycling and on the exhausted creep fatigue damage L. Further, the influence of mean stress σ_m on N_{fo} is taken into consideration by applying the Smith, Watson and Topper parameter [7] with the aid of a factor v_σ (Fig. 2). For the accumulation of the creep damage the rupture time t_{ro} is taken for the effective stress σ_{eff} and cyclic softening is considered. Further, a shifting of the rupture time t_{ro} due to creep fatigue interaction is considered by a factor τ_{cf} as a function of prior service type strain cycling [8,9].

The analysis of the single stage service type strain cycling tests with the above described method leads to a relative small scatter band with a mean relative creep fatigue life of $\bar{L} \approx 1$ (Fig. 3a). Microstructural analyses confirmed, that long term strain cycling and long term creep cause similar defects (Fig. 3b). Thus, the application of eq. 1 in the long term region is justified for the steels investigated.

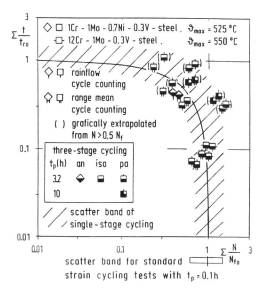

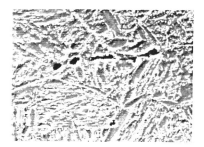

Fig. 3. Results of creep fatigue life analysis of service type strain cycling, scatter band for single stage testing and results of three stage testing (a) as well as microstructure after creep fatigue strain cycling during 26 000 h at 550 °C (b), 12Cr-1Mo-V-steel

THREE STAGE SERVICE TYPE STRAIN CYCLING

Multi-stage strain cycling is of high practical relevance at the heated surface of steam power plant components. Typical elementary cycles (Fig. 4) representing coldstart, warmstart and hotstart were arranged to a sequence (Fig. 5) presenting each three different ranges of temperature and total strain. The frequencies indicated for the elementary cycles are typical for a medium load power plant.

At first, anisothermal, isothermal and package type 3 000 h-tests were performed. Comparable results were observed for the different test types as earlier for single stage testing [6, 8, 9]. As a consequence, 30 000 h package type three stage tests were started which have exceed 15 000 h test duration, now. As a result, the softening (Fig. 6a) is relatively high at the coldstart cycle with its high total strain range $\Delta\varepsilon_C$. However, with decreasing total strain range of the elementary cycles, the softening is reduced. An arrangement of stress strain hysteresis loops of three subsequent elementary cycles (Fig. 6b) shows plastic deformation on the flank curves at coldstart with the high total strain range $\Delta\varepsilon_C$ but at the lowest total strain range $\Delta\varepsilon_H$ viscoelastic deformation dominates.

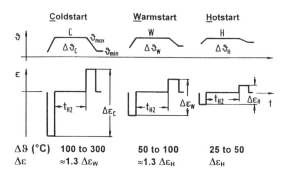

Fig. 4. Typical cycles of temperature and total strain at the heated surface of large components [10]

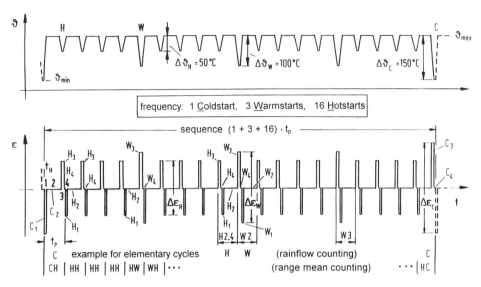

Fig. 5. Three stage service type strain cycling sequence and identification of individual cycles by rainflow cycle counting (cycles C, H and W) or by range mean cycle counting (cycles CH, HH, HW, WH and HC) [8]

A creep fatigue life analysis on the basis of the method described above shows similar relative life values as for single stage strain cycling (Fig. 3). The application of the rainflow cycle counting or the range mean cycle counting (Fig. 5) on this analysis gives only small differences. In the long term region, the proportions of the relative creep fatigue life exhausted are about 10 % for the coldstart cycles, 30 % for the warmstart cycles and 60 % for the hotstart cycles. On the basis of the methods demonstrated a PC programme SARA for life prediction of multi-stage strain cycling at the heated surface of large components is under development.

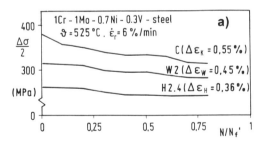

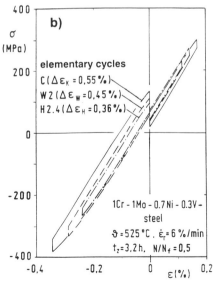

Fig. 6. Course of half stress range $\Delta\sigma/2$ versus number of cycles N normalized by the planned number of cycles to failure N_f' for the cycles C, W2 and H2.4, isothermal three stage service type strain cycling test planned for 10 000 h to failure (a) and stress strain hysteresis loop at mid-life selected with the aid of rainflow cycle counting (b), 1Cr-1Mo-0.7Ni-0.3V steel at 525 °C [8]

746

CONCLUSIONS

- The strain cycling at the heated surface of components can be simulated by service type creep fatigue tests.
- Isothermal tests on a bainitic and a martensitic rotor steel lead only to insignificantly different numbers of cycles to failure as comparable anisothermal tests.
- In the long term region, more economical package type tests can be carried out.
- A creep fatigue life analysis demonstrates the applicability of the generalized damage accumulation rule. As a basis, an analysis of cyclic deformation which considers internal stress, mean stress, stress relaxation and cyclic softening is needed. Preloading influences due to fatigue and creep have taken into consideration.
- Three stage cycling comprising coldstart, warmstart and hotstart cycles leads to similar values of relative creep fatigue life as single stage cycling.
- For failure times above several 1 000 h, three stage cycling can also be tested by isothermal package type cycles.
- For multi-stage strain cycles including coldstart and warmstarts the long term strain cycling is not negligible.
- A programme SARA for life prediction of multi-stage creep fatigue strain cycling is under development.

Thanks are due to the Bundesminister für Wirtschaft (AIF-No 8608), to the Verein Deutscher Eisenhüttenleute and to the Arbeitsgemeinschaft Warmfeste Stähle for their support of the work.

REFERENCES

1. Tremmel, D. and Mayer, K. H. (1984). *EPRI-Seminar "Life Assessment and Improvements of Turbo-Generator Rotors for Fossil-Plants"*, Raleigh.

2. Mühle, E.-E. and Gobrecht; E. (1993). *DVM-Vortragsveranstaltung "Betriebsfestigkeit"*, München.

3. Kloos, K. H., Granacher, J. and Rieth, P. (1980). In: *Proceedings of the Int. Conf. of Engineering Aspects of Creep*, 1, pp. 91-96; Sheffield.

4. Granacher, J. and Scholz, A. (1992). In: *Proceedings of the Third Int. Conf. on Low-Cycle Fatigue and Elasto-Plastic Behaviour of Materials*, pp. 235-241; K.-T. Rie (Ed.). Elsevier Science, London, New York.

5. Granacher, J. and Scholz, A. (1993). *Mat.-wiss. u. Werkstofftech.* 24, 409.

6. Granacher, J., Scholz, A. and Berger, C. (1997). In: *Proceedings of the of the Fourth Int. Charles Parsons Turbine Conf.*, pp. 592-602, A. Strang, W.M. Banks, R.D. Conroy, M.J. Goulette (Eds.). Ihe Institute of Materials, London.

7. Smith, K. N., Watson, P. and Topper, T. H. (1970). *Journal of Materials* 4, 767.

8. Kloos, K. H., Granacher, J. and Scholz, A. (1996). *Mat.-wiss. u. Werkstofftech.* 27, 331.

9. Granacher, J. and Scholz, A. (1996). In: *Fatigue under Thermal and Mechanical Loading*, pp. 209-214, J. Bresser and L. Remy (Eds.). Kluwer Academic Publishers, Dordrecht.

10. Wittich, R. (1981). *VGB Kraftwerkstechnik* 61, 383.

EFFECT OF SEQUENCE OF LOADING ON CUMULATIVE FATIGUE DAMAGE IN TITANIUM ALLOY

K. GOLOS

Warsaw University of Technology, Institute of Machine Design Fundamentals,
84 Narbutta St., 02-524 Warsaw, Poland

ABSTRACT

A method for evaluating the cumulative damage resulting of cyclic loading is presented. Both the crack initiation and propagation stages of the fatigue failure process are included in consideration. The experimental data for two stage loading tests of titanium alloy are presented. The comparison of the obtained experimental data for titanium alloy with the proposed cumulative damage theory is made, and is shown to be in fairly good agreement.

KEYWORDS

Cumulative damage, fatigue, sequence of loading, titanium alloy.

INTRODUCTION

Cumulative fatigue prediction is concerned with estimating the time length that material can serve its intended design function when it is subjected to cyclic loading. Because of the multide of possible loading patterns, it does not appear in general, that fatigue life results can be compiled for complex loading histories similar to that accumulated for pure sinusoidal loading. This indicates that for complex loading histories a certain amount of analysis must be resorted to in order to overcome the expected deficiency in experimental data directly relating to particular loading histories.

As yet no fatigue prediction method has been proposed for cyclic loading histories that does not require some sort of experimental data. Recognising this need, it has been the wish that no more than pure sinusoidal fatigue data would be necessary for fatigue prediction associated with any complex loading [1-4].

The main assumption adopted by most investigators is that operation at any given cyclic loading stage above fatigue limit will produce fatigue damage, the seriousness of which is related to the some damage parameters and history of loading.

748

It is further postulated that the damage incurred is permanent and operation at several different loading stages in sequence will result in an accumulation of total damage equal to the sum of the damage increments occurred at each individual loading level. When the total accumulated damage reaches a critical value fatigue failure occurs. Although the concept is simple in principle, much difficulty is encountered in practise, because the proper assessment of the amount of damage incurred by operation at any given load stage for a specified number of cycles ni is not straightforward. Many different cumulative damage theories have been proposed for the purposes of assessing fatigue damage caused by operation at any given stage load and the addition of damage increments to properly predict failure under conditions of spectrum loading.

In the present work the phenomenological theory is discussed. The results of the experimental study of the cumulative fatigue damage in titanium alloy are presented.

Tests has been performed for two-stage loading tests on solid specimen using MTS uniaxial servohydraulic testing machine.

A UNIFIED THEORY OF FATIGUE FAILURE

The phenomenon of the cumulative fatigue damage might be understood either from the point of view of the nucleation of critical crack and their subsequent propagation.

The amount of accumulated fatigue damage of steels can be associated either with a reduction of endurance limit or with the length of the propagating fatigue crack. These two approaches are equivalent on the basis of the nature of the nucleation mechanism of the critical crack size, and from the point of view of their subsequent propagation. One of the current topics of research in fatigue is the quantitative determination of the end of the initiation stage and the beginning of the propagation stage. We assume that the transition between these two stages can be expressed in terms of the critical damage curve which is associated with French's curve. This curve can be defined in the following manner. A specimen is cycled at a chosen stress level, Ds , for a given number of cycles, then the stress level is decreased to the fatigue limit and the cycling is continued. If the specimen fractures during the cycling at the fatigue limit, then the initial stress level $\Delta\sigma$ lies above the critical damage curve.

If the specimen does not fracture, the initial stress level was below the critical damage curve. Therefore, we can find the stress level associated with number of cycling at that damage is the same as that at the fatigue limit. This curve called critical damage curve separates the initiation and propagation stages and is associated with the critical crack length.

We postulate that for the crack propagation stage, the material has a "reduced" fatigue limit defined by σ^* and life time N^* , and is obtained through the fracture mechanics for notched specimen [5] or from the intersection of the extrapolated original σ - N curve with the critical curve.

It is assumed that the damage, d resulting from n cycles applied at a level loading described by damage parameter with an associated number of cycles to failure N is defined as

$$d = \left(\frac{n}{N}\right)^{f(\Psi;p)} \tag{1}$$

where $f(\psi, p)$ is a function of ψ, is damage parameter and p are material constants. This damage concept, it is noted, is specified as a nonlinear relationship for the cyclic loading for which the linear concept used in Palmgren-Miner's theory is a special case. As in the case of linear damage, the damage specification in (1) implies that failure occurs for pure sinusoidal history when n=N , that is when d=1.

In the analysis two-level sinusoidal history and the equivalent damage postulate will be used to present cumulative fatigue damage hypothesis. Now consider a specimen which is subjected to two-block cyclic loading. Let the damage parameter associated with the first loading block be denoted by ψ_1 at which is applied n_1 cycles. The damage curve can be expressed as:

$$d = \left(\frac{n_1}{N_1}\right)^{f(\Psi_1;p)} \qquad (2)$$

Changing the applied load and denoting the associated damage parameter by ψ_2 we continue cycling until failure occurs. Application of n_1 cycles at ψ_1 level will cause same damage in material which can be determined from the damage curve. We could find an equivalent number of cycles applied at the level ψ_2 which would cause the same amount of damage as the first loading block. Noting that

$$\left(\frac{n_{2eq}}{N_2}\right)^{f(\Psi_2;p)} = \left(1 - \frac{n_1}{N_1}\right)^{f(\Psi_1;p)} \qquad (3)$$

we get the cumulative damage low for the two-stage loading in the form of

$$\left(\frac{n_1}{N_1}\right)^{f(\Psi_1;p)/f(\Psi_2;p)} + \frac{n_2}{N_2} = 1 \qquad (4)$$

Putting as

$$f(\Psi_i;p) = \frac{1}{\lg(N_i/N^*)} \qquad (5)$$

we get the hypothesis presented by Golos in [3,4], i.e.

$$\left(\frac{n_1}{N_1}\right)^{\lg(N_2/N^*)/\lg(N_1/N^*)} + \frac{n_2}{N_2} = 1 \qquad (6)$$

We observe from the above that for the increasing value of "reduced" fatigue limit, N^*, the equation (6) is the same as the linear Palmagren-Miner rule [6,7]. Also in the case when the slope of the damage line is constant to that of the life curve we obtained the Palmgren-Miner rule.

COMPARISON WITH EXPERIMENTAL DATA

The material used in this investigation was titanium alloy. The chemical composition of the bar (wt.%) : (5,5-7.0)Al, (2.0-3.0)Mo, (1.0-2.5)Cr, (0.15-0.4)Si, (0.2-0.7)Fe, remainder titanium.

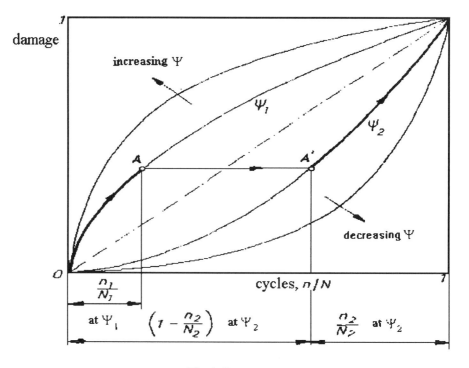

Fig. 1. Damage curves.

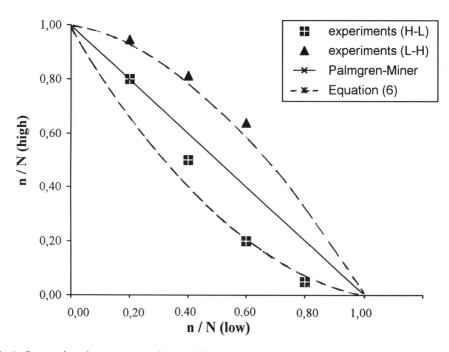

Fig.2. Comparison between experimental data and prediction for two-level loading.

Tests were conducted on solid specimens having 6 mm diameter. All tests were performed at ambient room temperature, in load controlled fully reversed cyclic mode. The testing program varied from low-to-high (L-H) and from high-to-low (H-L). The corresponding number of cycles to failure at the analysed stages ranges from $N = 800$ to $N = 40200$.

The results for the two-stages experiments are shown in Fig.2. In these figures, the ordinate represents the fraction of life spent at the high strain amplitude, and the abscissa is the fraction spent at the lover amplitude. The predictions of the proposed theory, equation (6), and the linear Miner-Palmgren rule are also shown in these figures. It is seen from the figures, that the investigated theory correctly predicts the trends of experimental results for both high and low strain amplitudes and given sequences.

REFERENCES

1. Fuchs, H.O. and Stephens, R.I., (1980) *Metal fatigue in engineering*, Wiley, New York.
2. Golos, K., (1988) Archiwum Budowy Maszyn, Vol.25, pp.5-16.
3. Golos, K., Ellyin, F., (1988) *Generalisation of cumulative damage criterion to multilevel cyclic loading*, Theoretical and Applied Fracture Mechanics, pp.169-176.
4. Golos, K., (1994) *Assessment of fatigue life of engineering structures*, Int. J. Pressure Vessel, Vol.59, pp. 307-311.
5. Lukas, P. and Klesnil, M., (1981) Mater. Sci. Engrg., Vol.47, pp.61-66.
6. Miner, M.A., J. Appl. Mech. Trans. ASME, Vol.12, (1945), A159-A164.
7. Palmgren, A., (1924) Z. Ver. Deutsch, Ing., Vol.68, P.339.

Low Cycle Fatigue and Elasto-Plastic Behaviour of Materials
K-T. Rie and P.D. Portella (Editors)
753

UNIVERSAL J-INTEGRAL FOR ASSESSING MULTIAXIAL LOW CYCLE FATIGUE LIVES OF VARIOUS MATERIALS

D. S. TCHANKOV[1] and M. SAKANE[2]

[1]Department of Strength of Materials, Technical University of Sofia,1000 Sofia, Bulgaria
[2]Department of Mechanical Engineering, Ritsumeikan University, 1-1-1 Nojihigashi,
Kusatsu-shi, Shiga 575-77, Japan

ABSTRACT

This paper proposes a unified correlation method of high temperature tension-torsion multiaxial low cycle fatigue lives of various kinds of materials. Many multiaxial strain/stress parameters have been proposed so far but the results of the correlation was material dependent. This paper proposes a new damage function based on J integral under biaxial stress state, termed universal J integral. The integral gave an excellent correlation of the multiaxial low cycle fatigue lives of SUS304, 1Cr-1Mo-1/4V, Inconel 738LC and 63Pb-37Sn solder. It is a suitable design criterion for engineering works.

INTRODUCTION

Multiaxial fatigue is the frequent cause of fracture of many components and several multiaxial parameters have been proposed [1-4]. The multiaxial parameters can be classified into three types by the physical background. They are the parameter based on yield theories, that based on strain energy and that based on crack behaviour or critical plane. Multiaxial parameters used in engineering design are mostly a simple extension of the static plasticity theories, such as the maximum principal stress and strain, the maximum shear stress and strain, the Mises (octahedral) strain and stress, etc. These parameters have an advantage of an apparent physical background and another advantage is that the evaluation of these parameters needs no material constants. However, recent studies are revealing that the suitable parameter for multiaxial low cycle fatigue (LCF) life is material dependent. Developing a new multiaxial parameter which is available to broad materials has been needed.

The objective of this paper is to propose a new multiaxial low cycle fatigue damage function which is universally applicable to various kinds of materials. The damage function is based on J integral under multiaxial stress states and has a small number of material constants. The brief derivation of the parameter will be stated and the application of the parameter to the multiaxial low cycle fatigue data of Type 304 stainless steel (SUS304) [5], 1Cr-1Mo-1/4V steel (CrMoV) [6], 63Pb-37Sn solder [7] and Inconel 738LC conventional cast Ni-base superalloy [8] will be discussed.

J-INTEGRAL BASED DAMAGE FUNCTION

Sakane et. al. [9] proposed an idea for the J-integral estimate under biaxial stress states and submitted two different J equations for small and large scale yielding bewlow.

$$J_{el} = \pi E a (\varepsilon^*)^2 \tag{1}$$

$$J_{pl} = 5.1 \times 10^{-3} \times f_1 \times f_2 \times f_3 \times (\varepsilon^*)^{1.25} \tag{2}$$

Equation (1) is for the small scale yielding (elastic J integral) and equation (2) is for the large scale yielding (plastic J integral). The elastic J integral is functions of Young's modulus, E, half crack length, a, and the equivalent strain based on crack opening displacement (COD strain)[3], ε^*. The plastic J integral is functions of strain biaxiality, f_1, crack geometry, f_2, material constant, f_3, and COD strain. The respective parameters are expressed as below.

$$f_1 = -33\phi^2 + 75, \tag{3}$$

where ϕ is the principal strain ratio defined as $\phi = \varepsilon_3 / \varepsilon_1$; ε_1 and ε_3 are the maximum and minimum principal strains, respectively. ϕ takes the unity value in torsion test while it has the value of –0.5 in tension test.

$$f_2 = \frac{w + 2a}{w - 2a} \sqrt{2aw} \tag{4}$$

In equation (2), a is a half crack length and w is a width of a cracked plate.

$$f_3 = \sigma_y^n A^{(1-n)} \left(\frac{A + \sigma_y}{E} \right)^{-0.3} \tag{5}$$

A and n are the strain the strain hardening coefficient and exponent expressed by the following equation.

$$\varepsilon_{pl} = \left(\frac{\sigma - \sigma_y}{A} \right)^{1/n} \tag{6}$$

ε_{pl} is plastic strain and σ_y is yield stress.

The COD strain is expressed as,

$$\varepsilon^* = \beta (2 - \phi)^{m'} \varepsilon_1, \quad \beta = 1.83, \quad m' = -0.66 \tag{7}$$

The COD strain physically expresses the amplitude of crack opening displacement for biaxially stressed mode I crack. The constants β and m' were derived by finite element analyses and do not depend on the material.

J integral is a fracture mechanics parameter which expresses the intensity of singularity of cracks in uniaxial elastic-plastic regimes. Many studies [10-15] reported that crack propagation rate in LCF is well correlated by J integral. Therefore, J integral may be an appropriate correlation parameter of multiaxial LCF lives if the major part of LCF life is the crack propagation life and J integral correlates well crack propagation rate under multiaxial stress states. This study employs these two assumptions and developed a damage function available to the multiaxial LCF data correlation.

For a perfectly elastic material, elastic J integral shown in equation (1) may be applicable to the date correlation and for a perfectly elastic material J integral in equation (2) to the data correlation. A combined parameter of equations (1) and (2) is presumably suitable parameter for elastic-plastic material, whereas the combination of the elastic and plastic parts loses its pure mathematical meaning as Dowling noted [12]. Weight function was introduced into the damage function, which determines the contribution of the elastic and plastic J integrals. Elastic constant was used as the weighting function of elastic J integral and yield stress as that of plastic J integral. Young's modulus is only the material constant which characterize the elastic property so that it is used as a weight function. The contribution of plastic J integral was weighted by the yield stress since yield stress mainly determines the contribution of plastic deformation to multiaxial parameter. Therefore, the multiaxial parameter termed universal J integral (UJI) is expressed by the following equation.

$$UJI = \frac{1}{BE^b}\frac{\Delta\varepsilon_{el}}{\Delta\varepsilon_{tot}}J_{el} + \frac{1}{C\sigma_y^c}\frac{\Delta\varepsilon_{pl}}{\Delta\varepsilon_{tot}}J_{pl} \qquad (8)$$

where J_{el} and J_{pl} are J-integral in small scale and large scale yielding. $\Delta\varepsilon_{el}$, $\Delta\varepsilon_{pl}$ and $\Delta\varepsilon_{tot}$ are the elastic, plastic and total strain ranges, respectively.

RESULTS AND DISCUSSION

Figure 1 shows the correlation of multiaxial LCF lives of the four kinds of materials with the Mises strain range. The four materials are SUS304, CrMoV, 63Sn-37Pb solder and Inconel 738LC conventional cast nickel-base superalloy. The specimen geometry is a hollow tube with 9mm I.D. and 11mm O.D. for the three former specimens and is a hollow tube with 5mm I.D. and 8mm O.D. Test temperature is shown in the figure.

Mises strain correlates the LCF lives of solder within a factor of two scatter band but it overestimates the shear dominant LCF lives for CrMoV and SUS304 steels. For these two materials, the torsion data are correlated out of a factor of two scatter band. Intermediate correlation is found in Inconel738LC between the solder and the two steels. Therefore, the suitable strain parameter changes by material. The correlation of LCF lives of the four materials with Mises stress range parameter is shown in Fig.2. The stress parameter correlates well only Inconel LCF data, it overestimates LCF live for other materials. This leads to the conclusion that the multiaxial stress parameter is material dependent as well as strain parameter.

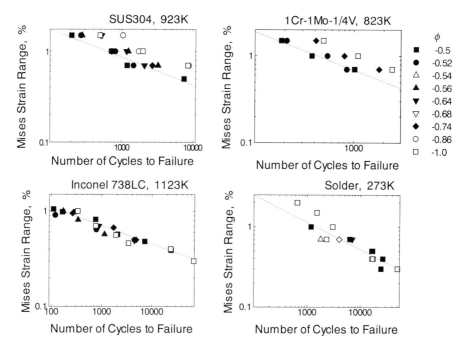

Fig.1. Variation of Mises strain with number of cycles to failure for different materials.

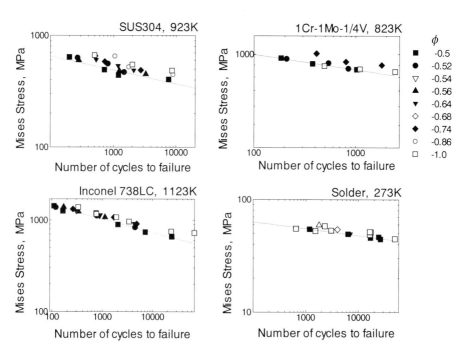

Fig.2. Variation of Mises stress with number of cycles to failure for different materials.

For the life prediction based on UJI, the following equation was used,

$$UJI = N_f^{\,k} \tag{9}$$

where k is a constant and N_f is number of cycles to failure. The constants B, b, C, c and k must be determined for LCF life prediction. They were determined by the least square method. The material constants determined are listed in Table 1. This paper uses a set of material constants for the prediction of the four kinds of materials.

TABLE 1
Material constants used for the life prediction

Parameter	B	b	C	c	k
Value	$6.6 \; 10^{-8}$	1.8535	122600	-1.1088	-0.46

Figure 3 shows the correlation of multiaxial LCF lives of the four kinds of materials with UJI parameter. The solid line in the figure represents the predicted fatigue lives estimated by equation (9) and dashed lines represent the scatter band of a factor of 3 based on the solid line. All the experimental data except few points are gathered within the scatter band independent on material. Therefore, tension-torsion multiaxial fatigue lives can be estimated by only knowing the Young's modulus and yield stress of the material. The constants listed in Table 1 are presumably a universal constant and their values are available to other materials.

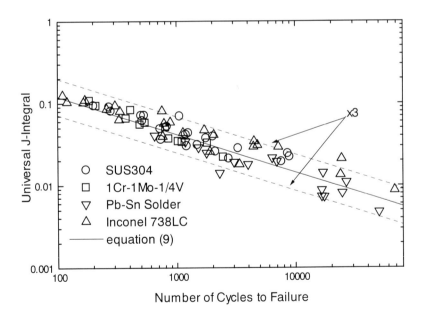

Fig.3 Variation of Universal J-Integral (UJI) with number of cycles to failure (Nf) for different materials and $-1 \le \phi \le -0.5$.

It should be noted that the results of the correlation in Fig.3 cover a wide range of the material. SUS 304 stainless and CrMoV exhibit the elastic-plastic deformation behaviour in LCF regime but Inconel 738 shows mostly elastic deformation behaviour. Plastic strain range is far larger than elastic strain in solder in LCF regime. The UJI parameter has a simple form and advantage of validity of its wide application with a small number of material constants.

CONCLUSIONS

The main conclusions of this paper are:

1. A new parameter entitled Universal J-Integral (UJI) for the fatigue life prediction is developed to correlate many LCF lives data for different materials in a unique curve: $UJI = N_f^k$. The UJI proposed here gives an excellent correlation of multiaxial LCF lives of materials and is a suitable design criterion for engineering works.

2. The UJI is based on the J-Integral estimates for the multiaxial LCF. The UJI consists of two components: the first corresponds to the elastic (or small scale yielding) and the second corresponds to the plastic strains (or large scale yielding). The plastic term is considered to depend on the material behaviour, while the elastic term depends only on the elastic modulus of the material. The calculation procedure is quite simple, and requires materials parameters from uniaxial tests only. The consideration of different crack propagation behaviour lead to introducing a weighting function which results in significantly improving the LCF life data correlation.

ACKNOWLEDGEMENT

A JSPS fellowship for the first author in the Ritsumeikan University is gratefully acknowledged.

REFERENCES

1. Socie D.F., J of Engng Mater. and Tech., Trans. ASME, 1988, vol. 110, pp. 380-388.
2. Ellyin, F., Valaire, B., SM Archives, vol. 10, 1985, pp. 45-85.
3. Itoh, T., SakaneM., Ohnami, M., Trans. ASME, JEMT, 1994, vol.116, pp90-97.
4. Vesselinov,K.V., Tchankov,D.S., Dimov,D.M., in Proc. XI Coll. Mech. Fatigue of Metals, Kiev, 1992, Vol.1, pp 14-19.
5. Sakane,M., Ohnami, M., Sawada,M., Trans. ASME, JEMT, 1987, vol.109, pp236-243.
6. Sakane,M., Ohnami,M., Sawada,M., Trans. ASME, JEMT, 1991, vol.113, pp241-251.
7. Yamamoto, T, et al., Proc. 31th High Temperature Strength of Materials, JSMS, 99, 1993.
8. Isobe, N.,et al., Proc. 30th High Temperature Strength of Materials, JSMS, 59, 1992.
9. Sakane, M., Itoh,T., Onami, M., in Proc. ICM 9, April 1997, Australia, pp.2071-2078.
10. Chow, T.J.Lu, Eng. Fract. Mech. 1991, (39), No1, pp1-20.
11. Dankert,M., Savadis, G., Seeger, T., Mat.-wiss. Und Werkstofftech, 1996, vol. 27, pp.1-8.
12. Dowling, W., Engng Fract. Mech., 1987, 223, pp333-348.
13. Hatanaka, K., Fujimitsu, T., in Proc. VI Int. Congr. On Exp. Mech., vol. 2, SEM, Portland, 1988, pp.1008-1013.
14. Savadis,G., Seeger,T., Fatigue&Fract. Eng. Mater.& Struct., 1996, vol.19, 8, pp.975-983.
15. Shikida,M., Sakane,M., Ohnami, M., JSME Int. J., Ser.A, 1995, vol. 38, 1, pp104-110.

Low Cycle Fatigue and Elasto-Plastic Behaviour of Materials
K-T. Rie and P.D. Portella (Editors)
759

CUMULATIVE DAMAGE MODELLING IN A LIFE PREDICTION UNDER RANDOM LOADING CONDITIONS

J. ČAČKO

Institute of Materials and Machine Mechanics, Slovak Academy of Sciences,
Račianska 75, P.O.Box 95, SK-830 08 Bratislava 38, Slovak Republic

ABSTRACT

An estimation of cumulative damage under service loading is significantly influenced by a time loading history. The problem of the history effect can be converted to the problem of investigation of loading with non-zero stress/strain mean value. Basis of the conversion method consists in specification of an equivalent amplitude of the harmonic cycle with zero stress and strain mean values. At the same time, a relative damage that is accumulated in the course of equivalent cycle must be the same as the relative damage of actual cycle that has been identified by the rainflow method. Because it is necessary to know both stress and strain values at any time, a conversion procedure is introduced in the paper. This procedure allows to determine the response value under either stress or strain control, without identification of the complete time loading history, using a rainflow decomposition. The method is also acceptable for structural materials with different tension and compression response properties. The submitted approach enables to utilize the knowledge of cumulative damage process and fatigue behaviour prediction under harmonic loading and to use it for any complex operational loading spectrum.

KEYWORDS

Rainflow decomposition, residual stress/strain, equivalent amplitude.

INTRODUCTION

The time loading history influence a fatigue behaviour of materials considerably. It is very difficult to predict a fatigue life and service reliability, especially under nonstationary operating loading. Moreover, an identification and/or specification of residual stresses is frequently a big problem in this case.

In order to predict a cumulative damage and residual life of structures, we must know stress-strain response properties of the material. If no residual stresses have been presented in a material, there is no problem to evaluate a cyclic degradation using some cumulative hypothesis or fatigue criteria. Otherwise, some recalculation procedure must be available and usual criteria must be altered.

The most of hypotheses are based on a dependence of a relative cumulative damage upon amplitudes of closed cycles with respect to corresponding mean values. Under residual stresses occurrence, an equivalent amplitude of a closed cycle must be calculated, so that an actual cycle increases a cumulative damage identically as a cycle with the equivalent amplitude without residual stresses consideration. The equivalent amplitude is calculated according to cyclic stress-strain response properties because it is dependent on a total plastic deformation that corresponds to the stress parameters.

For specification of all required stress and strain parameters of an actual closed cycle, we need to know the complete time loading history. But that is not very convenient, especially if high cycle fatigue properties would be investigated. Therefore, a suitable simplified interpretation of a time history is very desirable in this case. Such an interpretation can be made using a rainflow decomposition, which can decrease time and memory demands many times. The rainflow decomposition procedure is independent upon material properties, and it can be applied for any material and for any loading nature.

In order to create an effective method, the procedure must be optimally formulated. Therefore, stress-strain response relationships, e.g. a cyclic deformation curve, must be estimated and then analytically expressed and transformed, so that the procedure would be quick and performed on-line, in a real time. Such an approach enables a continuous monitoring of a fatigue damage and of a residual life with respect to a residual stresses influence.

CUMULATIVE DAMAGE EVALUATION

In order to evaluate the fatigue damage under dynamic loading, the cumulative representation of a relative damage of a closed cycle is accepted. Under harmonic loading with constant both amplitude and mean value of a cycle, stress and/or strain -life (S-N) diagrams can be experimentally obtained. Then it is possible to specify a relative damage of a closed loading cycle according to some hypothesis (e.g. Palmgren-Miner, Corten-Dolan, etc.). No theoretical problem could arise in the cumulative damage specification, if we have a macroblock that is composed of cycles with various amplitudes but with a constant mean value. The problem could arise, if the mean values of cycles are changing, because the relative damage of individual cycles cannot be simply added in this case, neither if the cycles have been identified according to the rainflow method [1]. The relative damage of a loading cycle with some definite amplitude and mean value under block loading need not correspond to the relative damage under pure harmonic loading. Moreover, the relative damage is significantly different in the case of a random loading.

The problem of the loading mean stress/strain effect is usually solved in such a way that instead of a cycle with the actual amplitude σ_a (stress control) or ε_a (strain control) and mean value σ_m (or ε_m respectively), we suppose the cycle with some equivalent amplitude σ_a^* (or ε_a^*) and zero mean value of σ (or ε). At the same time, of course, the damage effect of both cycles must be identical. The equivalent amplitude can be derived from the projection of parametric Haigh diagram [2].

The most serious difficulty is, that we must know complete load history until the actual loading cycle will occur, that is very inconvenient and frequently also practically uncontrollable. Therefore, it was necessary to find a new method to interpret the load history more properly. As a very perspective method seems to be the decomposition of closed hysteresis loops according to the rainflow method. For this purpose, the continual rainflow counting described in [3] can be very effectively used.

OPERATIONAL STRESS-STRAIN RESPONSE DIAGRAM UNDER RANDOM LOADING

In order to make an operational stress-strain response diagram, we must assume an ideal cyclic stable material, where hysteresis loops are entirely closing. At the same time, we also must assume that the material is in the stage before crack propagation, i.e. the hysteresis loops are not distorted.

In the case of a monotonous loading from the zero initial point (Fig. 1), we trace the operating stress-strain response curve *(OSSRC)* $\sigma = \Phi(\varepsilon)$ up to the reversal point σ_1 which is the origin of the first hysteresis loop. *OSSRC* is the curve that usually runs between the static monotonous curve and the cyclic strain curve for the saturated amplitudes. It is the real curve that characterizes a stress-strain response for the first dynamic cycle.

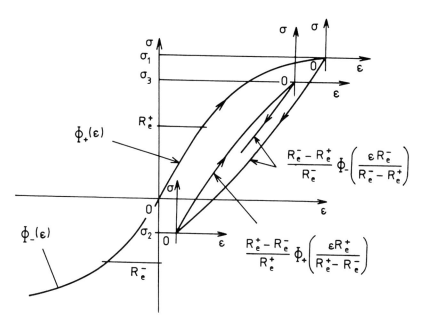

Fig. 1. Stress-strain response diagram and an analytical approximation of hysteresis loops.

Because the stress-strain responses in tension and in compression could differ (i.e. the yield stress in tension $R_e^+ > 0$ and the negative value of the yield stress in compression $R_e^- < 0$ is not equal), we generally suppose the *OSSRC* in the tension range as $\Phi_+ (\varepsilon)$, and in the compression range as $\Phi_- (\varepsilon)$. But we usually suppose that $R_e^+ = - R_e^-$ and then $\Phi_+ (\varepsilon) = -\Phi_-(-\varepsilon) \equiv \Phi(\varepsilon)$ and it holds

$$\sigma = \pm 2\ \Phi\!\left(\frac{|\varepsilon|}{2}\right) ; \quad \text{for } \varepsilon \gtrless 0 , \text{ respectively.} \tag{1}$$

The experimental results suggest that *OSSRC* can be expressed in the form: $\Phi(\varepsilon) = \pm K |\varepsilon - \varepsilon_e|^n$, for $\varepsilon \gtrless 0$, where ε_e is the elastic strain, and K and n are definite constants. Then, the relationship (1) can be adapted into the form

$$\sigma = \pm 2\, K \left(\frac{|\varepsilon - \varepsilon_e|}{2} \right)^n \; ; \quad \text{for } \varepsilon \gtrless 0 \text{, respectively.} \tag{2}$$

EQUIVALENT AMPLITUDE SPECIFICATION

The equivalent amplitude σ_a^* is such an amplitude of harmonic cycle which has zero mean value of both stress and strain, whereby a service life under such a loading is the same as that under repeating application of the actual cycle. Generally, it is possible to proceed from the projections of Haigh diagram $\sigma_A = f(\sigma_M)$ for different $\Delta\varepsilon_m$, where $\Delta\varepsilon_m$ denotes the shift of strain mean value of the actual cycle from the mean value of the corresponding cycle on *OSSRC*, e.g. $\Delta\varepsilon_m^{(1)} = \varepsilon_{m_2} - \varepsilon_{m_1}$; $\Delta\varepsilon_m^{(2)} = \varepsilon_{m_3} - \varepsilon_{m_1}$ according to Fig. 2. Such a diagram can be obtained like a classical Haigh diagram (for $\Delta\varepsilon_m = 0$) but for the material with a plastic prestraining. Then, we can construct the triaxial Haigh diagram using a composition of marginal dependencies (Fig. 3). Supposing that the actual cycle is equivalent to the cycle with $\sigma_m = 0$; $\Delta\varepsilon_m = 0$, we can express the equivalent amplitude as follows

$$\sigma_a^* = \sigma_a \frac{\sigma_P}{\sigma_H} \,. \tag{3}$$

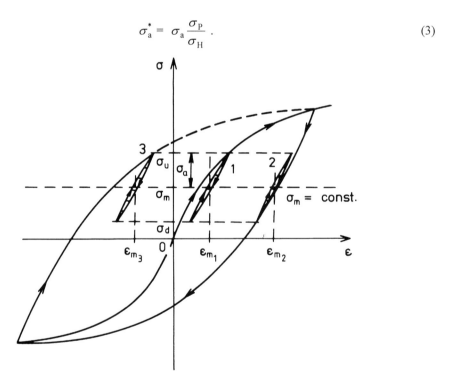

Fig. 2. Hysteresis loops with different residual strain.

For small σ_m and $\Delta\varepsilon_m$ values (mainly in the case of a narrow-band random loading process), we can consider the relevant part of the area in Fig. 3 as a plane (Fig. 4), and the relationship (3) can be linearized according to Fig. 5. Thus, we obtain the approximate relationship

$$\sigma_a^* = \sigma_a + \psi_\sigma \sigma_m + \psi_\varepsilon \Delta\varepsilon_m , \qquad (4)$$

where $\psi_\sigma = cotg\ \varphi_\sigma$ and $\psi_\varepsilon = cotg\ \varphi_\varepsilon$ (the dimension of ψ_σ is [1] and of ψ_ε is [MPa]).

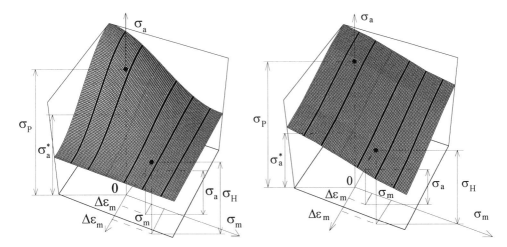

Fig. 3. Identification of an equivalent amplitude using the triaxial Haigh diagram

Fig. 4. Approximation of an equivalent amplitude using the linearized Haigh diagram

The equivalent amplitude can be effectively calculated using the description of the triaxial Haigh diagram according to Fig. 5.

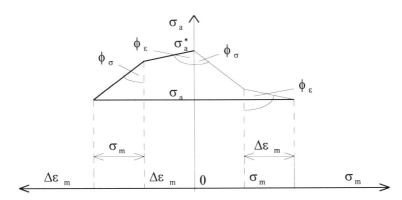

Fig. 5. Description of a linearized triaxial Haigh diagram to calculate the equivalent amplitude

Now we need an effective procedure, how to identify $\Delta\varepsilon_m$, without knowing the total time loading history.

RAINFLOW DECOMPOSITION OF A RANDOM LOADING SIGNAL

In order to identify the strain mean value of the actual hysteresis loop, we must be aware of the fact that in the case of an ideal cyclic stability of material, every hysteresis loop in the $\sigma - \varepsilon$ diagram is closing at the same point as it was starting to open, regardless the course of loading between the opening and closing instants. That means, we can dismiss already closed hysteresis loops during cyclic process in the $\sigma - \varepsilon$ diagram, and we can only consider hitherto unclosed loops and open branches.

If we indicate the origins of hitherto unclosed hysteresis loops as σ_1, σ_2, ... , σ_N , we could use the method which was described in [2]. Then we can express $\Delta\varepsilon_m$ as follows

$$\Delta\varepsilon_m = \Phi^{-1}(\sigma_1) - \Phi^{-1}(\sigma_N) + 2\sum_{i=2}^{N} \Phi^{-1}\left(\frac{\sigma_i - \sigma_{i-1}}{2}\right). \tag{5}$$

or if we know an analytical expression of *OSSRC*

$$\Delta\varepsilon_m = K^{-\frac{1}{n}}\left[\pm |\sigma_1|^{\frac{1}{n}} \mp |\sigma_N|^{\frac{1}{n}} \pm 2^{\frac{n-1}{n}} \sum_{i=2}^{N} |\sigma_i - \sigma_{i-1}|^{\frac{1}{n}}\right], \tag{6}$$

where for $\sigma_1 > 0$, $\sigma_N > 0$ and $\sigma_i > \sigma_{i-1}$, it holds the upper corresponding sign, otherwise the lower one.

CONCLUSIONS

The submitted method enables to evaluate the cumulative fatigue damage and to estimate the fatigue life under random loading on the basis of knowledge of the stress/strain life curve, parametric Haigh diagram and stress-strain response properties. According to the proposed procedure, the equivalent amplitude for any closed cycle in the loading history is specified, and the relative damage is calculated in the same way as in the case of operation under harmonic loading with zero mean value. The equivalent amplitude can be further calculated in order to respect the structure parameters (notches, surface treatment, welds, etc.), environmental effects (corrosion, radiation, high/low temperatures, etc.), loading mode (e.g. bending, torsion, complex and/or multiaxial loading) and other service conditions [3].

ACKNOWLEDGEMENT

This work was supported, in part, by the Slovak scientific grant agency, grant project No 95/5305/634.

REFERENCES

1. Čačko, J. et al. (1988). *Random Processes: Measurement, Analysis and Simulation*. Elsevier, Amsterdam.
2. Čačko, J. (1986). In: *Proceedings of ECRS-4*, ENSAM. Cluny en Bourgogne, pp 46-47.
3 Čačko, J. (1992) *Int. J. Fatigue* **14** , 183.

Low Cycle Fatigue and Elasto-Plastic Behaviour of Materials
K-T. Rie and P.D. Portella (Editors)

EFFECT OF THE DIFFERENT COUNTING METHODS OF CYCLIC LOADING SERVICE SPECTRUM ON CUMULATIVE FATIGUE DAMAGE

T. CALA and K. GOLOS

Warsaw University of Technology, Institute of Machine Design Fundamentals,
84 Narbutta St., 02-524 Warsaw, Poland

ABSTRACT

The non-deterministic feature of service loading as a sequence of variable amplitudes brings many difficulties and uncertainty in fatigue life evaluations. Practical implementation of cumulative fatigue rule in design against fatigue requires the analysis of cyclic loading spectrum. In the present paper two most often methods have been used: rainflow method and range branches method. Counting methods used to analysis cyclic loading with variable amplitude shows that fatigue lifetime depends on the treatment of the experimental loading spectrum. In the analysis the experimental loading spectrum of the cars elements have been used. All different step loading spectrum obtained from different above mentioned counting methods were transformed to constant mean value of stress patterns through Goodman's relation. Lifetimes were calculated according to Haibach's rule.

For the analysed experimental loading spectrum different results have been obtained, using different counting methods. However, some trends of these differences have been noted and they are reported in the paper. The calculation based on the rainflow method analysis has been found as the optimum counting procedure.

One aspect of the rainflow method needs some more consideration - it will always yield as largest range counted the load variation between the lowest trough and the highest peak. In other words, it is advisable to restrict the size of the load history on with the rainflow method can be applied at one time.

KEYWORDS

Counting methods, rainflow, range branches, loading spectrum, cumulative damage.

INTRODUCTION

Fatigue life calculation methods used by the industry are often based on the measurements of real changes in cyclic loading. The measurements are conducted in characteristic construction points using special measuring equipment. The results show the history of loading changes.

The real cyclic loading sequences, which are recorded for the characteristic construction points, e.g., in cars, can function as the basis for predicting fatigue life of other construction elements. Such procedure allows an acceleration of the designing process by means of limiting the amount of the required experimental research and results in a reduction of the costs of project's implementation [1-4].

However, the complexity of the events, which accompany fatigue processes, suggests the necessity to consider the influence of various factors on lifetime.

It is known that calculation of fatigue life of a construction element subjected to variable loads depends on the properties of the material used, nature of loading changes, external conditions, etc. Additionally, when recording random loads in low-cyclic fatigue, results may vary depending on the method of sequence processing. In other words, already in the initial stage of processing of the recorded signal a strong influence can be observed of an error of method on the process of fatigue life evaluation [3].

A frequently used procedure in the analysis of fatigue life based on real loading changes consists in summation of the cycles, their transformation into constant mean values, and implementation of a proper hypothesis of cumulative fatigue damage[4-7].

This paper presents the impact of implementation of various methods of cycle counting (the *rainflow* method and the *range branches* method) on the fatigue life of a construction element. Major differences can be noticed in the calculated fatigue life depending on the character and complexity of the loading sequence, as well as on the adopted method of cycle counting. The adoption of a proper method of cycle counting should be conditioned by the character of the analysed loading sequence.

DATA ANALYSIS

An analysis has been conducted of a sequence recorded as cyclic exchanges of stresses recorded for the car element. Sequence properties are shown in Table 1. Properties of investigated material are shown in Table 2.

The recorded sequence should be referred to the unnotched element with circular cross-section. The element underwent cyclical tension-compression only. Only sections of normal stresses were considered.

Table 1. Properties of loading sequence

Maximum loading peak [MPa]	Minimum loading peak [MPa]	Mean sequence stress [MPa]	Range of one loading level [MPa]
460	-51	217	17

Table 2. Properties of investigated material

Tensile strength	Yield strength	Number of cycles to failure	Fatigue limits	Wöhler curve exponent
R_m	$R_{p0,2}$	N_f	Z	m
[MPa]	[MPa]	[cycles]	[MPa]	
650	470	1.2E+06	210	8

DAMAGE CALCULATION

In element's fatigue life calculation the Haibach hypothesis [5] of damage cumulative has been used. Sequence processing is required for the purposes of an analysis of the loading sequence. The processing consists in a reduction of the sequence by means of a proper method of calculation and transformation of the obtained cycles into constant mean values. The implemented algorithm is shown in Fig. 1.

The calculation procedure enables an exchange of the loading sequence with an equivalent set of fatigue cycles. A fatigue cycle corresponds to a closed-loop hysteresis in the stress-strain system.

Each cycle obtained in that way has a different mean value and amplitude. In order for the discussed issue to have a two-dimensional character, each cycle requires implementation of a transformation procedure. Upon transformation each cycle will have a constant mean value (e.g., equalling zero). Thus, an equivalent set of cycles will have been received with correctly adjusted amplitudes and constant mean values. This paper uses the Goodman transformation equation, which has the following form:

$$\sigma_{Tai} = \frac{\sigma_{ai}}{\left(1 - \sigma_{mi} / R_m\right)} \tag{1}$$

where: σ_{Tai} - cycle stress amplitude after transformation, σ_{ai} - cycle stress amplitude, σ_{mi} – mean stress of cycle, R_m - tensile strength.

By sorting the transformed cycles according to their numbers on individual loading levels we can obtain the block loading spectrum.

The Haibach cumulating hypothesis is a modification of the Palmgren-Miner linear hypothesis [6,7]. The modification consists in adding an identical element to the Palmgren-Miner relation for the loads below the fatigue limit, but with the exponent of: $m' = 2m - 1$. The hypothesis allows, therefore, for an impact of the stresses below the fatigue limit, in accordance with the following equation:

$$N_c = N_{max} \left/ \left(\sum_{i=1}^{s} \left(\sigma_{ai} / \sigma_{a\_max}\right)^m \cdot t_{ai} + \sum_{i=s+1}^{L_p} \left(\sigma_{ai} / \sigma_{a\_max}\right)^{m'} \cdot t_{ai} \right) \right. \tag{2}$$

768

where: N_c - lifetime expressed as a number of cycles to failure, N_{max} - number of cycles to failure corresponding to the maximum loading amplitude, σ_{a_max} - maximum loading sequence value, t_{ai} - relative number of sequence cycles on a given loading level, s - number of loading levels above the fatigue limit, L_p - total number of loading levels, m - Wöhler curve exponent in the logarithmic system.

Fig. 1. Algorithm to estimate lifetime.

COMPARISON OF RESULTS

In accordance with the above described algorithm calculations have been made of fatigue life. The results of the analysis show that there were discrepancies concerning the obtained loading spectrum, depending on the employed method of cycle counting (Fig. 2).

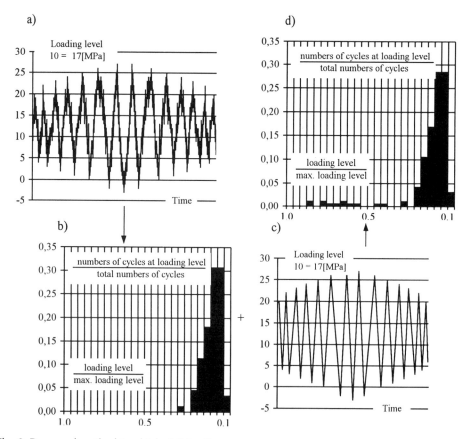

Fig. 2. Proposed method to obtain full loading spectrum.

For certain sequence fragments (Fig. 2a) the *rainflow* method does not produce acceptable results. The method results in cycles as well as in a certain remaining part of the sequence, which cannot be subject to further analysis using that method. Hence, the loading spectrum built on the basis of the obtained cycles (Fig. 2b) is not in its full form.

In order to supply the loading spectrum with the lacking cycles, the *range branches* method has been used in the remaining part of the sequence. Its use can be justified by the fact that it does not cause errors during simple loading sequences (i.e., such as in Fig. 2c) while retaining its quickness and simplicity. The final result of the analysis is a new loading spectrum obtained by summation of the cycles counted using the *rainflow* method and the cycles counted using the *range branches* method from the remaining part of the sequence.

The two loading spectra obtained using the presented methods (Fig. 2b and Fig. 2d) formed the basis for a fatigue evaluation using the Haibach fatigue damage cumulative hypothesis.

Table 3 presents lifetime results for a spectrum built using the above described analysis; the results were obtained using the following methods: the *rainflow method* alone, the *rainflow method* coupled with the *range branches* method, and the *range branches* method alone.

Fatigue lifetime results differ only slightly between those methods; this is due to the fact that the analysis, which uses the rainflow method alone, does not allow for all loading levels. Additionally, the omitted cycles are those from the levels with the highest amplitudes. Thus, they have a big influence on the evaluated fatigue life.

Table 3. Fatigue lifetimes for differences counting methods.

Counting method	Lifetime [cycles]
rainflow	3,16 E+09
rainflow + range branches	5,71 E+05
range branches	2,42 E+08

CONCLUSIONS

The paper covers a fatigue life analysis for the real loading sequence. Using the methods of cycle counting and the Goodman transformational equation, a loading spectrum has been built and lifetime has been evaluated using the Haibach hypothesis.

It has been observed that for certain loading sequences the *rainflow* method can not produce acceptable results. As a result of this method a part of the sequence remains which cannot be utilised by that method. The obtained loading spectrum does not include the cycles with highest loading amplitudes. As a result, lifetime is considerably higher than predicted.

In order to fill the remaining levels of the loading spectrum, the *range branches* method has been used in the part of the sequence remaining after the use of the *rainflow method*. The obtained fatigue life is considerably lower due to the fact that all sequence cycles have been allowed for. Simultaneous use of both discussed counting methods for sequence processing has been shown to produce the best results.

It has also been observed that the part of a sequence remaining after the *rainflow* method has been used would emerge only in specific loading sequences. Namely it is to be expected in steadily intensifying or steadily expiring sequence (Fig. 2a). In all other sequences, even those very complex, the *rainflow* method used alone seems to be the optimum choice.

REFERENCES

1. Cala, T., Golos, K., (1997) *Database integrated with system for fatigue analysis*, in: Proceedings Metody i srodki projektowania wspomaganego komputerowo, Warsaw University of Technology, pp. 69-75.
2. Downing, S.D., Socie, D.F. (1982) *Simple rainflow counting algorithms*, Int. J. Fatigue, pp. 31-40.
3. Golos, K., Cala, T.: (1997) *Computer aided analysis of loading spectrum*, in: Proceedings Poliootymalizacja 97, Koszalin, pp. 93-100.
4. Golos, K., Ellyin, F. (1988) *Generalisation of Cumulative Damage Criterion to Multilevel Cyclic Loading*, Theoretical and Applied Fracture Mechanics, pp. 169-176.
5. Haibach, E. (1970) In: *Modifizierate Lineare Schadensakkumulations-Hypothese zur Beruekssichtigung des Dauerfestigkeittsaballs mit Fortschreitender*, Darmstadt.
6. Miner, M.A. (1954) *Cumulative damage in fatigue*, J. Appl. Mech., pp. 159-164.
7. Palmgren, A. (1924) *VDI-Z*, **68**, 339.

Low Cycle Fatigue and Elasto-Plastic Behaviour of Materials
K-T. Rie and P.D. Portella (Editors)

LOW-CYCLE FATIGUE OF AROUND NOTCHED SPECIMENS OF ANISOTROPIC STEELS

KAZANTSEV A. G. and KORNILOVA G. Yu.
Department 32, Central Research Institute of Engineering Technologies,
Sharikopodshipnikovskaya str., 4
109088 Moscow, Russia.

ABSTRACT

Some results of experimental and theoretic low-cycle fatigue investigation of notched specimens produced from anisotropic steels are presented. For evaluating the magnitude of local stress and strain in notched specimens, cut out under different angles to the axes anisotropy, interpolation equation was used. It is based on energy relations and equations of Hill's theory of plasticity for anisotropic materials. The number of cycles before fracture was evaluated by using modified low-cycle fatigue equation of Coffin-Manson, taking into account orientation and values of normal strains and stress to the axes of anisotropy. Satisfactory coincidence of theoretic and experimental durability of notched specimens under low-cycle tension-compression loading was obtained.

KEY WORDS

Low-cycle fatigue, anisotropy of materials, notched specimens, stress-strain curve, durability.

Low-cycle tests of around smooth and notch ($\alpha_\sigma = 4,2$) specimens of anisotropic austenitic 16Cr-9Ni-2Mo and perlitic 2Cr-Ni-Mo-V steels were conducted. Steel 16Cr-9Ni-2Mo was delivered in a form of rolled plate 60 mm thick, steel 2Cr-Ni-Mo-V in a form of template , cut out from 160-mm forged shell. These materials are orthotropic, since it have three mutually perpendicular planes of symmetry of mechanical properties, and besides, as have shown results of tests, are circularly symmetric to the axis Z, which is perpendicular to the plane of rolling XY (axis X, Y – axis anisotropy) for steel 16Cr-9Ni-2Mo, and the axis X corresponds to the direction of rolling. For steel 2Cr-Ni-Mo-V, the axis Z corresponds to a radial direction of shell, axes X and Y can be accepted as meridional and tangential directions. Anisotropy of this materials can be characterized only by one parameter – an angle between axis Z and direction of cutting of specimen.

For these materials, low-cycle fatigue curve on specimens cutting out in X and Z directions were received under tension-compression (fig.1, 2). For a more detailed study of the influence of anisotropy on fatigue life cyclic tests of specimens, cutting under various angles $\varphi = 0$–$90°$ to the axis Z through $15°$ were conducted. The results of this tests ($\Delta\varepsilon = $ const) are shown on fig. 3, 4. For both materials at $\varphi < 45°$ a monotonic increase of durability with increasing angle φ takes place. At $\varphi > 45°$, experimental points have appeared, close to appropriate significance, received

772

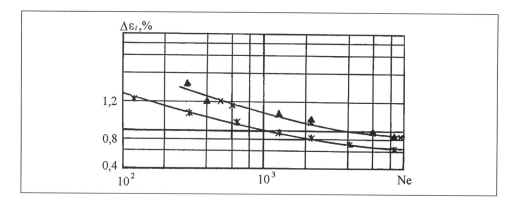

Fig. 1. Low-cycle fatigue curve 16Cr-9Ni-2Mo steel: ...– $\varphi = 0$ (z); ... – $\varphi = 90°(x)$; ✕ – $\varphi = 45°$

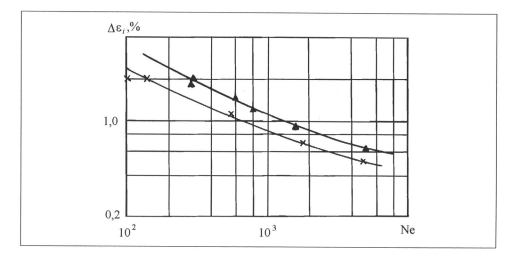

Fig. 2. Low-cycle fatigue curve 2Cr-Ni-Mo-V steel: ✕ – $\varphi = 0$ (z); ... – $\varphi = 90°(x)$;

in longitudinal specimen tests.

Thus, the minimal cyclic durability corresponds to the z-specimens. It decreases up to 2–4 times in comparison with x-specimens.

As follows from experimental data, stress-strain curves for16Cr-9Ni-2Mo steel depends on direction of cutting of specimen and minimal deformation resistance correspond to $\varphi = 45°$ (fig. 5). Behaviour of this steel is described by Hill's plasticity theory for anisotropic materials. Stress-strain curves for this steel are shown for static loading (K = 0) in σ–ε coordinates and for cyclic loading in S–ε coordinates (K = Ne). The latter the origin of coordinate coincides with the beginning of unloading in the cycle. Equivalent Hill's stress-strain curve for cyclic loading in coordinates S_{eq}–ε_{eq} calculated from S–ε curves is shown by dotted line on fig.5.

Cyclic stress-strain curves for 2Cr-Ni-Mo-V steel did not depend on the direction of cutting.

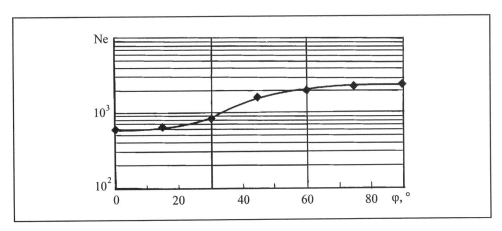

Fig. 3. Change in fatigue life with varied angle of cutting of specimens 16Cr-9Ni-2Mo steel.

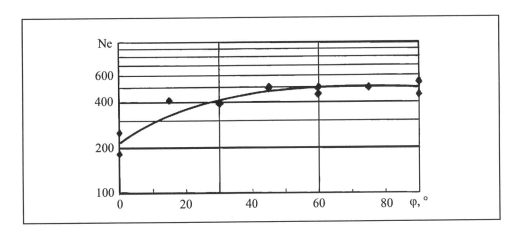

Fig. 4. Change in fatigue life with varied angle of cutting of specimens 2Cr-Ni-Mo-V steel.

The notched specimens of steel 16Cr-9Ni-2Mo were cut out in X and Z direction and at an angle $\varphi = 45°$ to the axis Z. The notched specimens of steel 2Cr-Ni-Mo-V were cut out under angles between $\varphi = 0–90°$ across 15°. All these specimens have around notch with 0.25 mm radius, 3 mm depth, apex angle 60°, and minimal diameter 10mm. The tests of notched specimens were carried out under tension-compression on symmetric cycle with load control. For 16Cr-9Ni-2Mo steel, the nominal stress range changed from 250 to 380 MPa, for 2Cr-Ni-Mo-V steel, the nominal stress range was constant: 510 MPa. As follows from our data, for steel 16Cr-9Ni-2Mo the decrease of durability was obtained for z–specimens and specimens cut out under $\varphi = 45°$ (table 1). For 2Cr-Ni-Mo-V steel, the durability of notched specimens is monotonously increasing with increase of the angle φ, as for smooth specimens. Minimal durability corresponds to z-specimens ($\varphi = 0°$) (table 2).

774

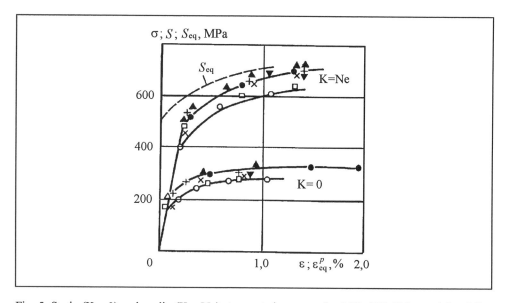

Fig. 5. Static (K = 0) and cyclic (K = Ne) stress-strain curves for 16Cr-9Ni-2Mo steel for different angles of cutting specimens: ...– $\varphi = 0$ (z); ● – $\varphi = 90°$(x); + – $\varphi = 75°$; ✕ – $\varphi = 60°$; □ – $\varphi = 30°$; O – $\varphi = 45°$; ▼ – $\varphi = 15°$.

Table 1. Experimental and calculated fatigue life of notched specimens for 16Cr-9Ni-2Mo steel

	z-specimens			x-specimens			specimens, cutting under $\varphi = 45°$			
$\Delta\sigma_n$	380	320	290	380	320	290	380	320	290	255
$\Delta\sigma_z$	851	792	775	0	0	0	513	480	465	451
$\Delta\varepsilon_z$	1,71	1,26	1,07	1,21	0,879	0,73	0,73	0,49	0,421	0,35
$\Delta\sigma_{z'}$	851	797	775	851	797	775	769	720	697	666
$\Delta\varepsilon_{z'}$	1,71	1,26	1,07	1,71	1,26	1,07	1,99	1,41	1,26	0,905
Nc	200	550	800	950	2200	4000	580	1400	2200	7000
Ne	90	310	1300	500	1300	6000	200	600	2000	6000

Note: In Tables 1 and 2 stress and strain are given in MPa and percent (%) for more dangerous points of notch; stress and strain with index z corresponds to Z-axis of anisotropy, those with index z' – longitudinal axis of specimen; Nc and Ne – calculated and experimental fatigue life.

For evaluating stress and strain at the top of notch, interpolation dependence for anisotropic materials was used in the next form [1]:

$$dA_l = \alpha_{\sigma_i}^2 \, dA_n$$

where dA_l and dA_n – increment of the energy of local and nominal strains; α_{σ_i} – stress intensity concentration factor.

Table 2. Experimental and calculated fatigue life of notched specimens for 2Cr-Ni-Mo-V steel

φ°	0	15	30	45	60	75	90
$\Delta\sigma_z$	0,27	0,123	0,28	0,83	1,38	1,78	1,93
$\Delta\varepsilon_z$	422	478	634	846	1060	1210	1270
$\Delta\sigma_{z'}$	1,93	1,93	1,93	1,93	1,93	1,93	1,93
$\Delta\varepsilon_{z'}$	1270	1270	1270	1270	1270	1270	1270
Nc	250	250	250	250	250	150	110
Ne	320	350	500	400	400	380	250

From this equation we derive the following equation:

$$2\sigma_{eq}d\varepsilon_{eq} + \sigma_i d\varepsilon_{ie} = \alpha_{\sigma_i}^2 \sigma_n d\varepsilon_n \tag{1}$$

where $d\varepsilon_{eq}$ – increment of equivalent Hill's strain; $d\varepsilon_{ie}$ –increment of intensity elastic strain; $d\varepsilon_n$ – increment of nominal elastic strain. For cyclic loading stress and strains in (1) change on corres- poding cyclic values. We assumed also, that the ratio of stresses at the top of notch is approxi- mately constant under elastic and elastoplastic loading.

According to (1), for 16Cr-9Ni-2Mo steel the values of stress and strain at the top of notch de-

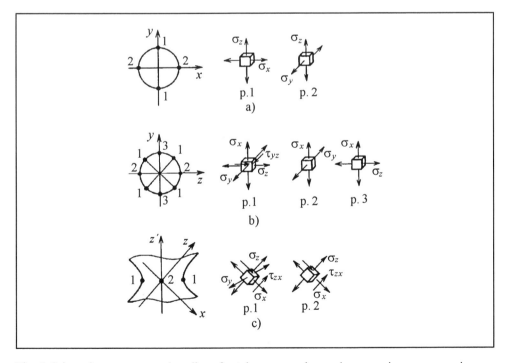

Fig. 6. Orientation stress around top line of notch: a – z-specimens, b – x-specimens, c – specimens, cut out under $\varphi = 45^\circ$ to the axis z.

pend on direction of cutting of specimens. Orientation of stress in some points at the top of notch is shown on fig. 6. For x-specimens and specimens cut out under $\varphi = 45°$ to the axis Z, the values of stress and strain change around the top line of notch. For z-specimens, stress and strain do not change due to around symmetry cyclic properties to the axis Z. For 2Cr-Ni-Mo-V steel, equation (1) correspond equation used in [2].

The number of cycles before fracture for anisotropic materials can be defined by the following equation [3]:

$$\frac{\sigma_j}{\sigma_{0j}} \Delta\varepsilon_j = C_{pj} N^{m_{pj}} + C_{ej} N^{m_{ej}} \tag{2}$$

The first and the second members of the right part (2) represent respective functions for the plastic and elastic components of strain. N – fatigue life (number cycles before crack initiation); C_{pj}, C_{ej} and m_{pj}, m_{ej} – coefficients and powers for plastic and elastic component of strain respectively. The index j in equation (2) specifies the direction of normal strain; σ_{0j} and σ_j – maximum normal stress on plane with normal j, acting, respectively in axial test and in studied type of stress state (at equal strain range). The most dangerous direction, which can be a line of acting maximum strain (specimen symmetry axis) or axis Z (if delamination takes place) is defined from the minimum number of cycles on (2).

Equations (1) and (2) were used for evaluation the number of cycles before fracture in nothed specimens of 16Cr-9Ni-2Mo and 2Cr-Ni-Mo-V steels. Data are shown in tables 1, 2 for the most dangerous points. Values of normal stress and strain corresponding to axial direction of specimens (axe Z) and direction of axis anisotropy Z are shown. The correlation of experimental and calculated data was satisfactory in all cases.

REFERENCE

1. Kazantsev A. G. (1994). *Evaluation of Low-Cycle Fatigue Durability of Anisotropic Materials in Concentration Stress Zones.*Tyazheloe mashinostroenie, **12**, 2.

2. Buczynski A., Glinka G. (1997). *Elastic-Plastic Stress-Strain Analysis of Notches under Non-proportional Loading.* In: *Proceedings of the 5th International Conference Biaxial/ Multiaxial Fatigue and Fracture*, Cracow, vol. 1, pp. 461–479.

3. Kazantsev A. G., Makhutov N. A. (1997). *Low-Cycle Fatigue of Anisitropic Steel under Non-proportional Loading.* In: *Proceedings of the 5th International Conference Biaxial/ Multiaxial Fatigue and Fracture*, Cracow, vol. 1, pp. 125–139.

PREDICTION OF CONSTANT-AMPLITUDE FATIGUE LIFE TO FAILURE UNDER PULSATING-TENSION (R≥0) BY USE OF THE LOCAL-STRAIN-APROACH

A. BUCH and V. GALPERIN
Faculty of Aerospace Engineering
Technion – Israel Institute of Technology
Haifa, Israel

ABSTRACT

The aim of the investigation was to compare LSA (Local Strain Approach) life predictions with pulsating-tension tests to failure. It was shown that, in the case of pulsating-tension, the ratio N_i/N_f does not change considerably and therefore the relative LSA can be used for predicting S-N lines for $R \geq 0$. The accuracy of the prediction is associated with choice of the K value in Neuber's relation $K_\sigma \times K_\varepsilon = K^2$. The K values which give the best life predictions were obtained close to those of the corresponding notch factors for pulsating-tension with $R = S_{min}/S_{max} = 0$. Because of the size and surface effects (which are not taken into account in LSA) and in view of the difficulty in knowing whether the considered specimen with a given K_T value has exactly the same cyclic parameters as those used in the LSA calculations, it is preferable to use a K-value proved to give a life prediction in agreement with a related pulsating-tension test.

INTRODUCTION

The fatigue life of material coupons and structural elements is divided into two phases – the crack initiation phase and the crack propagation phase. The Local Strain Approach (LSA) is conventionally applied for prediction of the life to crack initiation, N_i. This approach is based on the assumption that in notched and unnotched specimens with the same local strain history, cracking sets in at the same number of cycles to crack initiation.

However direct LSA predictions deviate considerably from the experimental crack initiation life, both for variable-amplitude and constant-amplitude loading. A scatter of N_{pred}/N_{exp} rations between 0.1 to 10 was observed in many investigations [1-4] and therefore some authors [5-8] recommended the relative method (method of correction factors) for improving the accuracy of predictions. It was observed in many investigations by the author [6], that the ratio N_{pred}/N_{exp} has close values for related aircraft loading cases. Accordingly use of N_{pred}/N_{exp} obtained for one loading case as a correction factor for another related loading case, yielded in an improved predictive accuracy with the scatter limited to 0.5 + 2.0. It was also shown, that in the case of typical aircraft loading distributions for the lower wing surface (which are predominantly of the tensile type), the relative method could be applied for more accurate prediction of the fatigue life to crack initiation, as well as of the life to failure [11], for cases where direct prediction resulted in large deviations from the actual life. It should be noticed that the scatter of N_{pred}/N_{exp} is for direct predictions evidently larger than scatter of N_i/N_f for different pulsating tension stress levels [11].

The aim of the reported investigation was to compare LSA predictions with pulsating-tension test results, with a view to better insight into application of direct and indirect prediction in the case of constant-amplitude tensile loading.

RESULTS FROM PREVIOUS INVESTIGATIONS

The Local Strain Approach has many well-known weak points, which, combined, are a source of inaccuracy in direct predictions [9]. LSA calculations do not make proper allowance for the loading sequence effect, since the used material memory rules are schematical and unreliable and as is well known, Miner's rule is not an exact one either. Another important weak point is the optional choice of the K-value in Neubers's relation $K_\sigma \times K_\varepsilon = K^2 = const$. Use of a specific value of K (e. g. K_T or K_F) may result in a good prediction for one loading case and in poor ones for others. The problem of poor allowance of the loading sequence effect and of the effect of Miner's rule inaccuracy exists, of course, only in the case of variable-amplitude loading. However, the effect of the chosen K-value on predictive accuracy exists also in the case of constant-amplitude loading.

It was observed in many investigations that in the case of pulsating tensile loading ($R \geq 0$), the crack propagation phase is short for massive parts and notched specimens [3, 10-12].

Table 1 presents Lowak's [12] pulsating-tension test results for AlCuMg2 (2024-T3) sheet specimens of different sizes with a central hole (d/w = 0.2, K_t = 2.5) of different diameters (d= 2, 4, 8 and 16 mm). As can be seen, the size has an effect on both the life to failure N_f and to crack initiation N_i, and also on the ratio N_i/N_f. However, the effect of the loading level on the ratio N_i/N_f of the specimens of the same size was not considerable (Table 1).

Table 1:   Life to crack initiation N_i and to failure N_f for geometrical similar 2024-T3 sheet specimens with central hole of different size (d/w = 0.2) under pulsating-tension (12), R = 0

D	S_{max} = 196 MPa		S_a = 98 Mpa	S_{max} = 252 MPa		S_a = 126 MPa
Mm	N_i in kc	N_f in kc	N_i/N_f	N_i in kc	N_f in kc	N_i/N_f
2	71.7	100.5	0.713	17.2	30.7	0.560
4	41.0	56.8	0.722	11.7	17.4	0.672
8	42.8	53.4	0.801	12.8	16.7	0.766
16	42.6	50.6	0.842	13.8	16.5	0.836

INVESTIGATED MATERIALS

The investigation covered 5 sheet materials: 2024-T3, 7075-T6, Ti6A14V, CK45 (carbon steel SAE 1045) and 42CrMo4 (alloy steel SAE 4140). For these materials, the cyclic material data needed for LSA calculation (P_{SWT}, $S'_{0.2}$, n and E) are available from several sources [3, 12-14], while S-N data for R = 0 are to be found in handbooks and publications [16-18]. In view of this diversity of sources, only data for chemically identical alloys with similar heat treatment and close properties (S_u and S_y) were taken into account. The analytical expression of the cyclic σ-ε diagram used in the LSA calculations was of the form

$$\varepsilon_a = \frac{\sigma_a}{E} + 0.002 \left(\frac{\sigma_a}{S'_{0.2}} \right)^n$$

TAE No. 649 [7] lists the cyclic yield limit $S'_{0.2}$, the exponent n and the modulus of elasticity E of the considered materials and the damage parameter values $P_{SWT} = (\sigma_{max}\varepsilon_a E)^{1/2} = f(N)$. The steel 42CrMo4 has the highest values of P_{SWT}, $S'_{0.2}$, $S_{0.2}$ and S_u, and the AlCuMg alloy 2024-T3 the lowest ones.

The P_{SWT}-N line used in the LSA calculation was constructed in a piecewise-linear form (in the log-log scale) for N = 10^2, 10^3 ... 10^7 cycles.

RESULTS OF THE LSA CALCULATIONS OF S-N DATA IN PULSATING-TENSION

As was mentioned before, measurements show that in pulsating tension variations of N_i/N_f are small when $S_{max} = 2 S_a$ varies between the fatigue limit value for N = 10^6 cycles and the yield limit of the material. Accordingly, when the ratios N_{pred}/N_i are close for two related loading cases, the ratios N_{pred}/N_f are also close.

Table 2 compares pulsating-tension test results from [18] for 2024-T3 and 7075-T6 notched sheet specimens (K_T = 4, 2 and 1.5) with the authors' LSA calculation results.

The calculations were performed for K = K_T and for K values giving the best approximation of the fatigue life to failure. As can be seen, the life estimates obtained for K = K_T deviated strongly from tests results, whereas for some chosen K values the calculated S-N data were close to the experimental. Titanium and steel specimens with K_T = 3.6 were considered in Table 3. The life approximation was here also quite accurate (both for N_f and N_i), when some K-values evidently smaller than K_T were used for life estimation.

Table 4 summarizes fatigue data for the considered five materials as well as the actual and predicted lives to failure for S_a = 140 MPa in the case of Al-alloys, and for S_a = 220 MPa in that of steel and Ti-alloy specimens.

For S_a = 220 MPa the best LSA life predictions were obtained for K values closer to the notch factor K_F in pulsating-tension than in tension-compression with the single exception of Ck 45. In the case of Al-alloys, the best K-value was always closer to the notch factor K_F for R = 0 than for R = -1. The notch factors for 2024-T3 and 7075-T6 were calculated using the two-parameter formula K_F/K_T = f (r, A, h) = $1/A \{1-(2x1h)/(r+r_o)\}$ where A = 1.1, h = 0.25 mm for R = 0 and A = 1.05, h. 0.25 mm for R = -1. The derivation of this formula is described in [19-20]. In the case of steel Ck 45 (with K_T = 3.6) the use of K = 2.4 instead of K = K_F = 2.08 results in smaller deviation of the predicted lives from the actual, indicating the better accuracy of the relative prediction method in the case in question.

Table 5 presents a comparison of pulsating-tension test results in the case $S_m = 30$ Ksi $= 210$ MPa $=$ const ($S_a < S_m$) with LSA life prediction results, using K-values giving the best approximation of the life. As can be seen the K-values used in calculations were mostly close to the corresponding K_F-values for the previously considered specimens with $K_T = 1.5$, 2 and 4 in the case $R = 0$ (Table 4). They were, however, evidently higher than the corresponding K_F values for the considered case of $S_m = 210$ MPa (Table 5).

Table 6 presents a comparison of best LSA predictions with pulsating-tension tests performed on 2024-T3 specimens (off-center hole, $K_T = 2.6$) for the loading cases $R = S_{min}/S_{max} = 0.25$, 0.4 and 0.6. The best approximation of lives $N = 10^6$ and $N = 10^5$ cycles was obtained for $K = 1.8$. However, in the case of low cycle fatigue life $N = 10^4$, the K-value which resulted in best approximation, was 2.4 and not 1.8 for all three above stress ratios. This can be explained by the effect of unsymmetrical distribution of residual stresses generated in the case of small cycle number $N = 10^4$ in the critical cross-section of specimens with a hole deplaced from the center line. For Al-specimens with central hole (Tables 2 and 4) the best K-value was identical for $N = 10^4$, 10^5 and 10^6. The different behaviour of specimens with a central hole ($K_T = 2$)and with an off-center one ($K_T = 2.6$) in the case of small fatigue life $N = 10^4$ is evident from the results for $R = 0.25$ presented in Tables 6 and 2.

CONCLUSIONS

1. In the case of pulsating-tension, the Local Strain Approach may also be used for prediction of life to total failure.
2. The best method of constant-amplitude life prediction is choice of a K-value which gives best agreement with some related pulsating-tension test. Agreement between LSA predictions and actual S-N data is then satisfactory for other stress levels.
3. The K-value which gives the best agreement between life prediction and test is often nearly equal to the notch factor for pulsating-tension ($R = 0$) at $N = 10^7$ cycles.
4. The value of $K = (K_\sigma \times K_\varepsilon)^{1/2}$ seems to be nearly constant for $S_y > S_{max} > S_{10}{}^6$.

REFERENCES
[1] Schütz, W. (1979) *Eng. Fracture Mech.* **11** , 405.
[2] Nowack, H. (1979). In: *ICAF Symposium Aeronautic Administration*, Brussels.
[3] Buxbaum, O., Opperman, H., Köbler, H.G., Schütz, D., Boller, C., Heuler, P., Seeger, T. (1983). *LBF-Bericht FB-169*, Darmstadt.
[4] Buch, A., Seeger, T., Vormwald, M. (1986) *Int. J. Fatigue* **3**, 175.
[5] Buch, A. (1986) *Materialprüfung* **28**, 315.
[6] Buch, A. (1988) *Materialprüfung* **30**, 67.
[7] Buch, A. (1988). *Fatigue Strength Calculations*. ttp.
[8] Buch, A. (1988). *ICAF Doc. No. 1672 – Technion TAE No. 624*
[9] Neuber, H. (1961) *J. Appl. Mech.* **28**, 544.
[10] Topper, T.H., Wetzel, R.H., Morrow, J.A. (1969) *J. Materials* **4**, 200.
[11] Conle, A., Nowack, H. (1977) *Experimental Mechanics* **2**, 57.
[12] Lowack, H. (1981). *LBF-Bericht FB-157*, Darmstadt.
[13] Boller, C., Seeger, T. (1987). *Materials Data for Cyclic Loading*. Elsevier, London.
[14] Bergmann, J. (1983). *Zur Betriebsfestigkeitsbemessung gekerbter Bauteile auf der Grundlage der örtlichen Beanspruchungen*. TH Darmstadt.
[15] Östermann, H., Matschke, C. (1971). *LBF-Mitteilung Nr. 62*. Darmstadt.
[16] Haibach, E., Matcshke, C. (1980). *LBF-Bericht FB-129*. Darmstadt.
[17] Haibach, E., Matcshke, C. (1980). *LBF-Bericht FB-153*. Darmstadt.
[18] AFSSC (1983). *Strength of Metal Aircraft Elements*. Military Handbook, Washington.

Table 2:   LSA life predictions for 2024-T3 and 7075-T6 sheet specimens under pulsating-tension (R = 0 and R = 0.25)

Material K_T and R Notch geometry	S_{max} MPa	$N_{exp} = N_f$ kilocycles	LSA predictions for dif. K kilocycles			Best K-value
2024-T3, $K_T = 4$			K = 4.0	K = 3.0	K = 2.8	2.8
R = 0	280	2.27	0.14	0.97	1.56	
Sym edge	238	4.56	0.43	2.91	4.53	
Notch	140	60.26	12.10	40.44	54.90	
$K_T = 2$			K = 2.0	K = 1.8	K = 1.7	1.7
R = 0	420	4.58	0.97	1.99	2.91	
Central	350	8.94	3.31	6.43	9.00	
Hole	280	27.53	12.10	18.70	23.73	
	210	120.70	40.44	64.78	84.62	
$K_T = 1.5$			K = 1.5	K = 1.4	K = 1.35	1.35
R = 0						
Sym edge	420	15.04	6.43	9.76	11.45	
Notch	288	82.80	40.40	54.50	64.78	
2024-T3, $K_T = 4$				K = 3.0	K = 2.8	2.8
R = 0.25	336	2.34		1.47	2.29	
Sym edge	224	14.47		14.05	18.40	
Notch	189	34.31		27.15	35.60	
$K_T = 2$				K = 1.80	K = 1.7	1.7
R = 0.25	294	51.9		21.1	26.4	
Central	224	241.5		131.0	268.0	
Hole	189	768.1		1260.0	3270.0	
7075-T6, $K_T = 4$			K = 4	K = 2.80	K = 2,75	2.8
R = 0	280	1.17	0.16	1.12	1.24	
Sym edge	210	5.65	-	5.48	-	
Notch	140	62.37	7.25	61.51	68.94	
$K_T = 2$			K = 2	K = 1.75	K = 1.70	1.7
R = 0	420	3.33	0.78	1.59	1.87	
Central	350	6.57	-	4.35	5.13	
Hole	280	16.87	7.25	15.76	18.69	
$K_T = 1.5$			K = 2	K = 1.75	K = 1.70	1.7
R = 0	420	3.33	0.78	1.59	1.87	
Sym. Edge	350	6.57	-	4.35	5.13	
Notch	280	16.87	7.25	15.76	18.69	
7075-T6, $K_T = 4$				K = 3	K = 2.8	2.8
R = 0	336	0.97		0.90	1.27	
Sym edge	224	9.01		7.23	10.49	
Notch	189	24.31		18.07	26.50	
$K_T = 2$				K = 1.8	K = 1.7	1.8
R = 0.25	294	28.26		12.7	17.4	
Central	224	150.50		166.0	305.0	
Hole	189	776.98		1093.8	4520.0	

Experimental data were taken from the Military Handbook [18].

Table 3: LSA predictions of S-N Data for Ti6Al4V, Ck 45 and 42CrMo4 flat specimens ($K_T = 3.6$) under pulsating-tension (R = 0)

Material	S_{max} MP$_a$	$N_{exp} = N_f$ cycles	LSA predictions Number of cycles			$N_{exp} = N_i$ cycles
Ti6Al4V			K = 3.6	K = 2.2	K = 2.15	
	560	4500	-	5282	5735	-
	410	20000	2700	20400	23100	-
	360	40000	4410	42040	47900	-
	320	65000	6700	83000	95600	-
			K = 3.6	K = 2.4	K = 2.2	
Steel	440	14000	3210	18300	29200	8000
Ck45	340	50000	8680	71000	113000	-
SAE1045	276	240000	25400	219000	347000	202000
			K = 3.6	K = 2.4	K = 2.68	
Steel	740	3000	715	3550	2209	2000
42CrMo4	570	10000	1900	10700	6652	6300
SAE4140	440	50000	5730	59100	29074	41000

Experimental data taken from [15, 16].

Table 4: Fatigue Properties of Investigated Materials (MPa)

Material and K_T	Fatigue limits for $K_T = 1$		Notch factor		K used in LSA.	N_{exp}	N_{pred}
	S_{pt}	S_{tc}	R = 0	R = -1	calcul.	kilocycles	
Ti6Al4V	580	380	Exp.	values		S_a =	220 MPa
3.6			2.10	2.03	2.2	16	16
Ck45	500	320					
3.6			2.08	2.20	2.4	14	18
42CrMo4	750	510					
3.6			2.34	2.68	2.4.	50	59
			Calcul.	values		S_a =	140 MPa
7075-T6	246	147					
4			2.88	3.02	2.8	1.2	1.1
2			1.70	1.78	1.7	17	16
1.5			1.33	1.39	1.35	76	77
2024-T3	234	147					
4			2.88	3.02	2.8	2.2	1.6
2.5			2.14	2.24	2.1	10	9
2			1.70	1.78	1.7	28	24
1.5			1.33	1.39	1.35	83	65

Table 5:   Comparison of Best LSA Predictions with Pulsating-Tension Test Results for Al-Alloy Specimens in the Case of S_m = const = 30 Ksi = 210 MPa, $S_a < S_m$

Notch Description	N_{exp} cycles*	2024-T3			7075-T6		
		S_{max}		N_{pred}	S_{max}		N_{pred}
		Ksi	MPa		Ksi	MPa	
central				for K = 1.7			for K = 1.9
hole d = 3 inch	10^4	53	371	1.53×10^4	50	350	0.61×10^4
d/w = 2/3	4×10^4	45	315	4.26×10^4	42	294	4.87×10^4
K_T = 2.04	10^5	42	294	0.97×10^5	40	280	1.00×10^5
K_F = 48/39 = 1.23*							
K_F = 49/38 = 1.25	10^6-10^7	39	273	1.24×10^6	38	266	0.43×10^6
				for K = 1.35			for K = 1.35
Edge notch	10^4	64	448	0.61×10^4	57.5	403	1.01×10^4
r = 0.76"	4×10^4	51	357	3.96×10^4	50	350	4.12×10^4
K_T = 1.5	10^5	47	329	0.81×10^5	46	322	1.16×10^5
K_F = 48/41.5 = 1.16*	10^6-10^7	41.5	290	2.75×10^6	40	280	1.46×10^6
K_F = 49/40 = 1.22							
				for K = 2.8			for K = 2.8
Edge notch	10^4	41	287	1.92×10^4	40	280	1.32×10^4
K_T = 4	4×10^4	38	266	5.09×10^4	37	259	4.94×10^4
K_F = 48/33.5 = 1.4*	10^5	37	259	0.76×10^5	36	252	0.86×10^5
K_F = 49/33.5 = 1.5	10^6-10^7	33.5	235	0.79×10^6	33.5	235	$4,71 \times 10^6$

*The experimental data were taken from [17].

The fatigue limits (N = 10^7) of unnotched specimens were S_{max} = 48 Ksi for 2024 and 49 Ksi for 7075 (S_m = 30 Ksi).

Table 6:   Comparison of Best LSA Predictions with Pulsating-Tension Test Results For 2024-T3 Al-Alloy Specimens in the Case of R = const

Notch Description and Mech. Properties	R= S_{min}/S_{max}	S_{max}		N_{exp}	N_{pred}
		Ksi	MPa	cycles	cycles
					for K = 1.8
Off-center					
hole d = 0.375 inch	0.25	26.7	187	10^6	1.57×10^6
w = 1.5 inch	0.25	33.0	231	10^5	0.96×10^5
d/w = 0.25	0.40	30.4	213	10^6	1.19×10^6
S_u = 67 Ksi	0.40	38.2	267	10^5	0.91×10^5
S_y = 57 Ksi	0.60	41.8	293	10^6	0.88×10^6
K_T = 2.6	0.60	46.8	328	10^5	2.61×10^5
					for K = 2.4
K_F = 2.08 for R = 0.25	0.25	41.3	289	10^4	1.24×10^4
K_F = 1.77 for R = 0.4	0.40	50.8	356	10^4	1.10×10^4
K_F = 1.73 for R = 0.6	0.60	63.0	441	10^4	1.65×10^4

Experimental data were taken from [17].

Chapter 11

CASE STUDIES; PRACTICAL EXPERIENCE

Low Cycle Fatigue and Elasto-Plastic Behaviour of Materials
K-T. Rie and P.D. Portella (Editors)
787

VALIDATION OF CREEP-FATIGUE ASSESSMENT PROCEDURES USING THE RESULTS OF COMPONENT-TARGETED FEATURE-SPECIMEN TESTS

S.R. HOLDSWORTH
ALSTOM Steam Turbine Group[‡]
Rugby CV21 2NH UK

ABSTRACT

The paper reviews the general principles and analysis options available for the assessment of engineering components subject to cyclic/hold loading at elevated temperatures. Particular attention is paid to the consideration of those circumstances in which damage accumulation occurs as a consequence of the combined effects of fatigue, and creep due to both primary and secondary loading. It also considers recent experience from a European collaborative activity which has shown that, irrespective of the complexity of the assessment procedure adopted, the precision of predicted endurances can be improved by using the results from component-targeted feature-specimen tests in conjunction with the selected category of material property input parameters.

KEYWORDS

Creep-fatigue, assessment, prediction, feature-specimen, validation.

INTRODUCTION

Life prediction methods applied to structures subject to creep-fatigue loading have advanced significantly in recent years. Throughout the power generation industry in particular, the main incentive has been to improve competitiveness through increased efficiency, performance and reliability; the manufacturers striving to provide equipment to a high technical specification at an economic price and the operators endeavouring to maximise on-line duty time by reducing the number of maintenance, inspection and unscheduled outages to a minimum. Published design [1-4] and assessment [5] codes are, by necessity, conservative and sometimes excessively so. For the reasons already given, in-house procedures have been developed wherever possible to predict creep-fatigue lifetimes with more precision; for example, by exploiting state-of-the-art methodologies, and/or feedback on effectiveness from service experience and the results of specific component-targeted feature-specimen tests [6].

[‡] Formerly GEC ALSTHOM Steam Turbine Group

788

Valuable experience with the use of component-targeted feature-specimen tests for assessment method validation has recently been gained in the BRITE-EURAM C-FAT project BE-5245. This is examined below following an overview of creep-fatigue assessment methodologies.

A simplified representation of the common steps in a number of creep-fatigue assessment procedures is shown in Fig. 1. The component is fully assessed in terms of the external forces imposed during the operating lifetime, and these are used in conjunction with material constitutive equations to define the distribution of stress and strain throughout the structure. Having established the stress/strain history at critical locations, cyclic and creep damage fractions (D_F and D_C)[1] are determined by reference to the appropriate material endurance property data. The D_F and D_C fractions are finally compared with the material characteristic crack initiation locus in a creep-fatigue damage assessment diagram, and the risk of damage evaluated.

Most creep-fatigue assessment procedures therefore comprise a series of stages for i) determining the state of stress and strain at critical locations, ii) calculating the damage due to cyclic and creep loading, and iii) summing the damage resulting from these two sources. There are various options available to the analyst at each of these stages, sometimes even within a single procedure, and these are reviewed in the following sections.

STATE OF STRESS-STRAIN

Analysis options for determining the state of stress and strain at critical locations range from simplified-elastic, for structures which can be shown to operate in shakedown (see below), to evolutionary full-inelastic requiring a detailed description of deformation as a function of temperature, strain rate, and any metallurgical changes leading to hardening or softening relative to monotonic behaviour. At one extreme, steady-state cyclic and creep deformation are assumed, with plasticity and creep being clearly distinguished. However, approaches range in increasing complexity to fully unified state variable models [7,8].

When global shakedown can be demonstrated, simplified methods of analysis may be used. The cyclic stress-strain behaviour at areas of local plasticity may be determined using an elastic-plastic $\Delta\sigma(\Delta\varepsilon)$ relationship for a given cycle number or for steady-state conditions, eg. [9,10], in conjunction with an appropriate strain concentration model, eg. [11-13], and an allowance for time dependent deformation. In such circumstances, periods of steady operation leading to load controlled creep, strain controlled stress relaxation or stress relaxation under the influence of elastic follow-up may be assessed using forward-creep $\varepsilon_c(\sigma,t)$ equations, eg. [14,15], with an appropriate hardening law and a correction for multi-axial loading eg. [16]. Alternatively, stress relaxation may be described directly, ie. using $\sigma_t(\sigma_o,t)$ expressions such as [17,18].

In the present context, shakedown refers to those circumstances where behaviour is essentially elastic at all points within the component throughout each cycle of loading in the steady cyclic state. Such a situation can develop after yielding during the first application of a particular load cycle.

---

[1] A list of nomenclature is given at the end of the paper

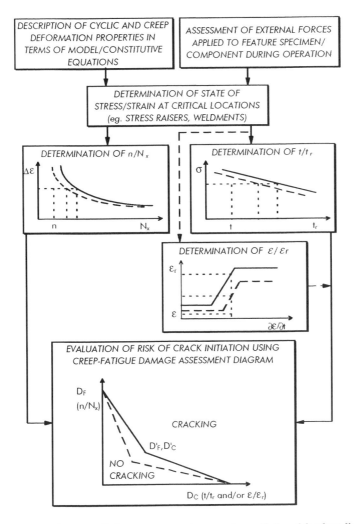

Fig. 1 Common steps in creep-fatigue assessment procedures (full and broken lines providing indication of assessment sensitivity to mean and lower bound material property data) [6]

If shakedown cannot be demonstrated, more detailed inelastic analysis may be necessary, which may be based on steady-state conditions or may incorporate evolutionary behaviour. Irrespective of the level of complexity, it is necessary to account for the effects of stress relaxation on cyclic stress-strain behaviour when the structure is deforming in the creep regime. The suite of constitutive equations making up the ORNL model [19] is reasonably successful in representing deformation at this level of complexity, but the approach is empirical.

In an attempt to overcome some of the difficulties associated with the preceding methods, models have been developed which use a state variable to describe the metallurgical condition of the material at any stage of the cyclic history. These variables evolve according to the time-temperature-strain history of the component. State variable models fall into two broad categories, namely non-unified and unified models.

Non-unified models retain separate sets of equations for plastic and creep deformation, *eg.* [20]. In this approach, the state variable reflects the cumulative effect of cyclic plasticity. This same variable is then incorporated into the creep equation to reflect the effect of cyclic hardening on subsequent creep. Conversely, the recovery processes which occur during creep moderate the effects of cyclic hardening, these changes reducing the extent of hardening in the plasticity equations.

Unified models include the Chaboche [7] and Interatom [8] models. Their principal feature is that they describe the stress-strain-time-temperature response of a material subjected to any combination of monotonic, cyclic and stationary loading. They are visco-plastic models based on internal variables evolving with time during the inelastic deformation process. The models are unified in that no partition or distinction is made between plastic and creep strain.

Stresses in high temperature components are generally categorised as being primary (directly applied) or secondary (self equilibrating), although not always in a consistent way in the published procedures (*eg.* in [1,2,5]). A potentially less confusing form of classification would be as load-controlled and displacement-controlled stresses, although this over simplifies the situation when secondary stresses are regarded as those which fully relax.

The methods used i) to determine cyclic and creep damage fraction and ii) to sum individual cyclic and creep damage components are reviewed in the following sub-sections.

CALCULATION OF DAMAGE

Creep-fatigue assessment procedures are broadly differentiated by the way components of fatigue and creep damage are determined. More specifically, they fall into two main categories, depending on how creep damage due to secondary loading is calculated, ie. by time-fraction or strain-fraction methods.

Time-fraction methods are currently the most widely used to determine the creep damage accumulated, *eg.* [1,2,4]. However, this approach leads to the anomolous situation, whereby a strong material which resists stress relaxation is attributed with a high creep damage fraction because it spends a long time under a large stress; even though in so doing it accumulates little or no creep strain [21]. Conversely, weak materials which rapidly shed the applied load, accumulating significant creep strain in the process, are attributed with a low creep damage fraction.

Cyclic Damage

Damage Fraction. Cyclic damage fraction is determined from a knowledge of $(\Delta\varepsilon)_i$ for each of i cycle types using $N_x(\Delta\varepsilon)$ endurance data, *eg.* [22], and the expression:

$$D_F = \sum_i \{n_i/N_x(\Delta\varepsilon)_i\} \tag{1}$$

N_x is the fatigue endurance to crack initiation, defined by the criterion, x. n_i is the number of cycles of type i.

Initiation Criteria. The accuracy of component life prediction can be significantly influenced by any discrepancy in the crack initiation criteria used to establish material property endurance data and to form the basis of the structure's performance in service. For example, the use of failure endurance data from 10mm diameter testpieces is likely to lead to non-conservative estimates of service cycles in 0.5mm thick nuclear reactor fuel cans. Similarly, data based on cycles to initiate a 0.5mm deep crack in the same laboratory testpiece can lead to unnecessarily conservative estimates of endurance in large components unless the cycles to grow such cracks to a detectable size are also considered in the life calculation. Consequently, it is now common practice to explicitly define the initiation criterion employed in both laboratory testpiece and component and to determine the number of cycles to extend from the smaller to the larger of the two crack sizes if they differ. Various approaches are available to determine inter-initiation criterion crack growth endurances [5b,23,24], including direct calculation by a fracture mechanics route using small crack $da/dN(\Delta\varepsilon, a)$ data (*eg.* [25]).

Creep Damage

Time Fraction. In time-fraction approaches, creep usage is determined as a function of time, ie.

$$D_C = \sum_i \{n_i [t_h/t_r(\sigma_P)]_i\} + \sum_i \{n_i [\int^{th} dt/t_r(\sigma_S)]_i\} \qquad (2)$$

The two terms on the right hand side of Eqn.2 respectively cover the damage accumulated due to primary and secondary loading respectively. The stress rupture times, $t_r(\sigma_P)$ and $t_r(\sigma_S)$, are determined using appropriate creep rupture data.

Some analysts account for creep damage due to secondary loading as part of the fatigue damage calculation by using a $N_x(\Delta\varepsilon)$ endurance model for cyclic/hold data in Eqn.1, *eg.* [26,27]. In such circumstances, creep damage fraction is estimated using only the first term in Eqn.2.

Strain Fraction

The best known procedure involving the determination of creep damage in terms of strain-fraction (or ductility exhaustion [21]) is the R5-Vol.3 procedure [5b]. Following on from the guidance given in the R5 procedure, fatigue damage fraction is determined according to Eqn.1 and creep damage fraction using an expression of the form:

$$D_C = \sum_i \{n_i[t_h/t_r(\sigma_P)]_i\} + \sum_i \{n_i z_i [\int^{th} \dot{\varepsilon}/\varepsilon_r(\dot{\varepsilon}).dt]_i\} \qquad (3)$$

where z_i is the elastic follow-up factor for the ith cycle type. The current guidance in R5 [5b] is to determine strain-fraction on the basis of a global (or testpiece reference length) based creep-rupture strain. This works well for austenitic steels but is excessively conservative for ferritic steels. The use of a local creep-rupture strain derived from Z_r measurements is more appropriate for such materials [28].

DAMAGE SUMMATION

The ways in which fatigue and creep damage may be summed are conveniently illustrated with reference to possible crack initiation locus options in a creep-fatigue damage summation diagram (Fig.2). Bi-linear damage loci, as used in [1,2], may be modelled using:

$$D_F = 1 - D_C (1 - D'_F)/D'_C \quad \text{for } D_C < D'_C \qquad (4)$$
$$D_F = [1 - D_C].D'_F/(1 - D'_C) \quad \text{for } D_C > D'_C$$

where D'_F and D'_C define the point of intersection [28]. For linear damage accumulation, as in [3,5], $D'_F, D'_C = 0.5$. An alternative is the L-shaped locus promoted in [29].

Using the strain range partitioning methodology, total damage is derived by the linear summation of individual damage components determined from up to four $N_x(\Delta\varepsilon)$ endurance curves for (i) plastic-plastic (PP), (ii) creep-plastic (CP), (iii) plastic-creep (PC) and (iv) creep-creep (CC) tension-compression cycles [30], ie.

$$1/N_x = f_{PP}/N_{PP,x}(\Delta\varepsilon_{PP}) + f_{CP}/N_{CP,x}(\Delta\varepsilon_{CP}) + f_{PC}/N_{PC,x}(\Delta\varepsilon_{PC}) + f_{CC}/N_{CC,x}(\Delta\varepsilon_{cc}) \qquad (5)$$

where $f_{PP} = \Delta\varepsilon_{PP}/\Delta\varepsilon$, etc., and $\Delta\varepsilon = \Delta\varepsilon_{PP} + \Delta\varepsilon_{CP} + \Delta\varepsilon_{PC} + \Delta\varepsilon_{CC}$.

The effectiveness of the strain range partitioning approach is very dependent on the scope of the endurance curves established for the creep component cycles, ie. (ii)-(iv). It is essential that the creep damage components inherent in the $N_{CP,x}$, $N_{PC,x}$, and $N_{CC,x}$ strain range functions adequately represent that for the target component operating cycle in service.

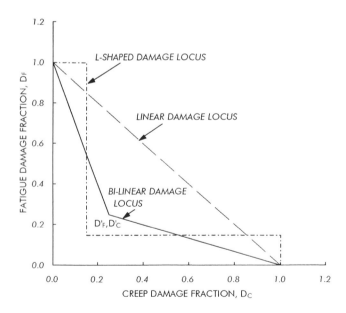

Fig. 2 Various locus options in creep-fatigue damage summation diagram

FEATURE TEST	SUMMARY DETAILS	COMPONENT(S)
i) LARGE SENB SPECIMEN [32] 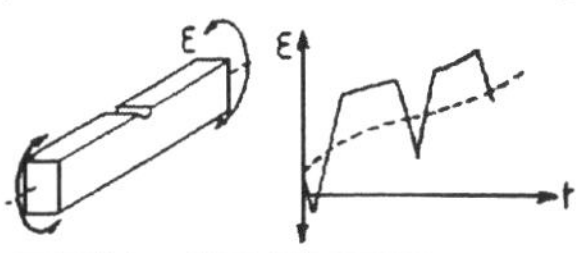	- Cast 1¼CrMoV, 550°C - 100 x 75 thick, with 6 rad notch - LCF with $\sigma_S + \sigma_P$ loading - HTSG, PD monitoring	HP/IP steam turbine chests, inner casings, diaphragms
ii) LARGE SCALE RATCHETING [33] 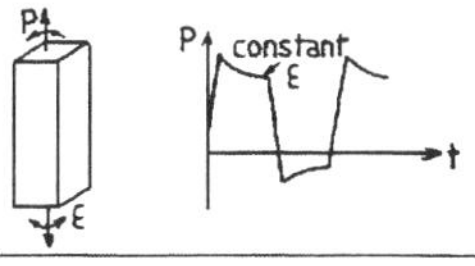	- Cast 1¼CrMoV, 550°C - 45 x 45 section - LCF with $\sigma_S + \sigma_P$ loading - HTSG, replica monitoring	Steam pipework
iii) LARGE CNRT SPECIMEN 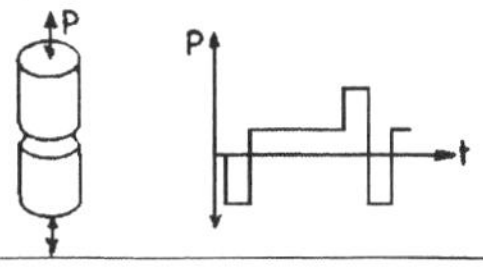	- 1CrMoV rotor, 525°C - 90 OD, with 10 rad notch - LCF with $\sigma_S + \sigma_P$ loading - HTSG, PD monitoring	HP/IP steam turbine rotors
iv) WELDED HEADER GEOMETRY 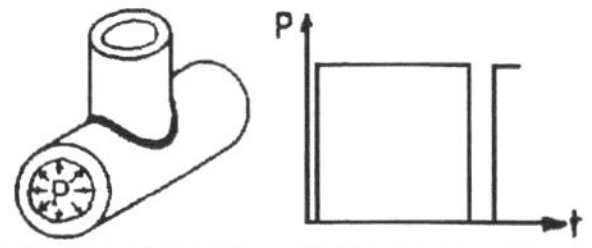	- Welded Mod9Cr, 585°C - OD x 23 thick T-joint - Cyclic pressure loading - HTSG, PD, US, replica monitoring	Parts for advanced boilers and turbines, header nozzles, stub geometries, pipework
v) WELDED WIDE PLATE [34] 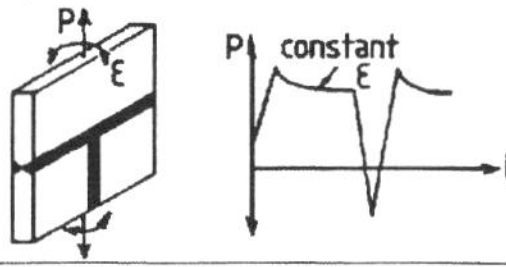	- Welded 18Cr11Ni, 570°C - 460 x 40 thick wide plate - LCF with $\sigma_S + \sigma_P$ loading - HTSG, replica monitoring	LMFBR fast breeder reactor structures, pipework

Fig. 3 Summary details and target components for C-FAT feature-specimen tests [6]

PREDICTION VALIDATION

General

The preceding review serves to highlight just some of the method options available to the analyst involved in predicting the lives of components subject to creep-fatigue loading. In addition to these, there are options concerning the material property input parameters.

The determined values of stress, strain and damage accumulated depend on the material property data used in the assessment. It is unusual for the analyst to have a complete data set of properties for the specific heat of material from which the target component has been manufactured. A conscious decision has therefore to be made as to whether to consistently use alloy-mean or alloy-lower-bound property data. Creep-fatigue endurances are particularly sensitive to the choice of scatterband parameter selected to characterise the creep-rupture properties of the material [31,32].

794

If the objective of the assessment is to avoid excessive conservatism and to predict component life with some precision, it is still necessary with the current state-of-the-art to validate the selected methodology with the results of specific component-targeted feature-specimen tests [6] and service experience.

Feature-Specimen Testing

The use of component-targeted feature-specimen tests to validate the effectiveness of assessment procedures employed to predict the lives of components subject to complex cycle creep-fatigue loading was exploited in the BRITE-EURAM C-FAT project [6]. In the C-FAT activity, five series of cyclic/hold loading tests were performed on large-size specimens representing critical features relating to a number of strategic power plant components (Fig. 3), *eg.* [32-34].

Two of the C-FAT test series provided endurance data for feature-specimens subject to displacement controlled cyclic/hold loading with superimposed primary loading, ie. [32,34]. These targeted power plant components which could be subject, in certain locations, to the combined influence of thermal fatigue and pressure loading; thereby providing a means of validating the application of assessment methods to circumstances in which cyclic damage accumulated in combination with creep damage due to both primary and secondary loading. In the former test, superimposed primary loading was achieved by computer simulation [32] whereas, in the latter case, the effect was produced directly using an advanced testing facility [34]. The direct approach was also adopted in a third series of tests concerned specifically with the development of deformation (ie. ii),Fig. 3) [33].

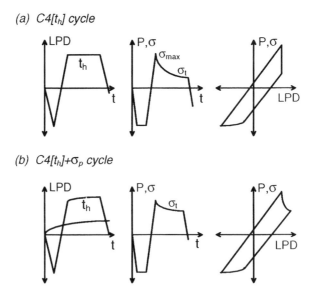

Fig. 4 Service type cycle (a) without and (b) with simulated superimposed primary loading

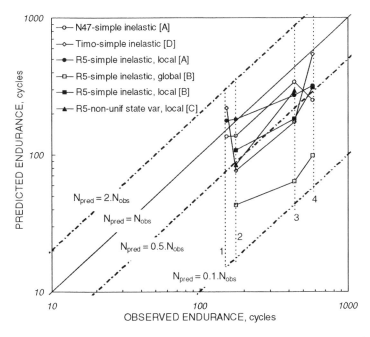

The legend refers to the damage assessment/summation approach, the stress/strain modelling approach, the fracture strain used (as appropriate), and an analyst reference

1 - C4[16h]+100MPa cycle; 2 - C4[16h] cycle;
3 - C4[0.5h] cycle; 4 - C4[0.5h]+200MPa cycle

Fig. 5 Comparison of predicted and observed endurances for SENB feature-specimen tests on cast 1¼CrMoV steel at 550°C (predictions made using heat specific material property data)

Large SENB Specimen Tests

The large SENB feature-specimen tests were performed on a 1¼CrMoV turbine casting steel at 550°C [25,32]. The testpieces were 100mm wide by 75mm thick, with a loading span of 400mm and a notch comprising a 75 x 6mm radius semi-circular groove located at mid-span. Local strain variations at the notch root were generated by controlling the load point displacement amplitude (LPD$_a$) using a service type cycle without and with primary load simulation (Fig. 4), and with hold times (t$_h$) of 0, 0.5 and 16h. In tests involving the C4[t$_h$]+σ$_P$ type cycle, primary loading was simulated by increasing the mean LPD along the creep deformation locus for the direct primary stress of interest (the specific heat of cast 1¼CrMoV steel under evaluation had already been characterised in terms of its creep deformation properties [31]).

The endurances of the tests performed were predicted using a variety of assessment procedures, *eg.* [32,35], and a sample of the predictions for various cycle types are summarised for heat specific and alloy scatterband material property input parameters in Fig. 5 and Fig. 6 respectively. The assessments performed included two time-fraction variants [1,26], with simplified inelastic stress analysis (open points), and three strain based analyses (filled points) [5]. Two of the strain based procedures employed simplified inelastic stress analyses, while the third incorporated a non-unified state variable evaluation (ie. a development of [20] for cyclic softening material [36]). The D'_F, D'_C values adopted in the N47 and R5 type assessments performed by analyst A had been previously optimised using the results of uniaxial tests on heat specific material [31].

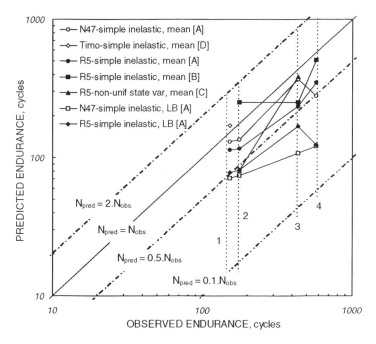

The legend refers to the damage assessment/summation approach, the
stress/strain modelling approach, the alloy scatterband property data
used (mean or lower bound), and an analyst reference

1 - C4[16h]+100MPa cycle;   2 - C4[16h] cycle;
3 - C4[0.5h] cycle;   4 - C4[0.5h]+200MPa cycle

Fig. 6 Comparison of predicted and observed endurances for SENB feature-specimen tests on cast 1¼CrMoV steel at 550°C (predictions made using alloy scatterband property data)

The heat specific data endurance predictions were generally conservative within a factor of ~2 of the observed behaviour (Fig. 5). The results of the R5 assessment using a global ductility model were overly conservative, and hence the move to a local ductility model for low alloy ferritic steels, consistent with [28]. The methodology represented by Eqn.3 was underpinned by experience with austenitic stainless steels [21], and the evidence to demonstrate that a Z_r based creep fracture strain was the most appropriate for low alloy ferritic steels in this context was an important deliverable from the analysis of this feature-specimen test.

The cast 1¼CrMoV steel used in the C-FAT study exhibited mean creep rupture properties [32], and the endurance predictions made using mean alloy scatterband data were similar to those determined with heat specific data (cf. Fig. 6, Fig. 5). The use of lower bound (LB) alloy scatterband data had a significant effect on predicted life (Fig. 6). This experience served to highlight the importance of 'tuning' the assessment procedure to the type of material properties being used as input data.

Welded Wide Plate Tests

The welded wide plate feature-specimens were manufactured from 18Cr11Ni austenitic stainless steel and contained a SAW-welded T-shaped seam [34]. Their overall dimensions were 750 x 460 x 40mm thick. The tests were performed at 570°C using a facility capable of directly applying the C4[40h]+10MPa cycle (Fig. 4). An additional metallurgical complication was that the weld metal was cyclic softening as the parent plate was cyclic hardening during the course of test [6].

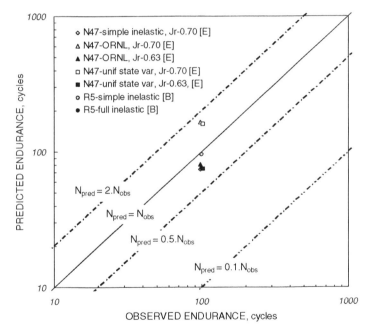

The legend refers to the damage assessment/summation approach, the stress/strain modelling approach, the weldment coefficient (as appropriate), and an analyst reference

Fig. 7 Comparison of predicted and observed endurances for welded 18Cr11Ni wide plate feature-specimen tests at 570°C (predictions made using alloy scatterband property data)

The endurances of the tests were predicted using an N47 approach with stress analyses at three levels of complexity, ie. employing i) simple inelastic calculation, ii) the ORNL model [19] and

iii) a fully unified state variable approach [8] (Fig. 7). In addition, two R5 assessments were performed, ie. with simplified and full inelastic stress analyses. The N47-ORNL and N47-fully unified state variable assessments conducted with a weldment coefficient of $J_r = 0.70$ gave non-conservative endurance predictions. However, the activity provided the evidence to enable J_r to be 'tuned' for these assessment procedures when used for this type of application.

Practical Implications

Three common stages have been identified in the assessment procedures commonly used to predict the life times of components subject to creep-fatigue loading, with a range of analysis options, in particular for determination of the state of stress (strain). The use of component-targeted feature-specimen tests provide an effective means of 'tuning' the appropriate procedure with the selected category of material property input parameters (*eg.* alloy scatterband mean/lower-bound).

From the foregoing, it would be tempting to conclude that the simplest analysis/data options, 'tuned' to the component category of interest, generally provide an effective solution. In practice, there are circumstances when such an approach would be technically inadequate. For example, conceptual designs of complex geometry components subject to relatively small numbers of critical cycles require a consideration of evolutionary behaviour which may only be adequately modelled by constitutive equation sets incorporating an appropriate yield surface (*eg.* [7,8,19,20]), both for physical realism and computational convenience. There are clearly 'horses-for-courses'.

CONCLUDING REMARKS

An overview of the general principles and analysis options available for the assessment of components subject to creep-fatigue loading has been presented. Particular attention has been paid to the analysis of those situations in which damage accumulation occurs as a consequence of the combined effects of fatigue, and creep due to primary and secondary loading.

The results from component-targeted feature-specimen tests have been shown to be essential to benchmark the predictions made using such procedures, in conjunction with the selected category of material property input parameters.

ACKNOWLEDGEMENTS

The author wishes to acknowledge the invaluable technical discussions with members of the C-FAT management team during the course of the project [6] on the issues covered in this paper.

REFERENCES

1 ASME Boiler and Pressure Vessel Code Case N47-17 (1992) *Class 1 Components in Elevated Temperature Service*, Section III, Division 1.
2 RCC-MR (1985) *Design and Construction Rules for Mechanical Components of FBR Nuclear Islands, Section I - Nuclear Island Components*, Edn. AFCEN.
3 TRD 301, Annex 1 - Design (1978) *Technical Rules for Steam Boilers*.
4 TRD 508 - Inspection and Testing (1978) *Technical Rules for Steam Boilers*.
5 R5 (1996) *Assessment Procedure for the High Temperature Response of Structures*, Iss. 2, Rev. 1, R.A. Ainsworth et al. (Eds), Nuclear Electric, (a) Vol.2 - *Analysis and Assessment Methods for Defect Free Structures*, (b) Vol.3 - *Creep-Fatigue Crack Initiation*.
6 BRITE-EURAM Contract BE5245 (1997) *C-FAT Synthesis Report*, S.R. Holdsworth (Ed).
7 Lemaitre, J. & Chaboche, J-L. (1990) *Mechanics of Solid Materials*, Cambridge University Press.
8 Bruhns, O.T. et al. (1987) In: *Proc. Int. Conf. on Constitutive Laws for Engineering Materials*, ICCLEM, Chongquing University.
9 Ramberg, W. & Osgood, W.R. (1943) *NACA Tech. Note No: 902*.
10 Masing, G. (1926) In: *Proc. 2nd Int. Congress of Appl. Mechanics*, 332.
11 Neuber, H. (1961) *Trans. ASME (Series E)* **28**, 544.
12 Hardrath, H.F. & Ohman, L. (1953) *NACA Report 1117*.
13 Polak, J. (1983) *Mat. Sci. Engng.* **61**, 195.
14 Garafolo, F. (1965) *Fundamentals of Creep and Creep Rupture in Metals*. Macmillan, New York.
15 Evans, R.W. & Wilshire, B. (1985) *Creep of Metals and Alloys*, Institute of Metals, London.
16 Cocks, A.C.F. & Ashby, M.R. (1982) *Progress in Metals Science*, 189.
17 Feltham, P. (1960) *J. Inst. Metals* **89**, 210.
18 Conway, J.B., Stentz, R.H. & Berling, J.T. (1975) *TID-26135*, US Atomic Energy Commission.
19 Pugh, C.E. et al. *ORNL-TM-3602*.
20 White, P.S., Hübel, H., Wordsworth, J. & Turbat, A. *Report of CEC contract RA1-0164-UK*.
21 Hales, R. (1987) In: *High Temperature Fatigue: Properties and Prediction*, R.P. Skelton (Ed), Elsevier App. Sci., 229.
22 Coffin, L.F. (1954) *Trans ASME (Series A)* **76**, 931.
23 Tomkins, B.T. (1968) *Phil. Mag.* **18** (115), 1041.
24 Pinneau, A. (1983) *Fatigue at High Temperatures*, Elsevier Appl. Sci., London, 305.
25 Holdsworth, S.R. (1998) In: *Proc. Intern. HIDA Conf. on Creep and Fatigue Crack Growth in High Temperature Plant*, Paris, April, Paper S5-34.
26 Timo, D.P. (1969) In: *Proc. Int. Conf. on Thermal Stresses and Fatigue*, Berkeley, 453.
27 Härkegard, G. (1984) In: *Proc. Int. Conf. on High Temperature Structural Design*, ESIS 12, Mech. Eng. Publ., London, 21.
28 Holdsworth, S.R. (1996) In: *Proc. Conf. on Life Assessment and Life Extension of Engineering Plant, Structures and Components*, Churchill College, Cambridge.

29 Bestwick, R.D.W. & Clayton, A. (1985) In: *Proc. 5th Int. Seminar on Inelastic Analysis and Life Predictions in High Temperature Environment*, Paris, 55.

30 Halford, G.R., Hirschberg, M.H. & Manson, S.S. (1973) In: *Fatigue at Elevated Temperatures*, ASTM STP 520, Philadelphia, 658.

31 Holdsworth, S.R. (1997) In: *Proc. SMiRT Post Conf. Seminar on Intelligent Software Systems in Inspection and Life Management of Power and Process Plants*, Paris.

32 Holdsworth, S.R. (1997) In: *Proc. HSE/I.Mech.E Seminar on Remanent Life Prediction*, I.Mech.E, London, I.Mech.E S539/007/98.

33 Maile, K., Bothe, K., Xu, H. & Mayer, K-H. (1997) In: *Proc. 20. Vortragsveranstaltung - Langzeitverhalten warmfester Stähle und Hochtemperaturwerkstoffe*, Düsseldorf.

34 Kussmaul, K. & Maile, K. (1998) In: *Proc. Intern. HIDA Conf. on Creep and Fatigue Crack Growth in High Temperature Plant*, Paris, Paper S3-20.

35 Booth, P., Budden, P.J., Bretherton, I., Bate, S.K. & Holdsworth, S.R. (1997) In: *Proc. SMiRT Conf.*, Lyon, Paper F-308.

36 Bretherton, I. et al. (1998) *to be published*.

NOTATION

a, Δa	crack length, crack extension
CNRT	circumferentially notched round tensile (testpiece)
da/dN	crack growth rate per cycle
D_C, D_F	creep damage, fatigue damage
D'_C, D'_F	damage fraction co-ordinates at point of intersection in bi-linear damage locus
HTSG	high temperature strain gauge
i	type of creep-fatigue cycle (*eg.* cold-start, warm start, hot start cycles)
J_r	weldment coefficient
LB	lower bound
LCF	low cycle fatigue
LPD, LPD_a	load point displacement, load point displacement amplitude
n_i	number of creep-fatigue cycles of type i
$N_x(\Delta\varepsilon)$	fatigue endurance to crack initiation criterion, x; expressed as a function of $\Delta\varepsilon$
N_{obs}, N_{pred}	observed endurance, predicted endurance (ie. for x = 0.5mm)
PD	potential drop (electrical crack monitoring)
SENB	single edge notched bend (testpiece)
t, t_h, $t_r(\sigma)$	time, hold time, time to rupture expressed as a function of stress
US	ultra sonic (inspection)
x	crack initiation criterion, *eg.* x = Δa (or percentage load drop in a conventional uniaxial specimen test)
z	elastic follow-up factor [5], ie. the ratio of the elastic strain recovered during the hold time to the total creep strain accumulated during the hold time
Z_r	reduction in area at creep-rupture
ε, $\dot{\varepsilon}$	strain, strain rate
ε_c, $\varepsilon_r(\dot{\varepsilon})$	creep strain, rupture ductility expressed as a function of strain rate
σ, σ_P, σ_S	stress, primary (directly applied) stress, secondary (self equilibrating) stress
σ_o, σ_t	stress at start of hold time, relaxed stress after time, t, also $\sigma_t(\sigma_o, t)$
$\Delta\sigma(\Delta\varepsilon)$	cyclic stress expressed as a function of cyclic strain

Low Cycle Fatigue and Elasto-Plastic Behaviour of Materials
K-T. Rie and P.D. Portella (Editors)

SIGNIFICANCE OF DWELL CRACKING FOR IN718 TURBINE DISCS

R.J.H. Wanhill
National Aerospace Laboratory NLR, Amsterdam, The Netherlands

ABSTRACT

The NLR investigated high temperature "dwell cracking" as part of a European project on IN718 turbine discs. Results of dwell and fatigue crack growth tests under simple loading conditions showed that dwell cracking would be unlikely to occur under actual flight conditions and that standard fracture mechanics specimens may be inappropriate for predicting crack growth in discs. Flight-by-flight loading tests by the NLR and others showed that dwell cracking was either absent or limited.

KEYWORDS

Nickel-base superalloy, gas turbine, dwell cracking

INTRODUCTION

A European cooperative technology project on lifing concepts for military aero-engine turbine discs was carried out from 1989 to 1997. The participants were two engine manufacturers, EGT (UK) and MTU (Germany), and four research institutes, IABG and RWTH (Germany), DERA (UK) and NLR. A review of the entire project is given in [1].

The project material was the nickel-base superalloy IN718, heat treated for optimum creep resistance [1]. A large test programme was done at 400 °C and 600 °C with standard LCF, creep, and fatigue crack growth specimens, "engineering" notched specimens simulating disc rim and bore locations, and model discs [1].

Part of the NLR's contribution, aided by other participants' tests, was investigation of "dwell cracking", the subject of this paper. Dwell cracking in nickel-base superalloys occurs under high temperature sustained loading. It is characterized by grain boundary oxidation and subsequent intergranular fracture [2-4], and depends on the load cycle frequency and waveform [4-6] and also the oxygen partial pressure of the environment [4, 7].

802

INVESTIGATION OF DWELL CRACKING

The following aspects of dwell cracking were investigated:

(1)   Effects of peak loads and underloads [8].
(2)   Characterization by fracture mechanics parameters [8].
(3)   Fatigue crack growth + dwells in standard and engineering specimens [9].
(4)   Crack growth under flight-by-flight loading [10-12].

Effects of Peak Loads and Underloads

Dwell crack growth tests were done at 600 °C using compact tension (CT) specimens and including peak loads and underloads representative for military aircraft turbine discs [8]. Peak loads strongly inhibited dwell cracking even when immediately followed by underloads, e.g. Fig. 1. The suppression of dwell cracking lasted for several hours, which is much longer than dwell periods in real flights. Suppression of dwell cracking was also obtained for fatigue cycling with repetitive load histories similar to that in Fig. 1 [1, 13].

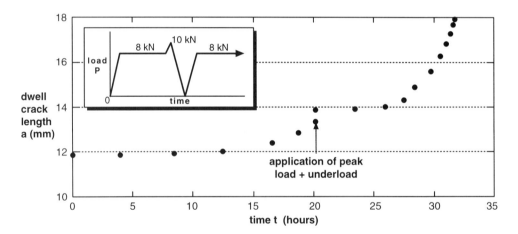

Fig. 1   Suppression of dwell cracking by a peak load, $P_{max}/P_{dwell} = 1.25$, for IN718 at 600°C [8]

Characterization by Fracture Mechanics Parameters

J and C^* were found unsuitable for predicting dwell crack growth in real components, if it occurs, because standard specimen tests showed the experimental and analytical values of each parameter to be very different [8]. In particular, the differences between experimental and analytical C^* values were enormous, Fig. 2. This is because the Norton creep law is inappropriate for dwell crack growth, which is controlled by environmental effects [2-4, 7, 8].

There remains the linear elastic parameter K_I, which actually provides good correlations of dwell crack growth rates, and also correlates data from different sources in a consistent manner, Fig. 3.

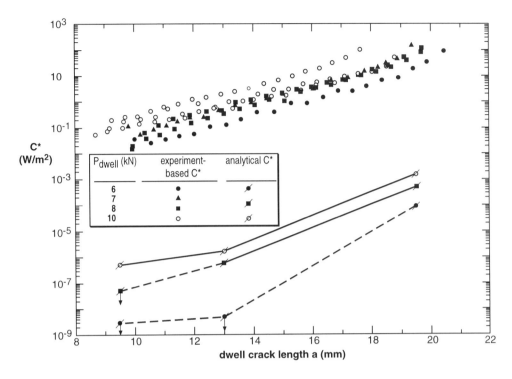

Fig. 2   Experimental and analytical C* for IN718 compact tension (CT) specimen dwell
         cracking at 600°C [8]

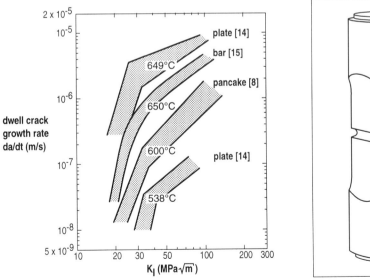

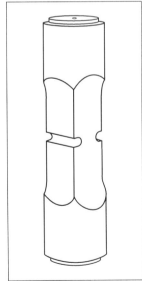

Fig. 3   IN718 dwell crack growth rates versus K_I [8]

Fig. 4   Engineering RIM
         specimen [1]

804

Fatigue Crack Growth + Dwells in Standard and Engineering Specimens

Figure 4 shows the engineering RIM specimen, whose notches generically simulate blade slots on the rim of a turbine disc. The NLR obtained fatigue crack growth rates fractographically for natural semi-elliptical cracks growing from the notches of RIM specimens tested by EGT [9]. The results are compared with those for standard (CT) specimens in Fig. 5. The very large frequency dependence for CT specimens is absent for the RIM specimens. The most probable reason is extensive monotonic yielding at the notch roots of the RIM specimens, followed by nominally elastic behaviour under subsequent fatigue loads: such initial yielding would result in a high slip density and homogeneous plastic deformation for a depth of millimetres ahead of the notch roots, and it is known that this metallurgical condition suppresses dwell cracking [3]. An alternative explanation, that the frequency independence of RIM specimen crack growth might be due to mean stress relaxation, is unlikely because the RIM specimen data in Fig. 5 are for different remote stress levels.

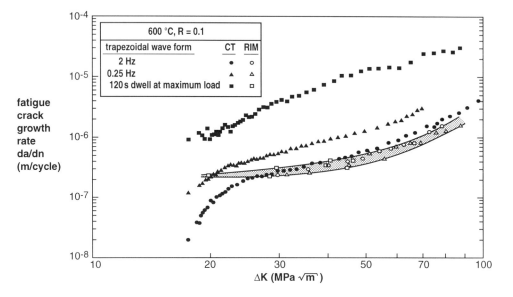

Fig. 5   Fatigue crack growth rates for CT and RIM specimens of IN718 at 600°C [9]

Crack Growth under Flight-by-Flight Loading

LCF, corner crack (CC) and engineering RIM specimens were tested at 600 °C by the NLR and IABG with the flight-by-flight loading sequence HOT TURBISTAN [16], e.g. Fig. 6. This is an isothermalized generic load history for military aircraft turbine discs, and all dwells are below the peak loads.

Fractographs of the specimens [10-12] show that dwell cracking was absent or limited. A similar conclusion was reached by Heuler and Bergmann [1] from analysis of the test data and the success of crack growth predictions that did not account for dwells.

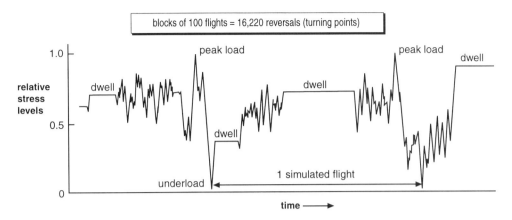

Fig. 6   Typical segment of the turbine disc standard load history HOT TURBISTAN [16]

CONCLUDING REMARKS

Peak loads and notch root plasticity are very effective in suppressing dwell crack growth, which therefore is unlikely to occur if turbine disc rim slots experience fatigue cracking under actual military flight conditions.

Dwell crack growth rates are well correlated by the linear elastic fracture mechanics parameter K_I, but not by the elastic-plastic parameters J and C^*. However, Fig. 5 shows that fatigue and dwell crack growth databases compiled from tests on standard fracture mechanics specimens may be inappropriate for predicting crack growth in discs (N.B.: corner crack (CC) specimens gave very similar results to the compact tension (CT) data in Fig. 5 [1]). In other words, the use of standard specimen data could greatly overestimate the frequency dependence of crack growth in turbine disc rim slots.

REFERENCES

1.   Heuler, P. and Bergmann, J.W. (1997). IEPG TA 31, A Research Programme into Lifing Concepts for Military Aero-Engine Components, Final Report, Industrieanlagen-Betriebsgesellschaft mbH (IABG), Ottobrunn.

2.   Andrieu, E., Ghonem, H. and Pineau, A. (1990). In: *Elevated Temperature Crack Growth, MD-Vol. 18*, pp. 25-29; S. Mall and T. Nicholas (Eds.). The American Society of Mechanical Engineers, New York.

3.   Ghonem, H., Nicholas, T. and Pineau, A. (1991). In: *Creep-Fatigue Interaction at High Temperature, AD-Vol. 21*, pp. 1-18; G.K. Haritos and O.O. Ochoa (Eds.). The American Society of Mechanical Engineers, New York.

4.   Gao, M., Chen, S.-F., Chen, G.S. and Wei, R.P. (1997). In: *Elevated Temperature Effects on Fatigue and Fracture, ASTM STP 1297*, pp. 74-84; R.S. Piascik, R.P. Gangloff and A. Saxena (Eds.). American Society for Testing and Materials, Philadelphia.

806

5. Clavel, M. and Pineau, A. (1978) *Metall. Trans. A* **9A**, 471.

6. Weerasooriya, T. (1988). In: *Fracture Mechanics: Nineteenth Symposium, ASTM STP 969*, pp. 907-923; T.A. Cruse (Ed.). American Society for Testing and Materials, Philadelphia.

7. Andrieu, E., Hochstetter, G., Molins, R. and Pineau, A. (1992). In: *Corrosion Deformation Interaction, CDI '92*, pp. 461-475; T. Magnin and J.M. Gras (Eds.). Les Éditions de Physique, Les Ulis.

8. Wanhill, R.J.H. and Boogers, J.A.M. (1993). The Effects of Peak Loads and Underloads on Dwell Crack Growth in Inconel 718 at 600 °C - Contribution to IEPG TA 31: Lifing Concepts for Military Aero-Engine Components, NLR TP 93300, National Aerospace Laboratory NLR, Amsterdam.

9. Wanhill, R.J.H., Hattenberg, T. and ten Hoeve, H.J. (1996). Fractographic Examination of Three EGT Rim Specimens, NLR CR 96451, National Aerospace Laboratory NLR, Amsterdam.

10. Wanhill, R.J.H., Boogers, J.A.M. and Kolkman, H.J. (1997). Completing NLR Contributions to the IEPG TA 31 Programme, Part III: Crack Growth Testing on RIM Specimens, NLR CR 97035, National Aerospace Laboratory NLR, Amsterdam.

11. Wanhill, R.J.H. and Boogers, J.A.M. (1997). Completing NLR Contributions to the IEPG TA 31 Programme, Part IV: Crack Growth Testing on CC Specimens, NLR CR 97035, National Aerospace Laboratory NLR, Amsterdam.

12. Wanhill, R.J.H. (1997). Fractographic Examination of Three IABG LCF Specimens, NLR CR 97083, National Aerospace Laboratory NLR, Amsterdam.

13. Wanhill, R.J.H. (1996). Fractographic and Metallographic Examination of IABG Corner Crack (CC) Specimens, NLR CR 96759, National Aerospace Laboratory NLR, Amsterdam.

14. Sadananda, K. and Shahinian, P. (1977) *Metall. Trans. A* **8A**, 439.

15. Liu, C.D., Han, Y.F. and Yan, M.G. (1992) *Engng. Fract. Mech.* **41**, 229.

16. Köbler, H.-G., Amelung, K. and Durwen, S. (1990). HOT TURBISTAN: LBF Contributions to the Experimental Part of the International Research Programme-Final Report, LBF Report No. 5092/6533, Fraunhofer-Institut für Betriebsfestigkeit, Darmstadt.

Low Cycle Fatigue and Elasto-Plastic Behaviour of Materials
K-T. Rie and P.D. Portella (Editors)
807

UNDERSTANDING OF CREEP, FATIGUE AND OXIDATION EFFECTS FOR AERO-ENGINE DISC APPLICATIONS

GERHARD W. KÖNIG [1)]

MTU, Motoren- und Turbinen- Union, München GmbH
Postfach 500640; D-80976 München, Germany

ABSTRACT

In service, aero-engine turbine discs are severely stressed by centrifugal forces, high temperatures and forces caused by stress gradients. As a result, fatigue is the life-limiting factor for most contemporary engine discs. However, the demand for increasing engine efficiency leads to increasing stresses and temperatures, so that creep and oxidation, in addition to fatigue, are becoming increasingly important for life consumption of discs. With the aim of extending the use of discs into the creep/fatigue/oxidation regime, a BRITE/EURAM project (BREU6021 „Development of an understanding of materials properties under the combined influence of creep, fatigue and oxidation") was performed by a consortium including European aero-engine designers and manufacturers (ALFA ROMEO AVIO, BMW/RR, MTU, ROLLS-ROYCE, SENER, TURBOMECA and VOLVO AERO CORPORATION), testing and evaluation laboratories (Defence Research Agency DRA and Istituto per la tecnologia dei materiali metallici non tradizionali, ITM) and a material supplier (THYSSEN REMSCHEID). The representatives of these partners are given in [1] to [10]. This paper describes the methodology and selected results of this project.

KEYWORDS

Creep, fatigue, oxidation, disc alloys, material models

PROJECT METHODOLOGY

The project consisted of the two major tasks: experimental and theoretical work.

Within the experimental effort, both specimen and component tests were performed. Two disc materials were investigated: the nickel base disc alloy U720LI optimised for disc applications [7] and the titanium alloy IMI834. Specimen tests were performed to investigate tensile, creep, fatigue (notched and unnotched) and crack propagation behaviour. A systematic study of the transition between creep and fatigue was undertaken with an approach illustrated schematically in Fig. 1. Pure fatigue was obtained by rapid cyclic loading and creep effects were introduced either by applying dwell times at maximum load (fatigue/dwell), reducing frequency by increasing the times for uploading or downloading (fatigue/frequency), applying pre-creep prior to fatigue testing (alternating creep/fatigue) or by superposition of tensile mean stresses (tensile mean stress/fatigue).

---

1) Coordinator of the BRITE/EURAM project BREU6021

808

For design, it is important to know the relevance of specimen test results to predict the behaviour of real components. Therefore, in addition to specimen tests also component tests were performed (using model discs representing critical features of discs used in engine service). Cyclic spinning tests on discs are very costly and time consuming and only five of these tests could be realised within the project (2 titanium [3] and 3 nickelbase model discs [1], [2] and [5]). Again, creep effects were investigated by comparing the results of test cycles with short and long dwell at maximum load. For illustration, Fig. 2 shows the titanium model disc and (schematically) the type of test cycle.

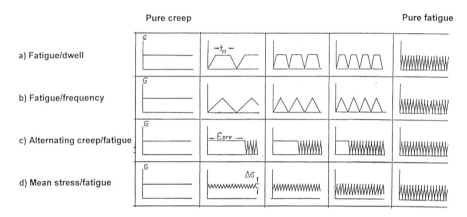

Fig. 1:   Scheme of testing for the transition from pure creep to pure fatigue

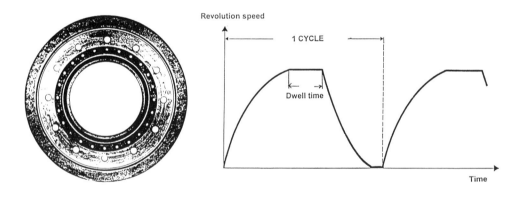

Fig. 2   Titanium model disc used for tests with short (1 second) and long (120 seconds) dwell at maximum load; critical areas for cracking are 12 circumferential holes

Within the theoretical part of the project, the emphasis was on the stress-strain analysis of the test pieces as well as on the assessment and application of material models. The models investigated cover the material properties of elasto-plastic behaviour, creep, crack propagation and creep-fatigue interaction, including simple engineering approaches as well as sophisticated unified constitutive models. For the assessment of the use of different models for design purposes, four main criteria were applied: the ability to accurately model different aspects of material behaviour, the calculation effort necessary to determine the model parameters, the amount of required material data and the suitability for implementation into (finite element) computer programs.

Besides a large number of conventional material models published in the literature, the two constitutive models of Bodner-Partom and Chaboche were selected to be studied in detail. The general advantage of constitutive models are their capability to model a large range of elasto-plastic material behaviour and their suitability for implementation in computer programs.

The Bodner-Partom model consists of a set of constitutive equations for viscoplastic material behaviour without yield criterion and is formulated on the basis of internal variables which depend on the loading history. The basic equations for uniaxial loading conditions and the method of determining the model parameters are given in [11], [14] and [15]. In contrast to the Bodner-Partom model, in the Chaboche model a yield limit is employed below which no inelastic deformation occurs and the material is assumed to be purely elastic. In this project modifications of the model as described in [13] were used.

RESULTS

In the following, some examples are given which demonstrate the capabilities and limitations of models to describe time-dependent material behaviour.

A) ELASTO-PLASTIC BEHAVIOUR

At low temperatures, the stress-strain behaviour of materials is virtually time-independent. With increasing temperature, time at temperature becomes a major variable for materials strength. This can be seen in tensile tests, where a very pronounced effect of strain rate on the flow stress can be observed (Fig. 3). High strain rates tend to produce strain hardening, while softening was found at low strain rates. This behaviour in tensile tests could satisfactorily be represented by both constitutive models.

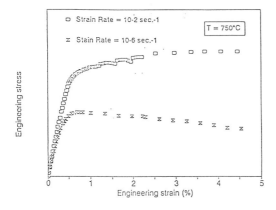

Fig. 3
Influence of strain rate on the flow stress, strain hardening and softening of U720LI
(ITM [4])

The situation is different in tensile tests where changes in strain rate were applied (Fig. 4). After each strain rate jump a transient effect occurs where, for a small amount of strain, softening or hardening can be experienced. In this type of test all models failed to describe the

810

material behaviour accurately. However, for the case of the constitutive models it is not clear whether this is due to limitations of the models or just a matter of inaccurate input data.

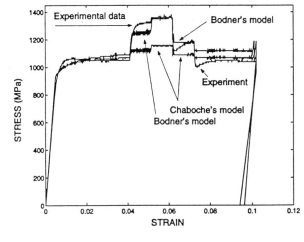

Fig. 4
Simulation of tensile tests with strain rate jumps using the models of Chaboche and Bodner-Partom (measurement and modelling by VOLVO AERO CORPORATION [9])

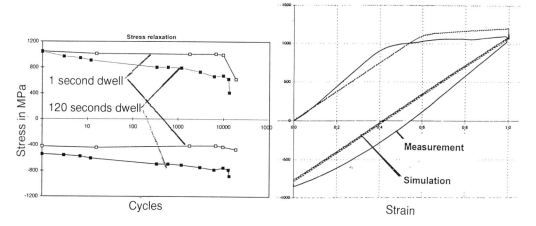

Fig. 5a: Cyclic stress relaxation of U720LI for 1 and 120 seconds dwell at 650 deg. C (MTU [10])

Fig. 5b: Comparison between measured and predicted (Bodner-Partom) hysteresis loops; U720LI, 1 second dwell (SENER [6])

Hysteresis loops, as measured during strain controlled testing of plain specimens, show strong time-dependent effects: the cyclic stress relaxation is much more pronounced at 120 seconds dwell, as compared to 1 second dwell (Fig. 5a). Both constitutive models were suitable to give satisfactory descriptions of these hysteresis loops with the connected effects of dwell and loading times (see example in Fig. 5b).

B) CREEP

A typical creep curve in terms of creep rate vs. creep strain is shown in Fig. 6. The three stages of the creep life consist of a (often very short) primary creep regime (deeply decreasing strain rate caused by hardening), a minimum creep rate (frequently called steady-state creep regime if it extends over a certain amount of strain) and the tertiary creep stage characterised by continuously increasing strain rates up to fracture. The major parts of these creep curves could be represented reasonably well by a number of conventional models (see example shown in Fig. 6).

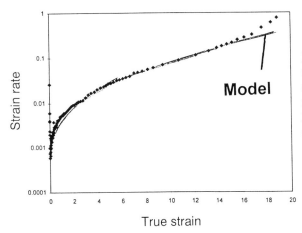

Fig. 6
Creep behaviour of U720LI at 700 deg. C; Comparison between measurement and simulation by ITM-creep model (ITM [4])

However, models which proved successful for monotonic loading were far less accurate in the case of load changes: the transient behaviour following these changes was not accurately modelled [4] (as in the situation of tensile strain rate jump tests; see above).

C) CRACK PROPAGATION

It is widely known that at high temperatures crack propagation rate is influenced by time-dependent as well as by time-independent contributions. Frequently it is assumed that these contributions add linearly and time-dependent crack growth can be obtained by an integration derived from the following type:

$$\Delta a = \int_{t1}^{t2} (da/dt)dt = \text{ const. } \int_{t1}^{t2} (K_{max}(t))^m dt \qquad \text{equation (1)}$$

However, the analysis of U720LI crack propagation data clearly showed discrepancies with this approach. The following table gives an overview of the influence of different wave forms on the crack propagation rate for U720LI at 650 deg. C.

Table 1:   Time-dependent crack growth (U720LI at 650 deg. C and 30 MPa$\sqrt{m}$)

Wave form *)	5Hz	1/1/30/1	1/1/1/1	1/120/1/1	30/1/30/1	30/1/1/1
Normalised da/dN **)	1	4,3	6,2	11	11	16

*) Time in seconds for up-loading/dwell at max. stress/down-loading/dwell at min. stress.
**) The crack growth rate da/dN is normalised with the da/dN at 5Hz.

812

As an effect of the additional time under load, all trapezoidal cycles show crack propagation rates which are significantly higher than under rapid cycling with 5 Hz. However, the analysis shows that the effects of time are rather complex ones. An up-loading time of 30 seconds increases da/dN stronger than a 120 seconds dwell at maximum load. Moreover, increasing the cycle time by introducing a down-loading time of 30 seconds obviously decreases da/dN. Both findings seem to be in conflict with the assumptions made in equation (1). Therefore it was necessary to modify existing crack propagation models by introducing additional terms accounting for the effects of the times of up-loading and down-loading [1]. With this engineering approach it was possible to get a reasonably good description of all data (Fig. 7).

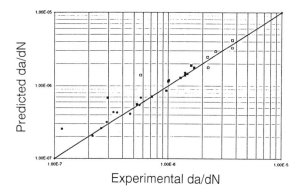

Fig. 7
Comparison between measured and predicted crack growth rates for the wave forms listed in Table 1; U720LI at 650 deg. C

(modelling: ALFA ROMEO AVIO [1])

Experimental da/dN

OXIDATION

The cyclic life is influenced by oxidation both with regard to crack initiation and propagation. Crack propagation rates under vacuum were found to be slower by more than one order of magnitude as compared to air both for U720LI [4] and IMI834 [3]. In the case of the titanium alloy, strong effects of oxidation were also observed on crack initiation life. At 630 deg. C, IMI834 reacts with oxygen resulting in a brittle surface layer. Under high strain cyclic loading this brittle surface cracks after only a few cycles leading to reduced crack initiation lives. The cracking can observed in the hysteresis loops of strain controlled tests [12]: At the beginning, minimum and maximum stresses develop in parallel up to the point where cracking of the surface layer occurs. Subsequently, the change of compliance due to cracking results in a change of the stress-strain slope.

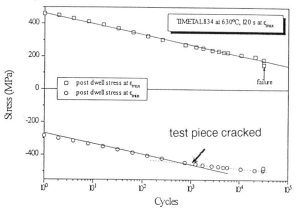

Fig. 8
Development of maximum and minimum stress in a fatigue test with 120 seconds dwell; influence of cracking of the brittle surface layer (DRA; [12])

CONCLUSIONS

The test results show that time-dependent effects at high temperatures make material behaviour much more complex than at low temperatures. Elasto-plastic, creep and crack propagation behaviour were found to be adequately described by a number of conventional models as well as by two advanced constitutive models to meet the requirements for many applications. The advantage of constitutive models are, in spite of the high effort for the required input material data, the broad applicability in design programs. However, the choice of the models to be used for specific applications will mainly depend on the requirements for accuracy as well as on the available data base and computer programs.

Some material properties such as the transient behaviour after load and strain rate changes or the time-dependent effects on crack growth rates, still require future refinements in modelling.

ACKNOWLEDGMENTS

This work was performed within the BRITE/EURAM project BREU6021. The author would like to thank the CEC for financial support and all partners for their contributions to the project.

REFERENCES

1. R. Festa (ALFA ROMEO AVIO): BREU6021 progress and final reports
2. U. Heßler, A. Fischersworring-Bunk (BMW/RR): BREU6021 progress and final reports
3. M. Hardy, G. Harrison (DERA, former DRA): BREU6021 progress and final reports
4. V. Lupinc et al. (TEMPE, former CNR-ITM): BREU6021 progress and final reports
5. S. Williams (ROLLS-ROYCE): BREU6021 progress and final reports
6. J. Vinals (SENER): BREU6021 progress and final reports
7. P. Janscheck (THYSSEN REMSCHEID): BREU6021 progress and final reports
8. F. Vogel (TUBOMECA): BREU6021 progress and final reports
9. N Järvstrat (VOLVO AERO CORPORATION): BREU6021 progress and final reports
10. G. König (MTU): progress and final reports
11. R. Sjöqvist, N. Järvstrat: „VpSim - interactive simulation of cyclic visco-plasticity" in proceedings of FATIGUE 96; eds. Lütjering and Novack, Elsevier Science, Oxford, 1996
12. M. C. Hardy: „The effects of oxidation and creep on the elevated temperature fatigue behaviour of a near-α titanium alloy"; submitted to LCF4 conference
13. J. Lemaitre, J. L. Chaboche: „Mechanics of solid materials"; Cambridge University Press, 1990
14. E. Cearra, J. J. Vinals: „Implementation of viscoplastic constitutive model in MARC computer code"; submitted to LCF4 conference
15. J.J. Vinals: „BP program - interactive evaluation of Bodner-Partom model material constants"; submitted to LCF4 conference

Low Cycle Fatigue and Elasto-Plastic Behaviour of Materials
K-T. Rie and P.D. Portella (Editors)
815

ON A FAILURE CRITERION FOR THERMOMECHANICAL MULTIAXIAL LOW CYCLE FATIGUE AND ITS INDUSTRIAL APPLICATION

E. CHARKALUK*,**, A. CONSTANTINESCU**, A. BIGNONNET*, K. DANG VAN**
* P.S.A. Peugeot-Citroën - Direction des Recherches et Affaires Scientifiques
Chemin de la Malmaison - 91570 BIEVRES - FRANCE
** Laboratoire de Mécanique des Solides (CNRS URA 317)
Ecole Polytechnique - 91128 PALAISEAU - FRANCE

ABSTRACT

The purpose of this paper is to assess the fatigue strength of complex structures submitted to low cycle thermomechanical fatigue. In a first step, the behaviour of the material in a high range of temperatures has been modelled using a simple elastoviscoplastic constitutive law. In a second step, a criterion based on the dissipated energy per cycle derived from the results of the numerical computations, allows the prediction of the fatigue strength of the structure.
Strain controlled isothermal low cycle fatigue tests were carried out on cast iron specimens for the identification of the constitutive and the fatigue laws. Temperature controlled tests on specimens and structures permitted the evaluation of the precision of the numerical computation under anisothermal loading. The lifetime prediction is showed to agree with the experiments both on specimens and structures.

KEYWORDS

Low cycle fatigue criterion, dissipated energy, constitutive law, thermal fatigue, structure computation.

INTRODUCTION

The prediction of fatigue and failure of structures submitted to thermal loading is one of the important problems in mechanical engineering. One can recognize two research directions in this field : one concentrated on the understanding and modelling of the physical phenomena underlying fatigue of materials and one concentrated on finding a robust prediction criterion to evaluate the fatigue strength of components from a structural computation. The prediction criterion should correspond to the expectation of the design engineer, i.e. to be identified from simple mechanical experiments on specimens and to be applicable on actual structures of complex geometry and complex thermomechanical loadings.
In spite of major advances in the understanding of fatigue phenomena at a local microscopic level and a large amount of experiments on specimens, robust criteria at a global macroscopic level to predict low cycle thermomechanical fatigue failure on structures are still missing. There are essentially two difficulties to attain this objective : a formulation of the material behaviour which gives access to an accurate description of the mechanical values which control the fatigue

phenomenon in a structure and a failure criterion, derived from simple tests, which provides a reasonable estimation of the lifetime of the structure.

The constitutive law should permit computer simulations in a acceptably small computation time of elastoviscoplastic structure subjected to complex thermomechanical loadings. The fatigue criterion should be capable of explaining isothermal and anisothermal LCF failure for a large range of temperatures and for multiaxial loadings.

From a material point of view , one can identify different steps in the fatigue life time (initiation of a crack, propagation, failure of a representative elementary volume, ...). The interpretation of these steps from the point of view of applications to industrial structure permits a distinction between different situations, for example :

- in the security assessment of a cracked structure, one can directly concentrate on the crack propagation rate and the critical size of the crack [1].
- in a particular design process, where no control is done after the delivery of the product to the customer, one can not accept crack growth leading to component disfunctionment before the targeted lifetime. Therefore one can consider in the design analysis the time to failure of a representative elementary volume (REV) as a failure criterion of the whole structure [2].

These considerations have suggested a certain number of assumptions in the present study :

- the material behaviour can be modeled by a simple elastoviscoplastic constitutive law, which permits the representation of the cyclic response of the structure under thermomechanical loading,
- the fatigue criterion can be based on an intrinsic quantity permitting the prediction of failure under isothermal and anisothermal multiaxial loading. The dissipated energy per cycle, seem to give good results.

This paper presents in a first section the identification of the constitutive law and in a second section the fatigue criterion and its applications to the fatigue assessment of actual structures using isothermal and anisothermal LCF experiments.

EXPERIMENTS AND COMPUTATIONS

The material studied is a spheroidal graphite cast-iron with silicium and molybden additions. Three types of test have been carried out on specimens and actual exhaust pipes of automotive engines :

- *Isothermal strain controlled tests* (TRV) were carried out in order to caracterize the mechanical behaviour of the material. The experiments have been done at different constant temperatures between 20°C and 800°C. The applied cycle corresponds to a tension-relaxation-recovery (TRV) test, with total strain rates applied to the specimen between $10^{-4} s^{-1}$ and $10^{-2} s^{-1}$ and maximal strain ranges between 2% and 5%.
- *Fully reversed isothermal low cycle fatigue tests* (LCF) ($R_\varepsilon = -1$) were carried out under strain control of the material. The experiments have been done at different constant temperatures between 200°C and 700°C, with a total strain rate of $10^{-3} s^{-1}$ and three different strain ranges : $\pm 0,25$ %, $\pm 0,5$ % and ± 1 %.
- *Thermal fatigue tests* (TMF) were conducted on clamped specimens heaten by the Joule effect. Maximum temperatures have been varied between 40 °C and 700 °C with an heating rate of 20°C/s. The maximum temperature has been obtained in a region of approximatively 10 mm in the center of the specimen. The maximum temperature gradient was 30-40°C/mm. The parameters of the test are the clamp value (183000 N/mm and 227000 N/mm) and the dwell time at 700 °C (60 seconds and 900 seconds).

- *Thermomechanical tests* were carried out on engine components. Maximum temperatures have been varied between 20°C and 700°C. It is important to remark that temperatures have not been distributed uniformly over the component. Some points never exceeded 200°C. The heating rate was approximately 10°C/s.

An elastoviscoplastic constitutive law with five temperature depending parameters has been identified for each temperature on the stabilized cycle obtained with the isothermal TRV. The rheological representation is shown in Figure 1. A comparison between computation and experimentation is given in the Figure 2 and it can be observed that relaxation is reasonnably well represented with a 10 % error on stresses. A difference of stress hardening was observed between compression and traction and similar behaviour for cast-iron has already been reported by Josefson et al. [3]. In the model an average of the two values of stress hardening has been used to minimize the error in tension and compression, which explains the differences between the computed and experimental curves of Figure 2.

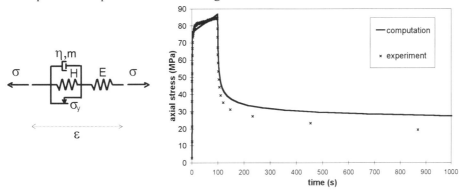

Figure 1 : a representation of the constitutive law   Figure 2 : strain controlled test at 700°C

Using this elastoviscoplastic constitutive law, FEM computation using the ABAQUS code has been performed in order to simulate the anisothermal fatigue tests. A typical result of computed axial stress in the specimen versus temperature is presented in Figure 3.

The observed difference between the experimental and the computed first cycle is actually due to the constitutive parameters of the elastoviscoplastic law. However one can remark that the stabilized cycle has been evaluated both in term of residual stress at room temperature and amplitude difference between loading and unloading. The dwell temperature has been considered constant during the computation, which differs from the experiment where a 20°C difference due to the thermal regulation can be observed.

The irregularities at 500°C are due to variations of the elastoviscoplastic parameters at this temperature, corresponding to a dynamic strain ageing phenomenon. A better control of the numerical convergence of the code can be achieved and is part of the study not reported here.

The anisothermal tests on components were also simulated using the same elastoviscoplastic constitutive law. A typical FEM model for an exhaust pipe containing about 20000 hexaedric volume elements is represented in Figure 4. The temperature computed distribution on the components has been checked with experimental values obtained by infrared thermography. The thermomechanical loading has been given by the temperature variation described before and stresses and strains distribution were obtained in all the structure.

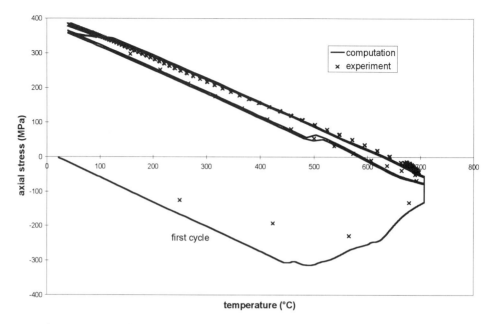

Figure 3 : comparison between the thermal fatigue test curve and the corresponding computation for a 40°C-700°C test with a 227000 N/mm clamp and a 60 s dwell time.

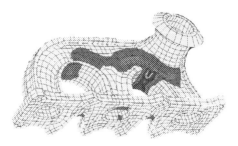

Figure 4 : FEM model of the exhaust pipe.

LIFE TIME PREDICTION

The defined objective was to determine a fatigue criterion in a multiaxial context permitting the reasonable assessment, i.e. in the standard deviation of the experimental results, of the number of cycles to failure of a representative elementary volume. The microscopic and fractographic analysis done after failure on the isothermal and anisothermal specimens showed little influence of parameters like oxydation or initial defects. This permitted a concentration on a criterion depending on mechanical state variables like stresses or strains.

A close inspection showed that the Manson-Coffin law [4, 5], the Strain-Range Partitionning (SRP) [6] or the Smith-Watson-Topper (SWT) function [7] expressed initially in an isothermal and uniaxial context are not appropriate for this problem.

The difficulties stem essentially from the multiaxial context and the high range of attained temperatures. Classical Manson-Coffin interpretation of isothermal LCF tests shows on one hand a high difference in lifetime for the same plastic strain-range at different temperatures, and on the other hand does not give a clear generalization of a multiaxial case. This makes any interpretation of cumulated plastic strain on an anisothermal stress-strain curve hazardous and unreliable.

SWT interpretation of the anisothermal experiments are difficult as the choice of σ_{max} is impossible in an anisothermal case (SWT function $= \sqrt{E.\sigma_{max}.\Delta\varepsilon}$). One can remark that for cast-iron, a maximum stress for the same strain can vary from 30 to 300 MPa depending on the temperature.

The principal criticism of the previously discussed criteria is their inapplicability for all types of loading case. A possible way to overcome this difficulty is an energy criterion. Different forms of energy criteria have already been proposed and discussed for example by Ostergren [8], Ellyin et al.[9] or Halford [10]. These papers present different expressions of macroscopic deformation energy including or excluding its elastic, plastic or work-hardening part [11].

An interesting interpretation of the cyclic behaviour of the material is presented in Skelton [12]. These results [12] suggest that the cumulated dissipated energy to the cyclic response stabilization can be used as a crack initiation criterion in low cycle fatigue. Denoting by N_S the number of cycles to stabilization, assumed to correspond to crack initiation, Skelton showed that the cumulated dissipated energy to stabilization: $Ns \times \Delta W_s$ is relatively constant for a given material, between 1 J.mm^{-3} and 10 J.mm^{-3}.

Our basic assumption is that the failure of the representative elementary volume corresponds to the failure criterion and in the case of a specimen, the failure of the representative elementary volume implies the complete failure. Using the ideas developed by Skelton, the present LCF and TMF tests on cast iron were analysed with a dissipated energy on the stabilized cycle.

The experimental versus predicted lifetime is presented in log-log plot in Figure 5.

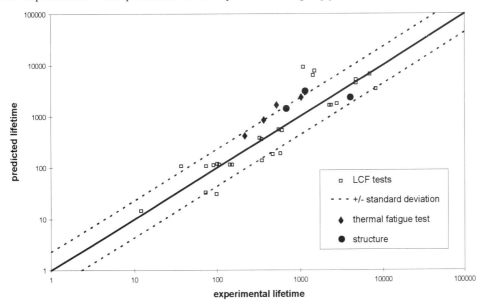

Figure 5 : comparison between estimated lifetime and predicted lifetime for LCF tests and thermal tests on specimens and anisothermal tests on real structures.

The criterion itself has been established by linear regression using *only* LCF tests. For these tests the dissipated energy has been evaluated by numerical integration from the stress-strain experimental curves. The dissipated energy for the TMF tests on specimens or structures has been obtained by FEM computations using the energy of the stabilized cycle.

For LCF tests, the difference between experimental and predicted lifetime represents actually only the deviation for each point with the identified criterion. The standard deviation has been represented by the two dotted lines on the graphics. For TMF tests on specimens or structures, experimental lifetime is obtained with the tests and predicted lifetime is obtained using FEM computations. The computed dissipated energy distribution over the structure indicates precisely the experimental cracked region.

The cumulated energy to failure on components and specimens was showed to be of the same order of magnitude as reported by Skelton [12], i.e. between 1 J.mm^{-3} and 10 J.mm^{-3}.

CONCLUSION

The aim of this paper was to investigate a possible low-cycle fatigue criterion and its applications to industrial structures subjected to thermomechanical loading.

It has been shown that a relatively simple elastoviscoplastic constitutive law can describe the mechanical behaviour of the structure in a large range of temperatures. The temperature dependance of the parameters permitted the emphasis on a plastic behaviour at low temperatures (20°C - 400°C) and a viscous behaviour at high temperatures (400°C - 800°C).

The dissipated energy per cycle, obtained directly from the computed stresses and strains demonstrated a robust fatigue criterion for the lifetime prediction of industrial structures. Moreover the total dissipated energy to failure has been found to be of the same order of magnitude as reported earlier by Skelton [12]. The applicability of an energy criteria to the lifetime prediction of structures subjected to complex low-cycle fatigue loadings has thus been proved.

However further efforts are necessary to improve the accuracy of life estimations of structures submitted to thermomechanical fatigue. A better quantitative understanding of the dissipation mechanisms and their relation to different types of failure will find a logical place in low-cycle fatigue lifetime predictions.

REFERENCES

1.   R5, BNL report (1990) Berkeley Nuclear Laboratories, Nuclear Electric plc, Berkeley.
2.   Dang Van, K. (1973) *Sciences Technique Armement*. Vol. 47, No 3.
3.   Josefson, B.L., Stigh, U. and Hjelm, H.E. (1995) *J. Engin. Mat. Tech*. Vol. 117, pp 145-150.
4.   Coffin, L. F. (1953) ASME, paper 53-A76, pp 931-950.
5.   Manson, S. S. (1953) NACA, TN 2933.
6.   Halford, G. R., and Manson, S. S. (1976) In : *Thermal fatigue of materials and components*, ASTM STP 612, pp 239-254.
7.   Smith, K.N., Watson, P. and Topper, T. H. (1970) *J. Mater*. Vol. 5., No 4, pp 767-778.
8.   Ostergren, W. J. (1976) *J. Test. Eval*. Vol. 4, No 5, pp 327-339.
9.   Ellyin, F., and Golos, K. (1988) *J. Eng. Mat. Tech*. Vol. 110, pp 63-68.
10.   Halford, G. R. (1966) *J. Mater*. Vol. 1, No. 1, pp 3-18.
11.   Lemaître, J. and Chaboche, J.L. (1985) In : *Mécanique des matériaux solides*, Dunod editions, Paris
12.   Skelton, R. P. (1991) *Mat. Sc. Tech*. Vol. 7, pp 427-439.

THE LOW-CYCLE FATIGUE OF FIBRE-METAL-HYBRID STRUCTURES

GEORG W. MAIR
Federal Institute for Materials Research and Testing, Section III.23
D-12200 Berlin, Germany

ABSTRACT

Searching for reliability restricted optima of carbon fibre reinforced steel pressure vessels for CNG-storage in local traffic busses a manageable mathematical tool was created for all appearing questions of life-time fatigue [1, 2]. This tool has been designed to work with the at most used material data sheets and has been made able to handle the different load numbers, stress ratios between the static tenacity and the fatigue strength for steel as well as for fibre reinforcement in reliability calculations and optimisations. This resulted in an analysing and optimisation tool to apply damage accumulation hypotheses without any additional simulating tests. This tool is useable to dimension structures by considering reliability as well as in cases of a high minimum probability too. The operation and results of working of this tool were shown by using the classic WÖHLER- and GOODMAN- diagrams.

KEYWORDS

CrMo4, carbon fibre, reliability, pressure vessel, liner, fibre reinforcement, stress ratios, fatigue strength

INTRODUCTION

Usually a lot of structures are still being dimensioned exclusively by aspects of static safety. Indeed most of these structures are subjected to a cycling demand, which can not be directly referenced to a safety factor based on static aspects. Especially in combination of metal and fibre reinforced plastics - so called hybrid structures - there are some differences and discrepancies between the static safety factor and the probability of fatigue or the total reliability. In this case there are also considerable increasing competitive disadvantages by using an unnecessary amount of material. The real fatigue strength of a part of a construction depends on the character and range of the local load as well as the local load is depending on the global structure load range.
Even in disregarding the aspects of total reliability often there are to guarantee safe-life reliabilities of some more as 90 % for the analysis of life-time fatigue strength.

To analyse such reliabilities in metal parts of constructions doesn't make any problem in consideration of 10^4 or more numbers of design load alternations. The mathematics used in such an analysis can be shown as a WÖHLER-diagram. Using this classic WÖHLER-tool, in most cases it isn't possible to analyse the reliabilities of fatigue caused by a lower load number. Considering the real upcoming loads in form of any load collective there are also to be taken into account load classes that have a load number which requires tools of low cycle fatigue. Especially using a linear damage accumulation method - like the one of MINER – makes it necessary to base the analysis of damage on the aspects of low-cycle fatigue.

These correlation is to be shown in the best way by the example of the liner of hoop raped hybrid cylinder vessels. The number of design loads at daily loading and design limit of 15 years can reach an amount of up to 5000 load cycles.

FATIGUE QUESTIONS OF HYBRID VESSELS

According to the analysis of the survival reliability especially the particularity of the low-cycle fatigue of steels has to be taken in account. In this case some deficiencies are appearing in the practised knowledge [3] which are to be investigated in a very clear and conservative way:

1st Comparing the different distributions of probabilities of the static fracture strength of metals it is to determine a uniform kind of distribution.

2nd A description model of the low-cycle-fatigue is to be derived from the data of the static fracture strength and the cycle fatigue strength.

3rd The borderline case of mean stress approximating the limit of the static fracture strength is to be described mathematically.

The multiple parametric calculation model for the young kind of reinforced plastics are as opposed to metals full of gaps in investigation. Therefor it is not safe assuming that in general accepted descriptions of strength are given. For the analysis of the life-time-fatigue of fibre reinforced pressure vessels - in conjunction with carrying metal liners - it is necessary to calculate the fibre reinforcement:

4th The statistician description of the fatigue strength of the exemplary investigated fibres;

5th A hypothesis about the influence of relative high mean stress.

FATIGUE OF METALS

Time-fatigue and cycle-fatigue strength of metallic materials e. g. CrMo4 can be described by using a WÖHLER-diagram like it is for the most part reproduced in literature (compare fig. 1). According to question 1 and 2 (see above) those diagrams are normally not able to take account of the low-cycle fatigue strength. But in this special case of concrete formulation of questions it might be adequate to interpret precisely the low-cycle fatigue as the transition of the yield strength to the static tenacity. This fact is visualised in the left of figure 1 and is the 1st result of investigation for a in [1] and [2] formulated mathematical model.

The main problem using such a transition is the question for a useful probability distribution. The static tenacity is normally described by using a WEIBULL or GAUSS distribution in a linear table. In opposite to the static load case the time fatigue is to be characterised in a log10 table. Linear extrapolating the log-scale and the used distribution of time fatigue up to the static tenacity it is possible to get a conservative (in the meaning of save) solution for this problem.

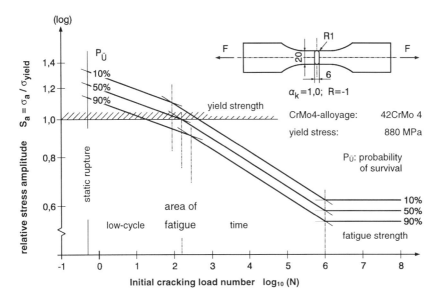

Fig. 1. Fatigue diagram based on a WÖHLER-line with a GAUSS distribution of hardened 42CrMo4-steel

After solving this problem for a constant mean stress ratio like it works by using a WÖHLER-diagram, it is necessary to get a model for using several mean stress levels. The load range of the metal liner shows a nearly constant maximum stress level.

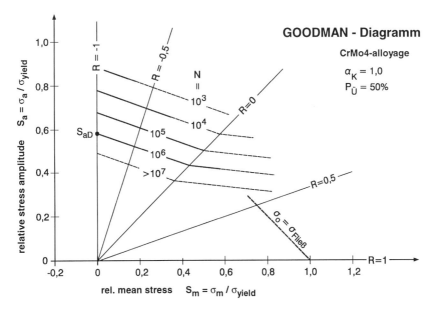

Fig. 2. GOODMAN-diagram of a CrMo4-alloyage

824

At the same time the minimum stress and therefor the stress ratio is alternating between the pure (negative) R = -1 inherent tension and the maximum stress level (R = 0).

Therefor it is necessary to build a GOODMAN-diagram based on the WÖHLER-diagram above for several load numbers of fatigue and a reliability of 50 % like fig. 2. There the lines of fatigue which are a derivative from the function shown in fig. 1 are crossing the yield strength limit. These and the affiliated questions of low-cycle fatigue above the yield strength obviously are in contradiction to the yield stress as one of the defined design limits of vessel optimisation. But in accordance to the required high reliabilities it is necessary to build such a calculation model as base for parallel lines of very low fatigue probability below the yield strength shown in fig. 3. This results in an area in this diagram in which the boundary of time fatigue disappears between the areas of low-cycle fatigue strength and fatigue limit strength. Therefor there are to analyse exclusively stresses below the yield limit but for quantifying the reliabilities it is the area above the yield strength that is to be taken in account.

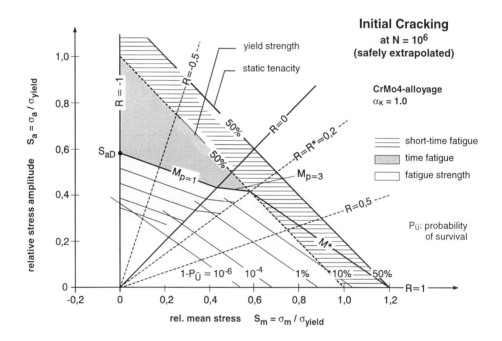

Fig. 3. Fatigue strength lines of steel in a GOODMAN-diagram with "Isoasfalen" as lines of constant reliability

The detailed description of the procedure and the formulation of the mathematical model can be read in more detail in [1].

FATIGUE OF FIBRES AND UNIDIRECTIONAL LAMINATES

To judge the reliability of fibre fatigue in hoop raped vessels first it is to be said that fatigue lines in half logarithm WÖHLER-diagrams are in most cases straight lines. So it is necessary to describe the distribution of reliability of the static tenacity on a linear scale. So the contradiction of different scales like occurring in the fig. 1 make no problem in analysing fibre fatigue.

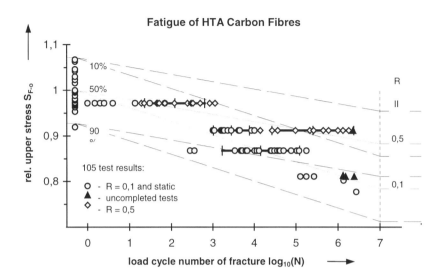

Fig. 4. Fatigue strength lines of the HTA- carbon fibre

Referring to the Tenax HTA fibre it is possible to describe all fatigue lines of different reliabilities and definite load ratios as a group of parallel lines as you can see it in figure 4.

The problem in justifying fibre fatigue by using those WÖHLER -lines of single stage load cycles is to judge the dependence of the stress ratio. For this HARRIS, REITER, ADAM, DICKSON and FERNANDO created a mathematical model ([4]; compare the ellipsoid lines in figure 5). According to the mean stress ratios of vessel reinforcements above R = 0 in figure 5 a triangle is marked which shows the appearing ratios. Comparing these limits with the hypothesis of HARRIS a. o. it is not helpful to use such ellipsoids. It is obvious that in some of these special cases of fibre loads the ellipsoids describe a higher allowable stress amplitude as there could work (static tenacity).

Comparing fig. 4 and the straight lines in fig. 5 another contradiction occurs between the used models. It is not possible to use formulations which result in straight fatigue lines as a function of load numbers in a WÖHLER-diagram and similar in a GOODMAN-diagram. Therefor it is to make an assessment in care of the danger of underestimation of the fatigue reliability. The proposition is to be made to calculate the gradients of the straight lines of fatigue in the WÖHLER-diagram for the several stress ratios by using the in fig. 5 shown relationship for a load number of e. g. $N = 10^8$.

Finally all the above shown figures and the correlating mathematical formulations result in a model that answers all the in front formulated questions of fatigue.

826

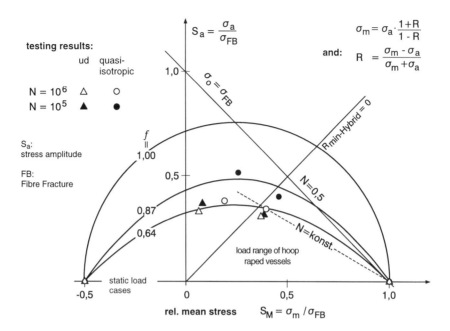

Fig. 5. Stress ratio and fatigue strength lines of carbon fibres in a GOODMAN- diagram

At last the discussed tools for justification of the reliability of liner and reinforcement of hybrid vessels are to be combined for use. By using this tool the results of some reliability restricted optimisation of fibre reinforced steel vessels show a possible area of development for saving up to 40 % of weight by keeping up the required reliability of $1 \cdot 10^{-6}$. Please compare this with [1,2,5].

REFERENCES

1. Mair, G. W. (1995). *Das Low-Cycle-Ermüdungsverhalten von Faser-Metall-Hybridstrukturen am Beispiel eines carbonfaserarmierten Druckgasspeichers im Nahverkehrsbus.* VDI-Verlag, Düsseldorf, Fortschritt- Berichte, Reihe 18, Nummer178.

2. Mair, G. W. (1995). *Das Low-Cycle-Ermüdungsverhalten von Stahl und Carbonfaser.* In: *DGLR-Jahrbuch 1995/II*, Bonn/Bad Godesberg, lecture [DGLR-JT95-116], pp. 665 - 674.

3. Haibach, E. (1989). *Betriebsfestigkeit: Verfahren und Daten zur Bauteilberechnung.* VDI-Verlag, Düsseldorf.

4. Harris, B., Reiter, H., Adam, T., Dickson, R.F., Fernando, G. (1990). *Fatigue behaviour of carbon fibre reinforced plastics.* In: *Composites* Vol. 21 No. 3; pp. 232 - 242.

5. Mair, G. W.(1996). *Zuverlässigkeitsrestringierte Optimierung faserteilarmierter Hybridbehälter unter Betriebslast am Beispiel eines CrMo4-Stahlbehälters mit Carbonfaserarmierung als Erdgasspeicher im Nahverkehrsbus.* VDI-Verlag, Düsseldorf, Fortschritt-Berichte, Reihe 18, Nummer 186, D83.

Low Cycle Fatigue and Elasto-Plastic Behaviour of Materials
K-T. Rie and P.D. Portella (Editors)

INADEQUATE DESIGN SPECIFICATIONS LEADING TO FATIGUE FAILURE IN TANK CONTAINER SHELLS

E. TERBLANCHE
*Department of Mechanical Engineering, University of Stellenbosch
Stellenbosch, South Africa*

ABSTRACT

Large tank containers have become an effective means of transporting liquid materials, both hazardous and other. The integrity of the shells of these containers is therefore important. Limited design specifications laid down by standard's organisations do not cover all eventualities of loading. Through shell cracks were observed on some containers. Finite element analyses of the various loading conditions normally prescribed predicted low stresses where the cracks were observed. This led to the re-assessment of the loading of the containers. Since road and rail transport was not responsible for the cracks, the only alternative was ship transport. Finite element analyses were used to investigate possible loading scenarios on board ship. A loading condition identified as being responsible for the fractures was a combination of stacking and the ship's rolling and/or pitching. To test this hypothesis a container was loaded cyclically with the calculated loads. The same modes of cracking were observed, as those in service. Slight design modifications led to the solution of the problem.

KEYWORDS

Tank container, finite element, fatigue failure, servo-hydraulic, dynamic load, container ship.

INTRODUCTION

Tank containers (See fig. 1) are used with increasing frequency to transport various liquid materials. It is essential that the structural integrity of these containers be maintained under all types of loading conditions. Some specifications exist that these tank containers must adhere to. These typically are stacking loads where loads are applied to a container to simulate a stack of filled containers above it, as well as longitudinal impact loads that a container is typically subjected to during rail transport. Various manufacturers of tank containers have developed their own in house tests [1] to fill the gaps in the current specifications.

The basis of these tests were normally derived from measurements on tank containers during typical transport conditions, especially by road and rail. Laboratory fatigue testing of the containers normally adhere to specifications derived from road and rail transport testing.

Fig. 1. Typical tank container with lagging removed in the test laboratory.

PROBLEMS ENCOUNTERED IN SERVICE

Certain containers were found with through shell cracks (See fig. 2). These containers had previously undergone rigorous testing in all the typical testing modes, without any ill effects in the area of the cracks observed in practice. Finite element analyses of the complete tank container employing all the different loading modes also showed very low stresses in the areas under question. It clearly was a case of applying the wrong loads, both experimentally as well as numerically. The approach now had to be to identify the loads responsible for the crack formation. The position of the cracks fortunately aided in identifying a possible load scenario.

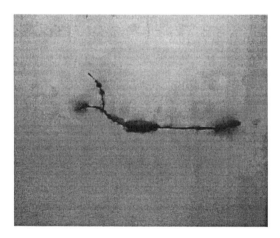

Fig. 2. Through shell crack observed on the inside of tank container in service.

On board a container ship the containers can be stored in two different modes. The first is that the container can be stored in the hold within rails to inhibit movement during the voyage. In a set of rails the containers can be stacked nine high. On deck the containers can be stacked about four or five high, depending on the type of container ship. In the latter case the containers are held on the deck with the aid of cross bracing.

Regardless of the method employed to inhibit the movement of the containers, some small movement is still possible, both in a lateral as well as a transverse direction. The reasoning was that it might be possible for containers in a stacked configuration to load the containers immediately below it in a longitudinal and transverse direction due to the ship's normal rolling and pitching. Figure 3 schematically depicts the effect on the tank containers.

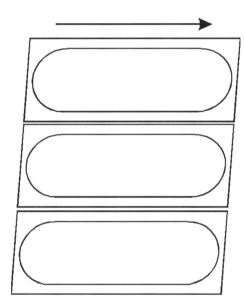

Fig. 3. Ship's rolling induced loads tend to deform tank containers as depicted.

With an absence of experimental data on the behaviour of these containers on ships the following approach seemed to be the one with the highest probability of success. Simple calculations by du Toit [2] were made to roughly determine the loads that one container can exert on the others as described above. Using typical rolling and pitching frequencies for these types of ships as well as typical friction values of steel on steel to simulate the shear load between any two containers, loads in the order of 4 tonnes per corner were arrived at. The assumptions made to arrive at these values were conservative. In practice the values may in fact be higher. A finite element analysis was then performed on a model of the tank container as shown in fig. 4.

The results proved that very high stresses were in fact set up in the area where cracks were observed. Figure 5 shows typical Von Mises stress contours on the inside of the tank in the region where the cracks had been observed.

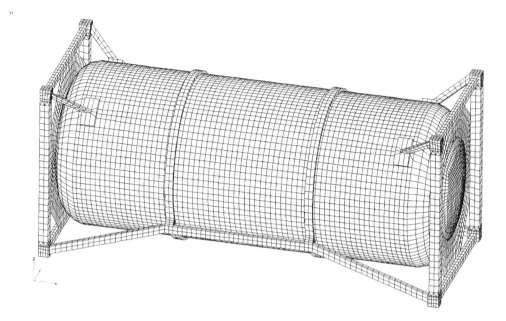

Fig. 4. Finite element model of a complete tank container.

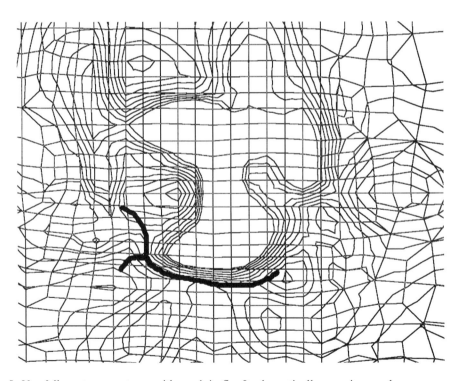

Fig. 5. Von Mises stress contours with crack in fig. 2 schematically superimposed.

SOLUTION OF THE FRACTURE PROBLEM

Once the postulated loading condition that caused the failures were numerically verified it was decided to test the container experimentally in the laboratory under the same conditions. Before this was done, it was decided to investigate what the effect would be when only one of the four corners of the container was loaded. Again use was made of the finite element model. The results proved that there was very little effect at the fourth attachment point to the shell due to the loads at the three other corners. With this in mind it was a lot easier to apply loading to the container, because it was only necessary to apply the load at one corner as shown in fig. 6. where use is made of a servo-hydraulic system under force control to apply a cyclic load.

Fig. 6. Servo-hydraulic loading frame to apply cyclic load at a top corner of the container.

It was possible to cause the formation of cracks at the same location where cracks had been observed in practice within about 1000 cycles with a cyclic load of amplitude 4 tonnes. The test was repeated at the other corners with the same results. This provided confidence in the identification of the cause of the crack formation.

The solution of the problem was fairly straightforward. The short top stays between the end frames and the tank shell attachment points had to be changed. Various possibilities were investigated. The final solution was by moving the attachment points on the shell further to the centre onto the top hat section that encircles the tank.

A comparison between the original and the final configurations can be seen in fig. 7 where only a quarter of the final finite element model is shown. In the areas immediately adjacent to the attachment points of the stays the finite element mesh was substantially refined to obtain better results. The original design is shown on the left in fig. 7 with the modified design on the right.

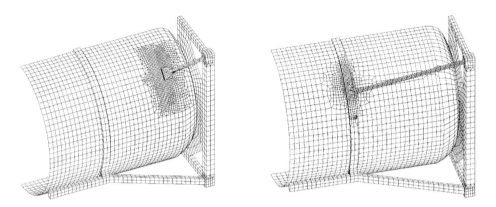

Fig. 7. Original (left) and modified (right) stay attachment positions.

CONCLUSION

The contours shown in fig. 5 were obtained from the analysis of the original structure as shown in fig. 7. The shifting of the attachment point of the stay as shown in the modified design reduced the Von Mises stress levels by a factor of approximately 2.5. The modified design was also tested in exactly the same fashion as the original design, but now an effective infinite life was the end result.

A further advantage of the modified design is that if cracks should occur it will be on the tophat section where there is no danger of propagating through the shell.

An additional advantage is that possible repairs in the latter case will be fairly simple, whereas with the original design it was a fairly involved process due to the relative inaccessibility of the cracked area.

REFERENCES

1. Terblanche, E. (1997) Longitudinal stacking induced loads on tank containers. *Internal report, Department of Mechanical Engineering, University of Stellenbosch.*
2. Du Toit, J.F. (1997) Finite element analysis of tank container structures. *Internal report, Department of Mechanical Engineering, University of Stellenbosch.*

Low Cycle Fatigue and Elasto-Plastic Behaviour of Materials
K-T. Rie and P.D. Portella (Editors)

SIMPLIFIED LIFE PREDICTION FOR NOTCHED COMPONENTS WITH HIGH LOCAL STRESSES

R. GERDES
ABB Power Generation Ltd., Steam Turbine Development
CH-5400 Baden, Switzerland

ABSTRACT

An improved prediction of lifetime of structures using the finite element method can be obtained if a unified material model is applied or a modified Neuber rule is used. A comparison between a unified model and the modified Neuber rule shows that the modified Neuber rule is a simple method to estimate the lifetime quickly. Both methods predict experimental data very well.

KEYWORDS

Life time prediction, notch, Neuber rule, plasticity, constitutive equations

INTRODUCTION

The need for lifetime prediction methods in the design of mechanical components is widely recognised. Using the finite element method detailed results of the state of stress can be obtained. For the estimation of the lifetime a precise method for notched components with high local stresses is necessary.

To evaluate the quality of lifetime prediction methods, a compact tension specimen with a slit under an angle of 57° was subjected to a pulsating force. For the first cycle 144% of the nominal load was applied, the following cycles were subjected to the nominal load (100%). The material used was a 1%-CrMoV-Steel. The geometry is shown in figure 1. The nominal force was determined by an elastic finite element calculation so that the maximum fictitious elastic stress in the notch was well above the yield stress. The number of cycles to crack initiation was measured by an electrical potential system.

For engineering applications the Neuber rule is often applied to predict the lifetime [1]. For technical purposes this rule is extended to general multiaxial state of stress and strain [2]

$$\sigma^* \cdot \frac{\sigma^*}{E} = \sigma \cdot \varepsilon(\sigma) \tag{1}$$

where σ^* denotes the fictitious elastic stress, σ the elastic-plastic stress, E the modulus of

834

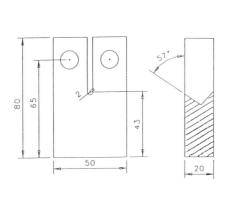

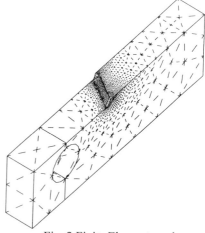

| Fig. 1: Geometry | Fig. 2 Finite Element mesh |

elasticity and $\varepsilon(\sigma)$ the stress-strain curve. The Neuber rule estimates the stress σ and strain $\varepsilon(\sigma)$ in a notch by the results σ^* and ε^* of a simple elastic calculation. The stress-strain curve $\varepsilon(\sigma)$ is described by a Ramberg-Osgood law up to a plastic strain of 0.2% followed by linear stress-strain hardening.

$$\varepsilon(\sigma) = \frac{\sigma}{E} + 0.002\left(\frac{\sigma}{\sigma_0}\right)^n \qquad \text{if } \sigma < \sigma_0$$

$$\varepsilon(\sigma) = \frac{\sigma}{E} + 0.002 + \kappa \cdot (\sigma - \sigma_0) \text{if } \sigma \geq \sigma_0 \tag{2}$$

The cyclic behaviour of the material can be described by applying the Masing-rule.

$$\frac{\Delta\varepsilon}{2} = f(\frac{\Delta\sigma}{2}) \tag{3}$$

Equations (1) through (3) can be used to model a cyclic stress-strain curve. Together with low-cycle fatigue data the lifetime of the specimen can be predicted. It is generally known that the Neuber rule agrees well with measurements in plane stress situations, the results are usually conservative for plane strain [3].

The mesh of a quarter of the specimen is shown in figure 2. The cyclic stress-strain curve was constructed from the elastic stresses in the notch. The first loading was calculated using the Neuber rule (eq. 1) and the stress-strain curve acccording to eq. 2. The following

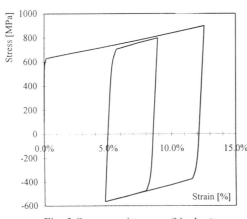

Fig. 3 Stress strain curve (Neuber)

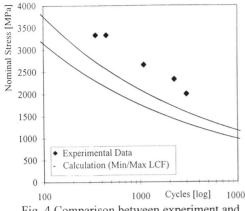

Fig. 4 Comparison between experiment and prediction of Neuber rule

cycles where calculated using the cyclic stress-strain curve (eq. 3) and also the Neuber rule (eq. 1). The resulting stress strain curve is shown in figure 3. To consider the influence of the mean stress, the Smith-Watson-Topper Parameter [4] was used to calculate the effective strain amplitude. To compare the theoretical results, experiments with similar specimens at different load levels were performed. The comparison between calculation and experimental results for five specimens is shown in figure 4. The calculated results are very conservative compared to the experimental data. Therefore a better approach for the prediction of the lifetime is desirable. In the following sections two methods will be introduced which allow a better prediction of lifetime. The stress-strain curve and the LCF data will be used for the calculation of lifetime only. The methods which will be analysed are:

- Modified Neuber rule. The Neuber rule gives good results for plane stress situations. For a general state of stress a modified Neuber rule can be used which considers the existing stress conditions in the notch.
- Constitutive equations. The stress-strain behaviour in the notch can be predicted by a material model which describes the cyclic behaviour of the material.

These methods are discussed in the following sections.

MODIFIED NEUBER RULE

The state of stress can be considered if a generalised Neuber rule is used [5]

$$\sigma^* \cdot \left(\frac{\sigma^*}{E} \right)^\kappa = \sigma \cdot \left(\varepsilon(\sigma) \right)^\kappa \tag{4}$$

The exponent κ describes the stress state where $\kappa=0$ implies stress control and $\kappa=\infty$ strain control. $\kappa=1$ is equivalent to the Neuber rule. For technical purposes the exponent κ has to be determined. It can be calculated by using the finite element method. Therefore an elastic calculation and an elastic-plastic calculation are performed. The stress result σ^* of the elastic calculation and the results of the elastic-plastic calculation σ_{elpl} and ε_{elpl} are used to determine κ by

$$\kappa = \log\left(\frac{\sigma_{elpl}}{\sigma^*} \right) / \log\left(\frac{\sigma^*}{E\varepsilon_{elpl}} \right) \tag{5}$$

For the specimen shown in figure 1 an exponent of κ=2 was calculated. Using this exponent κ the cyclic stress-strain curve can be constructed. After a cycle with 144% load using the original stress-strain curve (eq. 2), the following cycles were constructed using the cyclic stress-strain curve (eq. 3), both in combination with the Neuber rule (eq. 1). The resulting stress-strain curve is shown in figure 5. The strain amplitude can be determined from this cyclic stress strain curve. Together with the consideration of the mean stress according to the Smith-Watson-Topper rule and the LCF ata the possible number of cycles can be obtained.

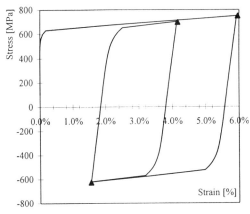

Fig. 5. Modified Neuber rule

UNIFIED MODEL

To calculate the cyclic behaviour of a material, a unified model can be used which describes kinematic and isotropic hardening. For these calculations the following model is applied [6]

$$\dot{\varepsilon}^{in} = C_1 \cdot \exp\left(-\frac{F_0}{kT}\right) \cdot \sinh\left(\frac{bA \cdot \sigma_{kin}}{kT}\right)$$

$$\dot{\sigma}_{kin} = H \cdot E \cdot \exp\left(-\frac{\delta \cdot bA \cdot \sigma_{kin} \cdot \text{sign}(\sigma - \sigma_{kin})}{kT}\right) \cdot \dot{\varepsilon}^{in} \qquad (6)$$

$$\dot{A} = -k_1 \cdot A^2 \cdot \dot{\varepsilon}^{in} + k_2 \cdot A \cdot \dot{\varepsilon}^{in}$$

The model uses 6 parameters(C_1, H, δ, k_1 and k_2 and the initial value of A) which have to be determined by comparison with experiments. The inelastic strain rate $\dot{\varepsilon}^{in}$ is temperature and stress dependent. Two internal variables, the kinematic back stress σ_{kin} and the activation area A are used. The kinematic back stress describes the cyclic behaviour, the activation area the cyclic softening of the material. Using cyclic stress-strain curves and information about cyclic softening, the values of these parameters are obtained. The model was adjusted to uniaxial data. The results of the adjustment for the uniaxial test are shown in figure 6 and 7.

The material model for these parameters is used for finite element analysis. For this calculation the finite element code ABAQUS was applied, for the implementation of the material model the user subroutine UMAT was programmed. The stress-strain behaviour in the notch is shown in figure 9. When reaching a quasi-static state, the strain amplitude is determined (Fig. 8). Again, the Smith-Watson-Topper rule was used to calculate the strain amplitude. The LCF data

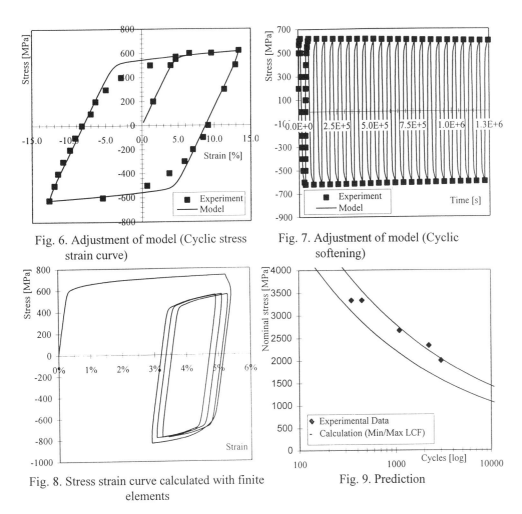

Fig. 6. Adjustment of model (Cyclic stress strain curve)

Fig. 7. Adjustment of model (Cyclic softening)

Fig. 8. Stress strain curve calculated with finite elements

Fig. 9. Prediction

allow an estimation of the maximum number of cycles. The stress-strain curve moves to higher mean strains for increasing number of cycles. This can be explained by the softening behaviour of the material.

RESULTS

The results of the different methods are summarised in table 1. Comparing the two methods, the results are nearly equivalent. This shows that the more accurate calculation with the unified material model can be estimated by the modified Neuber method with good accuracy.

	Unified model	Modified Neuber
Calculation steps	200	2
Strain amplitude	1.2%	1.1%
Maximum strain	5.5%	4.1%
LCF-Cycles	300	350

Tab. 1. Comparison of results

Therefore the modified Neuber rule is used to predict the lifetime for different stress levels. The comparison between experiment and calculation is shown in figure 9. To determine the possible number of cycles, the lower and upper scatter bands of the LCF data are used. The calculated lifetime and experimental results for 5 specimens subjected to different stresses are shown. The prediction with the modified Neuber rule estimates the lifetime of the specimens very well.

Therefore, if no detailed information about the local stress-strain behaviour is required, the modified Neuber rule allows a conservative estimation of the lifetime. For this estimation only an elastic and an elastic-plastic finite element calculation have to be performed. If more information about the local behaviour is necessary, a calculation with a constitutive model should be performed. This method is rather time consuming compared to the modified Neuber rule.

REFERENCES

1. Neuber, H. (1961) *J. Appl. Mech* **28**, 544.
2. Hoffman, M.; Seeger, A. (1985). *ASME Journal Engineering Materials and Technology* **107**, 250.
3. Fuchs, H., Stephens, R. I. (1980). *Metal Fatigue in Engineering.* Wiley-Interscience, Stanford.
4. Smith, K.N., Watson, P., Topper, T.H. (1970). *J. Mater.* **5**, 767.
5. Petrequin, P. et al. (1983) In: *Proceedings of Internation Conference on Advances in Life Prediction Methods.* ASME.
6. Gerdes, R. (1995). *Ein stochastisches Werkstoffmodell für das inelastische Materialverhalten metallischer Werkstoffe.* Braunschweiger Schriften Mechanik **20**.

839

THE ASSESSMENT OF REMAINING FATIGUE LIFE OF AIRFRAME DURING OPERATIONS

V. M. ADROV,

State Research Institute of Aircraft Maintenance and Repair

Lubertsy 3, Moscow distr., 140003, RUSSIA

ABSTRACT

The relationships between the assessments of fatigue life produced by two different approaches (as local strain and nominal stress) are derived. These relationships may be used both to correct the fatigue life assessments of airframe during maintenance and to convert these ones to each other. The results of this report are useful for right choose systems and devices that calculate remain airframe fatigue lifetime.

KEYWORDS

Airframe, fatigue life, remain lifetime

INTRODUCTION

One of the main condition for developing the progressive strategies of the service and repair of airframes is the accounting of the remain lifetime of its main structure elements. So, the task of choosing the methods and means for control the remain lifetime of airframe has a big value in present. Many systems (air-born and on-ground) that calculate (measure) airframe fatigue damage and remain life during operations are developed and used.

Let consider more interest systems that are united in two big groups. The first is group of systems and devices that treat information from special installed gages in terms of strains in some zones of structure elements. For example, FALC [1]. Other group is that ones input information is the signals from state gages that also used by state airplane systems. For example, OLMS [2].

Systems from first group calculate remain airframe lifetime using the main ideas of local strain approach. While devices from second group recalculate input gage signals to integral power factors (as bending mom) in concerning structure element for further using nominal stress approach to evaluate fatigue life.

The main objective of this report is by comparison two methods of predicting lifetime- one based on nominal stress analysis and another based on local stress-strain analysis - to evaluate the validation and verification remain lifetime assessments obtained by hardware systems, that used above mentioned methods.

To obtained the main conclusions two types of random loading are considered:

1) Gaussian loading with zero mean value and given root mean square σ^2.

2) loading that has straight-line distribution described by overall cumulative frequency

$H_+(S) = N_0 \exp(-aS)$, for peaks,

$H_-(S) = N_0 \exp(-bS)$, for valleys abs values,

where S is nominal stress, N_0, a, b are parameters.

Let calculate the mathematics averages of conventional fatigue damage for both loading types and both methods of fatigue analysis.

The calculations for nominal stress analysis are based on recent author's approach [3]. The assessments of average of number cycles to crack initiation are equal correspondent

$$\left(N_f\right)_1^{-1} = \frac{16\sigma^m}{C}; \qquad (1)$$

$$\left(N_f\right)_2^{-1} = \frac{40}{Ca^m}, a \equiv b \qquad (2)$$

Where C and m are parameters of S-N curve.

To obtain the assessments of average damage per loading cycle under local stress-strain analysis for each loading kind next expression is used

$$\xi_{1,2} = \int_0^{\infty} \frac{r_{1,2}(\Delta\varepsilon)}{N_f} d(\Delta\varepsilon), \tag{3}$$

where $r_{1,2}(\Delta\varepsilon)$ are density distribution functions of local strain range $\Delta\varepsilon$ for correspondent loading kind, $N_f(\Delta\varepsilon)$ is function of number of cycles to failure from $\Delta\varepsilon$ that is accepted in form proposed by S.S.Manson

$$N_f = \begin{cases} N_T \left(\Delta\varepsilon / \Delta\varepsilon_T\right)^{1/b} \exp\left\{\delta\left(\Delta\varepsilon/\Delta\varepsilon_T\right)^{\beta}\right\}, \Delta\varepsilon \le \Delta\varepsilon_T; \\ N_T \left(\Delta\varepsilon/\Delta\varepsilon_T\right)^{1/c} \exp\left\{\delta\left(\Delta\varepsilon/\Delta\varepsilon_T\right)^{\alpha}\right\}, \Delta\varepsilon > \Delta\varepsilon_T. \end{cases} \tag{4}$$

$$\beta = 0.5\frac{c}{b} - 0.3, \delta = -\frac{0.78}{c}\left(\frac{c}{b}\right)^{0.36}$$

Functions $r_{1,2}(\Delta\varepsilon)$ are calculated by known density distribution function of nominal stress range ΔS in form

$$f_S^1(\Delta S) \approx \frac{k^2 \Delta S}{\sigma^2} \exp\left(-\frac{k^2 \Delta S^2}{2\sigma^2}\right), \tag{5}$$

$$f_S^2(\Delta S) = a^2 \Delta S \exp(-a\Delta S), \tag{6}$$

where k is parameter of spectrum complexity and relationship between ΔS and $\Delta\varepsilon$ is accepted in form proposed by S.S.Manson

$$\Delta S = \begin{cases} B\Delta\varepsilon F_1, \Delta\varepsilon \le \Delta\varepsilon_T, \\ A(\Delta\varepsilon)^d F_2, \Delta\varepsilon > \Delta\varepsilon_T, \end{cases} \tag{7}$$

$$B = \left[2E\sigma_f^{'}\left(2N_T\right)^b / \left(k_f^2 \Delta\varepsilon_T\right)\right]^{\frac{1}{2}},$$

$$F_1 = \exp\left[b\delta/2\left(\frac{\Delta\varepsilon}{\Delta\varepsilon_T}\right)^{\beta}\right],$$

$$A = \left[2E\sigma_f^{'}\left(2N_T\right)^b / \left(k_f^2 \Delta\varepsilon_T^{b/c}\right)\right]^{\frac{1}{2}},$$

$$d = \left(b/c + 1\right), \alpha = -0.17 - 0.52\ln\left(c/b\right),$$

$$F_2 = \exp\left[b\delta/2\left(\frac{\Delta\varepsilon}{\Delta\varepsilon_T}\right)^{\alpha}\right],$$

842

where ΔS is range of nominal stress as an explicit function of $\Delta\varepsilon$.

Finally we get:

- for Gaussian loading kind

$$\xi_1 = A_1(z_1)\frac{1}{N_T}, \tag{8}$$

$$A_1(z_1) = \int_0^1 \frac{y^{1-\frac{1}{b}}}{z_1^2} e^{\delta(b-1)y^\beta}\left(1+\beta\frac{b\delta}{2}y^\beta\right)\exp\left(-\frac{y^2}{2z_1^2}e^{b\delta y^\beta}\right)dy +$$

$$+\int_1^\infty \frac{y^{2d-1-\frac{1}{c}}}{z_1^2} e^{\delta(b-1)y^\alpha}\left(d+\alpha\frac{b\delta}{2}y^\alpha\right)\exp\left(-\frac{y^2}{2z_1^2}e^{b\delta y^\alpha}\right)dy;$$

- for second loading kind

$$\xi_2 = A_2(z_2)\frac{1}{N_T}, \tag{9}$$

$$A_2(z_2) = \int_0^1 \frac{y^{1-\frac{1}{b}}}{z_2^2} e^{\delta(b-1)y^\beta}\left(1+\beta\frac{b\delta}{2}y^\beta\right)\exp\left(-\frac{y}{z_2}e^{b\delta y^\beta}\right)dy +$$

$$+\int_1^\infty \frac{y^{2d-1-\frac{1}{c}}}{z_1^2} e^{\delta(b-1)y^\alpha}\left(d+\alpha\frac{b\delta}{2}y^\alpha\right)\exp\left(-\frac{yd}{z_2^2}e^{b\delta y^\alpha}\right)dy;$$

$$z_1 = \frac{k_f\sigma}{k\Delta\varepsilon_T D}, z_2 = \frac{k_f}{a\Delta\varepsilon_T D}, D = \left[2E\sigma_f'/\Delta\varepsilon_T(2N_T)^b\right]^{\frac{1}{2}},$$

where $y = \dfrac{\Delta\varepsilon}{\Delta\varepsilon_T}$ is undimensional variable,

$z_1 = \dfrac{k_f\sigma}{k\Delta\varepsilon_T D}, z_2 = \dfrac{k_f}{a\Delta\varepsilon_T D}, D = \left[2E\sigma_f'/\Delta\varepsilon_T(2N_T)^b\right]^{\frac{1}{2}}$ are undimensional parameters.

The expressions (8) and (9) are shown in graphic form on Fig.1.

Let write the equations (1) and (2) in terms of equations (8):

$$(N_f)_1^{-1} \approx 6.5\frac{D^4\Delta\varepsilon_T^4}{k_f^4 C} z^{-4} \tag{10}$$

$$\left(N_f\right)_2^{-1} = 2.5\frac{D^4\Delta\varepsilon_T^4}{k_f^4 C} z^{-4} \tag{11}$$

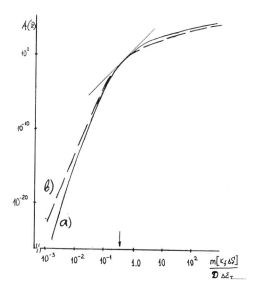

Fig.1. Fatigue damage as function of average range of undimensional nominal stresses: a) for expression (8); b) for expression (9).

Let place emphasis that slope in such curve is as that for S-N (m=4) curve only for a little interval of ranges of nominal stresses. This interval corresponds the transitive behavior of material deforming from elastic to plastic. For elastic behavior this parameter reaches values as 9...11, for plastic it decreases to 1,0.

CONCLUSIONS

The main conclusion of this report is that the relationships between assessments of fatigue damage obtained by different methods are developed. These ones allow to make corrections for assessments of fatigue life made by treatment of nominal loading histories and also mutual conversations of different assessments of remain airframe fatigue life. Obtained results shown are useful for aircraft specialists for evaluating and interpretations the assessments of remain aircraft life produced by different devices and for rational planning further airplane maintenance.

REFERENCES

1. Miodushevsky P., Podboronov B. (1991). In: *Proceeding of the Int.Conf. on aircraft damage assessnent and repair.* Inst. Eng., Austr., N 91/17, p.122 - 125.
2. H.- J. Meyer, V. Ladda.(1989). In: *Proceeding of the 15-th Symp. of the ICAF*, Jerusalem, Israel.
3. Adrov V.M. (1994) *Fat.&Fract.Eng.Mat.&Struct.*,**17**, 1397 - 1403.

Low Cycle Fatigue and Elasto-Plastic Behaviour of Materials
K-T. Rie and P.D. Portella (Editors)

LOW CYCLE FATIGUE CRACKING OF TITANIUM COMPRESSOR DISKS OF AIRCRAFT ENGINE D-30.

A.A.SHANIAVSKI and A.I.LOSEV
*State Centre of Flight Safety of Civil Aviation,
Airport Sheremetievo 103340, Moscow, Russia.*

ABSTRACT

Compressor disk from VT3-1 Ti-alloy has been investigated. Experiment was performed on disk having in-service developed fatigue crack. Block of two sequences of cyclic loads of the trapezium and triangle shapes was used to estimate correlation between a number of fatigue striations formed on fatigue fracture surface and a number of cycles produced by the block. Fractographic examinations revealed unified process of fatigue striations formation under both cyclic loads shapes. Based on the quantitative fractographic analysis, the five fatigue striations formation per flight has been shown.

KEYWORDS

Compressor disk, titanium, quantitative fractography, a number of striations per flight.

INTRODUCTION

It is well-known phenomenon of titanium compressor disks fatigue in service for various types of engines [1-5]. Fracture surfaces of disks may have two types of fracture features: (1) faceted pattern, which reflects bi-phase lamellar structure of titanium alloys, and (2) fatigue striations which can be seen on main or local areas of the surface.

Fatigue striations had been used to estimate fatigue crack growth life for many aircraft structure components [6]. This estimation is based on established relation between number of fatigue striations and number of load cycles which damage a component in flight. A number of striations per each flight depends on many factors for various aircraft structures.

Applicably to the titanium disks of various compressor stages and types of engines a number of striations per flight can be one, two, tree, and five [3,7]. It especially depends on a sequence of cyclic loads in flight for operated engine and, also, on a structure of titanium alloy which influence fracture surface patterns with or without fatigue striations [3,4]. The crack growth rate variation can be very large for different types of disks with the same mechanism of the fatigue fracture, and this variation exceeds 10 times or more for various mechanisms. Results of the fatigue crack growth life estimations performed on the basis of fractographic analyses depended on the established correlation between a number of fatigue striations and a number of aircraft flights.

The recent investigation was performed on the disk of aircraft engine D-30. This engine has the hardest regimes of cyclic loading in flights and the loads variation per flight is rather intensive for titanium disks of this engine. There were several cases of fatigue failures in flight of the disk the first stage of the low pressure compressor of these engines [3]. To introduce in practice non-destructive testing of these disks at the intervals established on the basis of the results of quantitative fractographic analyses, one of the damaged in service disks was used to discover correlation between fatigue striations number and a number of load cycles experienced by the disk in flight. Fractographic analyses of fatigue surface with the use the scanning electron microscope was essential in this case.

EXPERIMENTAL DETAILES

Material and Type of Disk.

A titanium disk of the first stage of the low pressure compressor of the engine D-30, used in test, had a small surface fatigue crack initiated in service, as shown in Fig.1. The crack discovered during maintenance by the ultra-sound non-destructive testing had the size near to 7mm on the disk surface. The place of the crack initiation was typical for the stress-state of the disk. The aircraft with this engine had flown 14,428 hours and had 8,656 flights at the moment of the crack detection.

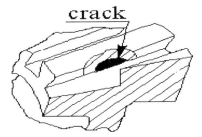

Fig.1 Schematic presentation of the disk rim area with the in-service initiated crack.

The disk was made from titanium alloy VT3-1 with the bi-phase lamellar structure being typical for the compressor disks. Investigation of the structure, material composition, and testing of mechanical properties, performed after the test, have shown that they fully conform the requirements for the industrial manufacture of disks from this type of Ti-alloy [4].

Test Procedure.

The disk test was performed on the special test rig [8] wich have been constructed to simulate the disk loading in service by rotation of the test spesimen at the full range of operating regimes.

The test specimen represented a part of the compressor rotor and consisted of the first stage disk assembled as in the real engine with the disks of two following stages.

The sequence, values of operating stresses and duration of their action in test cycle were chosen in conformity with those in real flight cycle in service, as shown in Fig.2a. The realistic period of test was achieved at the cost of the most protracted regimes durations of which were strongly reduced in proportion to their periods in flight cycle. As it is widely known, durations of these regimes have little effect on low cycle fatigue processes in compressor parts [4].

The whole test period consisted of 75 such test cycles which were divided to 5 identical blocks, as shown in Fig.2b. Each block included 15 test cycles in groups of 1, 2, 3, 4, and 5 cycles. Between these blocks and groups there was a number of socalled marking cycles of triangular shape used to fix a position of fronts of cracks developed in service and in test. A quantity of marking cycles was 40-35-30-25-20 and their stress level was 77% of maximum stress in the test cycle.

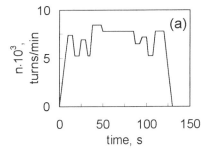 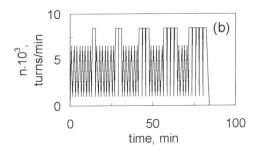

Fig.2 Scheme of cyclic loads sequences within (a) the test cycle and (b) one block of test cycles.

Fractographic Analysis.

After the end of the test the tested disk was inspected by the ultra-sound and then the crack was opened. It was found that the crack propagated under the test conditions and the revealed fatigue fracture had a semi-elliptyical shape.

Performed measurments have shown that the in-service developed crack had the surface half length c=3.5mm and the length a=2mm in the depth direction. During the test the crack front formation was not symmetrical. On the disk surface the crack has increased by 1.7mm in one direction from the position discovered during maintenance but in another one the increment was 1.3mm. In the depth direction the crack increment was 0.8mm under the test conditions. The revealed fatigue surface was investigated in scanning electron microscope with resolution not less than 0.009 µm.

Striations spacing was measured in the crack growth direction near the disk surface where the crack increment during the test was 1.1mm on the fatigue surface.

848

RESULTS AND DISCUSSION

Fatigue crack development in service and in the test was produced by the mechanism of fatigue striations formation. There were only two areas of the fatigue surface near the transition to the fast fracture developed by the crack opening after the test where the sequence of loads in the block was seen. The fatigue surface had striations in the region produced in the test with values of spacings in the range 0.26...1.25 μm. The values of striations spacing increased for the crack length increasing in the area produced by the test, Fig.3.

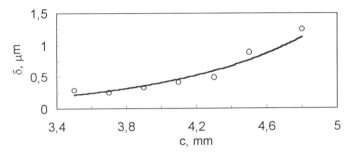

Fig.3 The dependence of the fatigue striation spacing, δ, on the crack length, c.

This dependence was used to estimate a number of striations formed in the fracture surface during test. There were approximately 1,600 striations in the crack growth direction.

As it can be seen on Fig.2, within the test cycle there are 5 sections of stress level increase and decrease which may be considered as the load cycles of trapezoidal shape. Both types of load cycle of triangular and trapezoidal shapes can produce fatigue striations. During test there were 5x75 = 375 and 150x5 = 750 cycles of trapezoidal and triangular shapes respectively or 1,125 cycles of both shapes.

A number of formed striations is nearly 1,600/1,125 = 1,43 times more than the number of cycles produced during the test by both cyclic load shapes. One can see that the only loads of the triangular shape are not enough to use for explanation fatigue crack development in the test. Cycles of the trapezoidal shape have formed fatigue striations, too.

The blocks of micro-lines were discovered in two areas of the fatigue surface. Spacings between the lines were in the range of 2-4.5 μm. It is nearly 10 times more than fatigue striations spacings. That is why these two areas were analysed more attentively.

The test conditions have shown one-to-one correspondence between the number of micro-lines in the areas and the number of test cycles, as it is shown in Fig.4. One can see within a micro-line the five fatigue striations developed by fatigue crack during the test cycle. The step between micro-lines correlated to the transition from maximum stress level to zero and, then, to the next maximum stress level. Therefore one micro-line correlated to one test cycle and 5 load cycles of the trapezium shape in each test cycle produced crack increment with formation of 5 fatigue striations. The striations which were formed under trapezoidal cycles had spacings values approximately two times more than the striations from the cycles of the triangular shape. Therefore a number of calculated striations should be diminished by the 375 because for measurment were used only small spacings from the cycles of triangular shape. Thus the calulated number of striations is nearly 1,225 and is approximately the same as summerized number of cycles of both shapes. One can see also that five alterations of engine operating regimes in flight damaged investigated disk in service and originated 5 fatigue striations per flight.

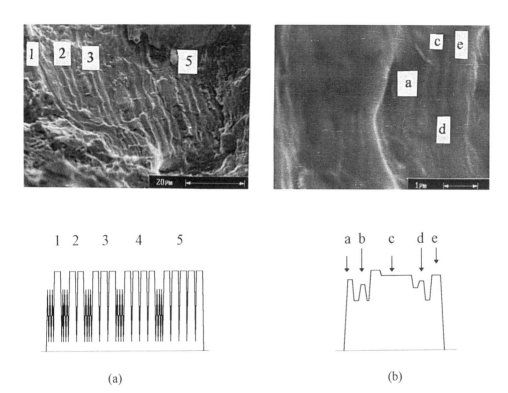

(a)   (b)

Fig.4 Correlation between the number of such fracture surface patterns as (a) micro-lines and (b) fatigue striations and the number of test cycles in one block of cyclic loads and the number of trapezoidal cycles in one test cycle.

Results of the investigation were used to estimate crack growth life for failured in service disks. After introducing inspection interval on the basis of the performed fractographic analyses no failure of the disk in service was seen.

CONCLUSION

The performed investigation of the crack growth in the disk of the first stage of the low pressure compressor of the engine D-30 have shown the five striations formation on the fatigue surface per flight. This value can only be applied to the fractographic estimation of the fatigue crack growth life for failure analysis when the fatigue striation formation mechanism is dominant for the titanium alloy VT3-1.

The test conditions which produced disk low cycle fatigue in test is not exactly the same as in operated disks. There exist small vibrations from the cycled blades placed on the titanium disk that influenced a stress state in the disk and accelerate crack growth rate during flight. This situation is to be checked in the next stage of the titanium disks investigations.

REFERENCES

1.   *NTSB Aircraft Acident Report PB90-910406, NTSB/AAR, 90/60,* US Governement.
2.   Shaniavski, A.A. and Losev, A.I. (1991) *Analysis methods of cyclic service durability of engine disks (overview).* Center of Scien. Techn. Inform. Civil Aviation. Moscow (in Russian).
3.   David H. (1991) *Avait. Week and Space Technol.* **20,**35.
4.   Shanyavsky, A.A. and Stepanov, N.V. (1995) *Fatigue Fract. Engng. Mater. Struct.* **5,** 539.
5.   Shaniavski, A.A. and Losev, A.I. (1996) In: *Proceedings of the ECF 11 on Mechanisms and mechanics of damage and failure,* ed. J.Petit and co-eds, Poitiers-Futuroscope, France,**II,**pp.1131-1136.
6.   Ivanova, V.S. and Shaniavski, A.A. (1988)   *Quantitative fractography. Fatigue fracture.* Metallurgia, Cheliabinsk (in Russian).
7.   Shaniavski, A.A. and Losev, A.I. (1998) Fatigue Fract. Engng. Mater. Struct. (to be published).
8.   Losev, A.I., Shaniavski, A.A. and Kashin, A.P. (1995) In: *Strength, Reliability and Flight Safety of Civil Aviation,* ed. A.Shaniavski,   Center of Scien. Techn. Inform. Civil Aviation, Moscow, **308,**pp. (in Russian).

Low Cycle Fatigue and Elasto-Plastic Behaviour of Materials
K-T. Rie and P.D. Portella (Editors)
© 1998 Elsevier Science Ltd. All rights reserved.

LOWER BOUND COLLAPSE LOAD AND FATIGUE OF MATERIALS

S. VEJVODA

VÍTKOVICE Institute of Applied Mechanics Brno, Ltd.
Veveří 95, 611 00 Brno, Czech Republic

ABSTRACT

The standards define the lower bound collapse load by criterion $\varepsilon_t = 2\varepsilon_e$. This condition is very strict in the stress concentration zone. The necessity of fulfilling of the criterion $\varepsilon_t = 2\varepsilon_e$ is different zones of vessels in analysed on the example of the pressure vessel. Results will used for the future revision of the standard A.M.E. Section III [1].

KEYWORDS

Lower bounds collapse load, stress category group, fatigue, strain, stress, limit.

INTRODUCTION

The Association of Mechanical Engineers of the Czech Republic published the A.M.E. Standard Technical Documentation „Strength Assessment of Equipment and Piping of Nuclear Power Plants of WWE Type" in the 1996 year [1]. This standard is based on the eastern standard [2] and its philosophy is conforming with ASME Code, Section III [3]. The criterion for lower bound collapse load is not sufficiently defined in relationship to fatigue of materials. Results of studies [5] will used for the future revision of the A.M.E. Standard [1].

LIMITS OF STRESES

The pressure vessel on the Fig. 1 is manufactured from steel with yield stress $S_y = 300$ MPa, tensile strength $S_u = 491$ MPa, Young's modulus $E = 192000$ MPa at 300°C. Number of load cycles $n = 2400$ during 30 years of the service life. Stresses and strains in elastic and elasto-plastic states were calculated by the FEM program SYSTUS. Relationship $\varepsilon_t = f(\sigma)$ for monotonic load :

$$\varepsilon_t = \varepsilon_e + \varepsilon_p = \frac{\sigma}{E} + \left(\frac{\sigma}{K}\right)^{1/n} \tag{1}$$

and $\Delta\varepsilon_t = f(\Delta\sigma)$ for cyclic loading :

$$\Delta\varepsilon_t = \Delta\varepsilon_e + \Delta\varepsilon_p = \frac{\Delta\sigma}{E} + \left(\frac{\Delta\sigma}{2^{1-n'}K'}\right)^{1/n'} \tag{2}$$

where is taken K = K' = 705,2 MPa, n = n' = 0,121, ε_e in elastic and ε_p plastic components of the total strain ε_t.

Limits for stress category groups in accordance with [1] :

$$(\sigma)_1 \leq [\sigma] \tag{3}$$

$$(\sigma)_2 \leq 1,3 \, [\sigma] \quad \text{or} \quad [p] = 0,67 \, p_C \tag{4}$$

$$(\sigma)_R \leq 2,5 \, [\sigma] \tag{5}$$

where allowed nominal stress

$$[\sigma] = \min\left\{\frac{S_y}{1,5}; \frac{S_u}{2,6}\right\} = \min\left\{\frac{300}{1,5}; \frac{491}{2,6}\right\} = 188,8 \; MPa \tag{6}$$

Stress category groups represent stress intensity σ_i (reduced stress) calculated according to the max. shear stress theory. The $(\sigma)_1$ is calculated from σ_m - general membrane stress; the $(\sigma)_2$ is calculated from σ_m or σ_{mL} - local membrane stress and σ_b - general bending stress; the range of stresses $(\sigma)_R$ is calculated from σ_m or σ_{mL}, σ_b, σ_{bL} - local bending stress and σ_T - general temperature stress. All stresses are nominal. The allowed pressure [p] is calculated from the lower bound collapse load p_C.

The criterion

$$\varepsilon_t = \varepsilon_e + \varepsilon_p = 2\varepsilon_e \tag{7}$$

is used for determination of the lower bound collapse load. It is question if this criterion $\varepsilon_t = 2\varepsilon_e$ or $\varepsilon_p = \varepsilon_e$ must be fulfilled in the stress concentration zone on the surface of the pressure vessel wall too. Relationship between criterion (7) and fatigue of materials was studied on the pressure vessel, Fig. 1.

STRES AND STRAIN ANALYSIS

Two tasks were analysed, the whole pressure vessel as the 1st task (Tab. 1) and the pressure vessel without the flat cover as the 2nd task (Tab. 2).

Table 1. Stress intensity for the 1st task

p [MPa]	σ_i [MPa]	Cross section N°					Limit k[σ] [MPa]
		1	2	5	6	8	
3,162	$(\sigma)_1$	17,2	47,4	48,0	62,7	43,9	188,8
	$(\sigma)_{2i}$	211,2	85,8	49,3	17,2	45,2	245,5
	$(\sigma)_{2o}$	**245,5**	85,8	47,9	17,2	46,2	245,5
	$(\sigma)_{Ri}$	211,2	554,7	49,3	101,4	45,2	566,7
	$(\sigma)_{Ro}$	245,5	263,7	47,9	51,1	46,2	566,7
3,230	$(\sigma)_1$	17,5	48,4	49,1	64,0	44,8	188,8
	$(\sigma)_{2i}$	215,2	87,7	50,4	17,5	46,2	245,5
	$(\sigma)_{2o}$	250,8	87,7	48,9	17,5	47,2	245,5
	$(\sigma)_{Ri}$	215,7	**566,7**	50,4	103,5	46,2	566,7
	$(\sigma)_{Ro}$	250,8	269,3	48,9	52,2	47,2	566,7

index: i - inside surface, σ - outside surface

Table 2. Stress intensity for the 2nd task

p [MPa]	σ_i [MPa]	Cross section N° 1	2	5	6	8	Limit k[σ] [MPa]
9,525	$(\sigma)_1$	-	-	144,7	**188,8**	132,1	188,8
	$(\sigma)_{2i}$	-	-	148,5	51,7	136,3	245,5
	$(\sigma)_{2o}$	-	-	144,3	51,7	139,1	245,5
	$(\sigma)_{Ri}$	-	-	148,5	305,3	136,3	566,7
	$(\sigma)_{Ro}$	-	-	144,3	154,0	139,1	566,7
17,679	$(\sigma)_1$	-	-	268,5	350,5	245,2	188,8
	$(\sigma)_{2i}$	-	-	275,6	95,9	252,9	245,5
	$(\sigma)_{2o}$	-	-	267,8	95,9	258,2	245,5
	$(\sigma)_{Ri}$	-	-	275,6	**566,7**	252,9	566,7
	$(\sigma)_{Ro}$	-	-	267,8	285,8	258,2	566,7

index: i - inside surface, σ - outside surface

The first plastic strains originated on the wall surface at the pressure p$_1$. The whole joints in the cross section originated at the pressure p$_p$, Tab. 3. Development of the plastic strain area in the cross section N° 2 is shown on the Fig. 2. Relationship $\varepsilon_t = 2\varepsilon_e$ on the inside radius surface of the cross section N° 2 is activated before than the plastic joint is formed in this cross section. However the plastic strain zones in the cross sections N° 5 and 8 are very large, when three plastic joints originate in the zone of the cylinder - cone connection, Fig. 3.

Results of the elasto-plastic calculation are shown in the Tab. 4. Allowed numbers of load cycles [N$_o$] are determined by program STATES [4] in accordance with standard [1].

Allowable numbers of cycles [N$_o$] are determined from following relationships :
  * Design fatigue curves of the Manson-Coffin type [4] are used for low cycle fatigue:

$$\frac{\varepsilon_{at}}{\varphi_w} = \frac{\varepsilon_f' - 0,35\varepsilon_{p,max}}{n_\sigma \left(2[N_o]_1\right)^{-c}} + \frac{\varepsilon_{ap,vr}}{n_\sigma} + \beta_\sigma \beta_n \frac{\sigma_f' - \sigma_m}{E n_\sigma} \left[\frac{1}{\left(2[N_o]_1\right)^{-c}} + \frac{\varepsilon_{ap,vr}}{\varepsilon_f'}\right]^{b/c} \tag{8}$$

$$\frac{\varepsilon_{at}}{\varphi_w} = \frac{\varepsilon_f' - 0,35\varepsilon_{p,max}}{\left(2n_N[N_o]_2\right)^{-c}} + \varepsilon_{ap,vr} + \beta_\sigma \beta_n \frac{\sigma_f' - \sigma_m}{E} \left[\frac{1}{\left(2n_N[N_o]_2\right)^{-c}} + \frac{\varepsilon_{ap,vr}}{\varepsilon_f'}\right]^{b/c} \tag{9}$$

  * Design fatigue curves of Wöhler type [4] are used for high cycle fatigue:

$$[N_o]_1 = N_C \left[\frac{\beta_n \beta_\sigma \sigma_C}{\beta n_\sigma \sigma_{a,nom}}\left(1 - \sigma_{m,nom}\frac{\sigma_C - \sigma_{AC}}{\sigma_C \sigma_{AC}}\right)\right]^m \tag{10}$$

$$[N_o]_2 = \frac{N_C}{n_N}\left[\frac{\beta_n \beta_\sigma \sigma_C}{\beta \sigma_{a,nom}}\left(1 - \sigma_{m,nom}\frac{\sigma_C - \sigma_{AC}}{\sigma_C \sigma_{AC}}\right)\right]^m \tag{11}$$

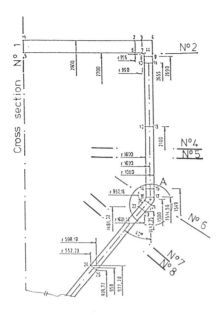

Figure 1. Scheme of cross sections

Figure 3. Plastic strain zone in the cross section No 6

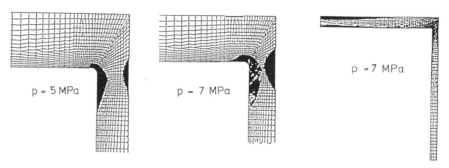

Figure 2. Plastic strain zone in the cross section No 1 and No 2

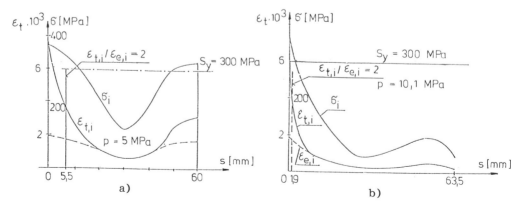

a)

b)

Figure 4. a) cross section No 2; b) cross section No 6

Allowable number of cycles:

$$\left[N_o \right] = \min \left\{ \left[N_o \right]_1 ; \left[N_o \right]_2 \right\}$$ (12)

For $n_\sigma\, \sigma_{a,nom} \leq \beta_n\, \beta_\sigma\, \sigma_C / \beta$ is $[N_o] = \infty$.

Used symbols in relations (8) to (11): E = Young's modulus; ε_{at} = amplitude of the total strain in the elasto-plastic state; σ_m = middle stress of the cycle; $\varepsilon_{p,max}$ = maximum of the plastic strain for all service time; $\sigma_{a,nom}$, $\sigma_{m,nom}$ = amplitude and middle values of the nominal stress of the cycle; n_σ, n_N = safety factors, usually $n_\sigma = 2$, $n_N = 10$; φ_w = reduction coefficient of the fatigue resistance by welding joints; ε_f' = fatigue ductility coefficient; σ_f' = fatigue strength coefficient; c = fatigue ductility exponent; b = fatigue strength exponent; $\varepsilon_{ap,vr}$ = amplitude of the returned plastic strain ($\approx 2.10^{-5}$); β_n = quality surface coefficient; β_σ = stress gradient coefficient (influence of the solid greatness); β = fatigue notch factor; σ_C = fatigue limit under symmetrical cycling; σ_{AC} = fatigue limit for pulsating stress cycle.

The design fatigue curve of Manson-Coffin type was used for calculation of the $[N_o]_{M-C}$ and the Langer type for calculation of the $[N_o]_L$, Tab. 4. Distribution of the σ and ε_t during thickness of the wall in cross section N$^\circ$ 2 and N$^\circ$ 6 is shown on the Fig. 4.

Table 3. The origin of plastic strains and plastic joints

Cross section N$^\circ$	1	2	4	5	6	7	8
p_1 [MPa]	4	1,75	18,63	> 18,63	5,9	17,73	> 17,73
p_p [MPa]	> 7,25	7,25	19,5	19,5	24,3	20,72	20,72

Table 4. Results of the elasto-plastic calculation

Vessel	whole vessel				without the flat cover				
cross section	N$^\circ$ 2, inside surface				N$^\circ$ 6, inside surface				
p [MPa]	4,0	5,0	6,0	7,0	5,9	10,1	15,1	17,34	24,3
σ_i [MPa]	350,8	377,1	406,4	430,5	313,9	362,9	406,4	427,9	504,6
$\varepsilon_{t,i}.10^3$	4,947	7,635	12,64	19,16	2,881	6,012	12,64	18,35	65,54
$\varepsilon_{e,i}.10^3$	1,827	1,964	2,117	2,242	1,635	1,890	2,117	2,229	2,628
$\varepsilon_{p,i}.10^3$	3,120	5,671	10,52	16,92	1,246	4,122	10,52	16,12	62,91
$\varepsilon_t/\varepsilon_e$	2,708	3,887	5,968	8,545	1,726	3,181	5,972	8,230	24,94
$\Delta\varepsilon_t.10^3$	3,304	4,806	7,679	11,47	2,207	3,896	7,679	10,98	38,41
$[N_o]_{M-C}$	13503	2417	613	142	1.E+6	5485	613	161	12
$[N_o]_L$	10027	2421	425	127	40842	5073	425	143	6
11 mm under wall surface (18,3 % of the wall thickness)									
σ_i [MPa]	304,3	315,7	347,2	378,2					
$\varepsilon_{t,i}.10^3$	2,548	2,949	4,508	7,775					
$\varepsilon_{e,i}.10^3$	1,584	1,644	1,808	1,970					
$\varepsilon_{p,i}.10^3$	0,963	1,305	2,863	5,806					
$\varepsilon_t/\varepsilon_e$	1,607	1,794	2,741	3,947					

ASSESSMENT OF LIMITS

The limits for stress category groups $(\sigma)_1$, $(\sigma)_2$ and $(\sigma)_R$ in accordance with (3) to (5) are fulfilled for the pressure p = 3,162 MPa, Tab. 1. Elastic calculation of stresses was made by the program SYSTUS.

The limit [p] = 0,67 p_C is possible to use when limit for $(\sigma)_2$ is not fulfilled. The limit $\varepsilon_t = 2\varepsilon_e$ for calculation of the lower bound collapse load-pressure p_C = 5,0 MPa is fulfilled of the 5,5 mm (9,2 % of the wall thickness) under the internal surface of the most stressed cross section N° 2, Fig. 1 and Fig. 4a). The allowed pressure [p] = 0,67 p_C = 0,67 . 5,0 = 3,35 MPa > 3,162 MPa. The limit of the $(\sigma)_1$ is fulfilled for the pressure [p] = 3,35 MPa too, see Tab. 1 and Tab. 2. The allowed number of cycles $[N_o]$ = 2417 is bigger then n = 2400, Tab. 4. The range of stresses $(\sigma)_{R1}$ = (3,35/3,162) . 566,7 = 600,4 MPa > 566,7 MPa. In accordance with [1] and [2] the limit of the $(\sigma)_R$ cannot be fulfilled, when local change of the vessel shape, due to plastic strain, has not influence on the function of this vessel. This local change of the vessel shape is possible.

CONSLUSION

Following limits are discussed for revision of the NTD A.S.I. Section III [1]. The limit of the $(\sigma)_1$ must be fulfilled. The condition $\varepsilon_t = 2\varepsilon_e$ for the lower bound collapse load is required in the membrane stress zone. The damage factor $D = \Sigma(n/[N_o]) \leq 1$ is required in the stress concentration zone and $\varepsilon_t > 2\varepsilon_e$ is possible in this case, but the length of the plastic strain zpne must be less then $0,7[D(s-c)]^{0,5}$. The distance of two similar zones must be bigger than $1,7[D(s-c)]^{0,5}$, where D - internal radius of the vessel, s - thickness of the wall, c - addition to the wall thickness. The condition $\varepsilon_t \leq 2\varepsilon_e$ is required in the depth 0,1 s under the wall surface.

REFERENCES

1. A.M.E. Standard Technical Documentation. Strength Assessment of Equipment and Piping of Nuclear Power Plants of WWER Type. Section III. Identification N° NTD ASI-III-Z-5/96. Prague, Brno 1996 (In Czech and English).
2. NTD SEV 4201-86 to 4214-86. Strength Calculation Standard of Equipment and Piping of Nuclear Power Plants, Interatomenergo Moskva, 1986 (In Russian).
3. ASME Boiler Pressure Vessel Code, Section III. Division 1. Rules for Construction of Nuclear Power Plant Components. Edition 1992. New York (In English).
4. Vejvoda, S. (1996). Assessment of Structures on Fatigue at the Varying Loading. Report of the IAM Brno, N° 2331/96 (In Czech).
5. Vejvoda, S., Černý, A. and Slováček, M. (1997). Criterion of the Limit State Load. Report of the IAM Brno, N° 2485/97 (In Czech).

SUBJECT INDEX

MATERIALS INDEX

xxviii

AUTHOR INDEX